MICROWAVE AND GEOMETRICAL OPTICS

Techniques of Physics

Editor

N. H. MARCH

Department of Theoretical Chemistry: University of Oxford, Oxford, England

Techniques of physics find wide application in biology, medicine, engineering and technology generally. This series is devoted to techniques which have found and are finding application. The aim is to clarify the principles of each technique, to emphasize and illustrate the applications, and to draw attention to new fields of possible employment.

1. D. C. Champeney: *Fourier Transforms and their Physical Applications*
2. J. B. Pendry: *Low Energy Electron Diffraction*
3. K. G. Beauchamp: *Walsh Functions and their Applications*
4. V. Cappellini, A. G. Constantinides and P. Emiliani: *Digital Filters and their Applications*
5. G. Rickayzen: *Green's Functions and Condensed Matter*
6. M. C. Huntley: *Diffraction Gratings*
7. J. F. Cornwell: *Group Theory in Physics, Vols I and II*
8. N. H. March and B. M. Deb: *The Single-Particle Density in Physics and Chemistry*
9. D. B. Pearson: *Quantum Scattering and Spectral Theory*
10. J. F. Cornwell: *Group Theory in Physics, Vol III: Supersymmetries and Infinite-Dimensional Algebras*
11. J. M. Blackledge: *Quantitative Coherent Imaging*
12. D. B. Holt and D. C. Joy: *SEM Microcharacterization of Semiconductors*
13. J. W. Orton and P. Blood: *The Electrical Characterization of Semiconductors: Measurement of Minority Carrier Properties*
14. P. Blood and J. W. Orton: *The Electrical Characterization of Semiconductors: Majority Carrier and Electron States*
15. S. Perkowitz: *Optical Characterization of Semiconductors: Infrared, Raman and Photoluminescence Spectroscopy*
16. S. Cornbleet: *Microwave and Geometrical Optics*

MICROWAVE AND GEOMETRICAL OPTICS

S. CORNBLEET

UNIVERSITY OF SURREY
GUILDFORD, UK

ACADEMIC PRESS
Harcourt Brace & Company, Publishers
London San Diego New York
Boston Sydney Toronto

This book is printed on acid-free paper

Academic Press Limited
24 - 28 Oval Road
London NW1 7DX

United States edition published by
Academic Press Inc.
San Diego, CA 92101

British Library Cataloguing in Publication Data is available

ISBN 0-12-189651-X

Printed in Great Britain by Hartnolls Ltd. Bodmin, Cornwall

CONTENTS

For

Joshua

Clare and Nicholas

PREFACE

This book is essentially a second edition of the original volume entitled "Microwave Optics" (Academic Press 1977). The advantage is taken, however, to include much of the material from a second book "Microwave and Optical Ray Geometry" (John Wiley & Sons 1984 now out of print), since receiving many requests for copies of that volume. In the interim an invited paper "Geometrical Optics Reviewed" (Proc IEEE April 1983 pp471-501) by the author gave an overview of the subject as a science in its own right. Since the publication of these books there have been many developments in the field of geometrical optics and practical antenna design and advances in my own field of theoretical study. The requirements of satellite antennas predominated during the decade of the eighties, shaped beam "footprints" and dual reflectors that produce them in particular, and a steady advance into the region of shorter and shorter wavelengths has made feasible the more optical lens devices that had hitherto been thought too unwieldy to consider at the longer wavelengths. This volume therefore gives these subjects greater consideration both with detailed studies and with surveys of current research. To do so means that parts of the original volumes have to be excluded in order to make way for the new material and a rearrangement made to make the subject more coherent.

Lest the title appears to be too comprehensive, it has to be stated at the outset that we deal mainly with design techniques for optical devices that can be used for microwave antennas and other microwave systems, polarization and the underlying theory that allows for a first order estimation of these effects, such as radiation patterns. There are several methods however that are applicable to the microwave range, are new in concept and capable of further development, which are also applicable to the field of light optics. These lead the subject into the field bordering on general theoretical physics, and are used as a demonstration of first order effects in other physics subjects. Such connections, it is hoped, make the subject matter of greater interest than simply a designer manual.

Other relevant subjects, such as fibre optics, the geometrical theories of diffraction and propagation in non-isotropic media, cannot be included for a variety of reasons, they all have an extensive literature of their own, such an inclusion would make the volume unduly unwieldy and this author cannot claim any expertise of his own in those subjects. It could be said that I have confined the subject to theories that depend essentially on the concepts of rays and phase fronts and on how the resultant devices can be rapidly assessed.

Thus the book retains its three original objectives. To the newcomer to the field of microwave optics it should provide a source of fundamental design ideas and a guide to the relevant literature for the entire subject. The more experienced designer, on the other hand, may find much familiar material cast in a new and hopefully interesting light and finally the theorist may find the connections of geometrical optics with general theoretical physics a source of deeper study and understanding.

Optics, and in particular geometrical optics, is amongst the oldest of the sciences, and the introductory chapters of any of the numerous texts on the subject would provide a suitable basis for discussion. Thus from the work of JL Synge ("Geometrical Optics" Cambridge University Press, 1937 p1)

> "Geometrical optics is an ideal theory and a useful one. The discovery that the propagation of light is an electromagnetic phenomenon made the subject of optics coextensive with electromagnetism. We may, however, study certain parts of the subject of optics without reference to electromagnetism, always understanding that there is a limit to the physical accuracy of the results so obtained. It is customary to use the name 'physical optics' for the more complex and physically accurate theory, and 'geometrical optics' for the simpler ideal theory with which we shall be concerned. It is possible to justify geometrical optics as the limiting case of physical optics, the wavelength of light in question tending to zero, but we shall be content with the development of geometrical optics on the basis of its own hypotheses."

What is not apparent in this quotation is that while the subjects of optics and electromagnetism are "coextensive", and the subject of my later chapters, their application lies in essentially different areas. If, in particular, we are concerned with the design of optical devices, then geometrical optics provides the only a priori method. This is, in fact, that certain part which can be studied without reference to electromagnetism. If, subsequently we need to investigate the precise properties of the design so obtained, then physical optics has to be applied. These will include such electromagnetic effects as polarization, fields at caustics and foci and the various geometrical theories of diffraction. The feed-back from the conclusions of the analytical treatment to the basic geometrical optics design, has not been thoroughly established. This leaves the geometrical design, as was said of optical lens designing, to be "more an art than a science".

In the context, this author considers the term "optical geometry" appropriate, defined as Euclidean geometry enhanced by Snell's laws of refraction and reflection. These include the approximations of geometrical optics, namely, that a point source of rays exists and that reflection or refraction at a curved surface is the same as that of the infinite plane surface tangential at the point of incidence. This definition itself shows certain limitations to the procedures. The radius of curvature of any such surface has to be large with respect to the wavelength and is no longer defined where a tangent surface cannot be defined, at corners specifically. Fortunately, the design of antennas with which we are mainly concerned can avoid such areas and this limitation is accepted throughout this volume.

The inclusion of "microwave" into the title alleviates considerably the limits of physical accuracy with which we need to be concerned. When, even at the highest frequencies envisaged, a small surface variation of a fraction of the wavelength is easy to realise physically, practical devices that are not geometrically perfect become permitted designs. Further differences with light optics exist, so much so, that one referee queried the right to allow the word "optics" in the title.

Optical antennas consist of systems rarely with more than two surfaces, designed essentially for the point focusing (or imaging as I was instructed to call it) of an incident plane wave with an additional criterion usually related to power distribution or aspect angle or suchlike. Lenses in light optics are designed to provide a wider field about the nominal focus, over which aberrations are minimised usually by adding further lens elements and hence a multiplicity of surfaces. The latter are customarily designed by ray tracing, a computerised technique, whereas the former can make use of "optics in the large" methods where the whole surface interacts with the ray pencil or phase front, a more "geometrical" concept. A prime example illustrating this is the difference of approach that will be found in the sections dealing with the design of the fundamental corrector for the spherical reflector, the Schmidt corrector.

Chapter 1 deals with the fundamentals of reflection and refraction applied to a single surface with a point source of illumination. This is presented in a manner that makes available several new concepts that will be required in later chapters, such as the numerous different ways that the basic laws of reflection and refraction can be presented, and in particular, the concept of the zero-distance phase front, its determination under several basic conditions and its relevance to the inversion theorem of Damien, and classical geometrical considerations such as catacaustics, evolutes and involutes.

Many of the methods to be outlined for the design of two surface systems are applicable to both dual reflector antennas and lenses and it was problematic whether to separate the subject into methods or devices. Since, as a reference work, a reader would be more likely to look for subject headings, it was decided to separate the subjects of the reflectors from the lenses at a possible cost of some repetition in designing procedures and analysis.

The optical principle on which the designs are based states that a degree of freedom is available for every independent surface contained in the design. Hence a single surface can only perform basic imaging. Two surface systems such as dual reflectors or a single lens, can accordingly perform an additional function, and there exists a variety of these for practical antennas. The basic ones are gone into in detail and methods suggested for the others.

Consequently dual reflector systems are considered in the second chapter, utilising the results from the first and including concepts such as caustic matching and coinvolution and applications of the inversion theorem, in the design process. This requires attention to the results of the design on aperture power distribution or off-axis performance for scanning or beam shaping these being the two major concerns of optical antennas although others are indicated. The chapter therefore includes a general survey of dual and shaped beam reflector antennas, their specific design methods and the necessary approximations that have to be made to enable a computational result to be obtained. The chapter concludes with a new application of the dual profile design process to a folded parallel plate line source lens with specified amplitude distribution or scanning capability.

Chapter 3 continues these methods into refraction and the design of lens systems, including waveguide lenses. The rapid advance of ultra-high frequency generators and receivers and a new generation of rigid dielectrics, brings about the need for lenses with similar properties of aperture power distribution and/or scanning ability that the dual reflectors perform. The latter requires the lens to obey the well-known sine condition, the application of which brings much of the designing procedure into the field of light optics. In addition there exists a whole range of intermediate devices, the phase corrected reflectors, relying for their action on both a refraction and a reflection. Lenses and phase corrected reflectors have an advantage not available to dual reflectors, that of stepping or zoning. The deterioration in performance that results can be minimized to keep the system within acceptable limits. With the thin devices that result, a further concept arises, that of the phase-only lens. These include for example Fresnel zone plates and axicons, designed on the principle of phase only, and without regard to ray directions. Because of their action in this manner, I have dubbed them Huygens' lenses.

Among the surprises is the part played by the spherical reflector in every facet of the design so far outlined. So much so that a separate chapter for the spherical reflector, its various approximations, caustic correction, phase correction and numerous forms of secondary sub-reflectors and their focusing properties, was considered. Unfortunately each of these topics occurs in a different section as design procedures and has to be illustrated in the spherical reflector separately. But it is an aspect the reader may care to look out for.

Once again at the end of the chapter, I presume on my author's licence and propose a lens transformation hypothesis, not yet subject to proof, that will transform any aplanatic lens (or dual reflector focused system) into a new one with the same imaging. This uses an extension to the inversion theory used in the design method, and although unproved has several cogent illustrative examples. This forms another of many ideas propounded that could be suitable subjects for further research.

The ensuing chapters follow broadly the lines of the earlier volumes expanded with more recent material in the fields of non-uniform lenses, scalar diffraction theory, pattern synthesis using zoned apertures and polarization.

Chapter 4 deals with non-uniform media and gives a general expansion of the ray equation for a large number of orthogonal curvilinear coordinate systems and expands the theory and design of geodesic lenses to surface guided wave devices. The plane stratified medium has separate attention due to its application as a transparent electromagnetic window, also achieving greater importance with the advent of shorter wavelengths of operation. The cylindrical medium is analysed and applied to the design of flat disc lenses and the general theory of the optical non-uniform fibre. Greatest attention is paid to the complete analysis of the non-uniform spherical medium and the lens and surface wave devices that can be derived from it. The spherical medium also plays a great part in the ensuing theoretical studies due to the seminal nature of the rays in the Maxwell fish-eye in comparison with other fundamental physical entities such as the harmonic oscillator and elliptical orbit trajectories. Rays in the general axisymmetric medium are shown to obey laws similar to geodesic flow on the sphere which are also stereographic projections of rays in practical lenses. This chapter now includes the major part of the non-uniform media section of the volume "Microwave and Optical Ray Geometry", mentioned earlier.

Chapter 5 deals with the scalar theory of diffraction by a (usually) circular aperture, maintaining the first order approximation that constitutes geometrical optics theory and shows that, as a design assessment procedure, it gives qualitative effects more simply and directly than the exact physical optics analysis. This even includes effects such as interference that one would expect only to arise from field theory. A particularly pleasing closed form of solution is obtained through the use of the circle polynomials of Zernike, and a Neumann series results. This combination was given its first airing in the original book where the classical nomenclature for the various series that arose was used. These as given in Watson's fundamental tome "Bessel Functions" were termed Neumann, Dini or Schlömilch series. The Zernike polynomials can be described as reduced Jacobi polynomials but to rename the standard series with a newly invented title to include that fact has not advanced the basic analysis in any material way so the subject has not had any major revision.

This treatment also permits the inclusion of the basic aberrations that occur in microwave systems and enables a correlation to be made between the near and far fields. The nature of the Neumann series and its companion the Dini series is such that the inverse problem, that of determining the near field required to produce a specific far field, can be considered. This illustrates the limitations that a continuous aperture distribution imposes on pattern shaping. Going to the discontinuous aperture created by zones with independent amplitude and phase control, theoretical results for shaped beams can be obtained, and the application to classical phase or amplitude zone plates is indicated.

Chapter 6 repeats the analysis of the polarization ellipse of the previous volume and uses the optical matrix technique for cascaded polarizing elements. It also introduces the practical use of quaternions, now in common use in polarized light optics, as rotation operators in preparation for their extended use in the final chapter as field transformation operators. Also added is an introduction to and a survey of the applications of polarization theory to radar polarimetry.

As indicated earlier, the final chapter is somewhat of a departure from applicable geometric optical concepts to the realm of the geometrical interpretation of physical theory. Several precedents will have been established in the text showing such relations, the rays in a transformed Maxwell fish-eye and the Kepler orbits of constant energy about a centre of gravitational attraction for example or the quaternion description of polarization with the Pauli spin matrices. Also included have been transformations of the optical system into other systems such as those that the inversion theorem of Damien induces.

In the field of the infinite isotropic non-uniform medium some properties peculiar to the geometry of the rays occur. These take the form of transformations that convert one system of rays into another. There are few transformations that transform the electromagnetic field covariantly, the Lorentz transformation for example. The "coexistence" between geometrical optics and the electromagnetic field mentioned earlier, must require that if such transformations exist with optical rays, they must also in some way occur through the conformal transformation of the electromagnetic field. The problem of finding those transformations which conformally transform the Poynting vectors of the electromagnetic field in the same way as the rays are transformed, and how the connection is made between them, is the concern of this final chapter. The study requires both a new algebraic formulation of conformal transformations using the theory of bi-quaternions, and a new potential description of the electromagnetic field that can be applied to rays more successfully. The former is based on the extension of the quaternion theory based on the study of polarization, to the complex bi-quaternions, and the latter on the use of potentials first defined by Bateman and hence termed Bateman potentials. The success of the theory in conformally transforming the spherically symmetric non-uniform medium is illustrated, but the application to the other ray geometric problems involving surfaces and boundary conditions has yet to be completed. Several such transformations are described throughout the work, nearly all involving some inversion, conformal transformation or reciprocity.

Problems such as those involving propagation in inhomogeneous media, are among the most difficult in all electromagnetic scattering theory. The attempt is made to simplify it to a geometrical ray situation with only a minor degree of success or mathematical rigour. However even this indicates a topic that could well deserve serious research consideration.

This essay into the conformality of ray systems and the electromagnetic field does show that there is a central position that geometry and optical geometry in particular can play in the study of theoretical physics. As one reviewer of the original book put it "It is only confined to the final chapter so no great harm is done".

Thus in most chapters of this book there are either new devices, new concepts or a new approach to the subject, many of which are open-ended or speculative, leaving much scope for any researcher willing to take the challenge of further investigating them.

In the first edition, references for each chapter were given at the end of the chapter and augmented by a considerable bibliography. To continue this procedure in the light of the recent upsurge in relevant publications and the extension of the material, would excessively increase the size of the book. Thus the bibliographies have been omitted. In these days of key words and electronic databases this will present no great difficulty to a dedicated investigator.

I have tried wherever possible to include references to sources outside the usual engineering fields, such as optics generally and mathematical physics. Inevitably, some references will have escaped the trawl net of library searches that have been undertaken in the attempt to make the book fully comprehensive, and I offer my apologies to those authors who feel that their work has not received the attention it deserves.

Leonardo da Vinci is said to have believed that "optics is the paradise of mathematicians". I can but hope to have made the subject as informative as possible but above all to have made it "enjoyable".

S Cornbleet
University of Surrey
Guildford

November 1993

ACKNOWLEDGEMENTS

As the volume has increased in size and scope from the original two books upon which it is based, so has the author's indebtedness to many others besides those to whom acknowledgements were originally given in those books. Thanks must go again to Dr RC Hansen of Tarzana, California who provided numerous reprints of articles and assessed my survey of them, to Professor Shane Cloude of Nantes University for a review of the theory of radar polarimetry and to Dr Bryan Westcott of the University of Southampton, who did the same for my study concerning the general theory of dual-reflector antennas. Also to Dr Westcott and to Research Studies Press, the publishers of his book "Shaped Reflector Antenna Design", my thanks for their permission to reproduce in full section 2.1 of that book, in an appendix to this one. All made valuable comments which I believe have enhanced the value of the book. Permission was also received with gratitude from Dover Publications for the reproduction of the formulae and geometry of curves given in Appendix III and from the editors of the Proceedings of the IEE for the use of parts of papers of mine, published in recent years in Part H.

Support also came from the Library, the Staff and Computing group of the Physics Department of the University of Surrey, and in particular, Mrs Joan Hilton who tolerated endless corrections of equations in mid-compute with efficiency and good humour.

The author is greatly indebted to James Gaussen and the staff of Academic Press for encouragement and assistance throughout the preparation of the work. The greatest debt of gratitude is owed to Mrs Gillian Smith of Quality Imaging Services Ltd for preparing the entire manuscript, as complicated as can be seen, for publication, with total competence and great patience, without which I would not have been able to complete the project.

1 THE SINGLE SURFACE OF REFLECTION OR REFRACTION

1.1 INTRODUCTION

The term optical geometry is taken to imply the entire body of classical geometry, with the single additional postulation that at some surfaces the optical laws of reflection and refraction are obeyed. These basic and well-known laws can be derived in precise terms from the boundary-value problem of the incidence of a plane wave at any angle to the plane interface between media with different refractive indices. In the approximation of geometrical optics it is assumed to hold in the locality of a non-planar surface at which a ray is incident. The normal to the surface the incident refracted and reflected rays are all coplanar. The surface is considered to be the infinite tangent plane at the point of contact of the ray, and reflection is considered to be a material with refractive index of -1. Different formulations of these laws are shown in Appendix I. Naturally these are not mutually exclusive and in principle any one can be derived from any other. However large differences can occur with the context with which any particular form is most useful, and can govern the form of solution and even the existence or uniqueness of that solution. This will become more relevant in the problems involving two or more surfaces. For the single surface we use the most basic form of the law of refraction, the bipolar differential form

$$\mathrm{d}r = \pm\eta\,\mathrm{d}\rho \qquad \eta = \eta_2/\eta_1 \tag{1.1.1}$$

where, as shown in Figure 1.1(a), r and ρ are radial coordinates from the source S and from the point of intersection F of the refracted rays respectively. For well-behaved surfaces this latter point exists in the limit of infinitesimal ray displacements. The ambiguity in sign stems from the manner in which r and ρ are either increasing together or increasing and decreasing respectively. This leads directly to the condition that either a real intersection of the refracted rays occurs and the negative sign is appropriate, or a virtual intersection occurs with the positive sign. Equation 1.1.1 derives directly from Fermat's principle for the variation of the optical path between the points S and F, for taking the two paths between them we have

$$\eta_1(r + \mathrm{d}r) + \eta_2\rho = \eta_1 r + \eta_2(\rho + \mathrm{d}\rho)$$

resulting in Equation 1.1.1.

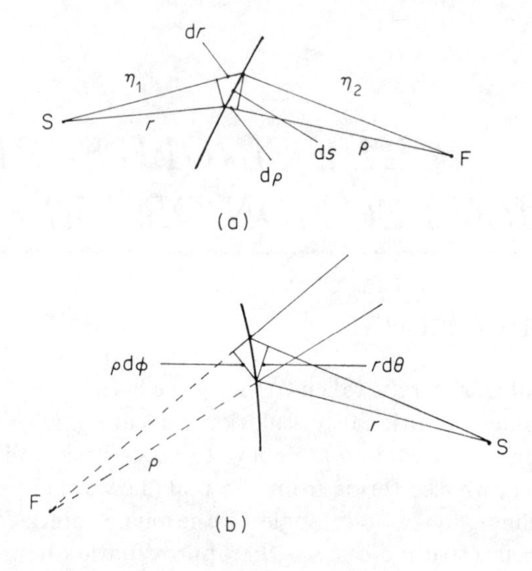

Figure 1.1: (a) The law of refraction $\eta_1\,\mathrm{d}r = \pm\eta_2\,\mathrm{d}\rho$
(b) The law of reflection $r\,\mathrm{d}\theta = \pm\rho\,\mathrm{d}\phi$

That this relation leads to the more commonly known form of Snell's law can be deduced immediately by dividing Equation 1.1.1 by the arc element $\mathrm{d}s$ from

$$\mathrm{d}s = \sqrt{\mathrm{d}r^2 + r^2\,\mathrm{d}\theta^2} = \sqrt{\mathrm{d}\rho^2 + \rho^2\,\mathrm{d}\theta^2}$$

with the result

$$\eta_1 \sin\theta = \eta_2 \sin\phi \tag{1.1.2}$$

In the case of reflection we put $\eta = -1$ and Equation 1.1.1 becomes the even more simple

$$\mathrm{d}r = \pm\,\mathrm{d}\rho \tag{1.1.3}$$

In this particular case a further simple relationship arises which can be seen from the congruence of the elemental triangle in Figure 1.1(b), that is

$$r\,\mathrm{d}\theta = \rho\,\mathrm{d}\phi \tag{1.1.4}$$

From these basic results the totality of geometrical optical design and ray-tracing procedures can be derived [1].

More importantly the forms given above can be cascaded to a system of reflecting and refracting surfaces by applying them to successive surfaces and using a multi-polar coordinate system. The boundary conditions applying to the integration of these equations will be the conditions applying to one known ray of the system. This will allow the reduction of the multi-polar coordinate system to a single chosen coordinate system.

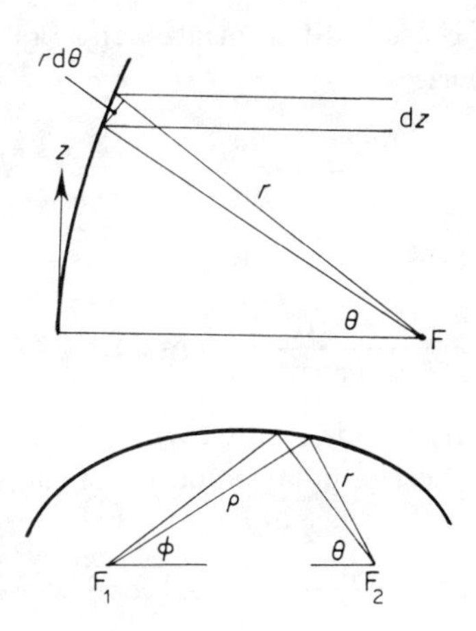

Figure 1.2: (a) Designing the paraboloid from $r\,\mathrm{d}\theta = \mathrm{d}z$
(b) Focal property of the ellipse, from $\mathrm{d}r = -\,\mathrm{d}\rho$

We illustrate these properties by applying the fundamental forms to "designing" conic reflectors. We require in the first instance an axially symmetric surface which will reflect all the rays from a source F into a direction parallel to the axis. Taking the axis of z perpendicular to the axis of symmetry we have (Figure 1.2(a)) from Equation 1.1.3.

$$r\,\mathrm{d}\theta = \mathrm{d}z \qquad z = r\sin\theta$$

Thus

$$\begin{aligned}
\mathrm{d}r/r &= \mathrm{d}\theta(1-\cos\theta)/\sin\theta \\
\log r &= \log(c\sec^2\theta/2) \\
r &= \frac{2c}{1+\cos\theta}
\end{aligned}$$

a parabola with focal length c.

A fundamental difference occurs between the use of Equations 1.1.3 and 1.1.4 in this manner. For a reflector which converts all the rays from a source at F_1 into a focused system at F_2 as in Figure 1.2(b) direct integration of Equation 1.1.3 gives, for a real focus

$$r = -\rho + \text{constant} \tag{1.1.5}$$

or $r+\rho = 2a$, the bipolar equation of the ellipse. This reduces to the polar equation by the application of the (obvious) geometry

$$\begin{aligned}
r\sin\theta &= \rho\sin\phi \\
r\cos\theta + \rho\cos\phi &= F_1F_2 = 2a\varepsilon
\end{aligned}$$

ε being the eccentricity.

However, if the second of these is differentiated and Equation 1.1.4 applied, we have the additional properties

$$\frac{d\theta}{d\phi} = -\frac{\sin\theta}{\sin\phi} \tag{1.1.6}$$

which integrates directly to give

$$\frac{\sin\theta}{1+\cos\theta}\frac{\sin\phi}{1+\cos\phi} = \text{const.} = A$$

Elimination of either θ or ϕ from this result by use of the triangle relation in Equation 1.1.4 leads directly to the polar equation of the ellipse

$$r = \frac{4Aa\varepsilon}{(1-A^2)-(1-A)^2\cos\theta}$$

from which we find that the constant of integration

$$A = \frac{1-\varepsilon}{1+\varepsilon} > 0$$

where ε is the eccentricity of the ellipse. Since the ellipse is described both by

$$r = \frac{l}{1-\varepsilon\cos\theta} \quad \text{and} \quad \rho = \frac{l}{1-\varepsilon\cos\phi}$$

Equation 1.1.3 gives us the further relation

$$\frac{d\theta}{d\phi} = -\frac{\sin\theta}{\sin\phi} = -\frac{1-\varepsilon\cos\theta}{1-\varepsilon\cos\phi}$$

For the hyperbola the related solution is

$$\frac{d\theta}{d\phi} = \frac{\sin\theta}{\sin\phi}$$

giving

$$\frac{\sin\theta}{1+\cos\theta}.\frac{1+\cos\phi}{\sin\phi} = \frac{1}{A} = \frac{1+\varepsilon}{1-\varepsilon}$$

We conclude furthermore that these (and the parabola of course) are the only single reflecting surfaces with the property of converting rays from a fixed point into rays converging onto a second fixed point. This agrees with the result of Brueggmann [1] for anastigmatic reflectors. For a single surface lens we require the definition that such a lens involves the change in direction of the ray through refraction at only one of the surfaces.

A lens will be defined to be a single surface lens if the change in ray direction through refraction occurs at only one of the surfaces of the element involved. The remaining surface or surfaces are then orthogonal to all the rays incident upon them. The analysis is confined, as in the previous section, to the cross-section of an axisymmetrical system or of a two dimensional system.

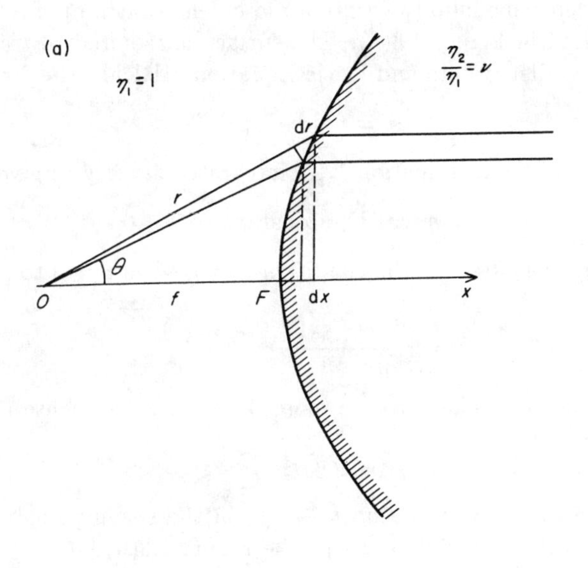

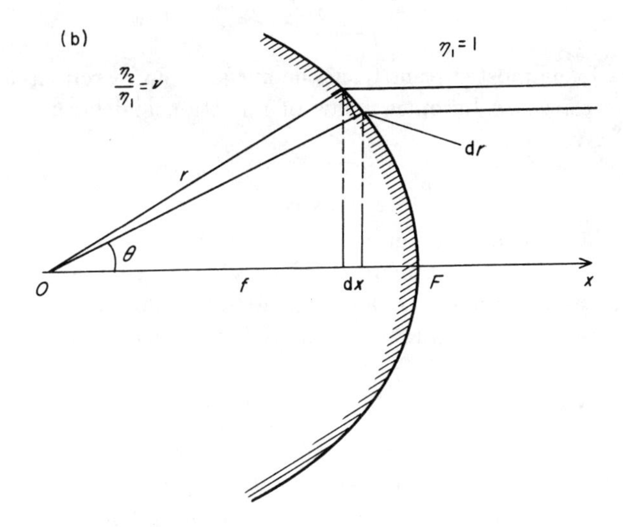

Figure 1.3: Single surface lenses. (a) Hyperbola; (b) Ellipse.

Consideration is first given to the refraction of a focused pencil of rays going from a less dense medium (air) into a more dense medium of refractive index η as illustrated in Figure 1.3(a). The surface is required to transform the pencil into a parallel system and hence Equation 1.1.1 has the form

$$\mathrm{d}r = \eta \, \mathrm{d}x \tag{1.1.7}$$

where $x = r\cos\theta$, and hence Equation 1.1.7 integrates directly to give

$$r = \eta r \cos\theta + \text{ constant}$$

The constant being evaluated by the condition $r = f$ when $\theta = 0$ to give the hyperbolic surface

$$r = \frac{f(\eta - 1)}{\eta\cos\theta - 1} \tag{1.1.8}$$

When the refraction takes place from a more dense to a less dense medium Equation 1.1.7 becomes

$$\eta \, \mathrm{d}r = \mathrm{d}x \tag{1.1.9}$$

This is identical to the transformation $\eta \to 1/\eta$ in the ensuing analysis and in the solution resulting in the elliptical profile Figure 1.3(b)

$$r = \frac{f(\eta - 1)}{\eta - \cos\theta} \tag{1.1.10}$$

with pole at the focus most distant from the surface. For a refractive index $\eta = -1$ the refractive law becomes a law of reflection and both Equations 1.1.8 and 1.1.10 give

$$r = \frac{2f}{1 + \cos\theta}$$

the parabola of the previous section.

These are the two commonest microwave lenses for focusing rays at infinity. An application of the same analysis can be made to the more general problem of refocusing the rays at a second, and possibly more convenient, point.

THE CARTESIAN OVALS

The Cartesian oval [2] is a single surface refracting lens which converts a perfectly focused pencil of rays into another perfectly focused pencil of rays at a general (finite) point as shown in Figure 1.4. This property of the surface will be termed aplanatic. These, in the terms of reference given, are rediscovered every ten years or so and have not found any useful application. They do have the capability however of projecting a focal point a considerable distance forward, if an exit surface, in this case purely spherical, is added (Figure 1.4(a)).

For the real focus the surface is subject to the relation

$$\mathrm{d}r = -\eta \, \mathrm{d}\rho \tag{1.1.11}$$

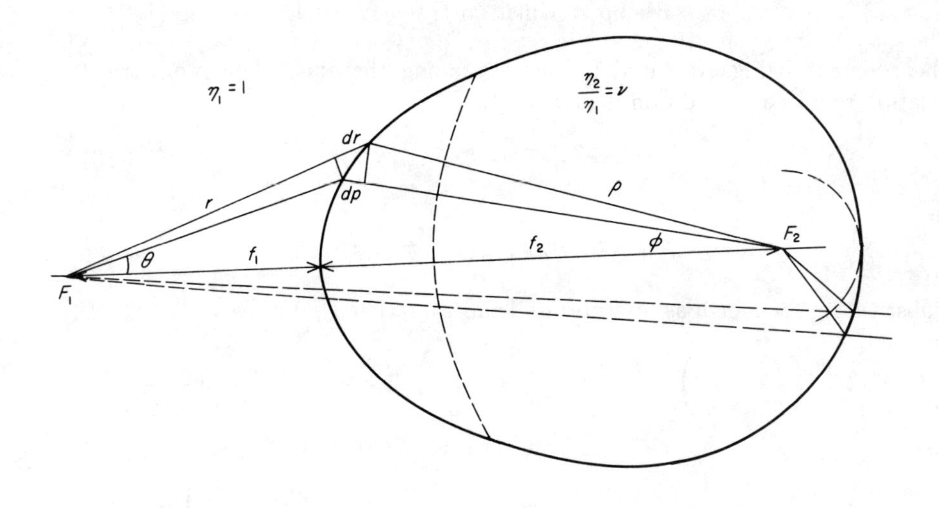

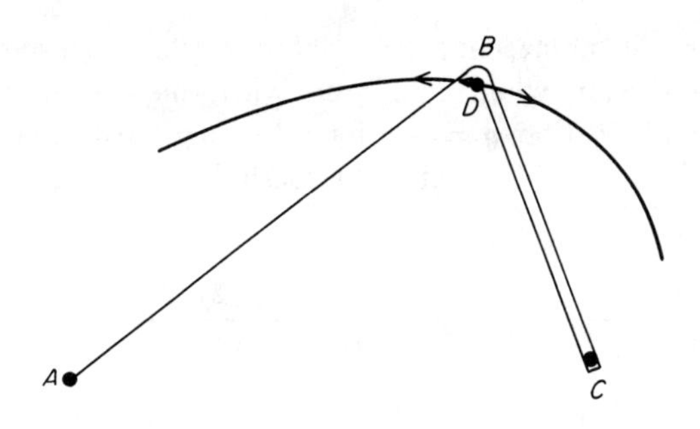

Figure 1.4: (a) Cartesian oval. The dashed circles centred on F_2 form possible second lens surfaces for focal projections.
(b) Drawing the Cartesian oval ABCD is a taut string

which integrates as usual to give the law of constancy of the optical path

$$r + \eta\rho = \text{ const } = f_1 + \eta f_2 \tag{1.1.12}$$

the constant being evaluated for the ray along the axis. The geometrical relation required in addition is one of either

$$\rho^2 = r^2 + F^2 - 2rF\cos\theta \tag{1.1.13}$$

or

$$r^2 = \rho^2 + F^2 - 2\rho F\cos\phi \qquad F = f_1 + f_2$$

Substituting for r or ρ as appropriate from Equation 1.1.12 gives

$$Fr\cos\theta = \left(\frac{f_1}{\eta^2} + \frac{f_2}{\eta}\right) + \frac{r^2}{2}\left(1 - \frac{1}{\eta^2}\right) - \left[f_1 f_2\left(\frac{1}{\eta} - 1\right) - \frac{f_1^2}{2}\left(1 - \frac{1}{\eta^2}\right)\right]$$

or

$$F\rho\cos\phi = (f_1\eta + f_2\eta^2)\rho + \frac{\rho^2}{2}(1-\eta^2) - \left[f_1 f_2(\eta - 1) - \frac{f_2^2}{2}(1-\eta^2)\right] \tag{1.1.14}$$

The final term in each of these relations is the constant of integration determined by the values of r and ρ at the boundary value $\theta = \phi = 0$ that is $r = f_1$; $\rho = f_2$.

Equations 1.1.14 are the polar equations defining the Cartesian ovals, a comparatively simple version when compared with the Cartesian equations themselves. The substitution $\eta = -1$ for reflecting surfaces in Equations 1.1.14 reproduces the conic, elliptical and hyperbolic mirrors (depending upon the relative values of f_1 and f_2).

$$r = \frac{2f_1 f_2}{(f_1 + f_2)\cos\theta - (f_1 - f_2)} \qquad \rho = \frac{2f_1 f_2}{(f_1 + f_2)\cos\phi - (f_2 - f_1)}$$

In the limit of either f_1 or f_2 becoming infinite these become the equations for parabolic mirrors.

If the same limit is taken in Equations 1.1.14 for general η the elliptical and hyperbolic lenses of Equations 1.1.8 and 1.1.10 result. It is also easily confirmed that Equations 1.1.14 transform into each other by the transformation

$$\eta \to \frac{1}{\eta} \qquad f_1 \leftrightarrow f_2$$

Equations 1.1.14 therefore embody the entire system of single reflecting and refracting surfaces with aplanatic property that have been dealt with so far. A method of drawing the Cartesian ovals, similar to that of the ellipse, with a taut string looped about the two "foci" is also shown.

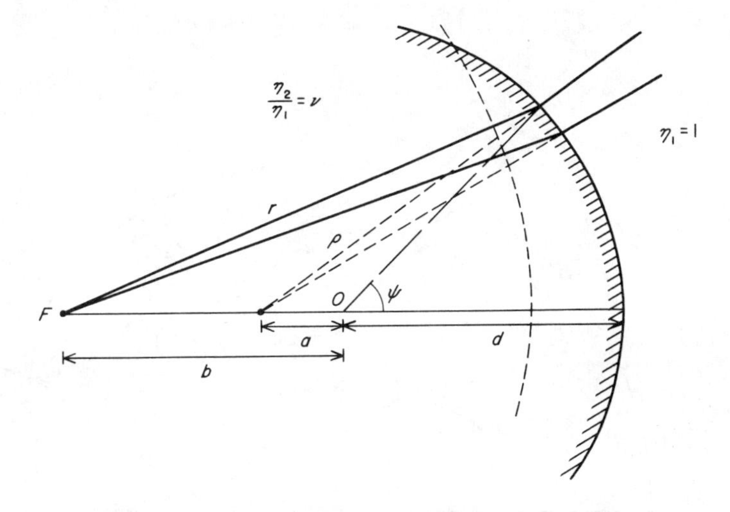

Figure 1.5: The aplanatic points of the sphere. The dashed circle centred on F is a possible second surface for a lens.

THE APLANATIC POINTS OF THE SPHERE

The Cartesian ovals develop in another manner to give a surface configuration with both optical and microwave applications. That is the condition for which the surface of refraction is purely spherical. The situation is illustrated in Figure 1.5 where a and b are the distances of perfect foci from the centre O of the sphere. We have then from the relation for a virtual image

$$\begin{aligned} \mathrm{d}r &= \eta \ \mathrm{d}\rho \\ \rho^2 &= d^2 + a^2 + 2ad\cos\psi \\ r^2 &= d^2 + b^2 + 2bd\cos\psi \end{aligned}$$

where d is the radius of the sphere. Differentiating the last two, eliminating ψ gives

$$b\rho \ \mathrm{d}\rho = ar \ \mathrm{d}r$$

and integrating the first gives

$$r = \eta\rho + \text{ constant}$$

We choose this constant to be zero and hence obtain the standard relationship for the aplanatic points of the sphere $b = \eta d$; $a = d/\eta$ [2].

THE LIMAÇON OF PASCAL

We require finally the convex refracting surface with the property of transforming a point focus into a second point focus. The second focus in this case

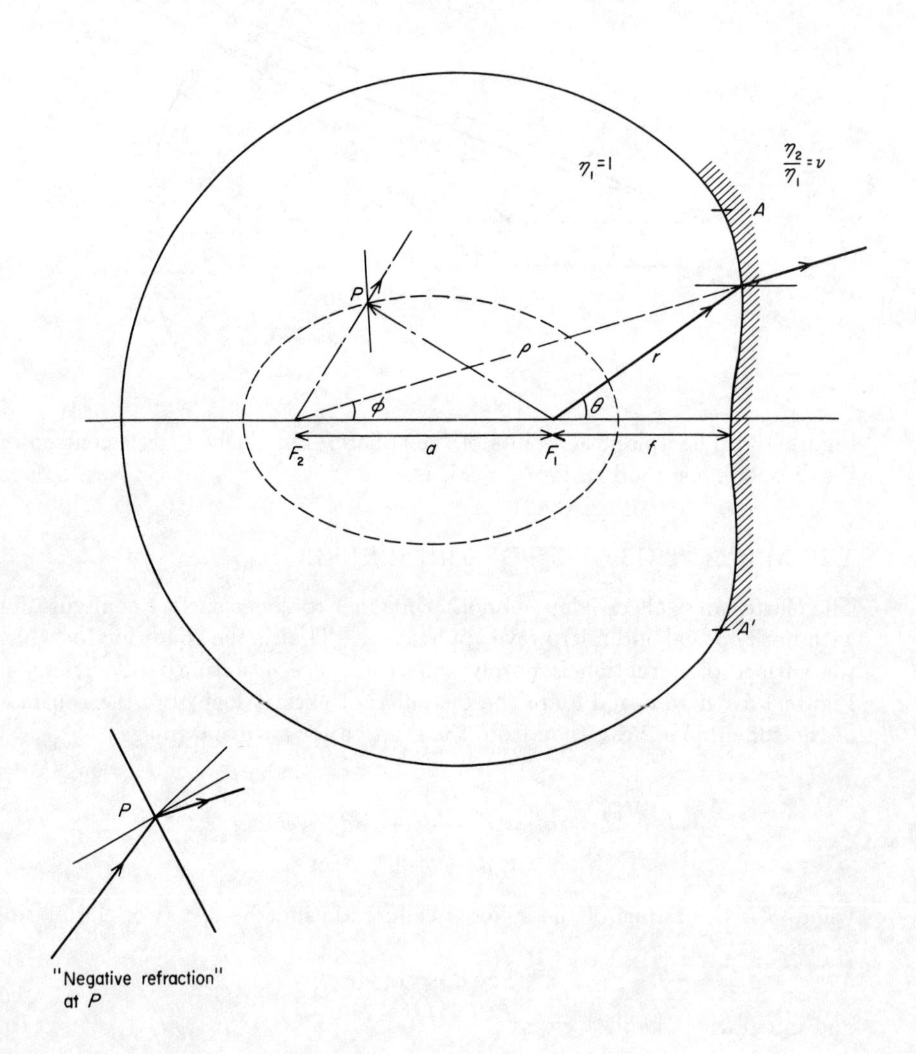

Figure 1.6: The limaçon of Pascal

will be at a fixed distance a from the first as in Figure 1.6 and will of necessity be a virtual focus. In this case the focus is displaced away from the surface and the action differs from that of the previous section. The two relations required are

$$\begin{aligned} \mathrm{d}r &= \eta \, \mathrm{d}\rho \\ \rho^2 &= a^2 + r^2 + 2ra\cos\theta \end{aligned} \tag{1.1.15}$$

or quite simply

$$r = \eta\sqrt{r^2 + a^2 + 2ra\cos\theta} + D \tag{1.1.16}$$

where D, the constant of integration is the value of $r = f$ when θ is zero, that is

$$D = f - \eta(a + f)$$

Re-arranging Equation 1.1.16 gives

$$(1-\eta^2)r = D+\eta^2 a\cos\theta+\sqrt{(D+\eta^2 a\cos\theta)^2 - (1-\eta^2)(D^2 - \eta^2 a^2)} \tag{1.1.17}$$

This curve is a particular form of the Cartesian oval known as the limaçon of Pascal as shown in Figure 1.6. It is more simply given by its direct bipolar form [5] which is the integral of equation 1.1.15

$$r = \eta\rho + D \tag{1.1.18}$$

The complete limaçon includes a second curve shown by the broken line in Figure 1.6. This derives from taking the negative root in Equation 1.1.17. It is the same as the Cartesian oval of the previous section but in the present context arises from taking a negative form of the law of refraction, that is the refracted ray is on the same side of the surface normal as the incident ray as shown in the inset in Figure 1.6.

1.2 THE ZERO-DISTANCE PHASE FRONT

A unique property is associated with the refraction or reflection of a pencil of rays at a single surface. To derive this we require firstly the definition of the phase front of zero distance.

Rays in an homogeneous isotropic medium, which originate from a point source, form a normal congruence. That is, if equal optical paths are measured along each ray from the source, the surface constructed by the end points will be normal to all of the rays in the congruence. It follows that a similar result will be obtained if the distances along each ray are measured from any one of the normal surfaces themselves. These surfaces are the phase fronts of the wave system, or orthotomics, for which the rays are the geometrical optics approximation. A fundamental theorem, the theorem of Malus and Dupin [9], states that a normal congruence will remain a normal congruence after any

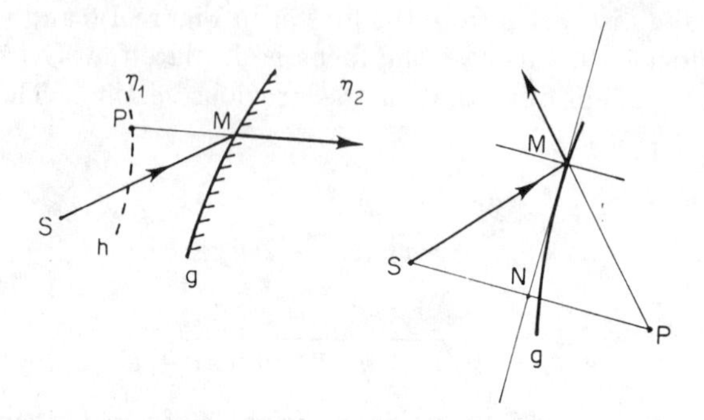

Figure 1.7: Derivation of the zero-distance phase front. The ray refracted at M is produced to P by a distance $-(\eta_2/\eta_1)$SM. The locus of P is the curve h, the zero-distance phase front. In the case of reflection $\eta_2/\eta_1 = -1$ and P lies on the opposite side to S and is the reflection of S in the tangent at M. The locus of N is the (first) pedal curve of g.

number of reflections or refractions. The phase front surfaces, however, will vary considerably at each intersection of the ray system with a reflecting or refracting surface. In an isotropic non-homogeneous medium the phase front is a continuously deformable surface normal to the now curving ray system.

We define a single special phase front relating to a specific combination of source and refracting surface [10]. This is constructed by the process illustrated in Figure 1.7. A general ray from the source S is incident upon the surface g, between two homogeneous media with refractive indices η_1, containing the source S, and η_2, at the point M. At M it undergoes refraction in accordance with Snell's law

$$\eta_1 \sin\theta = \eta_2 \sin\phi$$

where θ and ϕ are the incident and refracted ray angles with the normal to the surface at M. If the refracted ray is continued in its *reverse* direction to a point P, where

$$\eta_2 \mathrm{MP} = -\eta_1 \mathrm{SM}$$

the minus sign indicating the reversal of direction, the points P will all lie on a surface h, as M moves over the surface g. This surface h is defined to be the "phase front of zero distance". As a consequence of the theorem of Malus and Dupin, all the rays MP are normal to h, and the optical distance from the source S to P is zero. It is thus the virtual-*source* phase front for the image space to the right of g.

In most cases to be considered, the source will be in free space and the refracting medium uniform with refractive index η. Then with

$$\eta_1 = 1 \qquad \eta_2 = \eta$$

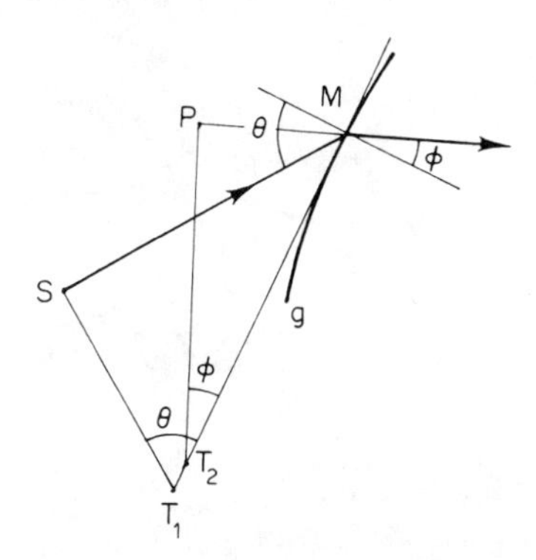

Figure 1.8: Construction to prove the coincidence of T_1 and T_2

the surface h will be defined by

$$\eta \mathrm{MP} + \mathrm{SM} = 0 \tag{1.2.1}$$

In principle we are replacing the source and first surface by a new source, the zero-distance phase front within an *infinite* medium with refractive index η. Propagation thereafter obeys Huygens' principle.

For the case of a reflector we put $\eta = -1$ in the last instance. P then lies on the *opposite* side of the reflector from S, and MP=SM. P is thus the reflected point of S in the tangent at M.

If now, as shown in Figure 1.8, we draw the tangent to g at M, then the perpendicular to SM at S will meet it at T_1 and the perpendicular to PM at P will meet it at T_2. The angle ST_1M then equals θ and the angle PT_2M equals ϕ as shown. Therefore by the construction for P

$$\frac{\mathrm{PM}}{\mathrm{T_2M}} = \sin\phi \qquad \frac{\mathrm{SM}}{\mathrm{T_1M}} = \sin\theta$$

But SM/PM $= \eta = \sin\theta/\sin\phi$; therefore $T_1M=T_2M$, and so T_1 and T_2 coincide at T, the result shown in Figure 1.9. This construction makes possible a general parametric description of the zero-distance phase front without recourse to point-by-point ray-tracing methods.

Taking the source at the origin of a plane coordinate system in which the refracting profile g is given parametrically by

$$x = f(t) \qquad y = g(t)$$

the tangent at $M(x, y)$ has equation

$$y - g(t) = \frac{g'}{f'}[x - f(t)] \tag{1.2.2}$$

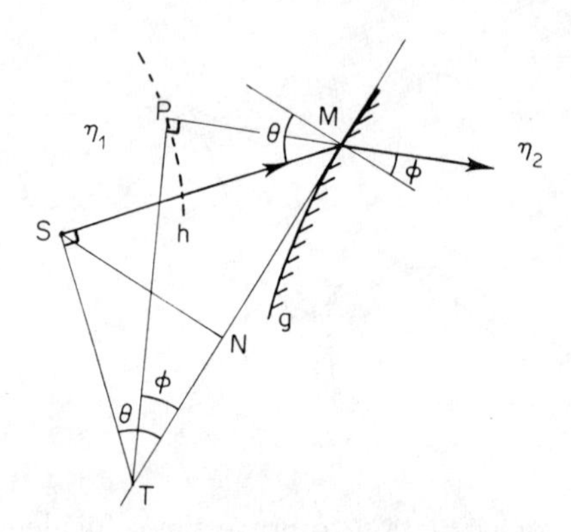

Figure 1.9: Geometrical construction of the locus of P. P is the intersection of the circle with diameter MT and the circle with centre M and radius (η_1/η_2)SM.

where primes refer to differentiation with respect to the parameter t (and in all following sections).

T is the intersection of this line with the perpendicular from the origin to SM, or

$$g(t)y = -f(t)x$$

Thus T has coordinates

$$X = \frac{fgg' - g^2 f'}{gg' + ff'} \qquad Y = \frac{fgf' - f^2 g'}{gg' + ff'} \tag{1.2.3}$$

P is now *one* of the intersections of the circle with centre M and radius PM=SM/η, and the circle with MT as diameter, that is the circles

$$\begin{aligned} (x-f)^2 + (y-g)^2 &= (f^2 + g^2)/\eta^2 \\ \{x - (X+f)/2\}^2 + \{y - (Y+g)/2\}^2 &= \{(X-f)^2 + (Y-g)^2\}/4 \end{aligned}$$

with the result, putting $N = gg' + ff'$

$$\begin{aligned} \eta^2(x-f)(f'^2 + g'^2) + Nf' &= \pm g' \sqrt{\eta^2(f^2+g^2)(f'^2+g'^2) - N^2} \\ \eta^2(y-g)(f'^2 + g'^2) + Ng' &= \mp f' \sqrt{\eta^2(f^2+g^2)(f'^2+g'^2) - N^2} \end{aligned} \tag{1.2.4}$$

The ambiguity in sign shows the existence of the second intersection of the two circles, and care has to be taken to obtain that intersection which agrees with the physical situation.

In general these equations are too complicated for the elimination of the parameter t, to enable the derivation of a functional relation between x and

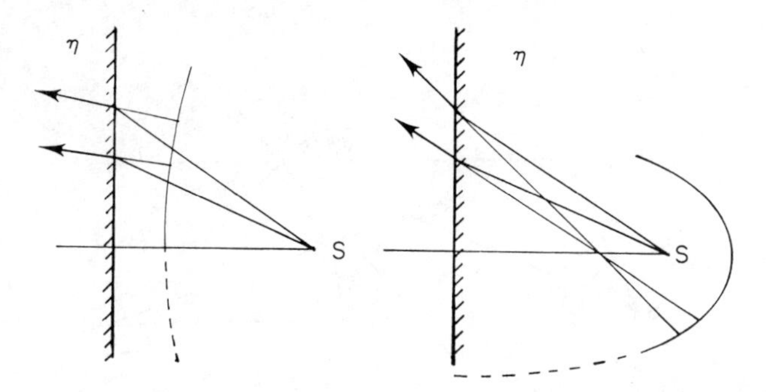

Figure 1.10: Construction of the zero-distance phase fronts for refraction in a plane interface

y alone. They are, however, quite suitable for computational methods. The result shows that a purely geometrical description of the zero-distance phase front is possible from a given geometrical surface without considering individual rays.

A simple illustration of this result is the zero-distance phase front of a plane interface. For this the plane is parametrized by

$$f(t) = c \qquad g(t) = t$$

where c is a constant. Equations 1.2.4 then give

$$\begin{aligned}\eta^2(x-c) &= \pm\sqrt{(c^2+t^2)\eta^2 - t^2} \\ \eta^2(y-t) &= -t\end{aligned}$$

which on eliminating t is

$$(x-c)^2 - \frac{y^2}{\eta^2 - 1} = \frac{c^2}{\eta^2} \tag{1.2.5}$$

This (Figure 1.10) is a hyperbola if $\eta > 1$ and the refraction is from a source in free space, or an ellipse if $\eta < 1$ and the source is within the refractive medium [11]. In either case the source is at the focus of the conic and the plane interface passes through the centre.

REFRACTION IN A CIRCULAR INTERFACE

For the all-important case of refraction in a circular interface we utilise the geometry of Figure 1.9 but with g a circle of radius r, whose centre O is at a distance k from the source S (Figure 1.11). Then equating the trigonometric identities for the triangles SPO, SMO and MPO, and with SM= ηPM, we

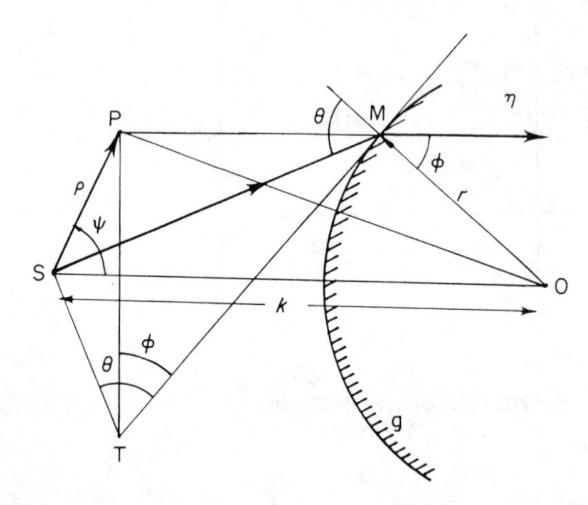

Figure 1.11: Construction of the zero-distance phase front for refraction in a circular interface

obtain respectively

$$\begin{aligned}
\text{SP}^2 = \rho^2 &= \text{SM}^2[1+\eta^2-2\eta\cos(\theta-\phi)]/\eta^2 \\
k^2 &= \text{SM}^2+r^2+2\text{SM}r\cos\theta \\
\text{PO}^2 &= k^2+\rho^2-2k\rho\cos\psi \\
&= r^2+\text{SM}^2/\eta^2+2\text{SM}r\cos\phi/\eta
\end{aligned}$$

It can readily be shown that $\rho = \pm\text{SM}(\cos\phi - \cos\theta/\eta)$ (by squaring and comparing with ρ^2 above), and hence SM and θ can be eliminated to give the polar (ρ, ψ) equation of P with respect to S; that is

$$\eta^2\rho^2 - 2\eta\rho(r+\eta k\cos\psi) + (\eta^2-1)(k^2-r^2) = 0 \tag{1.2.6}$$

The ambiguity in the sign of ρr has been resolved by the consideration of the case where g is a *reflecting* surface as shown in Section 1.2 REFLECTION IN A CIRCLE.

In performing this analysis we used only the fact that the normal to the surface at M goes through the centre O. Since the refraction of the ray at M is only a *local* effect, the same geometry results if g were taken to be the circle of curvature with double contact at M. This leaves the position of the tangent MT to the surface and to the circle of curvature, unaltered. Hence we can replace r and k in Equation 1.2.6 by R and p (Figure 1.12), where R is the radius of curvature and p the distance from S of the centre of curvature C at any point M on the now general surface g.

Thus the zero-distance phase front with respect to axes at S is given by (ρ, Ψ), where

$$\eta^2\rho^2 - 2\eta\rho(R+\eta p\cos(\Psi-\varepsilon)) + (\eta^2-1)(p^2-R^2) = 0 \tag{1.2.7}$$

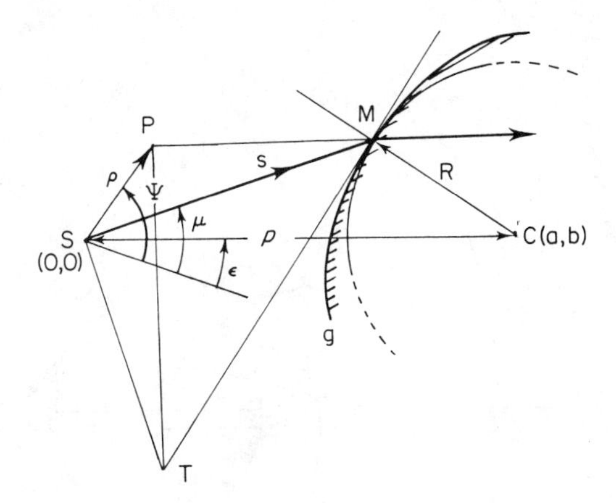

Figure 1.12: Construction for a general surface g. The circle with centre C is the circle of curvature of the point M

From the standard literature, if the surface g has polar coordinates (s, μ) with respect to S, then

$$R = \frac{\sqrt{s^2 + s'^2}}{s^2 + 2s'^2 - ss''} \qquad s' = \frac{\mathrm{d}s}{\mathrm{d}\mu} \qquad s'' = \frac{\mathrm{d}^2 s}{\mathrm{d}\mu^2}$$

and the coordinates of C are

$$\begin{aligned} a &= s\cos\mu - \frac{s^2 + s'^2 \mathrm{d}(s\sin\mu)}{s^2 + 2s'^2 - ss''} \\ b &= s\sin\mu + \frac{s^2 + s'^2 \mathrm{d}(s\cos\mu)}{s^2 + 2s'^2 - ss''} \\ \varepsilon &= \tan^{-1}(b/a) \end{aligned}$$

We can recover the condition for a plane interface by a method which has other applications. Equation 1.2.6 can be rewritten as

$$\eta^2\rho^2 - 2\eta\rho[(r-k) + k(1+\eta\cos\psi)] - (\eta^2-1)[2k(r-k) + (r-k)^2] = 0 \quad (1.2.8)$$

Then r and k can become infinitely large while the difference $(r-k)$ remains finite and equal to $-$C. In the limit

$$\rho = \frac{(\eta^2-1)C}{\eta(1+\eta\cos\psi)}$$

which is the polar equation of Equation 1.2.5.

We now consider the general conditions of the curve of Equation 1.2.6 for different positions of the source, that is for varying values of k and r, and for refraction both from the source interior to and exterior to the refracting circle.

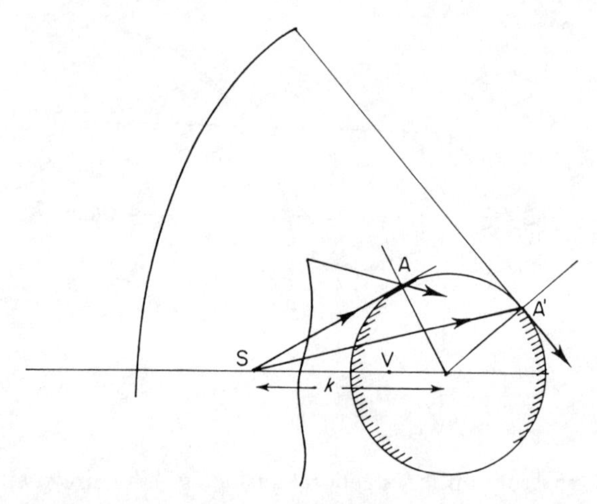

Figure 1.13: Zero-distance phase fronts for the convex and concave parts of a circular interface

In general the curve is double-branched, each branch being the result of the refraction in the convex or concave parts of the circle respectively (Figure 1.13). The source S is the focal point for the former and there is a second pole V relating to the concave refraction. The distance SV is given by

$$\mathrm{SV} = k(\eta^2 - 1)/\eta^2 \tag{1.2.9}$$

The bipolar equation to the curve in terms of radial coordinates r_1 from S and r_2 from V is

$$r_1 - \eta(r_2 \pm (\eta^2 - 1)/\eta^2) = 0$$

or

$$\eta r_2 - r_1 = \pm \eta r/k \tag{1.2.10}$$

Hence when $r = k$, that is the source is on the circle itself, the zero-distance phase front is composed of the isolated point S and the limaçon of Pascal [12].

$$\rho = 2r(1 - \eta \cos\psi)/\eta \tag{1.2.11}$$

For the refraction from a source in free space η is greater than unity. Then with the source within the circle, $0 < k < r$, the curve of Equation 1.2.6 is closed and entirely within the circle. It intersects the axis at points

$$(k - r)(1 - 1/\eta) \quad (k + r)(1 - 1/\eta)$$

With the source exterior to the circle the curve has two parts, one interior to the other, corresponding to the parts of the circle for which the surface is convex or concave towards S.

When $k = \eta r$ the interior curve degenerates to the point V.

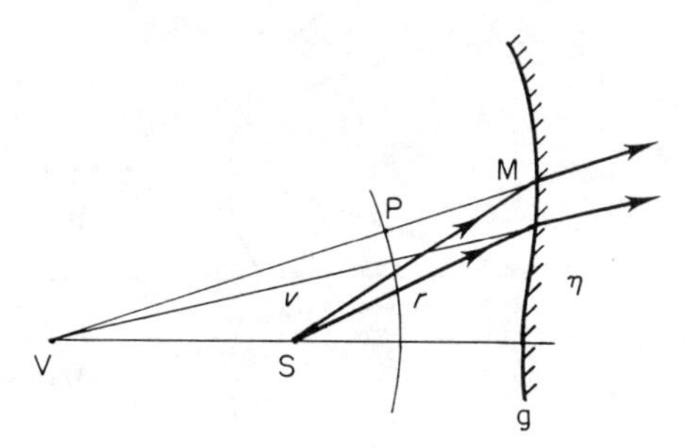

Figure 1.14: Surface required for a circular zero-distance phase front

For refraction from a source *inside* a circular refracting medium, η is less than unity and rays from the source are limited by the condition of total internal reflection to within an angle W such that

$$\sin W < r/\eta k$$

Thus if $k < r/\eta$ this condition will not occur and all rays will leave the circle.

If $k = r/\eta$ all the rays from S are concurrent at V which is at a distance ηr from the centre of the circle. This is the condition for the aplanatic points of the sphere.

Thus the form of the zero-distance phase front for refraction in a circular interface is seen to be intimately involved with the form taken by different parametric values of the limaçon of Pascal.

Since the zero-distance phase front replaces the source and the refracting surface with a new source that is the extended phase front in an infinite medium with the same refractive index, the zero-distance phase front for the Cartesian oval (Figure 1.4) must be a circle centred on the image point, for it to act as a focus of rays.

The determination of that surface profile that would produce a *circular* zero-distance phase front, by refraction from a source S in free space is obtained by the same method. We require the surface that would give a *virtual* focus for all the rays from V, which is then the centre of the phase front. This is identically the situation dealt with in Figure 1.6 that leads to the limaçon of Pascal. If then, as in Figure 1.14 the law of refraction at M is put into its differential form (Equation 1.1.1)

$$\mathrm{d}r = \eta\ \mathrm{d}v$$

where r and v are bipolar coordinates from S and V respectively, direct integration gives

$$r = \eta v + D \qquad D = f - \eta(a + f) \tag{1.2.12}$$

where the constant of integration is obtained by the condition for the central ray from S to V.

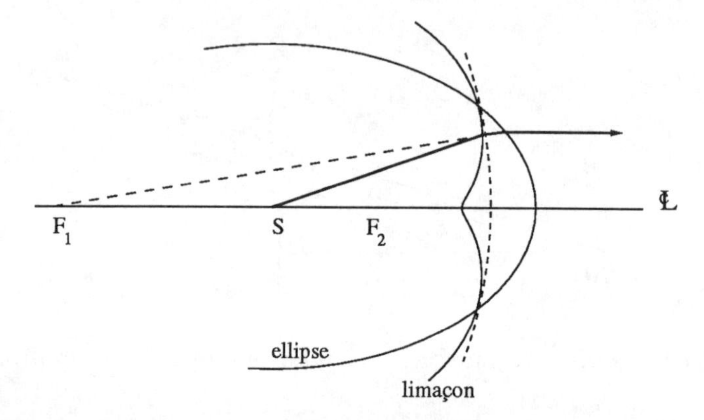

Figure 1.15: Limaçon/Elliptical lens. The ellipse with foci F_1 and F_2 collimates the rays from the focus at F_1. F_1 and S are aplanatic points of the limaçon, providing a new focus at S.

Equation 1.2.12 is the bipolar equation of the limaçon. Hence the situation is that refraction in a circle produces a zero-distance phase front which is a limaçon and refraction in a limaçon produces a phase front which is a circle. We have already met a similar situation in the limit of the plane surface refraction. Equation 1.2.5 showed this to be either elliptical or hyperbolic depending upon whether the source was inside the medium or in free space.

The limaçon of Figure 1.14 has an immediate application to the circular/elliptical lens. This suffers from having a long focal length since the source has to be situated at the remote focus of the ellipse. This can be considerably foreshortened by replacing the circle with its limaçon equivalent making the combination more practicable. As shown in Figure 1.15 a limaçon is required that will replace the circular phase front from the focus at F_1 with a circular phase front from the much closer point S. This is readily achieved by the process shown in Figure 1.14.

Now, as shown in Figure 1.16, if a source is embedded within a medium, then the ellipse is the single refracting surface that will produce parallel rays that is a plane phase front. Likewise a source in free space will be collimated to parallel rays by the single refraction in a hyperbolic surface. Hence a *plane* refractor gives a hyperbolic or elliptical zero-distance phase front, and the hyperbolic or elliptical refractor gives a plane phase front [Figure 1.16].

REFLECTION IN A CIRCLE

With the exception of the singular, and obvious, case of the plane surface, reflection can be dealt with by putting $\eta = -1$ in all of the foregoing analysis. Thus for the circular reflector, from Equation 1.2.6, we get immediately

$$\rho = 2(k \cos \psi - r) \tag{1.2.13}$$

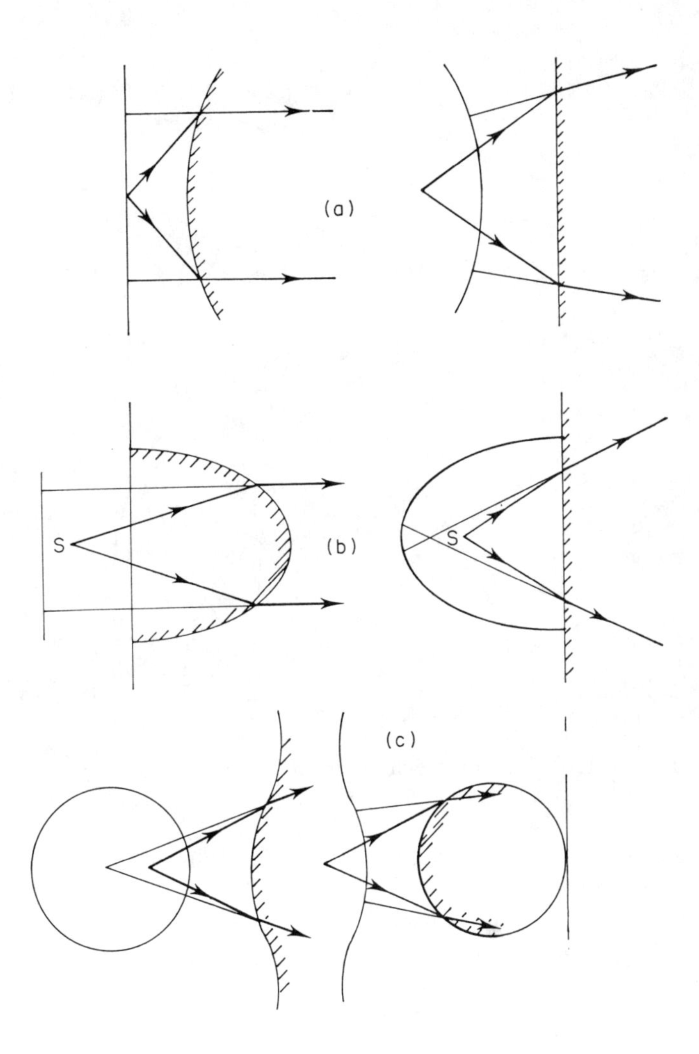

Figure 1.16: (a) Refraction in a hyperbola produces a plane phase front and refraction in a plane a hyperbolic phase front
(b) With the source in the interior, the same effect occurs with elliptical surfaces
(c) The symmetrical situation for the circle and limaçon, the inversion of (b) in the local source S
In (a) and (b) the two conditions shown are reflections, which are inversions with centre at infinity.

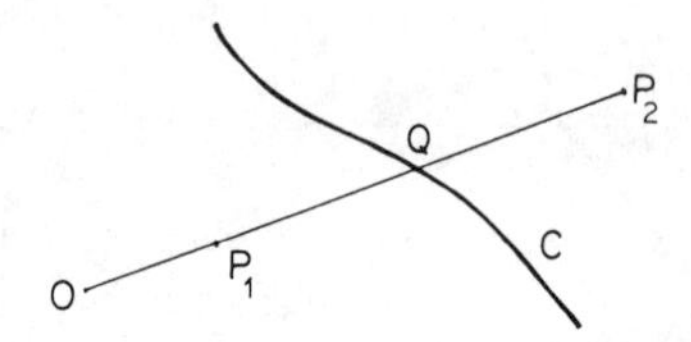

Figure 1.17: Construction of the conchoid of the curve C

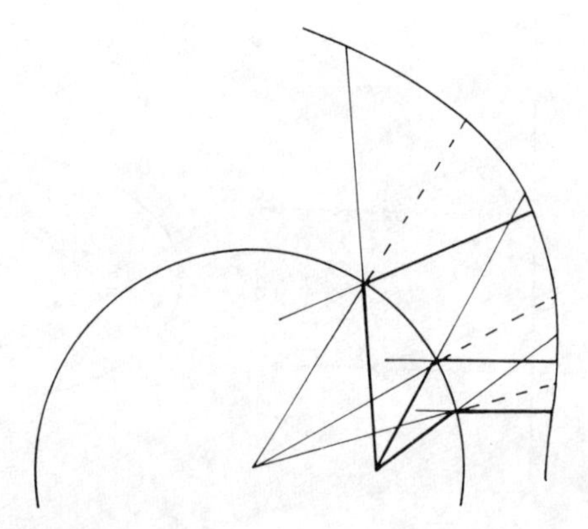

Figure 1.18: The epitrochoid as the conchoid of the circle

which again is the equation of a limaçon, also taking various forms as the ratio of k and r varies.

If, as in Figure 1.17, O is a fixed point and OP a line intersecting a curve C at a point Q, then the locus of points P_1 and P_2 such that

$$P_1Q = P_2Q = \text{constant}$$

are each a *conchoid* of C with respect to O. Equation 1.2.13 expresses the result that the zero-distance phase front of a source in a reflecting circle is the conchoid of one circle radius k with respect to a point on its circumference. It is also the *epitrochoid* generated by the fixed point on one circle, rolling without slipping upon another equal fixed circle (Figure 1.18). When $r = k$ the source is on the circumference and the curve becomes a *cardioid.* When $r = k/2$ it is a *trisectrix* (Figure 1.19). The nature of all these curves is described in the table of curves given in Appendix III.

If $r > k$ the source is inside the circle and the limaçon is a single loop. If $r < k$ the limaçon forms two loops which are the result of reflection in the respective parts of the circle, convex or concave towards the source.

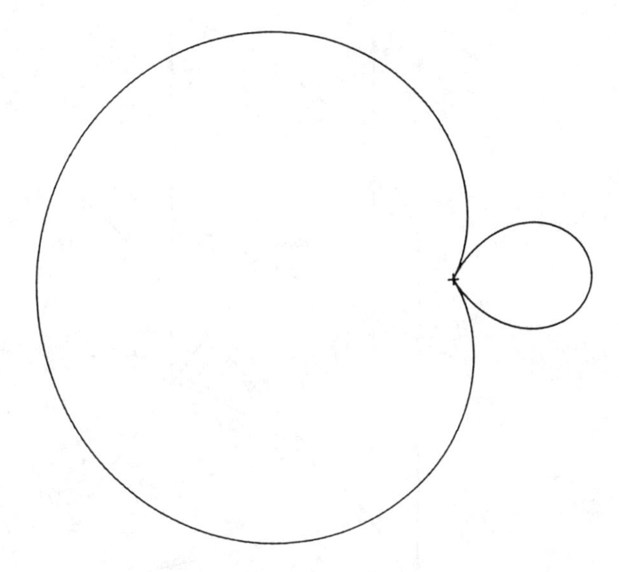

Figure 1.19: Trisectrix

REFLECTION IN A GENERAL CURVE

For a reflector of any shape whose profile is the curve $f(x,y) = 0$, we can obtain the zero-distance phase front for a source at any location (d,h) by direct means. We erect a cylindrical reflector with generators parallel to the z axis from the cross-section (Figure 1.20). Then a ray from the source S to the point (a,b,c) will, after reflection, intersect the plane $z = 0$ in a point P, on the zero-distance phase front, and independently of c. The direction of the ray can be derived directly from the vector form of the law of reflection (Appendix I.3)

$$\widehat{\mathbf{S}}_r = \widehat{\mathbf{S}}_i - 2\hat{\mathbf{n}}(\widehat{\mathbf{S}}_i \cdot \hat{\mathbf{n}}) \tag{1.2.14}$$

with unit vectors shown in Figure 1.20. Since the normal $\hat{\mathbf{n}}$ at (a,b,c) is given by ∇f, then P has coordinates

$$\begin{aligned} x &= d + \frac{\partial f}{\partial x} A/B \\ y &= h + 2\frac{\partial f}{\partial y} A/B \\ A &= \left[(a-d)\frac{\partial f}{\partial x} + (b-h)\frac{\partial f}{\partial y}\right] \\ B &= \left[\left(\frac{\partial f}{\partial x}\right)^2 + \left(\frac{\partial f}{\partial y}\right)^2\right] \end{aligned} \tag{1.2.15}$$

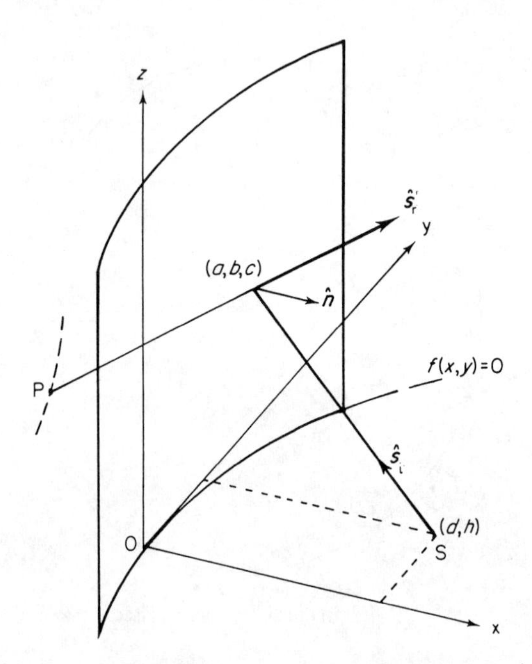

Figure 1.20: Zero-distance phase front of a general reflecting surface

where all the differentials are evaluated at the point (a, b). Since (a, b) also satisfics $F(a, b) = 0$, it is possible to eliminate them from Equation 1.2.15 and derive an implicit relation for the zero-distance phase front.

As an example we shall derive by this method the zero-distance phase front of a parabola with the extension required to deal with the possibility of having the source at an infinite distance. That is the zero-distance phase front for plane wave sources. We know, of course, that for the parabola with its focusing property, this will be a circle, strictly of infinite radius, centred on the focal point.

Using Equation 1.2.15 for the parabola with focal length a

$$f(x, y) = y^2 - 4a^2 + 4ax = 0$$

and a source on the straight line $x = 0, y = h$, the curve of intersection with the plane $z = 0$ is

$$F(x, y, h) = [y(x - 2a) - 2ha]^2 - x^2h^2 - (x - 2a)^2 = 0 \qquad (1.2.16)$$

This, incidentally, is the true curve of the phase front for an offset feed in a paraboloid and its digression from a straight line exhibits the cause of coma aberration in optical systems [4].

For an incident *plane* wave we require the envelope of this equation with varying h, that is we need to eliminate h between Equation 1.2.16 and the equation $\partial F/\partial h = 0$. This results in $h = 2ay/(2a+x)$ which, substituted back into Equation 1.2.16 gives the circle.

$$x^2 + y^2 = 4a^2$$

This radius is *not*, however, infinite, since we have not derived the zero-distance phase front but the phase front due to zero phase along the line $y = h$. For circular phase fronts, zero distance becomes a matter of defining the radius only.

The method illustrates that a phase front in general, and a zero-distance phase front in particular, can be obtained as the *envelope* of phase fronts from a moving point source. This is thus the process required to obtain the zero-distance phase fronts in those cases where the source is a *caustic*, or another proposed phase front. This envelope procedure, however, does *not* apply to the reflecting or refracting surfaces themselves.

We note also that Equations 1.2.6 take on a particularly simple form for a reflector when $\eta = -1$. Then if the reflector is given parametrically by

$$x = f(t) \qquad y = g(t)$$

the zero-distance phase front is

$$x = 2(fg'^2 - gf'g')/(f'^2 + g'^2)$$
$$y = 2(gf'^2 - ff'g')/(f'^2 + g'^2) \tag{1.2.17}$$

These coordinates are double in value to those of the point N in Figure 1.7, which is the foot of the perpendicular from S onto the tangent at M, as would be expected in the case of a reflector. The locus of N is defined to be the (first positive) *pedal* curve of g with respect to S.

Hence the zero-distance phase front of a reflector can be derived directly from the pedal curve of the reflecting profile. With a source at the point (d, h) and the parametric form of the reflector profile this construction, and Equation 1.2.6 give the zero-distance phase front

$$x = \frac{df'^2 + 2fg'^2 + 2(h-g)f'g' - dg'^2}{f'^2 + g'^2}$$
$$y = \frac{hg'^2 + 2gf'^2 + 2(d-f)f'g' - hf'^2}{f'^2 + g'^2} \tag{1.2.18}$$

the primes, as always, referring to differentiation with respect to the parameter t.

A table of pedal curves is included in the geometrical constructions given in Appendix III.

1.3 GEOMETRICAL CONSTRUCTIONS

We shall require, for future operations with zero-distance phase fronts, some geometrical results which are standard but are included for completeness.

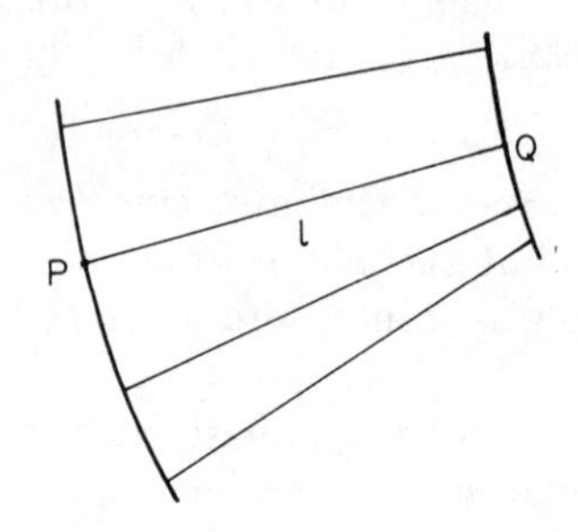

Figure 1.21: Construction of the parallel of a curve PQ = 1

Most of the details, and the tables in Appendix III are derived from Lawrence (1972) [5].

PARALLEL SURFACES

As already stated, the theorem of Malus and Dupin states that the propagating wave, or phase front in a uniform medium, is everywhere normal to the congruence of rays. Since the zero-distance phase front is a source of such a congruence, all ensuing phase fronts are "parallel" in the geometrical sense. That is, points on each surface are equidistant where the distance is measured along the common normal between points on each of the surfaces. If P is on a phase front whose profile is given parametrically by $f(t)$, $g(t)$, then $Q(x, y)$ will be on a parallel phase front at a distance l (Figure 1.20) given by

$$x = f(t) \pm lg'/\sqrt{f'^2 + g'^2}$$
$$y = g(t) \pm lf'/\sqrt{f'^2 + g'^2} \qquad (1.3.1)$$

We can see by inspection that the only form-invariant phase fronts are the sphere and the plane. The same relation applies when considering Huygens' construction for propagating wavefronts in a uniform medium.

INVERSION

The radius vector through the origin O, which is the *centre of inversion*, intersects a curve C at a point P (Figure 1.22). The point Q on OP such that

$$\mathrm{OP} \cdot \mathrm{OQ} = \kappa^2$$

is the inverse point of P with radius of inversion κ. The locus of Q as P moves on C is the inverse curve of C.

It is also important to define the negative inverse

$$\mathrm{OP} \cdot \mathrm{OQ} = -\kappa^2$$

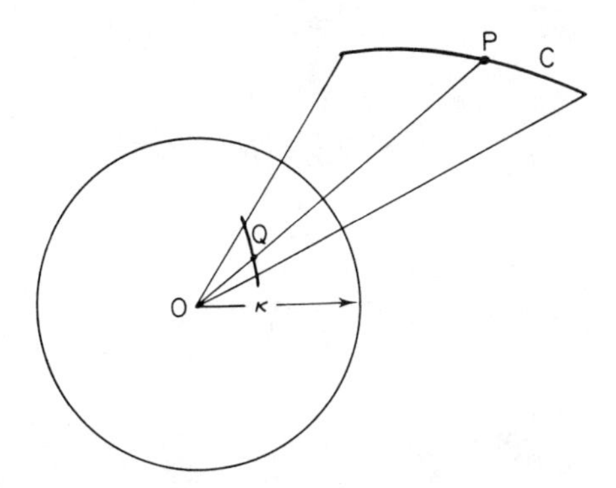

Figure 1.22: Construction of the inverse of a curve OP . OQ = κ^2

in which case P and Q are taken to be on opposite sides of O.

If O is the pole of a polar curve $r = f(\theta)$, the inverse with respect to the pole is given by

$$Rr = \kappa^2$$

or

$$R = \kappa^2 / f(\theta)$$

θ remaining invariant in this process.

If O is the point (a, b) and the curve C is given parametrically by $x = f(t)$, $y = g(t)$, the equation of the inverse curve $Q(x, y)$ is

$$\begin{aligned} x &= a + \frac{\kappa^2[f(t) - a]}{[f(t) - a]^2 + [g(t) - b]^2} \\ y &= b + \frac{\kappa^2[g(t) - b]}{[f(t) - a]^2 + [g(t) - b]^2} \end{aligned} \tag{1.3.2}$$

Under the inversion the angles between curves are unaltered and the mapping is *conformal.* Circles not passing through the centre of inversion invert into circles, and a circle passing through the centre inverts into a straight-line tangent to the circle at the centre of inversion. Similarly asymptotes invert into tangents of the inverse curve at the centre of inversion. Some curves such as the spirals are self-inverse and are termed *anallagmatic*. The inverse of an inverse in the same centre and with the same radius of inversion returns the original curve.

For inversion when the centre of inversion is at infinity, a requirement we shall later use, we adopt the convention that inversion is a reflection in *any* local plane whose normal is in the direction of the point at infinity specified as the centre of inversion.

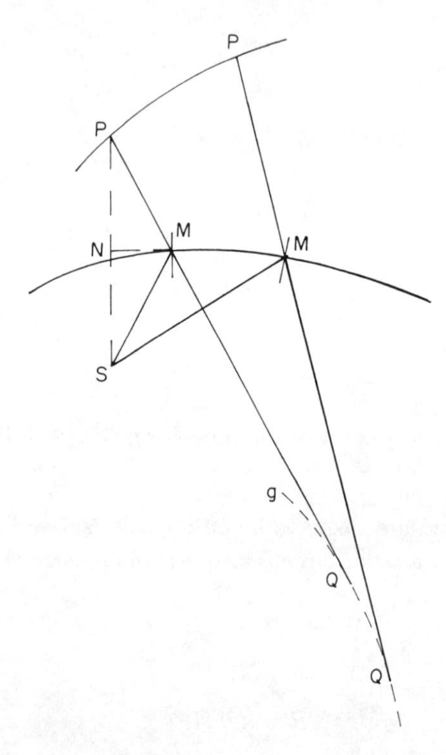

Figure 1.23: Construction of the catacaustic of a curve. Q is the centre of curvature of the curve at P

CAUSTICS EVOLUTES AND INVOLUTES

The caustic given by a curve C is the envelope of the rays emitted from the source S after reflection in the curve, the *catacaustic* or, after refraction, the *diacaustic*. From Figure 1.23, P is the reflection of S in the tangent to the surface at M for the case of a reflection, and MQ = MP. Since MQ is the reflected ray, it is normal to the locus of P and thus Q is the centre of curvature of the zero-distance phase front. The locus of the centre of curvature of a curve is the evolute of the curve. Thus Q describes the *caustic by reflection* in g, which at the same time is the evolute of the zero-distance phase front.

The evolute of a curve with parametric definition $x = f(t)$, $y = g(t)$ is

$$x = f(t) - (f'^2 + g'^2)g'/(f'g'' - f''g')$$
$$y = g(t) + (f'^2 + g'^2)f'/(f'g'' - f''g') \qquad (1.3.3)$$

A further derived curve from a given curve C is obtained from the motion of a point P fixed to the line L which rolls without slipping along the curve C. The same curve is obtained as if P were fixed to a taut inextensible string

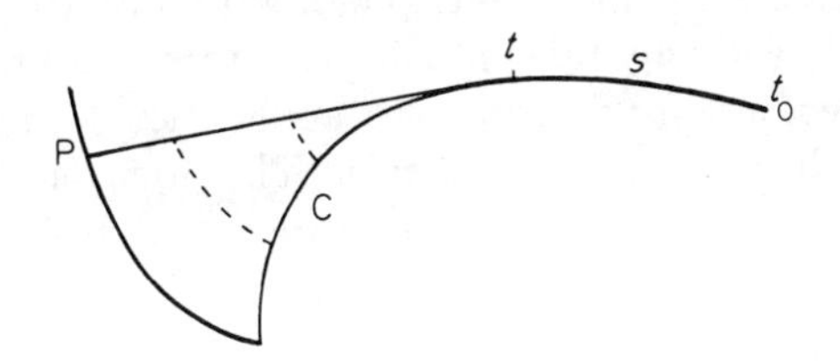

Figure 1.24: Construction of the involute of the curve C. Ptt_0 is a taut string attached to C at t_0 and wound about it to t

attached to a point on C and winding or unwinding over the curve, as shown in Figure 1.24. The part of the string forming the tangent to the curve replaces the line L of the first description. The locus of P is then the *involute* of the curve C. Different positions of the point P or different positions of the point of fixing of the string to the curve give involutes, every involute described in this manner being a parallel of every other involute. This indicates the connections to be found between involute curves and phase fronts.

If the fixed curve C has parametric definition $x = f(t)$, $y = g(t)$, and the tangent from P meets the curve at a point that has parameter t, then the equation of the involute is

$$X = f(t) - sf' \Big/ \sqrt{f'^2 + g'^2}$$

$$Y = g(t) - sg' \Big/ \sqrt{f'^2 + g'^2} \tag{1.3.4}$$

where s is the arc length of that part of the string lying along the curve from the fixed point to the point of tangency t. That is

$$s = \int_{t_0}^{t} \sqrt{f'^2 + g'^2}\, \mathrm{d}t \tag{1.3.5}$$

If a curve A is the involute of a curve B, then B is the evolute of A, and vice versa. Since the caustic is the evolute of the zero-distance phase front, the zero-distance phase front is the involute of the caustic. This gives a further method for its description, since caustics can be obtained as the envelope of reflected or refracted rays. Thus, since rays are tangential to a caustic, they are normal to the involute of the caustic, which is therefore a phase front.

The description of an involute to a caustic has to take into consideration the double-valuedness of the ray pattern on the illuminated side of the caustic. The commonest observed caustic is that of the plane wave incident upon a circular reflector (Appendix III). Since at each point on the illuminated side two tangents can be drawn to the caustic, each being a ray path, then any involute or phase front, which is nominally perpendicular to each ray, will likewise have two values at that point. The method of drawing the involute shown in Figure 1.24 will only take into account those rays which after reflection are single-valued, and for any practical application the aperture has to be limited to stop off the second pencil of rays. The point of attachment of the unwinding string then decides the nature of the local region of the involute.

EXAMPLES

To illustrate these points and the application of the table in Appendix III, we can derive the phase fronts for the illumination by a parallel beam of rays of some interesting curved reflectors.

First we shall consider a reflector with a profile given by the curve $y = \ln x$. The process for deriving the caustic is to obtain the envelope of the reflected rays. From Figure 1.25 the following relations apply

$$\tan\theta = \frac{dy}{dx} = \frac{1}{x}$$

The equation of the reflected ray

$$Y - y = \tan 2\theta (X - x)$$

then becomes

$$F(\theta) = Y - \log\cot\theta - \tan 2\theta (X - \cot\theta) = 0$$

and for the envelope

$$\frac{\partial F}{\partial \theta} = 1 - X\sin 2\theta = 0$$

Thus

$$\frac{dY}{dX} = \tan 2\theta = 1 \Big/ \sqrt{X^2 - 1}$$

so that

$$X = \cosh(Y + 1)$$

the constant being obtained from the reflection of the ray $y = 0$ into the ray $x = 1$.

Thus the caustic is the catenary and the phase front the involute of the catenary which is the tractrix. This is double-branched, the upper half showing a "real" phase front, that is after the rays have been tangential to the catenary, and the lower half a "virtual" phase front, occurring as it does before the ray has reached the caustic.

In Figure 1.26 the catacaustic of the cycloidal arch with parameter a is shown to be the two equal cycloidal arches as detailed in the tables in Appendix III. The involute of a cycloidal arch is an equal cycloidal arch, and hence the phase front is the latter arch as shown.

In the case of the logarithmic spiral both the catacaustic and its involute, the phase front, are equal logarithmic spirals, as illustrated in Figure 1.27.

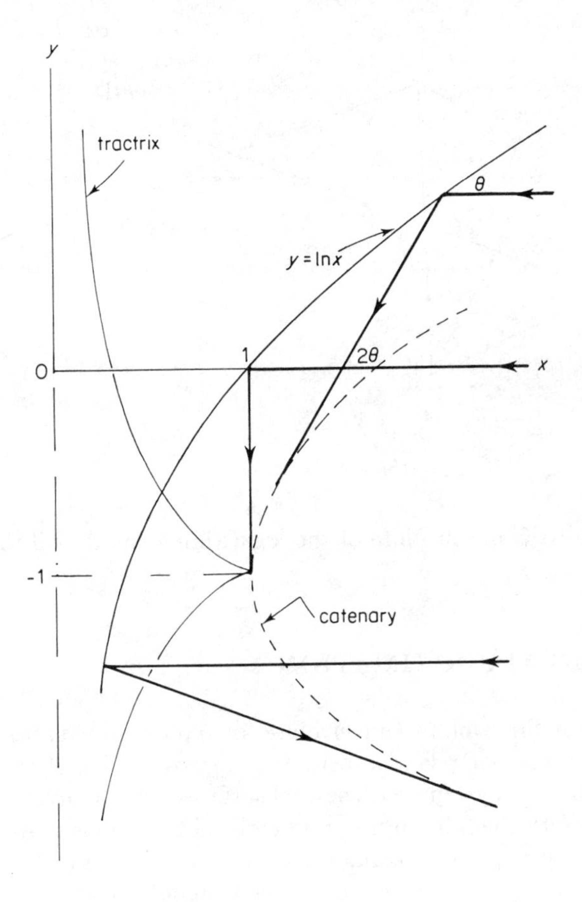

Figure 1.25: Caustic and involute for the curve $y = \ln x$

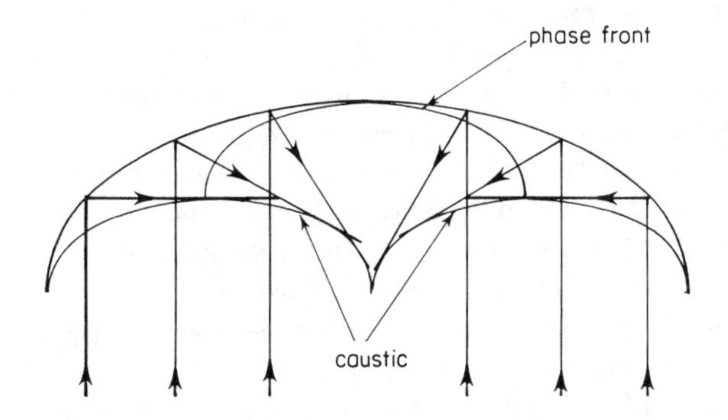

Figure 1.26: Caustic and involute of the cycloidal arch. Both caustic and involute are cycloidal arches with half parameter of the reflecting arch

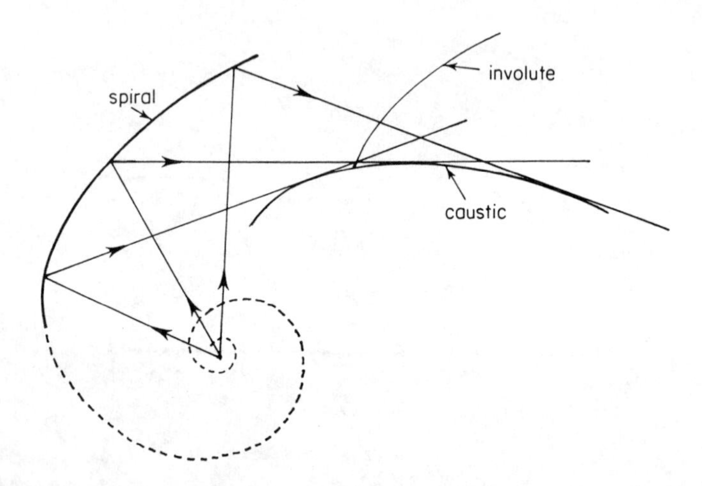

Figure 1.27: Caustic and involute of the logarithmic spiral. All three curves are equal spirals

1.4 CAUSTIC APPROXIMATION

The properties of the evolute and involute as applied to caustics and zero-distance phase fronts give rise to a method of approximation of great value in tracing fronts through systems with several surfaces. The analysis to date has been concerned with the full range of rays emitted by a point source. In practical instances only a specific angular range of rays will need to be considered, and in the paraxial case only a very small angle about the axis of the system. Hence only a small part of the curves concerned, phase fronts or caustics, will in general need to be used, and most often in the region about the axis of symmetry. Thus it will be possible to replace phase fronts over this region by simpler curves, curves that can be used to continue the process through to the next surface. Similarly, caustics can be replaced by simpler curves or even in some cases by a point source. In every case the approximation made by such replacements exactly specifies the amount of aberration *being introduced by the approximation.* The divergence of phase fronts from true circles (or planes if focused at infinity) or caustics from true points exactly specifies the amount of aberration inherent in the optical system.

The method for approximating a caustic to an equivalent point is, in fact, to derive the involute, that is the phase front, and approximate that to a circle. The centre of the circle will then be the best equivalent point for the caustic. Of course there are numerous ways to approximate the curve to the circle – a best mean-square fit, for example. In the examples chosen here we simply fit the circle to three arbitrary selected points on the phase front.

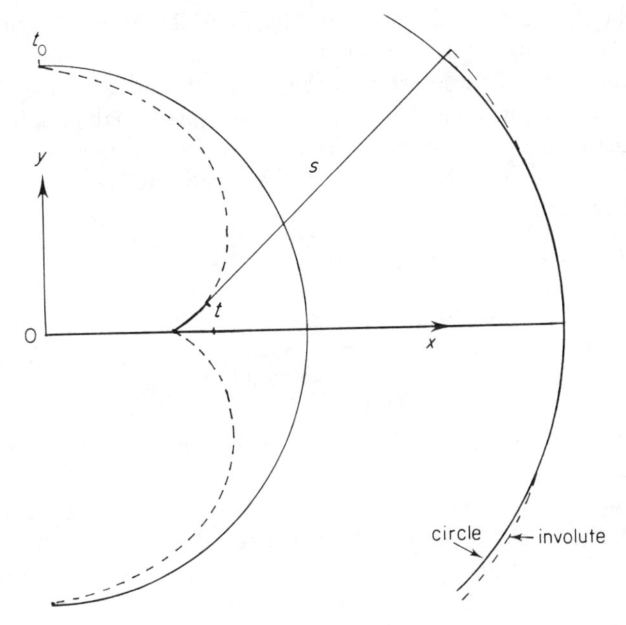

Figure 1.28: Approximation of the phase front, the involute of the caustic of a circular mirror by a circle

We take first the circular reflector illuminated by a parallel beam of rays (Figure 1.28). The caustic is the well-known cusp of the nephroid (Appendix III.1) with parametric equations

$$\begin{aligned} x &= a(3\cos t - \cos 3t) \equiv f(t) \\ y &= a(3\sin t - \sin 3t) \equiv g(t) \end{aligned} \tag{1.4.1}$$

Without loss of generality a can be taken to be unity; the radius of the circle is then 4. The cusp corresponds to $t = 0$ and the extrema on the y axis to $t = \pm\pi/2$, $y = 4$.

We then have from Equation 1.3.5

$$s = 6(\cos t_0 - \cos t)$$

where t_0 is the starting point for the unwinding string forming the involute. The parametric equation of the involute, that is of the phase front, is then

$$\begin{aligned} X &= 3\cos t - \cos 3t - 3(\sin 3t - \sin t)(\cos t_0 - \cos t)/\sin t \\ &= 3\cos t - \cos 3t + 6\cos 2t(\cos t - \cos t_0) \\ Y &= 3\sin t - \sin 3t - 3(\cos t - \cos 3t)(\cos t_0 - \cos t)/\sin t \\ &= 3\sin t - \sin 3t + 6\sin 2t(\cos t - \cos t_0) \\ & t_0 \geq t \end{aligned} \tag{1.4.2}$$

Care has to be taken with the signs of the final terms on the right-hand side owing to the ambiguity in the square-root of the denominators of Equations 1.3.4. The value of t_0 determines the range around the cusp for which the approximation is being sought. If we take the entire (half) nephroid then $t_0 = \pi/2$. Three points on the curve of Equation 1.4.2 can then be taken to be $t = 0$ and $\pm\pi/4$. Inserting these values we obtain the following three points on the true phase front

$$(8, 0) \quad (2\sqrt{2}, 4\sqrt{2}) \quad (2\sqrt{2}, -4\sqrt{2})$$

The centre of the circle through these points is on the x axis at the point

$$x = \frac{3\sqrt{2}}{2\sqrt{2}-1} \simeq 2.32$$

which is not the half radius usually chosen as the source point for a circular reflector.

The difference between the exact (computable) curve of Equation 1.4.2 and the circle *with this centre* expresses the spherical aberration of the reflector. Different circles are obtained if the range of the comparison is reduced by taking values of t closer to the central value $t = 0$.

We use the same method to determine the best position for a source point in a parabola to give a parallel beam of rays at some angle θ to the axis [6][7][14]. The caustic is Tschirnhausen's cubic (Figure 1.29)

$$\begin{aligned} f(t) &= -bt(t^2 - 3)/9 \\ g(t) &= bt^2/3 \end{aligned} \tag{1.4.3}$$

where $b = 9a \sin\theta$ and a is the focal length.

This curve has been referred to oblique axes which are the symmetry axes of the cubic (Figure 1.30). The illuminated portion of the paraboloid creates a region of this caustic around the parametric value $t = 0$. The value of t_0 depends on the geometry of the paraboloid; for a full focal plane paraboloid this is $t_0 = \pm 1$. For a paraboloid subtending a total angle 2Θ at the focus, t_0 is approximately $\pm\tan(\Theta/2)$.

The procedure given above leads to the parametric equation to the involute

$$\begin{aligned} X &= \frac{4bt^3 + bt_0(3 + t_0^2)(1 - t^2)}{9(1 + t^2)} \\ Y &= \frac{bt[t^3 - 3t + 2t_0(3 + t_0^2)]}{9(1 + t^2)} \end{aligned}$$

Taking a full focal plane paraboloid $t_0 = -1$, then values $t = 0$ and $t = \pm 1$ encompass the entire derived part of the caustic (not essentially the best choice but the simplest). The three points lie on a circle with centre (in the oblique coordinate system)

$$(b/21, -11b/126)$$

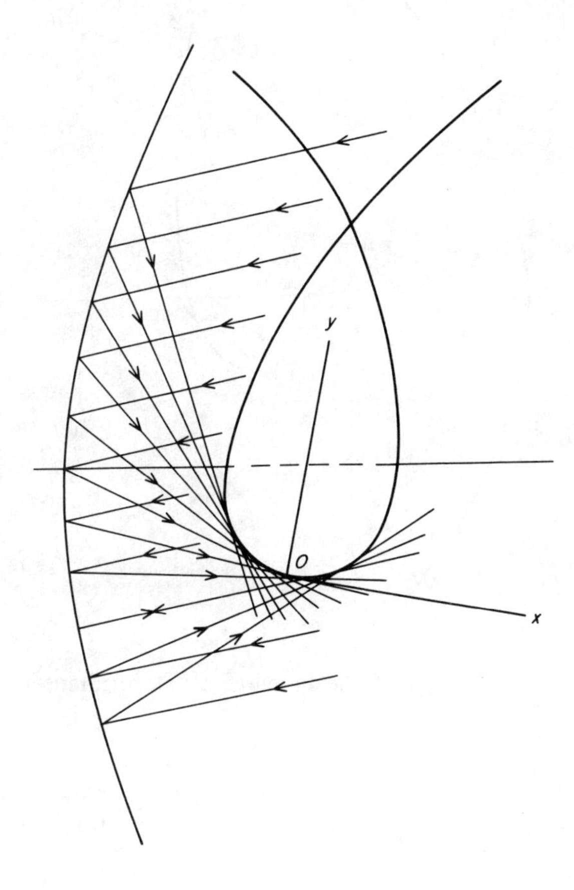

Figure 1.29: Caustic of asymmetrically illuminated parabola. Tschirnhausen's cubic

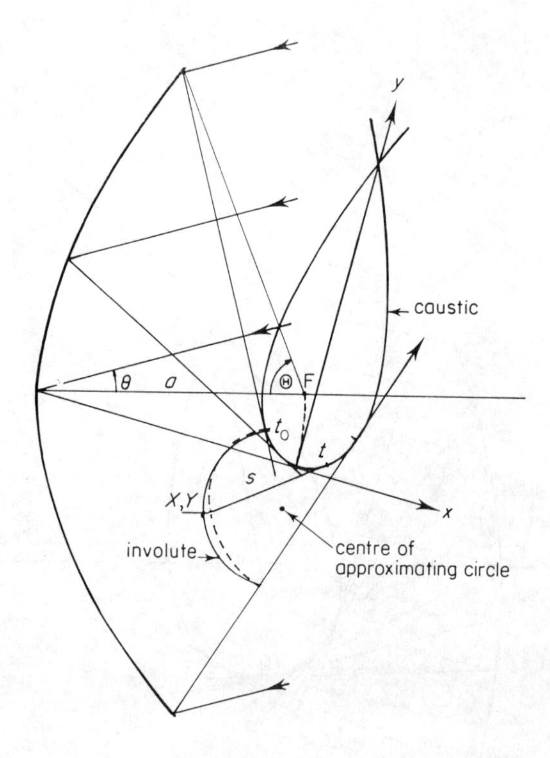

Figure 1.30: Approximation of the involute of Tschirnhausen's cubic by a circle

In these examples we have, of course, found a general phase front, but since involutes are parallel curves of the zero-distance phase front, the approximations to each will have the same centre.

1.5 THE INVERSION THEOREM OF DAMIEN

A major reason for establishing, in previous sections, a particular phase front, the phase front of zero distance, and examining its geometry, is that a unique property pertains to that phase front. This property was first explicitly enunciated by Damien [13] in a small monograph, although there are indications that a theorem of a similar nature was in the mind of Hamilton, and he refers to it as a continuation of a study by Cauchy. In its basic form the inversion theorem states (Figure 1.31)

> Given a surface g and its zero-distance phase front h with respect to a source S, then an inversion in a circle centred on S will invert g

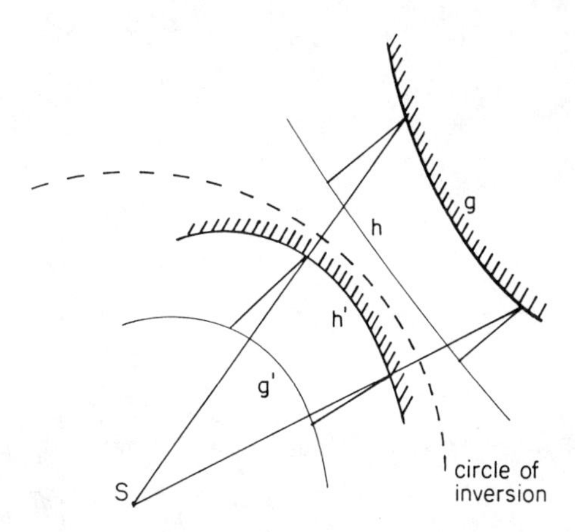

Figure 1.31: The inversion theorem of Damien. h is the zero-distance phase front of g. After inversion g′ is the zero-distance phase front of h′

> into g′ and h into h′, such that g′ is the zero-distance phase front of h′ with respect to a source at S and for the same refractive index.

It applies both to refracting and reflecting surfaces. It is a most unusual transformation in that it transforms a phase front, fixed in time by making it of zero-distance, into a refracting surface which is fixed in space. It is one of the very rare transformations that exist that completely transforms a given optical system, the first refraction, into a second optical system, purely geometrically.

Damien himself gives a purely geometrical proof based on the similarity of the quadrilaterals SPMT and SM′P′T′ in Figure 1.32. As can be seen the points P and M are transposed in this description making a refracting point M into a phase front point P. The proof given here is due to MC Jones [14]

Refer to Figure 1.32 and put $\eta_2/\eta_1 = 1/m > 1$. The construction of the zero-distance phase front gives

$$(1 - m^2)r^2 - 2rr' \cos(\gamma - \gamma') + r'^2 = 0 \tag{1.5.1}$$

Snell's law of refraction at M gives

$$(1 - m^2)r(\cos\gamma + \sin\gamma \tan\psi) = r'(\cos\gamma' + \sin\gamma' \tan\psi) \tag{1.5.2}$$

Also

$$\begin{aligned} \frac{\mathrm{d}r}{r\,\mathrm{d}\gamma} &= \frac{\cos\gamma + \sin\gamma\tan\psi}{\cos\gamma\tan\psi - \sin\gamma} \\ &= \frac{r'\sin(\gamma' - \gamma)}{r(1 + m^2) - r'\cos(\gamma - \gamma')} \end{aligned} \tag{1.5.3}$$

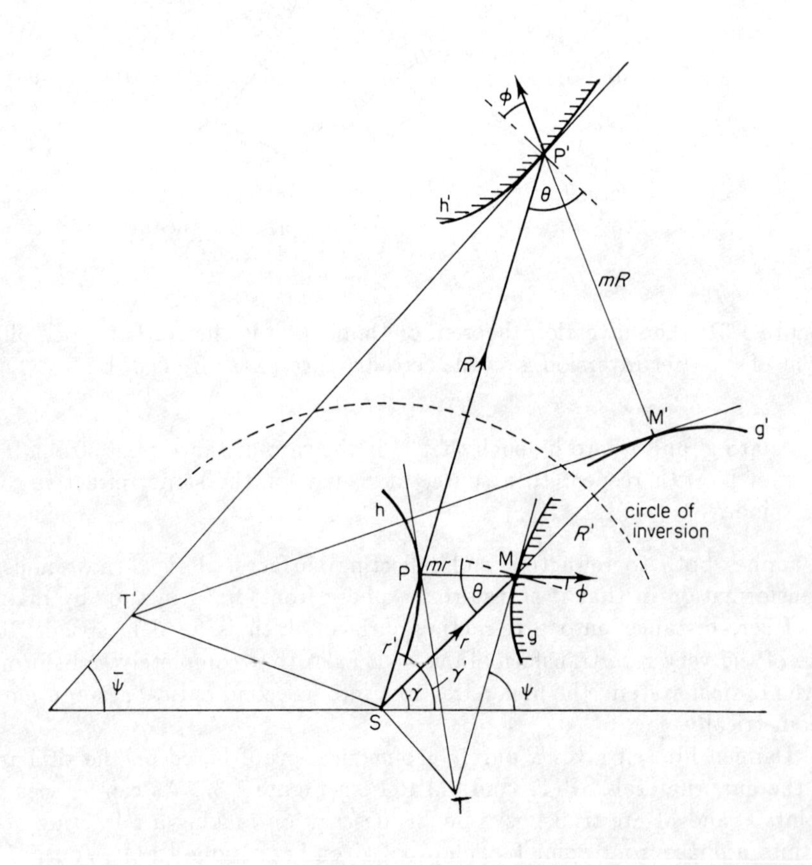

Figure 1.32: Proof of the inversion theorem. Note that the quadrilaterals SPMT and SM′P′T′ are similar

from Equation 1.5.2.

On differentiating Equation 1.5.1 with respect to γ and applying Equation 1.5.3 we obtain

$$[r\cos(\gamma-\gamma')-r']\frac{\mathrm{d}r'}{\mathrm{d}\gamma}+rr'\sin(\gamma-\gamma')=0 \tag{1.5.4}$$

Equations 1.5.1 and 1.5.4 are required to be invariant under the inversion

$$r\to\frac{1}{R'}\qquad r'\to\frac{1}{R}\qquad \gamma\leftrightarrow\gamma'$$

that is primed (phase front) coordinates are transformed into unprimed (refractive) coordinates, and vice versa. Without loss of generality the radius of inversion has been taken to be unity.

Equation 1.5.1 is invariant as it stands.

For the refraction at P′ we must have (cf. Equation 1.5.2)

$$(1-m^2)R(\cos\gamma'+\sin\gamma'\tan\bar{\psi})=R'(\cos\gamma+\sin\gamma\tan\bar{\psi}) \tag{1.5.5}$$

where

$$\begin{aligned}\tan\bar{\psi} &= \left(\cos\gamma'+\sin\gamma'\frac{\mathrm{d}R}{R\,\mathrm{d}\gamma'}\right)\Big/\left(-\sin\gamma'+\cos\gamma'\frac{\mathrm{d}R}{R\,\mathrm{d}\gamma'}\right)\\ &= \left(-\cos\gamma+\sin\gamma\frac{\mathrm{d}R'}{R'\,\mathrm{d}\gamma}\right)\Big/\left(\sin\gamma+\cos\gamma\frac{\mathrm{d}R'}{R'\,\mathrm{d}\gamma}\right)\end{aligned} \tag{1.5.6}$$

On substituting this result into Equation 1.5.5 and transforming the resultant equation, we have

$$(1-m^2)\frac{r\,\mathrm{d}r'}{r'\,\mathrm{d}\gamma'}=\sin(\gamma'-\gamma)+\cos(\gamma'-\gamma)\frac{1}{r'}\frac{\mathrm{d}r'}{\mathrm{d}\gamma'}$$

which, using Equation 1.5.1, reproduces Equation 1.5.4.

For the parabola of focus S and directrix d (Figure 1.33), rays from S reflected in the parabola are perpendicular to d which is thus the "wave front of emergence". Then the wave front of emergence for the inverse of d with respect to S, that is a circle passing through S, is the inverse of the parabola in the same circle. This is a cardioid with cusp at S. letting D be the foot of the perpendicular from S to d and putting SD $=2f$ the equation of the parabola is

$$r=\frac{2f}{1+\cos\theta}$$

and the equation of the vertical line is

$$r=\frac{2f}{\cos\theta}$$

Making an inversion of magnitude $4f$, the line d becomes the circle $r=2f\cos\theta$ and the parabola the cardioid $r=2f(1+\cos\theta)$. The cardioid is then the zero-distance wave surface for a source on the surface of a circular reflector.

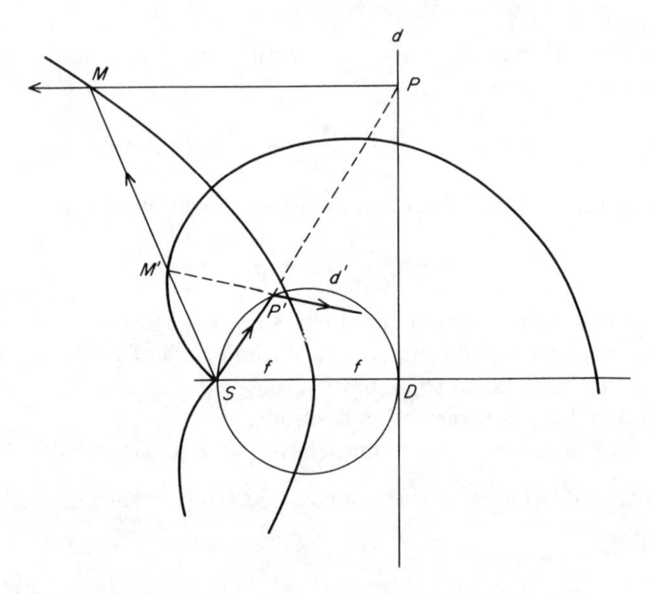

Figure 1.33: Damien's theorem for the parabola

Damien's theorem can, in an obvious way, be used to obtain the zero-distance phase front of a surface from the known zero-distance phase front of another surface. Thus for reflection in a circle, the situation considered in Section 1.2 REFLECTION IN A CIRCLE, we require a reflector that has a known circular zero-distance phase front. For this we take the ellipse with source at one focus (Figure 1.34), for which the zero-distance phase front is a circle centred on the second focus. The source is thus interior to the circle. The inversion in a circle centred on the source transforms the circular phase front into the circular reflector, and the elliptical reflector into the required phase front, which is the limaçon of Pascal without cusp, from the table in Appendix III. Variation of the radius of inversion or utilization of the concept of negative inversion provides for the other situations of source and reflector given in Section 1.2 REFLECTION IN A CIRCLE.

It is of interest to compare this result with the symmetry between surfaces and wavefronts illustrated at the end of Section 1.2 REFLECTION IN A CIRCULAR INTERFACE. The two instances, (a) and (b) of Figure 1.15, can be seen to be the result of an inversion, if the centre of inversion is taken to be the source at infinity, that is, a reflection. The limaçon and (eccentric) circle combination shown in Figure 1.15(c), is a direct application of Damien's theorem to the ellipse combination in Figure 1.15(b), with the centre of inversion this time at the source at the focus of the corresponding ellipse. Thus the ellipses in each case invert into limaçons of Pascal, and the plane surfaces into spheres.

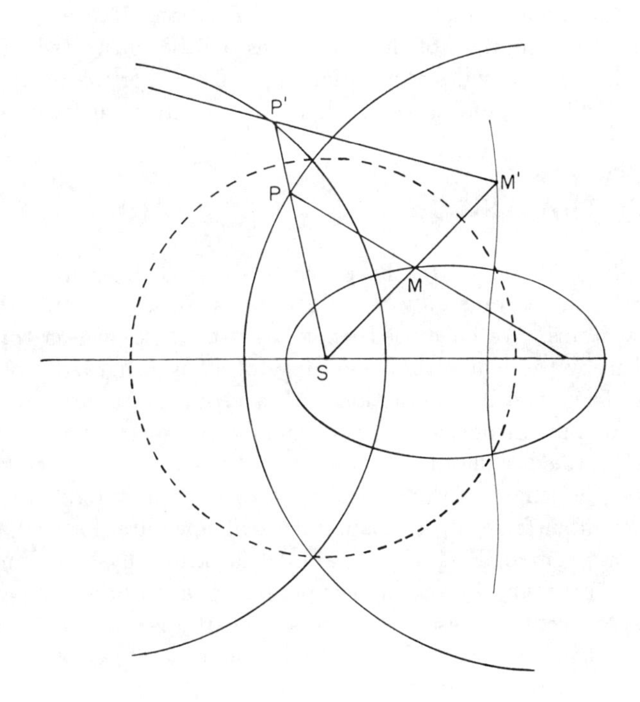

Figure 1.34: Application of the inversion theorem to derivation of zero-distance phase fronts

Therefore applying Damien's theorem to the Cartesian Oval, the circular phase front inverts into a (concave) circular refractive interface and the Cartesian oval inverts into a phase front which must therefore be a limaçon as was previously shown.

As has been stated already, the zero-distance phase front is the evolute of the caustic of the associated reflection or refraction. Hence for applications that require the definition of the caustic, as will be illustrated subsequently, Damien's theorem provides the initial step. The other major application of the theorem will be found in the design of two or more surface systems.

1.6 RAY TRACING

The procedures discussed in the previous sections differ in a fundamental manner from those commonly used in the design of light optical, generally focusing, systems. We have used what has become known as "optics in the large", whereby we deal with the entirety of a ray congruence or its associated phase front and its interaction with a geometrical surface. The design technique for optical devices consists mainly of ray tracing which will also play a part in the design of microwave optical devices. Ray tracing consists of establishing the length of the optical path through a system which can have any number of surfaces, up to a specified exit aperture, for each of the rays issuing from a source. Each ray is parametrized, usually by its angular position at the source, and by varying the parameter across the complete angle of rays at the source, the phase error across the exit aperture can be determined with respect to a base ray. This is usually the central ray of the system. In light optics this phase error determines the focal properties of the system, but commonly in the microwave optics of antennas, the rays are expected to be collimated and the phase error is the departure of the path length of the rays from that of the base ray, over the exit aperture. In optics the distribution of path lengths over the rays from the source is given by Hamilton's characteristic function or eikonal (see Appendix I). This is simplified by virtue of the collimation property required of microwave devices, but the general theory remains applicable. As with light optics, the aperture phase distribution is the required function to be entered into the aperture diffraction integral (Chapter 5) to give the first order approximation to the radiation pattern of the microwave antenna (or the distribution of the field about the nominal focus in light optics). The phase error in both cases is the function that requires minimisation, and the criteria differ for the two instances. Because of the considerably larger wavelength involved in the microwave field, phase errors up to a fair fraction of a wavelength can be tolerated depending on the specified purity of the ensuing radiation pattern. With such errors permitted, it means that at the aperture a proportion of the rays are travelling in a direction not exactly that required for the main beam of the antenna. These are the cause of the loss of forward power or loss of "aperture efficiency" and heightened sidelobe radiation.

In situations where these are not as important as having the definition given by a narrow confined angle for the main beam, systems may be devised where the ray direction plays only a secondary part to the derivation of a uniform phase. That is an aperture can be specified where the rays, although travelling in various directions, achieve across an exit surface an equality of phase. Such devices are known and will be demonstrated in the appropriate place (the Axicon for example). I have given them the generic title of Huygens' lenses, or reflectors as appropriate, since the propagation from the phase distribution created over the aperture is dependent on the principle going by that name.

We shall see in the next chapter the fundamental part this plays in the design of two surface systems. We can illustrate this procedure here once again in relation to the simple spherical reflector. As is well known, such a mirror suffers from the first order aberration namely spherical aberration. This is a quadratic function over the aperture and hence monotonically increases from the centre toward a maximum at the edge. However, given a fixed aperture angle at the source, the additional constraint of making the length of the edge ray equal to that of the central ray, can be applied. This in turn both fixes the position of the source and creates a phase distribution which is zero, both at the centre and at the edge thus thwarting the inherent spherical aberration. We take a spherical cap of radius 2 and depth d, as shown in Figure 1.35. The source S is a distance X from the apex of the reflector and the diameter containing it is taken to be the x axis. Then it is a simple geometrical exercise to calculate the length of any ray making the angle α with the axis from the source up to the exit aperture AA$'$.

The ray through S making the angle α with the axis meets the circle radius r at the point

$$x = r\cos^2\alpha + X\sin^2\alpha - \cos^2\alpha\sqrt{r^2 + 2rX\tan^2\alpha - X^2\tan^2}$$
$$y = (X - x)\tan\alpha \tag{1.6.1}$$

and after reflection, the aperture plane at the height

$$y' = y + (d - x)\tan(2\beta - \alpha)$$

where $\sin\beta = (2 - X)\sin\alpha/2$. The path length of each ray is then

$$E = \sqrt{(X - x)^2 + y^2} + \sqrt{(d - x)^2 + (y' - y)^2} \tag{1.6.2}$$

Taking $r = 2$ and a source angle of ± 45 degrees, the edge ray will equal the central ray when $X = d(3 + 2\sqrt{2})$ at which point the ray will have the length given in Equation 1.6.1 with $\alpha = 45^o$. The solution then is that $d = 0.164516$ and $X = 0.95887$. The error E with these values is shown in Figure 1.36. It can be seen that the curve is contained within ± 0.0002 of the

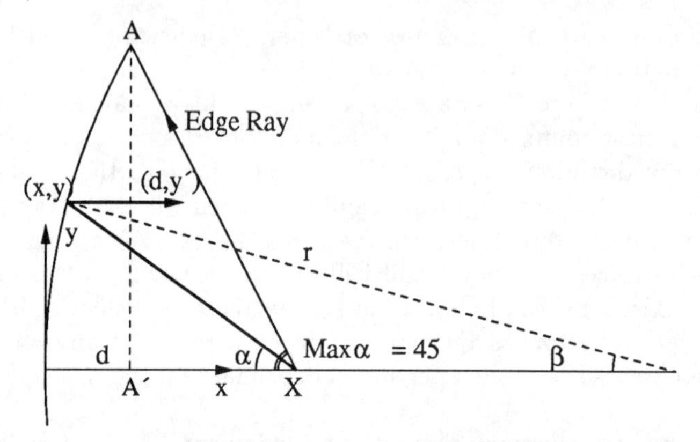

Figure 1.35: Spherical cap reflector

radius. Hence given a specified phase tolerance the size of reflector for which this is a suitable simplified antenna can be specified. It can be noted that the position of the source for this method is in close agreement with the position that a spherical cap of the same dimensions would require by the method of caustic approximation (Section 1.4) $X = 0.954$ in that case.

The mechanical advantage that such a cylindrically symmetrical reflector has in the way of beam rotation is obvious. The technique of creating equal path lengths for a central ray and two other edge rays, in the anticipation that the error for intermediate rays will be within a specified tolerance, is one that proves to be of great value in subsequent wide angle antenna designs.

Furthermore, by the construction alone, the incident and transmitted rays intersect on the circle of the reflector itself. As will be shown later (Appendix I) this is the Abbe condition for minimum coma when the source is moved transversely to the axis. Hence a degree of beam offsetting can be achieved from this combination.

1.7 POWER TRANSFORMATION

The second property that is required of microwave antennas is the ability to radiate the available power into a specified angular distribution. This takes the form of a transformation between the power available in a solid angle $d\Omega$ at the source and, after passing through the optical system, the power in the transmitted solid angle $d\Omega'$ (Figure 1.37) defined over a large "far-field" sphere with centre in the locality of the source position. If, as is commonly the case, it is the latter distribution which is specified, then this transformation constitutes the single available degree of freedom which specifies the reflector surface. It is therefore an alternative to the requirements of ray tracing as the specified condition.

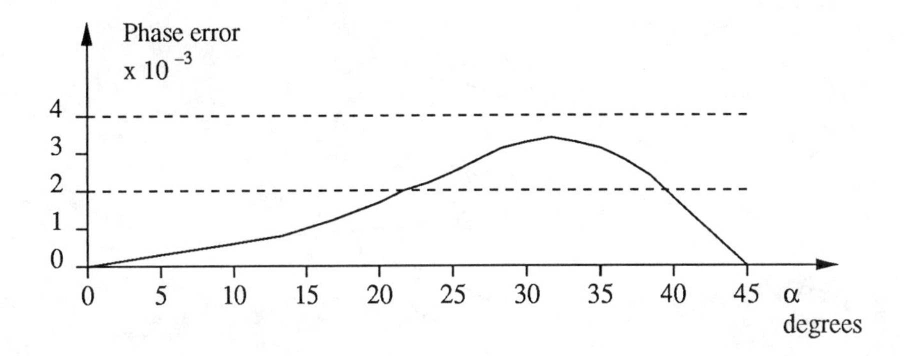

Figure 1.36: Phase error for spherical cap reflector

The transformation itself is an application of the principle of the conservation of energy, in this case in the form of the flux contained in a tube of rays undergoing geometrical optics reflection or refraction at the surface involved. The methods therefore apply to the problems of obtaining the aperture power distribution or the near field distribution from an already specified system of reflectors or lenses with a known source power distribution, or as a design degree of freedom applying to the derivation of the single surface that will convert a given source distribution into a specified far field power distribution.

We treat the three cases, the linear or cylindrical reflector, the axisymmetric or spherical reflector and the general surface as they increase in degree of complexity. In certain circumstances the linear case can be considered to be the cross-sectional profile of a more complex reflecting surface.

CYLINDRICAL AND SPHERICAL REFLECTORS

The standard method for obtaining the profile of a cylindrical reflector giving a specified shaped radiation pattern from a line source with a known characteristic has not been improved upon and is included here, yet again, for completeness [15][16]

If the line source has a radiation pattern per unit length given by $P_1(\psi)$ and the required secondary pattern is to be $P_2(\theta)$ then the assumption made is that all the energy within any angular range contained within the source distribution will be contained in a corresponding angular range within the

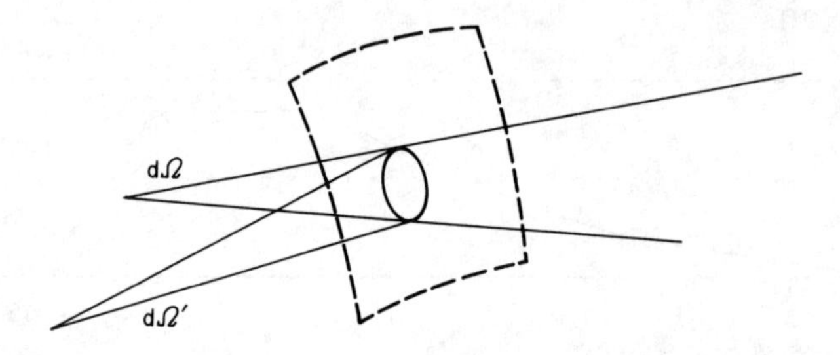

Figure 1.37: Source to far field transformation

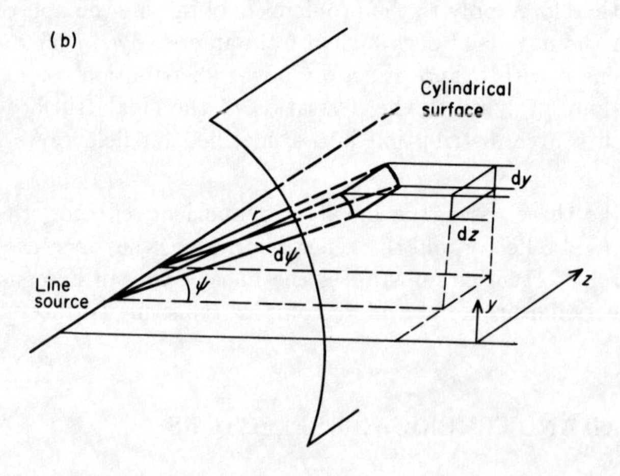

Figure 1.38: Conservation of flux for cylindrical systems

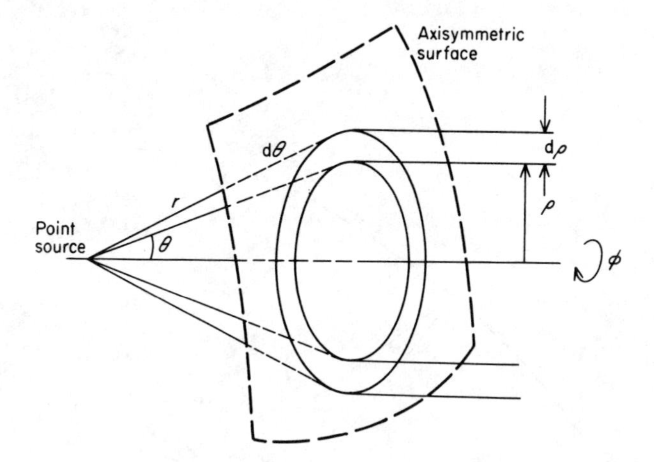

Figure 1.39: Conservation of flux for spherical systems

radiated distribution. This expresses itself as

$$\frac{\int_{\psi_0}^{\psi} P_1(\psi)\,\mathrm{d}\psi}{\int_{\psi_0}^{\psi_1} P_1(\psi)\,\mathrm{d}\psi} = \frac{\int_{\theta_0}^{\theta} P_2(\theta)\,\mathrm{d}\theta}{\int_{\theta_0}^{\theta_1} P_2(\theta)\,\mathrm{d}\theta} \tag{1.7.1}$$

where ψ and θ are any values in the range $[\psi_0, \psi_1]$ and $[\theta_0, \theta_1]$ respectively, which define the limits of the incident radiation and those of the radiated pattern. By graphical or other means [15] (Equation 1.7.1) can be used to derive a relationship between θ and ψ, say $\theta = g(\psi)$.

If the reflector has a curved profile shown as the curve AB in Figure 1.40(a) and is illuminated by rays from a source at the origin then from the geometry of the figure we have

$$\beta = \beta' - \pi/2$$
$$\beta' = \alpha + \psi$$

and

$$2\alpha = \theta - \psi$$

In terms of a Cartesian geometry the reflector will have a profile given by $y = f(x)$ then

$$\tan\beta = f'(x) = -\cot\left(\frac{\psi + \theta}{2}\right) \tag{1.7.2}$$

Alternatively, if the profile is given in polar coordinates (r, ψ) then from the law of reflection (Figure 1.40(b)) we have

$$\tan\alpha = \mathrm{d}r/r\mathrm{d}\psi$$

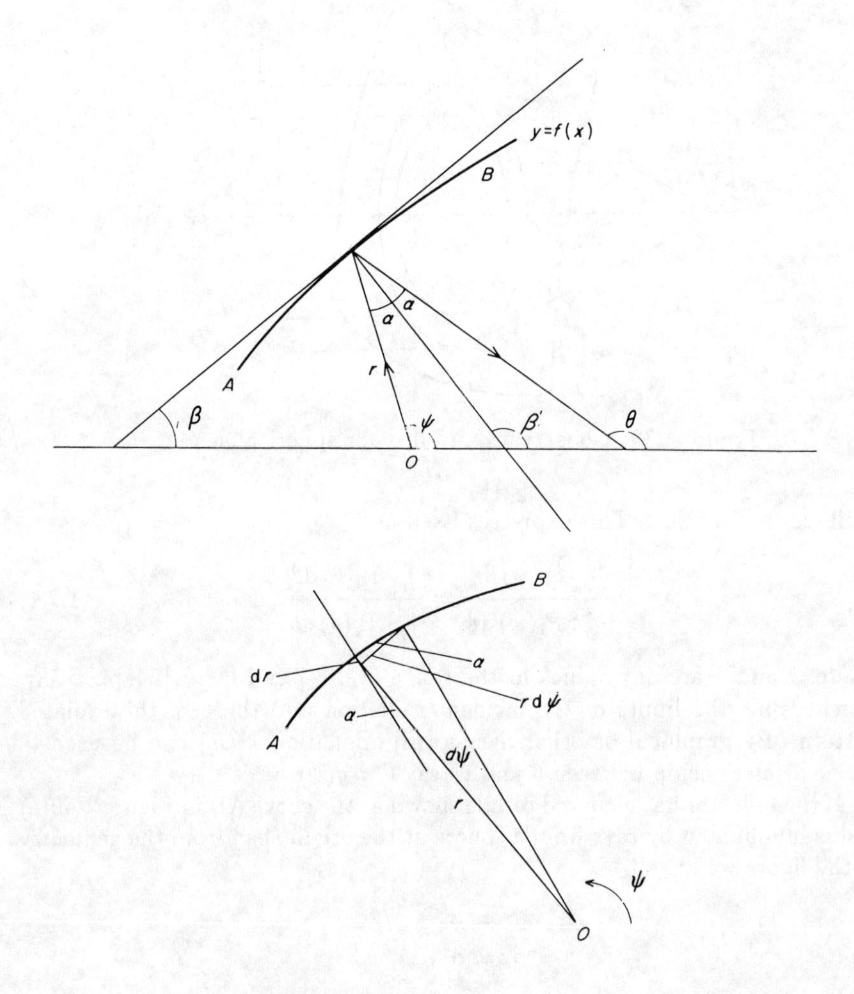

Figure 1.40: Differential geometry of a reflector

and hence

$$\frac{\mathrm{d}r}{r} = \tan\left(\frac{\theta - \psi}{2}\right) \mathrm{d}\psi \tag{1.7.3}$$

If θ can be obtained explicitly as a function $g(\psi)$ of ψ Equation 1.7.3 can be integrated numerically to give

$$\log(r/r_0) = \int_{\psi_0}^{\psi} \tan\frac{1}{2}(g(\psi) - \psi)\,\mathrm{d}\psi \tag{1.7.4}$$

where (r_0, ψ_0) refers to the starting point A of the process.

In the case of a cylindrical system we consider an element $\mathrm{d}z$ of the entire reflector and its line source then (Figure 1.38)

$$\mathrm{d}\Omega' = \mathrm{d}\psi\,\mathrm{d}z \quad \text{and} \quad \mathrm{d}\Omega = \mathrm{d}y\,\mathrm{d}z$$

Hence the power per unit length of the system in the plane aperture is

$$P_1(y) = \frac{\mathrm{d}\psi}{\mathrm{d}y} P_2(\psi) \tag{1.7.5}$$

where $P_2(\psi)$ is the power distribution per unit length of the line source. If the incident and refracted or reflected rays intersect on a surface given by $r = f(\psi)$ then since $y = r\sin\psi$

$$P_1(y) = P_2(\psi)/(f(\psi)\cos\psi + f'(\psi)\sin\psi) \tag{1.7.6}$$

For a circular profile this becomes

$$P_1(y) = P_2(\psi)/(a\cos\psi)$$

In the case of an axially symmetric system with near field collimation of the rays [16, p391] (Figure 1.39)

$$\begin{aligned} \mathrm{d}\Omega' &= \sin\theta\;\mathrm{d}\theta\,\mathrm{d}\phi \\ \mathrm{d}\Omega &= \rho\;\mathrm{d}\rho\,\mathrm{d}\phi \end{aligned}$$

and hence the final radial power distribution is given by

$$P_1(\rho) = \frac{P_2(\theta)\sin\theta\,\mathrm{d}\theta}{\rho\;\mathrm{d}\rho} \tag{1.7.7}$$

where $P_2(\theta)$ is the source distribution. Because of this factor $P_2(\theta)$, it is impossible to express $P_1(\rho)$ explicitly as a function of ρ. If the surface in which the transmitted rays meet their corresponding incident rays, that is the reflector surface in a single reflector system, has a cross-section given by $r = f(\theta)$, then since $\rho = r\sin\theta$ Equation 1.7.7 can be rewritten

$$P_1(\rho) = \frac{P_2(\theta)}{f^2(\theta)\cos\theta + f(\theta)f'(\theta)\sin\theta}$$

For a spherical cap reflector, $f(\theta)$ is constant $= a$ and the second term in the denominator is zero leaving

$$P_1(\rho) = \frac{P_2(\theta)}{a^2 \cos\theta} \tag{1.7.8}$$

For a paraboloid, $r = 2f/(1+\cos\theta)$; uniform illumination $P_1(\rho) =$ constant is obtained when $P(\theta) \propto (1+\cos\theta)^{-1}$ that is when the *amplitude* distribution $\sqrt{|P(\theta)|}$ is proportional to $\sec^2(\theta/2)$.

THE GENERAL SCATTERING OF WAVES BY SURFACES

The totally general case has been studied in both the optical and microwave antenna contexts. The connecting mathematical process is the use by both of the differential geometry of surfaces to connect the incident tube of rays or its intersection with a reflecting or refracting element with the radiated tube of rays or its intersection with a receiving surface.

A full survey of the principles of differential geometry involved is beyond the scope of this volume. The results of these studies will therefore be summarized so that their application to microwave antennas can be appreciated. Burkhard and Shealy [17] give the conservation principle in the form

$$F(\mathrm{d}S_1 \to \mathrm{d}S_2) = \rho\sigma \cos\phi_i \, \mathrm{d}S_1/\mathrm{d}S_2 \tag{1.7.9}$$

in which $F(\mathrm{d}S_1 \to \mathrm{d}S_2)$ is the flux per unit area on the receiving surface $\mathrm{d}S_2$ which corresponds to the reflector or refractor surface element $\mathrm{d}S_1$ (Figure 1.41), ρ is the reflectance or transmittance of the element (and could be a function of ϕ_i if required) and σ is the flux density of the incident beam. The directions of the tubes are obtained by the laws of reflection or refraction. The application of differential geometry then provides the method of obtaining the ratio $\mathrm{d}S_1/\mathrm{d}S_2$ required in Equation 1.7.9 and the results of this analysis are presented in the appendix.

It has two main applications to the requirements of microwave antenna design. Firstly if the "receiving surface" be the plane exit aperture of the optical system and σ the flux density of a non-isotopic source, aperture distributions appropriate to the scalar method could be obtained. Alternatively if the "receiving surface" be a surface through or near to the expected geometrical optics focus, the caustic of an incoming plane wave could be derived. A "shaped beam" can be constructed from a variety of such incoming plane waves and a composite feed system with sources at the caustics would then create a specific radiation pattern.

The law of reflection at P is, in terms of unit vectors (Appendix I.3)

$$\hat{\mathbf{y}} = \hat{\mathbf{r}} - 2\hat{\mathbf{y}}(\hat{\mathbf{r}}.\hat{\mathbf{n}}) \tag{1.7.10}$$

or

$$\mathbf{N} = r\hat{\mathbf{y}} - \mathbf{r} \tag{1.7.11}$$

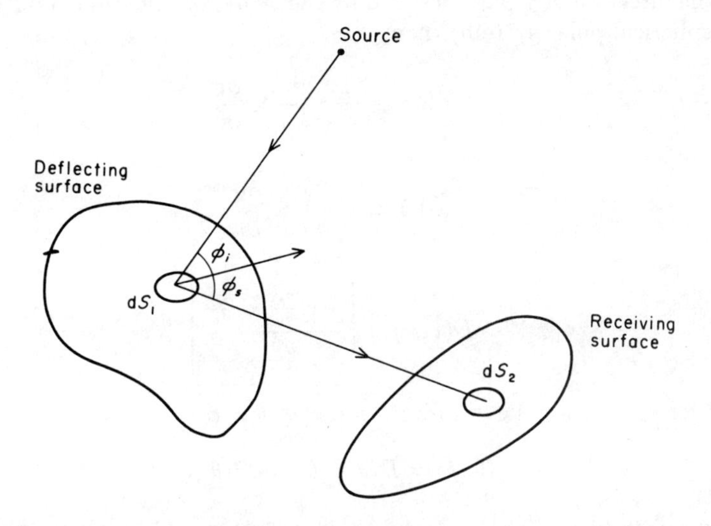

Figure 1.41: Geometry of general source deflector and receiver in scattering theory

where $\mathbf{r}$ is the position vector of the point P on the reflector with respect to the origin of rays at O, and $\mathbf{N}$ is a non-unit vector in the direction of the normal at P.

In the case of refraction

$$\hat{\mathbf{y}} = \frac{\eta_i}{\eta_t}\hat{\mathbf{r}} + \left(\cos\theta_t - \frac{\eta_i}{\eta_t}\cos\theta_i\right)\hat{\mathbf{n}} \tag{1.7.12}$$

If the surface is given parametrically by

$$\mathbf{r} = x(u,v)\hat{\imath} + y(u,v)\hat{\jmath} + z(u,v)\hat{\boldsymbol{k}}$$

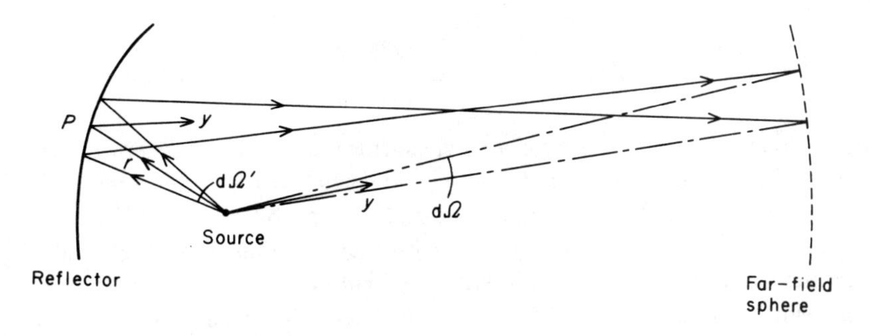

Figure 1.42: Geometry for far field of a reflector

and the direction of $\hat{\mathbf{y}}$ parametrized by the *same* coordinates (say $u = \theta$ $v = \phi$ in a spherical polar system) then

$$\mathrm{d}\Omega = -\frac{1}{r^3}\mathbf{r}.\frac{\partial \mathbf{r}}{\partial u} \times \frac{\partial \mathbf{r}}{\partial v} \tag{1.7.13}$$

and

$$\mathrm{d}\Omega' = \hat{\mathbf{y}}.\frac{\partial \hat{\mathbf{y}}}{\partial u} \times \frac{\partial \hat{\mathbf{y}}}{\partial v} \tag{1.7.14}$$

and hence

$$D(u, v) = \left| \frac{-\mathbf{r}.\frac{\partial \mathbf{r}}{\partial y} \times \frac{\partial \mathbf{r}}{\mathrm{d}v}}{r^3 \hat{\mathbf{y}}.\frac{\partial \hat{\mathbf{y}}}{\partial u} \times \frac{\partial \hat{\mathbf{y}}}{\partial v}} \right| \tag{1.7.15}$$

Then for spherical polar coordinates $u = \theta$, $v = \phi$

$$P(\theta, \phi) = D(\theta, \phi, \theta', \phi', P(\theta', \phi')) \tag{1.7.16}$$

The problem of synthesis is to find the surface $r(\theta, \phi)$ satisfying both Equations 1.7.11 and 1.7.16. The theory is obviously adaptable to analysis where $r(\theta, \phi)$ is known in advance.

For the reflector geometry shown in Figure 1.42, $\mathrm{d}\Omega$ in Equation 1.7.13 is given by $\mathrm{d}\Omega = \sin\theta$ and θ, ϕ are related to θ', ϕ' through the Jacobian of the transformation $J(\theta, \phi, \theta', \phi')$ or

$$J(\theta, \phi, \theta', \phi') = \begin{vmatrix} \partial\theta'/\partial\theta & \partial\theta'/\partial\phi \\ \partial\phi'/\partial\theta & \partial\phi'/\partial\phi \end{vmatrix} = \frac{\partial\theta'}{\partial\theta}\frac{\partial\phi'}{\partial\phi} - \frac{\partial\theta'}{\partial\phi}\frac{\partial\phi'}{\partial\theta} \tag{1.7.17}$$

Then

$$P(\theta', \phi')\sin\theta' \left[\frac{\partial\theta'}{\partial\theta}\frac{\partial\phi'}{\partial\phi} - \frac{\partial\theta'}{\partial\phi}\frac{\partial\phi'}{\partial\theta}\right] = P(\theta, \phi)\sin\theta \tag{1.7.18}$$

For reflectors limited in aperture angular size, $0 < \theta < \theta_M$, $0 < \phi < 2\pi$ and a similar limitation on the transmitted power distribution $0 < \theta' < \theta'_M$, $0 < \phi' < 2\pi$

$$\int_0^{2\pi}\int_0^{\theta'_M} P(\theta', \phi')\sin\theta'\,\mathrm{d}\theta'\,\mathrm{d}\phi' = \int_0^{2\pi}\int_0^{\theta_M} P(\theta, \phi)\sin\theta\,\mathrm{d}\theta\,\mathrm{d}\phi$$

The relations formulated here for a single surface and a specified power transformation become quite general for *any* incident field $P(\theta, \phi)$ and transmitted field $P(\theta', \phi')$ provided the intermediate surfaces can be connected by a mapping $\theta, \phi \Rightarrow \theta', \phi'$ through the interior geometry of the reflectors and the application of additional constraints in the form of required criteria.

In formal terms we will obtain θ' and ϕ' as a function of θ and ϕ plus all the available parameters, leading to highly complicated Jacobians. Even in the present single surface case, we have, because of the definition of the

(Equation 2.3.7 for example)

$$\theta' = \theta'\left(\theta, \phi, r, \frac{\partial r}{\partial \theta}, \frac{\partial r}{\partial \phi}\right)$$
$$\phi' = \phi'\left(\theta, \phi, r, \frac{\partial r}{\partial \theta}, \frac{\partial r}{\partial \phi}\right) \qquad (1.7.19)$$

Hence the Jacobian will require the complete partial differentials

$$\frac{\vartheta\theta'}{\vartheta\theta} = \frac{\partial\theta'}{\partial\theta} + \frac{\partial\theta'}{\partial r}\frac{\partial r}{\partial\theta} + \frac{\partial\theta'}{\partial(\frac{\partial r}{\partial\theta})}\frac{\partial^2 r}{\partial\theta^2} + \frac{\partial\theta'}{\partial(\frac{\partial r}{\partial\phi})}\frac{\partial^2 r}{\partial\theta\partial\phi}$$
$$\frac{\vartheta\theta'}{\vartheta\phi} = \frac{\partial\theta'}{\partial\phi} + \frac{\partial\theta'}{\partial r}\frac{\partial r}{\partial\phi} + \frac{\partial\theta'}{\partial(\frac{\partial r}{\partial\theta})}\frac{\partial^2 r}{\partial\theta\partial\phi} + \frac{\partial\theta'}{\partial(\frac{\partial r}{\partial\phi})}\frac{\partial^2 r}{\partial\phi^2}$$

and similar terms for $\vartheta\phi'/\vartheta\theta$ and $\vartheta\phi'/\vartheta\phi$. Substitution of this Jacobian in lieu of that in the bracketed term in Equation 1.7.18 leads to a second order non-linear differential equation of the form

$$A\frac{\partial^2 r}{\partial\theta^2} + B\frac{\partial^2 r}{\partial\theta\partial\phi} + C\frac{\partial^2 r}{\partial\phi^2} + D\left[\frac{\partial^2 r}{\partial\theta^2}\frac{\partial^2 r}{\partial\phi^2} - \left(\frac{\partial^2 r}{\partial\theta\partial\phi}\right)^2\right] + E = 0 \qquad (1.7.20)$$

with A B C D and E functions of θ ϕ r $\partial r/\partial\theta$ and $\partial r/\partial\phi$.

This is a (highly) elliptic Monge-Ampere equation [19] by virtue of the existence of the term in the brackets. The reason this solution is so much more complicated than previous expressions is of course that we are dealing here with the total surface and not simply the profile of a cross-section. However the solution of the equation by numerical means requires the establishing of a mesh of points over which the required surface is developed, which in principle is not much different from creating a total surface by cross-sections taken at closely angularly spaced "cuts". We shall be returning to this problem in our discussion of the general two surface system.

1.8 MAGNIFICATION AT A SPHERICAL SURFACE

We consider here the effect on a pencil of rays of reflection from a spherical surface. In subsequent applications, the results will be applied to a more generally curved surface replacing the radius of the spherical surface by the local radii of curvature of the more general one. The equations are therefore presented in terms that do not include the relative position of the source and the centre of the sphere. Figures 1.43(a) and (b) show a spherical surface illuminated by a pencil of rays from a source at the origin from the convex side and the concave side respectively. The infinitesimal pencil is a ray tube $\mathrm{d}\theta\,\mathrm{d}\phi$ which is reflected into a tube $\mathrm{d}\alpha\,\mathrm{d}\beta$ whose dimensions we wish to determine. The axial (ϕ = constant) and the transverse (θ = constant) situations are dealt with independently. Then as shown in the figures, for the axial case

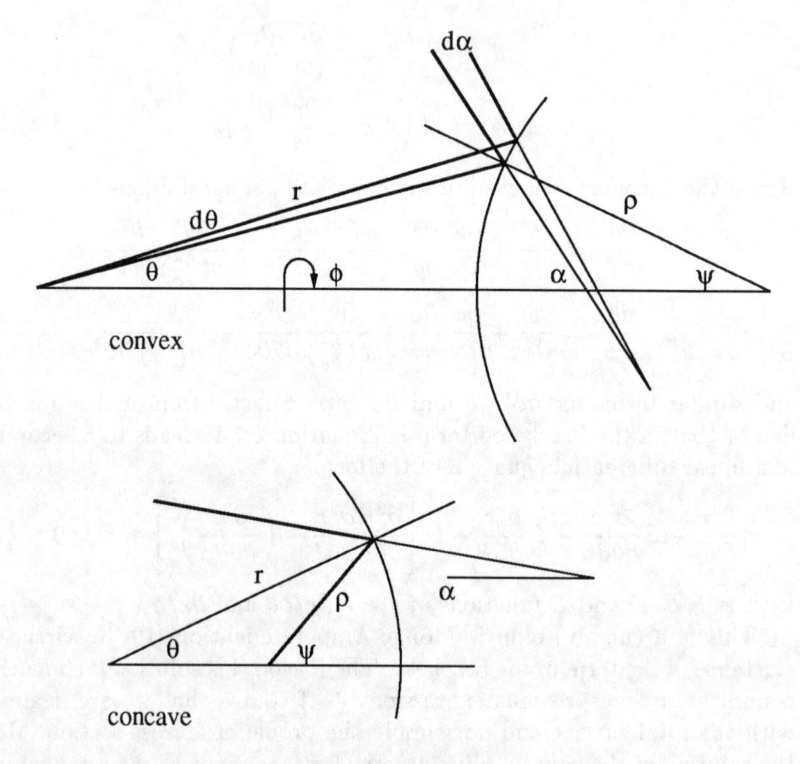

Figure 1.43: Reflection of rays from a sphere; lateral rays ϕ is constant

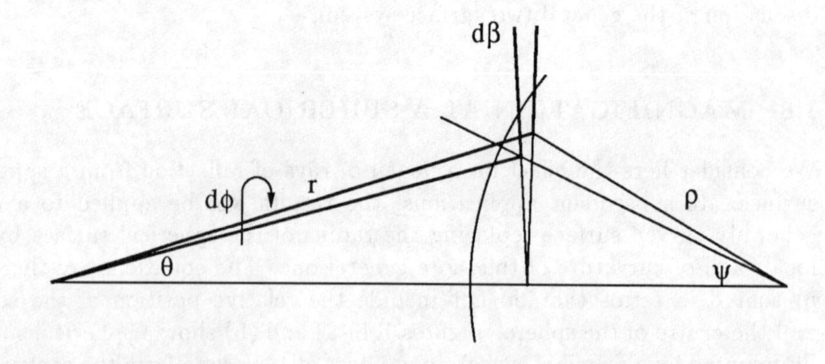

Figure 1.44: Reflection of rays from a sphere; transverse rays θ is constant

$$d\alpha/\,d\theta = 2\,d\psi/\,d\theta + 1$$

for the convex condition, and

$$d\alpha/\,d\theta = 2\,d\psi/\,d\theta - 1 \tag{1.8.1}$$

for the concave. The geometry gives

$$r/\sin\psi = \rho/\sin\theta$$

and

$$(\rho\,d\psi)^2 = (r\,d\theta)^2 + (\,dr)^2 \tag{1.8.2}$$

Solving for the ratio $d\psi/\,d\theta$ we obtain

$$\frac{d\psi}{d\theta} = \frac{\sqrt{k^2 - (1-k^2)\tan^2\theta} - \tan^2\theta}{k^2 - \tan^2\theta} \tag{1.8.3}$$

where $k = \rho/r$. In practical situations that permit a fair approximation to be made, we find that within 10% in the range of angles permitted (by a positive square root when appropriate)

$$d\psi/\,d\theta \approx 1/k \qquad 1/2 < k < 2 \text{ (exact when } k = 1) \tag{1.8.4}$$

Thus the "expansion" of the tube in a plane of constant ϕ is $2/k + 1$ or $2/k - 1$ for the convex and concave situations respectively. The degree of approximation is demonstrated for the value $k = 2$ which implies a source located at the node of the nephroid caustic (Appendix III) at the mid point of the radius in the concave situation. The resulting beam of rays is not a parallel beam as the approximation implies, but has, of course, spherical aberration and a small spread dx (≤ 0.1).

It is also to be observed that the two reflected rays constituting $d\alpha$ do *not* in general intersect on the line joining the origin of the rays to the centre of the sphere. On the other hand, the two transverse rays are in planes making the angle $d\phi$, which intersect along that axis. Consequently after reflection they must intersect at a point on that line. As the geometry of Figure 1.44 shows, the arc intercept on the sphere has length $\rho\sin\psi\,d\phi$ and hence $d\beta/\,d\phi = \sin(2\psi+\theta)$ for the convex reflector and $d\beta/\,d\phi = \sin(2\psi-\theta)$ for a concave one. In both cases

$$\sin\psi = \sin\theta/k \qquad k = \rho/r \tag{1.8.5}$$

The conclusion is that, even for a "simple" surface such as a sphere, a pencil of rays will, on reflection, be distorted in shape through differential magnification in the lateral and transverse planes and also no longer issue from a single point.

For more general curved surfaces, the same considerations apply. We can replace the sphere by the lines of curvature on the surface, one of which will apply to the lateral reflection and the other to the transverse. In this fashion an heuristic appreciation of the action of the surface on a pencil of rays can be made. However in going from the pure sphere to an irregular surface the angle θ has to have a more general interpretation. For the sphere it was the angle between the ray r, θ and the axis joining the ray source to the centre of the sphere. For the more general case, it has to be replaced by an angle, θ' say, between the ray and the line joining the source to the *centre of curvature*, and thus gives rise to a second angle, θ'' say, for the transverse (and different) radius and centre of curvature. More importantly this effect can be continued for more than one surface and can be applied incrementally to cover whole surfaces of quite general shape. For example a *local* condition in the transverse-concave reflection will allow $d\beta/d\phi$ to equal zero, and thus give a parallel beam. Similarly with two reflectors, in the plane $\phi =$ constant, $d\gamma/d\theta = (d\gamma/d\alpha)(d\alpha/d\theta) \approx (2/k_2 - 1)(2/k_1 - 1) = 0$; k_1 and k_2 referring to the curvatures ρ/r and P/R of the two reflectors (assumed to be concave as shown). There is no anomaly in the values k_1 or k_2 being equal to 2 since in the first instance the relation is an approximation only, and secondly it has to be remembered that the radii of curvature ρ and P are to be measured from the intersection of adjacent rays. Thus if one value is 2 the radius of curvature of the second surface will be infinite, (in the approximation considered here), a situation where one reflector is spherical and the second plane.

REFERENCES

1. Brueggemann HP (1968) *Conic Mirrors* The Focal Press.

2. Stavroudis ON (1972) *The Optics of Rays Wavefronts and Caustics,* p97. New York and London: Academic Press.

3. Sussmann MM (1966) Maxwell's ovals and the refraction of light. *Am Jour Phys* **34** p416.

4. Linfoot EH (1955) *Recent Advances in Optics* p51 Oxford University Press

5. Lawrence JD (1972) *A Catalog of Special Plane Curves,* p115 Dover (ISBN 0-486-60288-5).

6. Scarborough JB (1964) The caustic curve of an off-axis parabola *Applied Optics* **3** p1445

7. Stalzer HJ (1965) Comment on the caustic curve of a parabola *Applied Optics* **4** p1205

8. Cornbleet S (1979) Feed position for the parabolic reflector with offset pattern *Electron Lett* **15** p211

9. Herzberger M (1958) *Modern Geometrical Optics,* p152 NewYork: Interscience Publishers.

10. Eaton JD (1952) Zero phase fronts in microwave optics. *Trans IRE* **AP-1** p38.

11. Stavroudis ON (1969) Refraction of wavefronts: a special case. *Jour Opt Soc Amer* **59** p114.

12. Salmon G (1960) *Higher Plane Curves* New York: Chelsea Publishing.

13. Damien R (1955) *Théorème sur les Surfaces d'Onde en Optique Géometrique* Paris: Gauthier Villars.

14. Cornbleet S (1979) New geometrical method for the design of optical systems. *Proc IEE* **H3** p78.

15. de Size LK & Ramsay JF (1964) Chapter 2: Reflecting Systems. In Hansen RC (ed) *Microwave Scanning Antennas,* pp124-127, 161. London and New York: Academic Press.

16. Silver S (1949) *Microwave Antenna Theory and Design.* McGraw Hill MIT Radiation Laboratory Series **12** p402.

17. Burkhard DG & Shealy DL (1973) Flux density for ray propagation in geometrical optics. *Jour Opt Soc Amer* **63** no 3 p299.

18. Shealy DL & Burkhard DG (1973) Caustic surfaces and irradiance for reflection and refraction from an ellipsoid an elliptic paraboloid and an elliptic cone *Applied Optics* **12** no 12 p2955.

19. Westcott BS & Norriss AP (1975) Reflector synthesis for generalized far fields. *Jour Phys A: Maths & General* **8** no 4 p521.

20. Brickell F & Westcott BS (1976) Reflector design for two-variable beam shaping in the hyperbolic case. *Jour Phys A: Maths & General* **9** no 1 p113.

2
DOUBLE REFLECTOR ANTENNAS

2.1 INTRODUCTION

One major advance over the past decade has been in the design and production of dual reflector antennas, fulfilling a variety of functions, predominantly in the field of satellite communication.

The optical principle in respect to a system of two surfaces, states that there are now two a priori conditions that can be *exactly* fulfilled by the choice of two appropriate surfaces. This principle could be regarded as a consequence of the theorem of Malus and Dupin. If all the rays issue from a point source, then, after reflection from the first reflector, they still form a normal congruence and can therefore either be refocused to another point (at infinity if necessary) or can be given a specified phase front in accordance with Huygen's principle. For each specified first reflection this will be performed in a unique manner, and the two reflectors in combination will distribute the rays into a power distribution. Of all the combinations that can be created in this way, one only will have *exactly* the power distribution prescribed. Alternatives to the power distribution condition can be considered. The existence of an exact and unique solution in these circumstances, is a problem still to be proved.

The combination of two prescribed conditions may be made from a selection that depends essentially on the function of the antenna. Some possible criteria are shown in Table 2.1.

Antenna condition	Application
Aperture phase distribution	Pattern shaping
Aperture amplitude distribution	Apodisation, side lobe control
Abbe sine condition	Scan angle, pencil beam antennas
Specification of first surface	Paraboloid or sphere based antennas
Two symmetrically placed foci	Beam switching, direction finding
Two axially displaced foci	Zoom antennas
Reduction of cross polarisation	Communication antennas
Correction of a specific aberration	Elimination of manufacturing errors or gravity sag

Table 2.1:

Where the first condition given is that a specific first reflector has to be used, say in the case where a very large antenna is being modified for an alternative application, the problem reverts to the design of the single remaining surface to which the methods of Chapter 1 apply. For two general surfaces the problem becomes that of achieving a solution as exactly as *possible within specified tolerances.* From the nature of antennas at the frequency concerned, perfect geometrical optics is modified and it is always possible to have a certain degree of fulfilment of an additional condition. For example the basic paraboloid with a point source at the focus is a perfect geometrical optics design for a plane phase front, and the single condition is met by the single surface. It is well known however that a certain degree of amplitude shaping can be performed or that the beam can be offset in angle over a small range within specified tolerances. In rigorous geometrical optics this is not a function that can be analytically introduced. It has too great a dependence on the criteria chosen and the degree of approximation that can be tolerated. To the best of this author's knowledge, the only exact closed form solution that exists for a two surface antenna is for a specified first surface, that is, basically the single surface problem. Thus all other cases of the general two surface system require a computational procedure with the inevitable consequence of some form of approximation. This could arise from the termination of a series solution or of an iterative process or optimisation procedure, again all carried out until the design comes within the limits imposed by the wavelength tolerance. This is true even where precise differential equations are arrived at. These are of a form that also require computational solutions, and the existence and uniqueness of these solutions match the same considerations that concern the degrees of freedom of the two surface system. Relevant papers on the subject can, in fact, be found in the proceedings of conferences dealing with the solution of partial differential equations, without reference to the optical origin of the equations concerned. Of the many choices available for such processes, the better ones are those that can arrange a feed-back between the given criterion and the computational process, with the least computer complication or time of use.

The methodology of the limitation process also deserves consideration. In optical design a ray-to-ray stepping process is used, and to prevent the accumulation of errors, a feed-back corrective is applied at each step. This becomes analogous to point by point numerical integration. A whole surface description and specified phase front, on the other hand, is best fitted by a series of functionals such as a Fourier series or spline functions or generic curves with an iterative optimisation process for the coefficients. Similar methods govern the solution of the partial differential equations that arise in the exact analysis.

The success of the design in providing the required far field distribution is then decided, in the first approximation, by the scalar diffraction integral over the aperture as will be discussed in a later chapter. A more exact derivation can finally be obtained using the methods of physical optics which include diffraction effects and therefore the various geometrical theories of diffraction and other analytical procedures. These methods are the subject of a literature of their own and will not be included in this work.

Our consideration for the first part of the chapter will be for the *profiles* of the dual system. This then applies to the cylindrical reflector with a line source feed of which the cross-section is the profile concerned or to a rotationally symmetrical system (a spherical system). In some instances a more general surface can be constructed by these means by considering a regular array of profiles angularly displaced to form three-dimensional surfaces. Any design using this concept however, presupposes that each ray remains confined to a plane during its double reflection, which is a limitation on the generality of the design.

The fundamental optics of the system implies that the two criteria and corresponding surfaces are obtained by two applications of the laws of reflection for rays issuing from a point source with a given power distribution among them. There are however, numerous formalisms which the laws of reflection can take, and we give in Table 2.2 a wide selection of them. To avoid repetition, we include the laws of refraction, since many of the concepts considered apply equally to the refractive form of the two surface system, namely the optical lens.

Naturally, these formulae are not mutually exclusive and in principle any one can be derived from any other. However large differences arise in their application to the derivation of the equations to the surfaces concerned which are reflected in the suitability of the methods needed to be applied to their solution.

Thus every formalism eventually acquires a computational process with its limiting, iterative or optimisation procedure. The simplicity or time effectiveness of these computing programs then becomes a factor in deciding which of the formalisms is "best" for the designer. Discussion of computer techniques is beyond the scope of this volume, and of its author. There are, in any event, numerous articles in the learned journals and proceedings of conferences, for students and designers to investigate and seek their own conclusion. What is important is to observe the underlying principles that have led to the equations in the first place. The objective therefore is to trace this process through a selection of design procedures which are as general as possible. Some methods accordingly may not come under close scrutiny. This should not be taken to imply any lesser value for them, only that the limitations of space have to apply.

1.	Hamilton's differential form	see also Section 1.1
2.	Snell's law of sines	
3.	Fermat's principle and the eikonal function	
4.	Herzberger's fundamental optical invariant	
5.	Reflection and refraction dyadics	
6.	Reflection and refraction quaternions	
7.	Reflection and refraction matrices	
8.	Ray rotation operators	
9.	Classical vector formalisms	
10.	Sternberg's differential form	
11.	Projectivity of the refractive harmonic pencil	
12.	Electromagnetic boundary conditions — physical optics	
13.	Mapping in complex coordinates	
14.	Fourier optics — paraxial case only	
15.	Pure geometry of catacaustics and diacaustics	
16.	Refractive transformations between normal congruences	

Table 2.2: The laws of reflection and refraction
Derivations are given in Appendix I

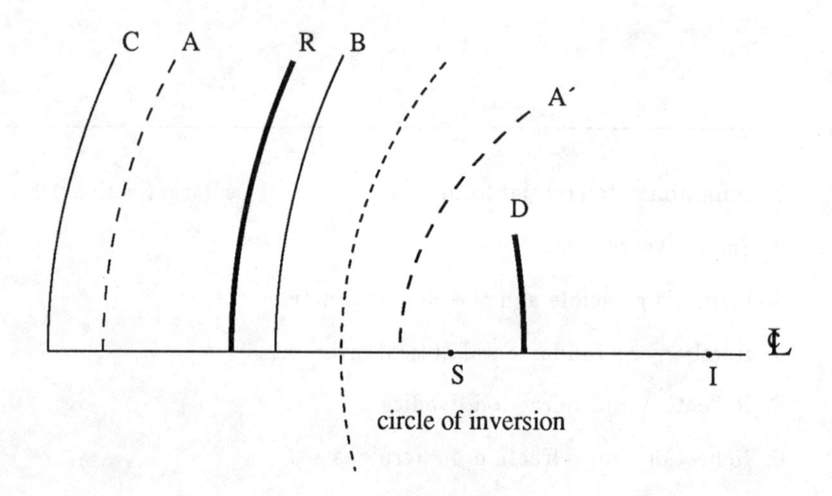

Figure 2.1: Derivation of the second surface of a two surface reflector antenna

2.2 APPLICATIONS OF DAMIEN'S THEOREM

We deal firstly with an optics-in-the-large procedure based on the theorem of Damien (p36) which demonstrates geometrically that, being given a first surface (profile) and a focusing condition, the second surface is uniquely and exactly determined (up to a scale factor). The method to be used, will be shown later (see Chapter 3), to apply equally to the refractive situation in determining the second surface of a lens, given the first surface and a specified image point. In fact, by virtue of the theorem of Malus and Dupin, the same method applies to a system of any number of reflecting or refracting surfaces, since the normal congruence of the ray corpus is maintained throughout the system. Then one can uniquely derive the final surface that will focus the rays at a specified point, given the phase front after the penultimate surface.

The proof of the method will be demonstrated for the more general case of refraction in Chapter 3, but we use it here for reflectors by considering reflective zero-distance phase fronts instead of refractive ones.

Thus, given a reflector R (Figure 2.1) and a specified image point I, the reflector being illuminated from a source S, first construct the zero-distance phase front of R with respect to S. This can be performed by any of the methods shown in the previous chapter. If the zero-distance phase front is the curve A as shown, we require a parallel to this curve, shown as A′. This can be arbitrary, but finally determines the (arbitrary) location of the second surface. Then proceed as follows:-

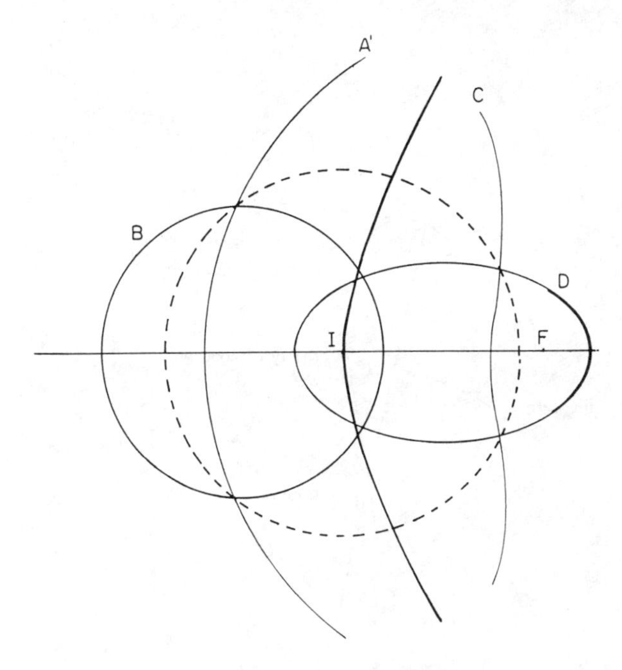

Figure 2.2: Application of the method to the Gregorian reflector, S at infinity

1. Invert the curve A′ in a circle centred on the specified image point I to give the curve B. The radius of the circle of inversion is also arbitrary since its application is to be reversed subsequently.

2. Obtain the zero-distance phase front of the *reflection* of B with respect to I to give curve C.

3. Invert C in the *same* circle as used in 1. to give the required curve D, the second surface.

All of this procedure can, of course, be detailed in terms of coordinate geometry if the exact closed form of solution is required.

We can illustrate the process by "designing" the paraboloid-ellipse combination that comprises the Gregorian telescope. For this the first reflector R in Figure 2.2 is the paraboloid, illuminated from a source S at infinity on its axis and thus by a uniform plane wave.

Hence the zero-distance phase front is a circle centred on the focus of the paraboloid F of indeterminate radius. Nevertheless all of its parallels are also circles with the same centre, and we can choose a local one A′. For simplicity we choose the apex of the paraboloid for our final image point I. Then inverting A′ in a circle centred on I will give the circle B. The reflective zero-phase front of B with respect to I is a limaçon, C (see p20) of one loop

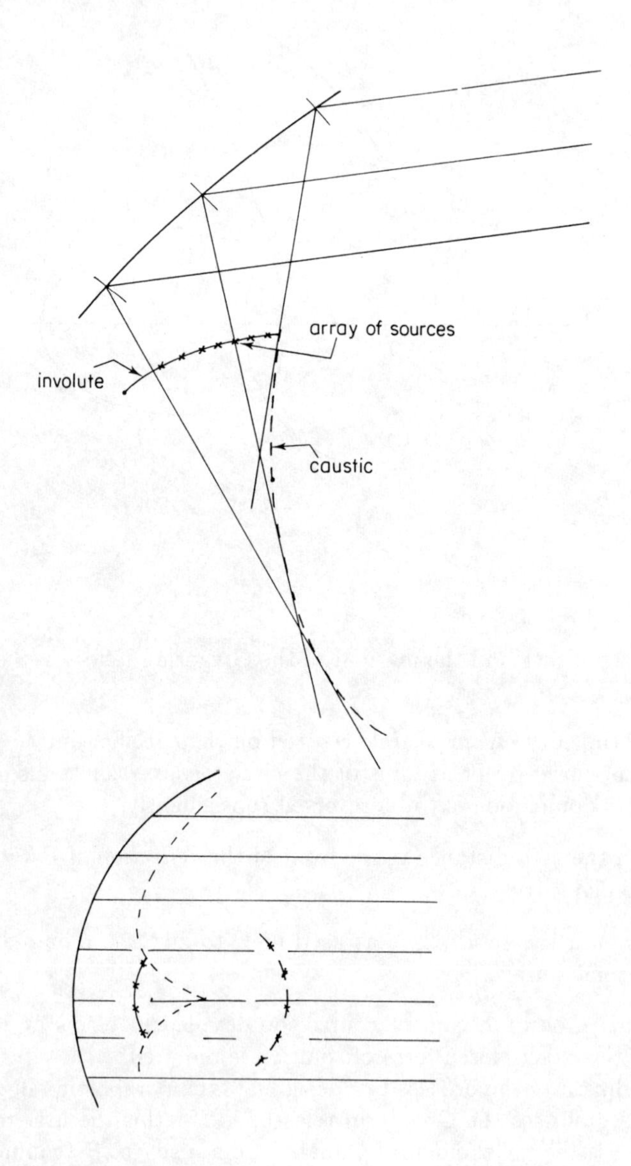

Figure 2.3: An array of sources on the involute of the caustic creates the ray pattern of the caustic

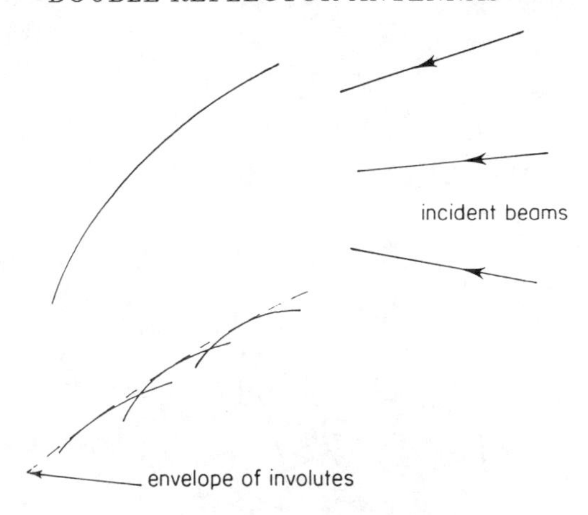

Figure 2.4: Base line for the array for many beams is the envelope of the involutes of the caustics of the individual beams

and finally the inverse of this limaçon in the same circle will give an ellipse with I and F as foci. The arbitrariness in the choice of A′ shows up as the choice from the infinite number of confocal ellipses that exist. Allowing the concept of *negative inversion* will enable the same method to produce the paraboloid-hyperboloid combination of the Cassegrain telescope. The same procedure can be used for more general non-symmetric surface *profiles* where offset dual reflectors are the objectives.

ARRAYS OF SOURCES

The reciprocity between the caustic created by a reflector and an incident beam of radiation, means that the creation of the caustic as a source would reproduce the beam as outgoing radiation. A means of creating such a source is to develop the involute of the caustic and regard it as the source phase front. Equiphased point sources at an appropriate spacing (usually less than $\lambda/2$ to avoid grating lobe effects) placed along this involute would then, by virtue of Huygens' construction, create the necessary parallels after reflection giving the caustic in one direction and the required radiating beam in the other (Figure 2.3).

If instead of a single beam and single caustic, several were to be used with radiation incident over an angular field, the separate involutes would have an envelope which would thus become the base line for an array of sources, or a moving single source, permitting the scanning of the beam over the angular range (Figure 2.4).

This method of inversion can thus be extended to multiple beams and hence to shaped beam dual reflector antennas with multiple sources. As shown in Figure 2.5 the shaped beam is constructed from individual narrow beams

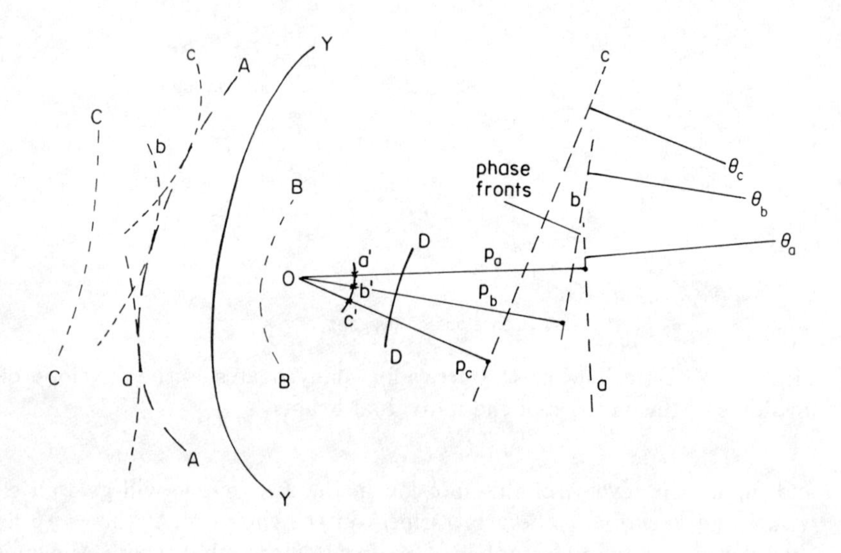

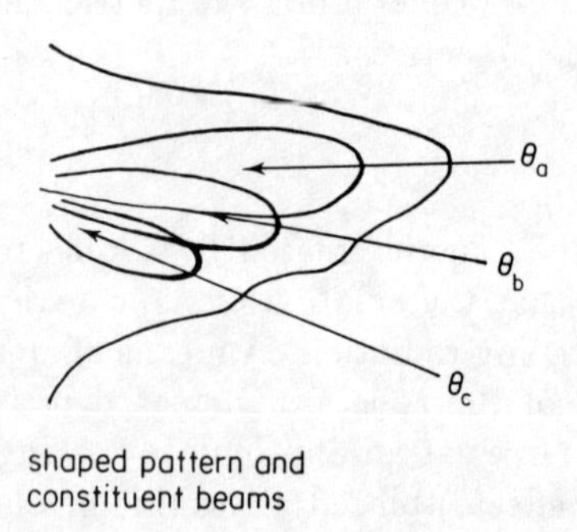

Figure 2.5: Shaped radiation pattern using the inversion construction with the envelope of the individual zero-distance phase fronts

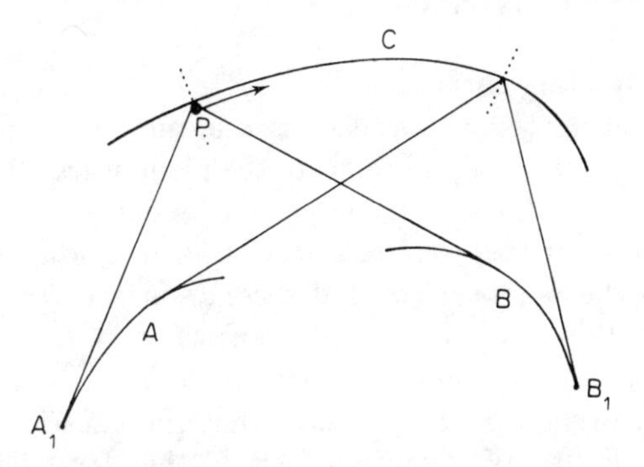

Figure 2.6: Co-involution of two caustics A and B. The point P is held in a taut string wound over the two curves

with appropriate phase and amplitude. This is reduced to a basic minimum of plane waves incident upon a *given* surface YY′ at different angles of incidence.

The zero-distance phase fronts for these waves are then derived by the process for an infinite source giving the curves a, b, c... each of which will be different in form owing to the difference in incident angle. Each will have a phase datum, say the distance of the perpendicular p, from a central point such as O (not necessarily a focus but generally some obvious mean position).

The envelope of these curves, if it exists, is the curve A. Now it can easily be seen that the phase of each wave can be individually adjusted by the length of the appropriate perpendicular from O. But this selection cannot be totally arbitrary as the requirement that the resultant set of curves must have an envelope has to be obeyed.

The curve shown in Figure 2.6 is that assumed to be given by a continuous decrease in p from a to b, and so on.

Curve A is now inverted to the image point O resulting in curve B as before. Its zero-distance reflective phase front is C, and the inversion of C gives the required shape of the second reflector D.

We now surround O with a small circle which will be the distribution of the fundamental sources a′, b′, c′.... To each of these sources a plane wave can be attributed through the geometry of the procedure. Hence, an amplitude and phase distribution can be imposed on the sources equal to that required by the original spectrum of waves.

This can be generalized in an obvious manner for sources distributed over other surfaces, say a planar phased array. If it is repeated in the other planes of a two-dimensional shaped pattern, the resulting curves construct a surface for full pattern shaping. A single-surface reflector can be obtained by the same method if YY′ is considered in the first instance to be a hypothetical infinite plane.

2.3 CAUSTIC MATCHING

If a reflector (or refracting surface) is illuminated by a plane wave, a catacaustic (diacaustic) is created, as was discussed in the first chapter. In reverse the caustic, used as a source, would recreate the plane wave after reflection at the surface. Accordingly, if a second source, say a point source, and a second reflector (or refracting surface) were to create the same, or approximately the same caustic, then the point source and second surface would become the appropriate combination to create the plane wave or system required. This matching of caustics can be achieved in more than one way.

The first of these relies on the property that the caustic itself is the involute of the phase front. We require a method therefore that gives us the necessary surface that will convert a given caustic, or point source which is a degenerate caustic, into the required caustic. The process has been termed (by the author) the "co-involution of two caustics".

The process of drawing the involute of a curve by a string wound about the curve was given in Section 1.3. We now propose an extension of this method involving two caustics (Figure 2.6), in which a taut inextensible string is fixed to points A_1 and B_1 on the caustics A and B enveloping the curves and held taut by the drawing point P as the string unwinds from one curve and winds onto the other. The locus of P is a curve C, the reflecting curve which will reflect all the rays creating caustic A into the envelope creating caustic B. If caustics A and B were to degenerate to perfect point foci, the method of drawing the reflector is the well-known method for drawing an ellipse with a string stretched over two pins at the foci. This construction was given by Leibnitz in a letter to John Bernoulli in 1704. The curves can be wound or unwound in the same or in contra directions, as shown in Figure 2.2. The proof that C is a reflector with this property arises from an application of Fermat's principle. If, as in Figure 2.7, A and B are points where the taut string is tangential to the caustics, then an incremental movement ds of the drawing point will "wind on" Δr on the A caustic and "unwind" $\Delta\rho$ on the B caustic, and obviously Δr will be equal in length to $\Delta\rho$. Hence from the figure we will have, measuring r and ρ from the intersection of the tangents from C_1 and C_2 respectively

$$r + \delta r + \rho - \delta\rho = \rho + \mathrm{d}\rho + \delta\rho + r + \mathrm{d}r - \delta r$$

Since $\Delta\rho = \Delta r$, and to the first order $\Delta\rho = 2\delta\rho$ and $\Delta r = 2\delta r$, we have

$$\mathrm{d}r = -\mathrm{d}\rho$$

which from Chapter 1 defines a reflector obeying Snell's law at C.

The term proposed by Leibniz for this operation was "co-ëvoluteo" but owing to its similarity to the drawing of involutions of curves, the term "co-involution" would seem more appropriate.

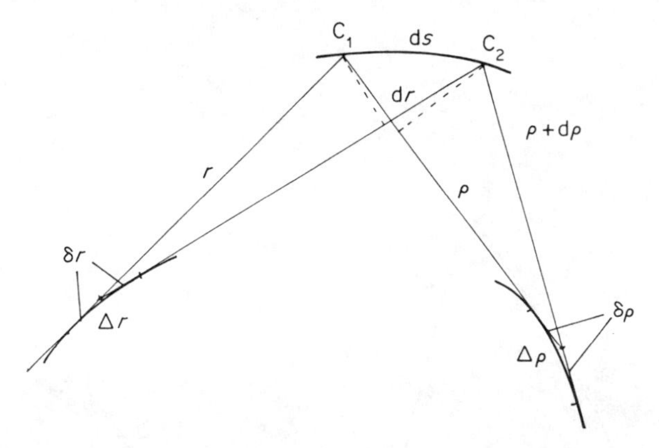

Figure 2.7: Proof that the co-involution drawing process gives rise to a reflecting surface C

We shall describe the method by the coordinate geometry of the caustics. Let caustic A have parametric descriptions $x = f_1(t)$, $y = g_1(t)$, and caustic B parametric description $x = f_2(s)$, $y = g_2(s)$; then a point (X,Y) on the curve C obeys

$$\sqrt{(X - f_1)^2 + (Y - g_1)^2} + \int_{t_0}^{t} \sqrt{f'^2_1 + g'^2_1}\,\mathrm{d}t + \sqrt{(X - f_2)^2 + (Y - g_2)^2} + \int_{s_0}^{s} \sqrt{f'^2_2 + g'^2_2}\,\mathrm{d}s = L \tag{2.3.1}$$

where L is the fixed length of string and t_0 and s_0 are the parameters of the points A_1 and B_1 where the string is attached to each caustic. Since the string is a tangent to A at the point with parameter t and to B at s, we have in addition

$$\frac{Y - g_1(t)}{X - f_1(t)} = \frac{\mathrm{d}f_1}{\mathrm{d}g_1} \tag{2.3.2}$$

$$\frac{Y - g_2(s)}{X - f_2(s)} = \frac{\mathrm{d}f_2}{\mathrm{d}g_2} \tag{2.3.3}$$

If the caustic B is reduced to a point focus at the point (a, b) the same equations apply with $f_2 = a$, $g_2 = b$ and the second line integral in Equation 2.3.1 and Equation 2.3.3 can be omitted.

EXAMPLES

1. Subreflector for parabola with offset beam

The caustic of a parabola $y^2 = 4ax$ receiving a beam of parallel rays making an angle θ with the axis is the cubic of Tschirnhausen (Figure 2.8). In the

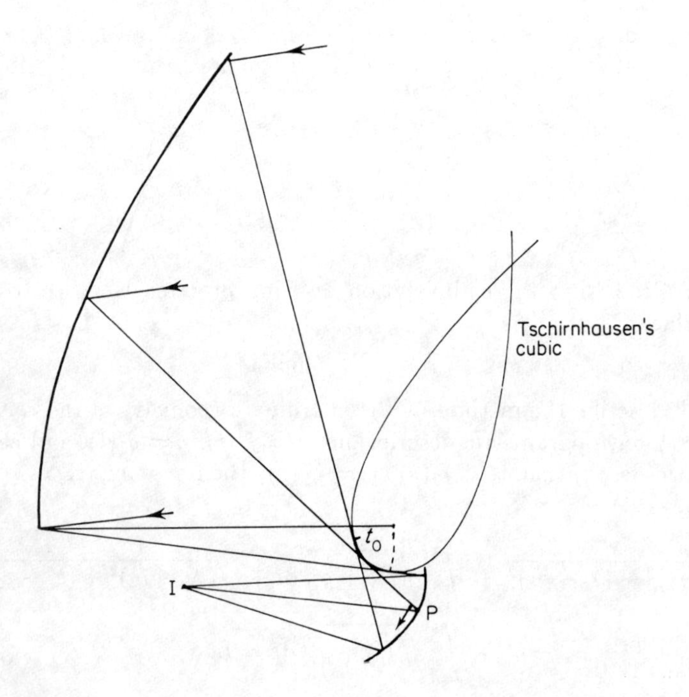

Figure 2.8: Drawing method for a sub-reflector in an asymmetrically illuminated parabola by a string wound about Tschirnhausen's cubic

coordinate system of the paraboloid itself this has equations

$$x = [-at\sin\theta(t^2-3)+a\cos\theta]\cos\theta + 3at^2\sin^2\theta \equiv f_1(t)$$
$$y = [at\sin\theta(t^2-3)-a\cos\theta]\sin\theta + 3at^2\cos^2\theta \equiv g_1(t)$$

To produce a focus at the origin, put $a = 0, b = 0$ in Equation 2.3.1 and (dropping the suffix since the second caustic is a point focus)

$$S(t) \equiv \int_{t_0}^{t} [f'^2 + g'^2]^{\frac{1}{2}} dt = at\sin\theta(t^2+3) - at_0\sin\theta(t_0^2+3) \qquad (2.3.4)$$

t_0 is the point of fixing on the caustic, which for a paraboloid subtending a total angle of 2Θ is given approximately by

$$t_0 = \tan\left(\frac{\Theta}{2}\right)$$

Substitution into Equations 2.3.1 gives the implicit parametric equations

$$S(t) + (Y-g)(1+\frac{g'^2}{f'^2})^{\frac{1}{2}} + \{[(Y-g)\frac{g'}{f'}+f]^2+Y^2\}^{\frac{1}{2}} = L$$
$$S(t) + (X-f)(1+\frac{f'^2}{g'^2})^{\frac{1}{2}} + \{[(X-f)\frac{f'}{g'}+g]^2+X^2\}^{\frac{1}{2}} = L$$

with

$$\frac{f'}{g'} = \frac{(1-t^2)\sin\theta\cos\theta + 2t\sin^2\theta}{(t^2-1)\sin^2\theta + 2t\cos^2\theta} \qquad (2.3.5)$$

The range of t in these equations is $\pm t_0$.

2. Correctors for spherical mirrors

The technique for obtaining the symmetrically placed subreflector for correcting the spherical aberration of a spherical mirror is illustrated in Figure 2.9. The caustic is the nephroid and the relevant equations are those given in Section 1.4 adapted and used in Equation 2.3.1. Hence

$$f(t) = a(3\cos t - \cos 3t)$$
$$g(t) = a(3\sin t - \sin 3t) \qquad -\pi < t < \pi$$

The cusp is at the value $t = -\pi$ as drawn in Figure 2.9. The integral of Equation 2.3.4 is then

$$S(t) = -6a\cos t \qquad f'/g' = -\tan 2t$$

These values can be substituted into Equations 2.3.1 to give, for example, a focus at the apex $(-4a, 0)$.

This case is of particular interest since an offset subreflector can be designed in an identical manner to give an axial beam not disturbed by the blockage of the centrally placed subreflector. This is shown in Figure 2.10.

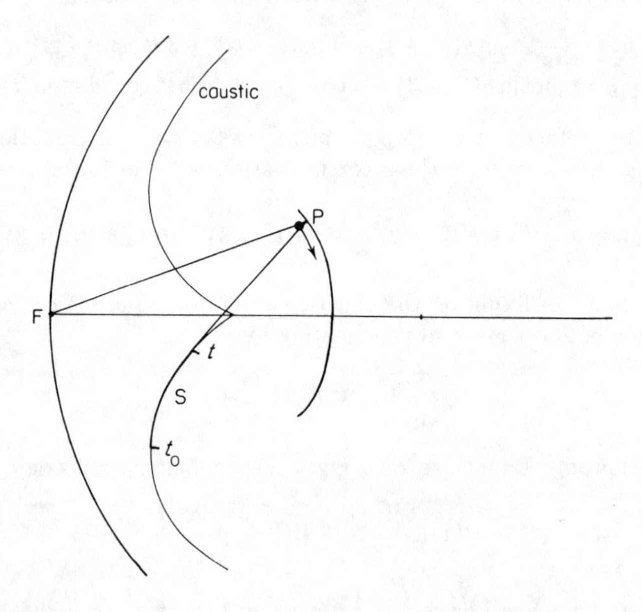

Figure 2.9: The corrector for a spherical mirror with a string wound about the nephroid caustic

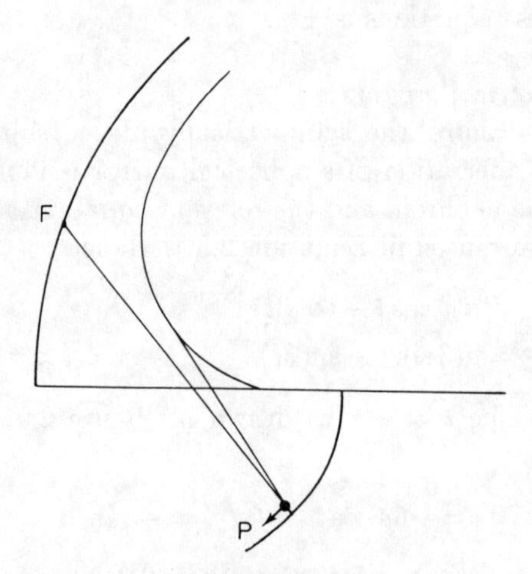

Figure 2.10: Arrangement for elimination of blockage in a circular reflector

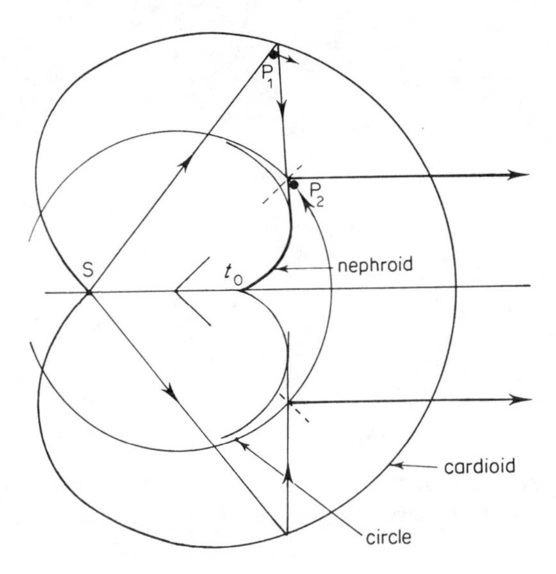

Figure 2.11: The co-involution of the caustic of the circle with the point at infinity gives the circle, and the co-involution of the source and the caustic gives the cardioid. This is then the Zeiss Cardioid combination

3. The Cardioid Reflector
The caustic of the circular reflector which has been used continuously in these studies appears again in the application of co-involution to the circle-cardioid combination known as the Zeiss Cardioid. By its construction the circular reflector is the co-involution of a nephroid caustic at a point at infinity. But, from the table in Appendix III the nephroid is also the caustic by reflection of a point source at the cusp of a cardioid; thus, as is shown in Figure 2.11 the cardioid will focus all the rays previously reflected by the circle at its cusp. If, instead of a point at infinity a local image were specified, a curve other than the nephroid would be its caustic in the circle resulting in a different reflector from the cardioid. The combination would thus have the properties of the Schwarzschild microscope objective.

A similar taut string technique applies to the description of a hyperbola when related to point foci. The image is then a virtual focus and the curve is the solution of Equation 1.1.1 with positive sign and $\eta = -1$, that is

$$r - \rho = \text{constant} \tag{2.3.6}$$

This curve can be described by a fixed point P on a string which is looped over the two foci F_1 and F_2 as shown in Figure 2.12, going on to a combined point T at a sufficient distance. If the combined strings move in the direction T,

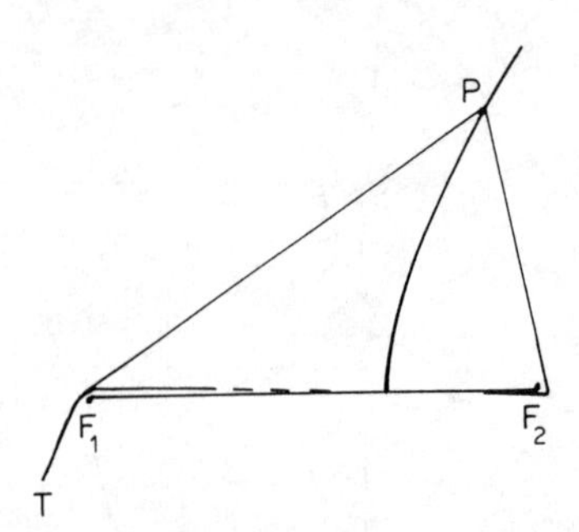

Figure 2.12: A string drawing method for the hyperbola. P is a fixed point on the string looped over the two foci and drawn together at T. F_2 is a virtual focus of F_1.

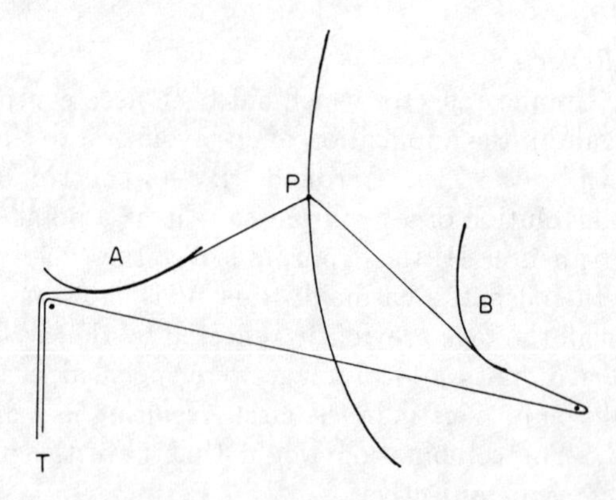

Figure 2.13: Extension of the co-involution process to a virtual and a real caustic

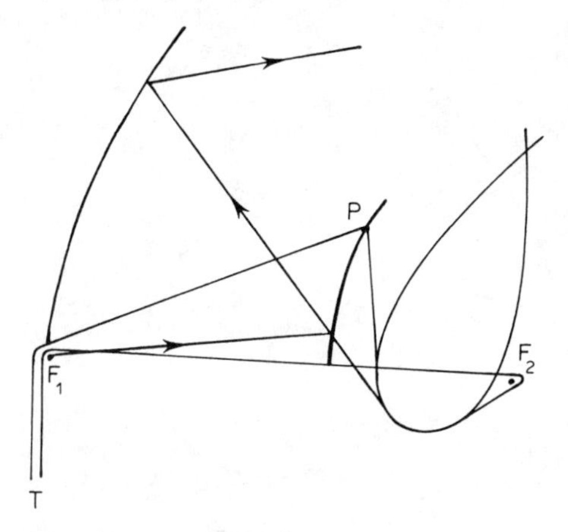

Figure 2.14: Application of the virtual caustic process to the design of a Cassegrain reflector with off-axis beam

both parts F_1P and F_2P are reduced at the same rate and hence the difference

$$F_1P - F_2P = \text{constant} = r - \rho$$

satisfies Equation 2.3.6.

By a similar analysis the point caustics F_1 and F_2 can be replaced by line caustics and the string overlaid in the same manner as in Figure 2.6. This gives a reflector that transforms the real caustic at A into a virtual caustic at B, as illustrated in Figure 2.13. This method could be used as shown in Figure 2.14 to obtain the Cassegrain design of subreflector that would convert the caustic, the Tschirnhausen cubic, to an arbitrarily situated focus. The string is combined over this required focus, one branch going to the fixed point P and the other winding about the caustic and looping over the second focus F_2 on that caustic. Note that only one part of the string at the focus F_2 has to be wound over the caustic to retain the validity of Fermat's principle.

THE GENERAL TWO-SURFACE REFLECTOR SYSTEM

Ideally the method could be made fully complete if there were a mechanical drawing method by which the original caustic could be derived. This lack can be avoided by a more general procedure. In the illustrative examples the common basic reflectors were taken, namely the circle and the parabola, each of which has a fairly complex caustic. However, it is obvious that we could instead choose a completely arbitrary caustic of simple definition, form a co-involution of it with one specified focus giving a primary reflector and a second co-involution with a second specified focus giving a subreflector. The two-reflector combination would then focus the first source into the second image. In this manner we trade simple reflectors with complex caustics for

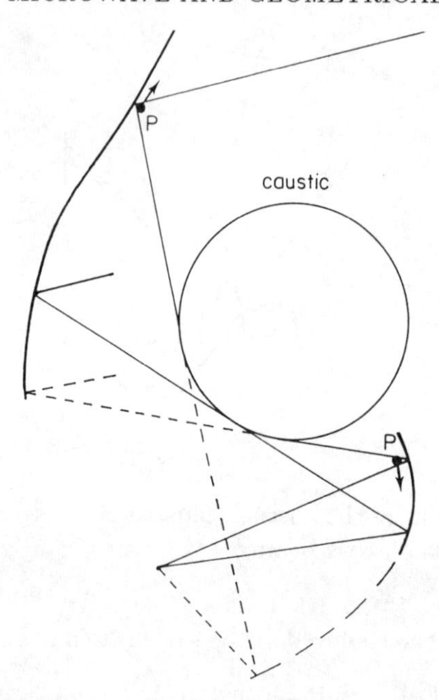

Figure 2.15: Simultaneous co-involution of a primary and a secondary reflector from an arbitrary mutual caustic

simple caustics with not too complex reflectors. Taking, for example, a circular caustic and a source at infinity, as in Figure 2.15, the two co-involutions performed simultaneously describe the main and subsidiary reflectors. The process can be termed simultaneous double co-involution. As is now obvious from this description the tangents to the caustic in both co-involution processes are contiguous. The parts of the tangents for which this is so could then be taken by a taut string contained within a rigid but extendable tube. As the tube rolls without slipping about the caustic, and the string is kept taut, the ends simultaneously describe both the main and subsidiary reflectors, a method illustrated in Figure 2.16. By rolling a link of this sort about a point caustic, the simultaneous drawing of the parabola and Gregorian elliptical subreflector can be shown to be an extension of the simple ellipse drawing method alone. It is possible to develop the method further and have several tubular links of the kind shown, each rolling without slipping on a specified caustic curve to give the profiles of a multiple reflector system with final end focusing requirements. Such a procedure would be a simultaneous multiple co-involution. The process extends naturally to the two-dimensional design of reflector surfaces by the rolling of a tubular link over a caustic surface, provided only that the parts of the string external to the tube and that part inside the tube are all kept coplanar.

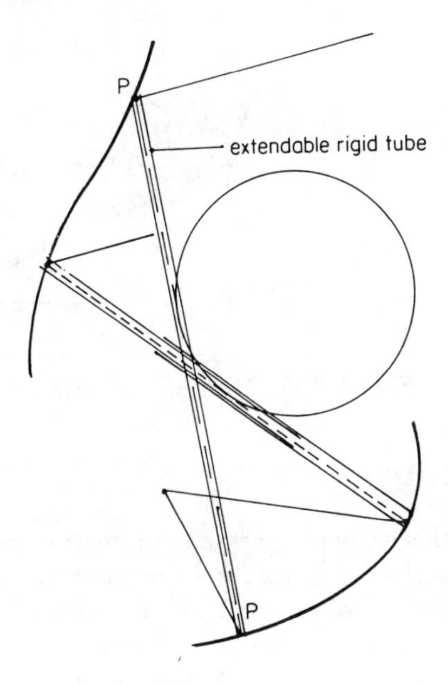

Figure 2.16: Simultaneous double co-involution, the tangential part of the string held in a rigid extendable tube

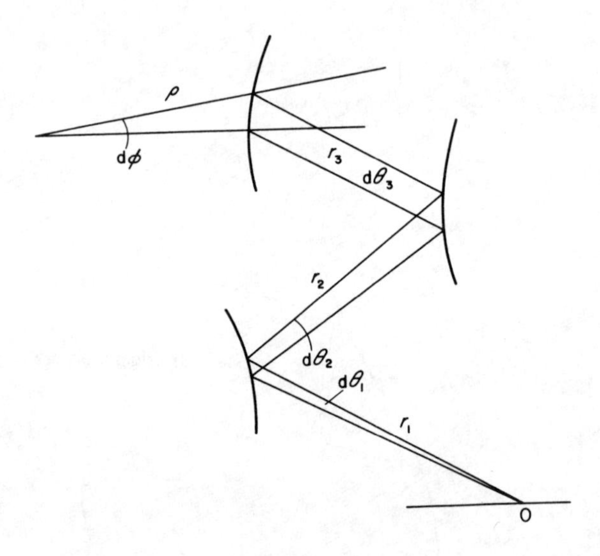

Figure 2.17: Multiple reflectors

2.4 RAY METHODS

Before we conclude this section it may be informative to compare the simple derivation of whole-surface reflectors from geometrical phase front principles with the geometrical optics of ray methods. It will also form an introduction to the method we shall use for more general two-surface systems. For two or more surfaces the differential form of Equation 1.1.1 is piecewise summable, and for the reflectors shown in Figure 2.17 takes the form

$$\mathrm{d}r_1 \pm \mathrm{d}r_2 \pm \ldots \pm \mathrm{d}r_n = \mathrm{d}\rho$$

and

$$r_1\,\mathrm{d}\theta_1 \pm r_2\,\mathrm{d}\theta_2 \pm \ldots \pm r_n\,\mathrm{d}\theta_n = \rho\,\mathrm{d}\phi \tag{2.4.1}$$

where the signs are determined by the same rules as for Equations 1.1.2 and 1.1.3 and r_1 is the distance from the fundamental source at the origin to the first reflector, r_2 the distance along the ray from the first to the second reflector and so on and (ρ, ϕ) are polar coordinates about the final focus of the rays. We then have the two relations equivalent to Equations 1.1.2 and 1.1.3 for n reflecting surfaces.

CASSEGRAIN AND GREGORIAN REFLECTORS

The simplest illustration of this principle is the combination of a paraboloid and a secondary reflector to give the same result as for the paraboloid alone. If

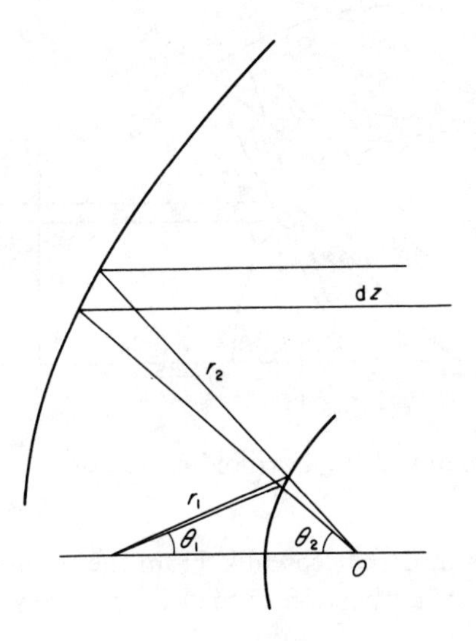

Figure 2.18: The Cassegrain antenna

the two surfaces in Figure 2.18 have a common origin O, then from Equation 2.4.1

$$r_1 \, d\theta_1 \pm r_2 \, d\theta_2 = dz \tag{2.4.2}$$

where the parabola is defined by $dz = r \, d\theta$.

If the polar equation of the secondary reflector is (ρ, ϕ) about the origin O, then $d\theta = d\theta_2$, and

$$r_1 \, d\theta = (r \pm r_2) \, d\theta_2 = \rho \, d\phi \tag{2.4.3}$$

But this is the defining equation of the conic mirror of the previous section and is thus a hyperbola (Cassegrain System) or ellipse (Gregorian System) depending on the sign.

ZEISS CARDIOID

In the previous example the two preconditions, the first surface and the required focusing have been met by two applications of Snell's law.

If we apply this to a spherical reflector there results two refocusing surfaces depending upon whether the reflection from the sphere is external (virtual focusing) or internal. The former of these as we now show, for a focus at infinity gives the cardioid reflector and sphere combination the cardioid of Zeiss as before.

We have again the relation referring now to Figure 2.19

$$r_1 \, d\theta_1 - r_2 \, d\theta_2 = dz \qquad r_1 = \text{FM} \qquad r_2 = \text{PM}$$

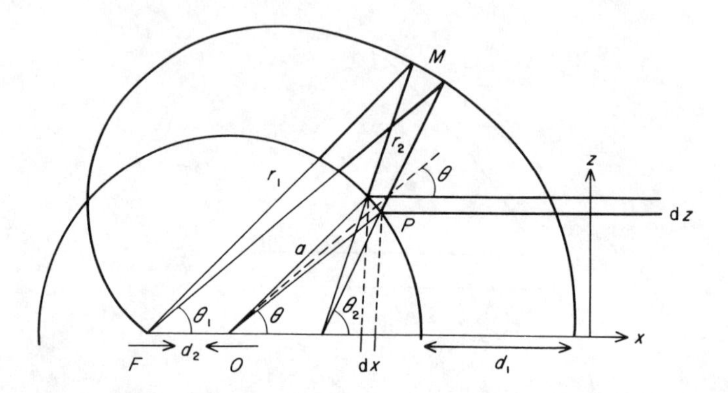

Figure 2.19: Cardioid corrector

where, for a circle radius a, $\mathrm{d}z = a\cos\theta \mathrm{d}\theta$. From the properties of the reflection at P, that is a further application of Snell's law , we see that

$$\theta_2 = 2\theta$$

and hence

$$r_1 \,\mathrm{d}\theta_1 = (a\cos\theta + 2r_2)\,\mathrm{d}\theta \tag{2.4.4}$$

The second statement of the principle gives

$$\mathrm{d}x = \mathrm{d}r_1 + \mathrm{d}r_2 = a\sin\theta\,\mathrm{d}\theta$$

which integrates to give the eikonal relation

$$r_1 + r_2 - a\cos\theta = \text{const} = 2d_1 + d_2 \tag{2.4.5}$$

A third relation is obtained from the geometry of the figure

$$r_1 = d_2\cos\theta_1 + a\cos(\theta - \theta_1) + r_2\cos(2\theta - \theta_1) \tag{2.4.6}$$

The solution of these equations is dependent on the arbitrary choice of d_2, the required position of the eventual focus, and d_1. It is found by inspection that the simplest of these is obtained by putting $\theta = \theta_1$ and $d_1 = d_2 = a/2$. The resulting solution has r_2 = constant = $a/2$, and

$$r_1 = a(1 + \cos\theta_1)$$

This is the equation to the cardioid.

When the above values are incorporated into Figure 2.19, we see as in Figure 2.20 that the quadrilateral FOPM is a trapezium with the three sides FO, OP and MP of fixed length. A parallelogram linkage can thus be constructed which will describe the cardioid by the movement of the point M while the point P is confined to the circle. The linkage however is not "pure" since it

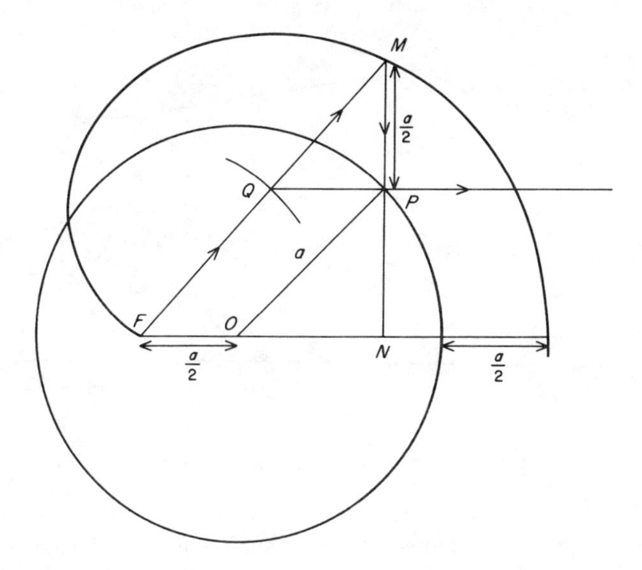

Figure 2.20: Zeiss cardioid

requires a sliding member at N. Further since the intersection of the initial and final rays are at the point Q and FOPQ is a parallelogram, Q will describe a circle centred at F. Hence this system automatically obeys the sine condition (see Appendix I). This is an additional and incidental advantage of the design. The cardioid is a roulette curve obtained by deriving the trajectory of a fixed point on a circle radius $a/2$ as it rolls without slipping around a circle radius $a/2$. Equation 2.4.5 can be rewritten as

$$r_1 = a\cos\theta_1 + 2d_1 + d_2 - r_2$$

Then other roulette curves are obtained when r_2 is kept constant and thus equal to d_1 leaving

$$r_1 = a\cos\theta_1 + d_1 + d_2$$

This is the general equation of the limaçon of Pascal again, obtained from the fixed point of a circle of radius different from $a/2$ rolling around the circle radius $a/2$. The curve will be re-entrant if the radius of the rolling circle is less than $a/2$ and smooth if greater. Thus the generalizations of the Zeiss sphere/cardioid combination are Zeiss sphere/limaçon systems. This result will prove to be of relevance in a later chapter (cf Example 3 in Section 2.2).

GENERAL SPECIFIED FIRST SURFACE

We conclude this series of studies involving the specified first reflecting surface (or in the case of the lens, refracting surface) by the study of systems where

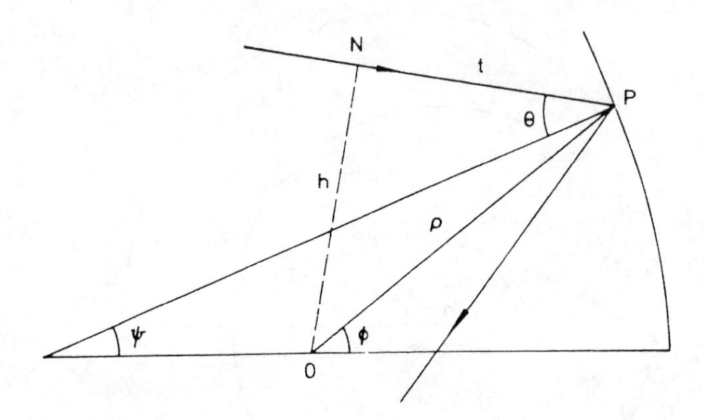

Figure 2.21: Geometry of primary reflector profile

the two preconditions, the shape of the first surface and the focusing property, are met *not* by two applications of Snell's law but by applying Snell's law of reflection at the first surface and defining an eikonal function, given by the path length of the rays through the system required for the focal property.

We consider a completely general specified main reflector surface with given polar equation $\rho = \rho(\phi)$ from an origin at O (Figure 2.21) and with the axis $\phi = 0$ corresponding to the x axis. The all important normal to the main reflector at a point $P(\rho, \phi)$ makes an angle ψ with the axis where

$$\tan\psi = \frac{\rho\sin\phi - \frac{d\rho}{d\phi}\cos\phi}{\rho\cos\phi + \frac{d\rho}{d\phi}\sin\phi} \tag{2.4.7}$$

from which can be derived the differential equation to the first reflector

$$\frac{d\rho}{\rho} = \tan(\phi - \psi)\, d\phi$$

Those cases for which this equation can be solved to give ϕ explicitly as a function of ψ are dealt with. This does in fact cover most of the common geometrical curves such as might form the primary reflector in a dual-reflector system.

Thus Equation 2.4.7 can be specified to be equivalent to its solution

$$\phi = F(\psi) \tag{2.4.8}$$

The incident (or transmitted) phase is obtained by establishing a datum, the point N, which is the foot of the perpendicular from the origin onto the incident ray. The path length of the intercept $NP = t$ depends on the nature of the illuminating pencil of which three forms are shown in Figure 2.22 for rays making angle θ with the *normal* at P.

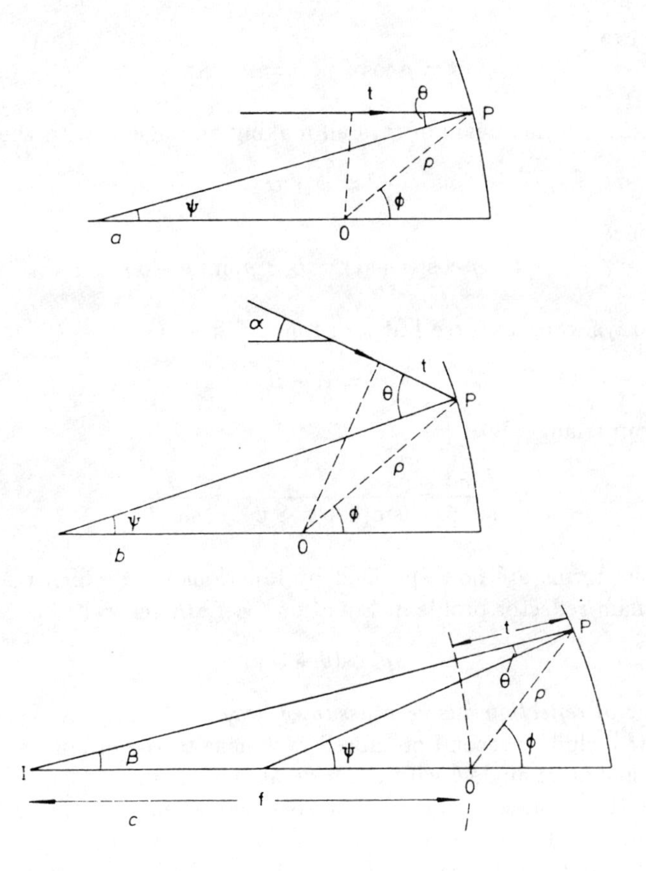

Figure 2.22: Illuminating ray systems
a) Parallel axial beam
b) Parallel beam at angle α
c) Distant focused beam

(a) with parallel axial rays

$$\theta = \psi$$

and hence

$$t = \rho \cos \phi \qquad h = \rho \sin \phi \tag{2.4.9}$$

(b) with a parallel beam of rays all making an angle α with the axis,

$$\theta = \psi + \alpha$$

and hence

$$t = \rho \cos(\phi + \alpha) \qquad h = \rho \sin(\phi + \alpha) \tag{2.4.10}$$

(c) for rays from a source I at a distance f from O

$$\theta = \psi - \beta$$

and from triangle IOP

$$\frac{f+t}{\sin \phi} = \frac{f}{\sin(\theta + \phi - \psi)} = \frac{\rho}{\sin \beta} \tag{2.4.11}$$

Thus all the terms are now specified by functions of ψ which itself derives from the main reflector profile in Equation 2.4.7. In general

$$t = \rho \cos(\theta + \phi - \psi)$$

and *the law of reflection has been assumed only at P.*

We now include a second or subreflector near to the origin and a general ray which makes an angle θ with the normal at P (Figure 2.23). The incident (or transmitted) phase surface is now specified when t in Figure 2.22 is an explicit function of any one of θ ϕ or ψ, as shown above. If, after reflection at P the ray meets the subreflector at Q(x, y) then the internal geometry gives

$$\begin{aligned} x &= \rho \cos \phi - r \cos(\theta + \psi) \\ y &= r \sin(\theta + \psi) - \rho \sin \phi \end{aligned} \tag{2.4.12}$$

where r is the intercept PQ.

We note here that these relations depend on having assumed that the ray crosses the axis between its reflection at P and incidence at Q. A second solution is available if the ray does not cross the axis in this manner. The point is not trivial when it comes to the solution of a differential equation for the surfaces.

With ϕ ψ and therefore θ given by the known shape of the first reflector, the shape of the second reflector can be obtained parametrically in terms of any one of them, by deriving a form for r given by the same parameter. This will depend on the specified nature of the focusing required, which will provide an eikonal function which all rays have to obey. It turns out that this function

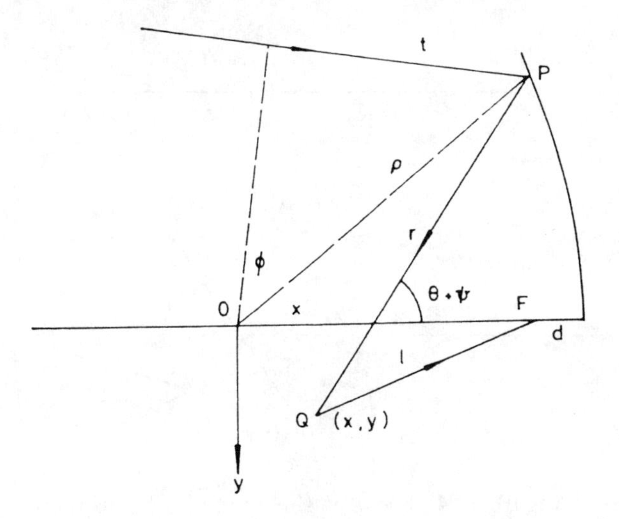

Figure 2.23: Geometry of subreflector profile

allows the explicit definition of the intercept r in terms of the parameter θ and hence the description of the second reflector surface profile. Even more importantly, the same consideration is found to apply when, instead of a given first surface, we have an unknown surface but a second criterion (Section 2.6).

Instead of using a second law of reflection at Q we use an equivalent, the constancy of an eikonal function from the incident phase surface to the image phase surface or focal point. For the latter this gives

$$t + r + \ell = \text{constant} = a \tag{2.4.13}$$

the constant being evaluated from a known basis ray, usually the central or axial ray of a rotationally symmetric system. Normalising the axial focal length to $\rho(0) = 1$ and for simplicity defining a focal point at the apex of the main reflector, then

$$\ell^2 = (1 - x)^2 + y^2 \tag{2.4.14}$$

and so after squaring Equations 2.4.12 and 2.4.13, r^2 is eliminated and the resultant linear relation in r gives

$$r = \frac{1 + \rho^2 - 2\cos\phi - (a - t)^2}{2[t - a - \cos(\theta + \psi) + \rho\cos(\theta + \psi - \phi)]} \tag{2.4.15}$$

All the terms in this relation are expressible in terms of ϕ alone (e.g. for a plane normal incident phase front $\theta = \psi$), and hence substitution back into Equation 2.4.12 gives the required reflector shape. Specification of the constant a determines the final overall geometry through locating the axial position of the centre of the subreflector.

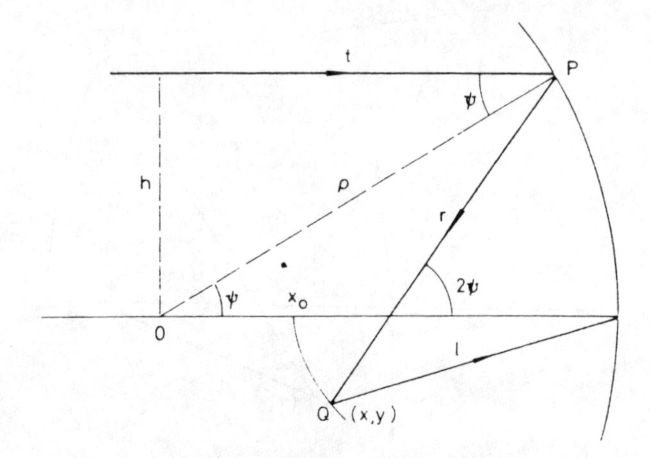

Figure 2.24: Focus at the apex of a hemisphere

SUBREFLECTORS IN THE CONCAVE HEMISPHERE

First the various forms of focusing possible with a subreflector in the interior of a concave hemisphere illuminated by an axial beam of parallel rays are discussed. Circular symmetry is assumed. The main reflector profile is thus $\rho = 1$ (say) for a parallel axial beam $\theta = \psi$ (Equation 2.4.9). Equation 2.4.10 gives (obviously) $\psi = \phi$ and thus $t = \rho \cos \psi$ and $h = \rho \sin \psi$. All subreflectors are therefore subject to the relations

$$x = \cos \psi - r \cos 2\psi$$
$$y = r \sin 2\psi - \sin \psi \qquad (2.4.16)$$

FOCUS AT THE APEX

For a focus at the apex of the hemisphere (Figure 2.24) Equations 2.4.13 and 2.4.17 become

$$\ell = a - r - t \qquad a = 3 - 2x_0$$

and $\ell^2 = (1 - x)^2 + y^2$ with the result, on eliminating ℓ

$$r = \frac{1}{2}\left[\frac{2 - 2\cos\psi - (a - \cos\psi)^2}{2\cos\psi - a - \cos 2\psi}\right] \qquad (2.4.17)$$

The parameter a specifies the axial position of the centre of the subreflector. Equations 2.4.12 with r given by Equation 2.4.17 are illustrated in Figure 2.25. As can be seen, the subreflector is a smooth curve when the centre is beyond the cusp of the well-known nephroid caustic of the hemisphere. Subreflectors in the illuminated region between the caustic and the hemisphere become double branched. This is because any point in this region is the intersection

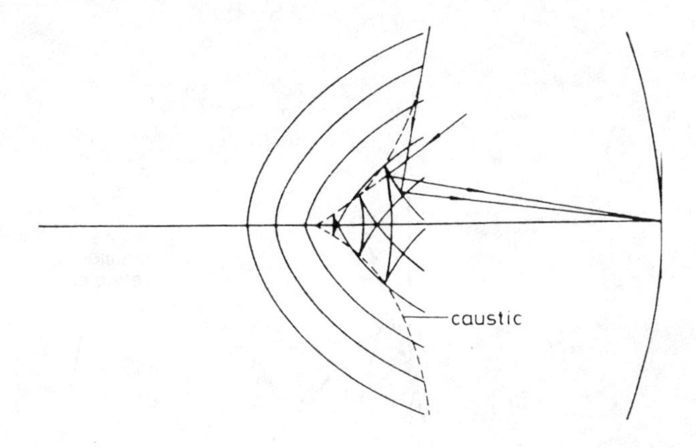

Figure 2.25: Profiles for focusing subreflectors in the hemisphere

of two rays which are the two tangents to the caustic from that point. One of the branches of the curve focuses one set of tangents, and the other branch focuses the second rays separately as shown in Figure 2.26. This is common to all main reflectors which have a caustic of reflection, and arises naturally from the parametric description derived.

TELESCOPIC SUBREFLECTOR

The telescopic configuration focuses the parallel beam into a smaller parallel beam as in Figure 2.27. The requirement is simply

$$r + \cos\psi - x = 2 - 2x_0 \tag{2.4.18}$$

or

$$r = \frac{2 - 2x_0}{1 + \cos 2\psi}$$

Subreflectors to this formula are shown in Figure 2.28. The required ray property in this case requires that $\mathrm{d}y/\mathrm{d}x = \cot\psi$ which can be deduced from the parametric equations.

CROSSOVER RETROREFLECTOR

If the rays on first reflection by the subreflector become perpendicular to the axis, then a symmetrical second reflection would give a crossover effect and the ray would return parallel to its incident direction. This is illustrated in Figures 2.29 and 2.30. For this to occur it is required that

$$r + y + \cos\psi = 2 - x_0 \tag{2.4.19}$$

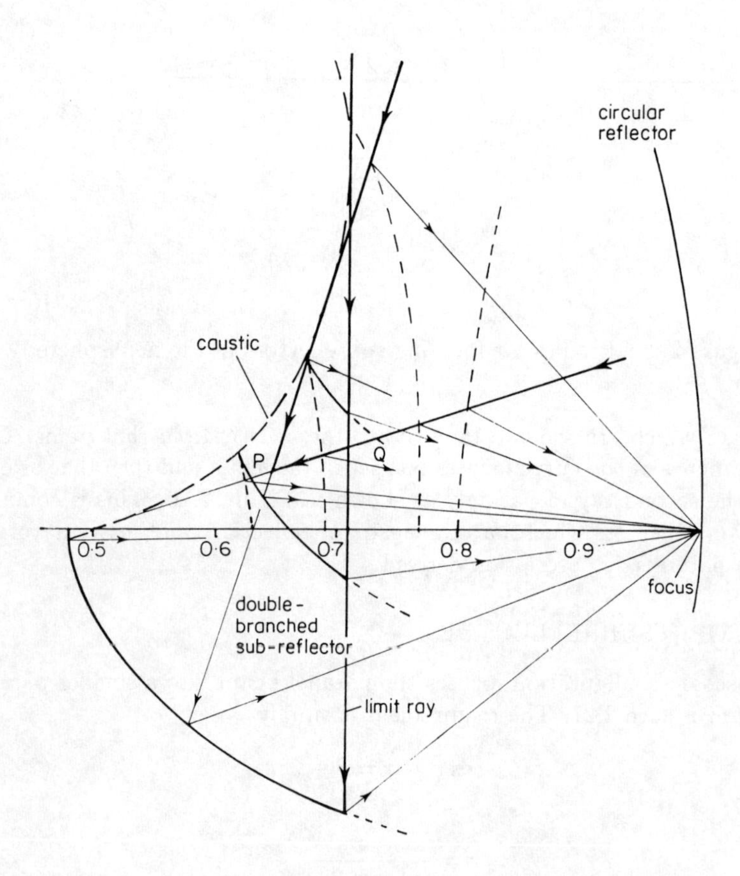

Figure 2.26: The development of the double-branched subreflector. The two rays through P reflect off different branches of the curve to focus at the apex.

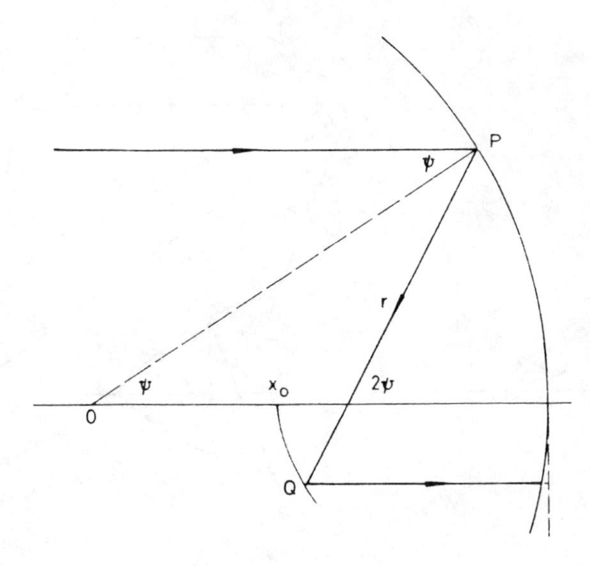

Figure 2.27: Geometry of telescopic subreflectors

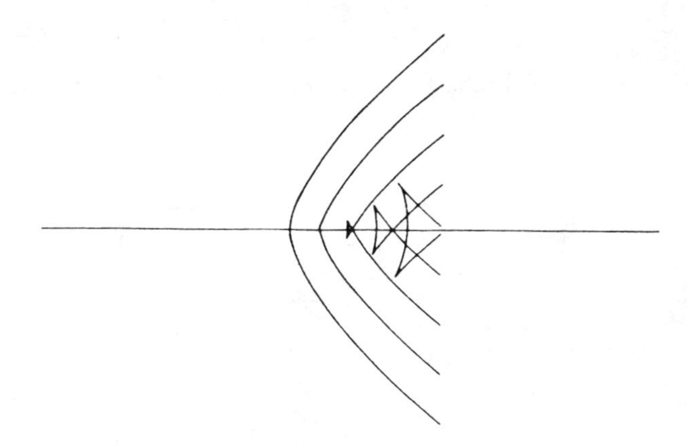

Figure 2.28: Subreflectors for telescopic system in the hemisphere

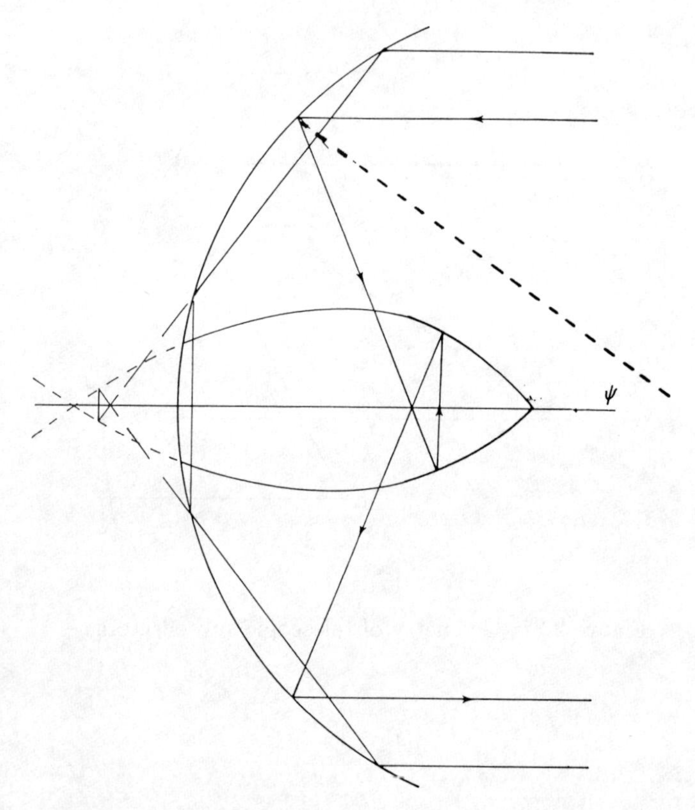

Figure 2.29: Subreflector for retro-reflection in the sphere

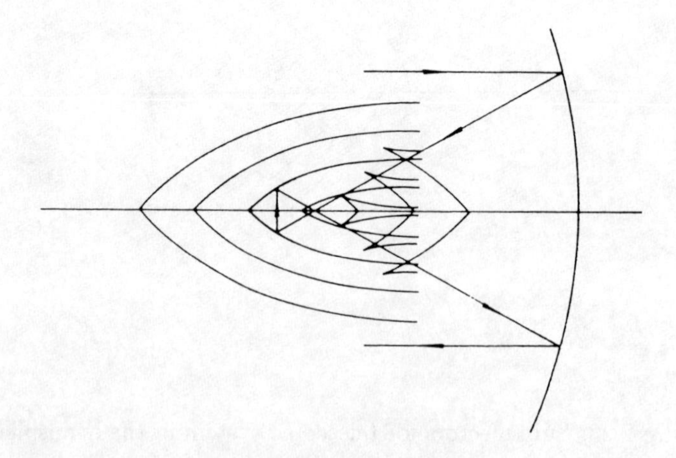

Figure 2.30: Retroreflectors in the hemisphere

from which

$$r = \frac{(2 - x_0) - \cos\psi + \sin\psi}{1 + \sin 2\psi} \tag{2.4.20}$$

giving, together with equations

$$x = \frac{1}{2\cos\psi} - \frac{\cos 2\psi[(4 - 2x_0)\cos\psi - \cos 2\psi - 2]}{2\cos\psi(1 + \sin 2\psi)}$$

$$y = \frac{\sin\psi}{1 + \sin 2\psi}[(4 - 2x_0)\cos\psi - \cos 2\psi - 2] \tag{2.4.21}$$

The reflection property here requires that

$$\frac{dy}{dx} = \frac{1 - \tan\psi}{1 + \tan\psi} \tag{2.4.22}$$

and, although this is difficult to derive from the differentiation of Equation 2.4.21, it can be integrated to derive the equation itself. As shown, the curve obeys the ray requirement geometrically, even for rays which can not physically reach it through being intercepted beforehand by the main spherical reflector.

In all these examples we have taken the position of the subreflector to be x_0 from the centre of the hemisphere.

RADIAL RAY RETROREFLECTOR

A subreflector which reflects every ray into a pencil with a virtual focus at the centre of the hemisphere, as shown in Figure 2.31 will, in combination with the hemisphere, reflect all rays back along their original direction. This property requires

$$r + \cos\psi + 1 - \ell = 3 - 2x_0 = a \text{ (say)}$$

and

$$\ell^2 = x^2 + y^2$$

From these

$$r = \frac{2a - a^2 - \cos^2\psi - 2(1 - a)\cos\psi}{2(2\cos\psi + 1 - a)} \tag{2.4.23}$$

is obtained with the resultant profiles shown in Figure 2.32.

INVOLUTE RETROREFLECTOR

A subreflector which is itself normal to all the incident rays will be the involute of the caustic, which is the tangential envelope of the rays. For this condition

$$r + \cos\psi = 2 - x_0 \tag{2.4.24}$$

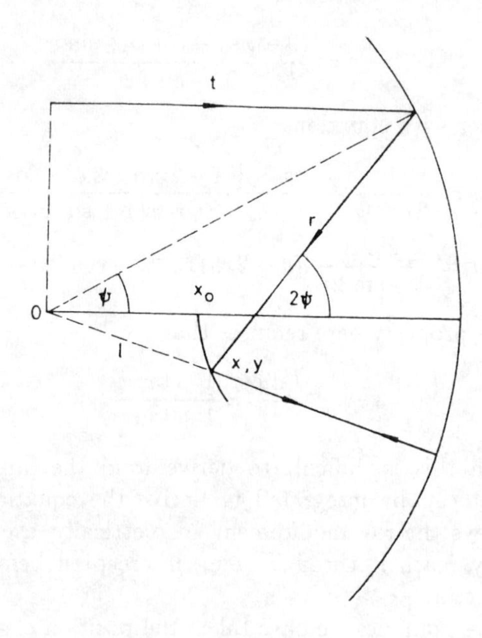

Figure 2.31: Radial ray retroreflector

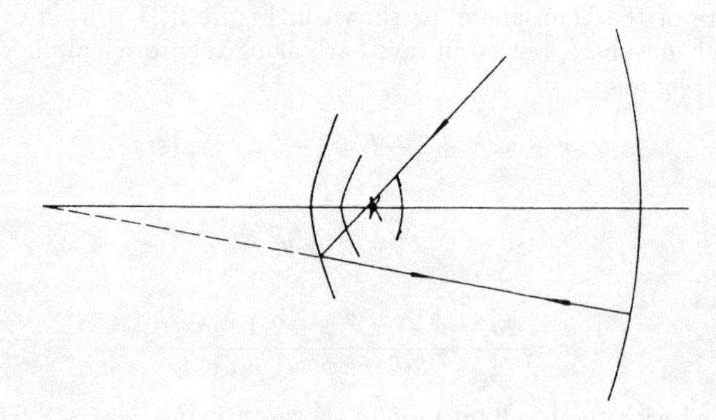

Figure 2.32: Profiles for Equations 2.4.23 and 2.4.16

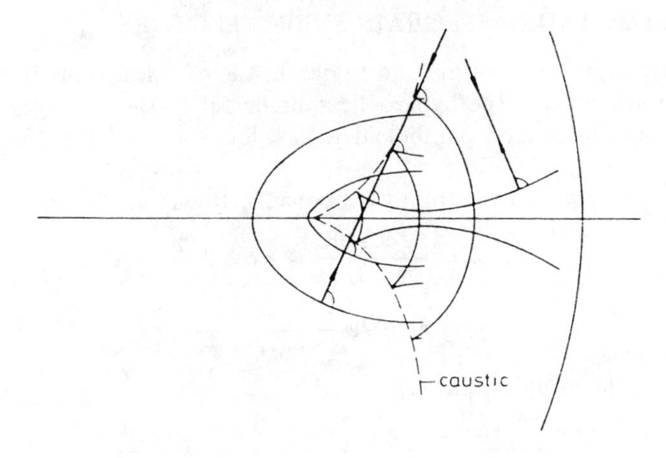

Figure 2.33: Involute of the caustic as a subreflector

This is a far simpler result than previous descriptions for the involute of this particular caustic. The subreflectors and the caustic are shown in Figure 2.33 from which this property can be confirmed.

Once a dual-reflector system is proposed for a given antenna requirement, there need be no compulsion to confine the main reflector to a paraboloid. From the foregoing a subreflector can be obtained for the focusing in any manner in quite general shapes of the main reflector. There may indeed be advantages to be gained in such shapes in the form of constructional simplicity, or, when assessed, in cross-polarization properties, amplitude distributions and multibeam systems. It is shown here that the necessity for Equation 2.4.7 to solve explicitly does occur for a variety of geometrical surfaces.

(a) the circle $\rho =$ constant, $\phi = \psi$

(b) the parabola

$$\rho = 2a(1 + \cos\phi) \qquad \phi = 2\psi \tag{2.4.25}$$

(c) the hyperbola

$$\rho = \frac{b^2}{a(1 + \varepsilon\cos\phi)}$$

$$\cos\phi = -\varepsilon\sin^2\psi + \cos\psi\sqrt{1 - \varepsilon^2\sin^2\psi}$$

where ε is the ellipticity.

(d) the cardioid

$$\rho = a(1 + \cos\phi) \qquad \phi = 2\psi/3 \tag{2.4.26}$$

GREGORIAN AND CASSEGRAIN SUBREFLECTORS

It is illustrative to use the method to obtain the elliptical subreflector of the Gregorian system and the hyperbolic subreflector of the Cassegrain system in an axially illuminated paraboloid with a focus at the apex (focal length unity).

The equations resulting from Equations 2.4.16 and 2.4.23 are

$$x = \frac{2\cos 2\psi}{1+\cos 2\psi} - r\cos 2\psi$$

$$y = r\sin 2\psi - \frac{2\sin 2\psi}{1+\cos 2\psi}$$

and from the focusing condition

$$r = \frac{(1-a^2)\cos^2\psi + 4\sin^2\psi - 2(1-a)\cos 2\psi}{2\cos^2\psi(2-a-\cos 2\psi)} \tag{2.4.27}$$

That these give ellipses and hyperbolas depending on the value of a has been confirmed. Although elaborate in a sense, it does have the advantage of being extendable to offset beams or other forms of focusing both on the incident or the focal sides, as shown below.

THE ZEISS CARDIOID

The cardioid reflector is represented by $\rho = a(1+\cos\phi)$ and hence (Equation 2.4.25) $\phi = 2\psi/3$.

With a focus at the cusp of the cardioid $\theta = \psi - \phi$ (Figure 2.34) and Equations 2.4.16 become

$$x = \rho\cos\phi - r\cos 2\phi$$

$$y = \rho\sin\phi - r\sin 2\phi$$

Solving for a reflected beam of rays parallel to the axis gives

$$r = \frac{4a - 2x_0 - a\sin^2\phi}{2\cos^2\phi} \tag{2.4.28}$$

which, for the special choice of $x_0 = 3a/2$ gives $r = a/2$, the circular profile of the Zeiss combination.

OFFSET BEAMS IN THE SPHERE AND PARABOLOID

Many situations exist which require non-axial beams mainly in the avoidance of subreflector blockage or for beam shaping applications. Considering an incident beam making an angle α with the axis of either a spherical or paraboloidal main reflector, from Equations 2.4.16 and 2.4.10

$$x = \rho\cos\psi - r\cos(2\psi+\alpha)$$

$$y = r\sin(2\psi+\alpha) - \rho\sin\psi$$

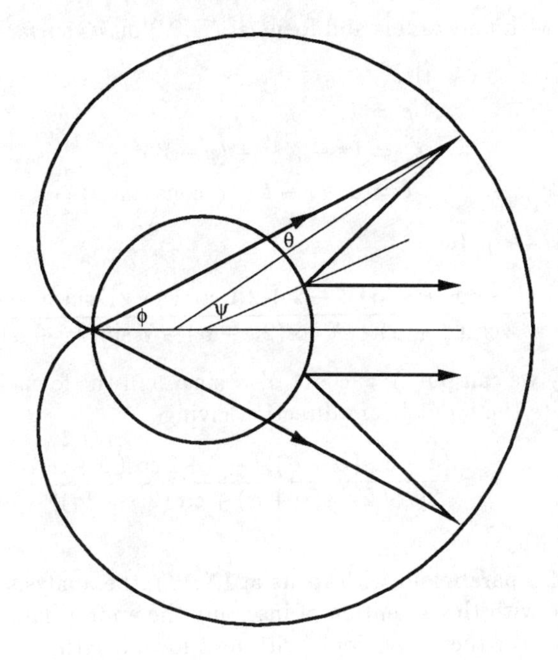

Figure 2.34: Zeiss cardioid

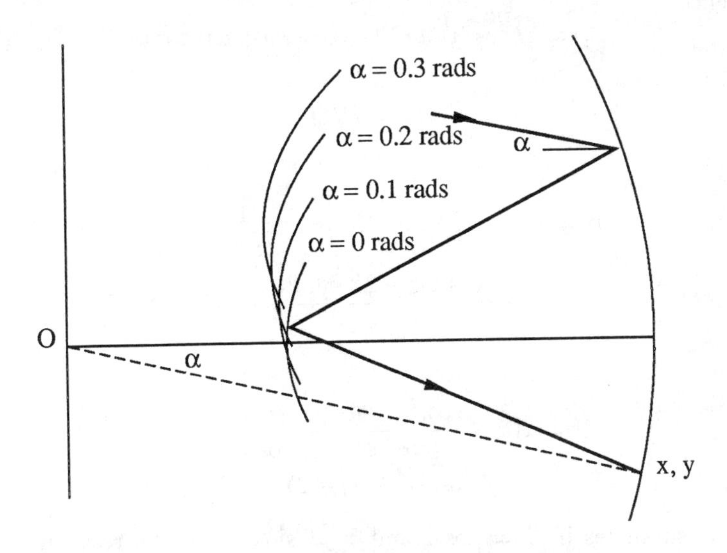

Figure 2.35: Offset reflectors in a hemisphere

For a sphere with unit radius and focus at (X, Y) on its surface (Figure 2.35), we have

$$\rho = 1$$
$$\ell^2 = (x - X)^2 + (y - Y)^2$$
$$\ell = a - r - t \qquad a \text{ constant}$$

and $t = \cos(\psi + \alpha)$. Hence

$$r = \frac{(a - \cos(\psi + \alpha))^2 - 2 + 2X\cos\psi - 2Y\sin\psi}{2[a - 2\cos(\psi + \alpha) + X\cos(2\psi + \alpha) - Y\sin(2\psi + \alpha)]} \tag{2.4.29}$$

For simplicity we can put $X = \cos\alpha$, $Y = \sin\alpha$ and the focus is then diametrically opposite the input beam direction giving

$$r = \frac{(a - \cos(\psi + \alpha))^2 - 2 + 2\cos(\psi + \alpha)}{2[a - 2\cos(\psi + \alpha) + \cos(2\psi + 2\alpha)]} \tag{2.4.30}$$

In the case of a paraboloid with focus at (X, Y), the analysis is the same as for the sphere with the exception of inserting the appropriate descriptions of ρ and hence t. For the paraboloid (with unit focal length)

$$\rho = \frac{2}{1 + \cos\phi} \qquad \phi = 2\psi$$

and hence

$$\rho = 1/\cos^2\psi \text{ and } t = \cos(2\psi + \alpha)\cos^2\psi \tag{2.4.31}$$

Thus

$$r = P/Q$$

where

$$P = \left(a - \frac{\cos(2\psi + \alpha)}{\cos^2\psi}\right)^2 - \frac{1}{\cos^4\psi}$$
$$+ \frac{2X\cos 2\psi - 2Y\sin 2\psi}{\cos^2\psi} - (X^2 + Y^2)$$

and

$$Q = 2(a - \frac{\cos(2\psi + \alpha)}{\cos^2\psi} - \frac{\cos\alpha}{\cos^2\psi}$$
$$+ 2X\cos(2\psi + \alpha) - 2Y\sin(2\psi + \alpha))$$

This too simplifies if $X = \cos\alpha$ and $Y = \sin\alpha$ with the resultant profiles shown in Figure 2.36.

The foregoing analysis has shown that the use of the parameter ψ, the angle between the normal at the surface of a main reflector and the specified axis,

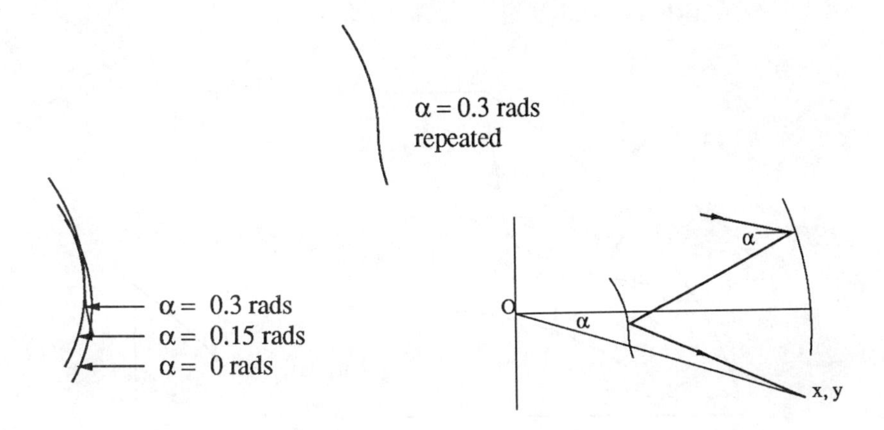

Figure 2.36: Offset reflectors in a paraboloid

leads to a very elementary description of the subreflector profile for any form of incident illumination and for any specified form of focusing. In particular the subreflectors for the sphere have a simple parametric equation. The example was chosen since constructionally the sphere has great advantages over the paraboloid. Whenever an offset geometry is specified, for aperture blockage or beam shaping purposes, the sections of a large offset paraboloid are complex doubly-curved surfaces each of an individual form, whereas those of the sphere are, of course, all of identical shape. This is particularly valuable in the case of deployable sectioned antennas and those requiring spares to be retained. In the offset geometry the second cross-section for the sphere too will be a circle (of different radius) but the design of the profile for that cross-section will retain the simplicity of the original design.

Some of the large variety of antennas based on the paraboloid or sphere as first reflecting surfaces, are shown in Figures 2.37 and 2.38.

As was stated earlier each will have for instance a different aperture amplitude distribution or possibly some other beneficial attribute that can only be discovered from a further investigation of the design. Those based on the paraboloid all consist of the judicious positioning of a second conic section surface (including the plane as a degenerate hyperboloid) with its focus at the same point as that of the paraboloid. Those based on the sphere are Rakovin antennas, so called after their designer, which use the symmetry of the sphere to perform wide angle scanning of the beam through simple rotations and motions of the source. Large radio astronomy antennas ($>$ 100 metres) could be constructed by this means. The cross-section of the subreflectors in this design can easily be seen to derive from a taut string co-involution process.

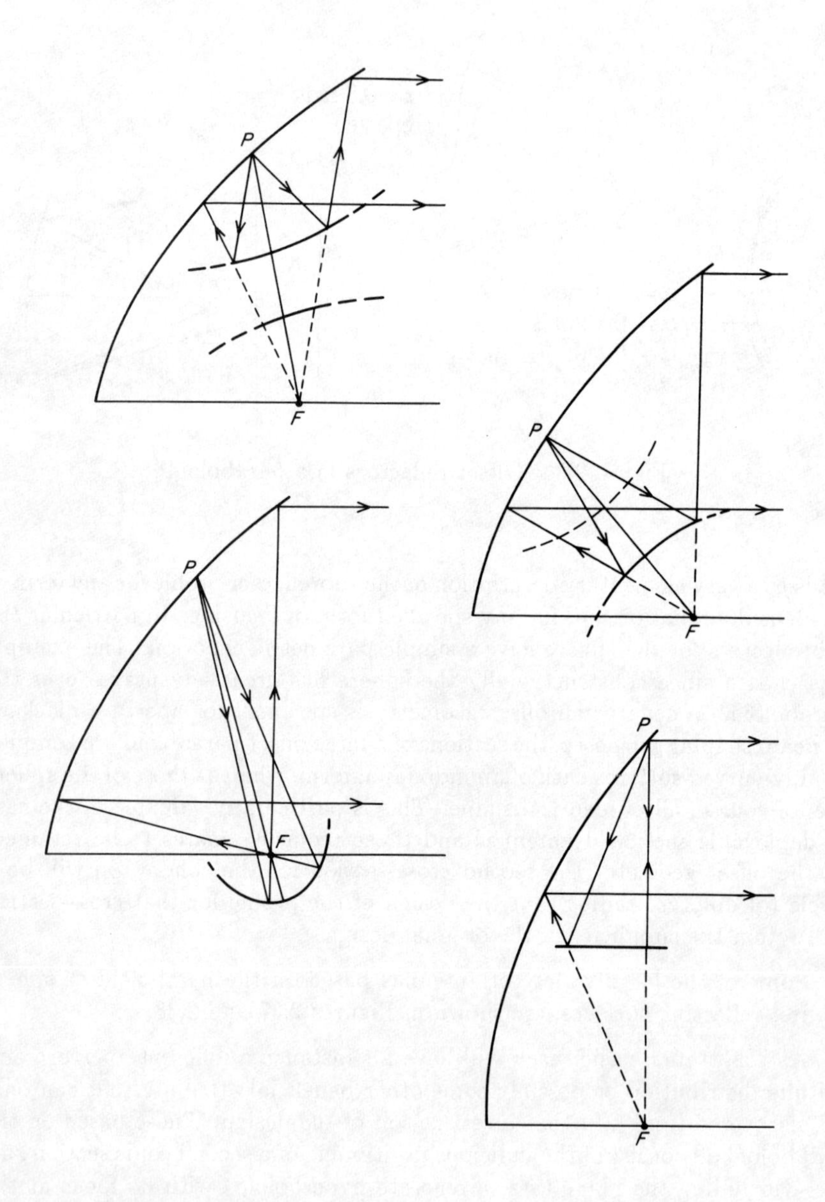

Figure 2.37: Subreflector geometries for offset feed in paraboloid. Subreflectors have rotational symmetry about line PF joining source point to focus

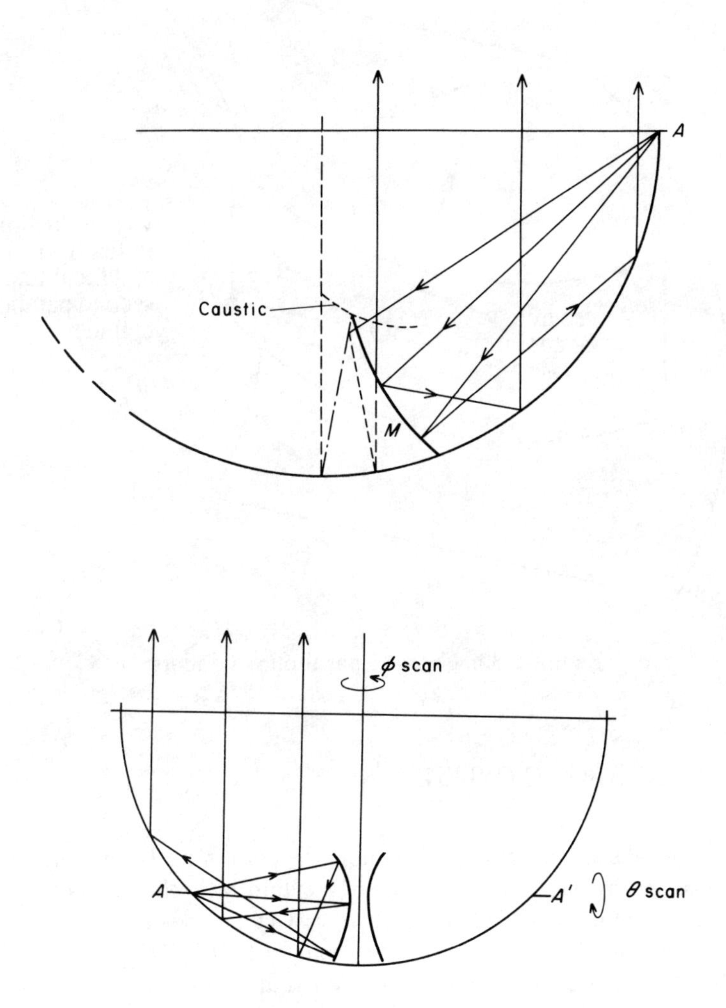

Figure 2.38: Subreflectors for offset feed in a sphere

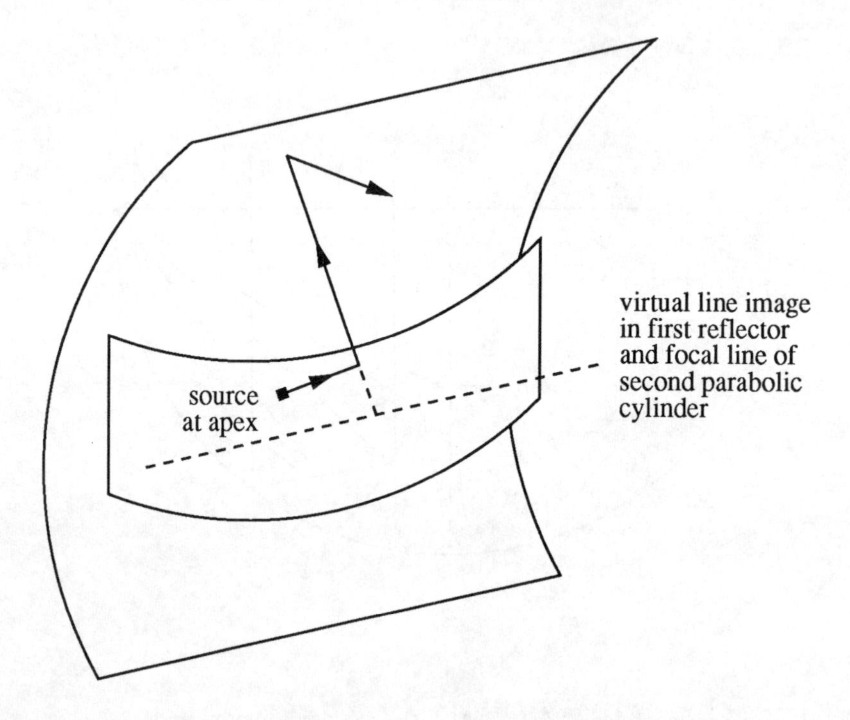

Figure 2.39: Crossed parabolic cylinders

CROSSED PARABOLIC CYLINDERS

A rather specialised construction but nonetheless a strictly two-surface system, is obtained by using the focal line of a cylindrical reflector as the source for a second cylindrical surface. Since the first cylinder is required to have a straight line virtual image, as seen in Section 1.2 REFLECTION IN A GENERAL CURVE, it has to have cross-section that is a conic section. We can take this to be parabolic. Likewise the second surface has to be such that it requires a line source for its illumination to give a collimated beam. It too therefore has to be parabolic in cross-section. Since the line image of the first reflector is (surprisingly?) at right angles to the generators of the cylinder, the generators of the two surfaces have to be perpendicular to each other as shown in Figure 2.39. One major advantage of the design is that, both surfaces being cylindrical and not doubly curved, no cross-polarization can be introduced. For some applications, notably space communications, this is a vital consideration. Within certain tolerances, a small degree of scanning can also be obtained in each of the basic directions, by rotating the appropriate cylindrical reflector.

2.5 METHOD OF GENERIC CURVES FOR ANTENNA PROFILES

The first method [1] we introduce is a continuation of the method given in Section 2.4 and uses the notation and figures given therein. In effect we preempt the necessity of having some form of series solution to an equation by creating a series representation of the first surface profile itself. We base this on the condition of Equation 2.4.8, where it is required to solve Equation 2.4.7 in the form

$$\frac{\mathrm{d}\rho}{\rho} = \tan(\phi - \psi)\ \mathrm{d}\phi \tag{2.5.1}$$

through a substitution

$$\psi = \psi(\phi) \tag{2.5.2}$$

Equation 2.5.1 becomes soluble for complete sets of reflector profiles each set having a free parameter which provides for a smooth transition between the curves. Obviously enough, we select profiles which can be expected to conform with the eventual shape anticipated, for example, sets of curves that are concave towards the source and second reflector. All curves are normalized to a unit focal length, that is, $\rho(0) = 1$. It has to be pointed out at this stage that the process forms an excellent exercise in integration for mathematics students.

At the appropriate stage of the analysis, which follows that of Section 2.4, a series of generic $\rho(\phi)$ curves of the form

$$\rho = a_1\rho_1(\phi, n_1, c) + a_2\rho_2(\phi, n_2) + \ldots + a_6\rho_6(\phi, n_6) \tag{2.5.3}$$

where $\rho_i(i = 1 \text{ to } 6)$ are the polar form of the generic functions with parameters n_i, will be introduced and the coefficients "manipulated" (that is computer optimised) to obtain agreement with a chosen second criterion.

We give here a selection of such profiles, and note that, in all cases, both ψ and ϕ can be derived explicitly in terms of the other, an essential point for all analysis.

1. $\psi = n\phi$

$$\frac{\mathrm{d}\rho}{\rho} = \frac{\sin(1-n)\phi}{\cos(1-n)\phi}\mathrm{d}\phi$$

and hence, putting $(n-1) \equiv m$,

$$\rho^m = \cos m\phi \tag{2.5.4}$$

These curves are the sinusoidal spirals (Appendix III) and include

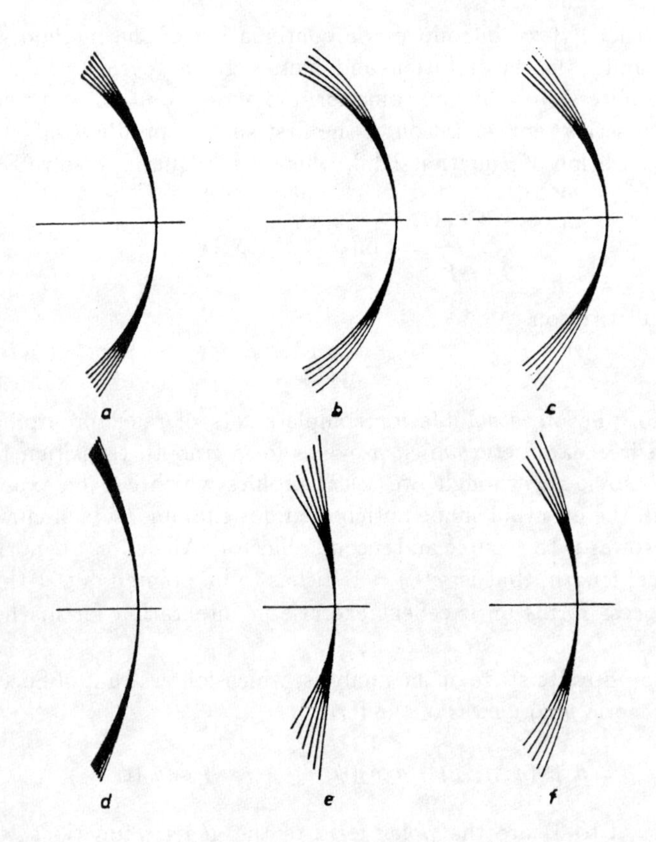

Figure 2.40: Generic curves for main reflector profiles
(a) $\rho^m = \cos(m\phi/c)$
(b) $\rho = (\cos^2\phi + n\sin^2\phi)^{-1/2}$
(c) Equation 2.5.7
(d) $\rho = (n+1)/(n+\cos\phi)$
(e) $\rho = 1 - n + n\sec\phi$
(f) $\rho = 1 - n + n\sec^2\phi$

$m = 0$ the logarithmic spiral
$= 1/3$ Cayley's sextic
$= 1/2$ the cardioid
$= 1$ the circle
$= 2$ lemniscate of Bernoulli
$= -1/3$ Tchirnhausen's cubic
$= -1/2$ the parabola
$= -1$ the straight line
$= -2$ the equiangular hyperbola (asymptotes at$\pm 45^o$)

As we are only to consider a finite angular aperture of the primary reflector, the variable ϕ is confined to $\pm 60^o$ (or less if necessary) in the ensuing analysis. Within this range, variations of m about any of the above values will give reflectors of variable curvature. For example, varying m between 0.8 and 1.2, say, will change the shape from the circle $m = 1$ toward the cardioid in one sense and the lemniscate in the other (Figure 2.40(a)).

These curves can be generalized further to $\rho^m = \cos(m\phi/c)$, in which form they are utilised subsequently.

We require for future analysis the limit, as ϕ and ψ tend to zero, of $\mathrm{d}\psi/\mathrm{d}\phi$ for each of the reflector systems under consideration. In assuming that both ϕ and ψ tend to zero on the axis we are stating that only curves that are smooth and normal to the axis at the apex are to be considered. This condition is mainly applicable to the Abbe sine condition and circularly symmetric systems.

In general

$$\lim_{\substack{\psi,\phi \to 0 \\ \rho(0) \to 1}} \frac{\mathrm{d}\psi}{\mathrm{d}\phi} = \frac{1 + 2\left(\frac{\mathrm{d}\rho}{\mathrm{d}\phi}\right)^2 - \frac{\mathrm{d}^2\rho}{\mathrm{d}\phi^2}}{1 + \left(\frac{\mathrm{d}\rho}{\mathrm{d}\phi}\right)^2} \tag{2.5.5}$$

For the sinusoidal spirals of this section

$$\frac{\mathrm{d}\psi}{\mathrm{d}\phi} \to 1 + \frac{m}{c^2}$$

2. $\tan\psi = n\tan\phi$
Hence

$$\frac{\mathrm{d}\rho}{\rho} = \frac{(1-n)\tan\phi\,\mathrm{d}\phi}{1 + n\tan^2\phi}$$

and thus

$$\rho = (\cos^2\phi + n\sin^2\phi)^{-1/2} \tag{2.5.6}$$

These are ellipses with major axis parallel to the x axis if $n > 1$, a circle if $n = 1$, and ellipses with major axis parallel to the y axis if $n < 1$. The profiles in the range $n = 0.8$ to 1.2 are shown in Figure 2.40(b).

We can also obtain

$$\lim_{\phi,\psi\to 0} \frac{d\psi}{d\phi} = n$$

3. $\tan\psi = n\sin\phi$ (looking innocent enough)
   Then, with $m^2 = 4n^2 + 1$, we obtain

$$\rho^{2m} = \frac{[\sqrt{m^2-1}+m][m+1-(m-1)\cos^2\phi]}{\sqrt{m^2-1}(1+cos^2\phi)+2m\cos\phi} \times \left[\frac{\sqrt{m^2-1}}{2}\sin^2\phi+\cos\phi\right]^{-m} \tag{2.5.7}$$

and

$$\lim_{\phi,\psi\to 0} \frac{d\psi}{d\phi} = n = \frac{1}{2}\sqrt{m^2-1}$$

4. $\tan\psi = n\sin\phi/(n+\cos\phi)$
   From which we obtain

$$\rho = \frac{n+1}{n+\cos\phi} \tag{2.5.8}$$

as shown in Figure 2.40(d), and

$$\lim_{\phi,\psi\to 0} \frac{d\psi}{d\phi} = \frac{n}{n+1}$$

These are the "near parabolic" hyperbolae or ellipses, that is, parabolic for $n = 1$, hyperbolae for $n < 1$ with asymptote at $\phi = \pi - \cos^{-1} n$ and ellipses with $n > 1$.

5. Other curves can be represented directly by their polar relation such as

$$\rho = (1-n) + n\sec\phi \tag{2.5.9}$$

These are the conchoids of Nicodemes and vary from the straight line $n = 1$ to the circle $n = 0$ as shown in Figure 2.40(e).

From Equation 2.5.1 we obtain

$$\cot\psi = \frac{n}{(1-n)\sin\phi\cos^2\phi} + \cot\phi \tag{2.5.10}$$

and hence

$$\lim_{\phi,\psi\to 0} \frac{d\psi}{d\phi} = 1-n$$

6.

$$\rho = (1-n) + n\sec^2(\phi/2) \tag{2.5.11}$$

Parabolae of varying curvature at the apex shown in Figure 2.40(f).

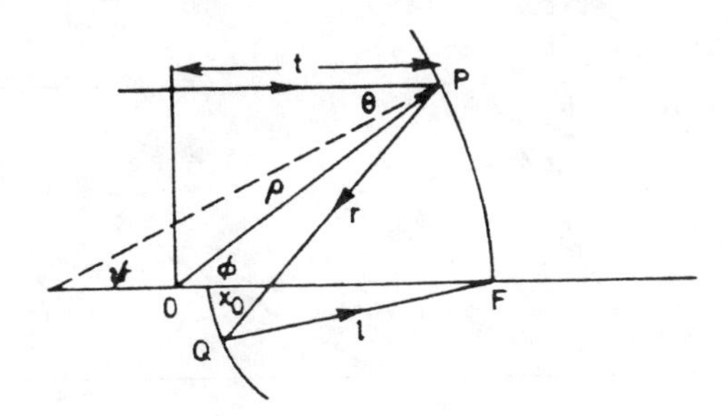

Figure 2.41: Parallel plane wave illumination

We obtain

$$\cot\psi = \frac{2n}{\sin\phi[2n + (1-n)(1+\cos\phi)^2]} + \cot\phi \tag{2.5.12}$$

and

$$\lim_{\phi,\psi\to 0} \frac{d\psi}{d\phi} = 1 - \frac{n}{2} \tag{2.5.13}$$

Numerous other curves of this kind, parametrized by a curvature variable n, can be considered, by integration of Equation 2.5.3.

The method is essentially the derivation of the intercept PQ $= r$ (Figure 2.41) for substitution into Equations 2.4.12 (repeated here for convenience).

$$\begin{aligned} x &= \rho\cos\phi - r\cos(\theta+\psi) \\ y &= r\sin(\theta+\psi) - \rho\sin\phi \end{aligned}$$

All terms are then given in parametric form, with parameter ϕ. For simplicity we take the plane wave, uniform phase condition shown in Figure 2.41 as the first criterion, although as illustrated previously, other phase forms or focus positions can be accommodated.

ABBE SINE CONDITION

The Abbe sine condition (Appendix I.14) for a symmetrical optical system with uniform phase in the aperture is met when the continuation of a ray from the source meets its eventual transmitted ray direction at a point which lies on a circle with centre at the source. When this condition is met the optical system suffers minimum aberration over a region near the focus transverse to the axis. It is thus suitable for antennas with a small angle of scan, or for multifeed applications with a view to beam shaping. As shown in Figure 2.42

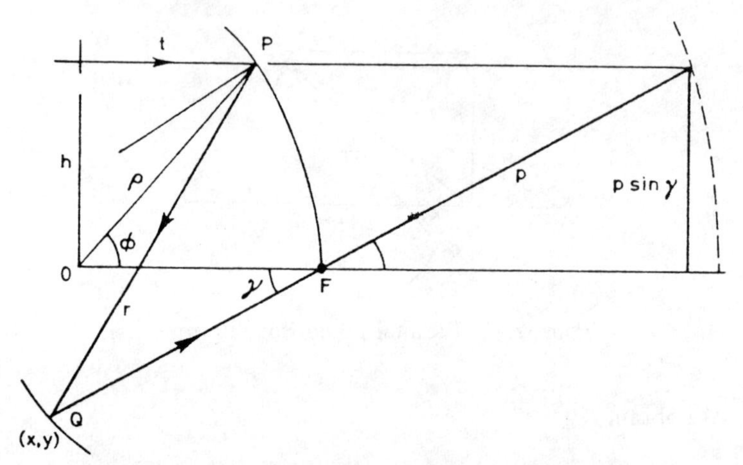

Figure 2.42: Abbe sine condition

we let the radius of this circle be p, then

$$h = p \sin \gamma = \rho \sin \phi$$

or

$$p = \rho \sin \phi \left[\frac{a - \rho \cos \phi - r}{r \sin 2\psi - \rho \sin \phi} \right]$$

This is valid over the entire range of ϕ, and so

$$p = \lim_{\phi, \psi \to 0} \left\{ \rho \sin \phi \left[\frac{a - \rho \cos \phi - r}{r \sin 2\psi - \rho \sin \phi} \right] \right\}$$

All $\rho = 1$ at $\phi = 0$ by design, and in the same limit

$$r \to 1 - x_0 = \frac{a-1}{2}$$

Hence

$$p = \frac{a-1}{2(a-1) \lim \frac{d\psi}{d\phi} - 2} \tag{2.5.14}$$

Solving for r and comparing the relation obtained with that for r in Equation 2.4.15 (repeated below),

$$r = \frac{1 + \rho^2 - 2 \cos \phi - (a-t)^2}{2[t - a - \cos(\theta + \psi) + \rho \cos(\theta + \psi - \phi)]}$$

we require

$$\frac{1 - a^2 + \rho^2 \sin^2 \phi - 2\rho(1-a) \cos \phi}{2[\rho \cos \phi - a - cos\, 2\psi + \rho \cos(2\psi - \phi)]} = \frac{\rho \sin \phi (a - \rho \cos \phi + p)}{p \sin 2\phi + \rho \sin \phi}$$

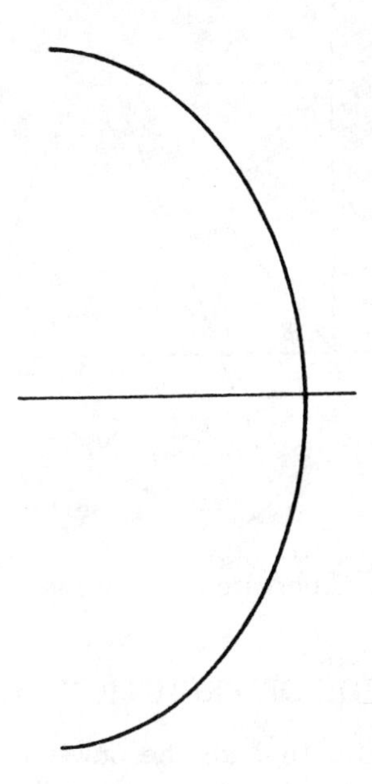

Figure 2.43: Profiles of antenna reflector for Abbe sine condition

We now substitute the series given in Equation 2.5.3.

In Equation 2.5.14 lim $d\psi/d\phi$ is given by Equation 2.5.5. The coefficients a_i and the parameters n_i are then optimised to make the difference between left- and right-hand sides of this equation minimum at selected values of ϕ in the range 0° to 60° and for given values of a. Rapid convergence occurs for general values of a in all except one instance. This turned out to be the case where the apex of the subreflector coincided with a cusp of the caustic of the main reflector when illuminated by a parallel beam of light. Different values of a imply different positions of the apex of the subreflector and hence determine factors such as maximum angle of ϕ, or aperture blockage or, generally, the overall geometry of the combination, although always allowing agreement with the Abbe condition. A typical main reflector to this design is shown in Figure 2.43 and is intermediate between a circle and a parabola.

The residual error is the amount by which the points of intersection of the source and aperture rays differ from the circle of radius p. The limit for this was set at the nominal tolerance on a reflector profile, say 10^{-3}. In the event this was improved upon by at least one order and over the paraxial region by two or more orders of magnitude. Obviously, limiting the aperture to smaller values of ϕ would improve even on this figure.

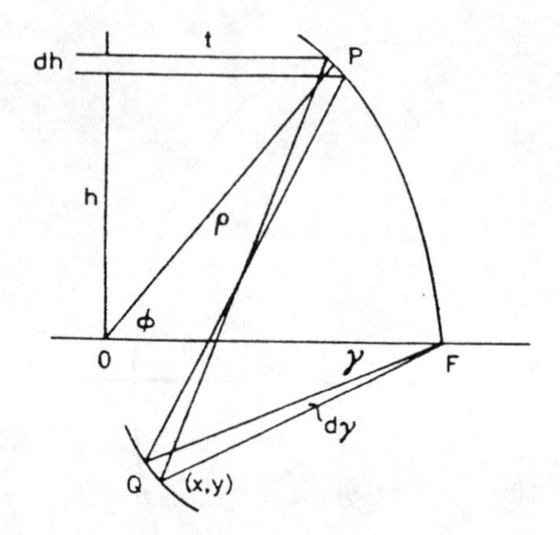

Figure 2.44: Aperture distribution - uniform phase

APERTURE AMPLITUDE DISTRIBUTION

Conservation theory states that all the power contained in the solid angle $2\pi \sin\gamma\, \mathrm{d}\gamma$ (Figure 2.44) is radiated into the solid angle $2\pi \sin\beta\, \mathrm{d}\beta$. Hence, if $P(\gamma)$ is the power distribution at the source and $P(\beta)$ that in the transmitted beam, these are related by

$$\frac{P(\gamma)}{P(\beta)} = \frac{|\sin\beta|\,\mathrm{d}\beta}{|\sin\gamma|\,\mathrm{d}\gamma}$$

In the case of uniform phase $\beta = 0$ and the power is radiated into an annulus of area $2\pi h dh$ where $h = \rho\sin\phi$. Then we obtain

$$\frac{P(\gamma)}{P(h)} = \frac{h\mathrm{d}h}{|\sin\gamma|\,\mathrm{d}\gamma} \tag{2.5.15}$$

and

$$\mathrm{d}h = -\rho\cos\phi\,\mathrm{d}\phi$$

The area of the annulus is therefore

$$2\pi h dh = \frac{2\pi\rho^2 \sin\phi\cos\phi\,\mathrm{d}\phi}{\cos(\psi - \phi)}$$

From Figure 2.44 and Equations 2.4.12 we have, for a source at the apex and $\rho(0) = 1$,

$$\cos\gamma = \frac{1 - x}{a - t - r} = \frac{1 - \cos\phi + r\cos 2\psi}{a - t - r} \tag{2.5.16}$$

For most cases the power distribution of the source can be expressed in the form

$$P(\gamma) = (\cos\gamma)^s \tag{2.5.17}$$

Then Equation 2.5.15 becomes

$$(\cos\gamma)^s \sin\gamma \, d\gamma = \pm P(h) h dh$$

with sign appropriate to the way h and γ vary, that is, whether two adjacent rays cross between the main and subreflectors or not. $P(h)$ can be expressed as any integrable function of h and Equation 2.5.17 integrates directly to give

$$(\cos\gamma)^{s+1} = -(s+1) \int P(h) h dh + \text{ const.} \tag{2.5.18}$$

From Equation 2.5.16 (point focus at the apex) the left-hand side is

$$\left[\frac{1 - \rho \cos\phi + r \cos 2\psi}{a - t - r} \right]^{s+1}$$

(For parallel axial rays $\theta = \psi$ and $t = \rho \cos\phi$ and other ray distributions give their appropriate forms.)

In Equation 2.5.18, r is given by Equation 2.4.15 and hence the right- and left-hand sides of Equation 2.5.18 can be "fitted" by a series of generic curves in the form of Equation 2.5.3, and optimised by the same computational routine.

For example, for a uniform in-phase aperture distribution $P(h) = 1$ and $h = \rho \sin\phi$ the right-hand side of Equation 2.5.18 is simply

$$1 - (1+s)\frac{h^2}{2} = 1 - (1+s)\frac{\rho^2 \sin^2\phi}{2}$$

where the constant of integration has been evaluated to give agreement at $\phi = \psi = 0$.

Substitution of the series for ρ given in Equation 2.5.3 can now be made and the coefficients manipulated to obtain agreement with Equation 2.5.19.

By virtue of the terms in ψ which occur in Equation 2.5.14 and in the expression for r itself, this equality is, in fact, a nonlinear differential equation but its relation to the exact nonlinear differential equations in Section 2.6 has not been determined.

The main reflector profile for this particular example with

$$P(\gamma) = (\cos\gamma)^4 \quad (a = 2.2)$$

is shown in Figure 2.45. The residual error is the difference between the right- and left-hand sides of Equation 2.5.15 after optimisation. Because this is a relation between power levels, the optimisation accuracy required can be considerably reduced, and, in the present case, the power distribution was uniform to within ± 0.05.

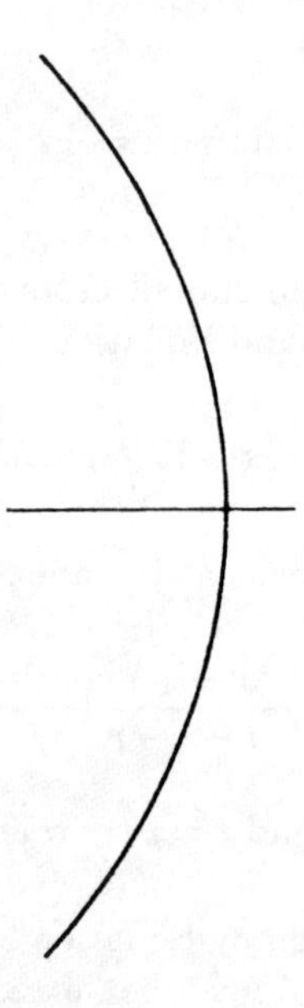

Figure 2.45: Profile of antenna reflector for uniform aperture distribution

OFFSET GEOMETRY

The true offset geometry will consist of two reflectors with an undefined axis, the source at any location and an illumination of any incidence. There will, however, be a single plane of symmetry in which the source will similarly be symmetrical or be a point in the plane. This, the offset plane, will intersect the reflectors in profiles which can be designed by the method given above. The same process can then be used for plane cross-sections angularly spaced about the nominal axis, the axis used in determining the reference phase constant a. The profiles obtained for each cross-section successively build up the three-dimensional surfaces of both reflectors being optimised as above at each stage. This procedure makes the assumption that the ray path is confined to the plane of the cross-section. It is discussed later (Section 2.7), where it is shown to be a good approximation in most practical cases.

It should be noted that the series in Equation 2.5.3 is not the most general that could be used. Each of the individual ρ_i could itself be a separate series of terms over the parameters n_i. This, plus the numerous forms which the ρ_i themselves can take, leads to an extremely wide selection for the accurate figuring of the main reflector profile. Furthermore, the series is not unique in any way, since some curves arise in each parametric form, the circle notably. There is therefore no possibility of solving for the coefficients by an orthogonality procedure.

Six examples of generic curves have been utilised and these are completely arbitrary. Other suitable curves can be considered. Their derivation and application then become part of the "art" of antenna design. For example, delicate figuring can be included through curves of the form

$$\rho = 1 - n + n \cos m\phi$$

where the value of m places a maximum or minimum at any required value of ϕ. The coefficients of the series, or any like series, can then be optimised by well-established numerical methods, including simple trial and error, to compel the profiles so obtained to obey the given second condition. The illustrations have been confined to symmetrical systems with a point source, but the method is completely general for offset systems both of source and/or reflectors and for general phase distributions. Consequently, provided a shaped beam can be specified by a relation between the incident angle of its constituent rays and the parameter ψ, then a relation between θ and ψ can be established and the method would then provide the required subreflector profile.

2.6 THE GEOMETRICAL THEORY

The purely formal theory is based on the coordinate geometry of surfaces and has all the elements that can be observed in the published treatments. We present here the fundamental geometric formulae governing the relations between two arbitrary surfaces connected by the laws of reflection and an eikonal function for the lengths of the ray paths. The derivation of the solutions for given a priori conditions, usually the power transformation between the source and the exit aperture and a prescribed phase surface at the aperture, is indicated. The problems that arise in deriving a solution can be compared with those arising from other methods of solution, notably those which give rise to nonlinear partial differential equations. The problem itself merits a complete study of the geometrical theory of surfaces of which only an outline can be given here.

Any surface can be parametrized in terms of two variables as

$$x(u,v) \quad y(u,v) \quad z(u,v) \tag{2.6.1}$$

then if l m and n are the direction cosines of the normal to the surface,

$$\frac{y_u z_v - z_u y_v}{l} = \frac{z_u x_v - x_u z_v}{m} = \frac{x_u y_v - y_u x_v}{n} = h \tag{2.6.2}$$

where $h^2 = eg - f^2$ and

$$\begin{aligned} e &= x_u^2 + y_u^2 + z_u^2 \\ f &= x_u x_v + y_u y_v + z_u z_v \\ g &= x_v^2 + y_v^2 + z_v^2 \end{aligned} \tag{2.6.3}$$

the suffices referring to partial derivatives.

It is also necessary to derive the principal radii of curvature of a surface for which we require

$$e' = \begin{vmatrix} x_{uu} & y_{uu} & z_{uu} \\ x_u & y_u & z_u \\ x_v & y_v & z_v \end{vmatrix}$$

$$f' = \begin{vmatrix} x_{uv} & y_{uv} & z_{uv} \\ x_u & y_u & z_u \\ x_v & y_v & z_v \end{vmatrix}$$

$$g' = \begin{vmatrix} x_{vv} & y_{vv} & z_{vv} \\ x_u & y_u & z_u \\ x_v & y_v & z_v \end{vmatrix} \tag{2.6.4}$$

The natural coordinates for the two surfaces which comprise a large or main reflector and a smaller subreflector are polar coordinates from a common origin (Figure 2.46). This origin is taken to be the source of a partial spherically symmetrical beam of rays with a known power distribution $P_s(\theta, \phi)$ about a central value $(\theta_0, 0)$. In the literature a variety of mixed coordinate systems are used, sometimes from independent origins, and a problem occurs in transforming between them. As we shall see, problems of a similar nature occur with a single origin as in the following. We shall designate all parameters and coordinates relating to the smaller reflector by lower case symbols and the main reflector by upper case symbols. The subreflector surface is therefore expressed in two parameter spherical polar coordinates by $x = r\sin\theta\cos\phi$; $y = r\sin\theta\sin\phi$; $z = r\cos\theta$ where, essentially, the surface is defined in shape by

$$r = r(\theta, \phi) \tag{2.6.5}$$

The main reflector is accordingly $X = R\sin\Theta\cos\Phi$; $Y = R\sin\Theta\sin\Phi$; $Z = R\cos\Theta$ and $R = R(\Theta, \Phi)$. Then with equivalent values in capital symbols

$$\begin{aligned}
x_\theta &= r\cos\theta\cos\phi + \frac{\partial r}{\partial\theta}\sin\theta\cos\phi \\
y_\theta &= r\cos\theta\sin\phi + \frac{\partial r}{\partial\theta}\sin\theta\sin\phi \\
z_\theta &= -r\sin\theta + \frac{\partial r}{\partial\theta}\cos\theta \\
x_\phi &= -r\sin\theta\sin\phi + \frac{\partial r}{\partial\phi}\sin\theta\cos\phi \\
y_\phi &= r\sin\theta\cos\phi + \frac{\partial r}{\partial\phi}\sin\theta\sin\phi \\
z_\phi &= \frac{\partial r}{\partial\phi}\cos\theta
\end{aligned} \tag{2.6.6}$$

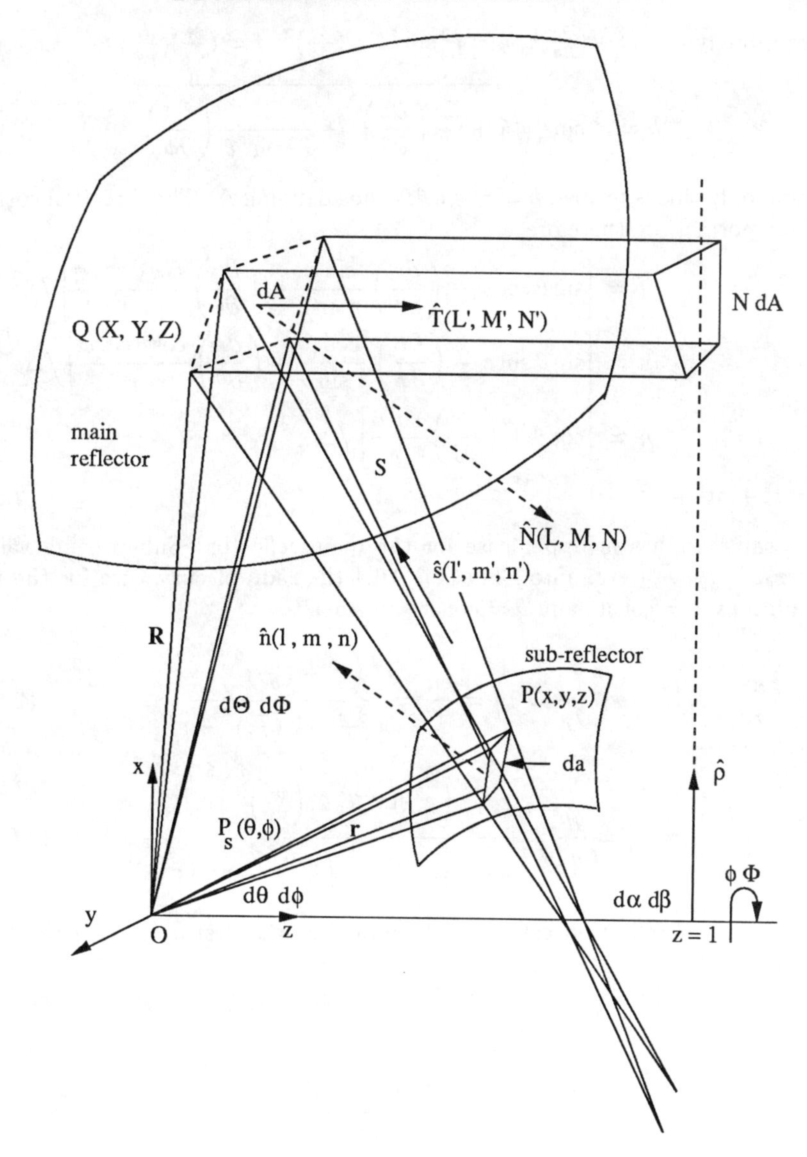

Figure 2.46: Reflection of rays from two reflectors. OPQ and PQN define different planes

Accordingly $e = r^2 + (\frac{\partial r}{\partial \theta})^2$; $g = r^2 \sin^2\theta + (\frac{\partial r}{\partial \phi})^2$; $f = (\frac{\partial r}{\partial \theta})(\frac{\partial r}{\partial \phi})$ and therefore

$$h = r^2 \sin\theta \sqrt{1 + \frac{1}{r^2}\left(\frac{\partial r}{\partial \theta}\right)^2 + \frac{1}{r^2 \sin^2\theta}\left(\frac{\partial r}{\partial \phi}\right)^2} \tag{2.6.7}$$

Commonly this is termed $h = r^2 \sin\theta\Delta$ thus defining Δ. The direction cosines of the normal are therefore

$$\begin{aligned} l &= \left[\sin\theta\cos\phi + \left(\frac{\partial r}{\partial \phi}\right)\frac{\sin\phi}{r\sin\theta} - \left(\frac{\partial r}{\partial \theta}\right)\frac{\cos\theta\cos\phi}{r}\right]/\Delta \\ m &= \left[\sin\theta\sin\phi - \left(\frac{\partial r}{\partial \phi}\right)\frac{\cos\phi}{r\sin\theta} - \left(\frac{\partial r}{\partial \theta}\right)\frac{\cos\theta\sin\phi}{r}\right]/\Delta \\ n &= \left[\cos\theta + \left(\frac{\partial r}{\partial \theta}\right)\frac{\sin\theta}{r}\right]/\Delta \\ l^2 + m^2 + n^2 &= 1 \end{aligned} \tag{2.6.8}$$

The same applies in upper case for the main reflector. Substituting values for $x_{\theta\theta}$, $x_{\phi\phi}$, $x_{\theta\phi}$ etc. into Equation 2.6.4 the radii of curvature for the two conditions $\phi =$ const. and $\theta =$ const. are then

$$\rho\left(\frac{\partial}{\partial\phi} = 0\right) = \frac{eh}{e'} = \frac{\left(r^2 + (\frac{\partial r}{\partial \theta})^2\right)^{3/2}}{r\frac{\partial^2 r}{\partial \theta^2} - 2\left(\frac{\partial r}{\partial \theta}\right)^2 - r^2} \tag{2.6.9}$$

$$\rho\left(\frac{\partial}{\partial\theta} = 0\right) = \frac{gh}{g'} = \frac{\left(r^2\sin^2\theta_c + \left(\frac{\partial r}{\partial \phi}\right)^2\right)^{3/2}}{\sin\theta_c\left[r\frac{\partial^2 r}{\partial \phi^2} - 2\left(\frac{\partial r}{\partial \phi}\right)^2 - r^2\sin^2\theta_c\right]} \tag{2.6.10}$$

Snell's law of reflection can now be applied to the first surface in the form (Figure 2.46)

$$\hat{\mathbf{s}} = \hat{\mathbf{r}} - 2\hat{\mathbf{n}}(\hat{\mathbf{r}} \cdot \hat{\mathbf{n}}) \tag{2.6.11}$$

with

$$\hat{\mathbf{r}} = \frac{x\hat{\imath} + y\hat{\jmath} + z\hat{k}}{\sqrt{x^2 + y^2 + z^2}}$$

$$\hat{\mathbf{n}} = l\hat{\imath} + m\hat{\jmath} + n\hat{k}$$

This form incorporates both the scalar and vector products of the incident ray vector and the normal vector as in Appendix I.3. The direction cosines of $\hat{\mathbf{s}}$ are therefore

$$\begin{aligned} l' &= [x - 2l(lx + my + nz)]/r &= x/r - 2l/\Delta \\ m' &= [y - 2m(lx + my + nz)]/r &= y/r - 2m/\Delta \\ n' &= [z - 2n(lx + my + nz)]/r &= z/r - 2n/\Delta \end{aligned} \tag{2.6.12}$$

since $lx + my + nz = r/\Delta$.

All these relations can be parametrized in terms of θ and ϕ. The ray meets the main reflector at the point $Q(X, Y, Z)$ and, with $PQ = S$,

$$\frac{X - x}{l'} = \frac{Y - y}{m'} = \frac{Z - z}{n'} = S \tag{2.6.13}$$

The differentials that form the direction cosines of the normals, have to be taken at the point of contact of the incident ray on the reflector, and while this is specified for the ray $\mathbf{r}$ from the origin, the point of application on the main reflector is as yet unknown. If this point has coordinates $X\ Y\ Z$ parametrised by $X = X(\Theta, \Phi)$ etc. then the incident ray is in the direction $\hat{\mathbf{s}}$ given by Equation 2.6.11 and the normal will be $\widehat{\mathbf{N}}$ with direction cosines LMN given by Equation 2.6.10 in upper case symbols. Thus the reflected ray will be given by

$$\widehat{\mathbf{T}} = \hat{\mathbf{s}} - 2\widehat{\mathbf{N}}(\hat{\mathbf{s}} \cdot \widehat{\mathbf{N}}) \tag{2.6.14}$$

The conditions that apply immediately for the plane phase front perpendicular to the axis are that $\widehat{\mathbf{T}}$ has zero components in the $\hat{\imath}$ and $\hat{\jmath}$ directions. Hence $\widehat{\mathbf{T}} = \hat{k}$. Therefore (compare with Equation 2.6.12)

$$\begin{aligned} L' &= l' - 2L(Ll' + Mm' + Nn') &= 0 \\ M' &= m' - 2M(Ll' + Mm' + Nn') &= 0 \\ \text{and } N' &= n' - 2N(Ll' + Mm' + Nn') &= 1 \end{aligned} \tag{2.6.15}$$

since all vectors are unit vectors.

Other ray directions appropriate to a non-planar phase front would still give three equations of the kind in 2.6.15 with ray direction cosines on the right-hand side. These can be solved, and for the planar case coupled with the conditions $L'^2 + M'^2 + N'^2 = 1$ and $l'^2 + m'^2 + n'^2 = 1$ give

$$L = \frac{l'}{\sqrt{2 - 2n'}} \quad M = \frac{m'}{\sqrt{2 - 2n'}} \quad N = \frac{n' - 1}{\sqrt{2 - 2n'}} \tag{2.6.16}$$

together with the intermediate relation $l'M = m'L$.

Written in full these expressions are complicated trigonometrical relations. The factors r and R in Equation 2.6.10 cancel and thus r and R only appear both explicitly and in the differentials in the expressions for $l\ m\ n\ L\ M\ N$ and Δ. They are, however, first order differentials only. Equations 2.6.16 constitute three functional relations between the five unknowns in the form

$$L(\Theta, \Phi, \frac{\partial R}{\partial \Theta}, \frac{\partial R}{\partial \Phi} \bigg| R) = (l', n')(\theta, \phi, \frac{\partial r}{\partial \theta}, \frac{\partial r}{\partial \phi} \bigg| r)$$

The remaining conditions are to be found in an eikonal relation for the value of the length of ray paths and in the relation governing the transformation of power from its distribution at the source to the required distribution in the aperture.

For the case of a plane aperture with a uniform phase front, we take the aperture to be the plane $z = Z = 1$ thus providing a scale factor for the antenna. The path length eikonal then is

$$r + S + (1 - Z) = \kappa \tag{2.6.17}$$

This we observe is a *first integral* of the differential form of Equation 2.4.1

$$\mathrm{d}r + \mathrm{d}S = \mathrm{d}Z$$

The use of the integral form precludes the necessity of differentiating functions which already contain partial derivatives. Perturbed or non-uniform phase fronts will have eikonal functions that depend on each ray individually, that is, κ will not be a universal constant but takes values according to the ray considered. It will still be applicable in this form.

Before continuing with the general analysis, we will consider the simpler problem of reflector systems that have a central plane of symmetry. In that plane $\phi = \Phi = 0$ $(Y = y = 0)$, the problem reduces to a profile one for which any of the previous methods are applicable. Here we reduce the foregoing equations by the substitutions above together with $\partial/\partial\phi = \partial/\partial\Phi = 0$. Then

$$\begin{aligned}
\Delta^2 &= 1 + \frac{1}{r^2}\left(\frac{\partial r}{\partial\theta}\right)^2 \\
l &= \left(\sin\theta - \frac{\cos\theta}{r}\frac{\partial r}{\partial\theta}\right)/\Delta \\
m &= 0 \\
n &= \left(\cos\theta + \frac{\sin\theta}{r}\frac{\partial r}{\partial\theta}\right)/\Delta
\end{aligned}$$

Hence

$$\begin{aligned}
l' &= \sin\theta - 2l^2\sin\theta - 2nl\cos\theta \\
n' &= \cos\theta - 2n^2\cos\theta - 2nl\sin\theta \\
l'^2 + n'^2 &= 1
\end{aligned}$$

From Equation 2.6.13 therefore

$$S = \frac{R\sin\Theta - r\sin\theta}{l'} = \frac{R\cos\Theta - r\cos\theta}{n'} \tag{2.6.18}$$

and from the eikonal function $S = x - r - (1 - R\cos\Theta)$. Thus R and Θ can be expressed explicitly in terms of r $\partial r/\partial\theta$ and θ.

Returning to the consideration of the total surfaces, we have now to include the remaining condition pertaining to the transformation of the power distribution at the source to the required distribution over the phase aperture $z = Z = 1$. The area illuminated by a pencil of rays included in the solid angular increment $\mathrm{d}\theta\,\mathrm{d}\phi$ at the first reflector is (Figure 2.46)

$$\mathrm{d}a = r^2\sin\theta\Delta\,\mathrm{d}\theta\ \mathrm{d}\phi = h\,\mathrm{d}\theta\ \mathrm{d}\phi \tag{2.6.19}$$

A surface area on the main reflector is correspondingly

$$\mathrm{d}A = H \mathrm{d}\Theta \,\mathrm{d}\Phi \tag{2.6.20}$$

This will project onto the plane $Z = 1$ into $N\,\mathrm{d}A$ where N is given by Equation 2.6.8 (in upper case symbols), or

$$N\,\mathrm{d}A = R^2 \sin\Theta \left(\cos\Theta + \frac{\partial R}{R\partial\Theta}\sin\Theta\right) \mathrm{d}\Theta\,\mathrm{d}\Phi \tag{2.6.21}$$

(the same as $\rho\,\mathrm{d}\rho\,\mathrm{d}\Phi$ in cylindrical coordinates in the aperture with $\rho = R\sin\Theta$). If the power distribution at the source is $P_S(\theta,\phi)$ and the power in the aperture is $P_A(\rho,\Phi)$, these are connected by

$$P_A N\,\mathrm{d}A = P_S \sin\theta\ \mathrm{d}\theta\ \mathrm{d}\phi \tag{2.6.22}$$

There is also the requirement that the total power from the source is converted into the total power in the aperture. That is, the appropriate integrals of the two sides of Equation 2.6.22 within their predefined boundaries, are equal.

What has yet to be established is the geometrical connection between the two areas $\mathrm{d}a$ and $\mathrm{d}A$, in order to make these last two equalities. As will be seen this is the major concern of all the published treatments of the subject and will be discussed in the ensuing section.

It is when full consideration is given to the total surfaces however, that problems arise. The first of these concerns the orientation of the central ray of the source distribution, θ_0 with respect to the direction of the radiated beam, the z axis. This can be met by a rotation of axes to make the central ray at the source parallel to the z axis and incline the output (plane) wave front by the angle θ_0. This introduces only a modest complication where the right-hand sides of Equations 2.6.15 and the eikonal relation have to incorporate the direction cosines of the inclined beam. The major problem is that the incident ray pencil of magnitude $\mathrm{d}\theta\,\mathrm{d}\phi$ will on reflection from the first reflector, have spread (or contracted) into a pencil $\mathrm{d}\alpha\mathrm{d}\beta$ (see Section 1.8 or the more exact derivation in Section 1.7), where the angles $\mathrm{d}\alpha$ and $\mathrm{d}\beta$ do not have a common point of intersection, and that as a consequence, the area illuminated on the main reflector will not conform to the regular orthogonal curvilinear coordinate section $\mathrm{d}\Theta\mathrm{d}\Phi$.

Consequently the projection of the element of the main reflector onto the exit aperture, needed to apply the power transformation law, remains to be resolved. This aspect is further complicated by the fact that, in general, the rays follow a skewed path. That is the plane containing the incident ray and its first reflection, is not the same as the plane containing the reflected ray from the subreflector and its reflection in the main reflector. It is in the manner with which these problems are formulated and dealt with that the various treatments of the subject differ. Finally there is the subject of the coordinate grids on which the geometrical or differential equations are satisfied, or for which the curvatures have been derived. These have to be fitted with a smooth (differentiable) surface. This is largely a matter of computing technique or of available computing power, and can only briefly have consideration.

2.7 METHODS AND SOLUTIONS

Numerous papers have been published in the past decade dealing with the subject of offset dual reflector antennas. It is worthwhile perhaps to recall the reasons why the design is so pursued. Originally the dual reflector system consisted of, usually, a main paraboloid and a subreflector with a conic profile. While these provided sufficient alternatives for some degree of variation of the amplitude distribution across the aperture, only a specialised form [3] gave a design specifically to create a uniform aperture distribution. This was obtained by a gradual variation from the basic conic pair retaining the phase, that is the constancy of the eikonal function, and monitoring the amplitude until the required uniformity (within a specified tolerance) was achieved. Being a rotationally symmetric design, only the cross-section profiles were considered, as they are in the alternative methods of previous sections and others [4].

The concept of offsetting arose as a method of overcoming the blockage caused by the centralized subreflector, and could be achieved by using the same conic section pair as with the centred system, obliquely illuminated by a cone of rays and matched by an oblique section of the larger paraboloid as in Figure 2.37. Being sections of a rotationally symmetrical system the ray paths were the same as for the complete paraboloid-conic pair, that is, the ray was confined to a single plane throughout its three linear increments from the source to the eventual aperture. Again, varying the position of the source, the angles of the oblique cross-sections and the shape and position of the conic section subreflector, gave a considerable variety of designs from which a best fit to the required aperture distributions could be made. These aperture distributions themselves fall into well-defined categories. For shaping the main beam, an in-phase aperture with variable amplitude distribution is singularly ineffective. That is to say, large variations in amplitude alone, are required to produce noticeable changes in the main beam shape. It follows that the tolerance on specified aperture distributions for this prescription can be very broad, allowing ample approximations to be made in whatever method of solution is being sought.

With the advent of satellite communications and ground base astronomy, a second category arose. This takes two forms, one with the need for the highest possible efficiency that a given aperture size can attain and a second with a grossly distorted angular radiation pattern, designed to fit a shaped "footprint" of ground coverage from a satellite antenna. For the latter it is obvious that the ray directions required can not be achieved from a uniform phase distribution in the aperture. Furthermore, as will be shown later (Chapter 5) the degree of non-uniformity attainable in a continuous field distribution is very limited. Consequently discontinuous phase distributions become necessary, which is reflected through the system to become discontinuous phase (and amplitude) distributions at the source. That is an array of individual feeds each with its own amplitude and phase fed from a distributive network. The radiation pattern is then the superposition of the individual patterns from each source and the reflector design can be comparatively conventional. To a certain extent, should the phase perturbation be smooth and small in degree (arbitrarily say, less than a fraction of a wavelength per wavelength of surface), a single reflector can provide a beam of quite distorted shape [5].

The production of high gain, low side-lobe antennas necessitates the removal of the central blockage and leads to the need for the theoretically most efficient system of offset dual reflectors. This too can be achieved in the manner of the centred system by taking an offset paraboloid-conic pair, and gradually changing the profiles to achieve the required aperture distribution [6]. The necessity in this method for the ray to be confined to a plane is acknowledged [8]. With this limitation, the method of generic curves (Section 2.5) applies or alternatively, a similar series of surface shaping $r_i(\theta, \phi)$ functions [7].

All designs consider the two conditions, the aperture phase and the aperture amplitude distribution, to be of paramount importance and we shall follow these criteria hereafter. Since there are two conditions and two surfaces involved the optical principle states that a unique solution exists, and we shall accept that conclusion even though it has given rise to much speculation regarding the solution of the analytical equations derived. To obtain that unique solution for the dual offset reflector antenna, precise and quite general analyses have been developed, which will be surveyed in the following. Broadly speaking, they differ mainly in the method by which they deal with the skewness of the ray paths and subsequently in the relationship between the two areas illuminated by an incremental pencil of rays on the two reflectors, that is da and dA in the analysis of the previous section. Where the transformation between them is described by a Jacobian, complications are caused by the fact that the differentials involved in the Jacobian imply relationships between the variables that can not have the degree of explicitness required. Most commonly this relates to the relation between (in the notation of Section 2.5) Φ and ϕ, and thus again to the skew behaviour of the ray. Much the same problem is posed in standard lens designs as can be seen from references to skew ray tracing in most classical texts, for example Born and Wolf [19, pp193-195]. Mostly the problem is similarly approximated. Further differences arise in the point at which simplifications have to be made or approximations resorted to. In all, it is essential to observe whether the problem being solved has inherent conditions, such as symmetry, that could simplify the general analysis at the outset.

The method most in keeping with the analysis of Section 2.6 is that of Kildal [9], which has the added attraction of a comprehensive list of references and a discussion of the salient points of their analyses. The reflectors are dealt with through their action on an incremental pencil of rays to give a wave front surface. These involve the local curvatures of the surfaces, in line with the method shown in Section 1.8 (but more exact). Since both surfaces are dealt with as reflecting surfaces, an eikonal function is not introduced and is, of course, unnecessary. The subreflector surface is illuminated by outgoing rays from the source, and the main reflector by incoming rays from the aperture. The problem is then to achieve a match of the wavefronts at an intermediate position, and then to derive the surfaces themselves from the principle curvatures. However the author includes the effects of the reflections on the polarization, thus attempting to use a two-surface system to optimise a three condition problem. Naturally it is found that an exact (computational in this case) solution can only be obtained in the planes of symmetry. The all important transformation between Φ and ϕ (our notation) is put in the form $\Phi = \phi + f(\theta, \phi)$ (Equation 13 of ref.) which is later represented by a Fourier series $f(\theta, \phi) = \sum f_i(\theta) \sin i\phi$, i=1, 2... (Equation 39) of which only the first term is used for a uniform plane aperture distribution. This ad hoc assumption forces constraints on the mapping which then no longer satisfies all the boundary conditions.

With the surfaces and phase fronts described by surface curvature parameters, a finite difference sequence can be arranged, which amounts in principle to ray tracing of closely adjacent rays, or dynamic ray tracing in the author's words. The method is justified by its successful outcome, but very little effect is obtained from the Φ, ϕ formulation and the grids show only minor distortion at the very edge. Where the dual reflector is used as the feeding arrangement for a much larger spherical reflector, the grid distortion is considerably larger and the solution not nearly as satisfactory. The central plane of symmetry is used as the starting boundary for the integration and limited by the specified boundaries of the aperture and the source fields, and a "floating grid" for the point by point solution. This agrees with optical ray tracing practice whereby the ray to ray integrations are checked by a circular argument to prevent the inevitable build-up of internal errors. In spite of the procedure being entirely numerical the authors claim it to be exact, as it is for those points on the grids and sufficiently exact for intermediate points provided they are close enough together. This aspect is also found in most other design methods. The conclusion has to be made that Φ and ϕ are sufficiently close to agreement over virtually the entire antenna surface and that this equality could have been made as a simplifying assumption at the outset [10].

An equivalent assumption is made in the design of a zenith-pointing telescope with minimum ground interference by von Hoerner. The antenna is dealt with as a receiver and the polar grid in the aperture is assumed to map into the coordinate system of the feed (receiving horn) orthogonally (ref [8,

Equation 16]). It is accepted, as we have shown, that the geometry itself provides that, once one surface is given the other is totally derivable. This assumption makes the power transformation Jacobian immediately applicable with, in the author's words, verification from the numerical results. The starting point is taken to be an asymmetric parabola-hyperbola combination and a relaxation procedure iterates the gradual distortion of these surfaces to the required shapes to give the aperture distribution required within the tolerances specified.

The most general methods involve the complete reduction of the sets of equations for the reflection from the first surface and the eikonal relation for the ray paths (usually but not essentially a constant) as the equivalent of the law of reflection at the second surface, into a single differential equation. As can be seen from Equations 2.6.12 2.6.13 and 2.6.16, were each successive relation substituted into the preceding one, elaborate and complicated coordinate geometry becomes involved [11, Equations23, pp39-42]. Only when these relations are conjoined with the power transformation condition, is the problem completed but with the consequence that second order differentials have had to be defined and included. The result is a second order nonlinear partial differential equation of the type known as a Monge-Ampere equation [12] [13]. There are two fundamental approaches to the derivation of the overall equation depending on the formulation used to describe the law of reflection, and the manner in which this definition confronts the problem of the skew behaviour of the rays and the consequent non-orthogonality of the projected areas. As we have stated, until that is resolved, the Jacobian of the transformation between source coordinates and aperture coordinates can not be precisely derived. In the terminology of these sections that is the relationship between (Θ, Φ) and (θ, ϕ) yet again.

Non-located ray directions can be specified as a point on the unit sphere and hence its direction after multiple reflections as successive points on the same sphere. A conical pencil of rays will accordingly describe a circle on the unit sphere. After a single reflection, this will be distorted into a non-uniform outline dependent on the action of the reflector. This distortion is as described in Section 1.8 between $d\theta\ d\phi$ and $d\alpha\ d\beta$. Locally, this transforms an incremental circle on the unit sphere into an ellipse, the equivalent of the reflection of an incremental pencil in the osculating sphere at the point of incidence. Thus it contains exactly the transfer of power between incident and reflected pencils and appears explicitly as a multiplying factor of the Jacobian [13, Equation 2.14]. The analysis is performed by stereographic projection of the unit sphere onto a tangent plane, a process first described in the design of illumination functions from a light source [14], and giving points in that plane complex coordinates [15] (Appendix I). The theory is extended to dual reflectors [16] by the inclusion of an eikonal relation. This requires a location to be defined for each ray in the aperture and a mapping (projection) of the aperture onto the second reflector. Reflected through to the first reflector, the problem of the non-location of the rays is resolved. (Otherwise all parallel rays from the aperture are designated by a single point direction on the unit sphere). The solution then requires the analytic solution of the elliptical form of the Monge-Ampere equation.

The problem of skewness then becomes the problem of matching coordinate grids of non-regular aspect by an iterative procedure involving finite differences with boundary conditions provided by the mapping of the source cone of rays into a defined area in the aperture and the arbitrary selection of a central ray and the position of its incident point at the first reflector. This is termed "anchoring" the solution and since the non-uniform coordinate grids are termed "floating" grids, the point appears to be to prevent the solution from "drifting" through the build up of computer errors. Mathematically, anchoring is necessary since the Monge-Ampere equation involves only derivatives of the dependent variable. It therefore classes as the provision of a constant of integration. However, this positioning of a central point, has to be taken with considerations that are not applicable to the equations as such. They stem more from an heuristic appreciation of the probable shapes and sizes of the final surfaces resulting from the solution. In the same way that a basic paraboloid-conic pair has an infinity of solutions depending on the position of the vertex of the subreflector conic, these will be concave or convex depending on whether the apex is between the main reflector and its focus or caustic. The solutions of the equations must show a similar dichotomy depending on which side of the eventual caustic, the starting point is taken. In fact there is a whole area in the region of the caustic node, which if used as the starting point for integration, the solutions will not properly integrate. Once again the distortion between the Φ and ϕ coordinate grids is very small [16, Figure 9], and the problem dealt with has a central plane of symmetry. Once the surfaces have been determined at the grid points the continuous surface is fitted in this instance by bicubic spline functions. Because of the symmetry involved and the finite difference method of solution, the system can be subjected to continuous linear coordinate deformation. This can be used to produce apertures of other shapes than the usual circular one, and is demonstrated for a very elongated elliptical aperture [17].

A more conventional treatment of Snell's law of reflection using the vector formula of Appendix I.3 in conjunction with an eikonal function for the second reflector together with the full coordinate geometry analysis, has also been presented [18] [20] and [11]. The equations again result in a non-linear partial differential equation. The earlier papers simplify the relation between Φ and ϕ by equating them exactly, [18, Equation 21a] and [20, Equation 14] and hence obtain a series of "radial" solutions which we have been calling profile solutions. These are shown only to differ from exact radial agreement in the outer regions of the reflectors as was noted previously. The result is that the power transformation is no longer exact but sufficiently accurate to be described as "excellent". However the problems dealt with have a circular aperture and conical ray source together with axial and, by virtue of the assumption, radial symmetry. The last of the papers [11] addresses the problem by asserting that a relation between Φ and ϕ can be chosen arbitrarily and corrected by a trial and error process during the computational procedure (reference at p22). This has, of course, to be tempered by practical considerations and is at first taken to be that arising from the geometrical optics analysis of the basic paraboloid-hyperboloid combination. In a later paper [21] (see also [22]) the Φ, ϕ relation is taken to be the first term of a Fourier cosine series (reference equation 10) with an adjustable coefficient for the cosine term.

An objective overview of the published treatments available suggests the following. The complete analysis sets up exact forms of the differential equations for each of the two surfaces. This was established by the very earliest analysis [23]. However it is becoming to be generally accepted that these equations do not have an exact solution in closed form, and hence all results depend on a form of numerical analysis. According to the optical principle, once a solution is reached, it is unique for the situation specified. The "exactness" then is dependent on the assumptions that have to be made to establish an iterative procedure or a series solution. This applies mainly to the method of deriving the Jacobian of the transformation involving the power distributions between source and aperture and concerns the non-planar path that a ray will, in general, have to follow.

The essential connection, describing the skewness of the ray in the most general situation, is inherent in the transformation described on the unit sphere and thence by complex coordinates as in Marder [12]. In all other instances this connection is dealt with on an ad hoc basis, and, quite commonly, simply reduces to the equality $\Phi = \phi$. In many situations this proves to be perfectly adequate (for example [24] [25]). The designs in those cases are invariably of a specialised nature. They have a central plane of symmetry, a uniform phase aperture distribution and circular boundaries. They have not been extended in any way to highly non-uniform distributions or aperture shapes, and it is questionable whether the analysis will permit this generalisation. In one instance where the dual reflector combination was designed to produce a wide angle beam to illuminate an even larger spherical reflector, [9] the Φ grid is greatly distorted away from the ϕ grid (reference Figure 10), requiring up to twelve terms of the approximating Fourier expansion. If, therefore, the problem itself is of a nature that can allow this eventual simplification, the adoption of an extremely complex analysis obscures the fact that the assumptions necessary to make the solutions of the differential equations possible, would have resulted in drastic simplifications had they been made at the outset of the design procedure. This particularly applies to designs that could be made, even if as a first approximation, by a series of $\Phi = \phi =$ constant profile cross-sections.

The inclusion of the distortion into the derivation by complex coordinates,

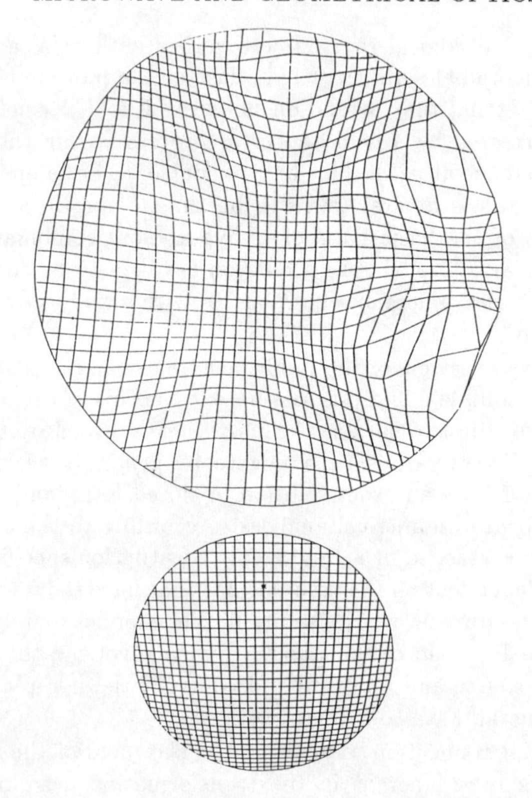

Figure 2.47: Synthesised main reflector and subreflector points projected onto a plane parallel to the aperture (from[24])

on the other hand, can allow for complete generality, and one such aperture grid [26] is shown in Figure 2.47. This was required for a main beam with different degrees of shaping and different non-uniform phase distributions in the major cross-sections of the transmitted beam. The mapping defined by the complex variable derivation of the Monge-Ampere equation contains no ad hoc assumptions regarding its form. Consequently when convergence is achieved the condition of the problem, aperture shape, power and phase distributions are met exactly.

The theories and analyses are now encapsulated in software programs, and with ultra high speed computing available, it becomes comparatively simple to establish a feed-back from the designed antenna surfaces to the radiated pattern and a correction loop for adjustment. This being the new form of trial and error. Because of this, other, more exact methods of obtaining the radiation pattern from reflector surfaces can be included, such as physical optics, mesh methods and the inclusion of edge diffraction effects [27]. In fact, many of these effects outweigh the approximations used in the solutions.

More practical considerations can be given to modified edge construction, and in fact edge diffraction can be utilised in cancelling other diffracted lobes [28] always with the proviso that cancellation in one direction implies enhancement in another. The cancellation of back-to-back cross-talk in land communication stations was effected by castellating the edge of the paraboloids as long ago as 1950.

This process of synthesis and optimisation with rapid computer turn round, has its logical conclusion in avoiding unnecessary complicated analysis of any sort by providing the computer with basic inputs, aperture size, source distribution, required radiation pattern and the laws of reflection plus a search program that will enable the computer to derive a close enough set of surface points that satisfy the criteria [29]. These are fitted in this instance by a general second degree polynomial in the aperture coordinates, "rounded off" with a Fourier series. Direct optimisation techniques are by their very nature time consuming, even on the most modern computers. Even where the process converges it may not yield a global minimisation of the error function but merely local ones. So the geometrical optics solutions are still needed in order to provide a good first approximation to a final solution using physical optics.

As stated at the beginning of the last two sections on dual reflector antennas, the number of designs, published papers and conference reports has grown steadily over the last decade. It would require a volume of its own to do justice to and acknowledge everyone who has contributed . What has been attempted here is to survey the major techniques in order that the relevant points at issue can be recognised and their treatment by other contributors appreciated.

Several interesting problems are outstanding. The still enigmatic existence of an exact solution and whether such exact solutions exist for criteria other than the standard amplitude-phase conditions usually applied. Such exact solutions are known for the offset aspheric reflector profiles with both input and output beams of parallel rays (the afocal telescope). The exact closed form solution for the circular symmetric, centred aspheric profiles, which obey phase and Abbe sine conditions is, of course, classical [30]. It thus appears that the amplitude distribution criterion does destroy the exactness of solution leading always to a dependence on numerical methods. The questions are, is that statement true and if so why. With that consideration we must leave the subject.

2.8 THE GRAPHICAL DESIGN OF TWO-SURFACE SYSTEMS

Approximate methods for the design of two-surface reflectors or lenses depend for their success upon the ease by which the criterion concerned with the available degree of freedom can be used to create an iterative stepwise procedure along the curves of the system profiles (assuming an axi-symmetric or cylindrical system). For sufficiently closely spaced points, the piecewise linear solution can be made as close to the exact curve as the usual $\lambda/16$ total phase discrepancy requires. In the case of lenses this is a differential phase shift between the ray in the precisely shaped lens and in the graphical approximation, and usually permits a greater physical surface tolerance than in the case of reflectors. In the reflector case the summation of the free space tolerances must be kept within the minimum permitted phase error and this has to be halved at each reflection. Thus a closer step procedure is desirable for reflectors, but with computing techniques this is easily met to within the tolerances required by engineering templates.

It is of interest to see how an otherwise indeterminate problem, the general two-surface optical device, becomes determined by the inclusion of a design criterion particularly for the case of the Abbe sine condition as applied to collimating systems. This requirement (Appendix I) is for every transmitted ray to intersect its corresponding incident ray at points which lie on a circle centred at the source of the rays. Using this criterion one can establish a connection between succeeding linear elements and their orientations for both surfaces. We give here the method of Ponomarev [31] for this design procedure.

The basis is the circle of intersection of the rays which is given (Figure 2.48). This is divided into small equal angular increments of a size determined by the eventual phase discrepancy. For the reflecting system shown a starting point A on the outer ray of the system is chosen. The horizontal ray through the intersection of the outer ray with the circle has then got to be the reflected ray from the second surface. The ultimate form of the system depends upon whether A is taken inside the circle or outside, and upon the initial slope of the increment at A. The choice of a tangential element at A then defines the starting point on the second reflector at B, and the tangential increment there. The three entities, the position of B and the incremental slopes at A and B are all determined once any one of them has been specified. The second ray and its horizontal counterpart then intersect the incremental elements at A and B respectively in points A_1 and B_1. Joining A_1B_1 new incremental elements and their orientations are obtained from which the intersections A_2 and B_2 can be derived and so on.

Two of the many diverse reflectors that can be obtained in this way are illustrated in Figure 2.40. Ponomarev gives the following recurrence formulae for deriving the coordinates of the points A_k and B_k from the preceding points A_{k-1} B_{k-1} using this procedure
for A coordinates

$$
\begin{aligned}
x_k &= x_{k-1}\frac{\tan[\theta-(k-1)\partial\theta]+\cot\{[\theta-(k-1)\partial\theta+\beta_{k-1}]/2\}}{\tan[\theta-k\partial\theta]+\cot\{[\theta-(k-1)\partial\theta+\beta_{k-1}]/2\}}\\
y_k &= x_k\tan(\theta-k\partial\theta)
\end{aligned}
$$

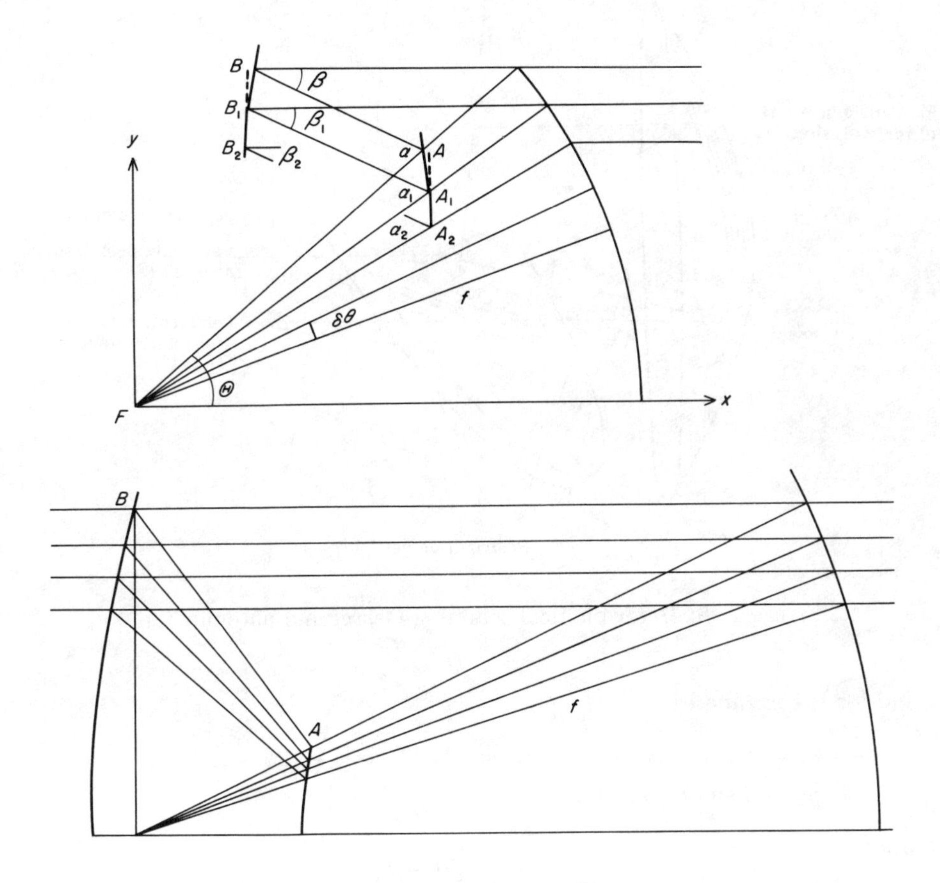

Figure 2.48: Graphical construction of two-surface reflector antennas obeying the sine condition (from Ponomarev[31])

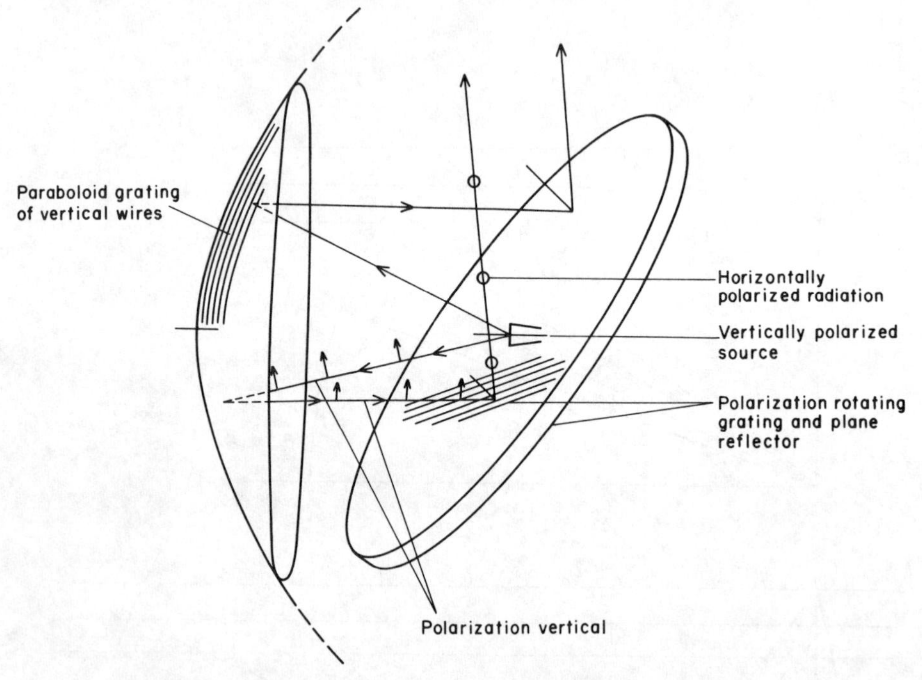

Figure 2.49: Hemispherical scanning Cassegrain antenna[33]

and for B coordinates

$$\begin{aligned} x_k &= x_{k-1} + 2f\tan(\beta_{k-1}/2)\sin(\partial\theta/2)\cos[\theta - (2k-1)\partial\theta/2] \\ y_k &= f\sin(\theta - k\partial\theta) \end{aligned}$$

where

$$\tan\beta_{k-1} = \frac{y_B - y_A}{x_B - x_A} \tag{2.8.1}$$

at $k-1$ points. This process has assumed the division of the Abbe circle into equal elements $\partial\theta$. Other similar procedures could be devised with a weighted division or by a ray to ray process [32].

2.9 SCANNING CASSEGRAIN ANTENNAS

Once the collimation of a beam has been achieved its deflection into any other direction can be simply performed by a plane mirror at a suitable orientation. Normally the directions into which such a beam is turned would have to avoid the original collimating device and the shadowing that this would create. This shortcoming can be overcome by using the polarization properties of the field

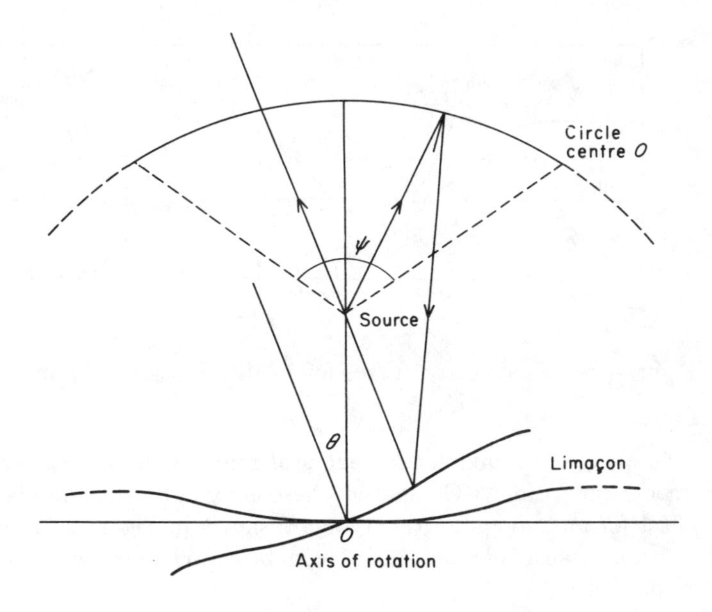

Figure 2.50: Modified scanning Cassegrain antenna

in an ingenious manner. The original collimating device can be made polarization sensitive by using for its reflecting surface a grating of closely spaced conductors, fine wires for example, which then appears as a continuous conducting surface to a wave polarized parallel to the direction of the grating elements. The plane mirror is then surfaced by a layer which besides reflecting the wave, rotates its polarization through 90°, a so-called twist reflector. The design of such elements is given in Chapter 6. Then, Figure 2.49 [33], the reflected wave is polarized perpendicularly to the grating of the original reflector and so is transmitted totally. Deflection of the beam can then be achieved by orientation of the plane twist-reflector with a magnitude of twice the angle of the rotation of the normal to its surface. Scan angles covering the entire hemisphere can be achieved in this way, since by intersecting a plane wave-front with a plane surface no further aberrations can be introduced.

In certain conditions where such a comprehensive angular coverage is not essential, other reflecting profiles can be considered which use the same polarization properties and scanning motion, but with possibly some advantage in geometry or in amplitude distribution over the aperture.

In such cases, the reflectors can be designed entirely by the basic theory. For example a spherical first reflector will have a limaçon profiled second twist-reflector as shown in Figure 2.50. The scan angle then permissible becomes limited by the onset of aberration effects, particularly coma, but since the

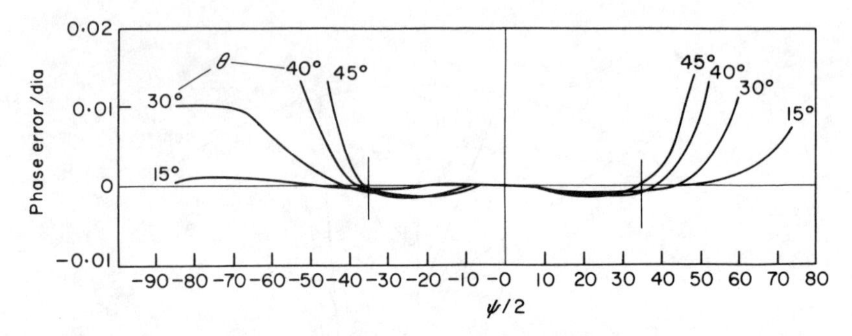

Figure 2.51: Phase error curves for modified Cassegrain antenna

two-to-one beam scan angle advantage still remains, most practical system requirements can be met. Computed phase error curves in the plane of the beam offset for an antenna of this type are shown in Figure 2.51. With this result a beam scan of θ up to $\pm 90^o$ can be anticipated, with an angular aperture of ψ of 70°.

DOUBLE CYLINDER SCANNING ANTENNAS[34]

Similar considerations apply to the periscopic antenna created by two cylindrical reflectors, Figure 2.52. The curves derived above for fully rotational systems would be the same for the orthogonal cross-section of such cylindrical reflectors. The use of two mirrors of equal curvature has already been proposed for this purpose in the field of optics and the reference given determines the degree to which a beam can be scanned by this method.

2.10 DUAL-PROFILE PARALLEL PLATE LINE SOURCE LENSES

A generalised dual-profile reflector antenna can be developed from the basic parallel plate cylindrical antenna to give a line source with specified phase and a secondary attribute, such as aperture amplitude distribution, scanning or multibeam property.

The original, and well-known, line source (Figure 2.53) is formed from a pair of parallel conducting planes and a parabolic cylinder, with the waveguide horn feed at the focus of the parabola. The polarization of the feed is such that the E vectors are perpendicular to the plates and the propagation is TEM, the equivalent of free space, provided only that the dimensions of the antenna are sufficiently large (say twenty or so wavelengths) that no resonant moding can occur. The device then operates purely optically in one dimension.

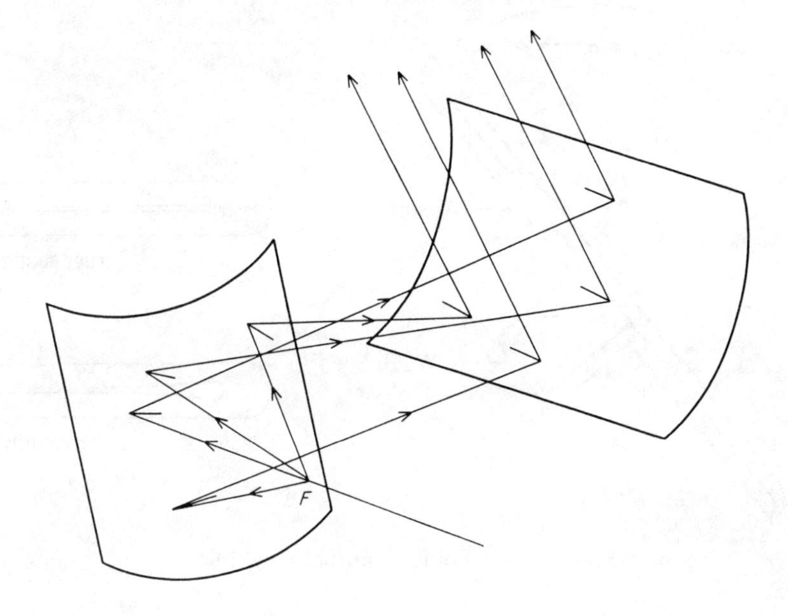

Figure 2.52: Double cylindrical scanning reflector

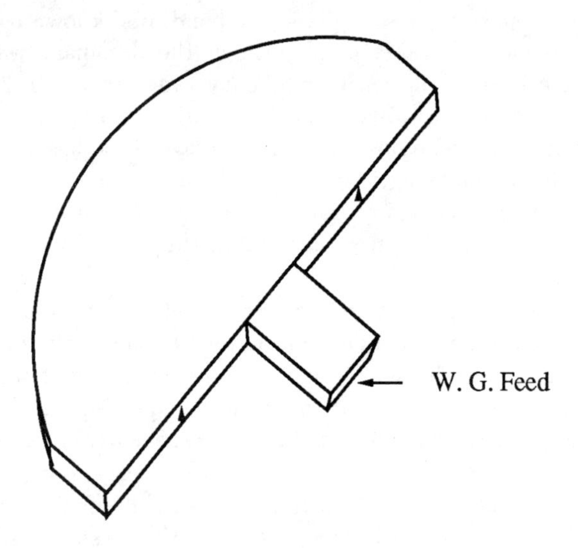

Figure 2.53: Flat parabolic cylinder

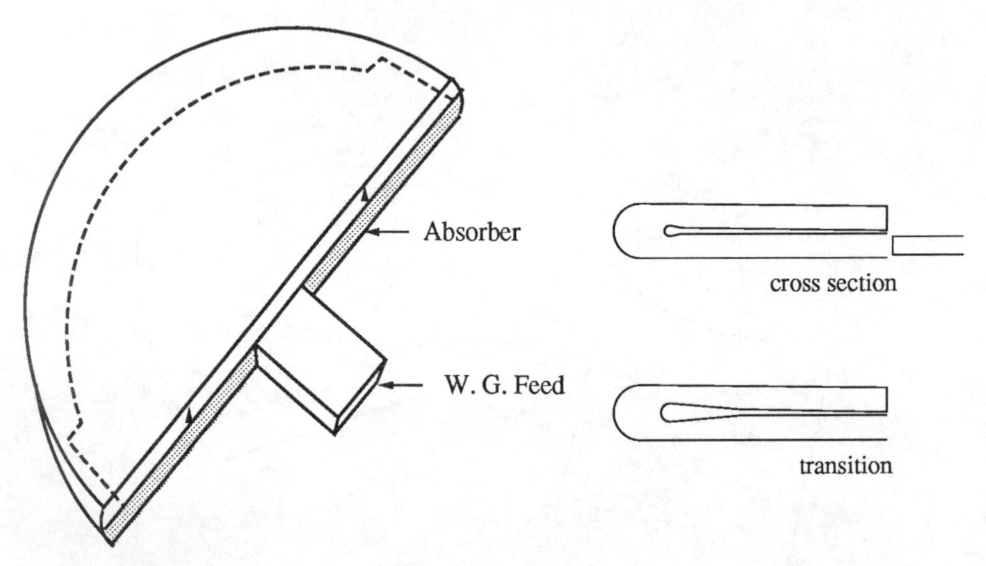

Figure 2.54: Folded parabolic cylinder

Retaining TEM propagation, a second parallel plate layer can be added as in Figure 2.54 where a "fold" is made between the two layers along the profile of the parabola, with a gap transition as shown in the cross-section. This transition is in the style of an E plane 180 bend, well-known to be impedance matched over a considerable bandwidth in the dominant waveguide mode and consequently over the entire frequency range in the TEM mode. The transition can be made through gradual tapering and forming a semicircular bend as indicated in the figure. This double layer then has the same focusing property as the original single layer, with the added advantage of the removal of the feed blockage. This is so purely by virtue of the fact that the rays will follow the Fermat least distance path within the two parallel plate regions, as they do in free space.

Now however a second "fold" can be introduced, with the same form of transition, which, if kept straight and parallel to the original aperture, introduces no error or any additional optical process to the antenna. This is shown in Figure 2.55. The feed is still at the focus of the parabolic "fold", and hence the ray paths obey the same law as for a free space parabola, but "folded" as indicated in Figure 2.56.

This double folding procedure thus provides the opportunity to develop and generalize the design. It is now possible, with free space ray optics, to fold between the successive layers of the antenna along curves that are no longer specifically parabolic/straight. In fact we can return to the basic problem of having two arbitrary profiles with which we can settle two a priori conditions. Inevitably one of these will be a condition of uniform phase in the final linear

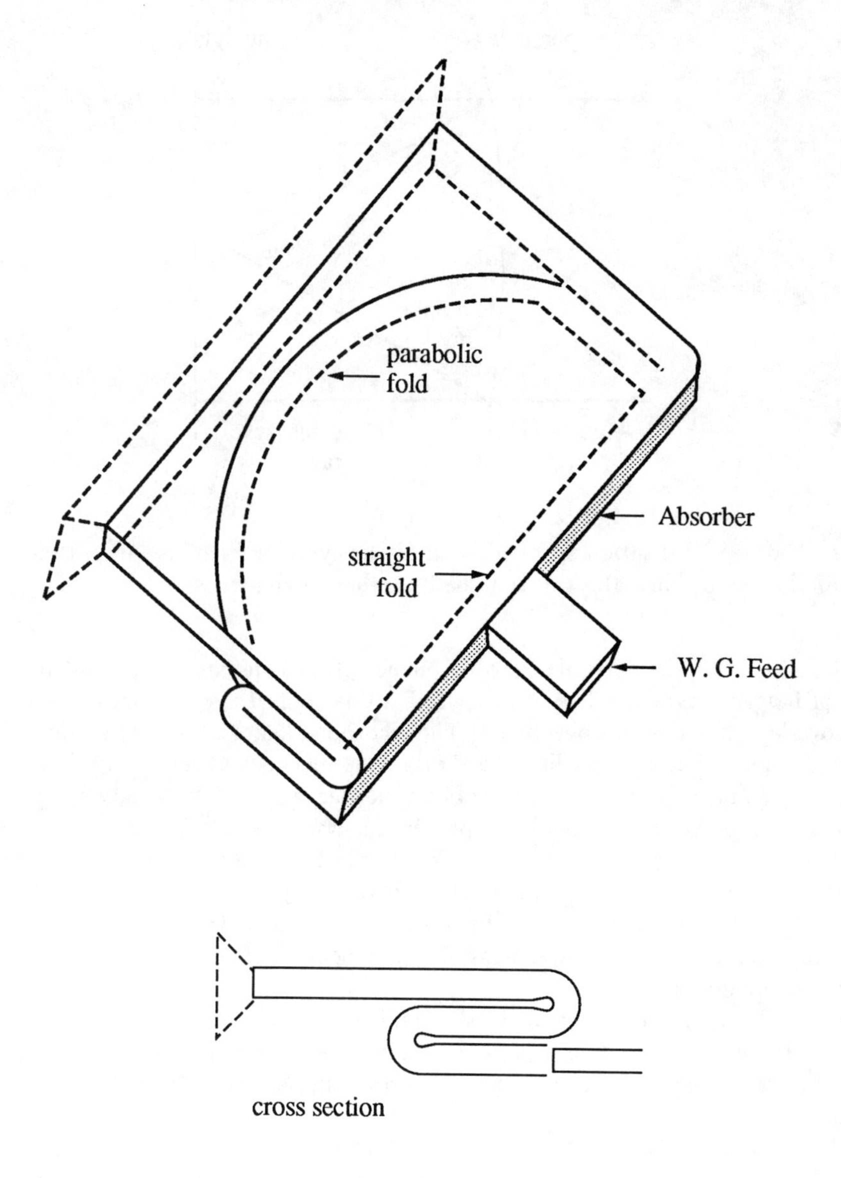

Figure 2.55: Re-folded parabolic cylinder

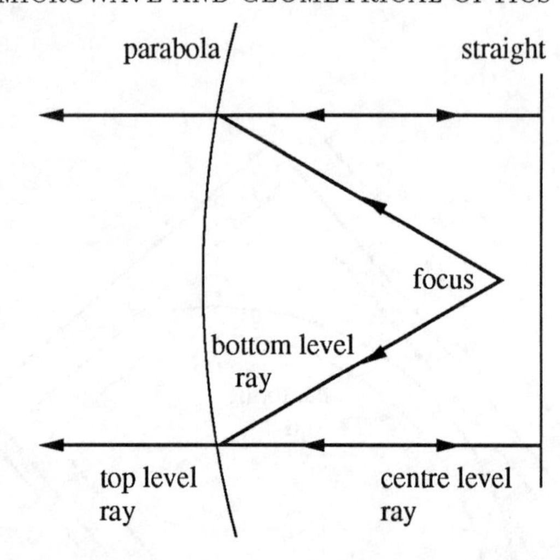

Figure 2.56: Generalization of refolded parabolic cylinder for phase amplitude or scanning conditions. Rays in re-folded parabolic cylinder

aperture. This will be derived as a consequence of the constancy of a "folded" eikonal function as if the lens consisted of a dual reflector combination but with overlaying of the rays permitted. The incorporation of a second function, proceeds along the identical lines to the various methods of designing dual-surface antennas in previous sections of this chapter, again with the advantage of the ray "overlay" that the folding process allows.

One example of this procedure is illustrated in Figure 2.57. Since the rays of the lower level intersect (in plan view) the exit rays of the upper level at points P which lie on a circle with the source at the centre, the lens obeys the Abbe sine condition which makes possible some measure of beam displacement or scanning capability.

Another combination that obeys the condition is the Zeiss cardioid. This can be adapted to the folding technique as illustrated in Figure 2.58 where the lower fold is made with a cardioid profile and the upper one with the circular profile.

More interestingly still, the method can be continued with the addition of further folds and ray path levels. Then in the spirit of optical lens design, every additional fold profile can be utilised to correct for an additional aberration or introduce some other lens function. So with additional folds to those already described, a lens can be created that is correct in phase, has a specified amplitude distribution and is in agreement with the Abbe sine condition. Of course, practical construction problems probably prevent this from ever becoming a viable proposition, but on theoretical optical grounds, this is the nearest to multi-element lens design that is arrived at in microwave practice.

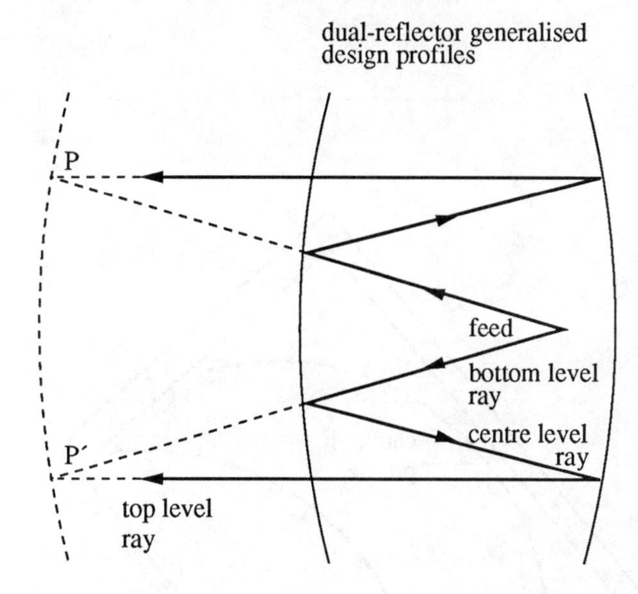

Figure 2.57: Generalization of folded dual reflector for phase amplitude or scanning conditions. Rays in folded dual reflector

2.11 TAUT STRING TECHNIQUE

Though regarded as somewhat arcane, the taut string method given in this section (Section 2.3) can provide an immediate assessment of quite complicated reflector designs including dual-reflectors, their offsetting properties, multiple feeding effects and multiple reflector designs. This arises from the fact that a taut string attached to the source point O and wrapped around the edge of a semi-infinite plane at P and with zero friction at the point of contact will obey Snell's law of reflection at P. From elementary statics, for the string to be in equilibrium under a tension T as shown, the angles θ on each side of the normal must be equal (Figure 2.59). If the string is of fixed length, the end point will describe the phase front of the reflected rays, in this case a circle centred on the image point O'.

The application of this fact to the known properties of the conic mirrors can be illustrated by wrapping a taut string attached to the foci F_1 and F_2 over the edge of an elliptically-shaped plane contour, where it will remain at rest in whatever position it is placed (Figure 2.60). If applied to a parabolic profile with the source end fixed at the focus, the end point of a string of fixed length will describe a straight line perpendicular to the axis, namely the plane phase front obtained from a focused parabola. It is obvious that the fixed length of string acts exactly as the eikonal of the ray system.

Making the obvious use of this last fact, fixed length eikonals can be ap-

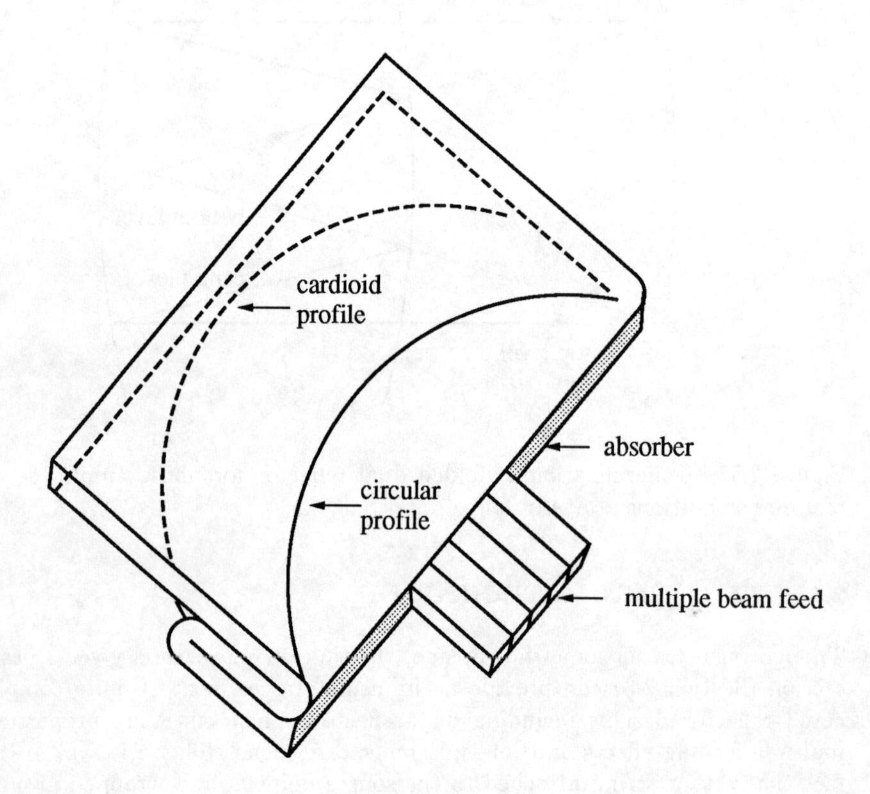

Figure 2.58: Parallel plate Zeiss cardioid

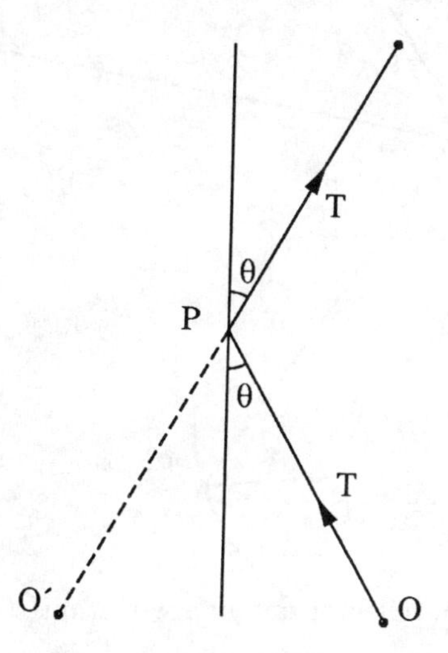

Figure 2.59: Snell's law of reflection obeyed by a taut string

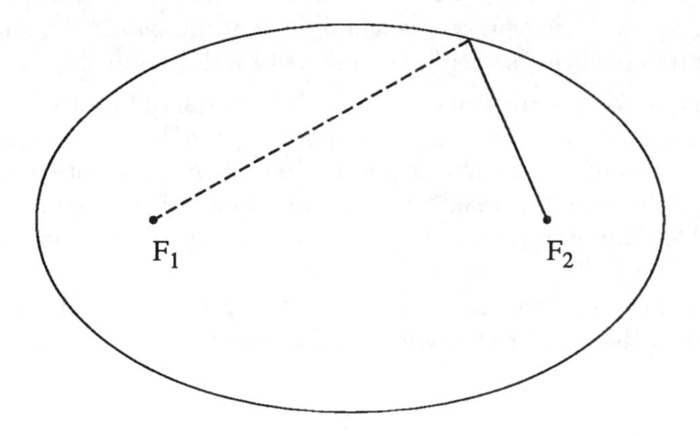

Figure 2.60: Taut string in equilibrium on an ellipse

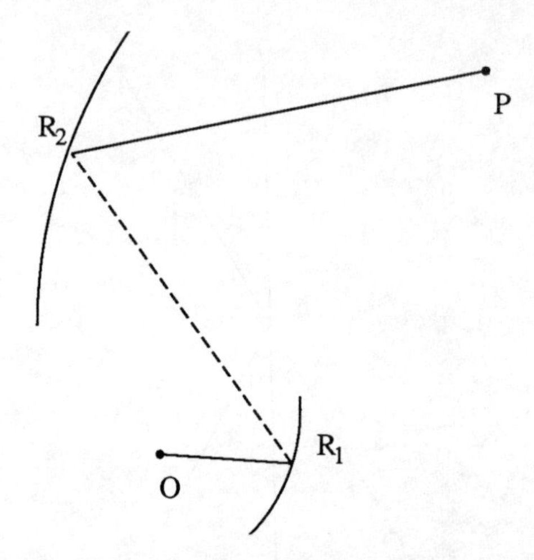

Figure 2.61: Dual-reflector antenna with taut string eikonal

plied to more than one reflecting profile, if the string is wrapped over each in turn. This is illustrated for a dual-reflector antenna in Figure 2.61. Several applications can be envisaged. There is the assessment of a dual-reflector antenna under conditions such as offsetting the feed or rotating the sub-reflector. Should a flexible curve be contrived that had the zero friction condition included, the design of shaped reflectors could be made in a first approximation. In a sense, the string process is simply a concomitant of the folded parallel plate system, which also could be designed by this method.

The zero friction condition is obviously essential to the process. One would consider a nylon type thread and a rounded-edge well-lubricated profile. Those of the readers of this volume engaged in teaching physics or engineering that might involve reflector design or an appreciation of Snell's laws, may find this a suitable laboratory exercise. It applies strictly to reflection alone. Even if a selective friction could be created that made the edge susceptible to the Snell law of refraction, the eikonal would have to be extended over the part included in the interior of any lens so designed.

REFERENCES

1. Cornbleet S & Smith G (1986) Dual reflector antennas with amplitude, phase or Abbe conditions. *Proc IEE* **H 133** no 3 pp221-226

2. Roberts S (1881) Historical note on Dr Graves' theorem on confocal conics. *Proc Lond Math Soc* **vol 12** p120

3. Williams WF (1965) High efficiency antenna reflector. *Microwave Jour* **8** pp79-82

4. Galindo-Israel V (1964) Design of dual reflector antennas with arbitrary phase and amplitude distribution. *IEEE Trans Antennas and Propagation* **AP-12** pp403-408

5. Jørgensen RJ & Balling P (1987) Generation of highly contoured beams by shaping the surface of an offset single reflector. In Clarricoats P (ed) *Advanced Antenna Technology* **2** MEPL

6. von Hoerner S (1976) The design of correcting secondary reflectors *IEEE Trans Antennas and Propagation* **AP-24** p336

7. Aoki K, Makino S, Katagi T & Kagoshima K (1993) Design method for an offset shaped dual reflector antenna with high efficiency and an elliptical beam. *IEE Proc* **H 140** no 2 p121

8. von Hoerner S (1978) Minimum noise maximum gain telescopes and relaxation method for shaped asymmetric surfaces. *IEEE Trans Antennas and Propagation* **AP-26** pp464-471

9. Kildal P-S (1990) Synthesis of multireflector antennas by kinematic and dynamic ray tracing. *IEEE Trans Antennas and Propagation* **AP-38** pp1587-1599

10. Oliker VI (1987) Near radially symmetric solutions of an inverse problem in geometric optics *Inverse Problems* **vol 3** pp743-756

11. Galindo-Israel V, Imbriale W & Mittra R (1987) On the theory of the synthesis of single and dual offset shaped reflector antennas. *IEEE Trans Antennas and Propagation* **AP-35** pp887-896

12. Marder L (1981) Uniqueness in reflector mappings and the Monge-Ampere equation *Proc Roy Soc London* **A378** pp529-537

13. Westcott BS (1983) *Shaped Reflector Antenna Design* Letchworth England: Research Studies Press

14. Schruben JS (1972) Formulation of a reflector-design problem for a lighting fixture. *Jour Opt Soc Amer* **62** no 12, pp1498-1501

15. Brickell F, Marder L & Westcott BS (1977) The geometrical optics design of reflectors using complex coordinates. *Journal of Physics A Maths & Gen* **10** no 2, pp245-260

16. Westcott BS, Stevens FA & Brickell F (1981) GO synthesis of offset dual reflectors. *IEE Proc* **H 133** no 1, pp57-64

17. Westcott BS, Graham RK & Wolton IC (1986) Synthesis of dual offset shaped reflectors for arbitrary aperture shapes using continuous domain deformation. *IEE Proc* **H 133** no 1, pp57-64

18. Galindo-Israel V, Mittra R & Cha AG (1979) Aperture amplitude and phase control of offset dual reflectors. *IEEE Trans Antennas and Progagation* **AP-27** pp154-164

19. Born M & Wolf LE (1954) *Principles of Optics,* pp193-195 Pergamon Press

20. Galindo-Israel V & Mittra R (1984) Synthesis of offset dual shaped subreflector antennas for control of Cassegrain aperture distributions. *IEEE Trans Antennas and Propagation* **AP-32** pp86-92

21. Galindo-Israel V, Imbriale WA, Mittra R & Shogen K (1991) On the theory of offset dual-shaped reflectors - case examples. *IEEE Trans Antennas and Propagation* **AP-39** pp620-626

22. Mittra R, Hyjazie F & Galindo-Israel V (1982) Synthesis of offset dual reflector antennas transforming a given feed illumination pattern into a specified aperture distribution. *IEEE Trans Antennas and Propagation* **AP-30** No 2, pp251-259

23. Kinber BY (1962) On two reflector antennas *Radio Eng Electron Phys* **6**

24. Lee JJ, Parad LI & Chu RS (1979) A shaped offset-fed dual reflector antenna *IEEE Trans Antennas and General* **AP-27** pp165-171

25. Oliker VI (1987) Near radially symmetric solutions of an inverse problem in geometric optics *Inverse Problems* **3** pp743-756

26. Westcott BS, Brickell F & Wolton IC (1987) Synthesis and design of shaped dual reflector antennas from specified aperture amplitude and phase distributions *Preprint series* **174** University of Southampton, Faculty of Mathematical Studies

27. Franceschetti G & Mohsen A (1986) Recent developments in the analysis of reflector antennas. A review (inc. 109 references). *IEE Proc* **H 133** no 1, pp65-76

28. Cwik TA & Kildal P-S (1989) A study of three techniques used in the diffraction analysis of shaped dual-reflector antennas. *IEEE Trans Antennas and Propagation* **AP-37** pp979-983

29. Bergmann J, Brown RC, Clarricoats PJB & Zhou H (1988) Synthesis of shaped-beam reflector antenna patterns *IEE Proc* **H 135** no 1, pp48-53

30. Schwarzschild KS (1905) *Untersuchen zur geometrischen optik. Abhandlungen der Königlichen* **Part II** Gesellschaft def Wisienschaften, Göttingen, Mathematische Physicallische Klasse IV

31. Ponomarev NG (1961) Graphical method for the design of profiles of aplanatic antennas. *Radio Eng & Electronics* **6** no2, p42

32. Luneburg RK (1964) *Mathematical Theory of Optics* University of California Press p208

33. Mariner PF & Cochrane CA (1953) British Patent No 716939 (August 1953 - October 1954)

34. White WD & de Size LK (1962) Scanning characteristics of two-reflector antenna systems. *Convention Record of the IRE* **I** p44

3 LENSES AND PHASE CORRECTED REFLECTORS

Microwave lenses have not been given too much consideration as antennas in the past because of their intrinsic bulk. For example to perform the same simple focusing as a thin metal sheet in the shape of a paraboloid, say a metre in diameter, requires a lens of the same diameter 20 cms thick at its centre. However more and more of the higher frequency bands are now being exploited and such figures can be scaled up to a factor of ten or more. This makes them far more of a practical proposition and their inherent advantages, zero blockage among them, should lead lenses and similar refractive elements to have a large potential for future development.

Since the design of lenses is governed by the same principles as reflectors, it should be expected that the methods of the previous chapter could in the main be repeated. Unfortunately the replacement of the law of reflection by the far more complex law of refraction at each surface seems to have rendered some of these processes extremely difficult. Only one attempt at the general dual surface refractive lens by the method used for the general dual surface antenna is known [1] and the additional complexity is obvious.

However, as a two-surface system, it can still perform two preconditional operations, and the general findings of the previous chapter apply. For example, if the first surface is specified then only a single condition, usually a focusing requirement, can be *exactly* obeyed. Skew or offset surfaces are rarely, if ever, met and so the design will concentrate on the cross-section of centred, rotationally symmetric systems or axially symmetric cylindrical ones. With the approximation of dealing with profiles alone, some methods are adaptable from the reflector profile technique.

3.1 APPLICATION OF DAMIEN'S THEOREM TO LENS PROFILES

We deal in the first instance with the design of the second surface of a lens for which the first surface the source and the required image are specified. This problem has to be exact but most methods involve numerical methods or approximate series solutions [2][3][4] and [5]. An extension of Damien's theorem gives the exact profile geometrically and its interpretation into coordinate geometry is standard, using the method of inversion [10].

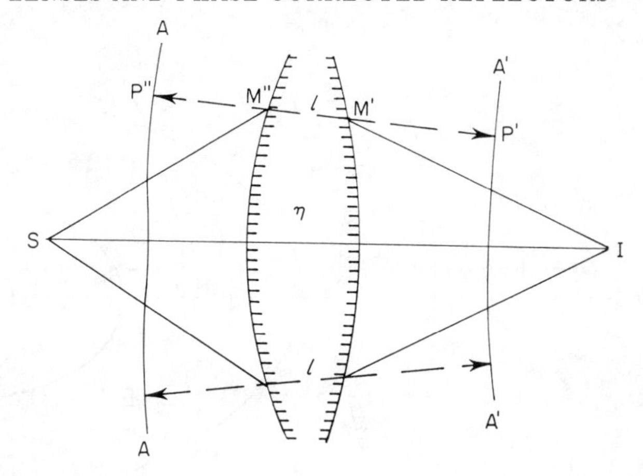

Figure 3.1: Transmission through a perfect lens. A is the zero-distance phase front of the source and first surface, A′ that of the image and second surface. A′ is parallel to A.

Then, as in Figure 3.1, the zero-distance phase front A of the source and first surface can be derived by any of the methods so far given. This phase front replaces completely both the source and the first surface and we regard it as a new source in the infinite medium with the given refractive index. We form at an arbitrary distance the parallel A′ of the curve A and we will show that this arbitrary choice is solely dependent on the so far arbitrary thickness of the lens at its centre.

In order to obtain the required focusing action the surface A′ must be the zero-distance phase front of some refracting surface with respect to the specified image I. For then, in the literal meaning of zero-phase, this occurs between the source and the surface A, and also between the image and the surface A′. Since A and A′ are parallel in a uniform medium, the phase along mutual normals is constant for all rays, and thus the total phase from source to image is constant along all rays. Hence the image will be a perfect focus.

What is required to complete the design is the process whereby the second surface is derived from the now known phase front A′. This process is carried out in three stages (refer to Figure 3.2 and Section 2.1).

1. Invert the surface A′ in a circle (of arbitrary but appropriate radius) centred on I to give curve B.

2. Obtain the zero-distance phase front of B with respect to I with the refractive index of the medium concerned to give curve C.

3. Invert curve C in the same circle as in (1) to give curve D.

Curve D is the profile of the second surface required. For, if A′ intersects the axis SI at P, curve B at P′, curve C at Q and curve D at Q′ as shown, then

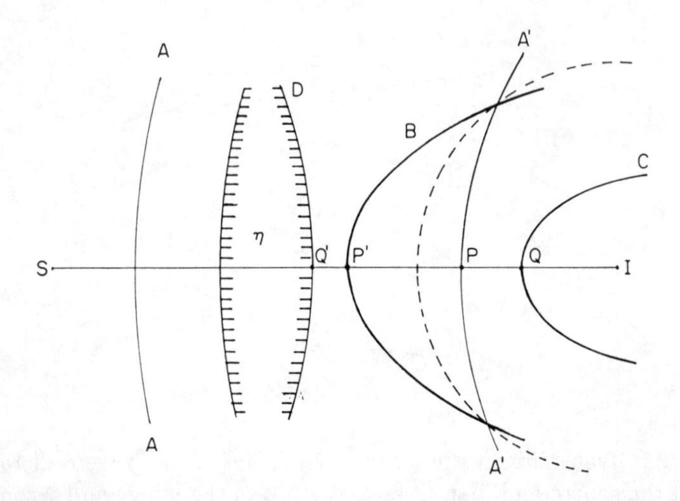

Figure 3.2: Construction of the second surface. A is the zero-distance phase front of the source and first surface, A′ is a parallel. B is the inverse of A′ in a circle centred on the image. C is the zero-distance phase front of B which inverts into D, the required second surface

the position of P is solely dependent on the chosen axial thickness of the lens and the process is independent of the radius of inversion. We have by the construction of zero-distance phase fronts

$$\mathrm{Q'P} = \mathrm{Q'I}/\eta$$

or

$$\mathrm{IP} = (1 - 1/\eta)\mathrm{IQ'} \tag{3.1.1}$$

P′ is derived by an inversion, so IP.IP′ $= \kappa^2$ (κ being the radius of inversion); Q is derived by the zero-distance phase front procedure

$$\mathrm{P'Q} = \mathrm{P'I}/\eta$$

or

$$\mathrm{IQ} = (1 - 1/\eta)\mathrm{P'I}$$

and Q′ is derived from Q by an inversion of the same radius IQ.IQ′ $= \kappa^2$. Hence

$$\mathrm{IQ'} = \frac{\kappa^2}{\mathrm{IQ}} = \frac{\kappa^2}{(1 - 1/\eta)\mathrm{IP'}} = \frac{\kappa^2}{(1 - 1/\eta)}\frac{\mathrm{IP}}{\kappa^2}$$

which is identical to Equation 3.1.1 and independent of κ^2.

The complex nature of the geometrical curves that arise in this procedure makes demonstrations of the methods difficult in all but a few well-known optical designs. Of these we take the aplanatic lens with hyperbolic surfaces.

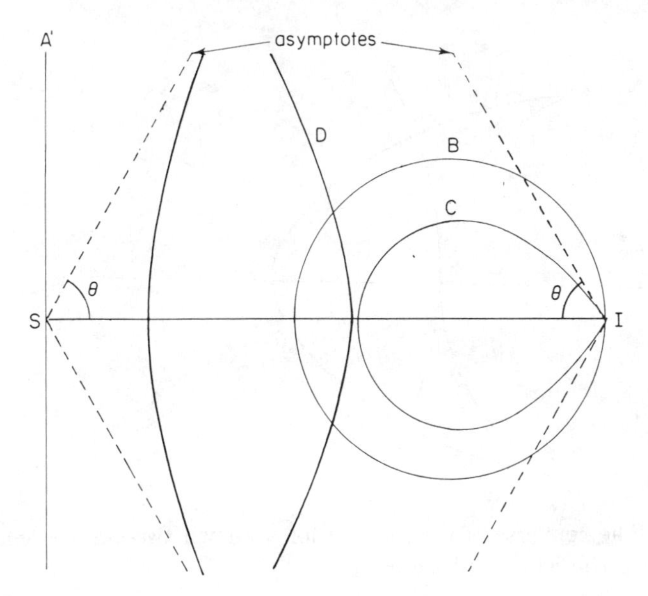

Figure 3.3: Application of the method to the exact lens with hyperbolic surfaces

The Gregorian reflector of the parabolic mirror was dealt with in Section 2.1. For these cases the zero-distance phase fronts of the first surface and source are plane and circular respectively.

For the former we are given the source S at the focus of a hyperbola, the interface with a medium of refractive index η (Figure 3.3). If the asymptotes of the hyperbola are at an angle $\pm\theta$ such that $\cos\theta = 1/\eta$, all the rays are refracted to become parallel to the axis. The zero-distance phase front is therefore a plane A, perpendicular to the axis and its parallel similarly a plane A$'$ perpendicular to the axis. The inversion of A$'$ in a circle centred at I is therefore the circle B passing through I in Figure 3.3. The refractive zero-distance phase front of a circle with a source on its perimeter ($r = k$ in Equation 1.2.6) is the interior loop of a limaçon with cusp at I on curve C. It is simple to show that tangents to the cusp make the same angle θ with the axis at I, and hence the inversion of the cusped curve will have asymptotes making the same angle with the axis. The inversion of a limaçon being a conic with asymptotes is thus the second hyperbolic surface D producing an exact focus at I.

The converse of the method also has applications for deriving the zero-distance phase fronts in situations where other methods might be too complicated. Thus if we have a two-surfaced system with known exact focusing properties, any one of the stages unknown in the procedure can be derived from the others which are known. For example, the two-surface lens with a

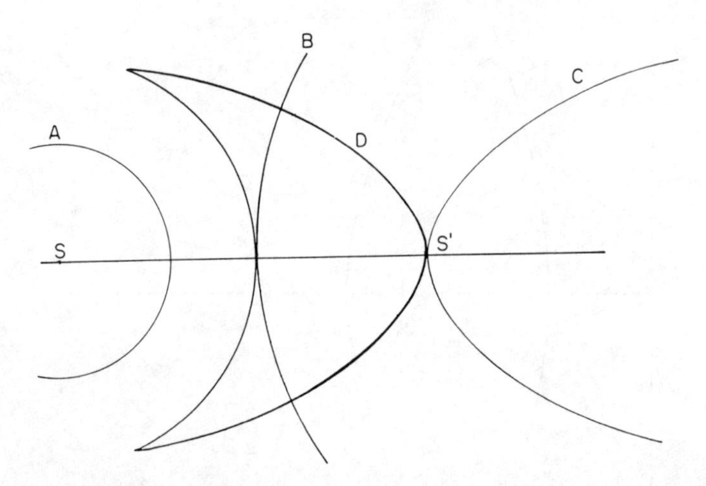

Figure 3.4: The converse of the method for a known two-surface lens, giving an unknown zero-distance phase front

circular first surface centred on the source and an elliptical second surface with focus at the source with eccentricity $\varepsilon = 1/\eta$ collimates all rays parallel to the axis (Figure 3.4). The zero-distance phase front of the first surface is the circle A centred on S, which we can take also to be its parallel A′ as an arbitrary choice of inversion arises later. The inversion of A′ in the point at infinity is a reflected circle B wherewith we take up the arbitrariness by making B tangential to the first surface of the lens and of radius SS′. The zero-distance phase front of B must be the curve C which will *reflect* (inverse at infinity) into the known second surface D. It is therefore itself the reflection of D. Hence the zero-distance phase front of internal refraction of parallel rays at a circular interface is the ellipse with eccentricity $\varepsilon = 1/\eta$.

PROGRESSION OF SURFACES - PARAXIAL APPROXIMATION

The zero-distance phase front of a source and surface combination becomes the source itself in the image space of the surface and behaves as if in the infinite medium of the image space. In the presence of a second surface there is, therefore, a new source and surface combination, the zero-distance phase front of which can be obtained by the envelope method for extended sources. This will give the zero-distance phase front in the image space of the second surface and the procedure can be continued, by obvious means, through an entire system of surfaces involving both refracting and reflecting elements. At any stage the actual caustic of the rays can be derived by taking the evolute of the phase front at that stage.

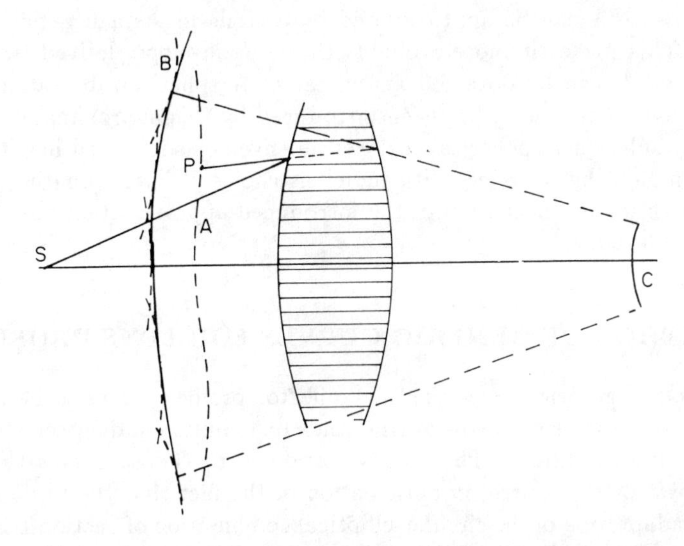

Figure 3.5: Phase front of a spherical lens. The zero-distance phase front of source and first surface is A. Points P on A give rise to zero-distance phase fronts of the second surface B. The envelope of B is produced to a parallel curve C. The evolute of C will be the caustic of the lens. In an approximation, A and the envelope of B can be made circular in a paraxial region

From the complicated form that a zero-distance phase front may have, even after a single refraction from a point source, it is quite obvious that geometrically a cascading process will rapidly become unwieldy. This is mainly because the procedure as previously stated applies to the wide angle of rays issuing from the source. In the paraxial approximation only those regions of the phase fronts about the axial direction would need to be considered. Approximations can then be made to those parts of the curve that would simplify the envelope method or provide a simpler method of progressing to the image space of the subsequent surface.

The process is illustrated in Figure 3.5 for a bi-convex spherical lens. For the first surface and source point S, the zero-distance phase front is a curve A, given by Equation 1.2.6 with the appropriate values of k η and r. The zero-distance phase front of A, now *interior* to the medium with the circular second surface, is obtained *exactly* as the envelope of zero-distance phase fronts B as a source point P moves over the curve A. Each separate curve B can be obtained from Equations 1.2.6 or 1.2.7 with a parameter introduced to define the position of P on A. Elimination of this parameter by the standard procedure gives the envelope of curve B. The exact phase front in image space is then a parallel surface C to the envelope B, and the caustic the evolute of C.

Now surface A can be approximated by a circle in a small region about the axis. This makes it more probable that the envelope derived from the varying point P can be obtained analytically. A spherical aberration (and higher symmetrical orders) has been introduced by this approximation.

For any subsequent spherical surface the envelope so derived has itself to be approximated by a circle. With judicious choice of approximating circles in both cases some cancellation of the introduced approximation aberrations could be achieved.

3.2 METHOD OF GENERIC CURVES FOR LENS PROFILES

The method of generic curves used for reflector profiles proves to be readily adaptable to the same problems of amplitude distribution and agreement with the Abbe sine condition. The only variation is to choose curves that are concave towards the source, in anticipation of the fact that the likeliest lens will be an adaptation of the circular-elliptical combination of Section 1.1. This is strictly a single surface lens, the refraction being completely obtained at the elliptical surface. The lens is completed by having a surface orthogonal to the rays and therefore non-refracting. This takes the form of a spherical surface with the source at its centre. The ellipse has the same source at the focus remote from its apex. This source can be projected towards the ellipse, by replacing the point source and circle combination with its aplanatic equivalent, the limaçon of Figure 1.6 as shown in Figure 1.15. This choice can be made in an infinite number of ways and a variety of lenses results, one of which may have the property required. Hence the choice of generic curves is made in a manner that suits this situation [11].

With the lens shown in Figure 3.6, the angle of the normal ψ is given by

$$\tan\psi = \frac{\rho\sin\phi - \frac{d\rho}{d\phi}\cos\phi}{\rho\cos\phi + \frac{d\rho}{d\phi}\sin\phi} \tag{3.2.1}$$

where $\rho = \rho(\phi)$ $\rho(0) = 1$ is the polar equation of the first surface.

The angle made with the horizontal by the ray after refraction is then

$$\alpha = \psi + \sin^{-1}(\sin(\phi - \psi)/\eta) \tag{3.2.2}$$

η being the refractive index of the lens material. This relation remains true for the alternative situation of incidence shown in Figure 3.7. This is the fundamental application of the Snell law at the first surface. For the second surface, we define the optical path length constant, the eikonal, which for a source at the origin is as follows

$$\rho + \eta r + D + 1 - x = 1 + \eta D \tag{3.2.3}$$

The internal geometrical relations are simply

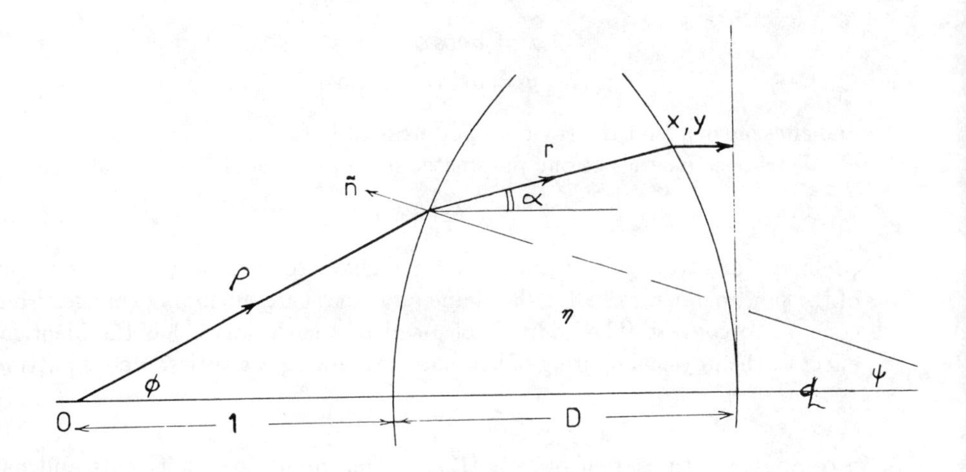

Figure 3.6: Ray geometry

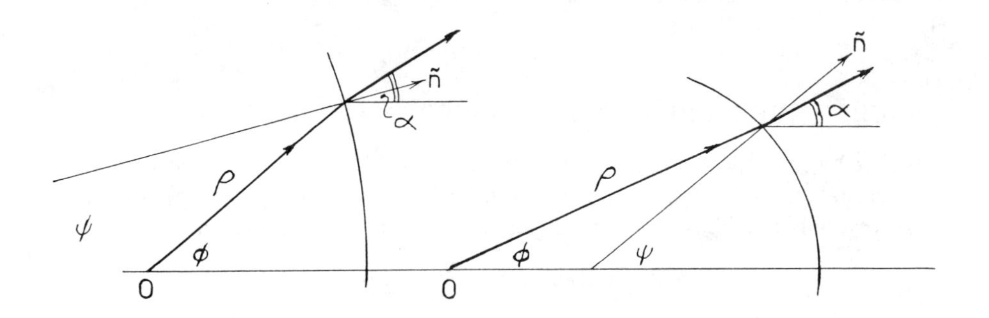

Figure 3.7: Alternative ray-incidence directions

$$x = \rho\cos\phi + r\cos\alpha$$
$$y = \rho\sin\phi + r\sin\alpha \tag{3.2.4}$$

and chosen such that the ray does not cross the axis between the two surfaces.

We choose a series of one parameter sets of curves of the general form

$$\rho_n = \rho(\phi, n) \tag{3.2.5}$$

which are arbitrary but of the anticipated shape for lens surfaces. Variations of the parameter n can alter the shape and curvature, including changes from concave to convex. The addition of a series of such curves has the identical effect to the physical figuring of lens surfaces, as in glass optics. Hence putting

$$\rho = \Sigma_n a_n \rho_n(\phi, n) \tag{3.2.6}$$

a computer optimisation process [12] can determine the coefficients and parameter values which give the best fit between the two sides of Equation 3.2.11. Some parametric generic curves are illustrated in Figure 3.8. It can be seen that some curves appear in more than one definition, so the series is not unique in any way. However, the resultant curve from the optimisation process does finally result in a defined curve which is the same whichever form of the constituent individual terms is taken. Additional curves can be obtained from any given set by inversion

$$\rho \rightarrow 1/\rho$$

SOURCE-APERTURE POWER TRANSFORMATION

If $P(\phi)$ is the power distribution at the source, which is at the origin of coordinates, and $P(y)$ is the power distribution in the (circular) aperture with radial coordinate y, then conservation of energy requires that

$$2\pi P(\phi)\sin\phi \; \mathrm{d}\phi = 2\pi P(y)y \; \mathrm{d}y \tag{3.2.7}$$

subject to

$$\int_0^{\Phi} P(\phi)\sin\phi \; \mathrm{d}\phi = \int_0^{R} P(y)y \; \mathrm{d}y \tag{3.2.8}$$

where Φ is the angular aperture at the source, and R is the radius of the lens. It is customary to assume that the source distribution can be put in the form of one or more terms of the cosine powers, that is

$$P(\phi) = A\cos^s\phi \tag{3.2.9}$$

This allows the left-hand side of Equation 3.2.8 to be integrated, with the result

$$y^2 = R^2(1 - \cos^{s+1}\phi)/(1 - \cos^{s+1}\Phi) \tag{3.2.10}$$

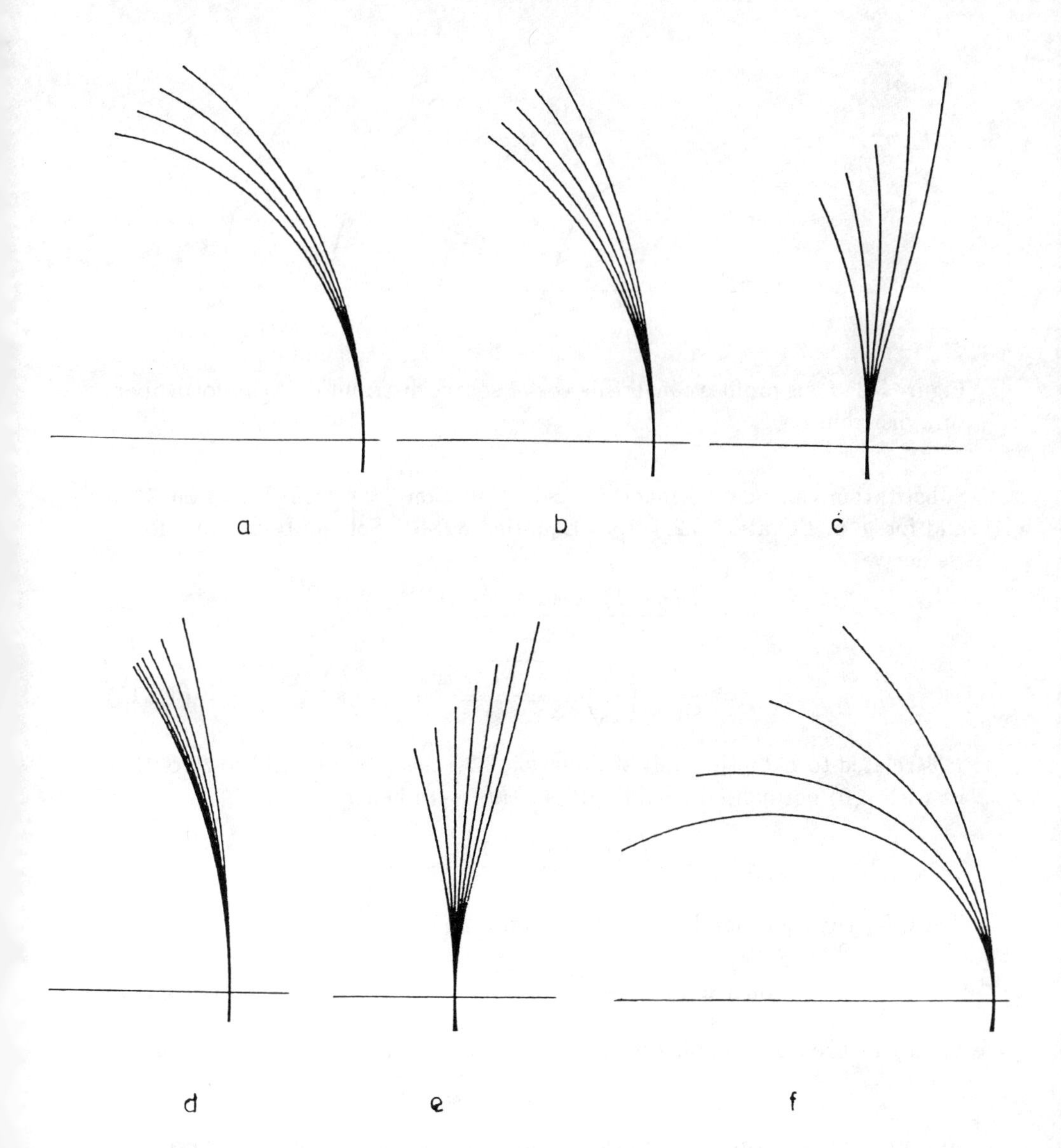

Figure 3.8: Systems of generic curves
(a) $\rho^n = \cos(n\phi/c) \qquad c = 1$
(b) $\rho = (\cos^2\phi + n\sin^2\phi)^{-1/2}$
(c) $\rho = (2 - \cos\phi)^n$
(d) $\rho = (n + 1)/(n + \cos\phi)$
(e) $\rho = 1 - n + n\sec\phi$
(f) $\rho = 1 - n + n\cos m\phi \qquad m = 2$

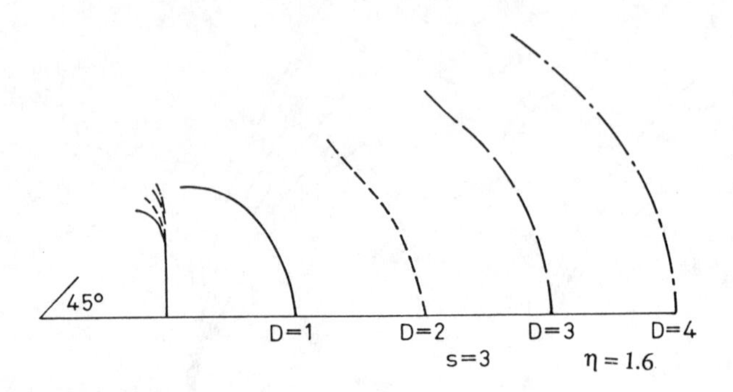

Figure 3.9: Lens profiles converting $\cos^s \phi$ source distribution to uniform aperture distribution

Substitution can now be made for x in Equation 3.2.4 from Equation 3.2.3 and for y, in Equation 3.2.4 from Equation 3.2.10. Solving both these for r, we derive

$$\begin{aligned} r &= \frac{D(\eta - 1) - \rho(1 - \cos\phi)}{\eta - \cos\alpha} \\ &= \frac{1}{\sin\alpha}\left(\frac{R\sqrt{1 - \cos^{s+1}\phi}}{\sqrt{1 - \cos^{s+1}\Phi}} - \rho\sin\phi\right) \end{aligned} \tag{3.2.11}$$

R is related to D by the limit of Equation 3.2.11 as ϕ and α tend to zero, that is, with $\rho(0)$ normalised to unity, $r \to D\,(\phi \to 0)$, hence

$$R = \sqrt{\frac{2(1 - \cos^{s+1}\Phi)}{s+1}}(1 + D/\eta)$$

Thus, for the right-hand side of Equation 3.2.11

$$r\sin\alpha = \sqrt{\frac{2(1 - \cos^{s+1}\phi)}{s+1}}(1 + D/\eta) - \rho\sin\phi \tag{3.2.12}$$

With this procedure applied to converting source distributions of the form

$$P(\phi) = A\cos^s\phi$$

to a uniform aperture distribution $P(y) = 1$, the lens profiles shown in Figure 3.9 ($s = 3$) were obtained. Equality between the two sides of Equation 3.2.11 depends upon the chosen values of the parameters s and D, and also on the angular aperture. With the focal length normalised to unity, this can be of the order 10^{-6}, as shown in Table 3.1.

The subtended angle at the source was chosen arbitrarily to be $\pm 45^o$ as a practical value for normal horn feeds, but the large improvement in performance that occurs for smaller angles is shown by the inclusion of the result for $\pm 30^o$.

s	D	Maximum Difference $\Phi = 45^o$	$\Phi = 30^o$
1	1	0.08	0.003
	2	0.014	$9\ 10^{-6}$
	3	0.005	$6\ 10^{-6}$
	4	0.002	$4\ 10^{-6}$
2	1	0.058	$1.2\ 10^{-5}$
	2	$2.6\ 10^{-5}$	$\leq 10^{-6}$
	3	$7\ 10^{-6}$	$\leq 10^{-6}$
	4	$3\ 10^{-6}$	$\leq 10^{-6}$
3	1	0.165	$2.8\ 10^{-6}$
	2	0.019	$\leq 10^{-6}$
	3	$1\ 10^{-6}$	$\leq 10^{-6}$
	4	$6\ 10^{-5}$	$\leq 10^{-6}$

Table 3.1: Difference between LHS and RHS of Equation 3.2.11 with $P(\phi) = \cos^s \phi$ and $\eta = 1.6$

Rather thick lenses result however, $D \geq 1$, depending on the physically allowed tolerance, that is variation in r, and bearing in mind probable machining capabilities for dielectric lenses at the wavelength concerned. Since the focal distance from the source to the first surface is normalised to unity, the errors shown can be interpreted as absolute phase errors when the wavelength of operation is known. The maximum error always occurs within the last few degrees of the incident angular aperture, approximately for angles $\phi \geq 26^o$. It can also be seen that the free parameter D, the central lens thickness, cannot be completely arbitrary, and attempts to make a thin lens, $D \ll 1$, fail on that account. This shows up in the ray segment r, which is equated for both sides of Equation 3.2.11 becoming negative at a value of ϕ within the specified range, gradually limiting the aperture to becoming more and more paraxial. Similar considerations govern the solution, or otherwise, of the partial nonlinear differential equation customarily derived.

Confining the incident angular aperture to 30° also results in a first surface that is virtually a plane for all thicknesses of D of the lens, and improves the residual error by several orders of magnitude as shown.

TAPERED APERTURE DISTRIBUTION

It is commonly required, for apodisation purposes, to have an aperture distribution that is reduced in magnitude at the periphery of the lens. For this we can select, for example

$$P(y) = 1 - 0.1y^2/R^2 \tag{3.2.13}$$

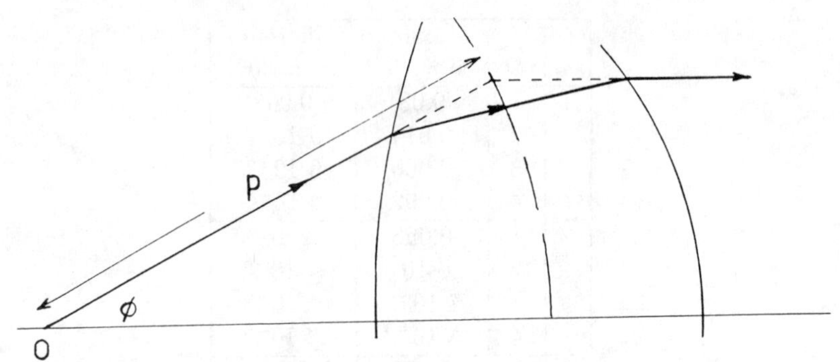

Figure 3.10: Abbe sine condition

where the factor 0.1 is chosen to give -10 dB at the edge. The same method of analysis applies, with additional complications, in deriving an explicit expression for y to replace that in Equation 3.2.10. Although elaborate, this presents no particular conceptual difficulty.

ABBE CONDITION LENS

This problem has been dealt with in a classic paper [13]. For the Abbe sine condition, we require that every ray from the source intersects its finally transmitted direction in points which lie on a circle with centre at the source (the origin in this case), as shown in Figure 3.10.

Taking this to be of, as yet undetermined, radius p, the condition translates into

$$y = p \sin \phi \tag{3.2.14}$$

The eikonal function is the same as for the previous example, as given in Equation 3.2.3. Thus substituting in Equation 3.2.4 for x from Equation 3.2.3 and for y from Equation 3.2.14

$$r = \frac{D(\eta - 1) - \rho(1 - \cos \phi)}{\eta - \cos \alpha} = (p - \rho)\frac{\sin \phi}{\sin \alpha} \tag{3.2.15}$$

From the limit, as ϕ and α tend to zero, we have $r \to \eta(p - 1)$ and hence

$$p = 1 + D/\eta \tag{3.2.16}$$

Equation 3.2.15 now becomes soluble for ρ but not explicitly in terms of ψ. It is here that a most unusual effect is observed. With this substitution in Equation 3.2.15, the right-hand side of the equation becomes identical to that of Equation 3.2.12, with the value $s = 1$. This can only mean that for this special set of conditions, namely a source distribution $P(\phi) = \cos \phi$, a dipole source, the lens will satisfy both the Abbe condition and the uniform amplitude condition simultaneously. Unfortunately, this source distribution, limited to the

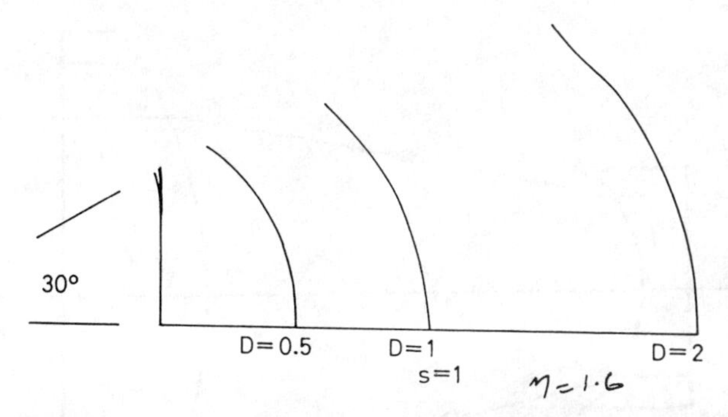

Figure 3.11: Lens profiles for sine condition with varying thickness

angular 30° aperture necessary to give an adequate approximation, is impractical. The residual error for this lens is therefore that of Table 3.1, for the values $s = 1$. For the thinnest lens attempted, $D = 0.5$, the maximum error is 0.0024. When D is reduced further, say to 0.25, the segment r becomes negative at the angle $\phi = 24^o$. The profiles for varying values of D are shown in Figure 3.11.

The results show that thick lenses give adequately accurate results, but for millimetre and submillimetre wavelengths, thinner lenses are practical. Indications are that, for amplitude shaping lenses, limiting the incident angular aperture to 30° is highly beneficial.

No attempt has been made as yet to improve on the results obtained by, for example, extending the variety of generic curves to include those with regions of higher curvature, and thus extend the angular aperture. The inclusion of such curves is a formality only. For the special distribution $P(\phi) = \cos\phi$, the profile formulae coincide, and both the amplitude-distribution condition and the sine condition can be satisfied simultaneously.

Alternative representations of the aspheric surface by a series include the standard power series taken up to the tenth order or a generalised aspheric series [6].

$$f(x,y) = \frac{c(x^2+y^2)}{1+\sqrt{1-c^2(1+k)(x^2+y^2)}} + \sum_{n=2} a_n(x^2+y^2)^n$$

The derivation of the coefficients for these series follows similar lines to the methods previously given.

COLLIMATED BEAM EXPANDER

An interesting "inside out" lens (glass to air / air to glass instead of the air to glass / glass to air of the usual lens (Figure 3.12)), which has applications to microwave antennas can be designed by these methods [7]. Applied to a laser

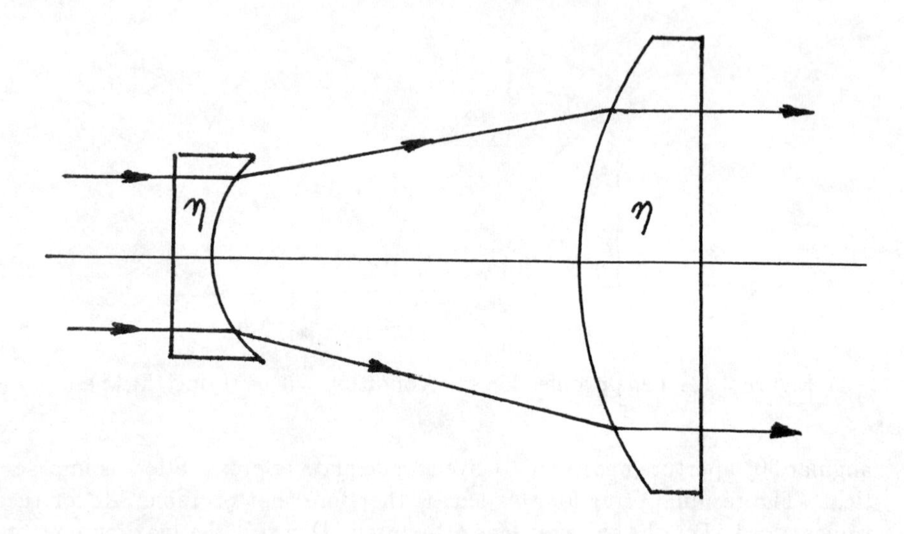

Figure 3.12: Beam expander

the requirement, typical of a microwave antenna, is for a uniform amplitude distribution over the output aperture.

In microwave practice the laser would be replaced by a corrugated horn [8], with its known illumination function to achieve a specified aperture phase and amplitude distribution.

3.3 MICROWAVE LENSES

In addition to the optical lenses which operate at microwave frequencies in the same way as with light, it is possible to use a refractive medium at microwave frequencies based upon the difference in phase velocity that exists between a wave in free space and a wave confined between parallel metal walls. Where the separation of the walls is between a half and one wavelength of the radiation being considered, the dominant TE_{01} mode alone propagates when the polarization is parallel to the walls themselves. In this situation a cylindrical lens fed by a line source is designed by the same principles as for the optical lens with the final result modified only by the fact that the apparent refractive index of the medium is less than unity. This refractive index is given by the relation

$$\eta = \sqrt{1 - \left(\frac{\lambda_0}{2a}\right)^2} \tag{3.3.1}$$

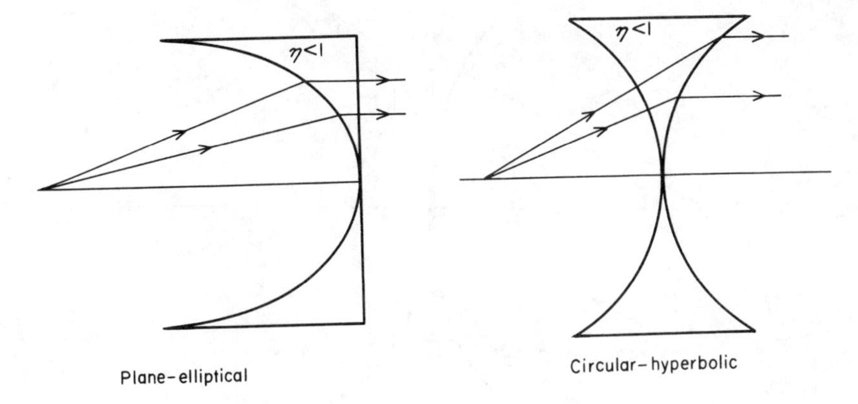

Figure 3.13: Metal plate lenses

where a is the spacing between the plates and λ_0 the operating wavelength in free space *. Substitution of a refractive index less than unity directly into the profiles given by Equations 1.1.8 and 1.1.9 converts them from hyperbolic and elliptical respectively into elliptical and hyperbolic (Figure 3.13). The second surface is then chosen to be normal to the incident or transmitted rays as appropriate. When a point source is used however this situation arises only in one plane of incidence, that containing the direction of the polarization of the source. In the orthogonal plane, the wave is constrained to change its direction to be parallel to the plates. A constraining action also occurs when the polarization is perpendicular to the set of parallel plates. These constraining actions give rise to pseudo refractive indices which are variable with the angle of incidence of the wave.

In order to make a lens which will affect all angles of incidence and polarization equally, as would be required by a rotationally symmetric device, a metal plate medium is constructed of two orthogonal sets of metal plates equally spaced (usually) to give a square waveguide medium with refractive index given by Equation 3.3.1. Such lenses have to be designed on the basis of optical path length because of the variability of the pseudo-refractive index with angle of incidence.

The principles of these two fundamental designs are well-established over many years [14][15].

It should be noted that the frequency dependence in Equations 3.3.1 gives a more highly dispersive medium than is found with natural dielectrics. Hence all designs based upon it will have an optimum frequency of operation and an associated bandwidth over which the performance will only be retained to within an acceptable limit. This means that the additional aberration of

*In practice the range of values for η is between 0.5 and 0.8.

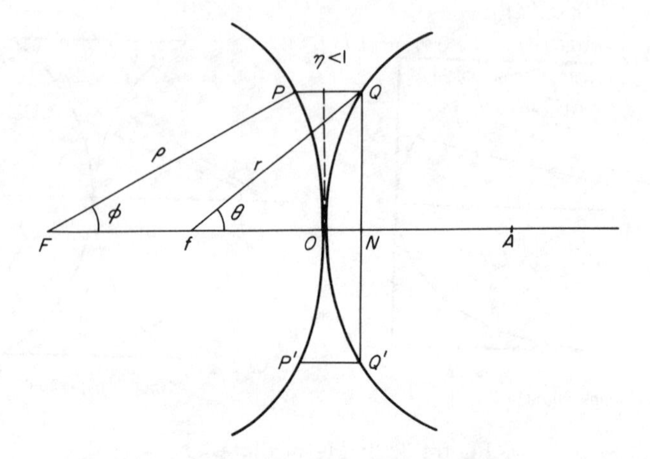

Figure 3.14: General two-surface waveguide lens

chromatism has now been added to the one or two normal ones associated with microwave optical devices.

Other forms of refractive index media of the waveguide type can be constructed from other shapes of waveguide notably the hexagonal medium the rectangular medium (for a single polarized source) and if the 10% loss can be tolerated, a medium of circular waveguides. A further valuable degree of freedom can also be obtained by making the medium variable in refractive index by altering the dimensions of the waveguides in the appropriate areas, subject to practical construction of course. The main distinction between microwave lenses of this type and optical lenses is that the latter have their maximum axial width at the centre. As can be seen from Figure 3.13, the reverse is the case for refractive index media less than unity, therefore in the following the thickness of the lens at the centre will be taken to be zero.

We now show that, because of the constraining action of the lens, the two main forms with elliptical and hyperbolic profiles respectively do not form two isolated instances as do the corresponding optical lenses but are two types of an infinitely continuous variation in form any of which can be chosen for an advantageous geometry or illumination distribution.

FOCUSING LENSES

The general two-surface lens with constant refractive index will consist of two curves OPP' and OQQ' (Figure 3.14) where the front face is a curve (ρ, ϕ) described from an origin at the source F, and the rear face of a curve (r, θ) described from an origin f. Without loss of generality the lens can have zero thickness at the centre, since in this case the addition of any constant optical length to the rays interior to the lens would produce a curve OQQ' parallel to the one with zero thickness as shown.

Letting for brevity $OF = F$ and $Of = f$ then equality of the optical path to produce a plane wave front QNQ' normal to the axis requires that for any point P and Q on the surfaces

$$FP + \eta PQ = FO + ON \tag{3.3.2}$$

or

$$\rho + \eta(F - \rho\cos\phi + r\cos\theta - f) = F + r\cos\theta - f$$

and hence

$$\rho(1 - \eta\cos\phi) - F(1 - \eta) = (1 - \eta)(r\cos\theta - f) \tag{3.3.3}$$

The principle of constraint states that the rays PQ are all parallel to the axis, that is

$$r\sin\theta = \rho\sin\phi \tag{3.3.4}$$

We can thus make a separation of variables between Equations 3.3.3 and 3.3.4 by making the (ρ, ϕ) sides of each functionally dependent and the (r, θ) sides the same functional dependence.

That is the two surfaces can be defined generally to be

$$\begin{aligned} \rho(1 - \eta\cos\phi) - F(1 - \eta) &= \text{ any function of } \rho\sin\phi \\ (1 - \eta)(r\cos\theta - f) &= \text{ same function of } r\sin\theta \end{aligned} \tag{3.3.5}$$

The most elementary solution is to make the arbitrary function simply a constant multiplier which we take to be $\alpha(1 - \eta)$; $0 < \alpha < \infty$. This gives

$$\rho(1 - \eta\cos\phi) - F(1 - \eta) = \alpha(1 - \eta)\rho\sin\phi$$

and

$$(1 - \eta)(r\cos\theta - f) = \alpha(1 - \eta)r\sin\theta$$

The resulting profiles are the ellipses

$$\rho = \frac{(1 - \eta)(F + \alpha\rho\sin\phi)}{1 - \eta\cos\phi}$$

or

$$\rho = \frac{F(1 - \eta)}{1 + \alpha\sin\phi + \eta(\alpha\sin\phi - \cos\phi)} \tag{3.3.6}$$

and the hyperbolae

$$r = \frac{f}{\cos\theta - \alpha\sin\theta} \tag{3.3.7}$$

When α is zero the elliptical plane profiles are obtained.

The asymptotes of the hyperbola make angles $\pm\tan^{-1}\alpha$ with the axis. Consequently a continuous variation of the single parameter α changes the shapes of both profiles causing the lens to shear in the axis direction retaining a changing elliptical profile on its front face and a corresponding changing hyperbolic profile on its rear face.

Thus a technique of "lens bending" becomes essentially simple to perform with this type of constrained medium.

Equation 3.3.5 may also be used to determine the second profile of the lens from any given first profile. Thus to design a lens whose second surface QOQ' is a circle of radius a centred at A we put its equation in the form

$$2a(r\cos\theta - f) - (r\cos\theta - f)^2 = r^2\sin^2\theta$$

which is a functional of the left-hand side of the second of Equations 3.3.5 in terms of another functional of the right-hand side.

The required profile for the (ρ, ϕ) face will then be obtained by repeating the functions in terms of the first of Equations 3.3.5 that is

$$2a\left[\frac{\rho(1-\eta\cos\phi)}{1-\eta} - F\right] - \left[\frac{\rho(1-\eta\cos\phi)}{1-\eta} - F\right]^2 = \rho^2\sin^2\phi$$

In general terms the functional relations of Equations 3.3.5 do not need to be explicit.

Some generalizations of this basic procedure can now be envisaged. For instance a non-uniform lens can be considered in which η is a function of distance from the axis. This makes it a function of both $r\sin\theta$ and $\rho\sin\phi$ and it can be included in its appropriate form in the respective equation of the two given in Equations 3.3.5. Another generalization would be to alter the law of constraint to make say $r\sin\theta = \beta\rho\sin\phi$ (β a second constant parameter) but structural complexity would doubtless outweigh most of the advantages that such designs may induce. On the theoretical side however the existence of these additional degrees of freedom does imply that beside the basic focusing property, consideration can be given to other properties such as amplitude distribution, bandwidth, additional points of focus or application of the sine condition (Appendix I).

WIDE ANGLE LENSES

Microwave lenses of the fully constrained type can use the available degree of freedom to provide exact focus at two symmetrically placed points transverse to the axis. The design due to Ruze[18] shows that as performed through the equality of the optical path from the source to each focus in turn, the shape of the front profile is established but only a single relation between the thickness of the lens and the refractive index is given, and thus the shape of the second surface, is not fully determined. A second condition is therefore required to make this relationship determinate and this allows for a third axial point to be made either a point of exact focus or at least of minimum residual aberration. The three points thus defined specify an arc along which the source can travel at each point of which the residual deviation from a plane wave front can be

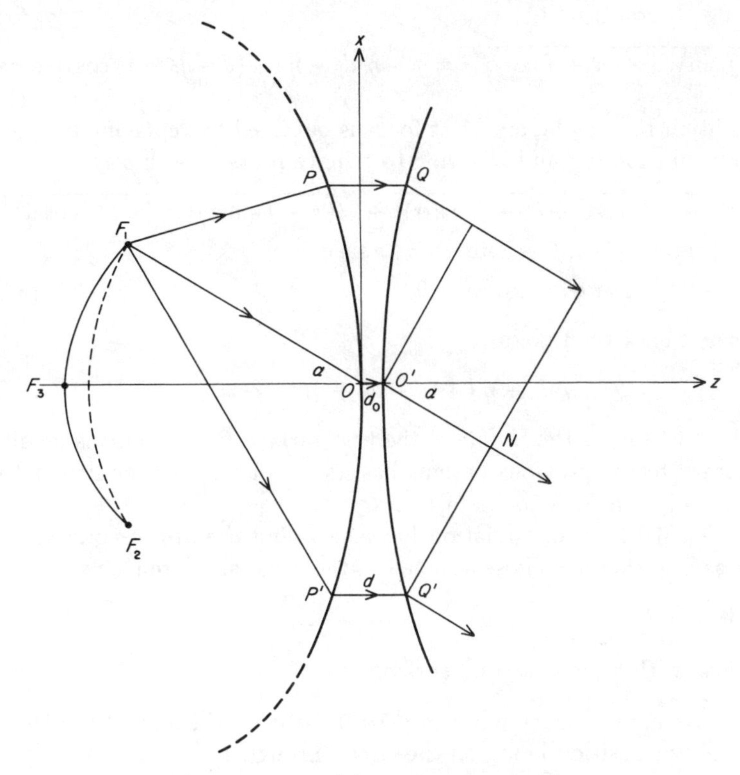

Figure 3.15: Wide-angle scanning lens

kept to within specified limits. This arc is usually taken for simplicity to be the circle containing the three points given above and the angle of travel is marginally greater then that subtended by the outer two perfect foci. The design of this lens formed something of a departure from the normal design of optical lenses and its main features will be summarized here for completeness.

The two curved surfaces required are POP' and $QO'Q'$ (Figure 3.15) where at this stage a finite thickness d_0 is assumed at the centre. F_1 and F_2 are the two chosen perfect foci subtending angles $\pm\alpha$ with the axis at O'. The refractive index η is a function of distance from the axis which we take to be the x coordinate axis. The axis of the lens is the z axis and the value of the refractive index at the centre is η_0.

Then since F_1 is a perfect focus

$$F_1P' + \eta P'Q' = F_1O + \eta_0 OO' + O'N \tag{3.3.8}$$

Calling the thickness $P'Q' = d$ and $F_1O = f$, Equation 3.3.8 in Cartesian coordinates becomes

$$\sqrt{(x+f\sin\alpha)^2+(z+f\cos\alpha)^2} = f+\eta_0 d_0 - \eta d + (d-d_0+z)\cos\alpha + x\sin\alpha \tag{3.3.9}$$

The condition that F_2 be a perfect focus is obtained by replacing α by $-\alpha$ in the above, subtracting and squaring to remove roots gives firstly

$$\sqrt{(x+f\sin\alpha)^2+(z+f\cos\alpha)^2} - \sqrt{(x-f\sin\alpha)^2+(z+f\cos\alpha)^2}$$
$$= x\sin\alpha + f - (f - x\sin\alpha) = 2x\sin\alpha$$
$$z^2 + 2xf\cos\alpha + x^2\cos^2\alpha = 0 \tag{3.3.10}$$

Substitution back then gives

$$\eta_0 d_0 + \eta d + (d - d_0 + z)\cos\alpha = 0 \tag{3.3.11}$$

Equation 3.3.10 gives the profile of the first surface POP'. This is an ellipse, but contrary to the previous designs has its major axis perpendicular to the lens axis, and with F_1F_2 as its foci.

Equation 3.3.11 is one relation between η and d and a second arbitrary relation can be chosen. Those selected in the original reference were

1. a lens of constant thickness
2. a lens with a plane second surface
3. a third and arbitrary point of perfect focus on the axis of the lens (the preferred position being on the circle through F_1F_2 centred at O), and
4. a constant refractive index

Each of 1. 2. and 4. then gave a preferred point of minimum phase distortion on the axis. The errors in phase from the plane wave front for a general position of the source along the circle through the three points of best focus can now be computed directly from the path lengths along the rays. If a level is then set for permissible deviations from the plane phase front, the scan angle is then determined as that angle subtended by the source at which these levels are attained. The levels are somewhat arbitrary and permit a deterioration up to a degree of the radiated pattern shape but once selected they serve as a basis of comparison for the different possibilities inherent in the design.

It was immediately apparent that even with the three points of focus so defined, the optimum curve for the source to travel, the so-called scanning arc, was not apart from 3. above, the circle through them. When the best scanning arc was achieved by a refocusing procedure for minimum phase error at all points along the path the total scan angle achieved was found to be

$$158\left(\frac{f^2\lambda}{a}\right)^{1/3} \quad \text{degrees}$$

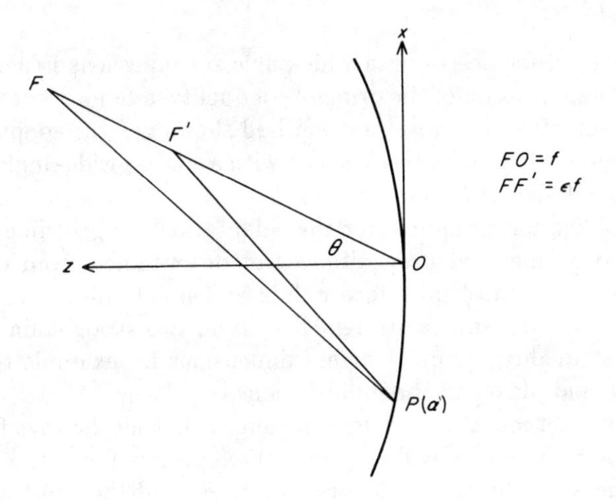

Figure 3.16: Refocusing

where λ is the operating wavelength and a the full aperture of the lens.

The refocusing procedure is essential for any optical system that requires an optimum point of focus over a wide range of angles. It consists of estimating to the first order the change in the optical path obtained by an incremental movement of the source along the radius vector to the centre of the lens. This particular motion is not the only one that could be considered but it is the most obvious practical choice.

Then if the feed is moved a small fraction εf (Figure 3.16) of the distance f from the centre of the lens the resulting path length change to the point $P(x, z)$ is given by [16]

$$\Delta = FP - F'P - \varepsilon f \simeq \frac{\varepsilon x^2}{2f}(1 - \theta x) \tag{3.3.12}$$

Such a term can thus be used to reduce both second and third order aberration coefficients, that is spherical aberration and coma. From the point of view of beam shape and beam position, where a choice has to be made it would be preferable to correct the third order term and leave the residual second order.

We note for future reference that the ellipse shape of the front surface of this lens can be given parametrically in the form

$$\begin{aligned} z &= f\cos\alpha(1 - \cos t) \\ x &= f\sin t \end{aligned} \tag{3.3.13}$$

where f and x are constants defining the polar coordinates of the foci F_1 and F_2.

THREE-RAY LENS

The second method for designing a wide-angle scanning lens is derived from the above by an application of the principle of duality, a fundamental principle of projective geometry. The resultant method shows greater adaptability for use in the design of natural dielectric lenses with a similar wide-angle property, as well as in the design of lens-reflectors.

In the simplified form required for this adaptation the principle of duality states that two geometrical propositions can be obtained from each other through the interchange of lines into points and points into lines [17]. For example as two points define a straight line so do two straight lines define a point. A similar duality occurs in higher dimensions for example the duality between points and planes in three dimensions.

In the design method above we have arranged that all the rays from three given directions converge individually into three perfect foci. The directions are $\pm\alpha$ and the axial direction and the foci F_1 F_2 and the third axial point F_3 as shown in Figure 3.15. We form the geometrical dual of this statement as follows: at every point along an arc F_1F_2 three of the rays of the complete pencil are correctly in phase.

These rays may be arbitrarily chosen but from the simplicity of subsequent analysis we choose

1. the main ray to the centre of the lens
2. the ray parallel to the axis of the lens, and
3. the ray meeting the lens at a point diametrically opposite that of the second ray.

This arbitrary choice receives further justification in its natural occurrence in the design of bi-cylindrical lenses.

The effect as the source moves over the arc is to spread the rays further apart making more of the surface of the lens conform at least in part to the required focusing property. The design is performed in exactly the same way, by assuring constancy of the optical length of the rays chosen. Then from Figure 3.17 we require

$$FP + \eta PQ + QN = FO + OM = FP' + \eta P'Q' \tag{3.3.14}$$

where a zero centre thickness has been assumed. By symmetry we must have $PQ = P'Q'$ hence the outer equation of 3.3.14 is simply

$$FP + QN = FP'$$

or squaring and rearranging

$$Q'N^2 = 2FP \cdot QN$$

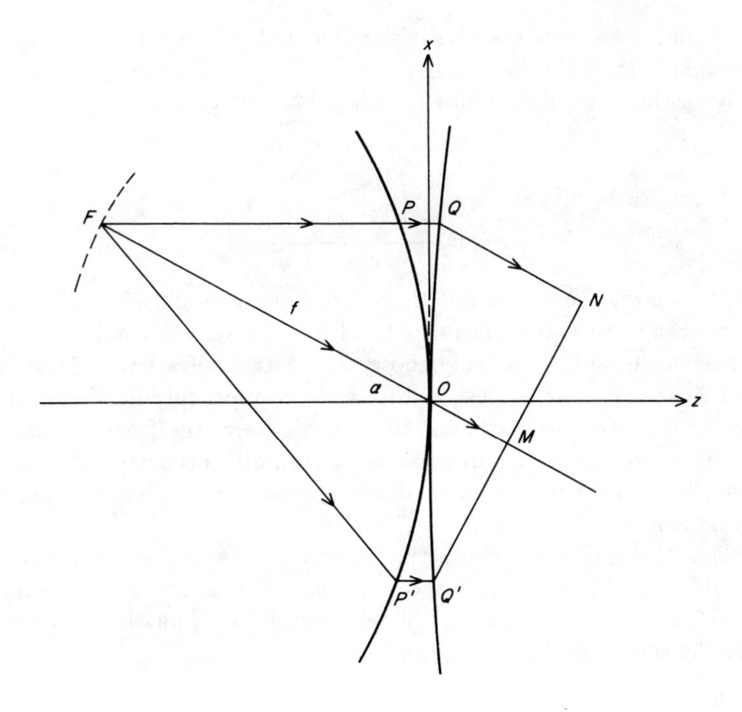

Figure 3.17: Three-ray lens. Rays from the variable position of F through variable points P P' are in phase

that is $4f^2 \sin^2 \alpha \cos^2 \alpha = 2(f \cos \alpha - z) \cdot 2f \sin^2 \alpha$ and hence in parametric form

$$\begin{aligned} z &= f \cos \alpha (1 - \cos \alpha) \\ x &= f \sin \alpha \end{aligned} \tag{3.3.15}$$

where f and α are now essentially variable and f as a function of α defines the scanning arc.

Solving the remaining equation in (3.3.14) now gives

$$\eta PQ = FT - FP$$

or

$$PQ = \frac{f \cos^2 \alpha (1 - \cos \alpha)}{\cos \alpha - \eta} \tag{3.3.16}$$

With f now given as a function of α Equations 3.3.15 and 3.3.16 give parametrically the two surfaces of the lens. (A second condition does not exist here since we have already chosen a constant refractive index.) Totally general lenses can now be designed from the a priori specification of any one of the three curves alone, that is the scanning arc, the front surface or the rear surface. Variation in refractive index with distance from the axis can be included by making η a function of $\sin \alpha$ and incorporating it directly into Equation 3.3.16.

Noting the similarity of description that exists between the parametric descriptions of the front surface of the three-ray design and the two focus design Equations 3.3.15 and 3.3.13 respectively, we find that if we were to specify the three rays to be (Figure 3.18)

1. the central ray FO

2. , 3. the rays through two *fixed* points P and P'

equality of the optical path then gives

$$QN + FP = FP'$$

or, if P is the point (z, x_p) where x_p is constant

$$\sqrt{(z + f \cos \alpha)^2 + (x_p + f \sin \alpha)^2} - \sqrt{(z + f \cos \alpha)^2 + (x_p - f \sin \alpha)^2} = 2x_p \sin \alpha$$

which is identically Equation 3.3.10(a) with x_p now a constant. The resulting profile in this case is therefore

$$z^2 + 2zf \cos \alpha + x_p^2 \cos^2 \alpha = 0$$

in analogy to Equation 3.3.10(b), but with f and α variable and x_p constant.

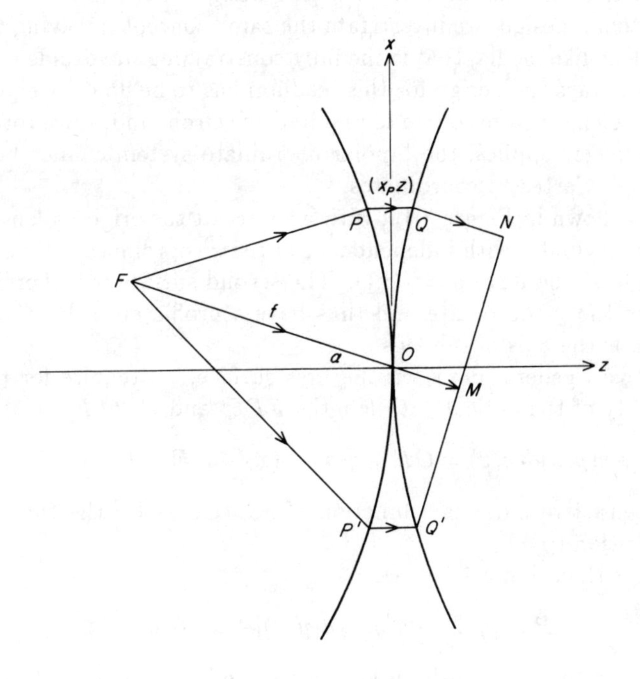

Figure 3.18: Three-ray lens (alternative design). Rays from the variable position of F through fixed points P P' are in phase

3.4 BI-CYLINDRICAL LENSES

In the same way that the reflective caustics of two cylindrical reflectors can be matched to create a bi-cylindrical reflector, two refractive caustics can be matched to create a bi-cylindrical lens. This principle is also well-established in optical practice and has been used in the design of spectacle lenses. In microwave antenna design, we investigate the same concept involving the use of microwave lens-like media, that is the fully constraining waveguide media of the lenses. As always the design for this medium has to be done by equalizing the optical path length from source to required wavefront and, since rotational symmetry no longer applies, the bipolar coordinate system cannot be used. We therefore use Cartesian coordinates.

We have as shown in Figure 3.19, with a source at the origin, a lens whose first surface is a cylinder with independence of the z coordinate, whose profile we assume to have the form $y = f(x_1)$. The second surface will therefore be independent of the y coordinate and thus have a profile given by $z = g(x_2)$ and the x axis is the axis of the lens.

Then if P is a general point on the first surface, we require for parallel rays the equality of the optical path lengths FPP' and $FOO'N$

$$FP + \eta(y,z)d(y,z) = OF + \eta_0 d_0 + (x_2 - OF - d_0) \tag{3.4.1}$$

where η the refractive index is a function of position and d the thickness of the lens at position (y, z).

Rearranging Equation 3.4.1 gives

$$FP = [\eta_0 d_0 - \eta(y,z)\, d(y,z)] + x_2 - d_0$$

The first factor occurs frequently in lens design and so for simplicity let

$$D(y,z) \equiv \eta_0 d_0 - \eta(y,z)\, d(y,z) \tag{3.4.2}$$

Squaring then gives

$$x_1^2 + y^2 + z^2 = (D + x_2 - d_0)^2 \tag{3.4.3}$$

but x_1 is independent of z and x_2 is independent of y. Therefore Equation 3.4.3 separates into two equations

$$\begin{aligned} x_1^2 + y^2 &= \text{constant} = A^2 \\ z^2 + A^2 &= (D + x_2 - d_0)^2 \end{aligned} \tag{3.4.4}$$

provided that $D(y, z)$ can be made to be a function of z only.

Solutions to Equation 3.4.4 can be freely chosen and depend upon the (almost) arbitrary choices of A and $D(z)$. The limiting consideration is not in the design mathematics but in the feasibility of the final construction, that is upon the limited range that exists for the refractive index in this type of medium.

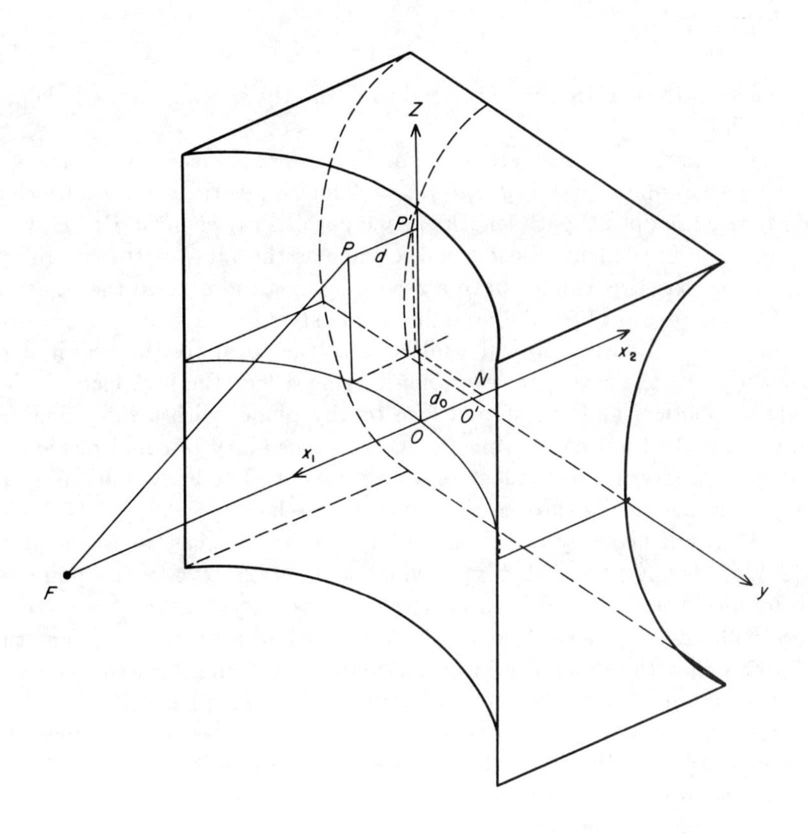

Figure 3.19: Bi-cylindrical focusing lens

The simplest solution is to make A the focal length OF and $D(z)$ equal to zero. The resulting profiles are then

$$x_1^2 + y^2 = F^2$$

and

$$z^2 + F^2 = (x_2 - d_0)^2 \tag{3.4.5}$$

The first of these is the circle centred at F and the second a hyperbola passing through O'.

There now arises the rather enigmatical consequence. The choice of $D(z)$ to be zero implies $\eta_0 d_0 - \eta(y,z)d(y,z) = 0$ for all positions of the surface point P. Hence the optical path length of each parallel ray element PP' is the same everywhere in the lens. Since it is the same as the value at the centre by this reasoning, the lens cannot have a zero thickness there as do the other lenses so far designed. This at first sight seems strange when one is accustomed to using the different optical widths of a lens to apply the required phase correction to the rays. However, considering a lens the first face of which is purely spherical and the second face totally plane, with a fully constrained waveguide medium connecting them, the plausibility of equal phase lengths for all rays internal to the lens establishes itself. The bi-cylindrical lens is in fact a symmetrically sheared version of such a lens.

The result given by Equation 3.4.5 further establishes the following statement: if for any lens of this type, which focuses the rays from a point source to infinity, the first surface has a circular cross-section, then the second surface will have a hyperbolic one and vice versa. In a practical design study it is found that the difference in profile between the circular and hyperbolic is very small and within the usual tolerance of $\lambda/16$ for phase differential. Thus there is no great error in making both profiles the same, say circular, and if necessary adjusting even this small error by a modification of the medium refractive index. We then have a lens which focuses exactly at infinity from a source on either side of the same focal length.

SCANNING BI-CYLINDRICAL LENSES

In accordance with the principle that the two surfaces can be used to a second effect beside that of ordinary focusing we establish a design for a feed point displaced from the point F, to a point with polar coordinates (ρ, α). That this has to take place in the x, y plane is apparent if one visualizes the cross-section of the centre of the lens in the other plane. This will be straight on the incident side and curved on the transmitted side. From Figure 3.20 it can be seen that for a displacement of the source through an angle α the beam position will be given by an angle β where

$$\tan\alpha = \tan\beta \sec\beta$$

Although this effect is small for small angles of scan and in practice when

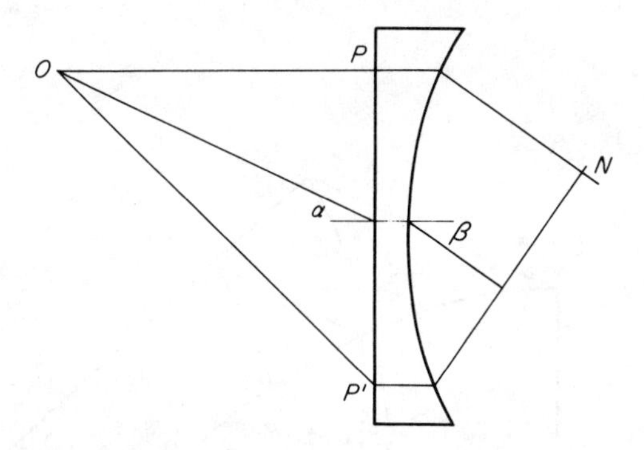

Figure 3.20: Scan in vertical plane

the remainder of the surface is brought into action hardly observable, we will nonetheless confine the scanning action to the plane specified above. Then as shown in Figure 3.21 we require

$$SP + \eta(y,0)PP' + P'N = SO + \eta_0 OO' + O'M = SQ + \eta(y,0)QQ' \quad (3.4.6)$$

where S is the displaced source at the point (ρ, α) with respect to the origin O.

From the outer equation in Equation 3.4.6 we obtain the parametric description of the curve PQ. In terms of coordinates (x_1, y, z) at O this is identically that derived for the three-ray lens given by Equation 3.3.15 namely

$$\begin{aligned} x_1 &= \rho(\alpha)\cos\alpha(1-\cos\alpha) \\ y &= \rho(\alpha)\sin\alpha \end{aligned} \quad (3.4.7)$$

where ρ as a function of α defines the scanning arc travelled by the source S.

Solving the remaining equation in (3.4.6) for the optical thickness, $\eta(y,0)$ $d(y,0)$, we obtain

$$\eta(y,0)d(y,0) - \eta_0 d_0 = 0$$

a condition in this case and not the free choice made in the previous section. Note that it only applies as yet to the section of the lens given by $z = 0$.

Thus we find that the bi-cylindrical lens in its capacity as a scanning lens automatically gives the principle of equal path length for each element of lens from the three-ray principle of lens design.

The second surface is obtained by considering (Figure 3.21) a general point R with coordinates referred to F given by

$$(F - \rho(\alpha)\cos\alpha(1-\cos\alpha), \rho(\alpha)\sin\alpha, z) \qquad (F = OF)$$

We require for a plane wavefront

$$FR + \eta(y,z)RR' = FO + \eta_0 d_0 + O'V \quad (3.4.8)$$

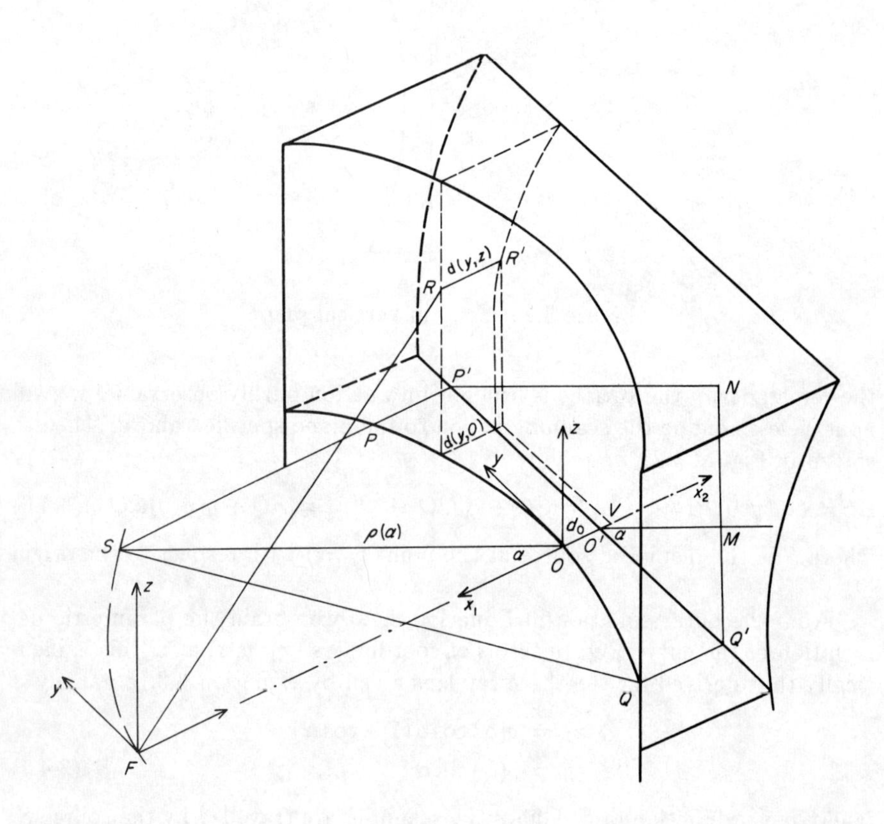

Figure 3.21: Scanning bi-cylindrical lens. Scanning motion is in the horizontal plane SPOQN

that is

$$FR = x_2 - d_0 + D$$

where

$$D = \eta_0 d_0 - \eta(y,z)d(y,z)$$

as before. Squaring Equation 3.4.8 and separating for x_2 and D dependent on z only we obtain

$$\begin{aligned}[F - \rho(\alpha)\cos\alpha(1-\cos\alpha)]^2 + [\rho(\alpha)]^2\sin^2\alpha &= \text{constant} = A^2 \\ (x_2 - d_0 + D)^2 - z^2 &= A^2 \end{aligned} \tag{3.4.9}$$

The first of these establishes the scanning arc by solving for ρ as a function of α. The second is the profile of the second face of the lens.

For example taking $A = F$ and $D = 0$ we recover the hyperbolic surface and the equality of the optical length for each element of the entire lens as was obtained for the simple focusing lens. This means that the profile derived from Equation 3.4.9 with this function of ρ inserted *must* be circular.

Thus from the first of Equation 3.4.9 with this choice we find

$$\rho = 2F\cos\alpha/(1+\cos\alpha+\cos^2\alpha-\cos^3\alpha) \tag{3.4.10}$$

and hence from Equation 3.4.7

$$\begin{aligned} x_1 &= 2F\cos^2\alpha(1-\cos\alpha)/(1+\cos\alpha+\cos^2\alpha-\cos^3\alpha) \\ y &= 2F\sin\alpha\cos\alpha/(1+\cos\alpha+\cos^2\alpha-\cos^3\alpha) \end{aligned}$$

That this is a circular arc with centre at F is readily confirmed since

$$(F - x_1)^2 + y^2 = F^2$$

but it is not a description of a complete circle. As the parameter α (*not* polar coordinate here) is varied this expression describes a limited circular arc traversing it once in each direction. This delimits the aperture of the lens and the scanning arc simultaneously. Being circular the profile also satisfies the requirement of the sine condition for this plane of scan.

Generalizations of this lens are possible through the various choices of the constant A and the function of thickness $D(z)$. Variation of A alters the hyperbolic profile of the second surface, hence the radius of the circular first surface and thus the scanning arc.

Variations in $D(z)$ would, as for the ordinary focusing lens, allow the scanning lens to have the same circular profiles on both faces. The symmetrical lens would then have the ability to focus rays from oblique angles within the range of the scanning arc, at individual points on either side of the plane of symmetry. This situation is indicated in Figure 3.22.

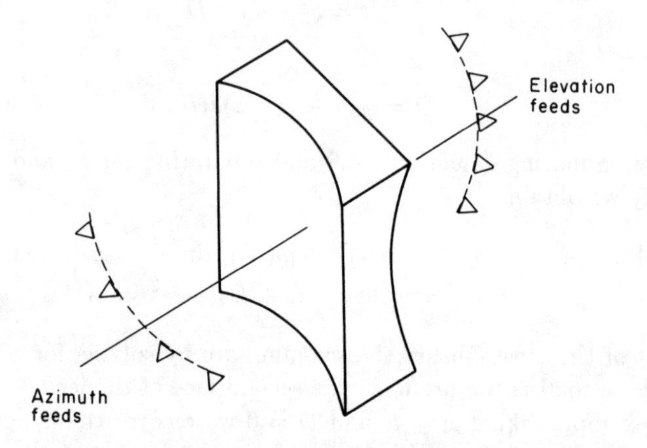

Figure 3.22: Multi-element lens

EXPERIMENTAL RESULTS

A scanning lens designed on these principles has been constructed using the waveguides of square cross-section and obtaining the variation in refractive index by filling each element with a dielectric of appropriate refractive index. It is found with the lossless dielectric foam materials that the refractive index is a linear function of the density and hence the required relation

$$\eta(x,y) = \sqrt{\kappa_e(x,y) - \lambda^2/4a^2} \tag{3.4.11}$$

which gives the refractive index as a function of position in the lens can be converted to a relation between the density of the expanded dielectric medium and position of the waveguide element. The range of refractive index values that can be achieved by this method lies between that of the lightest manufacturable foam and the solid dielectric (the unfilled waveguide being too fragile for the manufacturing process). The requirement that each element has the same *electrical* length places a limit on the maximum width of the lens at any point, unless wavelength stepping is introduced. This also defines the final aperture that can be achieved. The design, at a wavelength of 3 cm, of an unstepped lens was between 5 and 10 cm in thickness and gave an aperture (near circular) of 60 cm.

Radiation patterns of this lens when scanned in both E and H planes are shown in Figure 3.23. The E plane scan is the plane of offset as designed, the H plane is that facing the linear central section and thus subject to the possibility of the squint as described earlier. In practice both planes give a performance over $\pm 30^o$ much in line with the scanning patterns of spherical lenses designed by the alternative methods given previously. Thus the squint

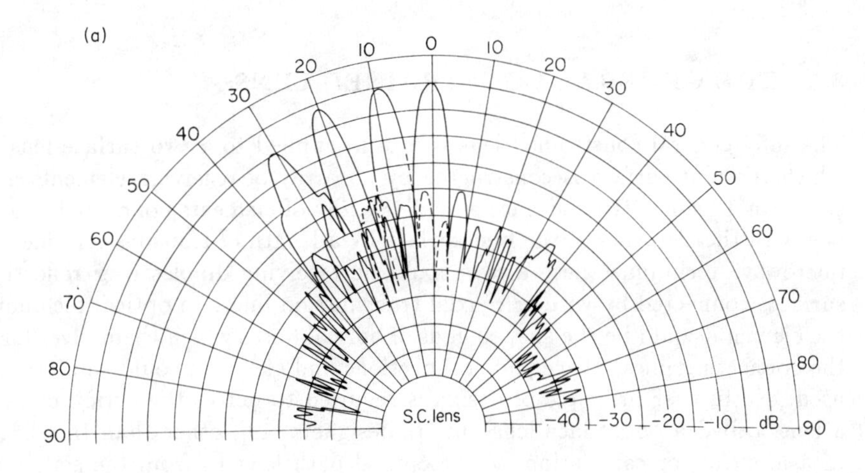

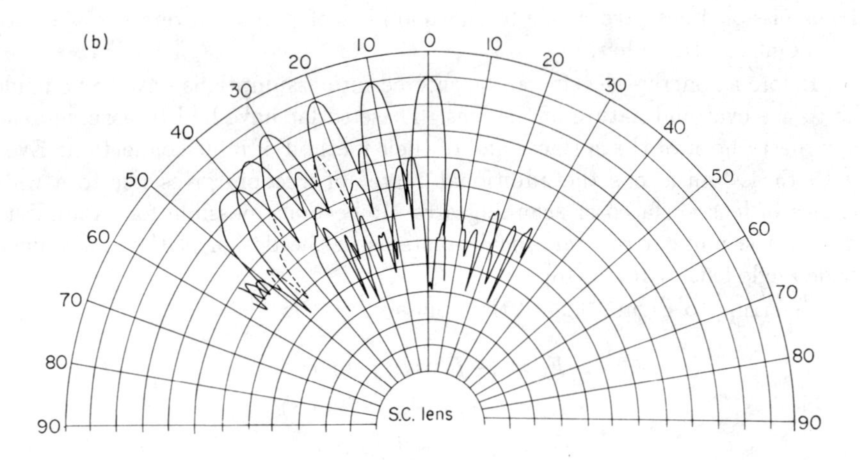

Figure 3.23: Scanning bi-cylindrical lens patterns (a) horizontal plane - E plane scan (b) vertical plane - H plane scan

effect predicted for a central cross-section is largely overcome when the entire aperture is brought into operation.

3.5 THE GENERAL CONSTRAINED LENS

The fully general constrained lens is a term applied to a two surface lens in which the first surface is constructed by an array of receiving elements and the second transmitting surface a similar array of elements connected one to one with the elements of the first surface by a length of transmission line. In microwave technique these arrays could be receiving dipoles over reflecting surfaces connected by wire or coaxial transmission lines. In optical technique the elements could be the shaped ends of optical fibres which themselves form the connecting lines. Waveguide connections and elements could similarly be used. No further principle of design is required to derive the action of such a lens, called a "boot-lace lens" by its designers [19], other than the direct measurement, or calculation, of the optical path lengths from the source to the required phase front under the various conditions to be considered.

In principle the two surfaces can be totally dissociated and the connecting transmission lines give a one-to-one mapping of points on one surface with the points on the other.

Before a concrete design can be entered into assumptions have to be made as to the eventual nature of the lens. These so far have had to assume axial symmetry both of the surfaces and of their method of inter-connection. Even with these limitations the additional degree of freedom gives rise to a wide choice of lens surfaces or scanning arcs. It becomes possible for example to double the number of *exact* focal points as in the design of the constrained wide angle lens of Ruze [18].

Referring to Figure 3.24, if the points

$$
\begin{aligned}
F_1 &= (-f\cos\alpha, f\sin\alpha)\\
F_2 &= (-f\cos\alpha, -f\sin\alpha)\\
G_1 &= (-g\cos\beta, g\sin\beta)\\
G_2 &= (-g\cos\beta, -g\sin\beta)
\end{aligned}
\qquad (3.5.1)
$$

are symmetrically placed (in pairs) points of exact foci which respectively give rise to plane phase fronts with normals at angles $\pm\alpha$ and $\pm\beta$ to the axis of symmetry then the optical path length conditions are

$$
\begin{aligned}
F_1P + \omega - \xi\cos\alpha + \eta\sin\alpha &= f + \omega_0 - a\cos\alpha\\
F_2P + \omega - \xi\cos\alpha - \eta\sin\alpha &= f + \omega_0 - a\cos\alpha\\
G_1P + \omega - \xi\cos\beta + \eta\sin\beta &= g + \omega_0 - a\cos\beta\\
G_2P + \omega - \xi\cos\beta - \eta\sin\beta &= g + \omega_0 - a\cos\beta
\end{aligned}
\qquad (3.5.2)
$$

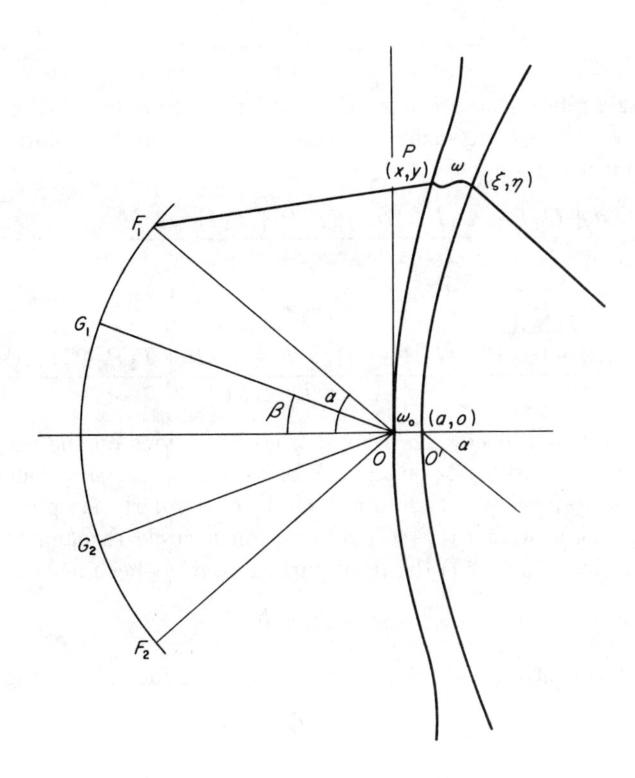

Figure 3.24: Boot-lace lens

in which (ξ, η) is the point on the second surface connected with $P(x, y)$ on the first surface through the transmission line of optical length $\omega(x, y)$ and $\omega_0 = \omega(O, O')$.

Subtracting the first pair and the second pair of these relations and dividing gives

$$\frac{F_2P - F_1P}{2\sin\alpha} = \frac{G_2P - G_1P}{2\sin\beta} \tag{3.5.3}$$

which is a single relation between x and y and thus describes the contour of the first surface. The second surface is then obtained from the solution of the remaining equations giving

$$\begin{aligned} \xi &= \frac{a + G_2P + G_1P - 2g - (F_2P + F_1P - 2f)}{2(\cos\beta - \cos\beta)} \\ \eta &= \frac{F_2P - F_1P}{2\sin\alpha} \\ \omega(x, y) &= \frac{\omega_0 + (G_2P + G_1P - 2g)\cos\alpha - (F_2P + F_1P - 2f)\cos\beta}{2(\cos\beta - \cos\alpha)} \end{aligned} \tag{3.5.4}$$

Various forms of lens can now be derived from conditions on the position of the points F_1 F_2 G_1 and G_2 or on $\omega(x, y)$ for example $\omega(x, y) =$ constant.

When the four focal points lie on a circle particularly simple forms of surface result. Thus with F_1 F_2 G_1 and G_2 on a circle through the point $(-h, 0)$ with radius $h/(1 - 2\mu)$ the front surface contour becomes the circle

$$x^2 + y^2 - xh = 0$$

of radius $h/2$ and centre $(-h/2, 0)$, and the second surface the ellipse

$$\xi = a + 2\mu(h - \sqrt{h^2 - \eta^2})$$

which has its centre at $(a + 2\mu h, 0)$ and axes $(2\mu h, h)$. The required path length delay between points (x, y) on the front surface and (ξ, η) on the second surface is given by

$$\omega = \omega_0 + (2\mu + 1)(h - \sqrt{h^2 - \eta^2})$$

Thus when $\mu = -1/2$ the focal arc is of radius $h/2$ and the connecting transmission lines are all of the same electric length $\omega = \omega_0$. A judicious choice of this length could be made to cancel the transmission line mismatch reflections. For a final assessment of the lens the phase front for the source at intermediate points of the focal arc require to be calculated and the errors minimized if necessary by refocusing. Other designs [19][20] can be made based upon the minimization of aberrations and on adaptations of the three-ray method. In the reference given three particular cases are discussed

1. a lens with a plane second surface with an almost circular scanning arc and low aberrations

2. a lens with near linear scanning arc

3. a frequency independent lens with zero geometrical aberrations on a circular scanning arc.

Some of the designs are illustrated in Figure 3.25.

The practical difficulties to be overcome are those facing the designers of most planar arrays. That is the consistency of the receiving and transmitting arrays of elements under conditions of varying wide angle incidence of plane waves, due to element interactions and impedance variations. Care has also to be taken that grating lobes due to element spacing do not occur in the field of operation.

Further generalizations can be considered such as the inclusion in the interior transmission lines of electronically controlled phase shifters or even of active elements amplifiers and frequency changers, but such a lens becomes an entire system in its own right and ceases to be simply an optically designed device.

The inclusion of short-circuits alone in the transmission lines creates a phase corrected reflector, but the limitation that only one surface is to be used both for incidence and transmission gives it no apparent advantage over the previous design. The reason for this is that with short-circuits the one-to-one correspondence between elements is the identity relation. If completely dissociated surfaces were to be considered with a general one-to-one correspondence between the elements on each then it is possible to imagine the second surface folded back over itself and into coincidence with the first surface. The array will then consist of a series of receiving and transmitting elements over a single reflecting surface connected in pairs by the bent over transmission lines. This forms the fundamental reflector array [23]. Applications of symmetry conditions can make this a monostatic reflector, for example if each element were connected to a diametrically opposite element (with respect to the centre of the array) by transmission lines all of equal length. Variations of the line lengths, the introduction of active elements and so on produce generalizations of this design. Fibre optical and acoustical versions based on the same principle are also known.

THE ROTMAN LENS

The principles of the "bootlace lens" provides one method of overcoming the limitation that the constancy of refractive index imposes on the Ruze type scanning lens (Section 3.3 Wide Angle Lenses). When constructed as a waveguide lens, natural mechanical requirements force the waveguides to be uniform throughout. The path length phase variation introduced with

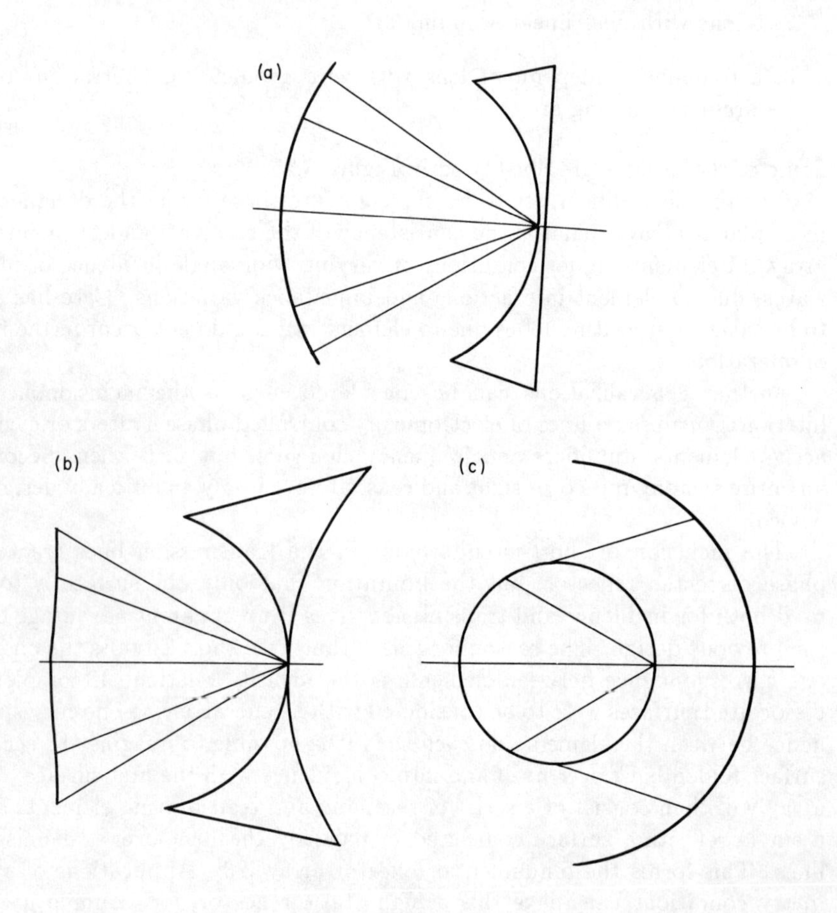

Figure 3.25: Boot-lace lens designs [20] (a) Lens with plane outer face, (b) lens with near linear scanning arc (c) Frequency independent lens with circular scanning arc

the bootlace technique is equivalent to a variation in refractive index. This requires the replacement of the waveguide array with an array of individual receivers connected to the output array of transmitters. Accordingly it also becomes possible to replace the single feed moving on its focal arc with an array of individual feeds serviced by a distributing network. This combination was first demonstrated by Rotman [25] and is consequently known as the Rotman lens. Ultimately, it is a sophisticated method of feeding a linear scanning array, without the limitation of requiring variable phase shifters. The radiation pattern itself can be scanned as far as linear arrays permit. The different forms that the focal arc and first receiving arrays can take provide differing illumination functions across the output array and an optimum in the linearity of the phase as the various feed positions are activated [26].

Modern introductions include the replacement of coaxial cable connections with stripline or printed circuits, and advantage can be taken of the introduced line circuits to add amplification to each output element.

Finally the progression to a three-dimensional configuration has proved possible [27] with transmitting elements covering a focal surface and the path length lens designed on the principles governing the bi-cylindrical lens of Section 3.4. A full survey of the designs available has been made [28].

3.6 PHASE CORRECTED REFLECTORS

The term "phase corrected reflectors" stands for the class of optical devices consisting of a reflector coated with a surface of refractive material either in the form of waveguide media or with natural dielectrics. Here we take a licence to include the correction of the spherical reflector by a detached lens, and start with that subject. Although the system manifestly consists of two designable surfaces, the fact that one of them has to be traversed twice by the rays, seems to prevent the *exact* application of two preconditional criteria. Most, therefore, are applied as wide angle scanning antennas for which the design has great advantages over the single lens or dual reflector.

CORRECTORS FOR THE SPHERICAL REFLECTOR

This a classical example of correction of aberration by the addition of a refracting element to a standard reflector [29]. The treatment by ray tracing leads to a series solution for the profile of the correcting element [35]. There are two alternatives based on optics-in-the-large procedures found in previous chapters.

The first is by caustic matching. To a first (and sometimes much higher) order, all simple cusped caustics are in agreement with each other. To that extent, the caustic of reflection in the sphere, the nephroid (Appendix III) and the caustic of refraction in a *plane* interface can be superimposed. Then

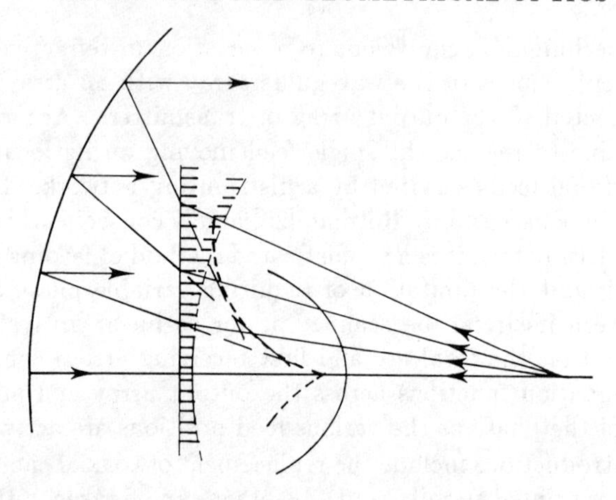

Figure 3.26: Matching the caustics of a reflector and a refracting surface. The caustic of the circular reflector can be matched approximately (paraxially) by the involute of the elliptical phase front of the refraction in a plane interface

illuminating the sphere by the caustic of refraction of the plane will produce a nearly aberration free beam. We derive this caustic of refraction from a source *within* the medium using the zero-distance phase front. For this form of refraction (see Section 1.2 Refraction in a Circular Interface) the zero-distance phase front is

$$(x-c)^2 - y^2/(\eta^2-1) = c^2/\eta^2 \tag{3.6.1}$$

The evolute of the ellipse $x = a\cos t$, $y = b\sin t$ is (Appendix III)

$$X = (a^2-b^2)(\cos^3 t)/a$$
$$Y = (b^2-a^2)(\sin^3 t)/a \qquad a > b$$

Applying this to the ellipse in 3.6.1 gives the caustic of refraction

$$X = c\eta\cos^3 t \qquad Y = c\eta\sin^3 t(1-\eta^2)^{-1/2} \tag{3.6.2}$$

with $\eta < 1$ through the choice of refraction from the medium into free space. This is a form of the "elliptical" astroid $x^{2/3}/c^{2/3} + y^{2/3}/d^{2/3} = 1$

The caustic of reflection in the circle is the nephroid

$$X = a(6\cos t - 4\cos^3 t) \qquad Y = 4a\sin^3 t$$

adapted from Equation 1.4.1. This has to be reflected and translated to match the refractive caustic giving

$$X = 4a - 6a\cos t + 4a\cos^3 t = 2a\cos^3 t$$
$$Y = 4a\sin^3 t \tag{3.6.3}$$

up to powers of t^4 or higher.

Thus by comparison with Equation 3.6.2, not only do we derive a good match between the two curves, but we find the necessary refractive index of the lens material. This is $\eta = \sqrt{3}/2$ inverted for real values to $2/\sqrt{3}$ since the source is embedded in the medium. It remains to obtain a method of removing the source into free space, and this is done by creating a second surface for which all the rays are not refracted. This is obviously the circle centred on the source as is shown in Figure 3.26. If however the position of the source is pre-specified as is usual in the literature this last spherical surface can be replaced by its aplanatic equivalent, the limaçon as before, which is the shape usually attributed to the Schmidt correcting element (Figure 3.27).

Thus the correction of spherical aberration in a spherical reflector, derives mostly from the *plane* surface of the corrector, with the figured surface being comparatively arbitrary. This figuring can, however, be adapted finally to take into account the approximations used in Equations 3.6.3 to give the "exact" result.

CO-INVOLUTION BY REFRACTION

A string drawing technique, suitable for adaptation to the process of co-involution, is known, for the point focusing by a single refraction as occurs with the Cartesian ovals given in Chapter 1.

As shown in Figure 3.28, if the string is looped one additional turn about one of the two foci, the point P, sliding inside the string, will be subject to the law

$$F_1P + 2F_2P = \text{constant} \;= L \text{ (the length of the string)} \tag{3.6.4}$$

This is the direct bipolar equation of a Cartesian oval from $r + 2\rho = \text{constant}$, implying a refractive index of two. Additional loops about the two foci apply in the same way to the derivation of the surface for any whole-number-fraction refractive index. That for $\eta = 1.5$ is shown in Figure 3.29.

We can extend this to the process of co-involution by winding or unwinding a part of the string about a caustic. Performing this for the nephroid caustic of a circular reflector, as shown in Figure 3.30 gives rise directly to the mechanical drawing of the Schmidt corrector for a spherical mirror [14]. Of course this produces a single surface refraction, and hence the second focus F_2 is embedded within the denser medium. It is "brought out" by any second refracting surface for which F_2 is an aplanatic point and most simply by a spherical surface with F_2 at its centre. The refractive index in the case illustrated is $\eta = 2$. Defining the surface in terms of algebraic geometry, however, allows the use of *any* given refractive index. The implicit equation to the surface states the fact that the string from (x_0, y_0) (Figure 3.30) to (x, y) and then tangentially to (X, Y) has to equal (the negative of) the distance from (X, Y) to the required focus $(a, 0)$ multiplied by the specified refractive index,

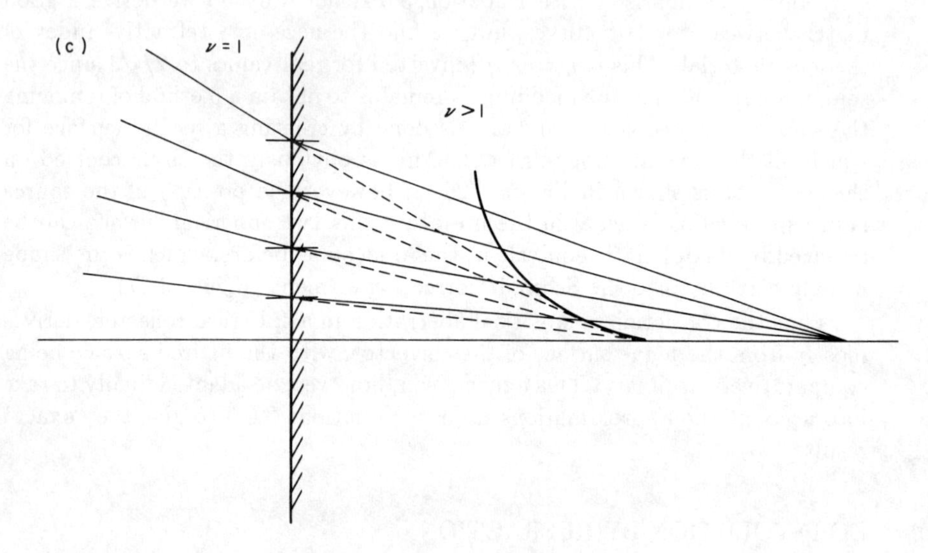

FIG. 1.13(c). Caustic of a plane refraction.

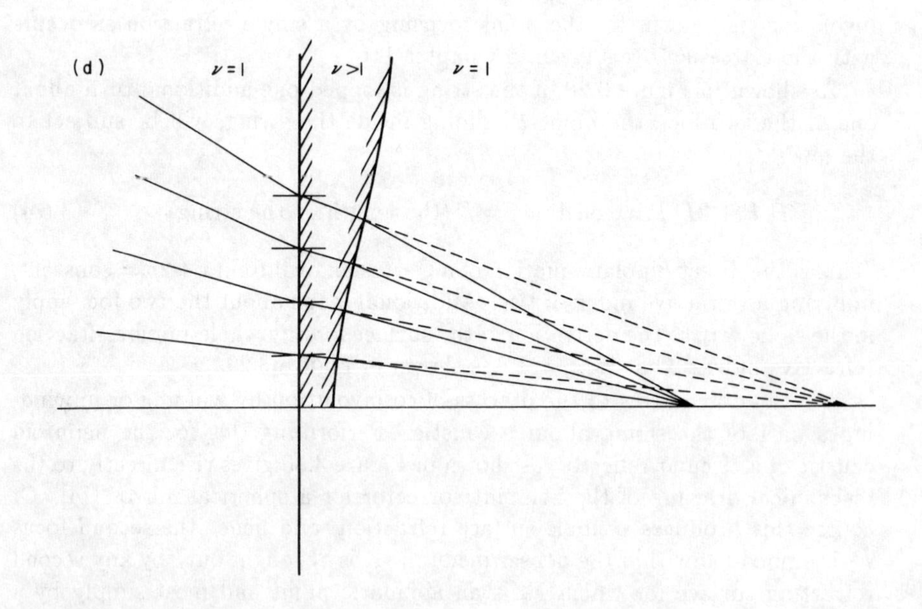

Figure 3.27: (a)Caustic of a plane refraction (b) Superposition of caustics and creation of external focus

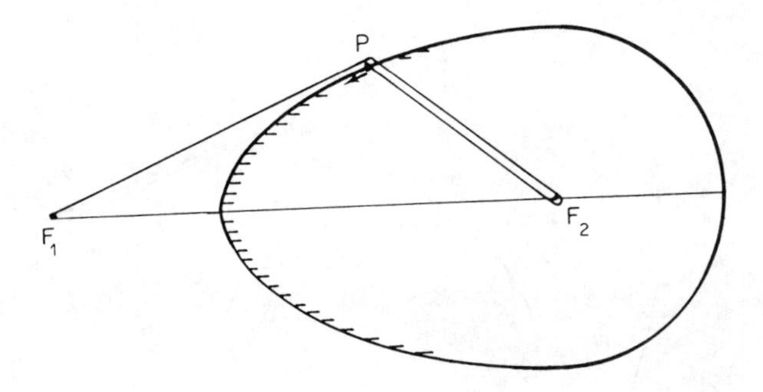

Figure 3.28: Taut string drawing process for a refraction with refractive index 2

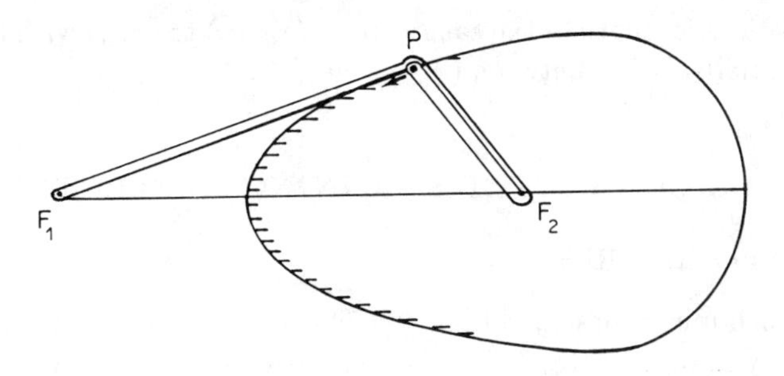

Figure 3.29: Drawing process for refractive index 3/2

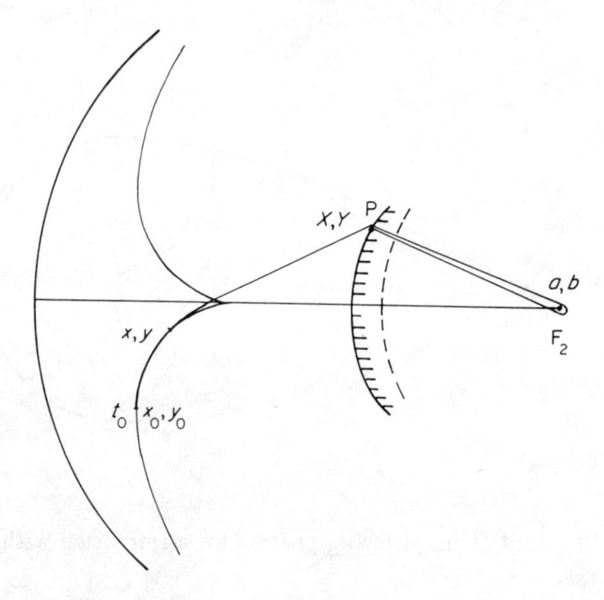

Figure 3.30: Co-involution with refraction giving the corrector profile for a circular reflector

to within a constant, the length of the string. Thus

$$S + \sqrt{(X-x)^2 + (Y-y)^2} = L - \eta\sqrt{(X-a)^2 + Y^2} \tag{3.6.5}$$

where S is the arc length of the caustic from (x_0, y_0) to (x, y) with the addition of a tangential relation between (x, y) and (X, Y).

3.7 PHASE CORRECTED SCANNING REFLECTORS

DESIGN PROCEDURES

The first such mirror designed to be free of coma and spherical aberration was described by Mangin in 1876 [31]. This consisted of two spherical surfaces the convex one being silvered. The application of such a mirror to microwave antennas has been fully investigated [17]. If it is required to use such a mirror as a mechanical scanner it is an obvious advantage to have the scanning arc on a circle. If further this circle was centred at the centre of the lens, a two to one advantage can be obtained by keeping the source fixed and rotating the mirror. Such a circular scanning arc is not a fundamental property of the original design of Mangin mirror. An improvement can best be sought by using the wide angle property of microwave lenses, and introducing a reflecting surface in the interior of the lens to convert it into a reflector with a correcting layer of refractive material. When this process is performed on the lenses designed

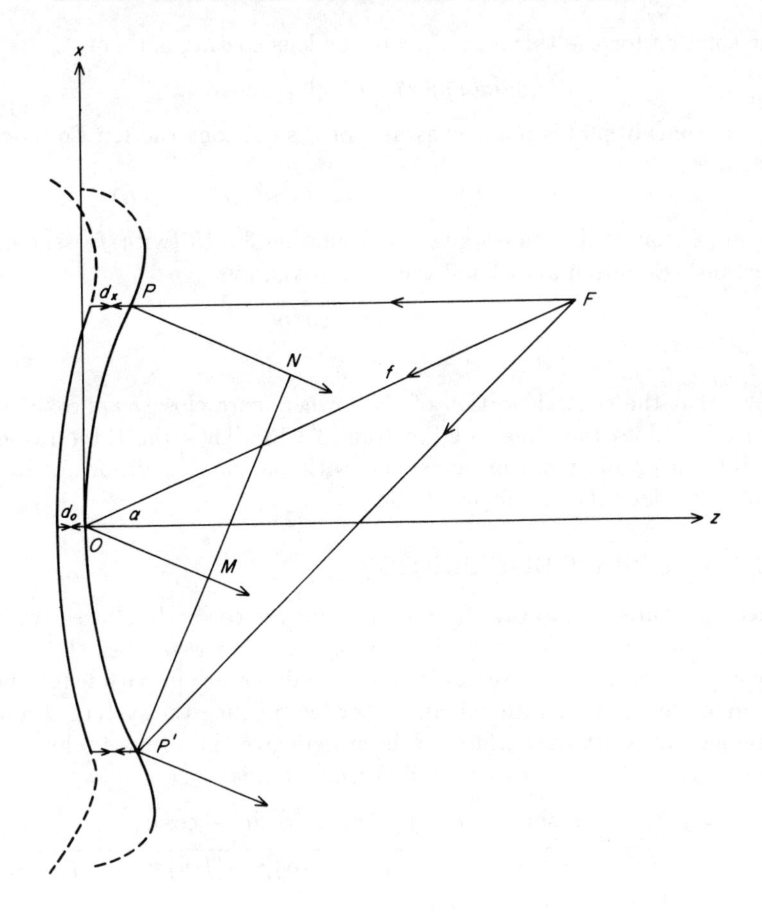

Figure 3.31: Phase corrected reflector

by the two focus method, the degree of freedom remaining that allowed for a third point on the scanning arc to be determined, is no longer available. This leaves the scanning arc undefined. This is not the case with the three-ray method. With the original choice of the "spreading" rays as shown in Figure 3.31, we require for equiphased points $P'MN$

$$FP + 2\eta(x)d(x) + PN = FO + 2\eta_0 d_0 + OM = FP' + 2\eta(x)d(x) \quad (3.7.1)$$

The outer equation gives

$$FP + PN = FP'$$

and the geometry is identical to that of the complete lens resulting in the same parametric form of profile as Equation 3.3.15.

$$\begin{aligned} z &= f \cos\alpha(1 - \cos\alpha) \\ x &= f \sin\alpha \end{aligned}$$

The solution for the thickness $d(x)$ of the lens coating is then

$$2[d(x)\eta(x) - d_0\eta_0] = z\cos\alpha \tag{3.7.2}$$

In particular if $\eta(x)$ is made constant for ease of construction Equation 3.7.2 becomes

$$2\eta[d(x) - d_0] = z\cos\alpha = f\cos^2\alpha(1 - \cos\alpha) \tag{3.7.3}$$

A comparison of the curves given by Equation 3.3.15 (with f assumed to be constant), Equation 3.3.13 and the circle given by

$$z = r(1 - \cos\alpha)$$
$$x = r\sin\alpha$$

shows that the central portion of the former more closely approximates the circle than does the ellipse of Equation 3.3.13. Thus the three ray method produces a profile more in agreement with the sine condition for minimum coma than does the two focus profile.

RESIDUAL PHASE DISTRIBUTION

Since only three of the rays from each point are correctly phased, we require expressions for the phase distribution of the other rays. Let the feed be in any angular position θ. We now require the difference in path length between the main ray to the centre and any other ray meeting the system at a point of parameter α. With the surfaces of the system given in terms of α by Equations 3.3.15 and 3.7.3 this path length difference, E_p, is

$$\begin{aligned} E_p = {} & f(0) - f(\alpha)\sin\alpha\sin\theta + f(\alpha)\cos\alpha\cos\theta(1 - \cos\alpha) \\ & - \sqrt{[f(\theta)\cos\theta - f(\alpha)\cos\alpha(1 - \cos\alpha)]^2 + [f(\theta)\sin\theta - f(\alpha)\sin\alpha]^2} \\ & - f(\alpha)\cos^2\alpha(1 - \cos\alpha) \end{aligned} \tag{3.7.4}$$

where the sign of α is the same as or opposite to that of θ, according as the surface point α is on the side adjacent to, or remote from, the displaced feed at θ. From this it can be seen that $E_p = 0$ when $\alpha = \pm\theta$ and when $\alpha = 0$, in accordance with the choice of the three focused rays.

SCANNING PROPERTIES OF PARTICULAR REFLECTORS

From the parametric expressions of Equations 3.3.15 and 3.7.3 (with constant refractive index) and the phase distribution given by Equation 3.7.4, we can analyse systems which are specified in one of three different ways:

1. Definition of f in terms of α. This specifies the scanning arc and hence the two profiles.

2. Specification of the refracting profile. Comparison of the equation to this curve with the parametric form required provides a definition of f in terms of α and hence the scanning arc and the reflecting surface.

3. Specification of the reflecting profile. When this profile is given, it is apparent that the shape of the refracting profile is dependent upon the refractive index of the surface coating. Thus since the scanning arc depends on the shape of the refracting profile, it too contains terms involving the refractive index. In this case, if the refractive index is given, the refracting profile can be obtained and the procedure is then the same as in (2).

SPECIFICATION OF SCANNING ARC

Two scanning arcs are of interest, the circle with centre at the reflector centre, and the straight line perpendicular to the axis. The former provides a scan motion which is mechanically simple to perform either for movement of the feed or tilt of the reflector. The latter focuses plane waves arriving at an angle to the axis into points of a plane, which, in pure optics, is a requirement for astronomical photographic processes.

CIRCULAR SCANNING ARC Until a later investigation revealed other refocused systems with a circular scanning arc, the first experiments were carried out with a reflector with f = constant = c. This gives

$$\begin{aligned} z &= c\cos\alpha(1-\cos\alpha) \\ x &= c\sin\alpha \\ d &= (x\cos\alpha)/(2\eta) \end{aligned} \tag{3.7.5}$$

the centre thickness d_0, being zero. The residual phase distribution is shown in Figure 3.32

STRAIGHT LINE SCANNING ARC To obtain a straight scanning arc at right angles to the axis, put $f = c/\cos\alpha$. The system is thus defined by

$$\begin{aligned} z &= c(1-\cos\alpha) \\ x &= c\tan\alpha \\ d &= (x\cos\alpha)/(2\eta) \end{aligned} \tag{3.7.6}$$

the centre thickness being zero.

The residual phase distribution is given in Figure 3.33. The high degree of asymmetry greatly limits the achievable angular aperture.

SPECIFICATION OF THE REFRACTING PROFILE

Let $x^2 = F(z)$ be the equation of the refracting profile. If this profile is also to be given by the parametric form of Equation 3.3.15

$$f^2\sin^2\alpha = F[f\cos\alpha(1-\cos\alpha)]$$

and thus f can be obtained as a function of α, defining the required scanning arc.

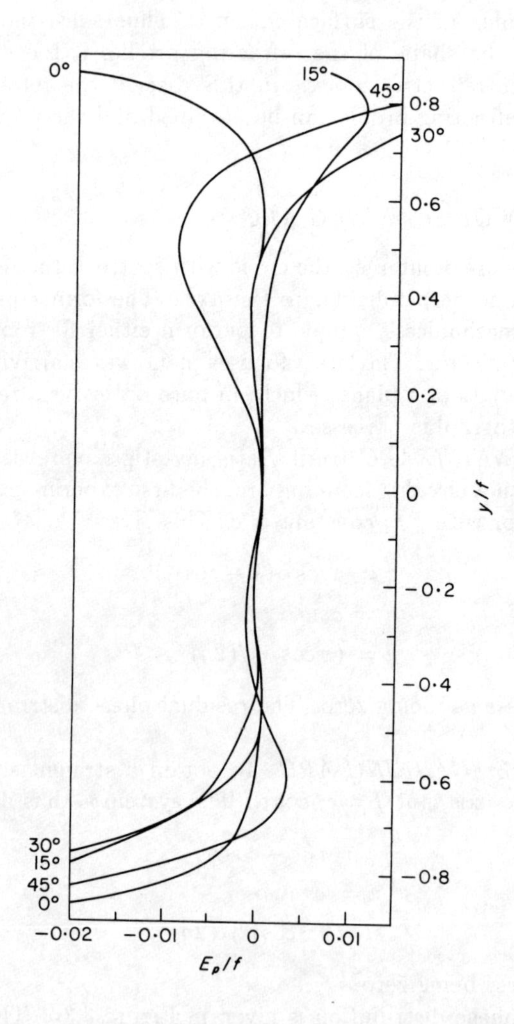

Figure 3.32: Phase error for constant-f scanning arc

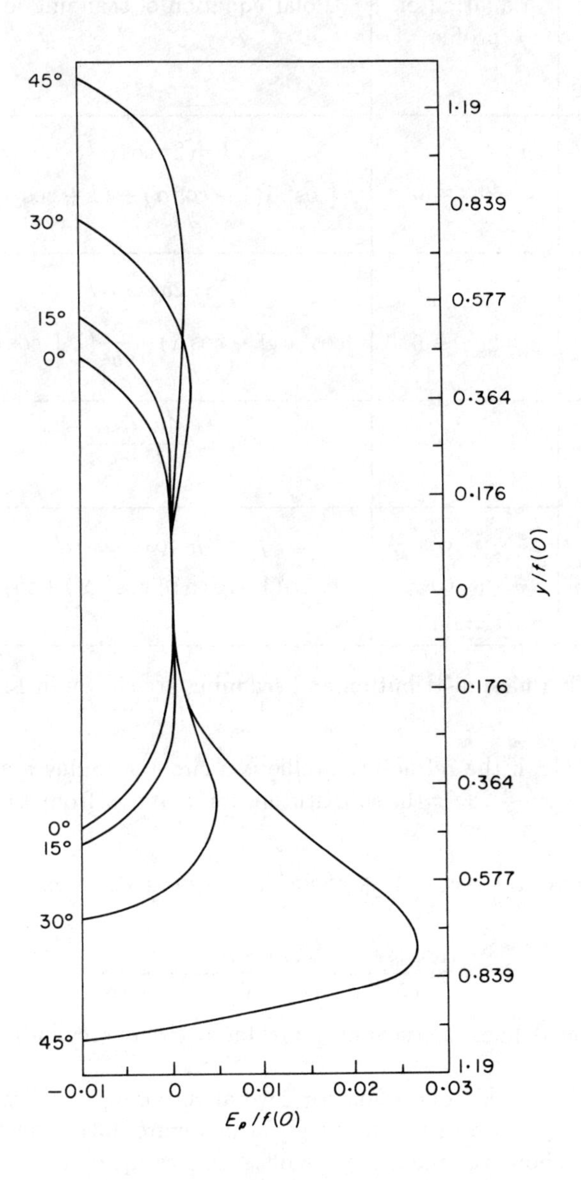

Figure 3.33: Phase error for straight line scanning arc

Refracting profile	Equation of profile	Polar equation of scanning arc	Axial focal length $f(0)$
Circle (radius r)	$z^2 + x^2 - 2rz = 0$	$f = 2r\cos\alpha / [\cos^2\alpha(1-\cos\alpha) + (1+\cos\alpha)]$	r
Ellipse (semi-axes a, b)	$z^2 + \frac{a^2}{b^2}x^2 - 2az = 0$	$f = 2a\cos\alpha / [\cos^2\alpha(1-\cos\alpha) + \frac{a^2}{b^2}(1+\cos\alpha)]$	$\frac{b^2}{a}$
Parabola (focal length a)	$x^2 = 4az$	$f = \frac{4a\cos\alpha}{1+\cos\alpha}$	$2a$
Two-point correction ellipse (foci f_0, $\pm\phi$)	$z^2 + x^2\cos^2\phi - 2f_0 z\cos\phi = 0$	$f = 2f_0\cos\phi\cos\alpha / [\cos^2\alpha(1-\cos\alpha)+\cos^2\phi(1+\cos\alpha)]$	$\frac{f_0}{\cos\phi}$

Table 3.2: Phase distribution and scanning arc shown in Figure 3.34

For example, if the refracting profile is a circle of radius r defined by the equation $z^2 + x^2 - 2zr = 0$, substituting for z and x from Equation 3.3.15 gives

$$f^2\cos^2\alpha(1-\cos\alpha)^2 + f^2\sin^2\alpha - 2rf\cos\alpha(1-\cos\alpha) = 0$$

i.e.

$$f = \frac{2r\cos\alpha}{\cos^2\alpha(1-\cos\alpha) + 1 + \cos\alpha}$$

This equation defines the scanning arc for a circular cylindrical refracting surface.

The results of this operation for several refracting profiles are given in Table 3.2. The phase distributions given in Figure 3.34 have a high degree of symmetry about the main ray for offset angles up to 45°. The similarity between these curves and the profiles of Schmidt correcting plates for spherical mirrors is marked since the main residual phase error is spherical aberration. Such systems may be refocused with advantage, and this will be dealt with in a later section.

The scanning arc for the two point corrected profile passes through the two correcting points as expected (Figure 3.34(d)). It is not, however, the anticipated circle centred on the reflector centre as can be seen.

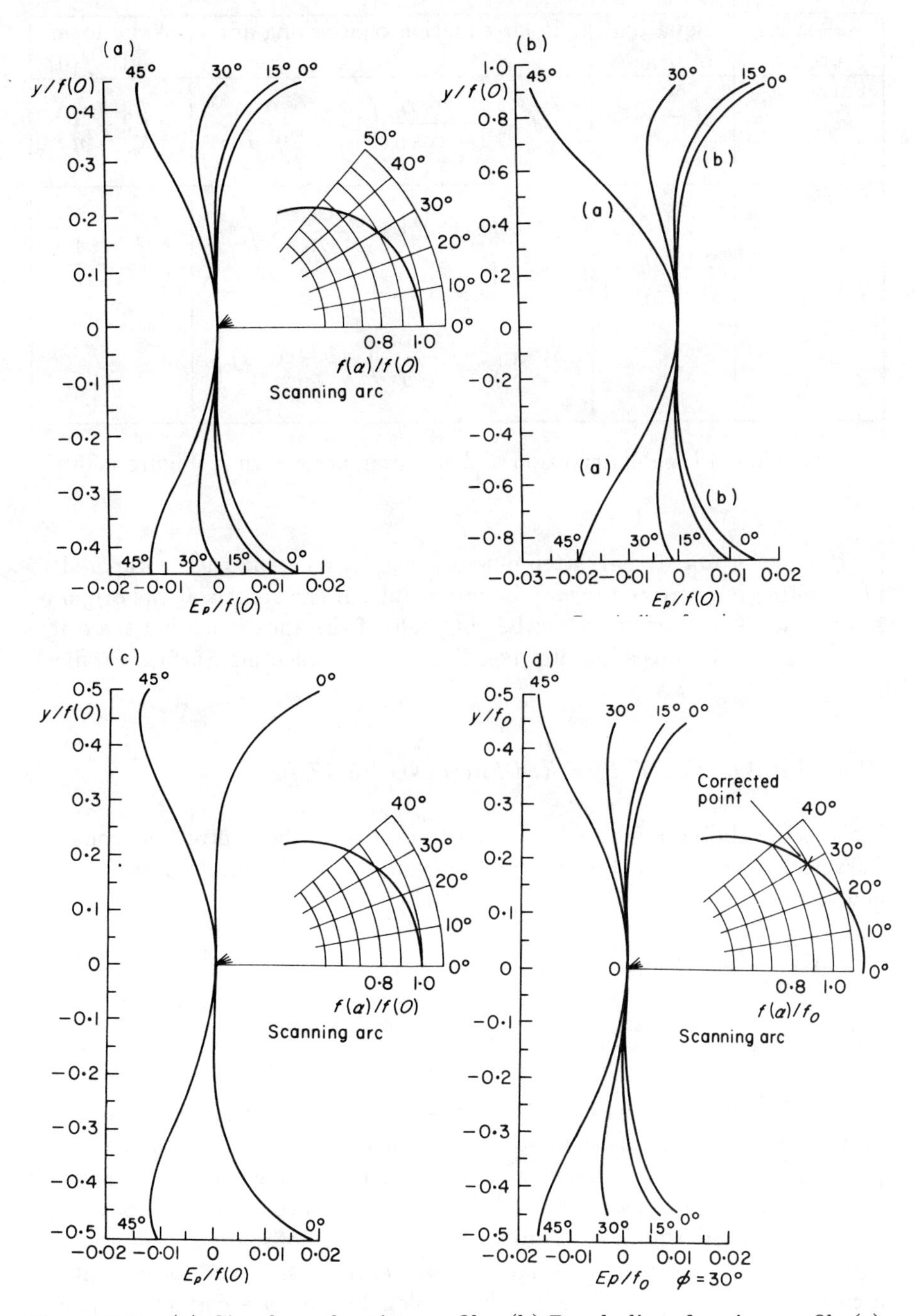

Figure 3.34: (a) Circular refracting profile. (b) Parabolic refracting profile (c) Elliptic refracting profile (d) Two-point corrected refracting profile

Reflecting profile	Equation of profile	Polar equation of scanning arc	Axial focal length $f(0)$
Parabola	$x_1^2 = 4az_1$	$f = \dfrac{4a\cos\alpha}{1+\cos\alpha}\left(1-\dfrac{\cos\alpha}{2\eta}\right)$	$a\left(2-\dfrac{1}{\eta}\right)$
Circle	$z_1^2 + x_1^2$ $-2az_1$ $= 0$	$f = 2a\cos\alpha\left(1-\dfrac{\cos\alpha}{2\eta}\right) \Big/ \left[1+\cos\alpha + \cos^2\alpha\left(1-\dfrac{\cos\alpha}{2\eta}\right)^2(1-\cos\alpha)\right]$	$\dfrac{a}{2}\left(2-\dfrac{1}{\eta}\right)$

Table 3.3: Phase distribution and scanning arc shown in Figure 3.35

It may be noted that attempts to design a reflector with a tapered or plane refracting surface prove to be impossible. If the apex is at the origin no solution for f in terms of α can be obtained. If the apex is behind the origin the distance f becomes infinite. (see Section 3.8 Reflecting Surface Profiles)

SPECIFICATION OF THE REFLECTING PROFILE

We assume the systems to have zero thickness at the centre and constant refractive index. From Equations 3.3.15 and 3.7.3 the reflecting surface has equations

$$x_1 = x = f\sin\alpha \tag{3.7.7}$$

$$z_1 = z - d = z\left(1-\frac{\cos\alpha}{2\eta}\right) = f\cos\alpha(1-\cos\alpha)\left(1-\frac{\cos\alpha}{2\eta}\right)$$

When this is expressed as $x_1^2 = F(z_1)$, the same analysis can be made as in the previous section. This method has been applied to the design of the corrected parabola and corrected circle, with the results as shown in Table 3.3 and Figure 3.35. In both cases the axial focal length $f(0)$ contains a factor $2-1/\eta$ which gives a lower limit of 0.5 to the possible range of refractive indices that can be used for the correcting layers. Furthermore, the axial focal length of a corrected parabola of focal length a is double that of a corrected circle of radius a, as is the case for uncorrected reflectors. In both cases a valid solution is obtained for $\eta = 1$. This means that the constraint alone enables phase correction to be obtained.

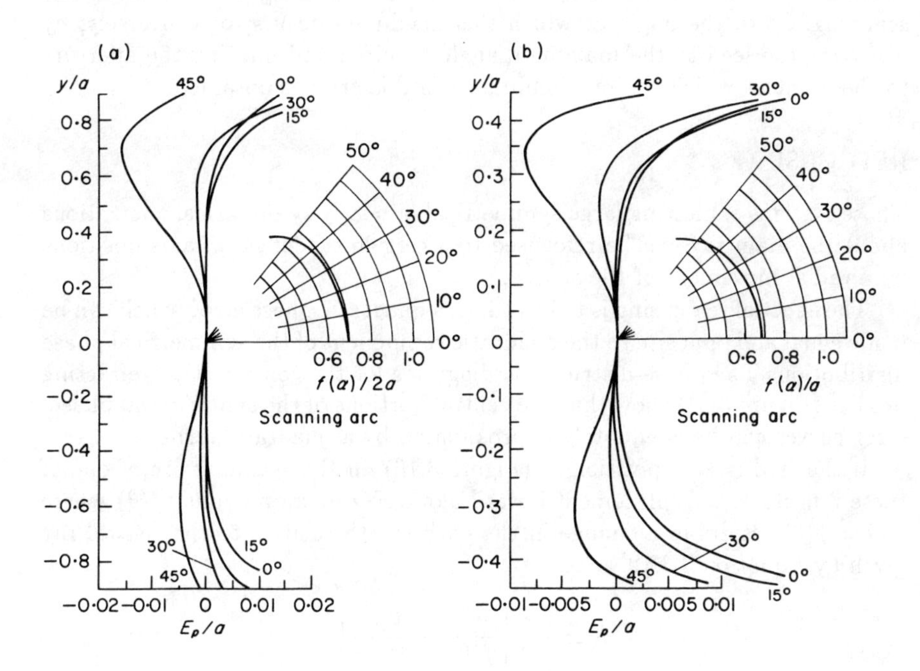

Figure 3.35: (a) Corrected parabola, $\eta = 1.6$ (b) corrected circle, $\eta = 1.6$

THE ANGULAR APERTURE

From the diagrams of the residual phase error it is possible to determine either the maximum angular displacement that a system with a given numerical aperture can permit or, for a maximum required angular displacement, the numerical aperture that will allow it. This can be done by defining the maximum permissible residual phase error and offsetting the feed along the scanning arc to the angle at which this maximum occurs; or conversely, by offsetting the feed to the maximum angle required and limiting the aperture to the point at which the maximum permissible error is obtained.

REFOCUSING

Those systems exhibiting large symmetrical or nearly symmetrical aberrations about the main ray can be refocused to a certain extent in off-axis positions by a radial movement of the source.

The effect of refocusing is to introduce a quadratic phase error which can be made equal and opposite to the quadratic component of the symmetrical phase distribution. The phase-distribution diagrams for the conic-section refracting profiles (Figure 3.34) show that the central portions of the symmetrical phase-error curves can be adequately approximated by a quadratic term.

If the feed is at a position F (Figure 3.16) on the scanning arc of one of these reflectors, a displacement inward along FO by an amount $\varepsilon f(\theta)$ where $FO = f(\theta)$ results in a change in the path length relative to the central ray given by Equation 3.3.12.

$$\Delta \simeq -\frac{\varepsilon}{2}\left[\frac{x^2}{f(\theta)} - \frac{\theta x^3}{f(\theta)^2}\right]$$

which is correct to the first order in ε and θ and to the third order in $x/f(\theta)$.

If Δ is then taken as the value of the quadratic approximation to the phase error at any intermediate value of α, the refocusing increment is given by

$$\Delta = -\frac{\varepsilon}{2f(\theta)}[f(\alpha)\sin\alpha]^2(1 - \theta\sin\alpha) \tag{3.7.8}$$

This can be calculated for several values of θ, with α and E_p given by the phase-error curve corresponding to the particular value of θ. With appropriate sign, this adjusts the scanning arc to a best-focal-position arc.

The cubic term introduces some asymmetry, particularly at the larger values of θ. The refocusing increments, calculated for the reflector with a circular refracting profile, are shown in Figure 3.36.

After removing the quadratic component a small residual phase error remains and a slight asymmetry, consisting mainly of the cubic component, is observed. The refocusing results in a best-focus arc which is circular with radius 0.94 of the original axial focal length.

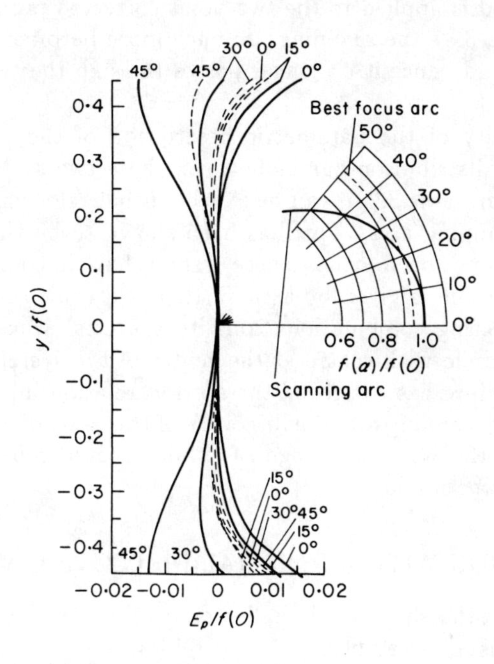

Figure 3.36: Effect of refocusing on reflector with circular refracting profile

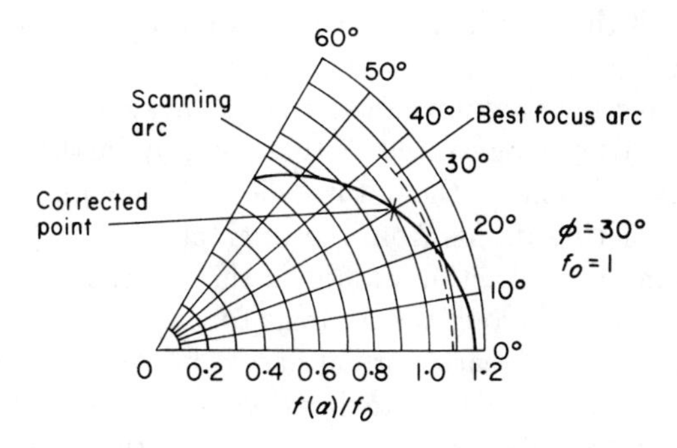

Figure 3.37: Refocusing scanning arc of two-point corrected refracting profile

When this method is applied to the two-point corrected profile and scanning arc as in Figure 3.37 the scanning arc once more becomes circular but with radius $1.09f_0$, and hence it no longer passes through the two correction points.

From the similarity of the parametric description of the scanning arcs and the phase error distribution curves between these two systems and the other systems given in Table 3.2, it can be expected that refocusing these also produces circular scanning arcs. This has been shown to be the case but a complete analysis has not been made. These systems have in common the fact that the refractive profile is given by a curve that is a conic section. Hence these refracting-reflecting combinations constitute a class whose best-focus scanning arc is the circle with centre at the centre of the system. Although this refocusing procedure has destroyed the a priori relationship between the refracting profile and scanning arc, the discovery of this class of related curves provides a basis for the iterative design of similar coated reflectors with a natural dielectric correcting layer.

PHASE CORRECTION WITH A NATURAL DIELECTRIC MEDIUM

It may be noted that the shape of the reflectors so far considered, for which the correcting layer is an array of metal waveguides, is, in general, the same as that of the original negative meniscus lens of Mangin, i.e. concavo-convex. This fact is contrary to the usual experience when comparing microwave with optical lenses. It means that, for these reflectors at least, the shape with a constrained refractive medium is in itself a satisfactory first approximation to a natural dielectric unconstrained correcting layer. This is particularly so in fully stepped systems, in which the thickness of the correcting layer need never be greater than half the wavelength in the medium used. The approximation involved in assuming that the rays are still parallel to the axis when they are within the correcting medium causes the quadratic error of the system to increase slightly. This can be cancelled, therefore, by further focusing.

A better approximation for the natural dielectric medium can be made by calculating the refracting profile from the original formula (Equation 3.3.15) and then calculating the reflecting profile on the assumption that the rays in the correcting layer travel along the normals to the refracting profile. In this case, the path length along each normal has to be the value of d calculated from the Equation 3.2.3. This approximation is particularly appropriate to the circular reflecting profile, where, in the axial position, the rays do, in fact, travel along these normals. The approximation here also results in an increase in the residual quadratic error and may be similarly corrected.

However, in all natural dielectric systems a further parameter is available in the thickness of the corrected reflector at its centre. It is found that variation of this parameter too gives rise mainly to a quadratic component of phase error which could be used for the cancellation of the above effects.

Finally the three-ray principle itself can be used directly for the natural dielectric case. Apart from the necessity of refocusing which has been demonstrated, the method by which the movement of the feed over a given arc describes both this arc and the refracting profile is a stepwise procedure suitable for programming a computer. To do this, one establishes the phase equality of the three rays concerned by considering the localities at which the outer two rays meet the reflector to be sections of thin reflecting prisms, similar in shape and symmetrical with respect to the axis. The three-ray principle, however, does not give a defined axial focal position. The scanning arc limits towards a point on the axis at which, of course, the three rays coincide. This point cannot therefore be taken as a starting point for the stepwise procedure. This means that a guess has to be made at the thickness at the outer edge, the guess being acceptable if, on reaching the axial position, a zero thickness has not been passed.

EXPERIMENTAL RESULTS

The analysis presented in the previous sections has dealt with two-dimensional or cylindrically symmetrical reflectors. The profiles concerned could, however, also be considered as cross-sections of a rotationally symmetrical reflector. Where such a reflector is required it is constructed by the rotation of the profiles about its axis. This, of course, introduces the further aberration of astigmatism which limits the angle of scan of rotationally symmetrical reflectors to a much lower value than cylindrically symmetrical ones. With the type of construction possible at microwave frequencies, namely a two-dimensional array of waveguides, systems with rectangular symmetry can be considered. In these the surfaces would be formed by the translation across a profile of a similar profile at right angles to it. The astigmatism of such systems, however, is still to be investigated.

Experimental work so far has been concerned with the construction of systems to test the elementary theory and to assess the limitations imposed on the scanning properties by the astigmatism of the rotationally symmetrical reflector and the dielectric approximation of the previous paragraph.

To test the basic theory a linear reflector (cylindrical system) was constructed for which the correcting layer was a parallel array of metal plates with effective refractive index of 0.6. The profiles were determined from Equations 3.3.15 and 3.7.3 with f constant: this gives the uncorrected scanning arc as the circle $f = c$. The reflector is 0.92 m long and has a focal length of 0.77 m operating at a frequency of 9.375 GHz. The f/D ratio is 0.83. From the residual phase-error curves it is to be expected that coma effects will become noticeable beyond 30° of angular displacement, because of the asymmetry in the phase-error curve. This is found to be the case, as shown by the scan patterns of Figure 3.38. A beam displacement of 30° (approximately 15 beam widths) is achieved without any observable deterioration in the beam shape,

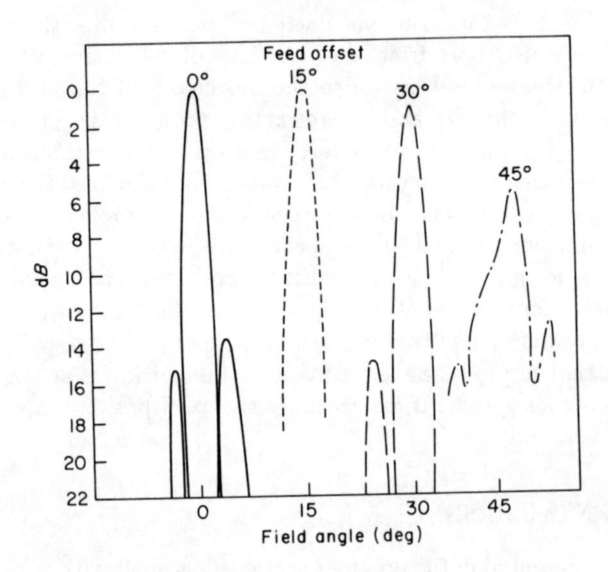

Figure 3.38: Polar diagrams of constant-f cylindrical reflector at different offset angles

and is followed quite suddenly by a decrease in gain and a large increase in the coma lobe.

A rotationally symmetrical system constructed from square-section metal tubes has scan patterns shown in Figure 3.39 demonstrating that the gain and beam shape deteriorate steadily on displacing the feed source along the scanning arc. Very similar results are achieved with a dielectric-coated reflector designed in accordance with the approximation given in the previous section.

Examination of the residual phase-error curves of Figure 3.36 shows that an expected scan angle of 45° should be achieved with a cylindrical reflector having a circular reflecting profile. A system for which $r = 20\lambda$ is possible with the maximum permissible phase error equal to $\lambda/10$ and with a refocused scanning arc which is circular.

A spherical natural dielectric system with the same aperture thus has an expected useful scan approaching 30°. A reflector of this type has been constructed. The scan patterns achieved give a useful scan of 30°, at which point the gain has decreased by 2 dB without any great deterioration in the main beam shape and before the coma lobe level has become pronounced.

Thus to summarize, the design procedure whereby three rays from every point of a specified scanning arc are kept equi-phased, gives rise to a class of phase-corrected reflectors with the following properties:

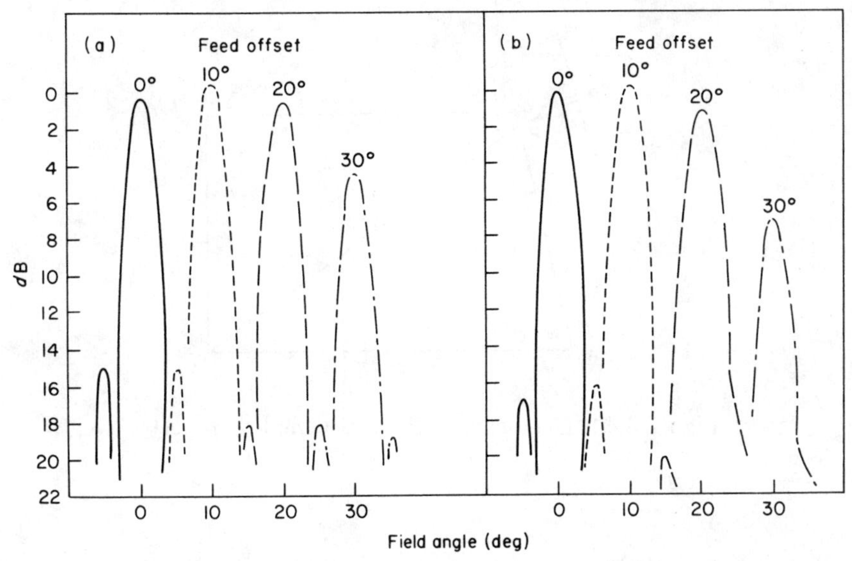

Figure 3.39: Polar diagrams of constant-f spherical reflector at different offset angles. a) H-plane scan b) E-plane scan

1. The refracting profile is a conic section and can, for simplicity, be circular.

2. The residual phase errors at angles of scan up to 45° are largely symmetrical about the main ray.

3. Refocusing the system at angles off-axis makes the scanning arc circular.

With such systems used as beam-scanning devices, i.e. with stationary feed and mechanical scan of the reflector, linear scans approaching $\pm 90^o$ can be achieved with cylindrical reflectors, and a cone of semi-angle 60° can be scanned with spherical reflectors.

3.8 THIN DIELECTRIC LENS REFLECTOR

Although not obvious at first sight, the design of a thin dielectric coated reflector turns out to be relevant to the correction of the spherical reflector. The thin lens reflector is formed from a proper cone dielectric surface, backed by a reflecting surface, the shape of which is determined by the usual combination of path length eikonal, the laws of reflection and refraction and the internal geometry, to give a uniform plane wave at the aperture plane perpendicular to the axis of the conical first surface.

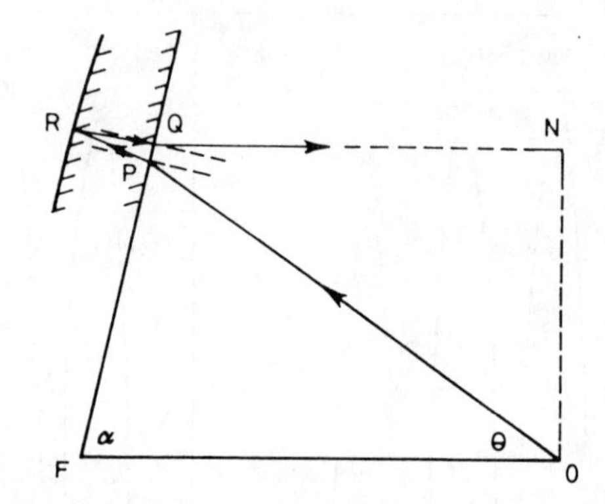

Figure 3.40: Geometry of the dielectric lens reflector

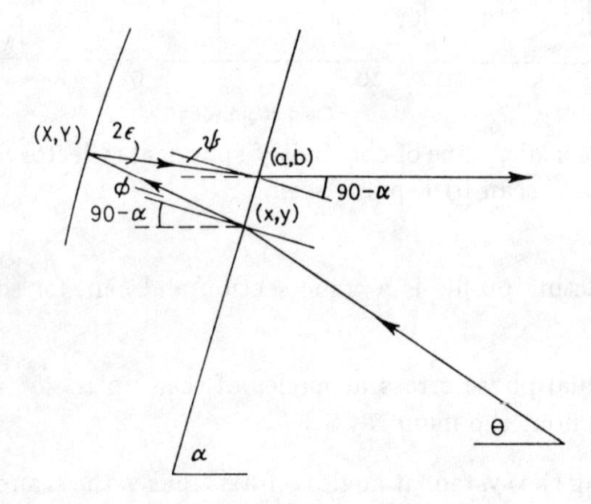

Figure 3.41: Enlargement of the refraction-reflection ray trace

RAY TRACE ANALYSIS

Figure 3.40 shows the elements of the necessary ray trace process. A ray from the origin meets the plane tilted front surface at $P(x, y)$ and is refracted there. After being reflected on the back surface at $R(X, Y)$ it is refracted a second time at $Q(a, b)$ on the plane front surface to become horizontal and parallel to the axis. An enlarged view of this refracting/reflecting process is shown in Figure 3.41 labelling all the angles involved in the analysis. It is required to find the relation between X and Y for the points R parametrically in terms of the angle of the ray θ, at the origin.

The refractive index of the material will be η and hence the following chain of ray equations results from Snell's law at each refraction or reflection. The focal distance OF (Figure 3.40) is taken to be unity.

$$\begin{aligned}
\eta \sin \psi &= \sin(90 - \alpha) = \cos \alpha \\
\eta \sin \phi &= \sin(-90 + \alpha + \phi) = -\cos(\alpha + \phi) \\
2\varepsilon &= \psi + \phi \\
x &= \frac{\tan \alpha}{\tan \alpha + \tan \theta} = \frac{\sin \alpha \cos \theta}{\sin(\alpha + \theta)} \\
y &= \frac{\tan \alpha \tan \theta}{\tan \alpha + \tan \theta} = \frac{\sin \alpha \sin \theta}{\sin(\alpha + \theta)} \\
OP &= \frac{\sin \alpha}{\sin(\alpha + \theta)} \\
\frac{Y - y}{X - x} &= \cot(\alpha - \phi) \\
\frac{Y - b}{X - a} &= \cot(\alpha + \psi) \\
\frac{(X - a)^2 + (Y - b)^2}{\cos^2 \phi} &= \frac{(X - x)^2 + (Y - y)^2}{\cos^2 \psi} \\
&= \frac{(a - x)^2 + (y - b)^2}{\sin^2(\phi + \psi)}
\end{aligned} \tag{3.8.1}$$

From the equality of phase length we require

$$OP + \eta RP + \eta RQ + QN = \text{ constant } = A \tag{3.8.2}$$

$$\frac{\sin \alpha}{\sin(\alpha + \theta)} + \eta\sqrt{(X - x)^2 + (Y - y)^2} + \eta\sqrt{(X - a)^2 + (Y - b)^2} + a = A \tag{3.8.3}$$

and hence from Equations 3.8.1 and 3.8.2

$$\frac{\sin \alpha}{\sin(\alpha + \theta)} + \frac{\eta(X - x)}{\sin(\alpha - \phi)}\left[1 + \frac{\cos \phi}{\cos \psi}\right] + a = A \tag{3.8.4}$$

From Equation 3.8.1

$$a = x - \frac{(X - x)\sin(\phi + \psi)\cos \alpha}{\cos \psi \sin(\alpha - \phi)} \tag{3.8.5}$$

Giving

$$\begin{aligned}
X &= \frac{[A - \sin \alpha(1 + \cos \theta)\sin(\alpha + \theta)]\sin(\alpha - \phi)}{\eta\,(1 + \cos \phi / \cos \psi) - \cos \alpha \sin(\phi + \psi)/\cos \psi} \\
&\quad + \frac{\sin \alpha \cos \theta}{\sin(\alpha + \theta)} \\
Y &= \frac{X + \sin \alpha \cos \theta / \sin(\alpha + \theta)}{\tan(\alpha - \phi)} + \frac{\sin \alpha \sin \theta}{\sin(\alpha + \theta)}
\end{aligned} \tag{3.8.6}$$

ϕ and ψ are derived in terms of α and θ from Equation 3.8.1.

Finally A is derived from the maximum value of θ to give

$$A = \frac{\sin\alpha(1+\cos\theta_m)}{\sin(\alpha+\theta_m)} \tag{3.8.7}$$

and the radius of the reflector will be (for unit focal length)

$$r = \frac{\sin\alpha\sin\theta_m}{\sin(\alpha+\theta_m)} \tag{3.8.8}$$

The fraction r in Equation 3.8.8 therefore converts the design from normalised focal length into normalised aperture radius. θ_m is the maximum value of θ subtended by the periphery at the origin.

REFLECTING SURFACE PROFILES

PLANE FRONT SURFACE

For a planar front surface $\alpha = 90^o$ and Equations 3.8.6 simplify considerably to

$$\begin{aligned} X &= 1 + (a - \sec\theta) \Big/ \left\{ \eta \left[\frac{\eta}{\sqrt{\eta^2 - \sin^2\theta}} + 1 \right] \right\} \\ Y &= \frac{a - \sec\theta}{\eta(\eta + \sqrt{\eta^2 - \sin^2\theta})} + \tan\theta \end{aligned} \tag{3.8.9}$$

This is shown in Figure 3.42 and the thickness, as previously noted, is approximately 0.065 of the diameter.

CONICAL FRONT SURFACE

With varying values of α, the front surface (on rotation) becomes a right circular cone and the central thickness reduces. This process can be continued until zero thickness is achieved at the apex (or even further if a central hole is allowed). This will occur for that value of α which makes A equal to twice the focal length, or, from Equation 3.8.7

$$\frac{\sin\alpha(1+\cos\theta_m)}{\sin(\alpha+\theta_m)} = 2 \tag{3.8.10}$$

These curves are also illustrated in Figure 3.42.

LENS REFLECTOR WITH CENTRAL HOLE ($\alpha = 70^o$)

The refractive index in the example illustrated was $\eta = 1.6$, but values in the range 1.6-2.0 showed the variation that would be anticipated. It can be seen from Figure 3.42 that, with zero thickness at the apex, the maximum thickness, taken perpendicular to the front surface, is only approximately 0.015 of the diameter, less than a quarter of the planar surface lens reflector. Even this can be reduced if the angle of the front surface, α, is made smaller and a hole is induced (or a negative thickness) at the apex.

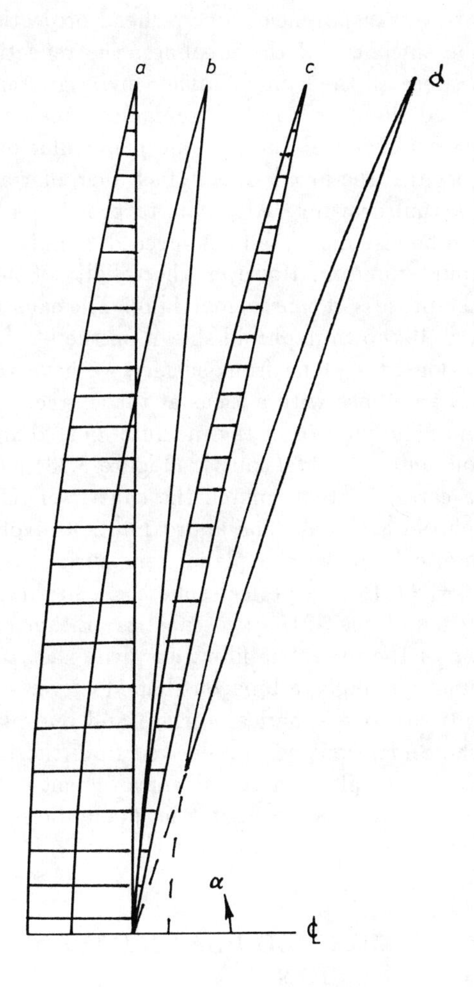

Figure 3.42: Lens cross-sections for unit aperture radius, $\theta_m = 45^o$
(a) Plane front surface lens ($\alpha = 90^o$)
(b) Intermediate conical front surface ($\alpha = 85^o$)
(c) Thin lens reflector with zero thickness at apex ($\alpha = 78.3^o$)
(d) Lens reflector with central hole ($\alpha = 70^o$)

APPROXIMATE REFLECTING SURFACE

In the course of preparing transparencies for overhead projection, it was observed that, within the thickness of the ink lines, the reflecting profile for fixed θ_m and diameter were all the same. Further investigation showed that, if the curve were displaced along its length, it remained constant and the profile of the thin lens was symmetrical about its perpendicular bisector. These properties could only occur if the profile was in fact a circular arc, within the approximation of the actual drawing. Attempts to prove this from the governing equations (as can be visualised) did not succeed, mainly because such a circle is only a close approximation. However, the validity of this assumption can be illustrated by the process of caustic matching. The caustic of reflection in a circular arc is the well-known nephroid shown on the left in Figure 3.43. For the ensuing refraction in a plane front surface we have that the phase front of zero distance is an ellipse with a focus at the source. Continuing the elliptical phase front into the interior of the medium and taking its involute, converts the phase front into a *virtual* caustic (Figure 3.43). This refractive caustic can then, to a certain extent, match the caustic of reflection in the empty sphere, the nephroid as usual. The approximation involved therefore, in using a "filled" spherical cap with a plane front surface is given by the difference of Equations 3.8.6 from a perfect sphere, or alternatively, the difference between exact definitions of the two caustics. Either can be used to figure the front surface of the dielectric filling to give exact correction, and this turns out to be, unsurprisingly, a limaçon shaped curve.

Analysing a system with one spherical surface and one plane surface is straightforward and the dimensions can be derived from the foregoing. The residual error function shows only a marginal improvement over the unfilled sphere however, but it can be used to figure the plane dielectric surface to give exact correction.

3.9 DIELECTRIC CORRECTED PLANE REFLECTOR: AN UNUSUAL SOLUTION

The three-ray method of the previous section does not apply to straight line surfaces such as the cone of the thin lens reflector or of the plane reflecting surface. Here, the problem of designing the correcting dielectric surface that will convert a plane reflector into a focusing system is approached by designing a refracting lens that collimates rays from a point source, but which is symmetrical about a central plane †. Thus when this plane is replaced by a reflecting surface, the combination is that of the dielectric corrected plane.

The problem of the symmetrical lens with asymmetrical foci, in this case one at infinity, is of considerable interest in itself, since its existence can only

†Problem first proposed by Dr RL Sternberg of ONR Boston, USA

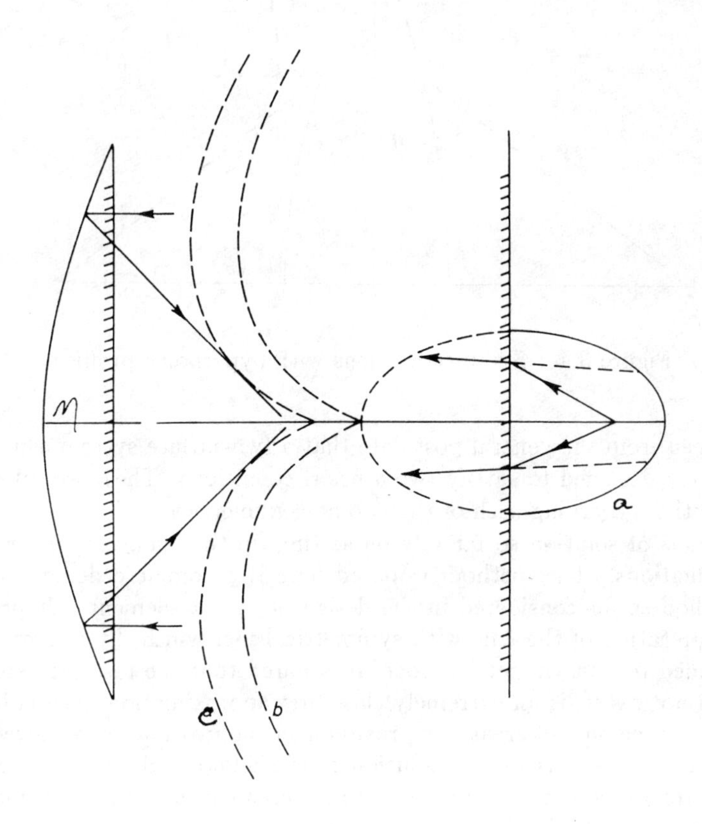

Figure 3.43: Caustic matching in a filled spherical cap
(a) is the zero-distance phase front of refraction in a plane interface
(b) is the involute of a) and is thus the caustic of the refraction
(c) is the caustic of reflection in the empty sphere
Over a paraxial region the two caustics are matched

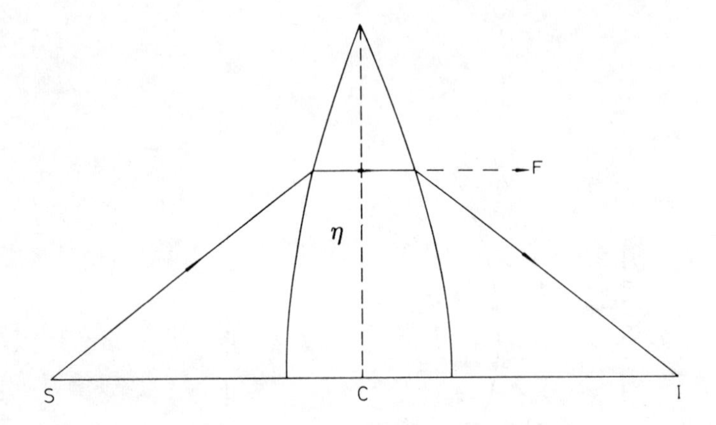

Figure 3.44: Symmetrical lens with hyperbolae profiles

be assumed from the general postulate that a two-surface system can always be uniquely designed to satisfy two a priori conditions. These we take to be the condition governing each of the two nonsymmetrical foci.

Methods of solution so far rely on setting up the usual nonlinear differential equations. The method proposed here is a complete departure from any method so far considered in the design of optical elements. It proposes an interpretation of the lens with symmetrical foci which, by inference, can be extended to nonsymmetrical foci. It is found that the resultant solution, although not exact, is an extremely close first approximation, particularly in the paraxial region and greatly improves on the approximation achieved after a number of iterations by the solution of the differential equation. Such a close approximation can then be used to set up a correcting process for exact agreement to be obtained.

GEOMETRICAL INTERPRETATION OF SYMMETRICAL LENS

The one refracting surface that will convert all rays from a point source into rays parallel to the axis (of circular symmetry) is the hyperbola with focus at the source point and with eccentricity equal to the refractive index. Hence, if this is "matched" by a second identical surface reflected about a central plane, as shown in Figure 3.44, the rays will refocus at the point which is the reflection of the first point focus in the same plane. A symmetrical aplanatic lens results. We now observe that the three points S C and I are homographic with respect to the point at infinity, and that this is the point F at which the rays would be focused if only half of the complete lens were to be operative.

The construction for homographically related points on a line is shown in Figure 3.45. S and I are any given points and C is a point between them, then if CP is any line, as shown, and SM and IN intersect on CP, where PMI and

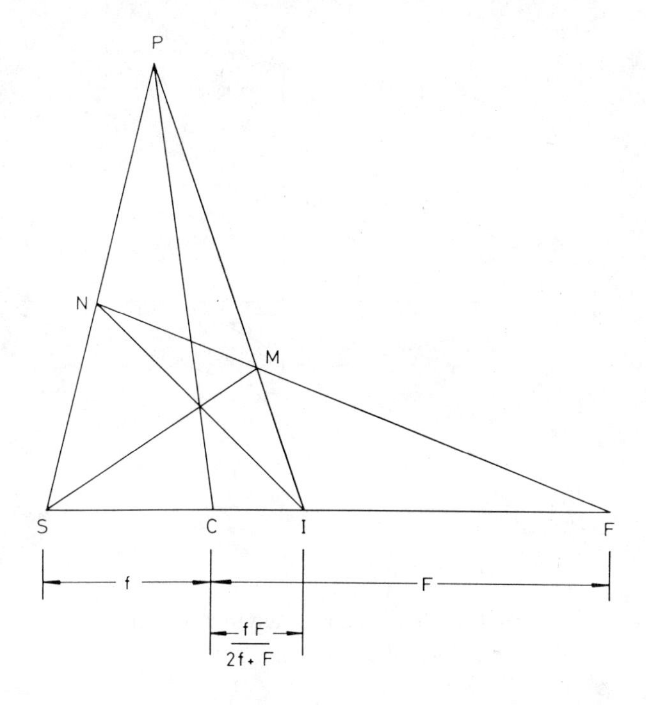

Figure 3.45: Homographically related points on line

PNS are collinear, then the intersection of MN with the axis gives the unique point F which is homographic with respect to S C and I. The construction is independent of the position of P and of the order of the points. That is, if F is homographically related to S C and I, I will be homographically related to C S and F and so on [33]. When C is the point midway between S and I then F will be at infinity. If the distance of the source S from the centre C is f and from C to the point F, F, then the point I is at

$$I = fF/(2f + F) \tag{3.9.1}$$

from C as shown.

It can now be seen that the symmetrical lens has the following properties:

1. The first surface converts the rays from the source S to rays which focus at infinity, the point F, that is the "half-lens" consisting of the first surface and the central plane focuses at infinity.

2. The second surface refocuses the rays to the point I.

3. The centre of the lens is the point C

4. The points F S C and I form a homography (or are in involution)

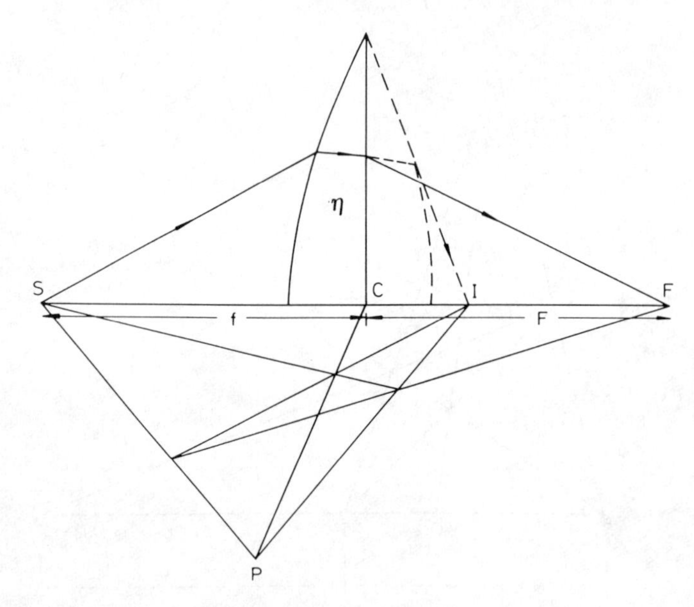

Figure 3.46: Construction for focus at asymmetrical point I. "Half-lens" focuses at F, full lens focuses at I.

EXTENSION TO NONSYMMETRICAL FOCI

We now make the assumption, which is only justified in the result, that, for a nonsymmetrical pair of foci, this homographic property should still pertain. That is, the first refracting surface and "half-lens" should focus the rays from the source at a point F, now at the finite distance from the centre C, (Figure 3.46) and consequently the complete symmetrical lens will focus the rays at the point I homographic with respect to S C and F. Hence all that is required is the design of a plano-convex "half-lens" to focus at the predetermined point F, a comparatively simple problem.

From Figure 3.47 we see that the ray trace equation for rays from the source S with the origin of coordinates at the centre of the lens C for refocusing at F is

$$SP + \eta PQ + QF = \text{ constant } = f - x' + \eta x' + F \equiv A \tag{3.9.2}$$

where η is the refractive index and x' is the chosen axial position of the first surface.

The planeness of the central surface makes this equation easily soluble giving the first profile in parametric form as the solution of the quadratic

$$\begin{aligned} x^2(\eta^2 - 1)\sec^2\phi - 2x[F\tan\theta\tan\phi - f \\ +\eta\sec\phi(A - F\sec\theta)] - f^2 - F^2\tan^2\theta + (F\sec\theta - A)^2 = 0 \end{aligned} \tag{3.9.3}$$

$$y = x\tan\phi + F\tan\theta \tag{3.9.4}$$

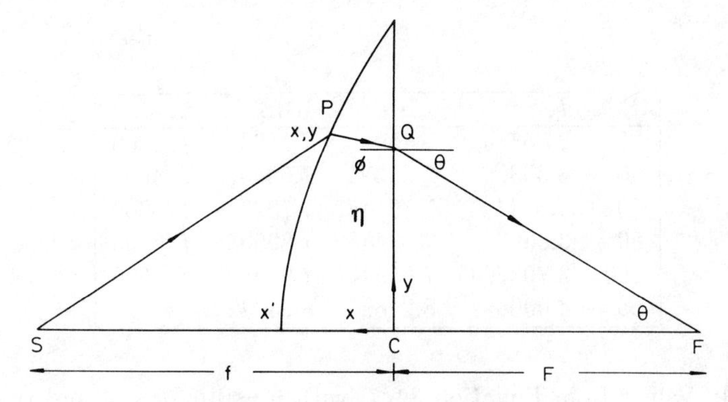

Figure 3.47: Design of "half-lens"

where $\sin\theta = \eta \sin\phi$. In these equations θ can be interpreted as an independent parameter. The maximum value of this parameter $\theta = \Phi$ gives the radius of the lens

$$R = F \tan\Phi$$

where

$$A + F\sec\Phi + \sqrt{f^2 + F^2\tan^2\Phi} \tag{3.9.5}$$

PARAXIAL APPROXIMATION

If this solution were exact, the path length for both the axial ray and an edge ray would be equal. The extent of disagreement is a measure of the approximation involved, and hence of its applicability to devices in the millimetre range of frequencies. Choosing as an example a lens with $\eta = 1.6$ and $f = 4$, the "half-lens" designed for an exact focus at F requires

$$0.6x' = F + 2.27689 - \sqrt{1 + F^2} \tag{3.9.6}$$

For the complete lens, on the other hand, we should have

$$\sqrt{1 + f^2} + \sqrt{1 + (fF/(2f + F))^2} = f + fF/(2f + F) + 1.2(f - x') \tag{3.9.7}$$

Substituting the value of x' from Equation 3.9.6 into Equation 3.9.7 we find that this equality does not occur, and that the difference is small and decreases rapidly with increase in F becoming exact as F tends to infinity, the symmetrical situation with hyperbolic profiles. These errors are tabulated in Table 3.4.

F	I	LHS	RHS	Error
20	2.85714	7.1502	7.1533	$f/1000$
40	3.333	7.60321	7.6045	$f/4000$
60	3.5294	7.79144	7.7923	$f/5000$
80	3.6363	7.89446	7.89508	$f/6000$
100	3.703704	7.95944	7.95992	$f/8000$
∞	4.000	8.24622	8.24622	

Table 3.4: Values from Equation 3.9.7, with $\eta = 1.6$, $f = 4$ and the semi-aperture angle $\Phi = 14^o$

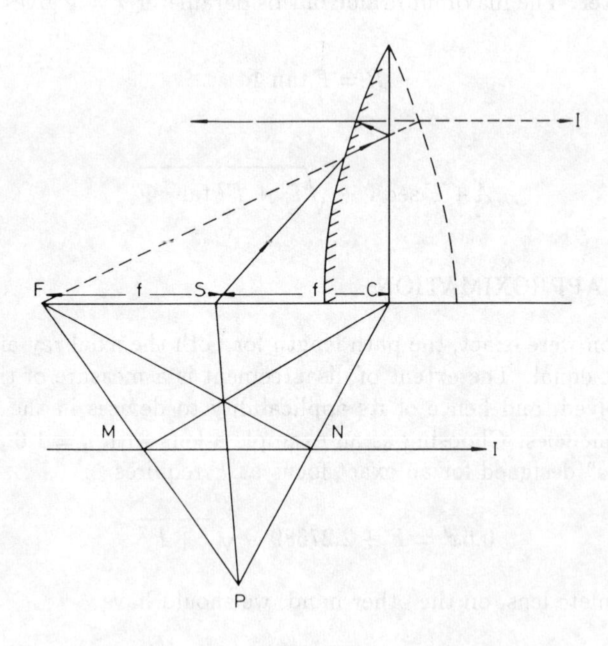

Figure 3.48: Homographic relation for focus at infinity. "Half-lens" has virtual focus at F

f	LHS	RHS	Error	Semiaperture angle
1	1.7111	1.6532		45°
2	2.4243	2.4142	$f/200$	26.6^o
3	3.2948	3.2915	$f/1000$	18.4^o
4	4.2245	4.2231	$f/3000$	14°

Table 3.5: Values from Equation 3.9.9

SYMMETRICAL LENS FOCUSED AT INFINITY

For this situation the "half-lens" needs to have a virtual focus at F on the same side as the source and equidistant from it; That is

$$SC \equiv SF$$

as shown in Figure 3.48. For a virtual source the ray path equation takes the form

$$SP + \eta PQ - QF = \text{ constant} \tag{3.9.8}$$

(cf. Equation 3.9.2). The result is that F in Equation 3.9.3 and subsequently in Equations 3.9.4 and 3.9.5 becomes negative, simply needing the replacement of F in those equations by $-F$. For the lens to focus at infinity the "half-lens" has to have a virtual focus at $F = -2f$, which value is to be inserted in Equation 3.9.4

We again look for equality between an edge ray and the axial ray, comparing the "half-lens" solution with that obtained from the complete lens. For the former with $\eta = 1.6$ we obtain

$$f - 0.6x' = \sqrt{1 + 4f^2} - \sqrt{1 + f^2} \tag{3.9.9}$$

whereas for the full lens

$$f - x' + 2\eta x' = \sqrt{1 + f^2} + x'$$

or

$$\sqrt{1 + f^2} = f + 1.2x'$$

The comparison is shown in Table 3.5 and the profiles are computed in Figure 3.49.

This shows clearly the increase in accuracy as the system approaches the paraxial situation.

The corrected plane reflector is thus the "half-lens" designed by the method given, with the plane centre surface covered by a conducting layer. A further example of the corrected plane, using a non-uniform refractive medium, will be given later in Chapter 4.

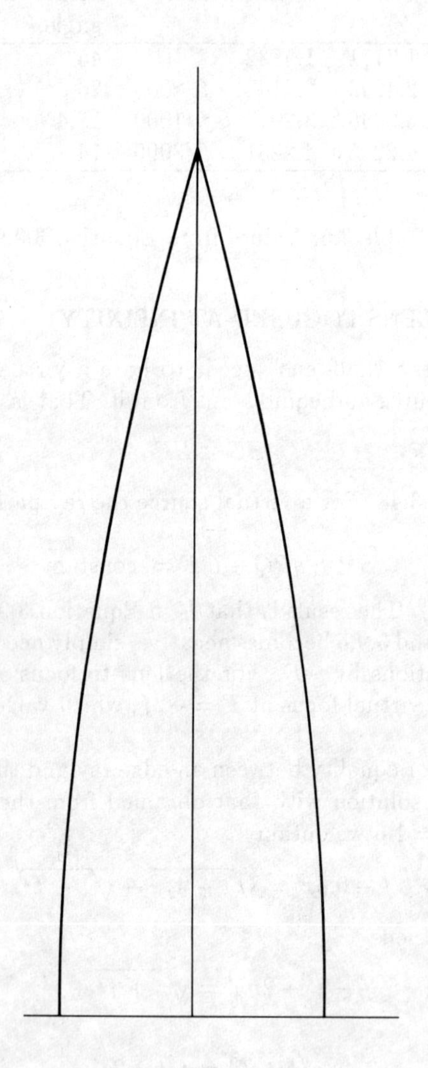

Figure 3.49: Symmetrical collimating lens

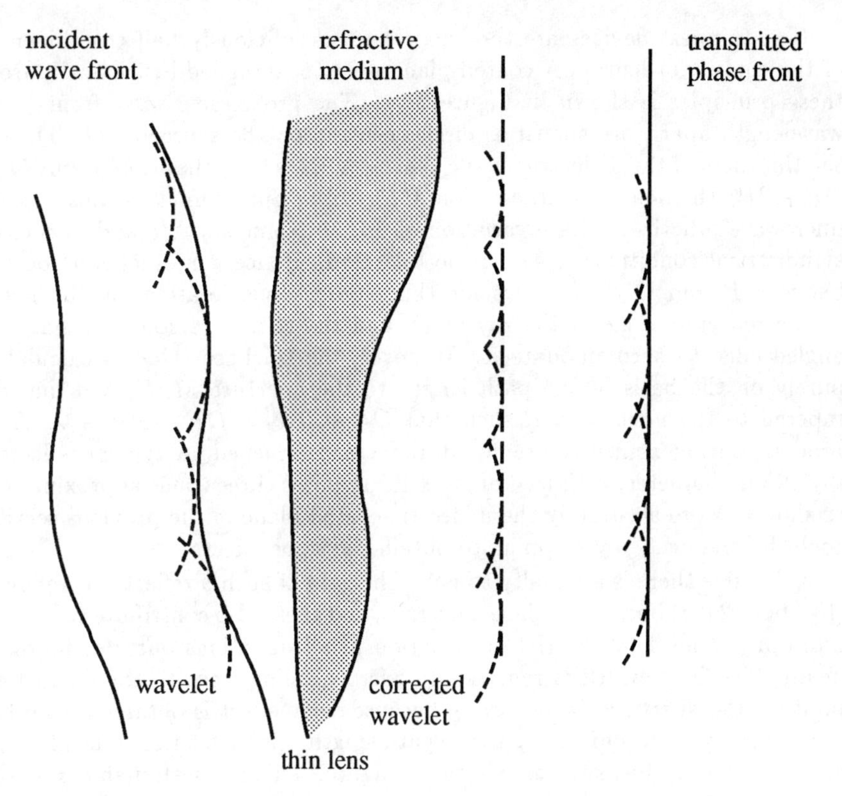

Figure 3.50: Huygens' lens

3.10 HUYGENS' LENSES AND LENS REFLECTORS

Wherever "thin" optical systems are being considered, thin being defined as an optical distance less than $\lambda/2\eta$, a case can be made for ignoring ray directions altogether along with Snell's law of refraction. We then consider the device to be a spatial phase transformer affecting only the Huygens' wavelets originating on a given wave front. This is illustrated in Figure 3.50. The incident wave is perturbed by its previous history in passing through the optical system and requires to be "regularised" by a thin layer of phase correcting dielectric medium. This, the wave is assumed to travel through, and being sufficiently thin not to affect greatly the direction of the rays themselves. Such a device is obviously of greater practicality at microwave frequencies than at optical ones. It has to be remembered that the phase correction required has to be calculated from the *differential* phase, that is the *difference* in phase between the incident wave surface and final surface with and without the refractive medium. That is the phase $\phi = 2\pi d/\lambda - 2\pi\eta d/\lambda$ for a thickness d of material.

The thinnest devices are the lens reflectors, obviously half the thickness of the equivalent lenses. A coated-plane reflector designed heuristically from these principles is shown in Figure 3.51. The progressive wave fronts, one wavelength apart, are shown as circles centred on the source at O. Then if the thickness of the dielectric at the centre, NM, is such that $OM+2\eta NM = OP+PP'$, the phase over the region MP' will be approximately constant. In microwave practice a phase variation of $\pm\lambda/10$ (some say $\lambda/8$ and in certain symmetrical conditions $\lambda/4$) can be tolerated. Hence a Fresnel zone occurs between P and Q that allows for this. Some consideration can be given to the reflected angle of the ray at Q, and the refractive zone can have an angled edge to accommodate it. However, the thickness QQ' is calculated purely on the basis of the path length to the aperture MM', and linearly tapered to the next zone R such that $OR + RR' = OP + PP' + \lambda$. The process continues until the required diameter is reached. A typical reflector, say 10λ in diameter, with feed angle $\pm 45^o$ requires three zones approximately as shown. More accurately the dielectric-coated plane of the previous section could be "reduced" by stepping to obtain the same effect.

Of course there is a penalty to pay. The gain of such a reflector is reduced (by about 2 dB) because of the errant rays, and these also contribute to a fairly uniform low level (< 20 dB) of radiation in all directions outside the main beam. The beamwidth is remarkably narrow, being close to the diffraction limit for the aperture size, possibly because some effect is obtained from the very edge rays still conforming to the path length principle outside the physical edge of the reflector, such as SS' in the figure. This is questionable, but the effect has been observed and is observable in other cone shaped surface devices such as the axicon (see Chapter 5). Finally, because of the near flatness and thinness of the device a fair degree of offset scanning is possible, $\pm 20^o$ for a loss of 1 dB in the one tested [34].

There are a number of different ways this form of zoning can be devised by this method, even when the use of light foam dielectrics is specified. The reflectors are particularly useful as low gain receiving antennas where the gain can be achieved by increasing the number of zones and the resulting narrow beamwidth used to contain cross-talk from adjacent transmitters.

3.11 THE SPHERE AS A LENS

Finally, we consider the simplest construction of all, that of a uniform sphere, and it is found by comparatively simple geometrical ray tracing, that over a limited range of paraxial rays, a caustic is created of a similar nature to that for a spherical mirror. This can be corrected in a similar way therefore by a Schmidt plate designed as before, as shown in Figure 3.52. The correcting plate has to be situated so that no ray crosses the axis between it and the sphere. In the notation of the figure we have $\sin\beta = (1/\eta)\sin\alpha$ from the law

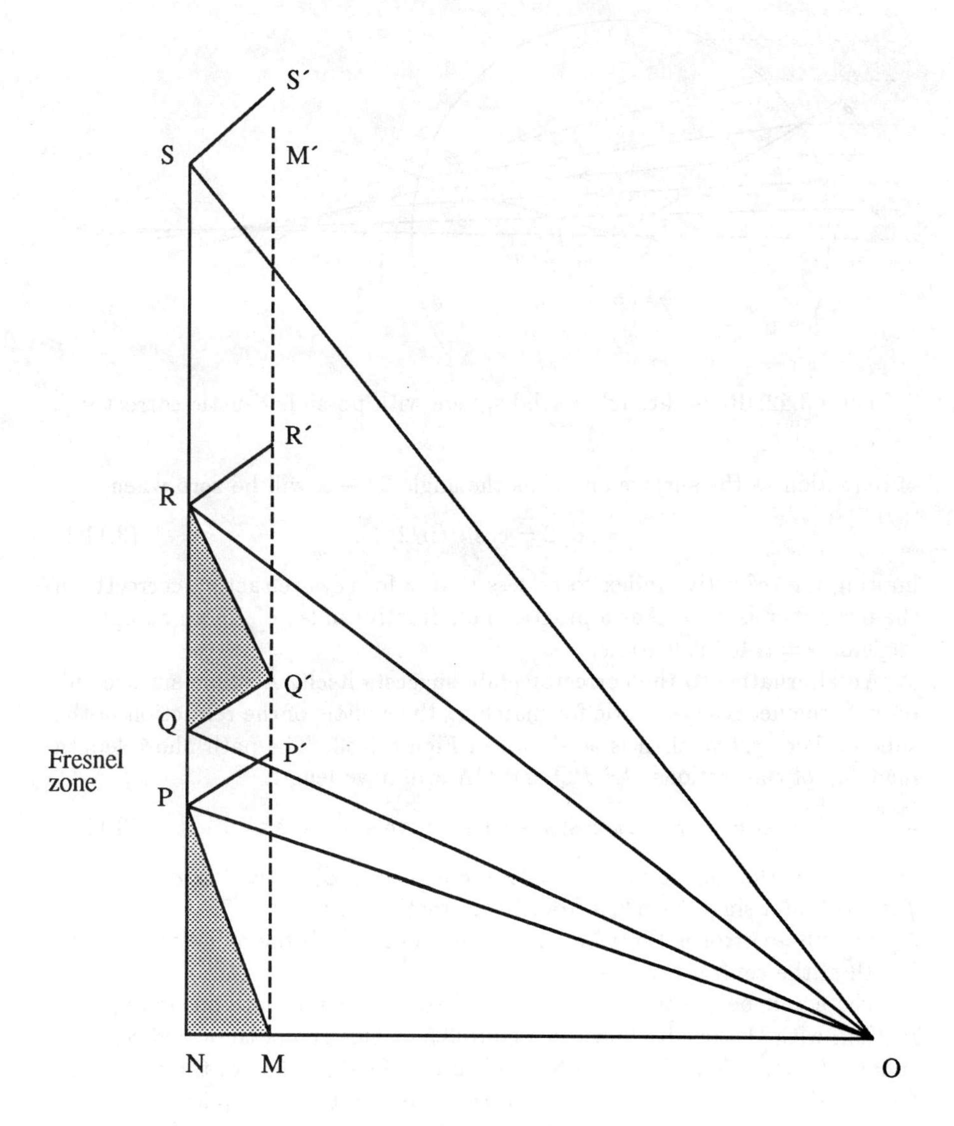

Figure 3.51: Huygens' lens reflector. All rays from O to MM' are within $\pm\lambda/10$ in path length

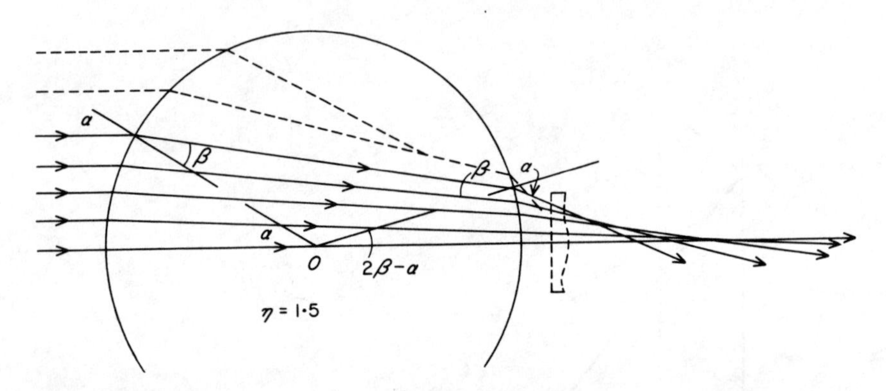

Figure 3.52: Rays through a solid sphere with possible caustic corrector

of refraction at the surface and thus the angle $2\beta - \alpha$ will be zero when

$$\alpha/2 = \cos^{-1}(\eta/2) \tag{3.11.1}$$

limiting the refractive index to be less than 2 for rays refracting correctly in the upper hemisphere. For a practicable refractive index of 1.5 an acceptable angle of $\alpha = \pm 45^o$ is feasible.

An alternative to the corrector plate suggests itself. A plane surface will provide the necessary caustic for matching the caustic of the refraction in the sphere. The system then is as shown in Figure 3.53. The path phase length made up of the sections OP PQ and ON will have length

$$f \sec\theta + \eta F \sec\psi \cos(\phi + \psi) + \eta \cos\delta + 1 - \cos\varepsilon \tag{3.11.2}$$

where from the geometry $\sin\delta = F \sec\psi \sin(\psi + \phi)$; $\varepsilon = \delta + \phi$; $\tan\psi = f \tan\theta/F$ and $\sin\phi = \sin\theta/\eta$ from the refraction at P.

The phase error is therefore Equation 3.11.2 minus the value of the path length of the central ray, $f + \eta(1 + F)$.

These can be optimised for f and F to obtain the best linear approximation, with the result shown in Figure 3.54. This shows the usual residual spherical aberration which can be applied to a final figuring of the plane surface, as shown in Figure 3.53. It will then take on the anticipated form of a limaçon. This final figuring can be averted with the use of a parallel layer of lower refractive index in the style of the oil immersion microscope objective [35].

3.12 A LENS TRANSFORMATION HYPOTHESIS

The application of inversion in Damien's theorem derives the second surface of a lens with known first surface and imaging property. This poses a question

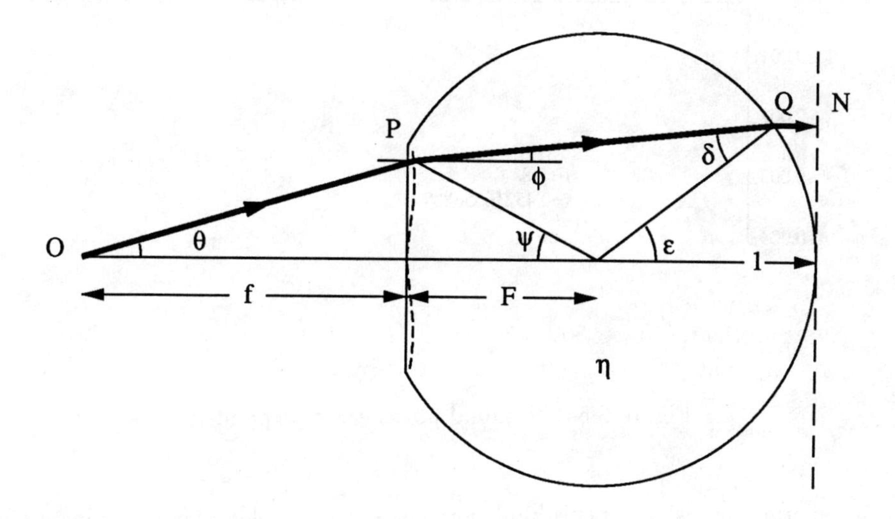

Figure 3.53: Correcting the sphere with a plane refracting surface (figuring shown greatly exagerated)

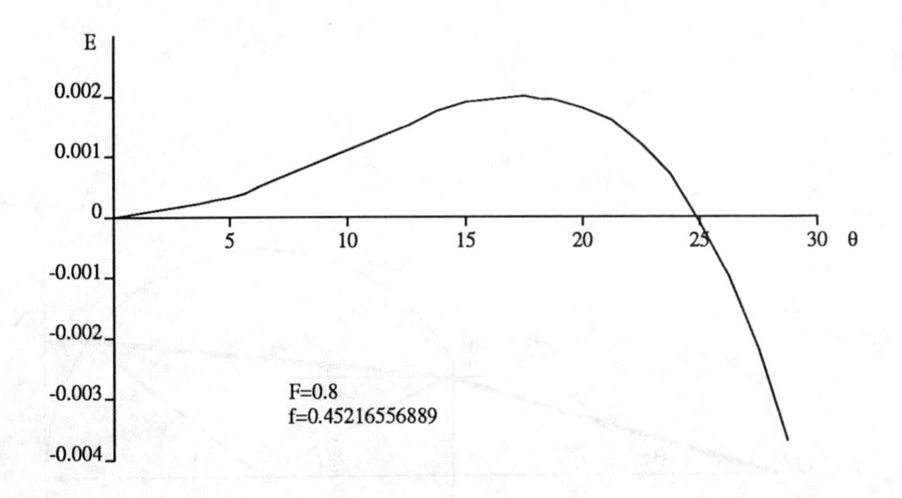

Figure 3.54: Residual phase error (expanded scale)

whether inversions of this kind can play a further role in transforming a given two surface lens with its imaging property into a different lens with identical imaging property. Any such theory will of necessity require that both the source and the image will be the centres of an inversion during the complete process of transformation. For inversion in a point at infinity, as for a collimated beam, the inversion will simply be a reflection in a plane perpendicular to the (supposed) axis of circular symmetry. This will entail no change in the shape of the curve being inverted. The inversion of both surfaces will therefore take place at some time, in circles centred on the source. For this to proceed in a symmetrical manner for both surfaces, each will therefore require to be inverted in both source and image point at some stage of the process. The procedure that does this is the grounds for the hypothesis. At the time of writing, this has not been given definitive proof. It is presented therefore as a conjecture, since it is to be feared that the time needed to obtain such a proof will exceed that available to the author to present the manuscript to the publisher. In doing so it should be observed that it is in full agreement with the current methods of theoretical physics, that of presenting hypotheses first and then establishing experiments to verify them.

The validity of this hypothesis can, however, be illustrated by its application to the many forms of aplanatic lenses and dual reflector combinations

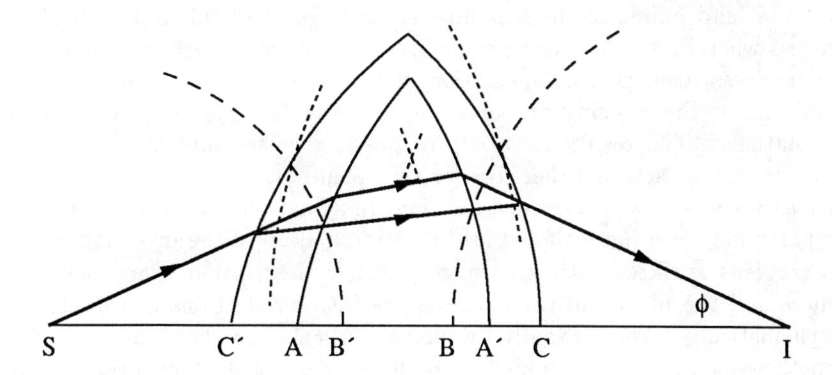

Figure 3.55: Transformation by inversion of an aplanatic lens. The circles of inversion are shown dotted

that have appeared in previous sections of this volume.

The process proposed is illustrated in Figure 3.55. The lens with source S and image I has two surfaces, A the "first"surface and A' the second. The hypothesis states that this will be converted into a new lens, with first surface C and second surface C' with the same source and image, by the following series of inversions:-

1. Invert the *first* surface A in a circle centred on the image I to give a surface B.

2. Invert the surface B in a circle centred on the source S to give the surface C (note here that C now forms the *second* surface of the transformed lens).

3. Invert the second surface A' in the *source* S to give the surface B'.

4. Invert B' in a circle centred on the image I to give the new *first* surface C'.

CC' is then a new lens with the same focusing property as the original lens AA'.

The radii of the circles of inversion are in each case arbitrary in the sense that the original lens can have arbitrary axial points. They have to be chosen such that the imperative of the process, that the first surface transforms into the second surface of the new lens and vice versa is obeyed. These radii of inversion (or position of reflecting plane), can also be arranged, as will be seen in the examples, to leave the axial points unaltered or, alternatively, the outer end points of the lens unaltered. Choices of this nature lead to a process whereby the lens transformation is a "lens bending" procedure. The arbitrariness itself is a manifestation of the principle that a single a priori requirement, the imaging property, for a two surface system has an infinity of solutions. The resultant variety of possible lenses that the process can produce, can be scanned therefore for the optimisation of a second condition, such as agreement with the sine condition. In a final version such a geometrical property may even be included in the determination of the inversion radii.

For lens surfaces with known geometrical description, particularly the sphere and the plane, all the above inversions or reflections can be detailed mathematically. These rapidly become complicated in the extreme, but are nonetheless amenable to computing methods. Each of the four inversions may be considered independently with regard to their respective radii or reflection plane to suit the practicality of the final result. In the absence of a general proof, we give an illustration of the process for several of the lenses that have been discussed in previous sections. Axial symmetry is presupposed, although it is possible that such a condition too is not a limiting factor. In most examples the system focuses at infinity, because that is the nature of most microwave lenses. There is no apparent reason though that if applicable to these systems it should not be equally valid for general imaging.

APLANATIC LENS WITH TWO HYPERBOLIC PROFILES

The simplest exact aplanatic lens is that shown in Section 3.9 consisting of two hyperbolic profiles, with asymptotes at the same angle to the axis given by $\cos\theta = \eta$. That is, *any* two hyperbolae of the same confocal net, but with one reversed. In the material of the lens, the ray is parallel to the axis. Since the hyperbolae have the same eccentricity, the arbitrariness in the design extends only to the central thickness of the lens. Each surface therefore will invert in each point, first into (the exterior loop of) a trisectrix, as shown, and then back into a hyperbola. Since the nodal angle of each trisectrix is the same as the asymptote angle of the hyperbola, and this is constant, the result is simply to derive a new lens with identical source and image points, with profiles two new hyperbolae from the same confocal net.

THE HYPERBOLIC/PLANE LENS

The very elementary nature of the first example above is deceptive. If the procedure is applied to the basic *single* surface lens, the hyperbola first surface is backed with a plane surface (see Figure 1.3). Then, as shown in Figure 3.56, we first reflect the hyperbola A into the hyperbola B since the inversion is at

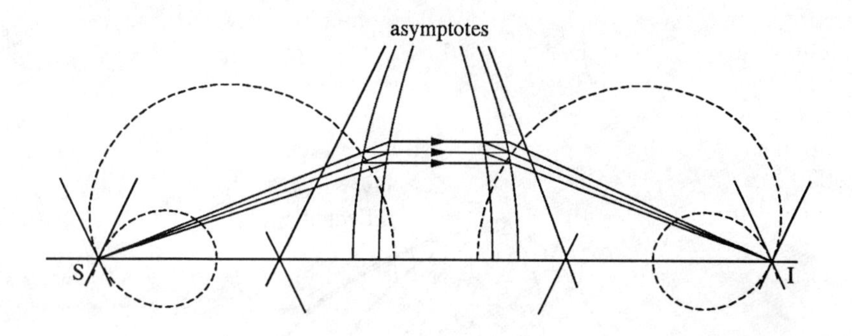

Figure 3.56: Imaging property of confocal hyperbolas and associated trisectrix inverses

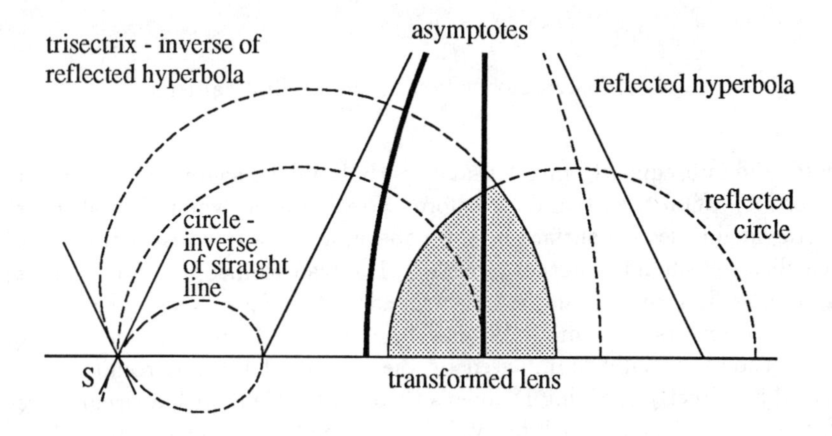

Figure 3.57: Transformation of hyperbolic/plane lens.

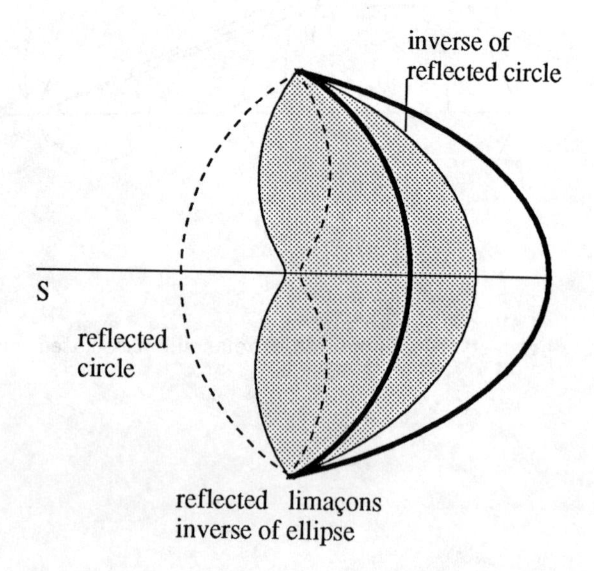

Figure 3.58: Transformation of circle/elliptical lens.

infinity, and subsequently invert that hyperbola in the source S to give (the exterior loop of) a trisectrix C as before. Now however, when we first invert the straight line second surface in S we obtain a circle B' passing through S which then reflects into another circle C'. The transformed lens now consists of a circular first surface and a trisectrix second surface. The exterior loop of the trisectrix (see Figure 1.19) has been chosen to match that area of the hyperbola normally illuminated by the source. The same result can be obtained by directly applying Damien's theorem to a lens with a circular first surface required to focus at infinity, if it is remembered that parallels of the limaçon progress through the cardioid to become the trisectrix.

THE CIRCULAR/ELLIPTICAL LENS

As was shown in Figure 1.15 the lens with an elliptical second surface can be conjoined with any circular first surface centred on the source at S, to give a

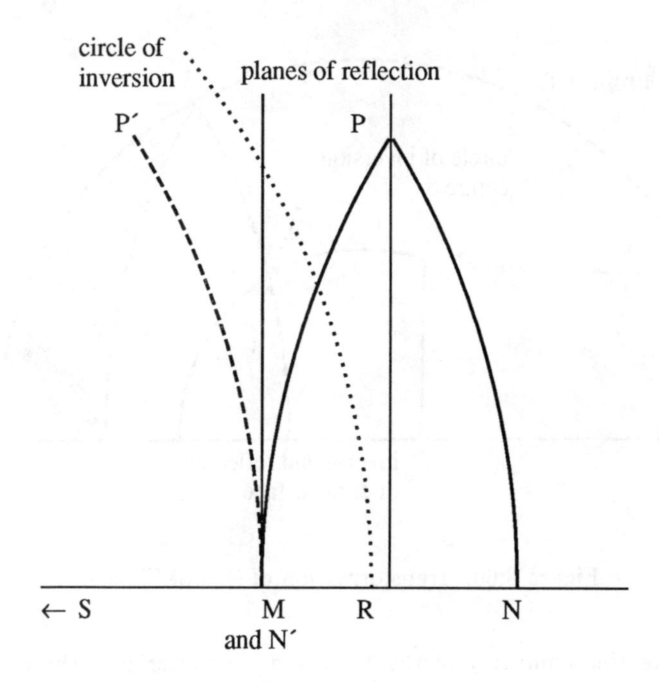

Figure 3.59: Transformation of the symmetrical lens into itself.

lens focused at infinity. The source is at the distant focus of the ellipse, and consequently its inverse in a circle centred on it will be a limaçon. This has to be reflected as shown in Figure 3.57, where the radius of inversion and the reflection have been arranged to leave the edge of the lens invariant. Similarly, the circular face is at first reflected and then inverted in a circle centred on S in the same way as to keep the edge intact, resulting in a new lens which has a front surface with a limaçon profile and a second circular surface, that still focuses rays from the source at infinity.

THE SYMMETRICAL LENS FOCUSED AT INFINITY

In contrast to the first example, this has asymmetrical source and image positions. It also obeys two a priori conditions, that of collimation and of being symmetrical about a central axis. Therefore it is unique and as such is compelled to invert into itself if it is to retain its focusing property, which is basic to the concept. Accordingly the radii of inversions are no longer arbitrary. The points M and N on the axis (Figure 3.58) must invert into each other and hence the radius of inversion SR must satisfy $SR^2 = SM \cdot SN$. Thus the first surface PM must reflect into the second surface, which it does

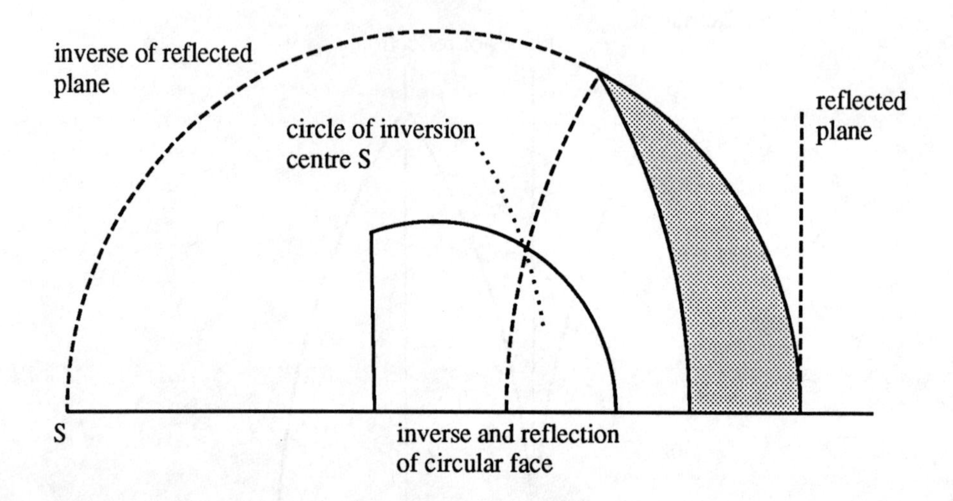

Figure 3.60: Transformation of lens of Section 3.

because of the symmetry of the lens, and thereafter into the curve $P'N'$, whereas the second surface inverts into $P'N'$ and thereafter reflects into PM. Thus we require a curve PN (or PM) with the property that its inversion in S with radius SR converts it into the *same* curve $P'N'$ which is a parallel shift through the distance MN along the axis. Of course, finding such a curve geometrically is probably as intractable as solving the original lens problem. Traditionally therefore such details are left as an exercise for the reader. The choice of radius of inversion to be the harmonic mean of the distances from the source to the axial points of the lens, is one method of retaining the same central thickness for the transformed lens.

THE PLANE/SPHERE LENS

It was shown in Section 3.11 that a plane section of a sphere could be used to reduce the spherical aberration of the complete sphere through the caustic match of the plane face refraction. Since planes and circles invert and reflect into planes and circles, the process can be applied as shown to derive concavo-convex lenses also with residual spherical aberration. This aberration will be different from that of the original lens since a point has been taken for the centre of inversion whereas in reality a caustic occurs. Whether this improves the aberration or worsens it is also part of the conjecture at this time.

DUAL REFLECTORS

One non-lens combination with an interesting result is the dual reflector system based on the paraboloid and conic section subreflector. These dual reflector combinations exhibit the same arbitrariness in choice of conic surface subreflectors, hyperboloid for Cassegrain style and ellipsoid for Gregorian style, with a plane subreflector as the dividing line and with totally arbitrary choice of source position, necessarily the second focus of the conic subreflector. Taking the simplest combination first, we have the paraboloid A' with a plane subreflector A midway between the source at the apex and the focus of the paraboloid. The "first" surface, the plane subreflector, inverts (reflects) into itself by inversion at infinity and then into a circle C which passes through the source at the apex of the paraboloid. The paraboloid A' inverts in the apex into a cardioid which then is reflected appropriately to give the curve C' and the Zeiss cardioid combination of Chapter 1. As was shown in the derivation given there, variation of the design parameters gave an infinite variety of cardioid/limaçon combinations that are alternatives to the cardioid circle pair of Zeiss. These would be derived by applying the inversion process to the various alternative conic surface subreflectors and source positions.

Applied to offset dual reflectors however, the process is limited in the same way that the Zeiss cardioid is, in that the final combination requires a subreflector (the cardioid) larger than the "main" reflector (the circle). The system would nontheless be a candidate for the folding process of the last section of Chapter 2.

REFERENCES

1. Westcott BS & Brickell F (1986) General dielectric lens shaping using complex coordinates *IEE Proc* **H 133** pp122-126

2. Born M & Wolf E (1954) *Principles of Optics,* pp196 Pergamon Press

3. Volosov DS (1947) A differential method for introducing non-spherical surfaces into calculation of optical systems *Jour Opt Soc Amer* **37** p342

4. Wolf E (1948) On the designing of aspheric surfaces *Proc Phys Soc* **61** p494

5. Lee JJ (1983) Dielectric lens shaping and coma correction zoning. *Trans IEE* **AP-31** p211

6. Rodgers JM (1984) Non-standard representations of aspheric surfaces in a telescope design *Applied Optics* **23** p520

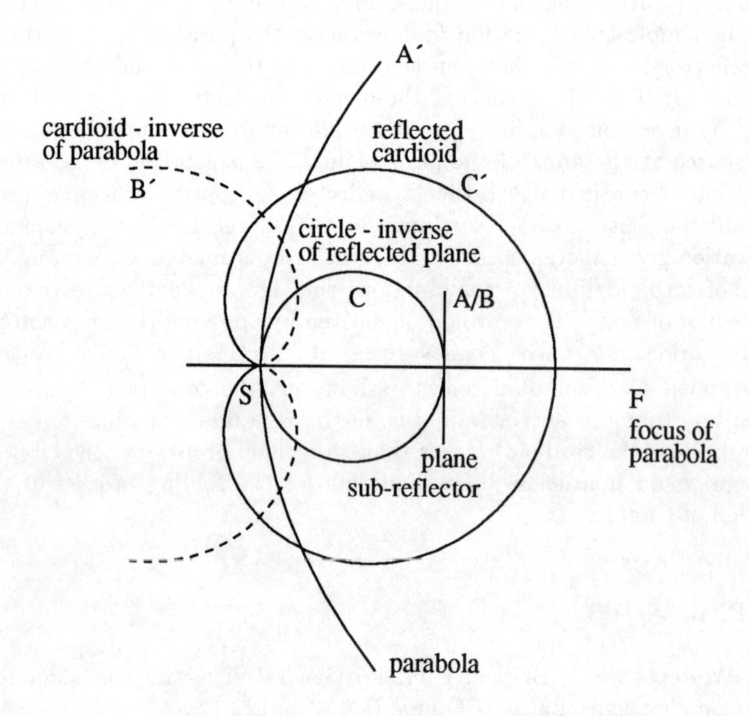

Figure 3.61: Transformation of parabola with plane subreflector to Zeiss cardioid and circular reflector

7. Rhodes PW & Shealy DL (1980) Refractive optical systems for irradiance redistribution of collimated radiation: their design and analysis *Applied Optics* **19** pp3545-53

8. Kay AF (1964) The scalar feed *AFCRC* TRG Report **19(604)-8057** March 30th

9. Lawrie RE & Peters L Jnr (1964) Modifications of horn antennas for low side lobe levels *Trans IEEE* **AP-14(5)** p1817

10. Cornbleet S (1979) Feed position for the parabolic reflector with offset pattern *Electron Lett* **15** p211

11. Bishay S, Cornbleet S & Hilton J (1989) Lens antennas with amplitude shaping or sine condition *Proc IEE* **136** pt H no 3 p276

12. Gill PE & Murray W Algorithms for the solution of the non-linear least squares problem *Siam Jour Numerical Analysis* **15** p977

13. Friedlander FG (1946) Dielectric lens aerial for wide-angle beam scanning *Jour IEE* **93** pt III, p658

14. Born M & Wolf E (1954) *Principles of Optics,* pp246 Pergamon Press

15. Silver S (1949) *Microwave Antenna Theory and Design* McGraw Hill: MIT Radiation Laboratory **12** p402

16. Brown J (1953) *Microwave Lenses* Methuen p33

17. Baker HF (1928) *Principles of Geometry* **I** Cambridge University Press p5

18. Ruze J (1950) Wide-angle metal plate optics, *Proc IRE* **38** p53

19. Gent H (1955) Lectures on development in microwave optics. Royal Radar Establishment UK (unpublished)

20. Gent H (1957) The bootlace aerial *RRE Journal*

21. Jones SSD, Gent H & Browne AAL (1961) British Patent No 860826

22. Pomot C, Sermet P, Munier J & Benoit J (1965) Lentilles et reflecteurs bidimensionelles a grand changes angulaire. from *Electromagnetic Wave Theory* Brown J (Ed) **2** Pergamon Press Proceedings of Symposium, Delft 1965 p685

23. Berry DG, Malech RG & Kennedy WA (1963) The reflectarray antenna *Trans IEEE* **APII** p645

24. Amitay N, Galindo V & Wu GP (1972) *Theory and Analysis of Phased Array Antennas* Wiley-Interscience

25. Rotman W & Turner RF (1963) Wide angle microwave lens for link source applications *Trans IEE Ant & Prop* **AP-11** p623

26. Hansen RC (1991) Design trades for Rotman lenses *Trans IEEE Ant & Prop* **AP-41(4)** p464-472

27. Sole GC & Smith MS (1987) Multiple beam forming for planar antenna arrays using a three-dimensional Rotman lens *Proc IEE* **134** pt H no 4 p375

28. Hall PS & Vetterlein SJ (1990) Review of radio frequency techniques for scanned and multiple beam antennas *Proc IEE* **137** part H no 5 p293

29. Schmidt B (1931) Central Zeitung fur Optik und Mechanik **52** heft 2

30. Rusch WVT & Potter PD (1970) *Analysis of Reflector Antennas* London & New York: Academic Press

31. Martin LC (1954) *Technical Optics* **II** Pitman p253

32. Gunter RC Jnr, Holt FS & Winter CF (1955) *The Mangin Mirror* AFCRC Report no TR-54-111

33. Baker HF (1928) *Principles of Geometry* **I** Cambridge University Press p118

34. Cornbleet S & Gunney B (1978) Corrected plane reflectors *IEE Confercncc Publication* **169** p201

35. Born M & Wolf E (1954) *Principles of Optics*, pp252 Pergamon Press

4
RAY-TRACING IN NON-UNIFORM MEDIA

The tracing of rays in an isotropic non-uniform refractive medium provides a rich field for the geometry of curves in space and the differential geometry of surfaces. This not only applies to the design of practical microwave optical devices but, being based upon a least-action integral, can be associated with the trajectories of particles in non-uniform potential fields and through this to many other branches of physical science. Recently techniques have become available which make possible the production of non-uniform optical glasses, leading to the design of optical components for fibre-optics or integrated optical systems.

As in the previous chapters we find that the practicability of the basic design methods is underlined by a transformation process of great theoretical complexity, and these too are beginning to find applications in the other least-action problems in physics. To demonstrate this equivalence, we use the derivation of Hamilton's canonical equations for rays given by Synge [1].

4.1 THE RAY-TRACING EQUATIONS

For any curve defined parametrically by

$$x = x(u) \qquad y = y(u) \qquad z = z(u)$$

to be a geometrical ray, it must obey Fermat's principle, that its optical length

$$V = \int_{u_1}^{u_2} \eta(x, y, z) \sqrt{\left(\frac{\mathrm{d}x}{\mathrm{d}u}\right)^2 + \left(\frac{\mathrm{d}y}{\mathrm{d}u}\right)^2 + \left(\frac{\mathrm{d}z}{\mathrm{d}u}\right)^2} \, \mathrm{d}u \tag{4.1.1}$$

is an extremum. We term the integrand W for brevity. The variation in passing from one curve to another near neighbour is

$$\delta V = \int_{u_1}^{u_2} \left(\sum \frac{\partial W}{\partial x} \delta \dot{x} + \sum \frac{\partial W}{\partial x} \delta x \right) \mathrm{d}u$$

Σ indicating the sum of similar terms with x y and z, and the dot representing differentiating with respect to u. Thus $\delta \dot{x} = \mathrm{d}/\mathrm{d}u \, (\delta x)$ and integration by parts gives

$$\delta V = \left[\Sigma \frac{\partial W}{\partial \dot{x}} \delta x \right]_{u_1}^{u_2} - \int_{u_1}^{u_2} \Sigma \left[\frac{\mathrm{d}}{\mathrm{d}u} \left(\frac{\partial W}{\partial \dot{x}} \right) - \frac{\partial W}{\partial x} \right] \delta x \, \mathrm{d}u = 0$$

The first part vanishes if the curves have common end-points, and thus the integrand in the second term must equal zero. For arbitrary displacements δx this implies the Euler-Lagrange equations for a ray.

$$\frac{\mathrm{d}}{\mathrm{du}}\left(\frac{\partial W}{\partial \dot{x}}\right) - \frac{\partial W}{\partial x} = 0$$
$$\frac{\mathrm{d}}{\mathrm{du}}\left(\frac{\partial W}{\partial \dot{y}}\right) - \frac{\partial W}{\partial y} = 0$$
$$\frac{\mathrm{d}}{\mathrm{du}}\left(\frac{\partial W}{\partial \dot{z}}\right) - \frac{\partial W}{\partial z} = 0 \qquad (4.1.2)$$

On substituting the expression for W we have

$$\frac{\mathrm{d}}{\mathrm{du}}\left[\frac{\eta\dot{x}}{\sqrt{\dot{x}^2+\dot{y}^2+\dot{z}^2}}\right] - \frac{\partial\eta}{\partial x}\sqrt{\dot{x}^2+\dot{y}^2+\dot{z}^2} = 0$$

The choice of u as parameter is arbitrary and we can make two specific selections:

1. $u = s$, the arc length along the curve. We then have three equations of the form

$$\frac{\mathrm{d}}{\mathrm{ds}}\left(\eta\frac{\mathrm{d}x}{\mathrm{d}s}\right) - \frac{\partial\eta}{\partial x} = 0$$

which collectively for terms in x y and z are

$$\frac{\mathrm{d}}{\mathrm{ds}}\left(\eta\frac{\mathrm{d}\mathbf{r}}{\mathrm{d}s}\right) = \nabla\eta \qquad (4.1.3)$$

where $\mathbf{r}$ is a position vector of a point on the ray from a given origin.

2. Alternatively selecting u to be $u = \int \mathrm{d}s/\eta$, then $\mathrm{d}u = \mathrm{d}s/\eta$ and the ray equations become

$$\frac{\mathrm{d}^2x}{\mathrm{d}u^2} = \frac{\partial}{\partial x}\left(\frac{1}{2}\eta^2\right)$$
$$\frac{\mathrm{d}^2y}{\mathrm{d}u^2} = \frac{\partial}{\partial y}\left(\frac{1}{2}\eta^2\right)$$
$$\frac{\mathrm{d}^2z}{\mathrm{d}u^2} = \frac{\partial}{\partial z}\left(\frac{1}{2}\eta^2\right) \qquad (4.1.4)$$

Equations 4.1.2 4.1.3 and 4.1.4 are different forms of the equations of rays, of which Equation 4.1.3 will be seen to be the most applicable.

We now consider a variation of V in which the end-points too are infinitesimally displaced, and we take the parameter u to equal s, the arc length along a ray. Then

$$\frac{\partial W}{\partial \dot{x}} = \eta\alpha \qquad \frac{\partial W}{\partial \dot{y}} = \eta\beta \qquad \frac{\partial W}{\partial \dot{z}} = \eta\gamma$$

where α β and γ are the direction cosines of the ray at its starting point. Using primes for similar entities for the ray at its end-point, e.g. $\partial W'/\partial \dot{x}' = \eta'\alpha'$, we have

$$\delta V = \sum_{\substack{x,y,z \\ \alpha,\beta,\gamma}} \eta\alpha - \sum_{\substack{x',y',z' \\ \alpha',\beta',\gamma'}} \eta'\alpha'$$

and thus

$$\frac{\partial V}{\partial x} = \eta\alpha \qquad \frac{\partial V}{\partial y} = \eta\beta \qquad \frac{\partial V}{\partial z} = \eta\gamma \qquad \text{etc.}$$

Since α β and γ are direction cosines, this results in

$$\left(\frac{\partial V}{\partial x}\right)^2 + \left(\frac{\partial V}{\partial y}\right)^2 + \left(\frac{\partial V}{\partial z}\right)^2 = \eta^2 \tag{4.1.5}$$

with a similar result for the primed coordinate system.

Finally, for an isotropic medium, rays are normal to wavefronts, and we can use Huygens' construction. If $S(x, y, z) = 0$ is the equation of a wavefront reaching the point (x, y, z) at time t, then the progression of the wavefront by Huygens' construction is given by

$$S(x, y, z) = c(x, y, z)t \tag{4.1.6}$$

the instantaneous positions being given by taking t = constant. In a non-uniform medium, the velocity of the wave v is a function of position, i.e. $v = v(x, y, z)$, where the refractive index $\eta(x, y, z) = c/v$.

The normals to this surface are the rays, and hence

$$\frac{\partial S}{\partial x} = \kappa\alpha \qquad \frac{\partial S}{\partial y} = \kappa\beta \qquad \frac{\partial S}{\partial z} = \kappa\gamma$$

where κ is a factor of proportionality. Hence, moving with the wave

$$c\ \mathrm{d}t = \mathrm{d}S = \sum_{x,y,z} \frac{\partial S}{\partial x}\,\mathrm{d}x = \left(\sum_{x,y,z} \alpha\frac{\partial S}{\partial x}\right)\mathrm{d}s = \kappa\,\mathrm{d}s \tag{4.1.7}$$

where $\mathrm{d}s$ is an increment of the ray path. Since $\mathrm{d}s = v\ \mathrm{d}t$ and $\eta = c/v$, then $\kappa = \eta$. Hence

$$\frac{\mathrm{d}S}{\mathrm{d}x} = \eta\alpha \qquad \frac{\mathrm{d}S}{\mathrm{d}y} = \eta\beta \qquad \frac{\mathrm{d}S}{\mathrm{d}z} = \eta\gamma$$

and therefore

$$\left(\frac{\mathrm{d}S}{\mathrm{d}x}\right)^2 + \left(\frac{\mathrm{d}S}{\mathrm{d}y}\right)^2 + \left(\frac{\mathrm{d}S}{\mathrm{d}z}\right)^2 = \eta^2 \tag{4.1.8}$$

that is, the same relation as is obeyed by the characteristic function V. To prove that these rays are identical with the rays of Fermat's principle, we have

$$\begin{aligned}
\frac{\mathrm{d}}{\mathrm{ds}}(\eta\alpha) &= \frac{\mathrm{d}}{\mathrm{ds}}\left(\frac{\partial S}{\partial x}\right) = \frac{\partial^2 S}{\partial x^2}\alpha + \frac{\partial^2 S}{\partial y^2}\beta + \frac{\partial^2 S}{\partial z^2}\gamma \\
&= \frac{1}{\eta}\left(\frac{\partial^2 S}{\partial x^2}\frac{\partial S}{\partial x} + \frac{\partial^2 S}{\partial x \partial y}\frac{\partial S}{\partial y} + \frac{\partial^2 S}{\partial x \partial z}\frac{\partial S}{\partial z}\right) \\
&= \frac{1}{2\eta}\frac{\partial}{\partial x}\left[\sum_{x,y,z}\left(\frac{\partial S}{\partial x}\right)^2\right] = \frac{1}{2\eta}\frac{\partial}{\partial x}\eta^2 \\
&= \frac{\partial \eta}{\partial x}
\end{aligned} \tag{4.1.9}$$

which follows directly from Equation 4.1.3.

Equation 4.1.7 shows the increment in S in passing from one wave surface to an adjacent one is the optical length of the ray; thus S is related to the characteristic function V through

$$S(x,y,z) - S(x',y',z') = V(x,y,z,x',y',z') \tag{4.1.10}$$

The light rays, being orthogonal to the surfaces of constant S, and by virtue of Equation 4.1.8, will obey the equation

$$\eta\frac{\mathrm{d}\mathbf{r}}{\mathrm{d}s} = \mathrm{grad}\ S \tag{4.1.11}$$

from which it follows that curl $\eta\mathbf{s} = 0$. ($\mathbf{s} = \mathrm{d}\mathbf{r}/\mathrm{d}s$ is the tangent to the ray; $\eta\mathbf{s}$ is termed the "ray vector".)

Hence, by applying Stokes' theorem to any closed curve $\int \eta\mathbf{s}\cdot\mathrm{d}\ell = 0$, and thus the optical path

$$\int_{P_1}^{P_2} \eta\mathbf{s}\cdot\mathrm{d}\ell$$

between any two points is independent of the path of integration. This is the "Lagrange integral invariant" [2, p246] for optical rays. Along a true ray $|\mathrm{d}\mathbf{r}| = \mathrm{d}s$, and thus the invariant is the optical length given by Equation 4.1.1.

From the ray-tracing point of view, the all-important result is that of Equation 4.1.3. It is therefore of some interest to derive the same result by considering the ray to be the geodesic of a "refractive space", a concept which will prove to be of value subsequently. We define the refractive space as a space with metric coefficients $G_{\mu\nu} = \eta^2 g_{\mu\nu}$, for μ, ν =1, 2, 3 where $\eta(x,y,z)$ is the medium refractive index and $g_{\mu\nu}$ the metric coefficients of the empty space.

The metric is

$$\mathrm{d}k^2 = G_{\mu\nu}\,\mathrm{d}x^\mu\,\mathrm{d}x^\nu \tag{4.1.12}$$

and the geodesics are given by [4]

$$\frac{\mathrm{d}^2 x^\mu}{\mathrm{d}k^2} + \left\{ {\mu \atop \nu\ \ \lambda} \right\} \frac{\mathrm{d}x^\nu}{\mathrm{d}k}\frac{\mathrm{d}x^\lambda}{\mathrm{d}k} = 0 \tag{4.1.13}$$

where the Christoffel symbols are defined by

$$\left\{ \begin{matrix} & \mu & \\ \alpha & & \beta \end{matrix} \right\} = \frac{1}{2} G^{\mu\sigma} \left[\frac{\partial G_{\alpha\sigma}}{\partial x^{\beta}} + \frac{\partial G_{\sigma\beta}}{\partial x^{\alpha}} - \frac{\partial G_{\alpha\beta}}{\partial x^{\sigma}} \right] \tag{4.1.14}$$

$$\alpha, \beta, \sigma = 1, 2, 3$$

This definition will form the basis for an inversion transformation which parallels that of Damien's theorem for the single refractive case.

Another curved-ray concept arises if the Snell construction for refraction (Appendix I) is extended to the situation where the ray is in a continuously variable medium. Thus, as we show in Figure 4.1(a) a parallel-sided system of thin layers has for a corresponding Snell diagram a system of concentric circles. The law of the refractive index in the single direction of variation is $\eta(x)$, and thus the law applied to the circle diagram is $\eta(x) \rightarrow \eta(r)$. As each transmitted ray from a given layer is the incident ray for the next, the construction proceeds as follows: $O'P'$ is parallel to the incident ray OP, Q' is obtained as the intersection of (the parallel to) the surface normal $\tilde{n}$ with the circle radius $r = \eta_1$, and PQ is parallel to $O'Q'$ (not $P'Q'$). Similarly QR is parallel to $O'R'$ and the ray PQR is traced by the successive determinations of $P'Q'R'$. Taking this procedure to the limit of infinitesimally thin layers, as in Figure 4.1(b), we obtain a smooth curve $P'Q'R'$. In this relation the tangents to the curve PQR are parallel to the radius vectors to the curve $P'Q'R'$. In the isotropic medium the tangents to the rays are perpendicular to the phase (or wave) front (cf Equations 4.1.6 and 4.1.7), the curve S in Figure 4.1(b). There is, therefore, a reciprocity between S and the curve $P'Q'R'$ in which the normals to S are parallel to the radius vectors to $P'Q'R'$ and the normals to $P'Q'R'$ are parallel to the radius vectors to corresponding points on S. In the simple case illustrated here, and for a monotonically varying index of refraction in the axial direction, $\eta = \eta(x)$ transforms into $\eta = \eta(r)$ in the Snell diagram and $P'Q'R'$ is the straight line parallel to the axis. In the medium the ray will be a curve $y = f(x)$, say, and in the Snell diagram $P'Q'R'$ is the line $r = h \operatorname{cosec} \theta$. Then $r = \eta$ and $\tan \theta = f'(x)$.

Consequently

$$f'(x)^2 = \frac{h^2}{\eta^2 - h^2} \qquad \eta = \eta(x)$$

and therefore

$$y = f(x) = \int \frac{h}{\sqrt{\eta^2 - h^2}} \, \mathrm{d}x \tag{4.1.15}$$

as will be found to be the case in the general expansion of Equation 4.1.3.

If we take the origin of the stratified system to be the point of intersection of the stratified layers, at infinity in the case presented, then the radius vectors to the wavefront at P are parallel to the normals to the curve $P'Q'R'$. That this is a general situation can be illustrated by making the stratification angular, as shown in Figure 4.2. The rotation of the normals to the stratified

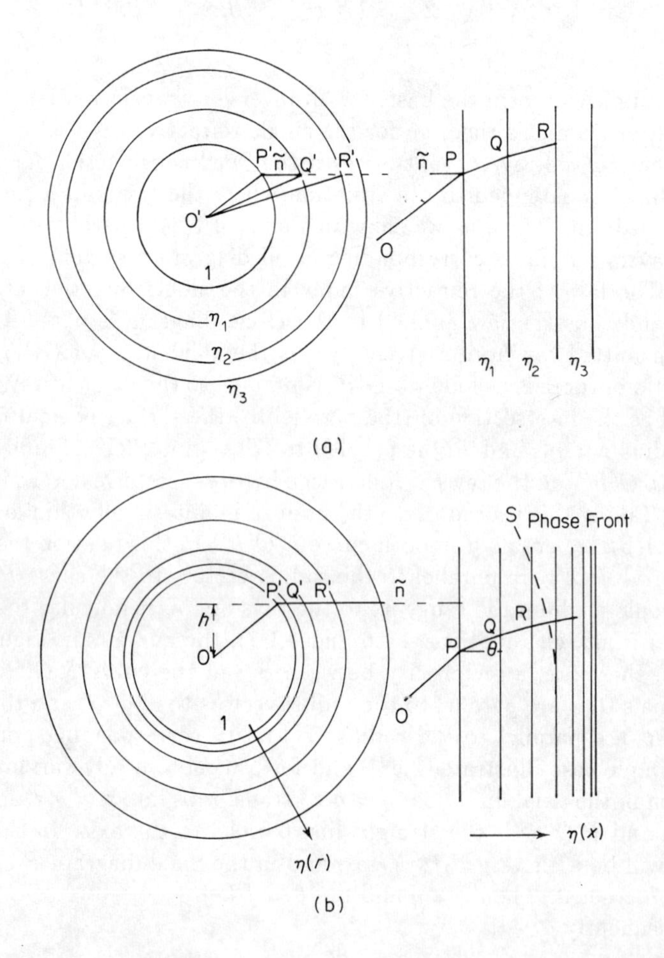

Figure 4.1: (a) Snell's construction for refraction for parallel sided layers of varying refractive index
(b) The construction continued for a continuously variable medium

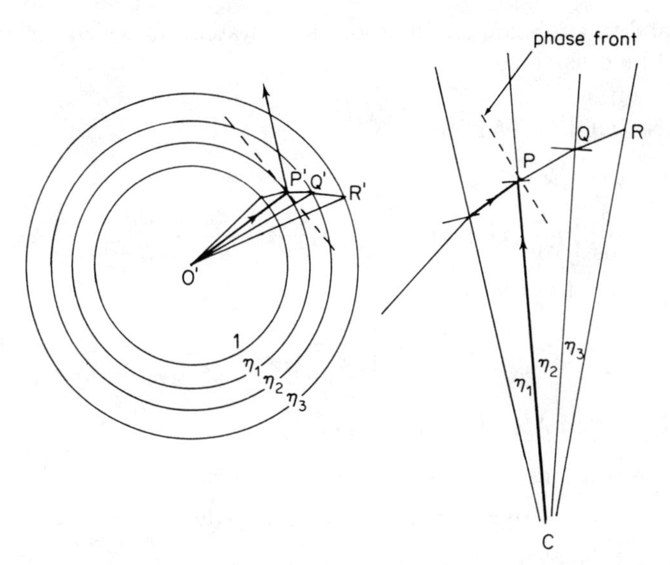

Figure 4.2: Reciprocity in an angularly varying medium

layers makes $P'Q'R'$ a curve as shown. As before $O'P'$ is parallel to the tangent at P and thus normal to the phase front at P, and CP is parallel to the normal to the refractive curve at P'.

If the non-uniformity extends to the remaining transverse direction the line $P'Q'R'$ becomes a surface which is called the "refractive index surface". The reciprocity we have illustrated between radius vectors and normals of the wavefront and refractive index surface is maintained even when the medium becomes anisotropic [3], and the ray tangent becomes different from the wavefront-normal.

It can be seen from the simple example that if the tangent at P (Figure 4.1(b)) has direction cosines cos θ and sin θ, the coordinates of P' are $\eta \cos \theta$ and $\eta \sin \theta$. Thus for a wave normal having direction cosines l m and n, the refractive index surface has direction cosines $(\eta l, \eta m, \eta n)$. This is then the surface of "normal slowness" and related to the construction of the Fresnel surface [5]. The reciprocity is that proposed by Hamilton based on a previous study by Cauchy [6] (see Budden's reciprocity, Chapter 7).

4.2 EXPANSION OF THE RAY EQUATIONS

The equation that has proved to be most applicable to practical problems is Equation 4.1.3, namely

$$\frac{\mathrm{d}}{\mathrm{ds}}\left(\eta \frac{\mathrm{d}\mathbf{r}}{\mathrm{d}s}\right) = \nabla \eta$$

and we expand this relation for the coordinate systems in which the ray trajectories will be considered

CARTESIAN COORDINATES

With $\mathbf{r} = x\hat{\imath} + y\hat{\jmath} + z\hat{k}$

$$\frac{\mathrm{d}}{\mathrm{ds}}\left[\eta\frac{\mathrm{d}}{\mathrm{ds}}(x\hat{\imath} + y\hat{\jmath} + z\hat{k})\right] = \frac{\partial\eta}{\partial x}\hat{\imath} + \frac{\partial\eta}{\partial y}\hat{\jmath} + \frac{\partial\eta}{\partial z}\hat{k}$$

giving three equations of identical form

$$\frac{\mathrm{d}}{\mathrm{ds}}\left(\eta\frac{\mathrm{d}x}{\mathrm{d}s}\right) = \frac{\partial\eta}{\partial x} \tag{4.2.1}$$

CYLINDRICAL POLAR COORDINATES

We require the relations

$$\mathrm{d}\hat{\boldsymbol{\rho}} = -\hat{\boldsymbol{\theta}}\,\mathrm{d}\theta \qquad \mathrm{d}\hat{\boldsymbol{\theta}} = \hat{\boldsymbol{\rho}}\,\mathrm{d}\theta$$

Then with $\mathbf{r} = \rho\hat{\boldsymbol{\rho}} + z\hat{k}$

$$\eta\frac{\mathrm{d}\mathbf{r}}{\mathrm{d}s} = \eta\frac{\mathrm{d}\rho}{\mathrm{d}s}\hat{\boldsymbol{\rho}} - \eta\rho\frac{\mathrm{d}\theta}{\mathrm{d}s}\hat{\boldsymbol{\theta}} + \eta\frac{\mathrm{d}z}{\mathrm{d}s}\hat{k}$$

and differentiation gives

$$\frac{\mathrm{d}}{\mathrm{ds}}\left(\eta\frac{\mathrm{d}\rho}{\mathrm{d}s}\right) - \eta\rho\left(\frac{\mathrm{d}\theta}{\mathrm{d}s}\right)^2 = \frac{\partial\eta}{\partial\rho}$$

$$\eta\frac{\mathrm{d}\rho}{\mathrm{d}s}\frac{\mathrm{d}\theta}{\mathrm{d}s} + \frac{\mathrm{d}}{\mathrm{ds}}\left(\eta\rho\frac{\mathrm{d}\theta}{\mathrm{d}s}\right) \equiv \frac{1}{\rho}\frac{\mathrm{d}}{\mathrm{ds}}\left(\eta\rho^2\frac{\mathrm{d}\theta}{\mathrm{d}s}\right) = \frac{1}{\rho}\frac{\partial\eta}{\partial\theta}$$

$$\frac{\mathrm{d}}{\mathrm{ds}}\left(\eta\frac{\mathrm{d}z}{\mathrm{d}s}\right) = \frac{\partial\eta}{\partial z} \tag{4.2.2}$$

SPHERICAL POLAR COORDINATES

We have

$$\begin{aligned}
\mathrm{d}\hat{\mathbf{r}} &= \mathrm{d}\theta\hat{\boldsymbol{\theta}} + \sin\theta\,\mathrm{d}\phi\hat{\boldsymbol{\phi}} \\
\mathrm{d}\hat{\boldsymbol{\theta}} &= -\,\mathrm{d}\theta\hat{\mathbf{r}} + \cos\theta\,\mathrm{d}\phi\hat{\boldsymbol{\theta}} \\
\mathrm{d}\hat{\boldsymbol{\phi}} &= -\sin\theta\,\mathrm{d}\phi\hat{\mathbf{r}} - \cos\theta\,\mathrm{d}\phi\hat{\boldsymbol{\theta}}
\end{aligned}$$

giving

$$\frac{\mathrm{d}}{\mathrm{ds}}\left(\eta\frac{\mathrm{d}r}{\mathrm{d}s}\right) - \eta r\sin^2\theta\left(\frac{\mathrm{d}\phi}{\mathrm{d}s}\right)^2 - \eta r\left(\frac{\mathrm{d}\theta}{\mathrm{d}s}\right)^2 = \frac{\partial\eta}{\partial r} \tag{4.2.3}$$

$$\frac{\mathrm{d}}{\mathrm{ds}}\left(\eta r\frac{\mathrm{d}\theta}{\mathrm{d}s}\right) - \eta r\sin\theta\cos\theta\left(\frac{\mathrm{d}\phi}{\mathrm{d}s}\right)^2 + \eta\frac{\mathrm{d}r}{\mathrm{d}s}\frac{\mathrm{d}\theta}{\mathrm{d}s} = \frac{1}{r}\frac{\partial\eta}{\partial\theta}$$

$$\frac{\mathrm{d}}{\mathrm{ds}}\left(\eta r\sin\theta\frac{\mathrm{d}\phi}{\mathrm{d}s}\right) + \eta r\cos\theta\frac{\mathrm{d}\theta}{\mathrm{d}s}\frac{\mathrm{d}\phi}{\mathrm{d}s} + \eta\sin\theta\frac{\mathrm{d}r}{\mathrm{d}s}\frac{\mathrm{d}\phi}{\mathrm{d}s} = \frac{1}{r\sin\theta}\frac{\partial\eta}{\partial\phi}$$

THE GENERAL CYLINDRICAL MEDIUM

We consider a generalized cylindrical system specified by the transformation

$$x = f(u, v) \qquad y = h(u, v) \qquad z = z \tag{4.2.4}$$

where u v are orthogonal curvilinear coordinates in a plane perpendicular to the z-axis.

The metric coefficients specified by

$$\mathrm{d}s^2 = g_{11}\,\mathrm{d}u^2 + g_{22}\,\mathrm{d}v^2 + g_{33}\,\mathrm{d}z^2$$

are therefore

$$g_{11} = f_u^2 + h_u^2 \qquad g_{22} = f_v^2 + h_v^2 \qquad g_{33} = 1$$

where the suffices denote partial differentiation. For orthogonality we have

$$f_u f_v + h_u h_v = 0 \tag{4.2.5}$$

the product of the Cauchy-Riemann equations. Unit vectors $\hat{\mathrm{e}}_u$ $\hat{\mathrm{e}}_v$ $\hat{\mathrm{e}}_z$ are given in terms of basis Cartesian vectors $\hat{\mathrm{e}}_x$ $\hat{\mathrm{e}}_y$ $\hat{\mathrm{e}}_z$ by

$$\begin{aligned}
\hat{\mathrm{e}}_u &= \frac{1}{\sqrt{g_{11}}}(f_u\hat{\mathrm{e}}_x + h_u\hat{\mathrm{e}}_y) \\
\hat{\mathrm{e}}_v &= \frac{1}{\sqrt{g_{22}}}(f_v\hat{\mathrm{e}}_x + h_v\hat{\mathrm{e}}_y) \qquad \hat{\mathrm{e}}_z = \hat{\mathrm{e}}_z
\end{aligned}$$

and consequently

$$\begin{aligned}
\mathrm{d}\hat{\mathrm{e}}_u &= \frac{1}{\Delta}\sqrt{\frac{g_{22}}{g_{11}}}[-h_u(f_{uu}\,\mathrm{d}u + f_{uv}\,\mathrm{d}v) + f_u(h_{uu}\,\mathrm{d}u + h_{uv}\,\mathrm{d}v)]\hat{\mathrm{e}}_v \\
\mathrm{d}\hat{\mathrm{e}}_v &= \frac{1}{\Delta}\sqrt{\frac{g_{11}}{g_{22}}}[-f_v(h_{vv}\,\mathrm{d}v + h_{uv}\,\mathrm{d}u) + h_v(f_{vv}\,\mathrm{d}v + f_{uv}\,\mathrm{d}u)]\hat{\mathrm{e}}_u
\end{aligned}$$

where Δ is the Jacobian

$$\Delta = \begin{vmatrix} f_u & h_u \\ f_v & h_v \end{vmatrix}$$

Substituting in the general ray equation for a non-uniform medium with

$$\frac{\mathrm{d}\mathbf{r}}{\mathrm{d}s} = \sqrt{g_{11}}\frac{\mathrm{d}u}{\mathrm{d}s}\hat{\mathrm{e}}_u + \sqrt{g_{22}}\frac{\mathrm{d}v}{\mathrm{d}s}\hat{\mathrm{e}}_v + \sqrt{g_{33}}\frac{\mathrm{d}z}{\mathrm{d}s}\hat{\mathrm{e}}_z \tag{4.2.6}$$

$$\nabla\eta = \frac{1}{\sqrt{g_{11}}}\frac{\partial\eta}{\partial u}\hat{\mathrm{e}}_u + \frac{1}{\sqrt{g_{22}}}\frac{\partial\eta}{\partial v}\hat{\mathrm{e}}_v + \frac{1}{\sqrt{g_{33}}}\frac{\partial\eta}{\partial z}\hat{\mathrm{e}}_z \tag{4.2.7}$$

gives the completely general result

$$\frac{\mathrm{d}}{\mathrm{d}s}\left(\eta\sqrt{g_{11}}\frac{\mathrm{d}u}{\mathrm{d}s}\right)+\frac{\eta\sqrt{g_{11}}}{\Delta}\frac{\mathrm{d}v}{\mathrm{d}s}\left[-f_v\left(h_{vv}\frac{\mathrm{d}v}{\mathrm{d}s}+h_{uv}\frac{\mathrm{d}u}{\mathrm{d}s}\right)\right.$$
$$\left.+h_v\left(f_{vv}\frac{\mathrm{d}v}{\mathrm{d}s}+f_{uv}\frac{\mathrm{d}u}{\mathrm{d}s}\right)\right]=\frac{1}{\sqrt{g_{11}}}\frac{\partial\eta}{\partial u}$$
$$\frac{\mathrm{d}}{\mathrm{d}s}\left(\eta\sqrt{g_{22}}\frac{\mathrm{d}v}{\mathrm{d}s}\right)+\frac{\eta\sqrt{g_{22}}}{\Delta}\frac{\mathrm{d}u}{\mathrm{d}s}\left[-h_u\left(f_{uu}\frac{\mathrm{d}u}{\mathrm{d}s}+f_{uv}\frac{\mathrm{d}v}{\mathrm{d}s}\right)\right.$$
$$\left.+f_u\left(h_{uu}\frac{\mathrm{d}u}{\mathrm{d}s}+h_{uv}\frac{\mathrm{d}v}{\mathrm{d}s}\right)\right]=\frac{1}{\sqrt{g_{22}}}\frac{\partial\eta}{\partial v}$$
$$\frac{\mathrm{d}}{\mathrm{d}s}\left(\eta\sqrt{g_{33}}\frac{\mathrm{d}z}{\mathrm{d}s}\right)=\frac{1}{\sqrt{g_{33}}}\frac{\partial\eta}{\partial z} \tag{4.2.8}$$

THE GENERAL AXISYMMETRIC MEDIUM

A generalized axisymmetric coordinate system (α, β, ψ) can be specified by the transformation

$$x = F(\alpha,\beta)\cos\psi$$
$$y = F(\alpha,\beta)\sin\psi$$
$$z = H(\alpha,\beta)$$

where ψ is the polar angle of rotation about the z-axis and $F(\alpha, \beta)$ and $H(\alpha, \beta)$ are quite general functions of the coordinates (α, β). For the system to be orthogonal we thus require (product of the Cauchy-Riemann equations)

$$F_\alpha F_\beta + H_\alpha H_\beta = 0$$

(suffices denoting partial differentiation). The metric coefficients, defined by

$$\mathrm{d}s^2 = \mathrm{d}x^2 + \mathrm{d}y^2 + \mathrm{d}z^2 = g_{11}\,\mathrm{d}\alpha^2 + g_{22}\,\mathrm{d}\beta^2 + g_{33}\,\mathrm{d}\psi^2$$

are thus

$$g_{11} = F_\alpha^2 + H_\alpha^2 \qquad g_{22} = F_\beta^2 + H_\beta^2 \qquad g_{33} = F^2 \tag{4.2.9}$$

The unit vectors $\hat{\mathrm{e}}_\alpha$ $\hat{\mathrm{e}}_\beta$ $\hat{\mathrm{e}}_\psi$ are then given in terms of constant Cartesian basis vectors $\hat{\mathrm{e}}_x$ $\hat{\mathrm{e}}_y$ $\hat{\mathrm{e}}_z$ by

$$\hat{\mathrm{e}}_\alpha = \frac{\sqrt{F_\alpha^2 + H_\alpha^2}}{H_\alpha F_\beta - F_\alpha H_\beta}[-H_\beta\cos\psi\hat{\mathrm{e}}_x - H_\beta\sin\psi\hat{\mathrm{e}}_y + F_\beta\hat{\mathrm{e}}_z]$$
$$\hat{\mathrm{e}}_\beta = \frac{\sqrt{F_\beta^2 + H_\beta^2}}{H_\alpha F_\beta - F_\alpha H_\beta}[H_\alpha\cos\psi\hat{\mathrm{e}}_x + H_\alpha\sin\psi\hat{\mathrm{e}}_y - F_\alpha\hat{\mathrm{e}}_z]$$
$$\hat{\mathrm{e}}_\psi = -\sin\psi\hat{\mathrm{e}}_x + \cos\psi\hat{\mathrm{e}}_y$$

The factor $H_\alpha F_\beta - F_\alpha H_\beta$ occurs frequently in the ensuing analysis and we define for brevity

$$H_\alpha F_\beta - F_\alpha H_\beta \equiv K(\alpha,\beta)$$

Differentiation then gives

$$\mathrm{d}\hat{\mathrm{e}}_\alpha = -\frac{K}{H_\alpha\sqrt{F_\beta^2+H_\beta^2}}\mathrm{d}\left[\frac{H_\beta\sqrt{F_\alpha^2+H_\alpha^2}}{K}\right]\hat{\mathrm{e}}_\beta + \frac{H_\beta\sqrt{F_\alpha^2+H_\alpha^2}}{K}\mathrm{d}\psi\hat{\mathrm{e}}_\psi$$

$$\mathrm{d}\hat{\mathrm{e}}_\beta = -\frac{K}{H_\beta\sqrt{F_\alpha^2+H_\alpha^2}}\mathrm{d}\left[\frac{H_\alpha\sqrt{F_\beta^2+H_\beta^2}}{K}\right]\hat{\mathrm{e}}_\alpha + \frac{H_\alpha\sqrt{F_\beta^2+H_\beta^2}}{K}\mathrm{d}\psi\hat{\mathrm{e}}_\psi$$

$$\mathrm{d}\hat{\mathrm{e}}_\psi = \left[-\frac{F_\alpha}{\sqrt{F_\alpha^2+H_\alpha^2}}\hat{\mathrm{e}}_\alpha - \frac{F_\beta}{\sqrt{F_\beta^2+H_\beta^2}}\hat{\mathrm{e}}_\beta\right]\mathrm{d}\psi \tag{4.2.10}$$

Hence from Equation 4.2.6

$$\frac{\mathrm{d}\mathbf{r}}{\mathrm{d}s} = \sqrt{F_\alpha^2+H_\alpha^2}\frac{\mathrm{d}\alpha}{\mathrm{d}s}\hat{\mathrm{e}}_\alpha + \sqrt{F_\beta^2+H_\beta^2}\frac{\mathrm{d}\beta}{\mathrm{d}s}\hat{\mathrm{e}}_\beta + F\frac{\mathrm{d}\psi}{\mathrm{d}s}\hat{\mathrm{e}}_\psi \tag{4.2.11}$$

and from Equation 4.2.7

$$\nabla\eta = \frac{1}{\sqrt{F_\alpha^2+H_\alpha^2}}\frac{\partial\eta}{\partial\alpha}\hat{\mathrm{e}}_\alpha + \frac{1}{\sqrt{F_\beta^2+H_\beta^2}}\frac{\partial\eta}{\partial\beta}\hat{\mathrm{e}}_\beta + \frac{1}{F}\frac{\partial\eta}{\partial\psi}\hat{\mathrm{e}}_\psi \tag{4.2.12}$$

Differentiation of Equation 4.2.11 gives the final general result

$$\frac{\mathrm{d}}{\mathrm{d}s}\left[\eta\sqrt{F_\alpha^2+H_\alpha^2}\frac{\mathrm{d}\alpha}{\mathrm{d}s}\right] - \eta\frac{K}{H_\beta}\sqrt{\frac{F_\beta^2+H_\beta^2}{F_\alpha^2+H_\alpha^2}}\frac{\mathrm{d}}{\mathrm{d}s}\left[\frac{H_\alpha\sqrt{F_\beta^2+H_\beta^2}}{K}\right]\frac{\mathrm{d}\beta}{\mathrm{d}s}$$

$$-\eta\frac{FF_\alpha}{\sqrt{F_\alpha^2+H_\alpha^2}}\left(\frac{\mathrm{d}\psi}{\mathrm{d}s}\right)^2 = \frac{1}{\sqrt{F_\alpha^2+H_\alpha^2}}\frac{\partial\eta}{\partial\alpha}$$

$$\frac{\mathrm{d}}{\mathrm{d}s}\left[\eta\sqrt{F_\beta^2+H_\beta^2}\frac{\mathrm{d}\beta}{\mathrm{d}s}\right] - \eta\frac{K}{H_\alpha}\sqrt{\frac{F_\alpha^2+H_\alpha^2}{F_\beta^2+H_\beta^2}}\frac{\mathrm{d}}{\mathrm{d}s}\left[\frac{H_\beta\sqrt{F_\alpha^2+H_\alpha^2}}{K}\right]\frac{\mathrm{d}\alpha}{\mathrm{d}s}$$

$$-\eta\frac{FF_\beta}{\sqrt{F_\beta^2+H_\beta^2}}\left(\frac{\mathrm{d}\psi}{\mathrm{d}s}\right)^2 = \frac{1}{\sqrt{F_\beta^2+H_\beta^2}}\frac{\partial\eta}{\partial\beta}$$

$$\frac{\mathrm{d}}{\mathrm{d}s}\left(\eta F\frac{\mathrm{d}\psi}{\mathrm{d}s}\right) - \eta\frac{H_\beta(F_\alpha^2+H_\alpha^2)}{K}\frac{\mathrm{d}\alpha}{\mathrm{d}s}\frac{\mathrm{d}\psi}{\mathrm{d}s}$$

$$+\eta\frac{H_\alpha(F_\beta^2+H_\beta^2)}{K}\frac{\mathrm{d}\beta}{\mathrm{d}s}\frac{\mathrm{d}\psi}{\mathrm{d}s} = \frac{1}{F}\frac{\partial\eta}{\partial\psi} \tag{4.2.13}$$

From the orthogonality condition and the definition of K we find also that

$$F_\alpha^2+H_\alpha^2 = \frac{H_\alpha K}{F_\beta} \qquad F_\beta^2+H_\beta^2 = \frac{F_\beta K}{H_\alpha}$$

With circular symmetry, $\partial\eta/\partial\psi = 0$ and the last in Equation 4.2.13 reduces to

$$\frac{\mathrm{d}}{\mathrm{d}s}\left(\eta F\frac{\mathrm{d}\psi}{\mathrm{d}s}\right) + \eta F_\alpha \frac{\mathrm{d}\alpha}{\mathrm{d}s}\frac{\mathrm{d}\psi}{\mathrm{d}s} + \eta F_\beta \frac{\mathrm{d}\beta}{\mathrm{d}s}\frac{\mathrm{d}\psi}{\mathrm{d}s} = 0 \tag{4.2.14}$$

or

$$\frac{1}{F}\frac{\mathrm{d}}{\mathrm{d}s}\left(\eta F^2\frac{\mathrm{d}\psi}{\mathrm{d}s}\right) = 0$$

$\eta F^2(\mathrm{d}\psi/\mathrm{d}s) = \kappa$ is a ray constant called the "skew invariant" of the ray.

4.3 RAYS IN LINEAR MEDIA

We can now deal with those solutions of the ray equations which give rise to practical focusing devices or systems in which we need to derive the ray trajectory within a given medium. In the main our attention will be concentrated on the linear cylindrical and spherical coordinate systems, and on refractive indices in a realistic range, that is real and greater than unity (less than a value only a few times this). In microwave practice this range can be extended somewhat, although the densities required and the bulk of material rapidly becomes excessive in all but the very short wavelength range.

However, for purely theoretical purposes it is of great interest to consider the entire range of (real) values of refractive index, including values less than unity, zero and infinite. Values less than unity occur in practice in the microwave waveguide analogue of propagation of a non-TEM mode between parallel conducting plates as in Section 3.3. With such a range of values we can retain the concept of the "refractive space" which has particular ray properties which are subject to transformations and subsequently from which "real" regions can be selected to provide practical focusing devices. By retaining the complete refractive space, other theoretical considerations lead to strong interconnections with analogous problems in the other least-action subject, namely, particle trajectories in non-uniform potentials.

It is found, in general, that solutions of the ray equations are most easy to derive when the ray is considered to be confined to a surface for which one coordinate of the system has a constant value - that is, on meridional planes, planes containing the axis of symmetry in a cylindrical system or, of course, cylinders coaxial with the refractive system.

Some general laws can be enunciated at the outset. A ray can, and often does, at some point of its trajectory, become parallel to the general stratification of the medium. A ray can only achieve a direction perpendicular to the stratification when the refractive index is infinite. A ray which is perpendicular to the stratification at its source will thus remain so for its total trajectory. At an interface between two non-uniform media, the ray will obey Snell's law of refraction at each point, as though the media were infinite and with values obtaining at that point, the so-called "geometrical optics" approximation.

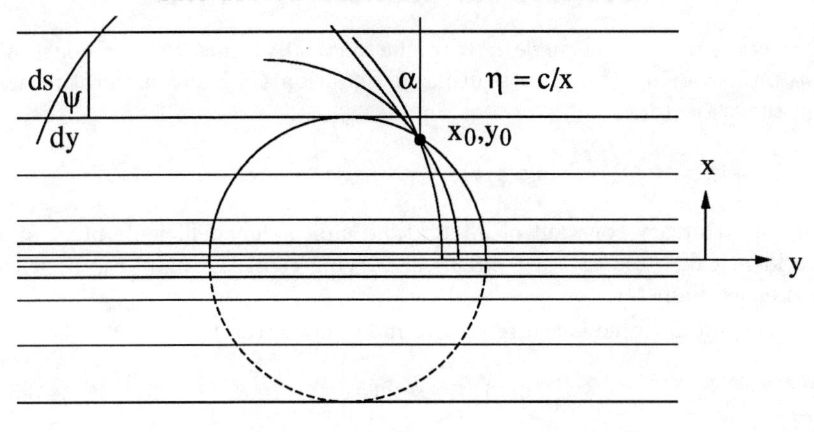

Figure 4.3: Circular rays in a medium with refractive index varying as c/x

Reflections of rays at such surfaces as is the common practice in optics are ignored.

THE LINEAR STRATIFIED MEDIUM

We consider first the stratification to depend upon the single coordinate x and the ray to lie in the (x, y) plane. This reproduces the conditions of Figure 4.3. Then $\mathrm{d}s^2 = \mathrm{d}x^2 + \mathrm{d}y^2$ and Equations 4.2.1 become, since $\eta = \eta(x)$

$$\frac{\mathrm{d}}{\mathrm{d}s}\left(\eta\frac{\mathrm{d}x}{\mathrm{d}s}\right) = \frac{\mathrm{d}\eta}{\mathrm{d}x}$$

$$\frac{\mathrm{d}}{\mathrm{d}s}\left(\eta\frac{\mathrm{d}y}{\mathrm{d}s}\right) = 0 \tag{4.3.1}$$

Both equations lead to the same result, on substituting $\mathrm{d}/\mathrm{d}s = \cos\psi\ (\mathrm{d}/\mathrm{d}x)$, namely

$$\eta\frac{\mathrm{d}y}{\mathrm{d}s} = \text{ constant } = h \text{ along a ray}$$

Since $\mathrm{d}y/\mathrm{d}s = \sin\psi$ (Figure 4.3) this is simply the continuation of Snell's law, $\eta\sin\psi = h$. Then

$$\mathrm{d}y = \frac{h\,\mathrm{d}x}{\sqrt{\eta^2 - h^2}} \quad \text{or} \quad y = \int\frac{h\,\mathrm{d}x}{\sqrt{\eta^2 - h^2}} \tag{4.3.2}$$

as was shown by Equation 4.1.15.

This trajectory can be treated on two levels, one given the refractive index to determine the ray path. For example, consider the law $\eta(x) = c/x$ and a ray originating at the point $x = x_0$, $y = 0$; at which point the refractive index is $\eta_0 = c/x_0$, making an angle α with the vertical. Hence the ray constant h is $\eta_0 \sin\alpha = c\sin\alpha/x_0$. Substituting in Equation 4.3.2 and integrating we obtain the trajectories

$$(y\eta_0 \sin\alpha + b)^2 = c^2 - \eta_0^2 x^2 \sin^2\alpha$$

b being an arbitrary constant of integration. Since the coefficients of x^2 and y^2 are identical, these rays are circles of varying radii through (x_0, y_0) with the y-axis as diameter.

The second example is the refractive index law given by

$$\eta(x) = \eta_0 \operatorname{sech}(ax) \tag{4.3.3}$$

giving

$$\mathrm{d}z = \frac{A\,\mathrm{d}x}{\sqrt{\eta_0^2 \operatorname{sech}^2(ax) - A^2}}$$

where, for rays originating from the origin making an angle α with the vertical

$$A = \eta_0 \sin\alpha$$

Using the trigonometric and hyperbolic identities converts this equation to

$$\mathrm{d}z = \frac{\sin\alpha \cosh(ax)\,\mathrm{d}x}{\sqrt{\cos^2\alpha - \sin^2\alpha \sinh^2(ax)}}$$

which is directly integrable to give

$$\sin(az) = \tan\alpha \sinh(ax)$$

the constant of integration being made zero by a choice of the position of the origin $z = 0$.

The rays in this case are continuous and repeatedly focus at points along the z-axis with separation $2f = \pi/a$ (Figure 4.4(a)).

The equation to the wave fronts is given as

$$\cos az = p \cosh ax$$

p being a variable parameter for the different wave fronts

The illustration here is of a second type of focusing which is not of the closed loop type. In this case it is periodic and at an infinite number of points along a straight line. If the origin of rays were to be displaced from the axis of symmetry two lines of images are formed as shown in Figure 4.4(b) on opposite sides of the axis.

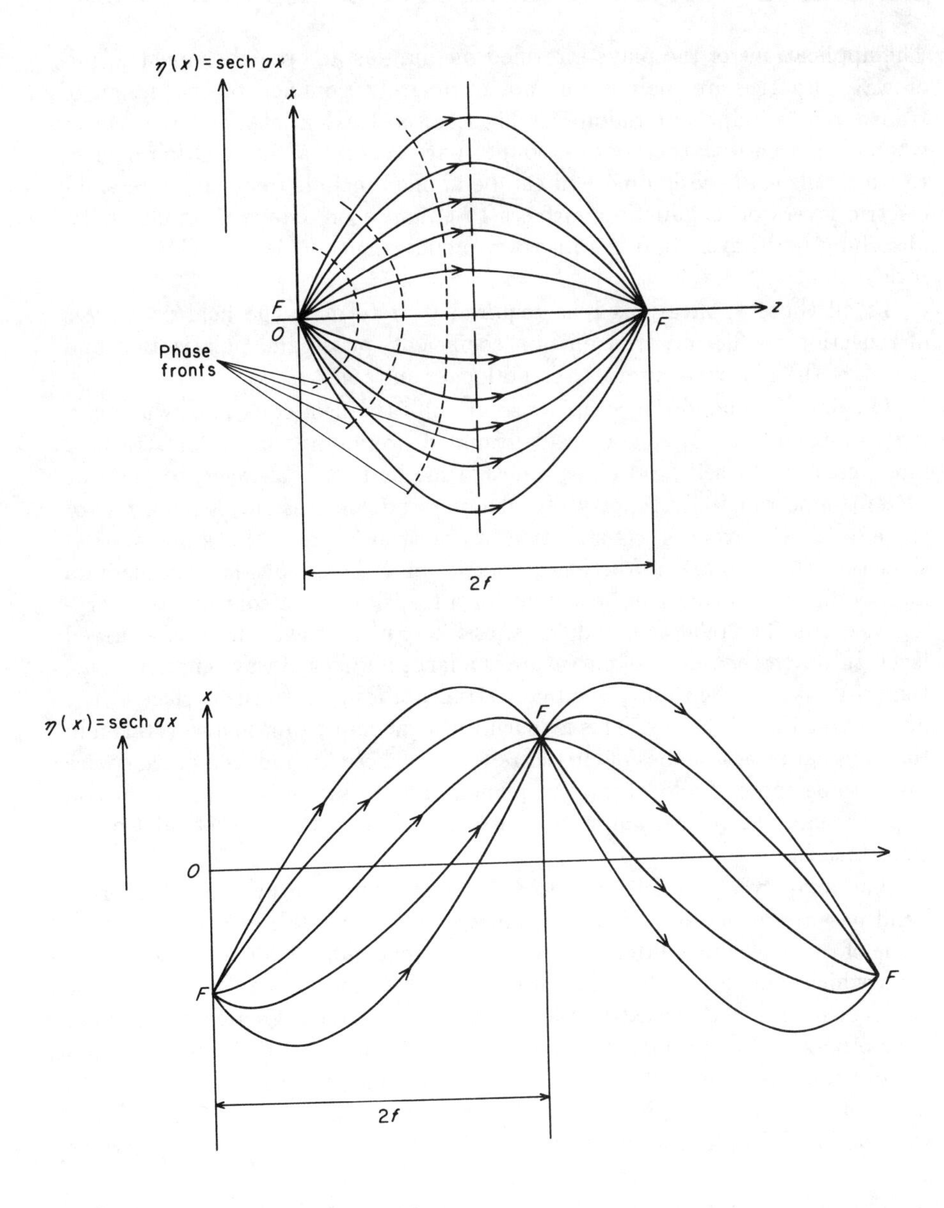

Figure 4.4: (a) Ray paths and phase fronts in a medium with sech law of refractive index [6] (b) Ray paths in a medium with sech law of refractive index focus displaced from axis [7]

4.4 PLANE MULTILAYER STRUCTURES AND RADOMES

The applications of the plane stratified medium as an optical element in microwave practice are mainly, but not exclusively, confined to the design of transparent windows or radomes. These consist essentially of few surfaces when compared with their optical counterpart the optical filter. Other devices to which the same design method can be applied include frequency filters, dielectric layers on conducting surfaces to enhance or reduce the reflectivity, absorbing multilayers and the microwave equivalent of the half-silvered mirror.

In all these applications it is required to determine the field properties of reflection coefficient, transmission coefficient, phase and polarization and therefore the ray optic approach is no longer suitable.

On the other hand the complete solution of the field equations, while giving the exact solution for a pre-determined non-homogeneous or stratified layer, cannot of itself be used as a design method for that layer, to give the system some required property of reflection or transmission. Having regard therefore to the eventual construction of practical layers of this kind, we consider the reflection and refraction properties of a plane multi-layered medium and modifications that can be sought for a transition to a continuously varying medium. The reflection and transmission of plane waves through a curved layer, in which the radius of curvature is a large number of wavelengths is customarily taken to be that of the plane structure tangent to the surface at the point of incidence of a ray. This is much the same approximation as is taken in the physical optics studies of curved reflectors where the induced currents are taken to be those of a local tangent plane. The validity of this approximation in the case of curved radomes has not been tested either theoretically or by experiment.

The analogy of a microwave multilayer radome with an optical filter as band-pass frequency devices is well-known. The optical filter is also an analogue of the wave guide filter which itself is a frequency analogue of the spatial form which compares diffraction gratings with linear arrays. Some of these subjects have been treated in the literature to a much greater depth than have others and a considerable body of mathematical method has developed around some of them, which by virtue of the analogies, could be applied to many of the others. Readers may enjoy discovering for themselves those areas where a large fund of theory can be readily interchanged between the various problems.

The method to be employed in the following is the standard procedure in optics [2, pp57-65] of multiplying the transmission matrices of the individual layers that go to make up the multilayered structure. It is considered to be well-known that complete transmission does not occur unless the multi-layer is fundamentally symmetrical (unless each layer is separately reflectionless). Advantage can be taken of this symmetry to simplify certain of the procedures.

The degrees of freedom available then increase by one for each symmetrically placed pair of surfaces. Since the three layer structure is then the simplest to consider, we illustrate the different uses to which the additional degree of freedom can be put, with variations on that system.

To make the transmission to a smoothly varying medium the same transmission matrix method can be used. However, it can be shown that the usual procedure whereby the smoothly varying medium is approximated by a series of infinitesimally small steps, is only amenable to this treatment if the resulting "infinitesimal" matrices are correctly multiplied. The approximate multiplication of the matrices gives a result which is only valid for a layer whose *entire* overall optical width is only a small fraction of a wavelength. This is the Born approximation. The proper multiplication of the matrices is highly complex and greatly reduces the benefit that is to be derived from the process.

An improvement of this matrix method is presented which gives an exact result in the limiting process and is applicable to a layer of any width of a plane stratified medium. This proves to be analogous to the WJKB approximation.

The procedure whereby losses of the materials and conductivity can be included in the analysis is standard even if mathematically complex, and some new forms of radome structure can be assessed by these general methods.

SINGLE LAYER TRANSMISSION MATRIX

UNIFORM PLANE LAYER

The transmission matrix for a uniform layer of finite width of homogeneous dielectric material, is determined by the solution for the continuity of the field components at the two surfaces of the layer [2, pp57-65]. This method is standard in most text books on electromagnetic theory. In terms of this matrix the fields on the incident side of the layer, for a normally incident plane wave in free space, are given by

$$\begin{pmatrix} E_{inc} \\ H_{inc} \end{pmatrix} = \begin{pmatrix} \cos\beta d & \frac{i\omega\mu}{\beta}\sin\beta d \\ \frac{i\beta}{\omega\mu}\sin\beta d & \cos\beta d \end{pmatrix} \begin{pmatrix} 1 \\ \frac{\gamma'}{i\omega\mu'} \end{pmatrix} \tag{4.4.1}$$

where $\beta = \omega\sqrt{\mu\varepsilon}$ is the propagation constant for a plane wave in an infinite medium of the same material as the layer and for increased generality the medium on the transmission side (Figure 4.5) has been made complex with medium constants ε' μ' and propagation constant γ'.

For oblique incidence we have to distinguish between the two polarization states of the wave, perpendicular when the electric vector is perpendicular to the plane containing the wave and the surface normals (the plane of incidence) and parallel when the magnetic vector is perpendicular to this plane and hence has the electric vector parallel to it.

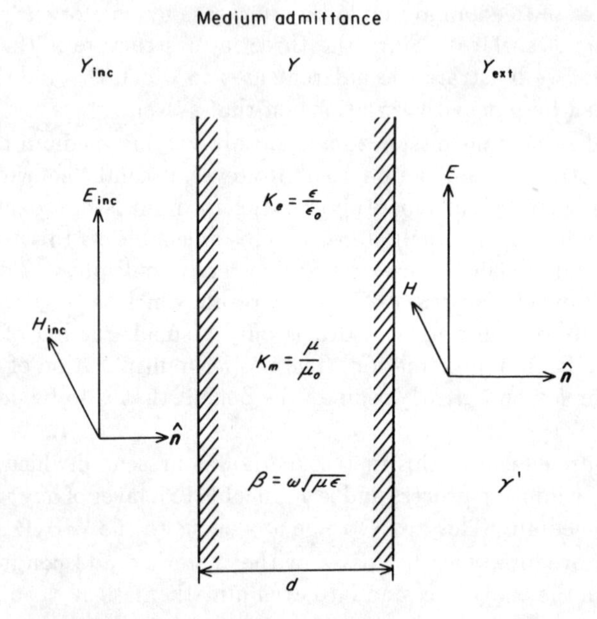

Figure 4.5: Plane wave normally incident upon a uniform layer

The relative permeability K_e and permeability K_m are defined by $K_e = \varepsilon/\varepsilon_0$ $K_m = \mu/\mu_0$ and the media admittance by $Y = \sqrt{\varepsilon/\mu}$; $Y_0 = \sqrt{\varepsilon_0/\mu_0}$ where ε_0 and μ_0 are the free space constants.

Then Equation 4.4.1 can be written

$$\begin{pmatrix} E_{inc} \\ H_{inc} \end{pmatrix} = \begin{pmatrix} \cos A & \frac{i}{Y}\sin A \\ iY\sin A & \cos A \end{pmatrix} \begin{pmatrix} 1 \\ Y_{ext} \end{pmatrix} \tag{4.4.2}$$

where A is the phase angle

$$A = \frac{2\pi d}{\lambda}\sqrt{K_e K_m - \sin^2\theta}$$

For a perpendicularly polarized incident wave Y becomes

$$Y_\perp = \frac{Y_0\sqrt{K_e/K_m - \sin^2\theta}}{\cos\theta}$$

and for a parallel polarized wave Y becomes

$$Y_\| = \frac{Y_0(K_e/K_m)\cos\theta}{\sqrt{(K_e/K_m) - \sin^2\theta}} \tag{4.4.3}$$

Y_{ext} is the admittance of the exterior medium on the transmission side.

Assuming temporarily for complete generality that the admittance of the medium on the incident side is Y_{inc} then the reflection coefficient [10] referred to the incident surface is

$$\rho = \frac{Y_{inc}E_{inc} - H_{inc}}{Y_{inc}E_{inc} + H_{inc}} \tag{4.4.4}$$

and the transmission coefficient referred to the *same* surface is

$$\tau = \frac{2Y_{inc}\sqrt{Y_{ext} + Y_{ext}^*}}{(Y_{inc}E_{inc} + H_{inc})\sqrt{Y_{inc} + Y_{inc}^*}} \tag{4.4.5}$$

where $*$ refers to complex conjugate.

In most of the cases to be considered, the incident and external media will both be free space and the relations 4.4.4 and 4.4.5 simplify to

$$\rho = \frac{Y_0E_{inc} - H_{inc}}{Y_0E_{inc} + H_{inc}} \qquad \tau = \frac{2Y_0}{Y_0E_{inc} + H_{inc}} \tag{4.4.6}$$

This result can be simplified further since the magnitude of the reflection and transmission coefficients are obviously only dependent upon the *relative* admittance between the media. Thus defining the relative admittance of the homogeneous layer by

$$Y' = Y/Y_0$$

the *same* result as in Equation 4.4.1 can be obtained from the transmission matrix

$$\begin{pmatrix} E_{inc} \\ H_{inc} \end{pmatrix} = \begin{pmatrix} \cos A & \frac{i}{Y'}\sin A \\ iY'\sin A & \cos A \end{pmatrix} \begin{pmatrix} 1 \\ 1 \end{pmatrix} \tag{4.4.7}$$

and the coefficients

$$\rho = \frac{E_{inc} - H_{inc}}{E_{inc} + H_{inc}} \qquad \tau = \frac{2}{E_{inc} + H_{inc}}$$

Throughout the foregoing the medium has been considered to be lossless. Such a loss can be introduced by the modification of the various parameters to give K_e complex values that is

$$\begin{aligned} K_e = K_e' - iK_e'' &= K_e(1 - i\tan\delta_e) \\ &= \frac{K_m}{\varepsilon_0}\left(\varepsilon - \frac{i\sigma}{\omega}\right) \end{aligned} \tag{4.4.8}$$

where σ is the conductivity of the medium and $\tan\delta_e$ the dielectric loss factor. For a highly conductive medium the approximation for large σ may be made in Equation 4.4.8 to give

$$\sqrt{K_e} = \sqrt{\frac{\mu\sigma}{2\omega\mu_0\varepsilon_0}}(1 - i) \tag{4.4.9}$$

In this case $Y_{\parallel}$ and $Y_{\perp}$ become complex, as do the propagation constants. The trigonometric functions in the transmission matrix then have to be replaced by hyperbolic functions. An application of this result is given later in this section for the radome with central conducting layer (see page 261).

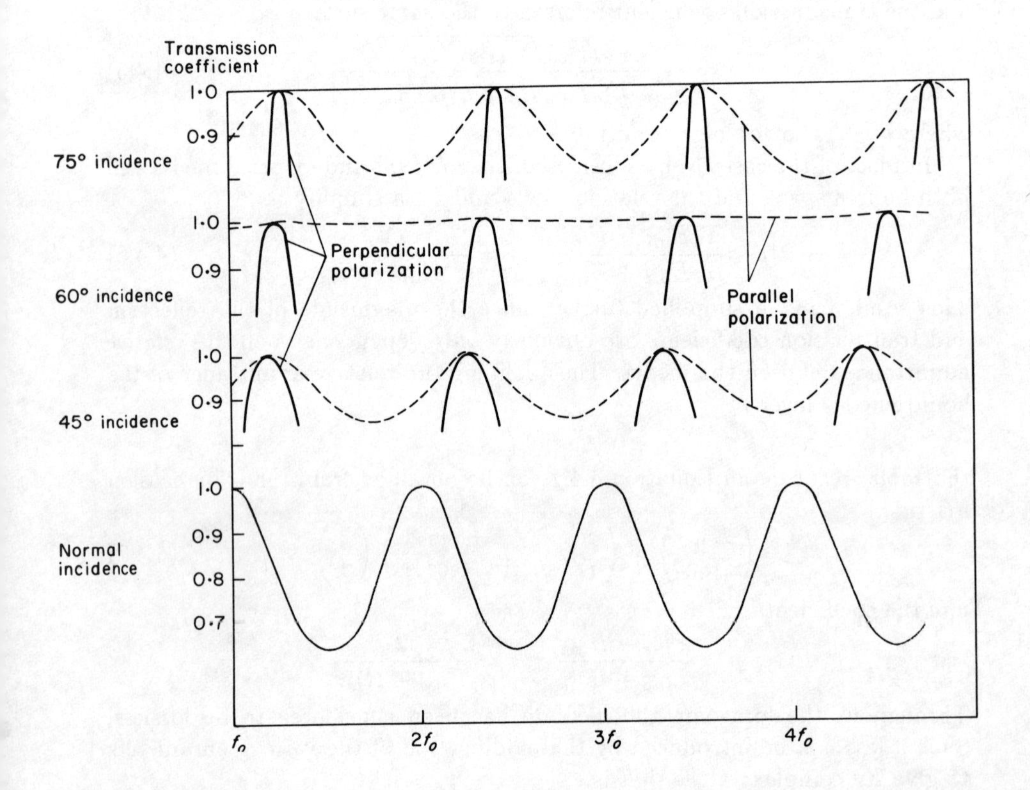

Figure 4.6: Periodicity of transmission maxima for uniform layer

THE REFLECTION COEFFICIENT OF A SINGLE HOMOGENEOUS LAYER

The properties of reflection and transmission through a single homogeneous layer of width d are summarized here for future reference when comparing the limiting effects in otherwise non-uniform layers. From Equations 4.4.6 or 4.4.7

$$\rho = \frac{i \sin A[(1/Y') - Y']}{2\cos A + i \sin A[(1/Y') + Y']}$$

$$\tau = \frac{2}{2\cos A + i \sin A[(1/Y') + Y']} \tag{4.4.10}$$

These values are periodic with frequency, being respectively zero and unity for those values of A for which

$$A = \frac{2\pi d}{\lambda}\sqrt{K_e - \sin^2\theta} = n\pi$$

The lowest order for a reflectionless layer is thus the value of d for $n = 1$ namely

$$d = \frac{\lambda}{2\sqrt{K_e - \sin^2\theta}} \tag{4.4.11}$$

It is however, the effect of oblique radiation that limits the operational frequency band of a panel of this type. If such a radome is to accommodate all angles of incidence then both polarization states have to be considered. It is then found that the frequency of maximum transmission for the two states become separate from that for normal incidence, as illustrated in Figure 4.6. It thus becomes impossible to obtain a single frequency with maximum transmission over anything but a narrow range of incidence angles.

This effect can be demonstrated by plotting

$$|\tau|^2 = \frac{4}{4\cos^2 A + [Y' + (1/Y')]^2 \sin^2 A}$$

in a polar (τ, A) coordinate system (Figure 4.7).

For normal incidence the graph is an ellipse. With oblique incidence both the minor axis and the radius vector are scaled by a factor $Y' + (1/Y')$ and the periodicity of A changes. Thus the ellipse separates into two systems, one of increasing eccentricity and the other with decreasing. The former corresponds to perpendicular polarization and the latter to parallel polarization, and which becomes circular at the Brewster angle of the medium. At the same time, because of the change of periodicity, successive orbits of the ellipse with change of frequency rotates the major axis giving an effect akin to the advance of the perihelion in a planetary orbit. This causes the maxima of the transmission coefficients to disperse from their original first order agreement with the normal incidence ellipse.

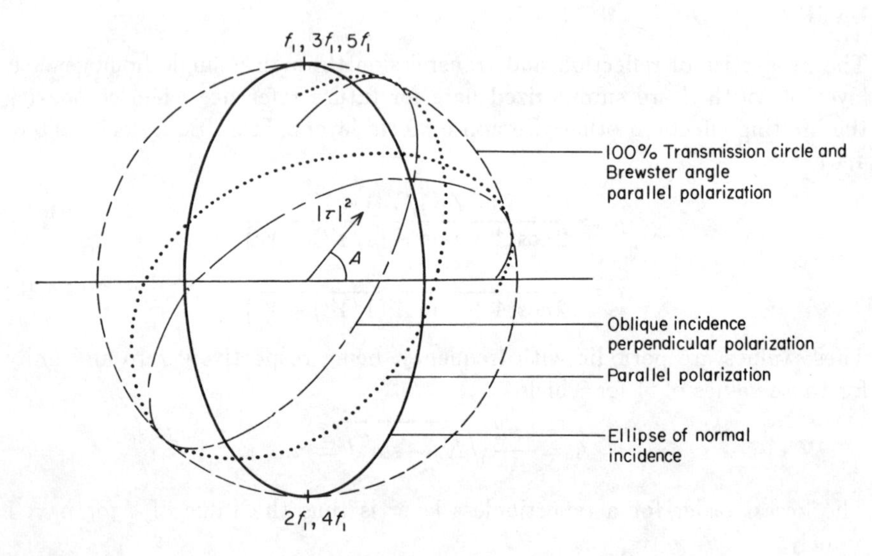

Figure 4.7: Polar coordinate representation of power transmission through uniform layer

THE THREE LAYER RADOME

The properties of multilayer radomes can be obtained directly by the product of the respective transmission matrices of each layer. Outside the basic half-wave homogeneous layer, the first extension giving a degree of freedom that can be applied to improvement of performance, is the symmetrical structure of three layers. The basic three layer radome, called a sandwich radome, is designed for complete transmission at a specified angle of incidence in one of the polarization states, usually the perpendicular, since the parallel state still gives a useful Brewster angle effect and can usually be left to take care of itself [9].

For the structure shown in Figure 4.8, M and K are the relative admittances of the layers whose refractive indices are $\sqrt{K_{e1}}$ and $\sqrt{K_{e2}}$ respectively. Then from Equation 4.4.7 the fields on the incident side are given by

$$\begin{pmatrix} E_{inc} \\ H_{inc} \end{pmatrix} = \begin{pmatrix} \cos\alpha & \frac{i}{K}\sin\alpha \\ iK\sin\alpha & \cos\alpha \end{pmatrix}$$
$$\begin{pmatrix} \cos\beta & \frac{i}{M}\sin\beta \\ iM\sin\beta & \cos\beta \end{pmatrix} \begin{pmatrix} \cos\alpha & \frac{i}{K}\sin\alpha \\ iK\sin\alpha & \cos\alpha \end{pmatrix} \begin{pmatrix} 1 \\ 1 \end{pmatrix} \qquad (4.4.12)$$

where

$$\alpha = \frac{2\pi d_1}{\lambda}\sqrt{K_{e1} - \sin^2\theta} \quad \text{and} \quad \beta = \frac{2\pi d_2}{\lambda}\sqrt{K_{e2} - \sin^2\theta}$$

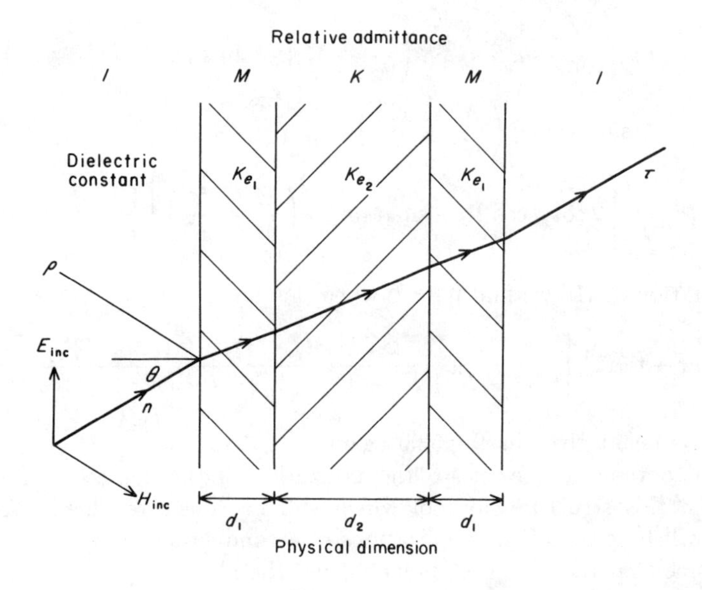

Figure 4.8: Three layer symmetrical sandwich

For perpendicularly polarized waves, $K_\perp = \sqrt{K_{e1} - \sin^2\theta}/\cos\theta$ and for parallel polarization, $K_\| = K_{e1}\cos\theta/\sqrt{K_{e1} - \sin^2\theta}$ with similar relations for $M_\perp$ and $M_\|$.

The reflection coefficient is then found to be

$$\begin{aligned}|\rho|^2 = {} & \left[\sin\beta\cos^2\alpha\left(\frac{1}{M} - M\right) + \sin 2\alpha\cos\beta\left(\frac{1}{K} - K\right)\right. \\ & \left. - \sin^2\alpha\sin\beta\left(\frac{M}{K^2} - \frac{K^2}{M}\right)\right]^2 \Big/ \\ & \left[\left\{2\cos\beta\cos 2\alpha - \sin\beta\sin 2\alpha\left(\frac{K}{M} + \frac{M}{K}\right)\right\}^2\right. \\ & + \left\{\sin\beta\cos^2\alpha\left(\frac{1}{M} + M\right) + \sin 2\alpha\cos\beta\left(\frac{1}{K} + K\right)\right. \\ & \left.\left. - \sin^2\alpha\sin\beta\left(\frac{M}{K^2} + \frac{K^2}{M}\right)\right\}^2\right] \end{aligned} \tag{4.4.13}$$

and the phase of the transmission coefficient referred to the incident face

$$\arg \tau = \tan^{-1} \left\{ -\sin\beta\cos^2\alpha \left(\frac{1}{M} + M \right) - \sin 2\alpha \cos\beta \left(\frac{1}{K} + K \right) \right.$$
$$+ \sin^2\alpha \sin\beta \left(\frac{M}{K^2} + \frac{K^2}{M} \right)$$
$$\left. \Big/ \left[2\cos\beta\cos 2\alpha - \sin\beta\sin 2\alpha \left(\frac{K}{M} + \frac{M}{K} \right) \right] \right\} \tag{4.4.14}$$

From Equation 4.4.13 we find $|\rho| = 0$ when

$$\beta = n\pi - \tan^{-1} \left[\frac{2KM(K^2-1)\sin 2\alpha}{(M^2-K^2)(K^2+1) + (M^2+K^2)(K^2-1)\cos 2\alpha} \right] \tag{4.4.15}$$

which is the result given in [9, p533] again.

Several options now exist for the utilization of the degrees of freedom inherent in this structure, four of which will be considered here. A final example will be given of the applications of a conducting centre layer which modify somewhat the above relations but use the identical process.

We consider the following cases

1. A three-layer structure which is reflectionless at normal incidence for two frequencies f_0 and nf_0 and with $K_{e1} > K_{e2}$ (this type of structure is commonly called an A sandwich).
2. A structure reflectionless at two frequencies and for which $K_{e2} = K_{e1}^2$ (a B sandwich).
3. A structure which at a given frequency is reflectionless both for normal incidence and for oblique incidence of both polarizations.
4. A structure which is reflectionless at normal incidence and has zero phase delay at a given frequency.

Finally we will consider a three layer structure in which the centre layer is an extremely thin film of conducting material. These cases, as will be seen subsequently, do not exhaust all the possibilities inherent in even a three layer structure.

THE A SANDWICH

A double frequency design can be simply effected by choosing d_1 and K_{e1} to be those values for which the layer in isolation would be a half-wave reflectionless sheet at the higher frequency and then, with a value of K_{e2} sufficiently near to unity say $1 < K_{e2} < 1.2$, the three layer structure can be made reflectionless at the second frequency by the use of Equation 4.4.15. With a low value of

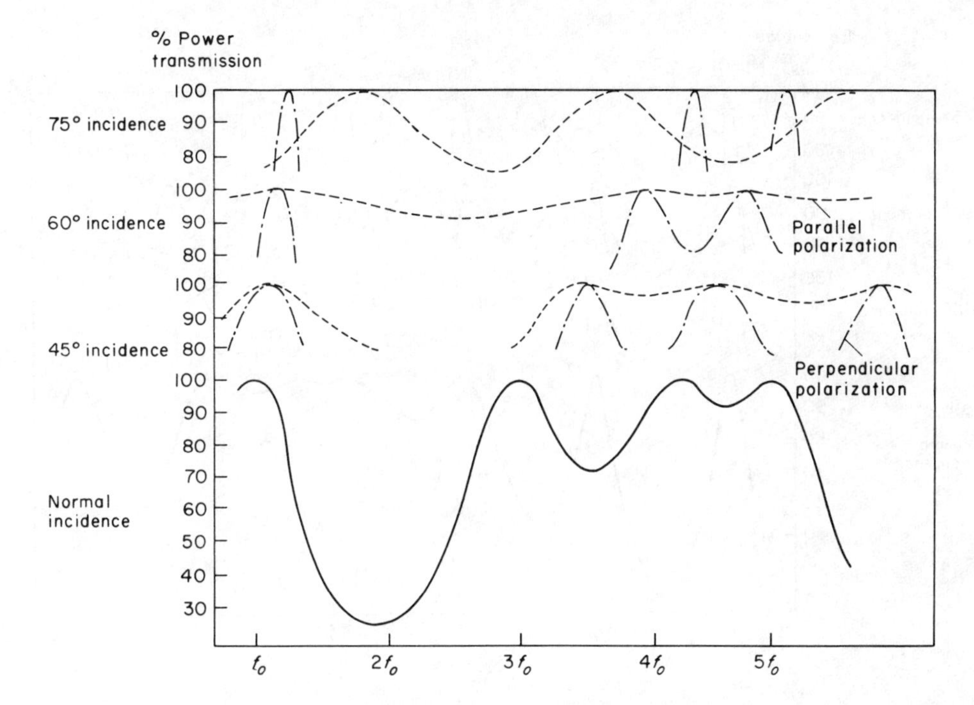

Figure 4.9: Three layer A sandwich computed transmission curves. The sandwich designed for 100% transmission at f_0 and $5f_0$

K_{e2} little or no redesign of the outer layers is found to be required. The results for two frequencies f_0 and $5f_0$ in which $K_{e1} = 4$ and $K_{e2} = 1.1$ are shown in Figure 4.9. Typical effects can be noted, that is additional "windows" in the frequency bands at $3f_0$ and $4f_0$, but an unexplained destructive interference at $2f_0$. A Brewster angle effect exists at the angle appropriate to the material of the exterior layers, and a similar separation of the frequencies of the maximum transmission peaks, with angle of incidence occurs as in the homogeneous layer. This latter effect makes it extremely difficult to obtain reflectionless operation over a wide range of incidence angles at any one of the selected frequencies.

THE B SANDWICH

Choosing the same two frequencies as in the previous example, we make the outer layers equivalent to 1/4 wavelength matching surfaces to the high refractive index centre layer [2, pp57-65]. That is $K_{e2} = K_{e1}^2$ and each surface of the centre layer is made to be independently reflectionless at the higher frequency. The centre thickness can then be chosen quite arbitrarily and can

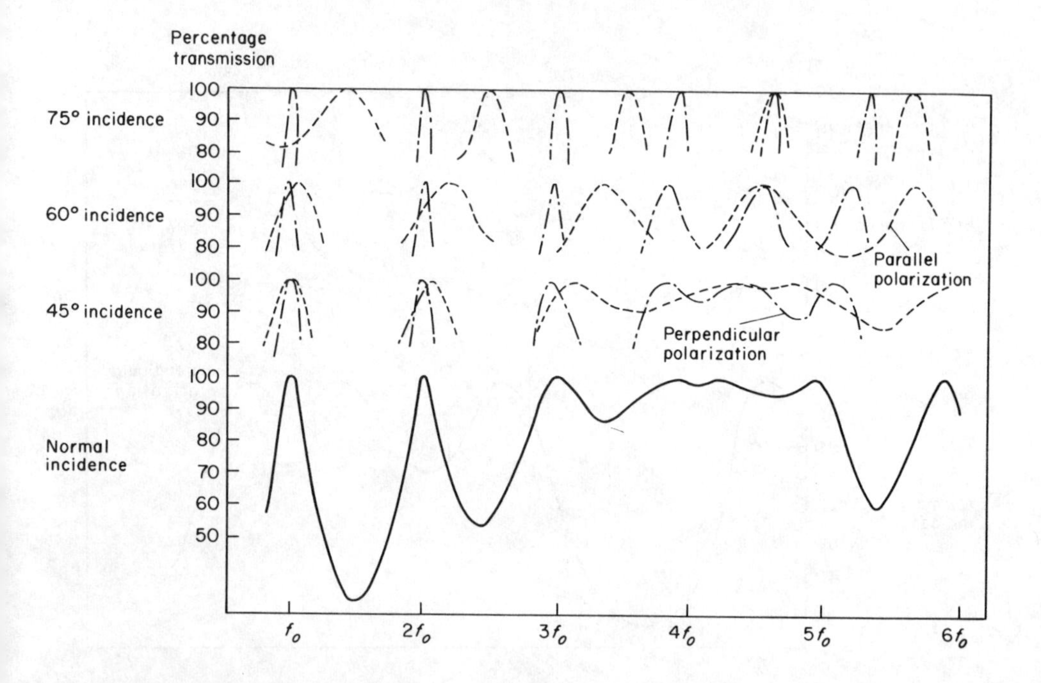

Figure 4.10: Theoretical transmission curves for three layer B sandwich designed for 100% transmission at f_0 and $5f_0$

thus be chosen so that the overall *equivalent* layer is a half-wave layer at the lower frequency. For parameters $K_{e2} = 16$, $K_{e1} = 4$ the transmission peaks are shown in Figure 4.10. Notable differences have now arisen between this result and the previous one. Transmission bands now occur at many more multiples of f_0 (up to six were encountered in the range investigated) and at the higher frequency chosen two of these coalesce to give a broad band of operation. The peaks in themselves although narrower than for the A sandwich, remain fixed in frequency with change of incidence angle for a wide range of the latter, and there is absence of any Brewster angle effect.

RADOME WITH ZERO INSERTION PHASE DELAY [9]

At any given angle of incidence the insertion phase delay is the phase difference at the transmission surface between an unimpeded plane wave and the wave in the presence of the layer. This then is

$$\psi = \arg \tau - \frac{2\pi}{\lambda}(2d_1 + d_2)\cos\theta \tag{4.4.16}$$

where arg τ is given in Equation 4.4.14.

For this to be made zero it is intuitively obvious that the centre medium must be a "phase-advance" material if the outer layers of dielectric are "phase-retarding". Such a material exists in the form of the two-dimensional array of square waveguides such as is used for the microwave waveguide lenses discussed in Chapter 3. The obvious advantage of such a construction is in those applications where the insertion phase of the radome is a major cause of beam distortion or displacement. The use of a waveguide medium changes the description of the relative admittance since for the constrained field the admittance will be the same both for perpendicular and parallel polarization. That is for symmetrical waveguides of square or circular cross-section and approximately for hexagonal cross-section

$$M_{\parallel} = M_{\perp} = \sqrt{1 - (\lambda/\lambda_c)^2}$$

and

$$\beta = \frac{2\pi d_2}{\lambda}\sqrt{1 - (\lambda/\lambda_c)^2}$$

where λ is the cut-off frequency for the dominant mode of the waveguide. Typical values for M in this case would be in the range 0.5 to 0.75 and for K_{e1} approximately 4. Simultaneous solution of Equation 4.4.16 with $\psi = 0$ and Equation 4.4.15 is now possible. With the choice of $M = 0.6$ $K_{e1} = 4$ this is found to give identical values for α and β namely $\alpha = \beta = 28.6^o$. This gives $d_1 = 0.1192$ cm $d_2 = 0.3986$ cm, at a design wavelength of $\lambda = 3$ cm. The result (Figure 4.11) shows that a very broad band radome results and that over the larger part of this band and for a wide range of incidence angles the insertion phase delay can be kept within ± 1 radian (Figure 4.12).

The applicability of a medium with relative admittance less than unity gives rise to a further interesting phenomenon. We note that with M greater than unity the denominator in Equation 4.4.15 has a real root at which point an integral multiple of π has to be added to obtain a solution for β. Physically this means that as the surface layers become infinitely thin the three layer A sandwich structure tends towards a half wave wall of the material of the centre layer. If this refractive index is very close to unity, this half-wave width can be nearly as large as a half wavelength of the radiation itself, and hence does not tend to zero, which would be the most logical result for a vanishing reflectivity overall.

With the waveguide medium, M is less than unity, and this condition no longer arises (Figure 4.13). With increasingly thin surface layers, the central core becomes equally thin and there arises a system of very thin structures which are reflectionless. We then observe that the frequency band of these structures can be extended beyond the cut-off frequency for the waveguide medium. This makes both M and β in Equation 4.4.17 complex, but since $\sin\beta$, and either M or $1/M$ always appear as a product, the result remains real with some changes of sign. Then for wavelengths far beyond cut-off β tends to a limit value, M tends to infinity and the reflection coefficient tends

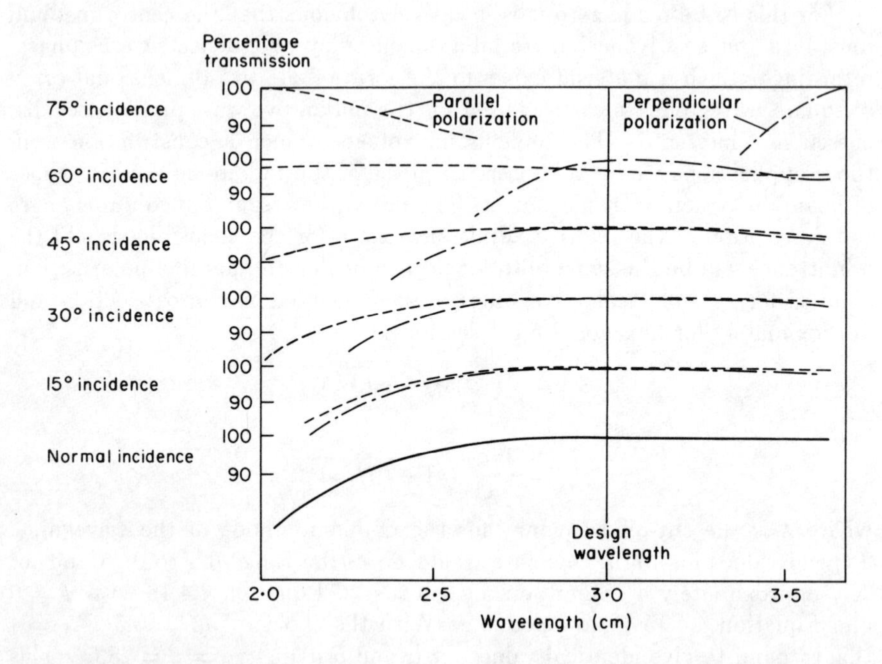

Figure 4.11: Theoretical transmission curves for waveguide sandwich $K_{e1} = 4$ $M = 0.6$ $d_1 = 0.12$ cm $d_2 = 0.4$ cm

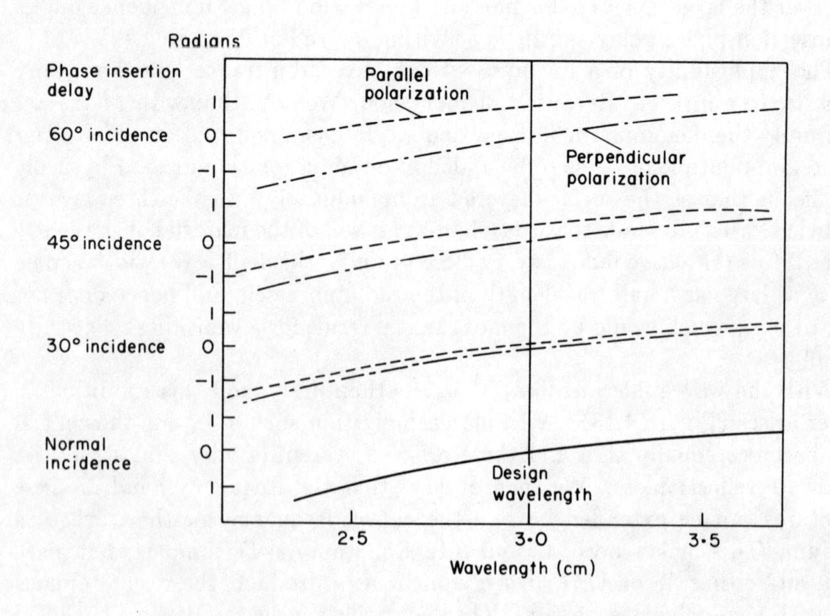

Figure 4.12: Phase insertion of waveguide sandwich

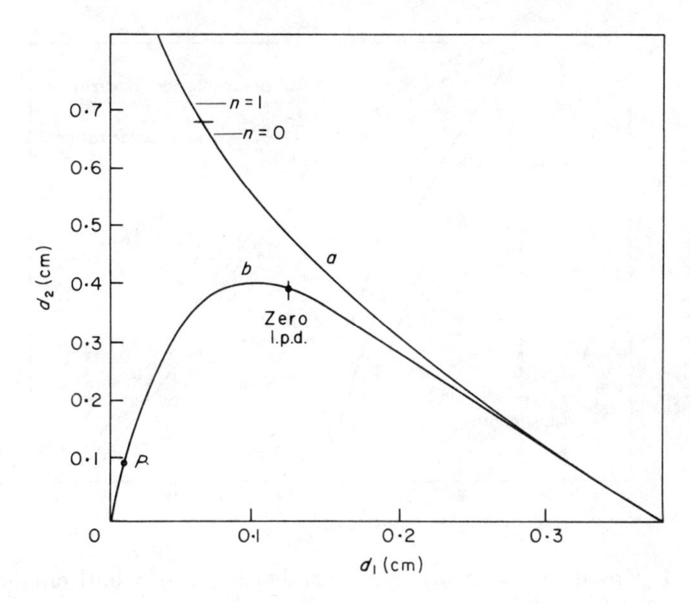

Figure 4.13: Dimensions for reflectionless sandwich constructions. (a) A sandwich $K_{e1} = 4$ $K_{e2} = 1.2$ (b) waveguide sandwich $K_{e1} = 4$ $M = 0.6$

to unity. The rate at which it does so, is determined naturally by the thickness of the cut-off medium, and this as has been shown, can be made as small as required.

Computation shows that for a sandwich of very small dimension the reflection loss can be kept at less than 10% over a 10:1 frequency band. This requires, at a frequency of 3 cm, a core thickness of 1 mm and skin thickness of 0.1 mm P in Figure 4.13. A radome of double these dimensions has approximately half the bandwidth. The theoretical transmission is shown in Figure 4.14.

THE TWO ANGLE RADOME

Equation 4.4.10 gives the condition for zero reflectivity in terms of the four parameters α β M and K each of which appears in the relation in the form appropriate to a given angle of incidence and prescribed polarization.

The equation is soluble for these parameters under a variety of a priori conditions. We choose for this example a radome which besides being reflectionless at normal incidence is reflectionless for *both* polarizations at one other angle of incidence at a single given frequency. This could be generalized to be a reflectionless condition for both polarizations at two prescribed angles of incidence.

Using the subscripts 0 $\perp$ and $||$ to indicate normal incidence, oblique in-

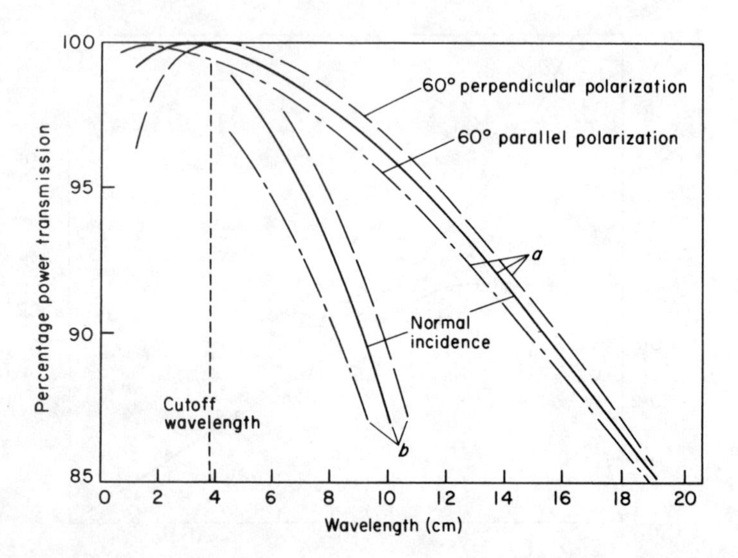

Figure 4.14: Broadband thin waveguide sandwich (a) $d_1 = 0.01$ cm $d_2 = 0.10$ cm (b) $d_1 = 0.02$ cm $d_2 = 0.185$ cm

cidence with perpendicular polarization and oblique incidence with parallel polarization the latter at a fixed incidence angle θ, we require to satisfy three conditions

$$\tan\beta_0 = \frac{-2K_0 M_0(K_0^2 - 1)\sin 2\alpha_0}{(M_0^2 - K_0^2)(K_0^2 + 1) + (M_0^2 - K_0^2)(K_0^2 - 1)\cos 2\alpha_0} \tag{4.4.17}$$

$$\tan\beta_\theta = \frac{-2K_\parallel M_\parallel(K_\parallel^2 - 1)\sin 2\alpha_\theta}{(M_\parallel^2 - K_\parallel^2)(K_\parallel^2 + 1) + (M_\parallel^2 + K_\parallel^2)(K_\parallel^2 - 1)\cos 2\alpha_\theta} \tag{4.4.18}$$

and

$$\tan\beta_\theta = \frac{-2K_\perp M_\perp(M_\perp^2 - 1)\sin 2\alpha_\theta}{(M_\perp^2 - K_\perp^2)(K_\perp^2 + 1) + (M_\perp^2 + K_\perp^2)(K_\perp^2 - 1)\cos 2\alpha_\theta} \tag{4.4.19}$$

From Equation 4.4.3 we have $Y_\parallel = K_e/Y_\perp$ for *all* angles of incidence. Using this relation for $M_\parallel$ and $K_\parallel$ the equality of Equations 4.4.20 and 4.4.21 gives by division

$$\begin{aligned}\cos 2\alpha_\theta = {} & (K_\perp^2 - 1)(K_{e2}^2 K_\perp^2 - K_{e1}^2 M_\perp^2)(K_{e1}^2 + K_\perp^2) \\ & -K_{e1}K_{e2}(K_{e1}^2 - K_\perp^2)(M_\perp^2 - K_\perp^2)(K_\perp^2 + 1) \\ & /[(K_\perp^2 - 1)(K_{e1}^2 - K_\perp^2)(K_{e1} - K_{e2})(K_{e1}M_\perp^2 - K_{e2}K_\perp^2)]\end{aligned} \tag{4.4.20}$$

Since we also have the relations

$$K_\perp = \sqrt{K_{e1} - \sin^2\theta}/\cos\theta \quad K_0 = \sqrt{K_{e1}}$$

and

$$M_{\perp} = \sqrt{K_{e2} - \sin^2\theta}/\cos\theta \qquad M_0 = \sqrt{K_{e2}}$$

Equation 4.4.20 is a single relation between K_{e1} and K_{e2} for a given angle α_θ and Equation 4.4.19 is then a second relation between these two parameters, but with α_0 determined by being the zero value of α_θ and β_0 by β_θ. An iterative procedure thus suggests itself which need only be continued up to the point where the transmission band required has been achieved. The results of a partial solution to this problem are shown in Figure 4.15 for a reflectionless wall at zero and 60° angles of incidence at a wavelength of 3.2 cm. It is found that better results are obtained with high dielectric constant values in the outer layers and the examples have constants appropriate to ceramic materials. The computed results show an optimum design at 3.1 cm based on an average value of the ambiguity d_2.

There exists in radome theory an angle/frequency relationship whereby any example involving a wide range of incidence angle could *instead* apply to a wide range of frequencies. We see that this last example is an angle version of the A sandwich, in the form of a generalized two frequency design.

RADOME WITH CENTRAL CONDUCTING LAYER

The identical procedure for a three layer radome in which the centre layer is a very thin sheet of highly conducting material requires that the central matrix in Equation 4.4.12 be replaced by the matrix

$$\begin{pmatrix} \cosh\beta & (1/M)\sinh\beta \\ M\sinh\beta & \cosh\beta \end{pmatrix} \tag{4.4.21}$$

where β by virtue of Equation 4.4.9 is

$$\beta = \frac{2\pi d_2}{\lambda}\sqrt{\sqrt{\frac{\mu\sigma}{2\omega\mu_0\varepsilon_0}} - i\sqrt{\frac{\mu\sigma}{2\omega\mu_0\varepsilon_0}} - \sin^2\theta} \tag{4.4.22}$$

with corresponding complex forms for M and $M_{\parallel}$. The condition for zero reflection then becomes

$$\tanh\beta = \frac{[(1/K) - K]\sin 2\alpha}{[K^2/M - M/K^2]\sin^2\alpha + [1/M - M]\cos^2\alpha} \tag{4.4.23}$$

with M complex.

Using

$$\beta = \tanh^{-1} ix = \frac{1}{2}\log\left(\frac{1-x^2}{1+x^2} + \frac{i2x}{1+x^2}\right)$$

and equating real and imaginary parts, this can be solved for given realistic values of K_{e1} and σ to obtain d_1 and d_2 the thickness of the respective layers. The theoretical centre thickness for highly conducting materials such as tin or

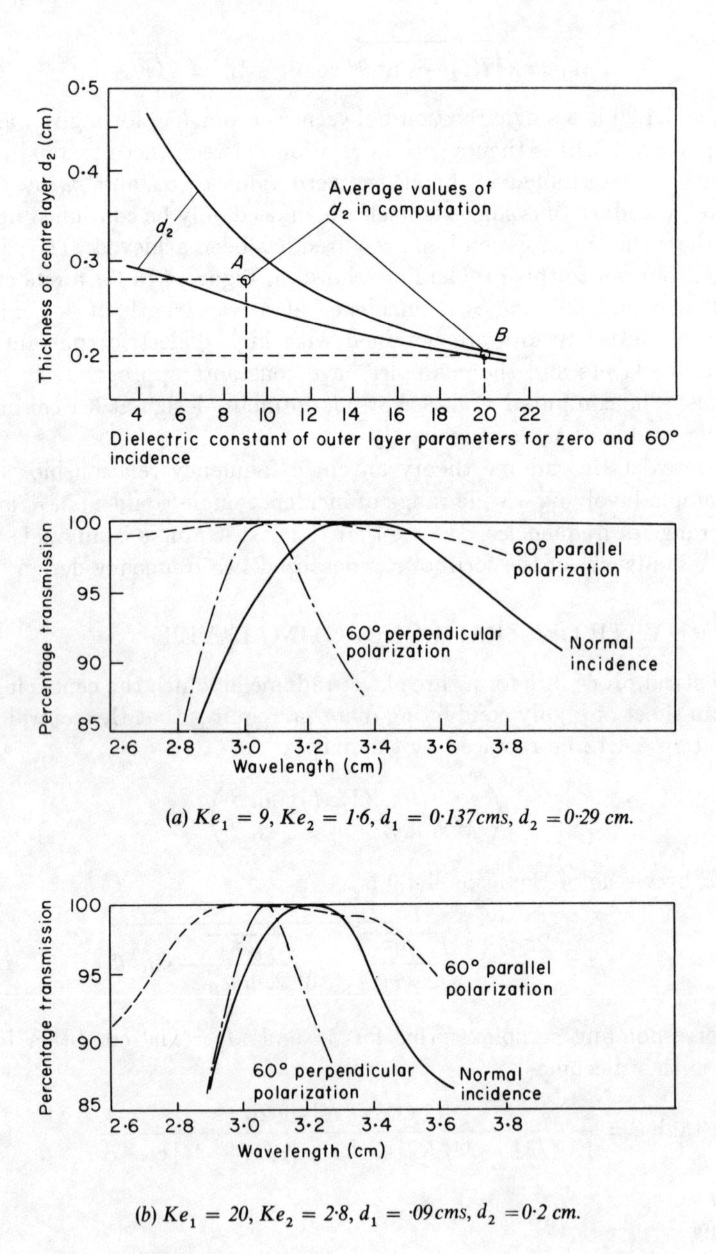

(a) $Ke_1 = 9, Ke_2 = 1{\cdot}6, d_1 = 0{\cdot}137$ cms, $d_2 = 0{\cdot}29$ cm.

(b) $Ke_1 = 20, Ke_2 = 2{\cdot}8, d_1 = {\cdot}09$ cms, $d_2 = 0{\cdot}2$ cm.

Figure 4.15: Reflectionless wall at zero and 60° incidence

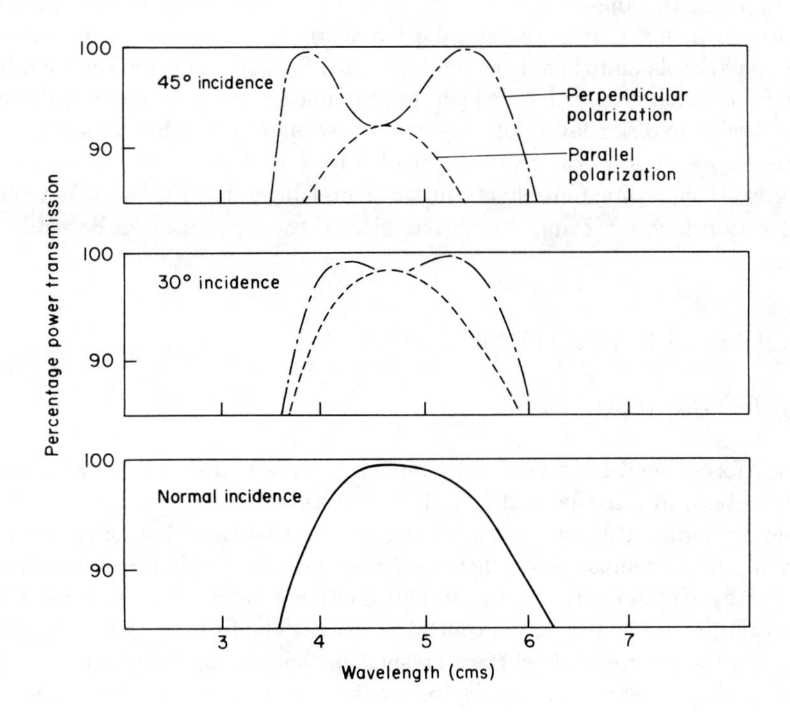

Figure 4.16: Transmission characteristics of three layer sandwich with central conducting layer $K_{e1} = 9$ $d_1 = 0.3$ cm $d_2 = 0.2442 \times 10^{-7}$ cm Conductivity $\sigma = 0.87 \times 10^7$ mho/m

aluminium are of the order of 10-20 angstrom units, the higher value relating to exterior layers of higher dielectric constant. The theoretical reflectivity for a design at $\lambda = 4.5$ cm is shown in Figure 4.16. The exterior layers are of ceramic with $K_{e1} = 9$ and the centre layer tin for which $\sigma = 0.87 \times 10^7$ mho/m. The computed thicknesses are then $d_1 = 3 \times 10^{-3}$m and $d_2 = 24.42 \times 10^{-9}$m. The change in form from the standard reflectivity curves shown in Figure 4.16 is remarkable and there is no evidence of a Brewster angle effect. Similar curves with much greater bandwidth, approximately 2:1 can theoretically be obtained with exterior layers of dielectric constant $K_{e1} = 4$ but this requires the even more impractical centre layer of 9.15×10^{-9}m thickness. At these thicknesses other thin film effects in the metal layer have to be taken into consideration before a complete appreciation of the structure can be made.

MULTILAYERED AND STRATIFIED MEDIA

MULTILAYERED MEDIA

From the foregoing it can be seen that numerous possibilities will be created by the utilization, in a similar fashion, of the additional degrees of freedom that become available in using five and seven layer structures. Formulae for the reflection and transmission coefficients for symmetrical radomes of this kind are given by Kaplun [11]. The examples include a single layer radome with wire grating reactive grid in the central position. The formulae are presented in terms of the intermediate surface Fresnel coefficients. In Chapter 6 we shall be giving the transmission matrix for reactive wire gratings. The inclusion of such gratings in radome walls can then be carried out by the same matrix procedure as in the previous sections.

Theories relating to multilayer structures which appear in the literature of optics usually have to do with structures which, translated into the microwave region, would till now be of impractical size. There has been developed in this theory methods applicable to the multiplication of large numbers of transmission matrices of the symmetrical kind that is considered here. This theory is therefore applicable to microwave practice, where the multilayer can be considered to be constructed, as it would be in practice, from a large number of very thin layers. The process then lends itself to the design as well as the analysis of such radome walls.

There is firstly a theorem by Herpin which states that any symmetrical multilayer can always be represented by a single equivalent transmission matrix. The additional statement that any non-symmetrical multilayer can be represented by at most two transmission matrices is also valid but not of our immediate concern. The former has been given an elegant proof by Young [12].

Transmission matrices are of the form

$$M = \begin{pmatrix} c & is/n \\ isn & c \end{pmatrix}$$

and can be transformed to a symmetric matrix

$$M' = \begin{pmatrix} isn & c \\ c & is/n \end{pmatrix}$$

by the product $M' = AM$ where A is the spin matrix

$$\begin{pmatrix} 0 & 1 \\ 1 & 0 \end{pmatrix}$$

A symmetric multilayer whose matrix is

$$S = M_1 M_2 \ldots M_N \ldots M_2 M_1$$

becomes

$$\begin{aligned} S &= AM_1' AM_2' \ldots AM_2' AM_1' \\ &= A(M_1' AM_2' \ldots AM_2' AM_1') = AS' \end{aligned}$$

Now S' is symmetric since, because of the symmetry of M' and A, $S'^T = S'$. Therefore S is the equivalent matrix of the *single* layer that would represent the entire multilayer.

Symmetric periodic multilayers have been studied by many authors through the method of multiplication of a repetitive transmission matrix. In particular Mielenz [13] shows that for a symmetrical periodic structure of $2n$ or $2n+1$ layers the unitary matrix representing the repeating *pair* of matrices has to be raised to the nth power. This is done as follows. If M and N are the individual transmission matrices of the repeating pair and

$$U = MN = \begin{pmatrix} a_{11} & a_{12} \\ a_{21} & a_{22} \end{pmatrix}$$

we require to evaluate U^n. Other more complex forms of symmetry lead to a similar mathematical requirement.

Since U is unitary it can be put in the form

$$U = a\sigma_0 + b\sigma_1 + c\sigma_2 + d\sigma_3$$

where

$$\sigma_0 = \begin{pmatrix} 1 & 0 \\ 0 & 1 \end{pmatrix} \quad \sigma_1 = \begin{pmatrix} 0 & 1 \\ 1 & 0 \end{pmatrix} \quad \sigma_2 = \begin{pmatrix} 0 & -i \\ i & 0 \end{pmatrix} \quad \sigma_3 = \begin{pmatrix} 1 & 0 \\ 0 & -1 \end{pmatrix}$$

are the Pauli spin matrices, and $a^2 - b^2 - c^2 - d^2 = 1$.*

*Since a b c and d may be complex U can, with identical analysis and final results, be expanded in quaternion form discussed in Appendix IV

An inductive proof then shows that

$$U^n = \begin{pmatrix} a_{11}S_{n-1}(x) - S_{n-2}(x) & a_{12}S_{n-1}(x) \\ a_{21}S_{n-1}(x) & a_{22}S_{n-1}(x) - S_{n-2}(x) \end{pmatrix} \tag{4.4.24}$$

where $S_n(x)$ are Chebychev polynomials (the same as appear in Herpin [14] given by

$$S_n(x) = \frac{\sin(n+1)\psi}{\sin\psi} \qquad x = 2\cos\psi \qquad x \leq 2 \tag{4.4.25}$$

or

$$S_n(x) = \frac{\sinh(n+1)\phi}{\sinh\phi} \qquad x = 2\cosh\phi \qquad x \geq 2 \tag{4.4.26}$$

and

$$x = a_{11} + a_{22}$$

These results simplify a great deal under conditions appropriate to optical work such as a multilayer of individual 1/4 wave layers, but which become impracticable at microwave frequencies. However, they have not been fully assessed recently, with the possibility of the use of computing techniques for millimetre wave applications.

The method provides a link between optics and other branches of microwave activity. The same analysis applies to the design of quarter wave coupled filters and loaded line synthesis [17] and non-uniform waveguide transmission. In these fields Chebychev polynomials and spin matrices make their appearances. Chebychev polynomials are also a key design method for linear arrays of sources [19] in which problem spin matrices, or their equivalent the quaternions, have not yet made an appearance. Other methods in filter analysis, for example Walsh functions [20] would presumably have applications in array theory and hence in optical filter theory.

STRATIFIED MEDIA

There are two methods by which the matrix theory of a multilayered medium can be extended to a medium with a continuously variable refractive index. The first is to consider the continuous variation to be made up of discrete infinitesimal steps and to multiply the corresponding "infinitesimal" matrices. This in principle is no great advance since the multiplication of such matrices is a highly complex procedure. If an approximate method is employed for this *multiplication* process, the degree of approximation is similarly multiplied until in the end result the solution can only be applied to a very small overall (in optical wavelengths) thickness of the medium. As such it is analogous to the Born approximation, as it would be applied to the actual differential equation governing the propagation in such a medium [21]. We include this method for illustrative purposes.

For a layer of thickness δx the transmission matrix is

$$\Delta M = \begin{pmatrix} \cos(\frac{2\pi\delta x\alpha}{\lambda}) & \frac{i}{Y(x)}\sin(\frac{2\pi\delta x\alpha}{\lambda}) \\ iY(x)\sin(\frac{2\pi\delta x\alpha}{\lambda}) & \cos(\frac{2\pi\delta x\alpha}{\lambda}) \end{pmatrix} \tag{4.4.27}$$

where α is the phase angle $\sqrt{K_e(x) - \sin^2\theta}$ and $Y(x)$ the variable admittance of the medium.

Then if the small angle approximation is made

$$\Delta M \simeq \begin{pmatrix} 1 & \frac{i}{Y(x)}\frac{2\pi\delta x}{\lambda}\alpha \\ iY(x)\frac{2\pi\delta x}{\lambda}\alpha & 1 \end{pmatrix} \tag{4.4.28}$$

and multiplication of a large number of such matrices results in the complete scattering matrix of the layer

$$M \simeq \begin{pmatrix} 1 & \frac{2\pi i}{\lambda}\sum\frac{\alpha\delta x}{Y(x)} \\ \frac{2\pi i}{\lambda}\sum\alpha Y(x)\delta x & 1 \end{pmatrix} \tag{4.4.29}$$

where powers of δx greater than the first have been ignored at each matrix multiplication.

Proceeding to the limit this becomes

$$M \simeq \begin{pmatrix} 1 & \frac{2\pi i}{\lambda}\int_0^x \frac{\alpha}{Y(x)}\,\mathrm{d}x \\ \frac{2\pi i}{\lambda}\int_0^x \frac{\alpha}{Y(x)}\,\mathrm{d}x & 1 \end{pmatrix} \tag{4.4.30}$$

The error in this result can be illustrated by applying it to a *uniform* layer of thickness D. Then with α and $Y(x)$ constant M in Equation 4.4.30 becomes

$$M \simeq \begin{pmatrix} 1 & \frac{i}{Y}\frac{2\pi D\alpha}{\lambda} \\ iY\frac{2\pi D\alpha}{\lambda} & 1 \end{pmatrix}$$

which has now lost any connection with the periodic result of Equation 4.4.7

$$M = \begin{pmatrix} \cos(2\pi D\alpha/\lambda) & \frac{i}{Y}\sin(2\pi D\alpha/\lambda) \\ iY\sin(2\pi D\alpha/\lambda) & \cos(2\pi D\alpha/\lambda) \end{pmatrix}$$

In fact the two results only compare for values of D for which

$$\cos(2\pi D\alpha/\lambda) \simeq 1 \quad \text{that is} \quad D\alpha \ll \lambda$$

Hence an improved method is required that could apply to a layer of any width. We proceed by diagonalizing the matrix of a single thin uniform layer. The transmission matrix of Equations 4.4.7 and 4.4.3 then separate into surface "admittance" matrices separated by a spacing or phase angle matrix as follows

$$M = \frac{1}{2Y}\begin{pmatrix} 1 & -1 \\ Y & Y \end{pmatrix}\begin{pmatrix} e^{iA} & 0 \\ 0 & e^{-iA} \end{pmatrix}\begin{pmatrix} Y & 1 \\ -Y & 1 \end{pmatrix}$$

$$A = 2\pi d\alpha/\lambda \tag{4.4.31}$$

Multiplication of a number N of such layers to produce a layer of thickness D is achieved through

$$M_D = M_1 M_2 \ldots M_N$$

or

$$\begin{aligned} M_D = & \begin{pmatrix} 1 & -1 \\ Y_1 & Y_1 \end{pmatrix} \begin{pmatrix} e^{iA_1} & 0 \\ 0 & e^{-iA_1} \end{pmatrix} \begin{pmatrix} Y_1 & 1 \\ -Y_1 & 1 \end{pmatrix} \frac{1}{2Y_1} \\ \times & \begin{pmatrix} 1 & -1 \\ Y_2 & Y_2 \end{pmatrix} \begin{pmatrix} e^{iA_2} & 0 \\ 0 & e^{-iA_2} \end{pmatrix} \begin{pmatrix} Y_2 & 1 \\ -Y_2 & 1 \end{pmatrix} \frac{1}{2Y_2} \\ & \ldots \text{etc} \end{aligned} \tag{4.4.32}$$

The internal products of the "admittance" matrices are of the form

$$\begin{pmatrix} Y_1 & 1 \\ -Y_1 & 1 \end{pmatrix} \frac{1}{2Y_1} \begin{pmatrix} 1 & -1 \\ Y_2 & Y_2 \end{pmatrix} = \frac{1}{2Y_1} \begin{pmatrix} Y_1 + Y_2 & Y_2 - Y_1 \\ Y_2 - Y_1 & Y_1 + Y_2 \end{pmatrix} \tag{4.4.33}$$

If now Y_2 is incrementally different from Y_1 and A_2 likewise incrementally different from A_1 Equation 4.4.33 becomes

$$\frac{1}{2Y_1} \begin{pmatrix} Y_1 + Y_1 + \delta Y_1 & Y_1 + \delta Y_1 - Y_1 \\ Y_1 + \delta Y_1 - Y_1 & Y_1 + Y_1 + \delta Y_1 \end{pmatrix} = \begin{pmatrix} 1 + \frac{\delta Y_1}{2Y_1} & \frac{\delta Y_1}{2Y_1} \\ \frac{\delta Y_1}{2Y_1} & 1 + \frac{\delta Y_1}{2Y_1} \end{pmatrix} \tag{4.4.34}$$

In the limit $\delta Y_i \to 0$ and all the internal products tend to the unit matrix. Equation 4.4.32 then becomes

$$\begin{aligned} M_D = & \begin{pmatrix} 1 & -1 \\ Y_1 & Y_1 \end{pmatrix} \begin{pmatrix} e^{iA_1} & 0 \\ 0 & e^{-iA_1} \end{pmatrix} \begin{pmatrix} e^{i(A_1 + dA_1)} & 0 \\ 0 & e^{-i(A_1 + dA_1)} \end{pmatrix} \cdots \\ & \begin{pmatrix} Y_N & 1 \\ -Y_N & 1 \end{pmatrix} \frac{1}{2Y_N} \\ = & \begin{pmatrix} 1 & -1 \\ Y_1 & Y_1 \end{pmatrix} \begin{pmatrix} e^{i\int_0^D A\,dx} & 0 \\ 0 & e^{-i\int_0^D A\,dx} \end{pmatrix} \begin{pmatrix} Y_N & 1 \\ -Y_N & 1 \end{pmatrix} \frac{1}{2Y_N} \end{aligned} \tag{4.4.35}$$

Reconstituting M_D gives the final result

$$M_D = \begin{pmatrix} \cos\Phi & i/Y_N \sin\Phi \\ iY_1 \sin\Phi & Y_1/Y_N \cos\Phi \end{pmatrix} \tag{4.4.36}$$

where Φ is now the phase integral [22]

$$\Phi = \int_0^D \frac{2\pi}{\lambda} \alpha\, dx = \frac{2\pi}{\lambda} \int_0^D \sqrt{K_e(x) - \sin^2\theta}\, dx \tag{4.4.37}$$

Multiplication of two such matrices produces a third of the identical form, and it can be seen that a matrix of this kind implies an infinitesimal layer in which the medium properties vary in a *linear* manner from the incident to the transmission sides. Such a layer would have characteristic matrix

$$\Delta M = \begin{pmatrix} \cos 2\pi\delta x\alpha/\lambda & i/(Y_2)\sin 2\pi\delta x\alpha/\lambda \\ iY_1 \sin 2\pi\delta x\alpha/\lambda & Y_1/(Y_2)\cos 2\pi\delta x\alpha/\lambda \end{pmatrix} \qquad (4.4.38)$$

Multiplication of a string of such matrices and proceeding to the limit produces the same result as in Equation 4.4.36. Thus the result is in effect that of making the refractive index profile up by incremental linear sections with no discontinuities at the surfaces between the layers.

These phase integrals were introduced by Eckersley and Budden into their solution for the propagation of radio waves in a variable atmosphere [23] but the introduction appears to be somewhat arbitrary. Jacobsson [25] arrives at the same result and shows that it can be explained as an averaging process of the Fresnel reflection coefficients of the discontinuously stepped approximation. The derivation is from the differential equation governing the propagation and the observation is made that the approximation concerned is the WKJB approximation [27].

It is clear from the description of the phase integral in Equation 4.4.38 that there exist conditions on the thickness of the layer D and the law of the refractive index $K_e(x)$ for the case when the integrand becomes complex. This physically would imply a total internal reflection in a large depth of material, or at very oblique incidence. These problems are more pertinent to the study of radio wave propagation and hence to all the theory that has been applied to it. The matrix analysis here forms a positive link between these theories and the other branches of microwave discussed previously. It can be applied particularly to the theory of non-uniform waveguides.

As stated in [25], "In principle inhomogeneous films, possibly combined with homogeneous films, can be used to realise most of the spectral requirements of modern thin film optics". The additional degrees of freedom obtained by dissociating the reflections at interfaces from the phase length between them, and in addition by having different values of Fresnel coefficients at these interfaces, permits a large potential of radome designs to be considered. Some of the many possible variations are shown in Figure 4.17.

4.5 THE CYLINDRICAL NON-UNIFORM MEDIUM

THE CIRCULARLY SYMMETRICAL MEDIUM

The most commonly found of all non-uniform materials are the glass fibres currently used for optical transmission. For a rotationally symmetrical medium uniform in the longitudinal z direction, we put in Equation 4.2.2

$$\frac{\partial \eta}{\partial \phi} = 0 \qquad \frac{\partial \eta}{\partial z} = 0$$

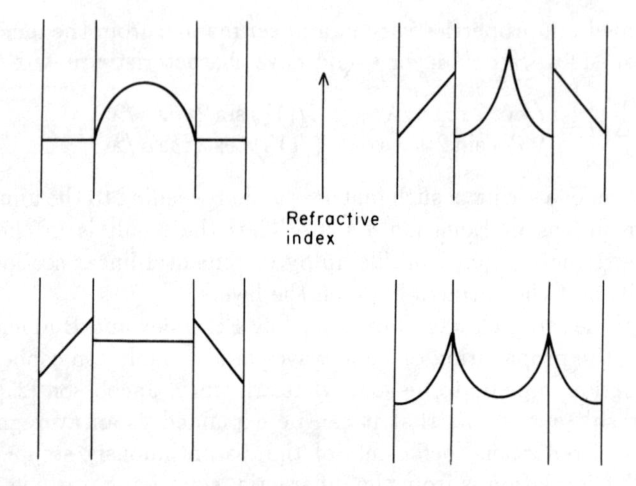

Figure 4.17: Non-uniform three layer sandwich designs

We obtain $(\mathrm{d}/\mathrm{d}s)[\eta(\mathrm{d}z/\mathrm{d}s)] = 0$ identical to Equation 4.3.1 and from 4.2.2

$$\frac{\mathrm{d}}{\mathrm{d}s}\left(\eta\rho^2\frac{\mathrm{d}\theta}{\mathrm{d}s}\right) = 0$$

Thus if the rays are confined to meridional planes and $\mathrm{d}\theta/\,\mathrm{d}s = 0$ as a consequence, the system is the equivalent of the rotation about the axis $x = 0$ of the linear system in Section 4.3. Thus the cylindrical form with refractive index

$$\eta(\rho) = \eta_0 \mathrm{sech}\left(\frac{\pi\rho}{2a}\right) \qquad \text{see Equation 4.3.3}$$

will have the same trajectories, in the meridional planes only, shown in Figure 4.4

At any position midway between two perfect foci, the rays by symmetry alone will be parallel to the stratification; thus if the medium were to be terminated there with a plane surface perpendicular to the axis, the rays would continue into free space as a parallel beam. This design of lens was first proposed by Brown [7, p89] and termed the "short-focus horn" (Figure 4.18)

The medium could, of course, be terminated by any other shaped surface to provide a local perfect focus, or indeed any specified caustic, symmetric or asymmetric.

If the rays are no longer confined to the meridional plane, Equation 4.5.1 gives a second constant, applicable to the now skew ray. That is

$$\eta\rho^2\frac{\mathrm{d}\theta}{\mathrm{d}s} = \text{constant } = h \tag{4.5.1}$$

the Herzberger "skew invariant" [28].

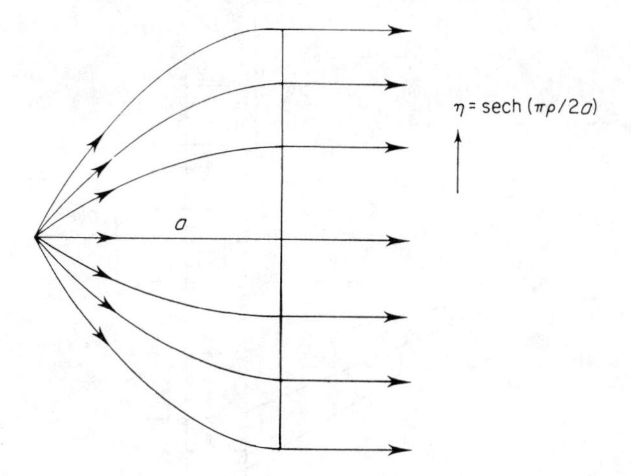

Figure 4.18: The short-focus horn

The implication is that when h is zero the rays are once again confined to meridional planes, and thus Equation 4.5.1 contains a measure of the skew behaviour of these rays. This is borne out by the analysis given by Luneburg. Recasting Equation 4.5.1 as a dependent function of the axial variable z ie substituting $\mathrm{d}s^2 = \mathrm{d}\rho^2 + \rho^2\,\mathrm{d}\phi^2 + \mathrm{d}z^2$, one obtains Luneburg's equation

$$\frac{\eta\rho^2 \frac{\mathrm{d}\phi}{\mathrm{d}z}}{\sqrt{1 + (\frac{\mathrm{d}\rho}{\mathrm{d}z})^2 + \rho^2(\frac{\mathrm{d}\phi}{\mathrm{d}z})^2}} = h$$

Solving this for $\mathrm{d}\phi/\mathrm{d}z$ and integrating gives

$$\phi_1 - \phi_0 = h \int_{z_1}^{z_0} \frac{1}{\rho}\sqrt{\frac{1 + (\mathrm{d}\rho/\mathrm{d}z)^2}{\eta^2\rho^2 - h^2}}\,\mathrm{d}z \qquad (4.5.2)$$

where (ϕ_0, z_0) and (ϕ_1, z_1) are the *angular* positions at the end-points of a ray.

We now separate the optical path of a ray from a point in the z_0 plane to a point in the z_1 plane, into an equivalent "radial" path and an "angular" optical path, the latter given by $h(\mathrm{d}\phi/\mathrm{d}z)$. Then the "radial" path will be given by the total path minus the "angular" path, that is

$$\int_{z_0}^{z_1} \left[\eta\sqrt{1 + \left(\frac{\mathrm{d}\rho}{\mathrm{d}z}\right)^2 + \rho^2\left(\frac{\mathrm{d}\phi}{\mathrm{d}z}\right)^2} - h\frac{\mathrm{d}\phi}{\mathrm{d}z}\right]\mathrm{d}z \qquad (4.5.3)$$

which equals, after substituting for $\mathrm{d}\phi/\mathrm{d}z$

$$\int_{z_0}^{z_1} \sqrt{\left(\eta^2 - \frac{h^2}{\rho^2}\right)\left(1 + \frac{\mathrm{d}\rho}{\mathrm{d}z}\right)}\,\mathrm{d}z \qquad (4.5.4)$$

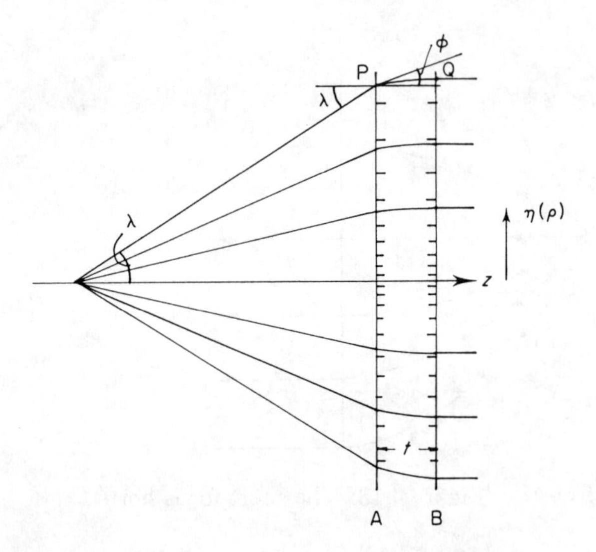

Figure 4.19: Non-uniform flat-disc lens

which is (ρ, z) dependent only; hence the definition of "radial" optical path.

This is equivalent to the path of a ray in a *meridional* plane in a medium with refractive index

$$\eta(\rho, z) = \sqrt{\eta^2 - \frac{h^2}{\rho^2}} \tag{4.5.5}$$

Thus for a skew ray we obtain the radial movement, with respect to distance along the axis, from Equation 4.5.4 as for a Cartesian geometry but using the refractive index $\eta(\rho, z)$. and the angular motion from Equation 4.5.2.

FLAT-DISC LENSES

A collimating lens similar to the short focus horn can be obtained in a considerably reduced body of material if the rays are allowed their natural spread in free space up to some plane surface A, as shown in Figure 4.19, and curve thereafter to become horizontal at a second plane surface B, assisted by the refraction that occurs at the surface A. This creates a flat-disc radially non-homogeneous lens with an external source.

A lens of this type first described by Wood [29] made use of a paraxial approximation and the assumption that, within the body of the lens, the rays were arcs of circles. As will be seen, this assumption gives a very good indication of the form of refractive index law required. The result here is for a wide-angle lens $f/D \simeq 1$, and consequently for exact ray paths.

Once established the method is adaptable to lenses with other focal requirements and to multi-lens systems.

We consider, as in Figure 4.19, a ray from a source on the axis at *unit* focal distance OP from the plane surface A perpendicular to the axis. This ray makes an angle θ with the axis, and at P undergoes a refraction in accordance with Snell's law. We assume this refraction to occur as though the medium at P were infinite, and of uniform refractive index of value $\eta(P)$. This is the usual assumption for non-uniform surfaces or media in physical optics.

At the point P, therefore, we have

$$\eta(P)\sin\phi = \sin\theta$$

and hence

$$\sin\phi = \frac{\rho}{\eta\sqrt{1-\rho^2}} \tag{4.5.6}$$

At P, $\eta\cos\phi = A$, and therefore the ray constant for the ray OPQ is

$$A = \sqrt{\eta^2 - \frac{\rho^2}{1+\rho^2}} \tag{4.5.7}$$

evaluated at P. This function will be designated $A(P)$. Within the medium itself we use Equation 4.3.2

$$z = \int_{\rho(P)}^{\rho(Q)} \frac{A(P)\,\mathrm{d}\rho}{\sqrt{\eta^2(\rho) - A^2(P)}} \tag{4.5.8}$$

In order to obtain a second plane surface where the rays have become horizontal, we require a function $\eta(\rho)$ such that Equation 4.5.8 results in a constant value $z = t$, the thickness of the lens, for all values of θ and hence for all values of ϕ as given by Equation 4.5.6.

Since the value of $\rho(Q)$ is then that at which the ray is horizontal, or $\mathrm{d}z/\mathrm{d}s \equiv 1$, then at Q

$$A(Q) \equiv A(P) = \eta(Q)$$

In other words $\rho(Q)$ has the value that makes the denominator in Equation 4.5.8 equal to zero, a fact which will be made some use of in the ensuing analysis. It is, of course, understood that the objective in making the second surface B plane, is not to have to consider a further refraction there. Other lenses of this kind, with non-planar surfaces, or with different focal properties, can be designed by a similar analysis extended by including a refraction at the second surface.

The solutions of Equation 4.5.8 for the complete medium, that is the short-focus horn, has $\rho(P)$ zero and η can be obtained by the standard method of converting Equation 4.5.8 to an integral of Abel's form. With the integral now incomplete these methods no longer apply and the approach used is to insert such trial functions of $\eta(\rho)$ as make Equation 4.5.8 integrable by analytic methods. It is soon found that quadratic and quartic functions of both $\eta^2(\rho)$ and $1/\eta^2(\rho)$ can be dealt with by the use of incomplete elliptic integrals of the first kind [30].

The closest approximation to a plane-surfaced lens has come from modifications to the parabolic law of refractive index $\eta = a - b\rho^2$. Besides being the most commonly used law in gradient index fibres, it is the result obtained by Wood himself in the approximate solution previously mentioned.

This can be extended by applying a law of the form

$$\eta^2(\rho) = a^2 - 2b^2\rho^2 + c^2\rho^4 \tag{4.5.9}$$

Then the constant lens thickness is obtained from

$$\begin{aligned} t &= \int_{\rho(P)}^{\rho(Q)} \frac{A\,\mathrm{d}\rho}{\sqrt{a^2 - A^2 - 2b^2\rho^2 + c^2\rho^4}} \\ &= \int_{\rho(P)}^{\rho(Q)} \frac{A\,\mathrm{d}\rho}{c\sqrt{(\alpha^2 - \rho^2)(\beta^2 - \rho^2)}} = \frac{A}{c}I \end{aligned} \tag{4.5.10}$$

where

$$\alpha^2 = [b^2 + \sqrt{b^4 - c^2(a^2 - A^2)}]/c^2$$
$$\beta^2 = [b^2 - \sqrt{b^4 - c^2(a^2 - A^2)}]/c^2$$

and from Equation 4.5.7

$$A^2 = a^2 - 2b^2\rho^2 + c^2\rho^4 - \rho^2/(1 + \rho^2)$$

at $\rho = \rho(P)$. Then by virtue of the fact that $\rho(Q)$ is a root of the denominator, it must equal β, and hence with

$$\alpha > \beta > \rho(P) > 0$$

Equation 4.5.10 fulfils the conditions for the incomplete elliptic integral of the first kind [30] again and

$$I = gF(\psi, k)$$

where

$$g = \frac{1}{\alpha} \qquad \psi = \sin^{-1}\frac{\alpha^2\sqrt{\beta^2 - \rho^2(P)}}{\beta^2\sqrt{\alpha^2 - \rho^2(P)}} \tag{4.5.11}$$

and $k^2 = \beta^2/\alpha^2$. In the limit $\rho \to 0$, $a \to A$ and $k \to 0$, and therefore

$$\psi \to \psi_0 - \sin^{-1}\left(\sqrt{\frac{1}{1 + 2b^2}}\right)$$

This gives the axial thickness of the lens

$$t_0 = \frac{a}{\sqrt{2b^2}}F(\psi_0, 0)$$

Since, with $k = 0$, $F(\psi_0, 0) \simeq \sin\psi_0$ over the range usually encountered

$$t_0 \simeq \frac{a}{\sqrt{2b^2 + 4b^4}}$$

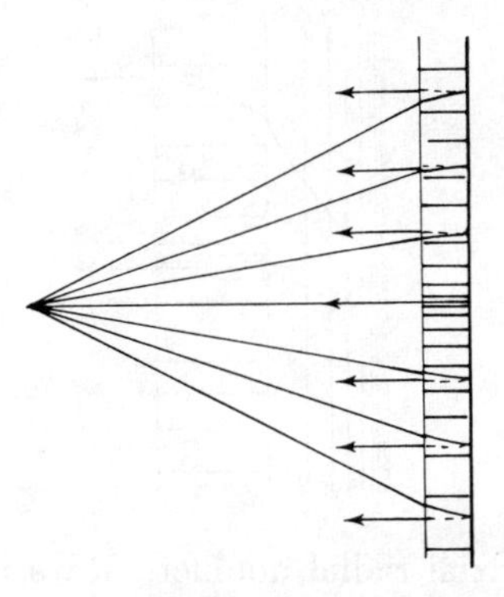

Figure 4.20: The bisected disc lens as a coated plane reflector

The requirement is therefore for t of Equation 4.5.10 to be made constant in value for all values of $\rho(P)$ and equal to t_0, by adjusting the arbitrary constants a, b and c. This can be achieved by optimization processes such as minimizing the value of $\sqrt{t^2 - t_0^2}$.

Subsequent ray-tracing computation shows that for $\eta^2 = 16 - 24\rho^2 + 12.85\rho^4$, the rays are horizontal (to 1:10^{-6}) over a surface which is flat to within 0.0004 (OF $= 1$) over the range $0 < \rho < 0.5$, that is for an f/D ratio of unity. The thickness is approximately $f/6$.

The high degree of symmetry arising from the plane surfaces of the flat-disc lens opens up many opportunities. The lens focuses identically from sources on either side and therefore solves the problem of the symmetrical lens dealt with in Section 3. In the same way, the insertion of a mid-plane reflector converts it into a corrected plane reflector (Figure 4.20). Owing to the simplicity with which refraction can be dealt with at plane surfaces, multi-lens systems are conceivable for such effects as variable beam shaping, zoom focusing and aberration correction. Other applications could include focusing end-plates for laser cavities, matching junctions for gradient index fibres and large reflector feeding systems. The method is, of course, particularly applicable to acoustic lenses with similar focusing requirements.

FLAT LENS DOUBLETS

A two layer cylindrical lens can be designed such that the first layer produces an upward curvature of the rays which are then curved backwards to the horizontal by the second layer. No further principle of action is involved and the ray paths are soluble with a quartic description for the refractive index as Equations 4.5.11 using at most, elliptic integrals of the second kind. The upward curvature of the rays in the first region can be obtained in two different ways: first by a radially varying refractive index which *increases* with radius

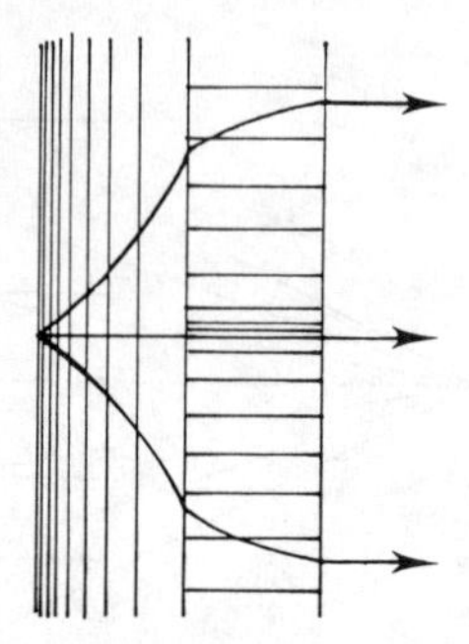

Figure 4.21: Flat-lens axial radial doublet. Rays in the first region acquire their upward curvature from the decrease in refractive index in the axial direction

or alternatively by an axial variation of refractive index *decreasing* in the axial direction (Figure 4.21). Also shown is a telescopic lens to this design (Figure 4.22). A theoretical evaluation showed that the wave front transmitted was plane to within ±0.0002 (for a unit diameter) with a magnification of 2.5:1. This ability to vary the degree to which the rays curve obviously leads to the possibility of adjusting the power distribution between the source and the aperture to give a prescribed aperture distribution.

4.6 THE GENERAL CYLINDRICAL MEDIUM

The coordinate system for the general cylindrical system is given in Section 4.2, where $f(u, v)$ and $h(u, v)$ are generalized orthogonal coordinates in a plane. Such coordinate systems can be derived directly from the conformal mapping

$$\begin{aligned} z \equiv x + iy &= F(u + iv) \\ &= f(u, v) + ih(u, v) \equiv F(w) \end{aligned} \tag{4.6.1}$$

although conformality in addition to orthogonality, Equation 4.2.5, is not essential and would prove an unnecessary restraint on the selection of cross-sectional geometries. This can be seen by comparing transformations 1 and 3 in Table 4.1, where the first is orthogonal but not conformal whereas the third is both.

Solutions to these equations can be found as before for rays confined to a constant coordinate surface. Therefore putting $u =$ constant, that is $\mathrm{d}u/\mathrm{d}s = 0$ in Equation 4.2.8 and postulating that the medium is uniform in the z direction, that is, with g_{33} =constant = 1, Equation 4.2.8 becomes

$$\eta \frac{\mathrm{d}z}{\mathrm{d}s} = \text{constant } = A \tag{4.6.2}$$

1. Circular polar coordinates: rays on cylinder

$$x = u\cos v \qquad y = u\sin v \qquad g_{22} = u^2$$

2. Circular polar coordinates: rays in meridional planes

$$x = v\cos u \qquad y = v\sin u \qquad g_{22} = 1$$

3. †Exponential coordinates: rays on cylinders

$$z = \exp^w \qquad x = \exp^u \cos v \qquad y = \exp^u \sin v \qquad g_{22} = \exp^{2u}$$

4. Exponential coordinates: rays in meridional planes

$$z = \exp^{iw} \qquad x = \exp^v \cos u \qquad y = \exp^v \sin u \qquad g_{22} = \exp^{2v}$$

5. Elliptic coordinates

$$z = a\cosh w \qquad x = a\cosh u\cos v \qquad y = a\sinh u\sin v$$

$$g_{22} = \sinh^2 u + \sin^2 v$$

6. Logarithmic coordinates

$$z = a\log w \quad x = a\log(u^2+v^2) \quad y = 2a\tan^{-1}\frac{v}{u} \quad g_{22} = 4a^2$$

7. Tangent coordinates

$$z = \tan w \qquad x = \frac{\sin u\cos u}{\cos^2 u + \sinh^2 v}$$

$$y = \frac{\sinh v\cosh v}{\cos^2 u + \sinh^2 v} \qquad g_{22} = \frac{1}{\cos^2 u + \sinh^2 v}$$

8. †$z \tan(iw) = -i\tanh w$ as for (7) with $u \leftrightarrow v$

9. Inverse coordinates

$$z = 1/w \qquad x = \frac{u}{u^2+v^2} \qquad y = \frac{v}{u^2+v^2} \qquad g_{22} = \frac{1}{u^2+v^2}$$

†The transformation $w \to iw$ is equivalent to the transformation $u \leftrightarrow v$

Table 4.1: Orthogonal cylindrical coordinates

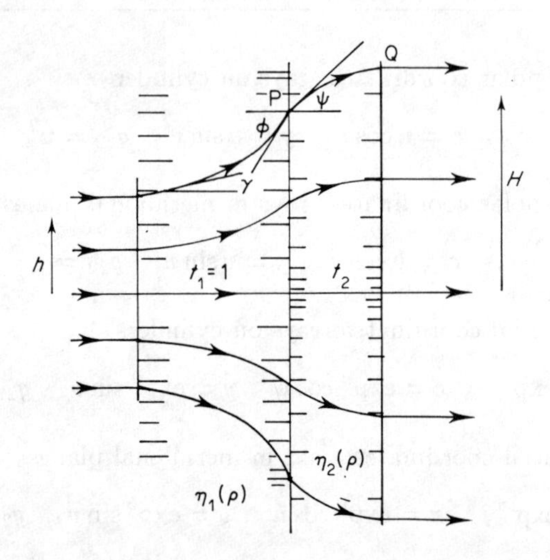

Figure 4.22: The flat-lens doublet as a telescopic lens

This is the generalization of Snell's law for a medium of this type. The constant A is specific to a given single ray and thus has different values for different rays in a pencil of rays. The precise parametric dependence of A is vital to the interpretation of the eventual mathematical results. Equations 4.28 reduce to

$$\frac{\eta\sqrt{g^{11}}}{\Delta}\left(\frac{\mathrm{d}v}{\mathrm{d}s}\right)^2(-f_v h_{vv} + h_v f_{vv}) = \frac{1}{\sqrt{g_{11}}}\frac{\partial \eta}{\partial u} \tag{4.6.3}$$

and

$$\frac{\mathrm{d}}{\mathrm{d}s}\left(\eta\sqrt{g_{22}}\frac{\mathrm{d}v}{\mathrm{d}s}\right) = \frac{1}{\sqrt{g_{22}}}\frac{\partial \eta}{\partial v} \tag{4.6.4}$$

In addition, we now have

$$\mathrm{d}s^2 = g_{22}\,\mathrm{d}v^2 + \mathrm{d}z^2 \qquad \frac{\mathrm{d}}{\mathrm{d}s} = \frac{\partial}{\partial v}\frac{\mathrm{d}v}{\mathrm{d}s} + \frac{\partial}{\partial z}\frac{\mathrm{d}z}{\mathrm{d}s}$$

The ray constant A is now u-dependent.

We shall require the following identity

$$\frac{g_{11}}{\Delta g_{22}}(-f_v h_{vv} + h_v f_{vv}) = -\frac{1}{2}\frac{\partial}{\partial u}\log g_{22} \tag{4.6.5}$$

where

$$g_{11} = f_u^2 + h_u^2 \qquad g_{22} = f_v^2 + h_v^2 \qquad \Delta = f_u h_v - f_v h_u$$

This result is obtained by applying the orthogonality relation of Equation 4.2.5 and its derivative

$$f_u f_{vv} + f_v f_{uv} + h_u h_{vv} + h_v h_{uv} = 0$$

in turn to the left-hand side of Equation 4.6.4.

Consider first Equation 4.6.4. Multiplication by the factor $\eta\sqrt{g_{22}}(\mathrm{d}v/\mathrm{d}s)$ gives, since $\partial\eta/\partial z = 0$

$$\eta\sqrt{g_{22}}\frac{\mathrm{d}v}{\mathrm{d}s}\frac{\mathrm{d}}{\mathrm{d}s}\left(\eta\sqrt{g_{22}}\frac{\mathrm{d}v}{\mathrm{d}s}\right) = \frac{\partial\eta}{\partial v}\frac{\mathrm{d}v}{\mathrm{d}s}$$

and thus

$$\left(\eta\sqrt{g_{22}}\frac{\mathrm{d}v}{\mathrm{d}s}\right)^2 = \eta^2 + \text{constant} \tag{4.6.6}$$

the constant being u-dependent. From the definition of $\mathrm{d}s$ and Equation 4.6.2 this constant is $-[A(u)]^2$ and consequently $|A(u)| < \eta$.

We now multiply Equation 4.6.3 by g_{22} and use the identity in Equation 4.6.5 and the result in Equation 4.6.6 to get

$$-[A(u)]^2\frac{1}{g_{22}}\frac{\partial g_{22}}{\partial u} = -2\eta\frac{\partial\eta}{\partial u}$$

or

$$\eta^2 = \frac{1}{g_{22}}\int [A(u)]^2\frac{\partial g_{22}}{\partial u}\,\mathrm{d}u + B(v) \tag{4.6.7}$$

with $B(v)$ as an arbitrary function of the integration. This is the general form of the refractive index law for which a ray on a surface u = constant is constrained to remain there.

The ray path can be obtained by integration of the ray equations separately; that is, from Equation 4.6.2

$$\frac{\mathrm{d}z}{\mathrm{d}s} = \frac{A(u)}{\eta}$$

and from Equation 4.6.6

$$g_{22}\frac{\mathrm{d}v}{\mathrm{d}s} = \frac{\sqrt{\eta^2 - [A(u)]^2}}{\eta} \tag{4.6.8}$$

Combining these gives

$$z = \int A(u)\sqrt{\frac{f_v^2 + h_v^2}{\eta^2 - [A(u)]^2}}\,\mathrm{d}v \tag{4.6.9}$$

The ray is parallel to the z-axis at points where $\mathrm{d}v/\mathrm{d}s = 0$ (i.e. where $A(u) = \eta$), and transverse where $\mathrm{d}z/\mathrm{d}s = 0$ (where $\eta \to \infty$).

If the ray makes an angle ψ with a z-directed generator (Figure 4.23) then $\mathrm{d}z/\mathrm{d}s = \cos\psi$ and $\sqrt{g_{22}}(\mathrm{d}v/\mathrm{d}s) = \sin\psi$, giving the "starting" conditions for a ray from a source in the medium.

Although the analysis has been derived for the situation in which a ray is confined to a surface u = constant, the simple substitution $u \leftrightarrow v$ in the transformation specifying the system (and in all subsequent results) gives the

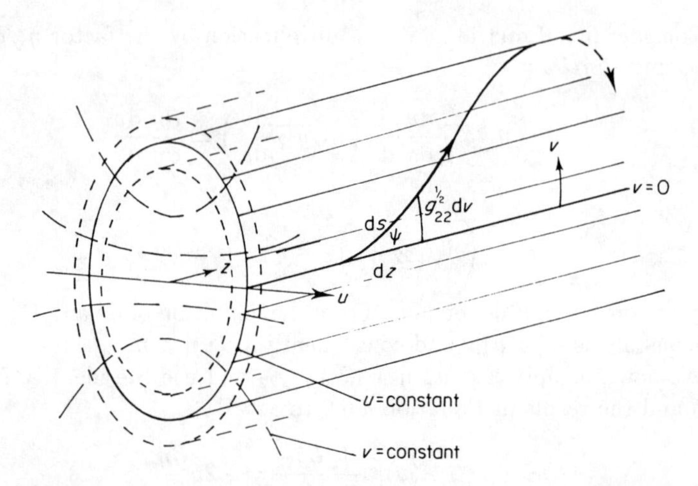

Figure 4.23: Rays in a general cylindrical coordinate system, confined to the surface u =constant

ray trajectory in the orthogonal system. Hence two orthogonal cross-sections of a ray pencil can be dealt with by the analysis given. Skew rays, of course, require the complete solution of the general equations.

Equations 4.6.7 and 4.6.9 can be solved for specific conditions of ray behaviour. The most important is coherence, that is, the phase length along each individual pencil of rays must be equal. When the surface concerned is a closed surface enclosing the axis, the ray paths will, in general, be spirals, encircling the axis, or they will be confined to a zone parallel to the axis. Coherence is maintained when rays of a particular pencil spiral with the same pitch.

For non-spiralling ray pencils, coherence is maintained by continuous refocusing along the axis. Both these conditions are well known in the case of the circular polar coordinate system: here it is extended to refractive index variations in both the radial and angular coordinates.

Any function $A(u)$ will give rays which spiral on the cylinder, in a system for which g_{22} is a function of u only. This can be seen by immediate substitution into Equation 4.6.7. Subsequently $\mathrm{d}v/\,\mathrm{d}s$ can only become zero in the limit of a completely uniform medium.

For example, in the circular system (1) of Table 4.1 with $g_{22} = u^2$, the function $A^2 = a - bu^2$ gives the commonly used "parabolic" gradient of refractive index $\eta^2 = a - bu^2/2$ $(B(v) = 0)$. The helix angle of the spiral is then $\tan^{-1}\{2[(a/bu^2) - 1]\}$ which is u-dependent. A pencil of such rays will thus form a diffuse system, and coherence will not in general be retained. For weakly guiding media (small b) and small-diameter cylinders, the spiral pitch becomes large and diffusion of rays can be kept small.

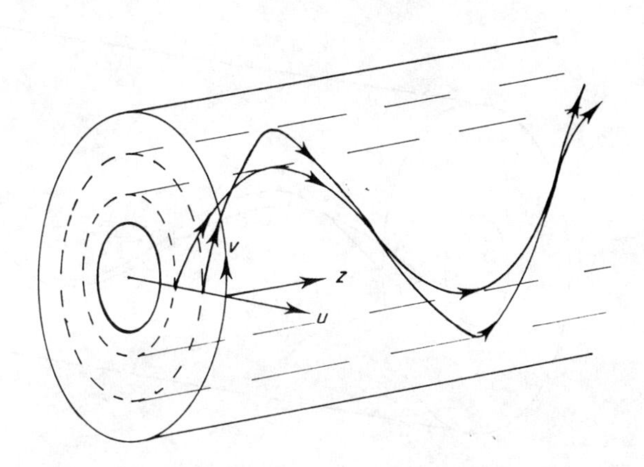

Figure 4.24: Spiralling band of rays on a tube with radially varying refractive index

If, however, we dispense with the condition of regularity along the axis on the assumption that the guiding medium is a hollow tube, we can, in such a case, establish exact conditions for spiral rays from a flat pencil to retain coherence by making the pitch independent of the parameter u (but not, obviously, independent of ψ). This implies that the rays from an extended radial-line source would spiral as a ribbon with constant pitch, as shown in Figure 4.24.

Retaining $B(v) = 0$ in the first instance, then, for constant helix angle $p = \tan\psi$, the ratio of the two expressions in Equations 4.6.4 becomes

$$\sqrt{\frac{\eta^2 - [A(u)]^2}{A(u)}} = p \tag{4.6.10}$$

and thus

$$\eta^2 = (p^2 + 1)[A(u)]^2 \quad p > 0$$

Substitution into Equation 4.6.7 gives for the circular cylindrical system with $g_{22} = u^2$

$$(p^2 + 1)[A(u)]^2 = \frac{1}{u^2}\int 2[A(u)]^2 u \, du$$

or

$$A(u) = au^{-p^2/(p^2+1)}$$

Such a solution is possible only where g_{22} is a function of u alone. Thus

$$\eta^2 = a^2(1 + p^2)u^{-2p^2/(p^2+1)}$$

The limit of a uniform medium and straight rays is attained as p tends to zero.

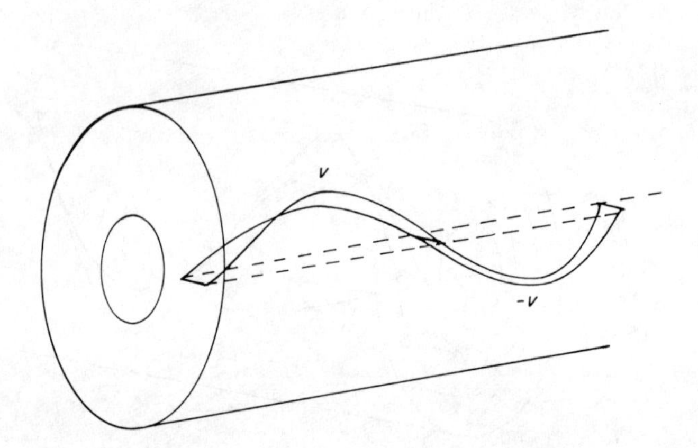

Figure 4.25: Frustrated spiral forms a coherent band within angular limits $\pm v$

General functions of $B(v)$ can now be included. This has the effect of preventing the complete spiral from forming as the numerator in Equation 4.6.9 has a zero value at some value of v. The ribbon of rays will fluctuate in a zone confined between these v values. Figure 4.25 shows the effect for arbitrarily chosen $B(v) = -2\sin^2 v$.

With such an included function $B(v)$, the refractive index law becomes

$$\eta^2 = a^2(1+p^2)u^{-2p^2/(p^2+1)} + B(v)/u^2 \tag{4.6.11}$$

In this manner, isolated channels of rays can be formed on a single circular tubular fibre. In particular, for $p = 1$

$$\eta^2 = 2a^2/u + B(v)/u^2$$

It is now apparent that the same procedure can be adopted for any coordinate system for which g_{22} is a function of u alone, and where the surfaces $u =$ constant are closed continuous cylinders containing the axis.

This generalizes the result of Equation 4.6.11

$$\eta^2 = a^2(1+p^2)[g_{22}(u)]^{-p^2/(p^2+1)} + B(v)/g_{22} \tag{4.6.12}$$

Equation 4.6.9 is generally integrable by the method of Appendix III for those coordinate systems for which $g_{22} = f_v^2 + h_v^2$ is a constant or is a function of v only.

For rays not performing continuous spirals, the trajectory will in general fluctuate about a line generator of the cylinder. If we consider a pencil of rays from a source on the generator $v = 0$, then the angle made by the ray there is given by

$$\eta(u, v = 0) = A(u)\sec\psi$$

Also from Equation 4.6.9, the ray will be parallel to the generators again at a value $v = v_1$, where

$$\eta(u, v = v_1) = A(u)$$

Since these occur on the surface $u =$ constant $= c$, we can designate them

$$\eta_{c,0} = A(c)\sec\psi$$

and

$$\eta_{c,1} = A(c)$$

On the surface itself, $\eta = \eta(c, v)$.

For the continuous coherent refocusing of a flat pencil of rays (flat in the sense of all rays being confined to the same surface), we require Equation 4.6.9 to be independent of ψ on the surface $u =$ constant, i.e.

$$z = \int_0^{v_1} A(c)\sqrt{\frac{g_{22}(u = c)}{\eta^2(c, v) - A^2(c)}}\,\mathrm{d}v = \kappa \tag{4.6.13}$$

where

$$\eta^2(c, v) = \frac{1}{g_{22}}\left[\int A^2(u)\frac{\partial g_{22}}{\partial u}\,\mathrm{d}u + B(v)\right]$$

evaluated at $u = c$.

For simplicity, we can write $\eta^2(c, v) = D^2(v)/g_{22}(u = c)$ and make the appropriate substitutions into Equations 4.6.7 to give

$$\int_0^{v_1} \frac{\eta(c, 0)\cos\psi\sqrt{g_{22}(u = c)}\,\mathrm{d}v}{\sqrt{D^2(v)/g_{22}(u = c) - \eta^2(c, 0)\cos^2\psi}} = \kappa$$

or

$$\int_0^{v_1} \frac{D(0)\cos\psi[g_{22}(u = c)]}{\sqrt{D^2(v) - D^2(0)\cos^2\psi}}\,\mathrm{d}v = \kappa \tag{4.6.14}$$

and where v_1 is the value of v for which

$$\cos\psi = \frac{D(v)}{D(0)}$$

that is, the ray is parallel to the axis.

Equation 4.6.14 can be integrated by the formal procedure of Appendix III giving

$$D(v) = \operatorname{sech}\left\{\frac{\pi}{2\kappa}\int [g_{22}(u = c)]\,\mathrm{d}v\right\}$$

and consequently

$$\eta(c, v) = \frac{1}{\sqrt{g_{22}(u = c)}}\operatorname{sech}\left\{\frac{\pi}{2\kappa}\int [g_{22}(u = c)]\,\mathrm{d}v\right\}$$

and thus generally

$$\eta^2(u,v) = \frac{1}{g_{22}}\left(\int [A(u)]^2 \frac{\partial g_{22}}{\partial u}\,\mathrm{d}u + \mathrm{sech}^2\left\{\frac{\pi}{2\kappa}\int [g_{22}(u=c)]\,\mathrm{d}v\right\}\right) \quad (4.6.15)$$

The same analysis can be carried through for an orthogonal surface ($v =$ constant) by first transforming the surfaces by the substitution $u \leftrightarrow v$, deriving the results in Equation 4.6.14 and then transforming back. Thus, two orthogonal-plane pencils of refocusing rays can exist in the cylindrical medium, not necessarily confined to the axis, since the surface $u = c$ and the source can be arbitrarily positioned. In this case $A(u)$ can be derived to symmetrize Equation 4.6.15 with the result

$$\eta^2(u,v) = \frac{1}{g_{22}}\left\{\mathrm{sech}^2\left[\frac{\pi}{2\kappa}\int g_{22}(u=c_1)\,\mathrm{d}v\right]\right\} + \frac{1}{g_{11}}\left\{\mathrm{sech}^2\left[\frac{\pi}{2\kappa}\int g_{11}(v=c_2)\,\mathrm{d}u\right]\right\} \quad (4.6.16)$$

for a source located on the intersection of the two surfaces $u = c_1$, $v = c_2$.

Coordinate systems can be chosen, which for small values of one coordinate can approximate to a rectangular strip, or even to a thin film guide, for which the above law, in its small-value approximation, would give the refractive index profile for continuous self-focusing rays.

In certain cases, where the field can be expressed by a single component E vector, such as the perpendicular component of the field over a conducting ground plane, this refractive index law can be converted to an equivalent height profile for a dielectric guiding strip.

4.7 RAYS IN SPHERICAL AND AXISYMMETRIC MEDIA

Ray-tracing in a non-uniform spherical medium provides a design method for many practical optical devices, mainly in the field of microwaves where transparent non-uniform refractive media can be constructed from foamed or loaded plastic materials. Early studies derived mainly from the work of biologists [31] who found that the focusing properties of the eyes of marine animals required such ray tracing methods. New technologies applying to glass manufacture make it possible to design optical components with varying refractive index, and if a diametral cross-section of the spherical medium alone is considered then integrated optical components with specified focusing properties can be designed by ray tracing methods.

However, besides being another example of the method for solving Equations 4.2.3 and 4.2.13 some other theoretical consequences have been discovered similar to the inversion method given in previous chapters. There is a well-known analogy between optical rays and the dynamics of particles, both being describable by a similar form of least-action principle. The theoretical discoveries applying to optical rays therefore have an immediate connection with gauge transformation theory of the phase space of the dynamical system. In this chapter we shall deal with the practical design methods and transformations that are allowed and we shall consider the consequences of the transformation theories in the final chapter.

THE SPHERICALLY SYMMETRIC MEDIUM

The procedure is similar to that of the previous chapter. The ray equations become integrable when the ray is considered to be confined to a surface with a constant value of one coordinate, and for the first we take the spherically symmetrical medium with rays confined to the equatorial plane. The same analysis would be applicable to the *transverse* cross-section of the cylindrical systems of the previous chapter.

We therefore put $\mathrm{d}\theta/\mathrm{d}s = 0$, $\theta = \text{constant} = \pi/2$, and $\partial\eta/\partial\phi = 0$, and $\partial\eta/\partial\theta = 0$ in Equations 4.2.3 reducing them to

$$\frac{\mathrm{d}}{\mathrm{d}s}\left(\eta\frac{\mathrm{d}r}{\mathrm{d}s}\right) - \eta r\left(\frac{\mathrm{d}\phi}{\mathrm{d}s}\right)^2 = \frac{\partial\eta}{\partial r}$$

and

$$\frac{\mathrm{d}}{\mathrm{d}s}\left(\eta r\frac{\mathrm{d}\phi}{\mathrm{d}s}\right) + \eta\frac{\mathrm{d}r}{\mathrm{d}s}\frac{\mathrm{d}\phi}{\mathrm{d}s} = \frac{1}{r}\frac{\partial\eta}{\partial\phi} = 0 \qquad (4.7.1)$$

The ray is now given by $r = r(\phi)$, and

$$\mathrm{d}s^2 = \mathrm{d}r^2 + r^2\,\mathrm{d}\phi^2 \qquad (4.7.2)$$

The consequence of Equation 4.1.11

$$\operatorname{curl}(\eta\hat{\mathrm{s}}) = 0 \qquad (4.7.3)$$

is that, in the complete refractive space, the rays are all closed loops. If we include the line at infinity, then closed loops in the projective plane includes curves such as parabolas and hyperbolas. This too exemplifies the dynamical analogue since Equation 4.7.3 describes a conservative field, and hence one that derives from a potential function.

With circular symmetry Equation 4.7.1 reduces to

$$\frac{\mathrm{d}}{\mathrm{d}s}\left(\eta r^2\frac{\mathrm{d}\phi}{\mathrm{d}s}\right) = 0$$

or

$$\eta r^2\frac{\mathrm{d}\phi}{\mathrm{d}s} = \text{constant} = A \text{ along a given ray} \qquad (4.7.4)$$

Writing this as $\eta r r\,\mathrm{d}\phi/\mathrm{d}s = A$, we have

$$\eta r\sin\psi = A \qquad (4.7.5)$$

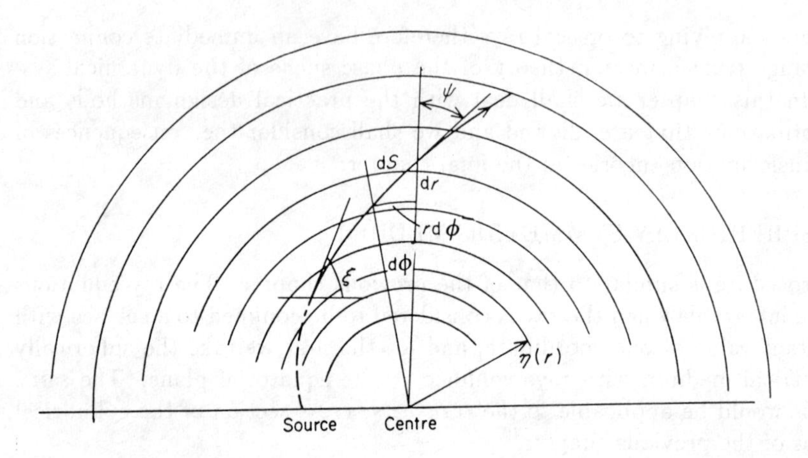

Figure 4.26: Ray path in a spherical medium of varying refractive index. Bouguer's theorem

where ψ is the angle between the radius vector to a point on the ray whose ray constant is A, and the tangent to the ray at that point (Figure 4.26). This generalizes Snell's law for a spherical medium and Equation 4.7.5 is known as Bouguer's theorem.

Substitution of this result, and applying Equation 4.7.2 to Equation 4.7.1, gives

$$\mathrm{d}\phi = \frac{A\,\mathrm{d}r}{r\sqrt{r^2\eta^2(r) - A^2}}$$

and consequently

$$\phi_1 - \phi_2 = \int_{r_1}^{r_2} \frac{A\,\mathrm{d}r}{r\sqrt{r^2\eta^2 - A^2}} \tag{4.7.6}$$

This is the polar equation of the ray trajectory as shown in Figure 4.27. Due to the fact that r may be variously an increasing or decreasing function of ϕ, the range of integration has to be split at a value r_m where this changes. That occurs whenever the ray is at either its closest to the origin or the furthest away, and at both these points the ray will be parallel to the stratification. Hence ψ in Equation 4.7.5 will be $\pi/2$ at these turning values, $r\eta(r)$ will equal A and the denominator in Equation 4.7.6 will be zero. Thus the limit r_m in the integral will be that value which makes the denominator zero.

If the boundary conditions are chosen such that $\phi_1 = 0$ when $r_1 = 1$, the polar equation of the ray is

$$\phi = \int_1^r \frac{A\,\mathrm{d}r}{r\sqrt{r^2\eta^2(r) - A^2}} \tag{4.7.7}$$

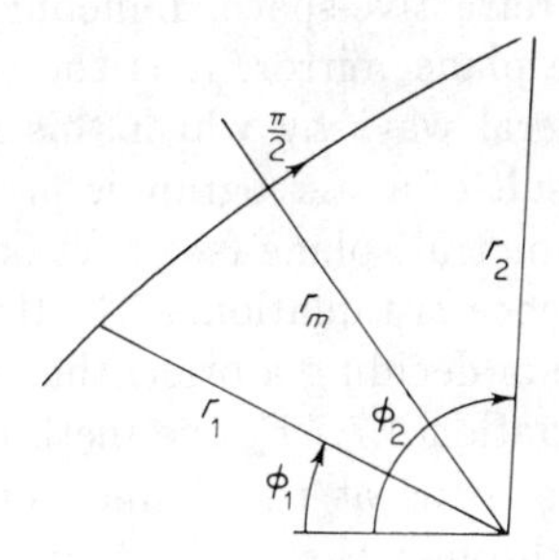

Figure 4.27: Turning value of the ray in the spherical medium occurs at the point where the ray tangent is perpendicular to the radius vector

The solution of this equation for ϕ given as a function of r for the refractive index law $\eta(r)$ is again the method using Abel's integral given in Appendix III.

An interesting transformation can be effected if, as in [32], we take the angular variable to be the angle between the tangent to the ray and a line parallel to the central diameter containing the source, angle ξ in Figure 4.26.

Then using the relation given in the reference [32]

$$\mathrm{d}\xi = \frac{\tan \psi}{\eta}\,\mathrm{d}\eta$$

and Equation 4.7.5, we obtain

$$\mathrm{d}\xi = \frac{A\,\mathrm{d}\eta}{\eta\sqrt{\eta^2 r^2 - A^2}}$$

This can be obtained from Equation 4.7.7 by the direct transformation

$$r \leftrightarrow \eta \qquad \phi \leftrightarrow \xi$$

This integration of Equation 4.7.1 takes on two forms depending upon whether the refractive index $\eta(r)$ is specified in advance or whether the refractive index has to be derived from the required geometry of the rays as given by their (θ, r) variation. In the latter case a standard procedure exists in which the integration is performed through the use of Abel's theorem (Appendix III).

THE SPHERICAL SYMMETRIC LENSES

The archetypal non-uniform spherical lens is considered to be Maxwell's "fish-eye". In it all the rays from a source on the perimeter of the unit sphere ($\phi = 0$ when $r = 1$) and at which point the refractive index is unity, are imaged into the diametrically opposite point. The result is the "ideal lens" which images perfectly every point in the refractive space. Luneburg [33] proves that, apart from the trivial case of the plane mirror, it is the only optical device with this ability. There are several ways by which this result can be obtained. Luneburg gives it as the result of a stereographic projection of the geodesics of the sphere on to the (diametral) plane (see Section 4.11). Alternatively it can be derived as a consequence of Equation 4.7.3. Here, and in what follows, we use the general principle of deciding a priori the ray path as a function of r and ϕ and integrating Equation 4.7.7 by the method of Appendix III.

Since all rays come to a focus at the diametrically opposite point, the radius vector has to turn through the angle $\phi = \pi$ for all values of A in Equation 4.7.7. Hence

$$\pi = 2\int_1^{r_m} \frac{\sin\alpha\, dr}{r\sqrt{r^2\eta^2(r) - \sin^2\alpha}} \tag{4.7.8}$$

where, as shown in Figure 4.28, r_m is the radius at which the ray becomes horizontal. From the solution in Appendix III, Equation 7

$$r\eta(r) = \text{sech}(\log r)$$

and hence

$$\eta(r) = \frac{2}{1 + r^2} \tag{4.7.9}$$

Consider first the entire space for which the law of Equation 4.7.9 applies, that is extending it throughout the region in which the refractive index has a fictitious value less than unity.

The ray paths are then completely circular and all the rays through a given fixed point, the source, intersect again at a second common point the focus. The focusing is thus of the continuous closed loop kind. If the source is at a point on the unit circle where $\eta = 1$, the focus is at the diametrically opposite point on the same circle. The rays then form a system of circles (Figure 4.28) with centres on the diameter intersecting the source-focus diameter at right angles. If α is the angle made by an individual ray with this diameter the equations to these circles are given parametrically by substituting this result into Equation 4.7.7. The ray trajectories are the coaxal system of circles, shown in Figure 4.28, with equations

$$r = \sqrt{1 + \sin^2\phi\cot^2\alpha} + \sin\phi\cot\alpha \tag{4.7.10}$$

Luneburg shows that this ideal focusing effect continues into the space external to the unit sphere and hence to regions where the refractive index is less than unity. Since the rays are a coaxal system of circles, the orthogonal system is also circular, and hence the phase fronts form circles with centres on the diameter containing the source. Thus if the lens is terminated by a surface coinciding with a spherical phase front, the ray will continue normal to that

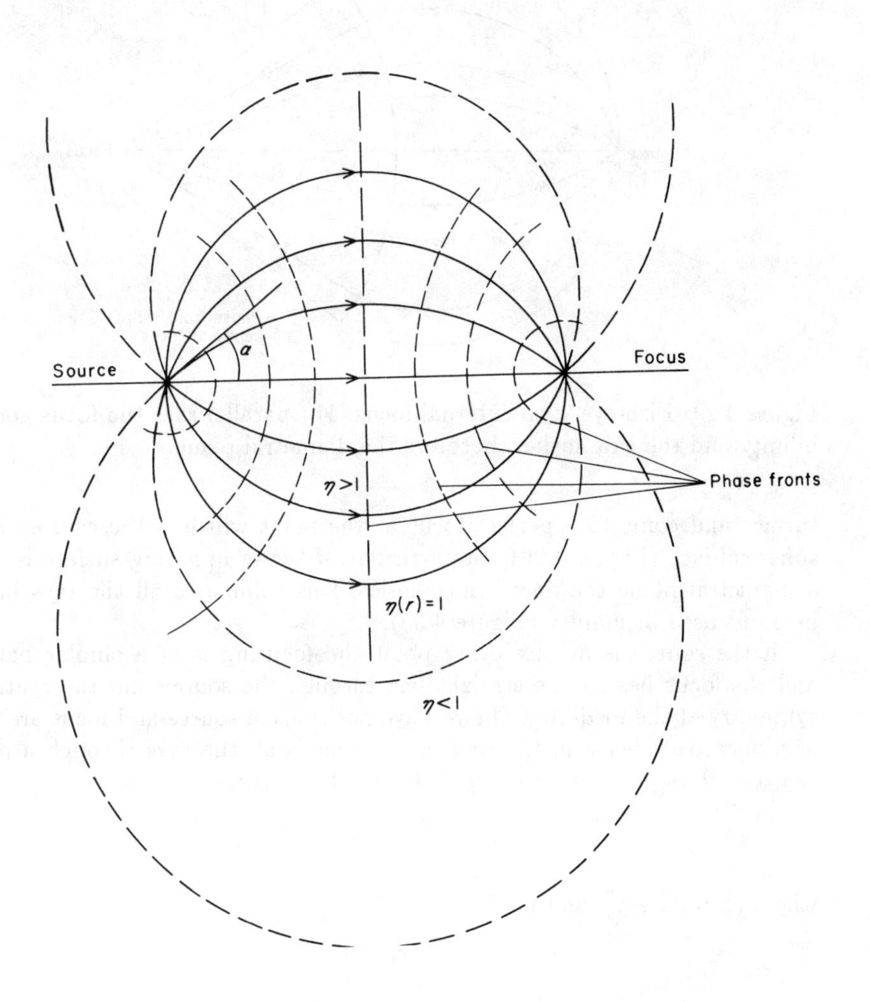

Figure 4.28: Rays in Maxwell's fish-eye lens

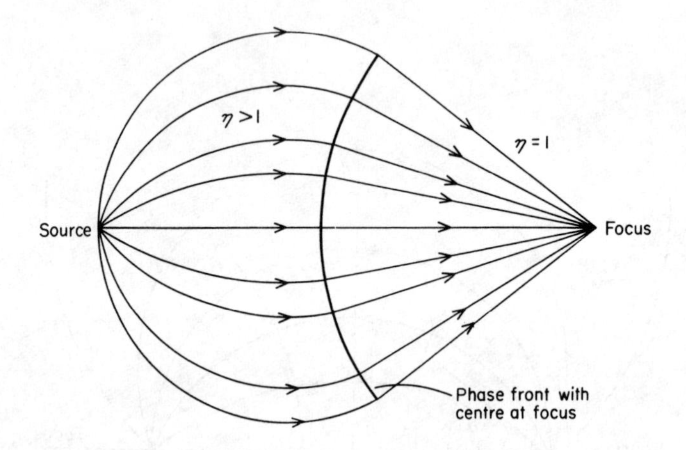

Figure 4.29: Fish-eye with external focus. For parallel rays the focus goes to infinity and the exit surface becomes the diametral plane

surface and come to a perfect focus at the point which is the centre of the spherical face (Figure 4.29). In particular, if the terminating surface is itself a diametral plane the now hemispherical lens collimates all the rays into a beam focused at infinity (Figure 4.30).

If the source is at any other point the focusing is of a similar nature, and the focus lies on the straight line through the source and the centre of symmetry of the medium. The relative positions of source and focus are that of a negative inverse in the unit circle, that is all the rays through a point (x_0, y_0) intersect again at (x_1, y_1) (Figure 4.31) where

$$x_1 = \frac{-x_0}{r_0^2} \qquad y_1 = \frac{-y_0}{r_0^2}$$

where $r_0^2 = x_0^2 + y_0^2$, and if $r_1^2 = x_1^2 + y_1^2$,

$$r_0 r_1 = 1$$

In all cases the wave fronts are the orthogonal system of circles.

One generalization of the Maxwell fish-eye needs to be noted at this stage. If instead of focusing all the rays to the diametrically opposite point as in Figure 4.28 the rays are all focused at a point $\phi = \pi/a$, then the left-hand side of Equation III.7 in Appendix III becomes $2\pi/a$ and hence the refractive index law

$$r\eta(r) = \text{sech}\left(\frac{a}{2}\log r\right)$$

or

$$\eta(r) = \frac{2r^{a-1}}{1 + r^{2a}} \tag{4.7.11}$$

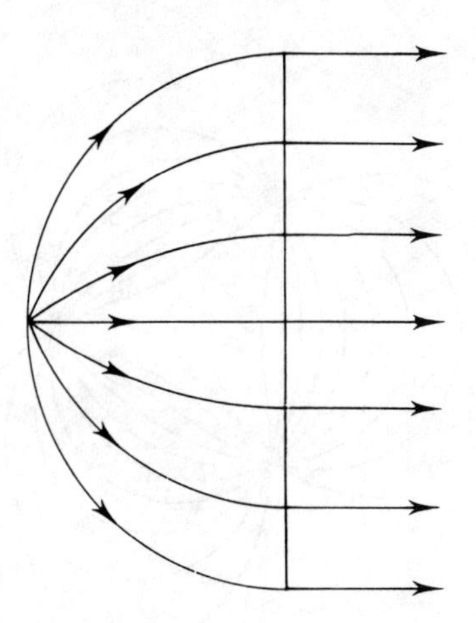

Figure 4.30: The hemispherical fish-eye lens focused at infinity

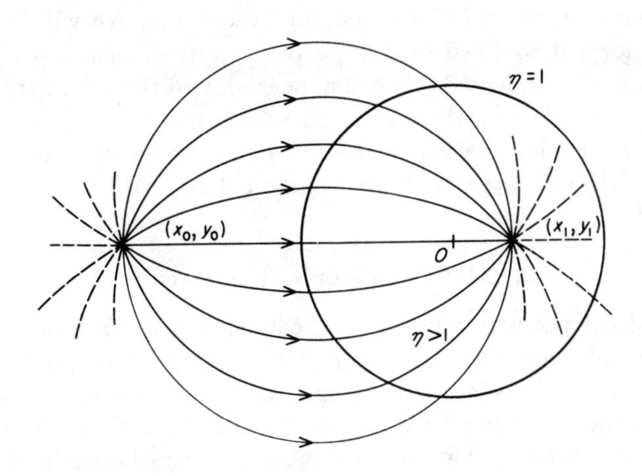

Figure 4.31: Maxwell fish-eye with displaced source

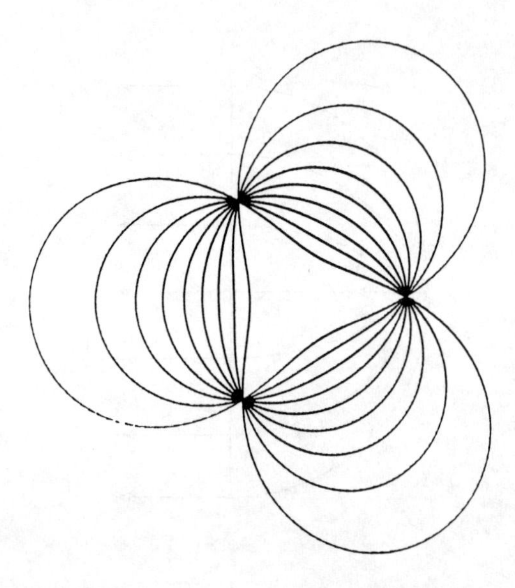

Figure 4.32: The generalized fish-eye refocusing at points on the unit circle - the ring resonator

Obviously the system is no longer spherically symmetrical, and hence the ray pattern applies to diametral planes only otherwise a source will be focused into a ring of foci. If confined to a single plane section, then if a is a whole number fraction of π, repeated foci will appear around the periphery and one such focus will coincide with the source. The result would be a ring resonator. The ray pattern for this situation can be derived by the substitution of the law of Equation 4.7.11 into the integral of Equation 4.7.7, with the result shown in Figure 4.32

$$r^{-a} = \sqrt{1 + \sin^2 a\phi \cot^2 \alpha} + \sin a\phi \cot \alpha \tag{4.7.12}$$

The second class of non-uniform spherical lenses is again based upon the interior of the unit circle with, in the complete refractive space, perfect focusing of the rays at a point diametrically opposite to the source. In this second case, however, the rays are all parallel to the axis when they first meet the unit circle, at which point the refractive index is unity. Hence in real space, the rays all continue to be parallel to this direction after leaving the lens. The refractive law for such a lens was first derived by Luneburg [34] again and we can label this class of lens as the generalized Luneburg lens. For the general pattern of rays shown in Figure 4.33, all the rays from the source are required to arrive again at the unit circle where their directions will all be parallel to the angle $p\pi/2$ as shown. The similarity of this effect to the manner in which the fish-eye was generalized can be noted. Again ray spherical symmetry is

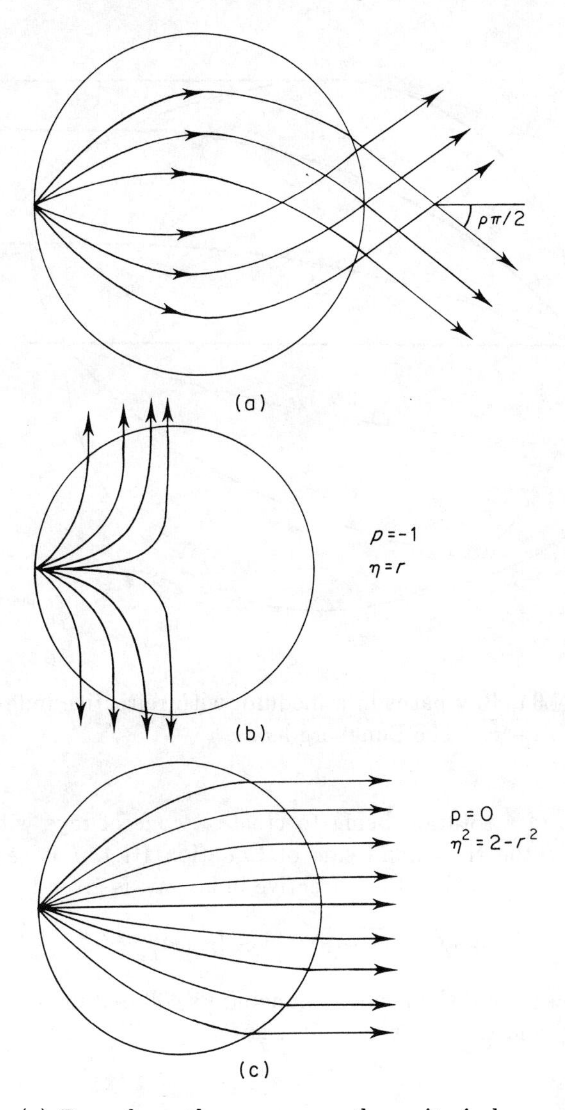

Figure 4.33: (a) Rays from the source on the unit circle separated into two beams
at angles $\pm p\pi/2$. The refractive law is obtained from

$$[1+\sqrt{1-\eta^2 r^2}\,]^{p+1} = [\eta r]^{p+2}/r^2$$

(b) With $p = -1$ all rays are perpendicular to the axis and $\eta = r$ (fictitious since $\eta < 1$).
(c) With $p = 0$ all rays are parallel to the axis - the Luneberg lens with refractive law $\eta^2 = 2 - r$

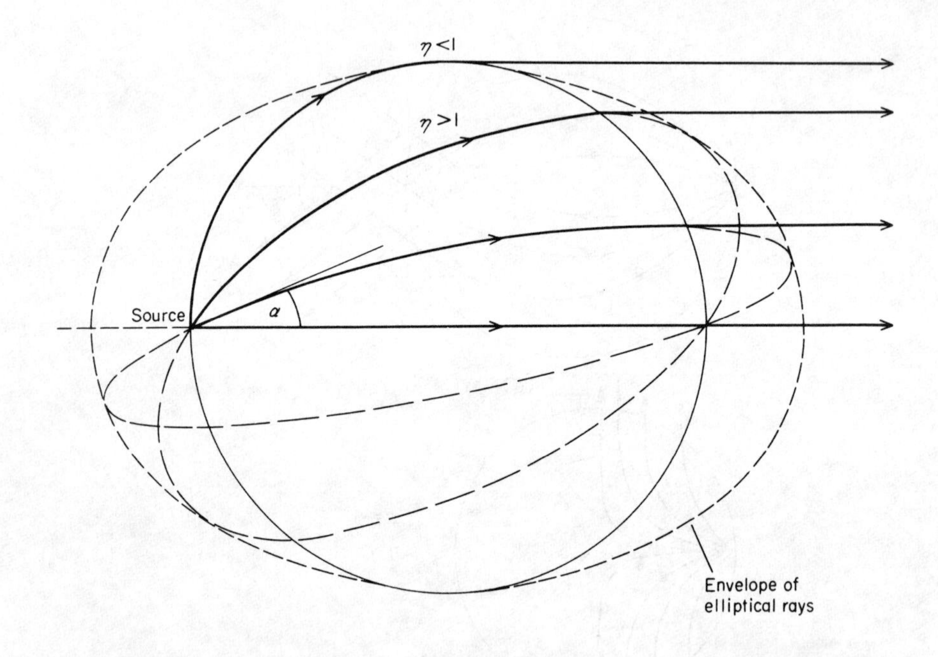

Figure 4.34: Ray paths in a medium with refractive index law $\eta(r) = \sqrt{2 - r^2}$. The Luneberg lens

lost, the result of a rotation being to create a cone of rays with half-angle $p\pi/2$. Hence in the right-hand side of Equation III.4, I is required to be $\pi/2 + p\pi/4 - \alpha/2$, and thus the refractive index law is given by

$$[1 + \sqrt{1 - r^2\eta^2(r)}\,]^{p+1} = [r\eta(r)]^{p+2}/r^2 \tag{4.7.13}$$

If p is put equal to zero the result is Luneburg's collimating lens having both total symmetry and the refractive index law

$$\eta^2(r) = 2 - r^2 \tag{4.7.14}$$

The ray paths, from Equation 4.7.7 using the substitution $r^2 = 1/v$ are all ellipses (Figure 4.34) passing through the two diametrically opposite points with equations

$$r^2 = \sin^2\alpha/[1 - \cos\alpha\cos(2\phi + \alpha)] \tag{4.7.15}$$

The major axes are inclined at $\alpha/2$ to the diameter containing the source and of length $2\sqrt{2}\cos(\alpha/2)$, the minor axes are of length $2\sqrt{2}\sin(\alpha/2)$ and the foci are at points $r = \sqrt{\cos\alpha}$. The medium only extends to the maximum value $r^2 = 2$ from Equation 4.7.14 and the refractive index is only larger than unity within the unit circle.

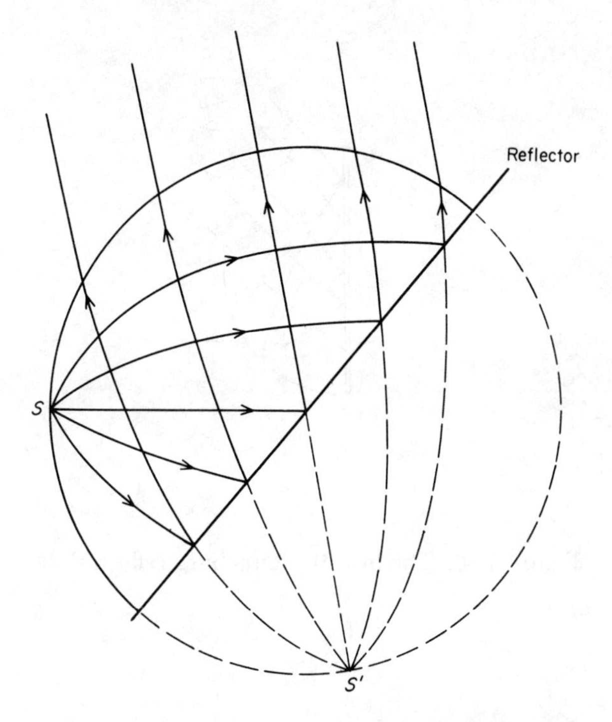

Figure 4.35: Virtual source reflecting Luneberg lens

The envelope of this system of ellipses is another ellipse

$$x^2/2 + y^2 = 1$$

If the focusing is in the extended medium and thus of closed loops (Figure 4.34) in this case the exit surface normally used is that of the unit sphere itself. The tangents to the ellipses are all parallel to the x axis at the points where they intersect the exit surface and, since the refractive index is unity at that surface, no further refraction occurs, and the entire ray system is then parallel to the axis as required. Adaptations of the Luneburg lens include the reflecting hemispherical lens and the retro-reflector (Figures 4.36 & 4.37)

EXTENSIONS OF THE LUNEBURG LENS

There are two fundamental extensions to Luneburg's analysis that have been made, both with the practical application of the concept to microwave antennas in mind.

The first of these is by Morgan [35] who continues the original system analysed by Luneburg, that of the spherical region with external source and focus, to that where the refractive index law is piecewise continuous. That is the sphere is constructed of shells in each of which the refractive index has

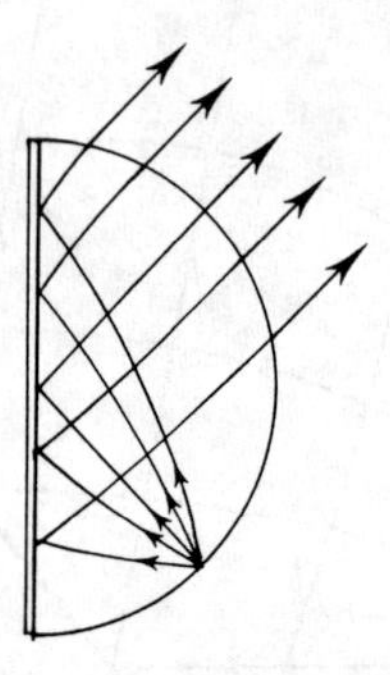

Figure 4.36: The bisected Luneburg reflector

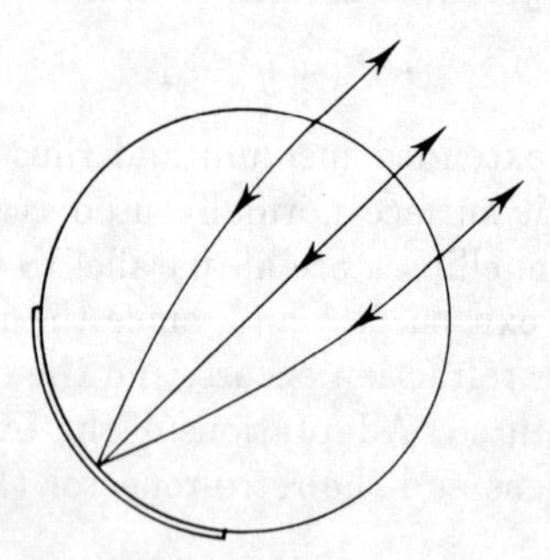

Figure 4.37: The Luneburg lens retro-reflector

a particular law. The most important of these is the lens of two regions in which the outer shell has a constant refractive index and the core then has a derived refractive index law. It is the derivation of these laws for particular outer layer parameters that is the concern of the reference given.

The method used is that of the appendix as before, but as basic study of the entire subject of spherical non-homogeneous lenses, the analysis and conclusions of the reference are fundamental reading to anyone intending the design of such an antenna.

Specifically once the refractive lens for the outer layer is specified, the central region law is determined. The different lenses that are obtained from different specified outer layers give rise to different aperture amplitude distributions, when non-isotropic sources are used and thus affect the radiation patterns eventually achieved.

Considering the case where the outer layer is of a constant refractive index, then since this cannot be unity a discontinuity in refractive index now occurs at the surface and a refraction takes place. This lens as shown in Figure 4.38(a) then strictly only applies to the rays contained between the rays tangent to the inner core as shown. As such it is applicable to a microwave source with pattern width less than $\pm 90^o$. The design no longer falls within the present theoretical study of perfectly focusing systems, but is an approximation, which however remains perfectly adequate for most practical designs as will be shown later. We note in this respect the conclusion reached in the treatment, namely that a uniform outer layer results in a refractive index law for the central core that has a *higher* value at the centre than did the original Luneburg lens, but that the *overall* variation in refractive index becomes smaller. Some typical results are shown in Figure 4.38(b). Some practically designed lenses omit to take these conclusions into account, when including a surface protection layer.

THE GUTMAN LENS

The second extension to the Luneburg result is the refractive index law obtained when the source is within the unit sphere itself. For a focus at infinity, the solution has been obtained by Gutman [36] and the law derived has unity value at the surface of the unit sphere, which then becomes the exit surface. The rays thus have to have tangents, at their intersections with the sphere, which are parallel to the x axis, as in the original Luneburg design. Gutman derives these ray paths by yet another method which, although given by Luneburg [8] again was not in fact used in designing his lenses. The principle is to consider a light ray as the path of a particle in a potential field. For a refractive index η this potential field is given by

$$\phi = -\frac{1}{2}\eta^2 \qquad \text{cf Equation 4.1.4}$$

By this association it can be seen that a great deal of the theory of particle dynamics, can be applied to the theory of optics and the whole of the

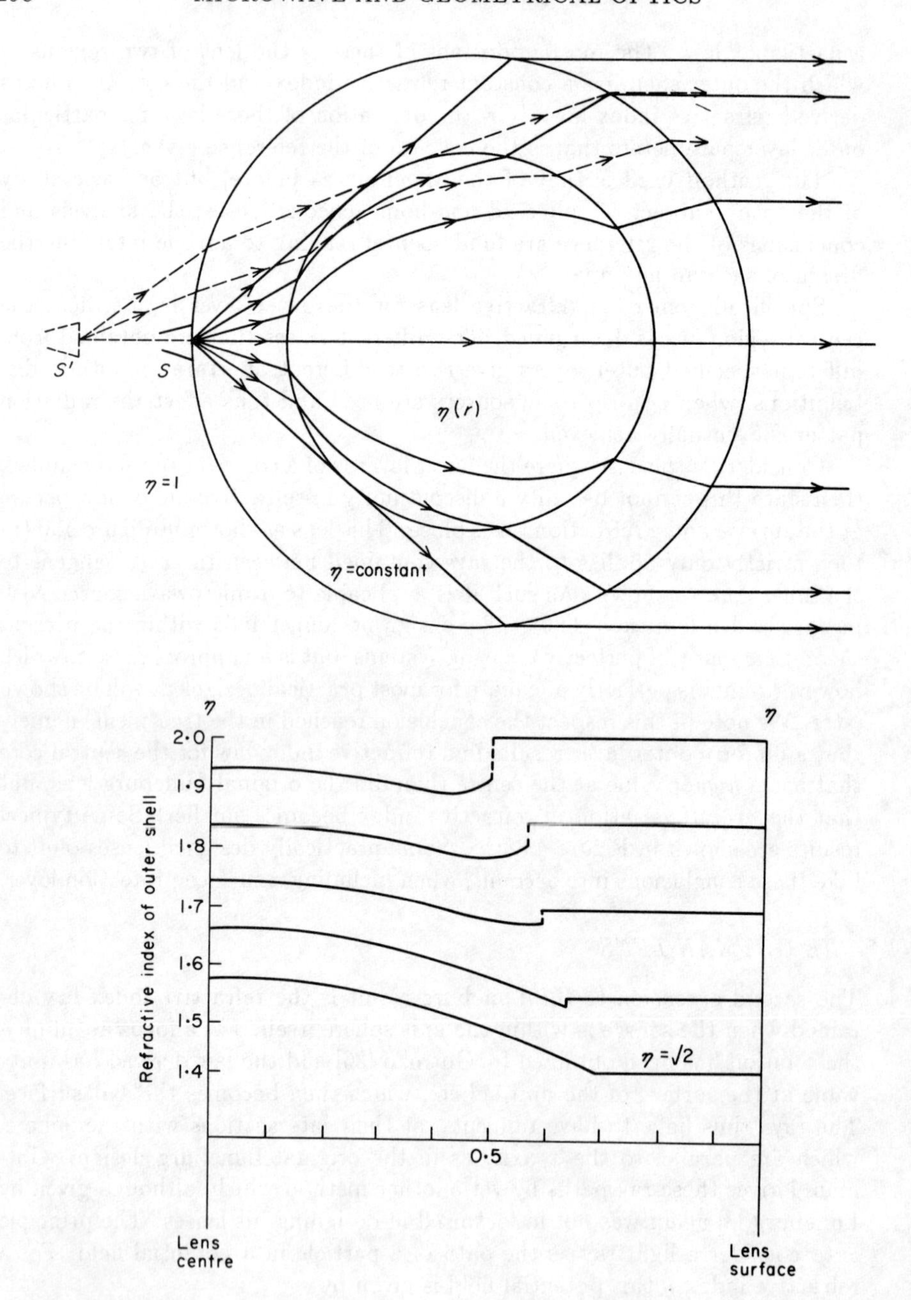

Figure 4.38: (a) Ray paths for lens surrounded by a shell of constant refractive index. The source has pattern width less than $\pm 90^o$. S' is an alternative design for a source not on the actual surface of the lens.
(b) Refractive index of lens with constant outer shell

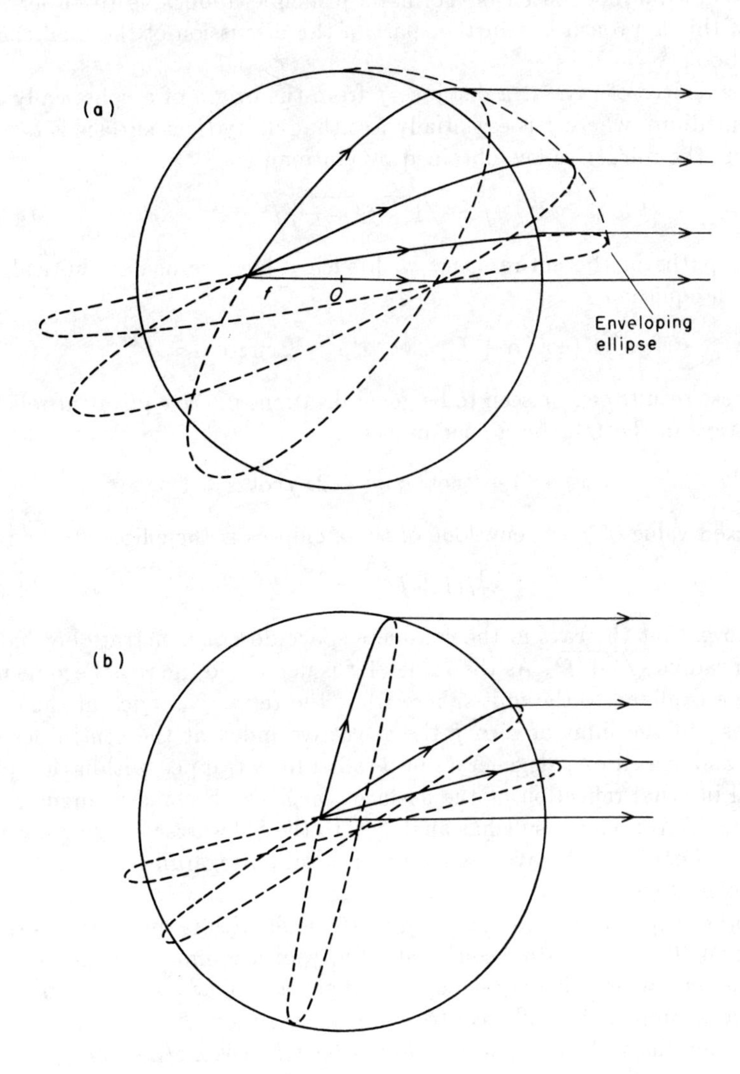

Figure 4.39: Ray paths in a Gutman lens (a) $f = 1/2$ (b) small f

Hamiltonian method based on Fermat's principle applicable to either. The value of this approach is a further part of the discussion of the final chapter of this book.

For a source of rays at a distance f from the origin of a spherically symmetric medium, where f is essentially less than unity (lens surface is taken to be $r = 1$) the refractive law obtained by Gutman is

$$\eta = \sqrt{1 + f^2 - r^2}/f \tag{4.7.16}$$

The ray paths in the entire space with such a law are again elliptical, this time with equation

$$x^2 + (\cot^2\alpha + f^2\operatorname{cosec}^2\alpha)y^2 - 2xy\cot\alpha = f^2 \tag{4.7.17}$$

Both these results can be seen to be generalizations of the equivalent relations for the rays in the Luneburg lens, namely

$$x^2 + (1 + 2\cot^2\alpha)y^2 - 2xy\cot\alpha = 1$$

For a fixed value of f the envelope of these ellipses is the ellipse

$$x^2/(1 + f^2) + y^2 = 1 \tag{4.7.18}$$

This shows that the rays in the complete space do not penetrate beyond the circle of radius $\sqrt{1 + f^2}$. As the value of f is decreased the rays become more and more confined to the unit sphere while the refractive index at the centre increases. In the limit of zero f the refractive index at the centre becomes infinite and the ellipses degenerate into radial lines trapped within the sphere by total internal reflection at the surface which has become a singularity of Equation 4.7.16. This result has analogies to the Schwarzschild singularity in relativity theory, which gave rise to the proposition regarding the existence of black holes (Figure 4.39)

Another ray pattern that can be directly dealt with is the situation shown in Figure 4.40, that is with a source at infinity, or a collimated beam, incident upon the lens which thereupon separates into a similar conical beam to the foregoing example. For all rays to make an angle $q\pi/2$ as shown, the substitution for the right-hand side of Equation III.12 is $q\pi/2 - 2\alpha$, giving the refractive index law

$$[1 + \sqrt{1 - r^2\eta^2(r)}]^{q-2} = [r\eta(r)]^2/r^2 \tag{4.7.19}$$

When $q = 2$ the rays continue as straight lines through the lens since $\eta = 1$. When $q = 0$ the rays enter the lens, perform one orbit about the origin and are transmitted parallel to the incident direction. Equation 4.7.19 gives for this result

$$\eta^2 = \frac{2}{r} - 1 \tag{4.7.20}$$

This also results from the value $q = 4$.

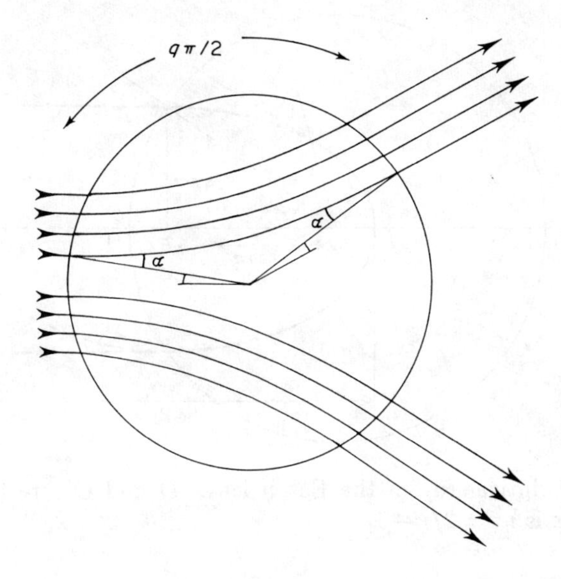

Figure 4.40: The radially variable beam divider. Power division can be achieved by displacing the input beam transversely from the axis

Substitution of this law into Equation 4.7.9 shows the rays in the complete refractive medium to be ellipses all with a focus at the origin and all with major axis of length 2 lying along the diameter parallel to the direction of incidence (Figure 4.41). Their equations are

$$r = \sin^2\alpha/(1 - \cos\alpha\cos\phi) \tag{4.7.21}$$

This lens was first derived by Eaton [37] as a perfect 360° retro-reflector. Unfortunately the innermost ellipses require such high eccentricity that the principle of minimum curvature (the radius of curvature of a ray cannot be smaller than the wavelength of the radiation) is violated and the resultant design is highly inefficient.

4.8 THE GENERAL THEORY OF FOCUSED RAYS IN A SPHERICALLY SYMMETRICAL MEDIUM

As can be seen from the foregoing, the spherically symmetrical non-homogeneous medium has been the focus of attention with regard to applications to microwave lenses. We consider as before the properties of a total space in which the refractive index law is spherically symmetric with regard to a fixed origin of coordinates, and we allow in this space the entire range of fictitious refractive indices including values that are less than unity, negative, zero and

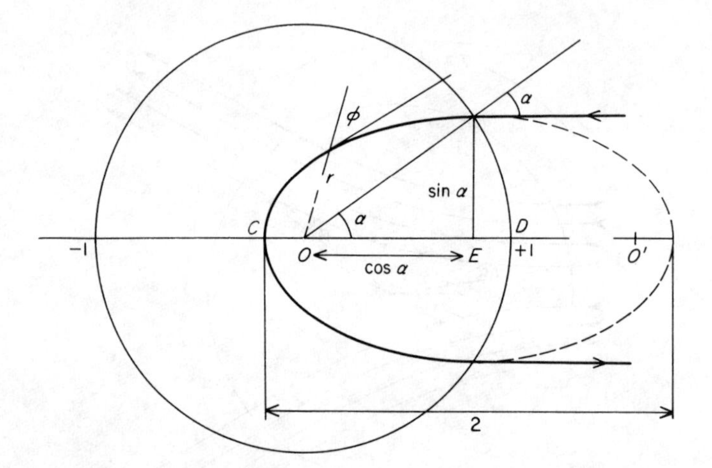

Figure 4.41: Elliptical ray in the Eaton lens. O and O' are foci. Law of refractive index is $\eta^2 = 2/r - 1$

infinite (but *not* complex). The refractive index is a continuous function of the radial coordinate r and has a continuous first derivative. Real lenses can be constructed in this space by defining a region with "realistic" refractive index. If, in such a medium perfect focusing is possible, then all the rays intersecting at a given point, the source, will intersect again at a second point, the focus and hence all the rays must form closed loops. If we include in this space (or the plane cross-section we need only consider) the circle at infinity then curves such as hyperbolas and parabolas can be considered also as closed loops. The plane is then the projective plane.

The optical distance from the source to the focus along any one such closed loop must be the same for both paths around the loop. In the sense of the dynamical analogue of particle trajectories, the field is conservative and hence describable by a potential function. This too can be seen from the fact that the curl of Equation 4.1.3 is identically zero, implying that the entire optical path around a loop in one direction of travel is zero.

We consider points P and Q on a loop to be image points. Then, by virtue of the equivalence of the *optical* path from P to Q, for the infinitesimally small triangle at P imaged in the small triangle at Q (Figure 4.42) we have

$$\eta_P PP' = \eta_Q QQ' \quad \text{and} \quad \eta_P PP'' = \eta_Q QQ''$$

that is

$$\eta_P r_P \, \mathrm{d}\theta = \eta_Q r_Q \, \mathrm{d}\theta \tag{4.8.1}$$

and

$$\eta_P \, \mathrm{d}r_P = \mp \eta_Q \, \mathrm{d}r_Q \tag{4.8.2}$$

the ambiguity in sign referring as in Chapter 1 to real or virtual focusing.

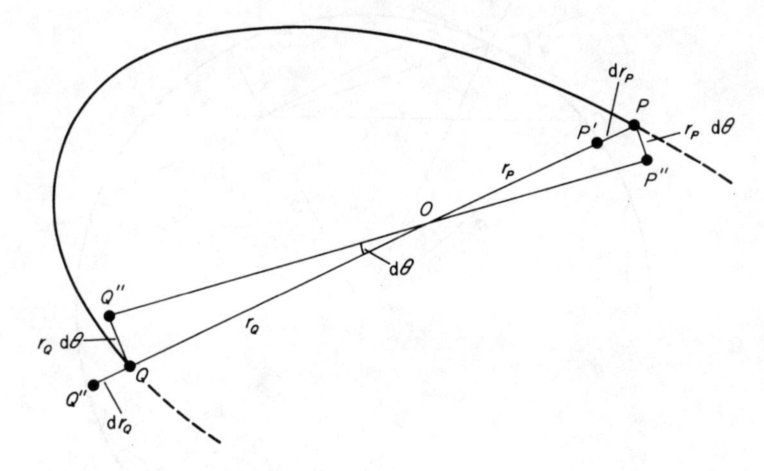

Figure 4.42: Source and image on a closed ray path

Division of these relations gives

$$\mathrm{d}\,[\log r_P] = \mp \mathrm{d}\,[\log r_Q]$$

and hence either

$$r_P r_Q = a^2 \quad \text{or} \quad r_P/r_Q = b^2 \tag{4.8.3}$$

where a and b are arbitrary constants.

If each source has only a single focus then P and Q can be interchanged hence $r_Q/r_P = b^2$ and thus for this situation $b^2 = 1$.

We now show that the first relation in Equation 4.8.3 gives the law for Maxwell's fish-eye and the second the Luneburg and Gutman lenses.

Ray paths obeying the law $r_P r_Q = a^2$ where P and Q are any two points on the same radius vector through the origin are inverse curves with respect to a circle radius a. We wish to derive the refractive index law for which the ray paths are circular. Let one such ray be the circle radius R centre M Figure 4.43 where OM is the length D. Then $2a$ is the length of the chord at right angles to OM and thus $r_P r_Q = a^2$. For a general point such as A on the ray we have

$$D^2 = R^2 + r^2 - 2Rr\sin\alpha \tag{4.8.4}$$

From Equations 4.8.1 and 4.8.2 we can derive Bouguer's law namely

$$\eta r \sin\alpha = \text{constant along a ray} = C$$

Then substituting into Equation 4.8.4

$$D^2 = R^2 + r^2 - 2RrC/\eta$$

or

$$\eta = \frac{2RC}{r^2 + R^2 - D^2} \tag{4.8.5}$$

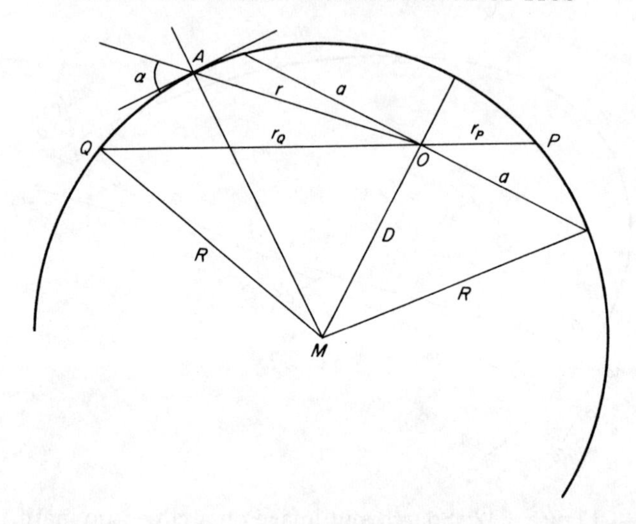

Figure 4.43: Circular ray in spherically symmetrical medium

This is the law of refraction for Maxwell's fish-eye with constants chosen so that $RC = 1$, $R^2 - D^2 = 1$, that is for a refractive index that is unity at the unit circle.

In the second situation the ray paths satisfy $r_P = r_Q$ where P and Q are on opposite sides of a radius vector through the centre of spherical symmetry. The simplest curves for this case are thus the central conics. We choose those ellipses that pass through the points ± 1 on a diameter which we choose as the x axis. We further require the tangents to these ellipses to be parallel to the x axis at the points where they intersect the unit circle. The ray paths can be recognised as those of the Luneburg lens. These ellipses, with the conditions prescribed above must have equations

$$x^2 + a^2y^2 + 2bxy = 1 \tag{4.8.6}$$

If the ellipse makes the angle α with the x axis at the point $x = -1$ and hence $\pi + \alpha$ at the point $x = 1$, and has the tangent property required the constants in this equation can be evaluated to give

$$x^2 + y^2(1 + 2\cot^2\alpha) - 2xy\cot\alpha = 1$$

If the tangent at any point P with coordinate r on this ellipse makes an angle ϕ (Figure 4.44) with the radius vector, then it can be shown, with some fairly complicated algebra, that the equation to the ellipse can be expressed as

$$r^2 = 1 \pm \sqrt{1 - \sin^2\alpha / \sin^2\phi} \tag{4.8.7}$$

the sign depending upon the parts of the ellipse internal or external to the unit circle.

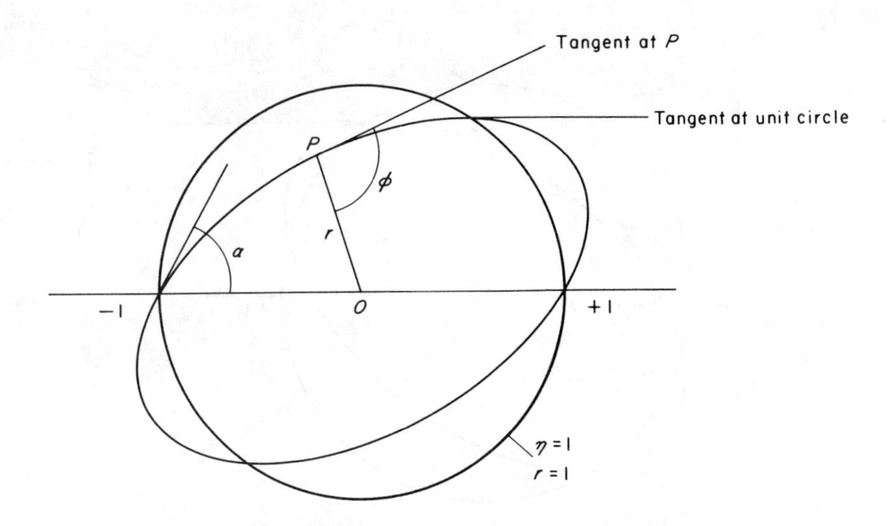

Figure 4.44: Geometry of elliptical ray Luneburg lens

If then $\eta r \sin\phi$ =constant for the ray as Bouguer's theorem states, $\eta r \sin\phi = \sin\alpha$ since $\eta = 1$ when $r = 1$ at which point $\phi \equiv \alpha$.

Substituting for $\sin\alpha/\sin\phi$ in Equation 4.8.7 gives the refractive index law

$$\eta^2 = 2 - r^2$$

required for the Luneburg lens.

The Gutman modification can be obtained in a similar manner. If the source of rays be placed at a point F a distance $f < 1$ from the origin, and the same conditions of tangency made for the ellipse, the equation for the ellipse becomes

$$x^2 + y^2\left(\frac{f^2 + \cos^2\alpha}{\sin^2\alpha}\right) - 2xy\cot\alpha = f^2 \tag{4.8.8}$$

from which can be derived after considerably more complex algebra the relation (Figure 4.45)

$$2r^2 = 1 + f^2 \pm \sqrt{(1 + f^2)^2 - 4f^2\sin^2\alpha/\sin^2\phi} \tag{4.8.9}$$

Substituting $\eta(r)r\sin\phi = \eta(f)f\sin\alpha = \sin\alpha$ the first equation relating to the ray at the source and the second to the ray at its intersection with the unit circle, result in the law

$$\eta^2 = (1 + f^2 - r^2)/f^2 \text{ of the Gutman lens}$$

Thus particular solutions of Equation 4.8.3 give the lenses with which we are now familiar.

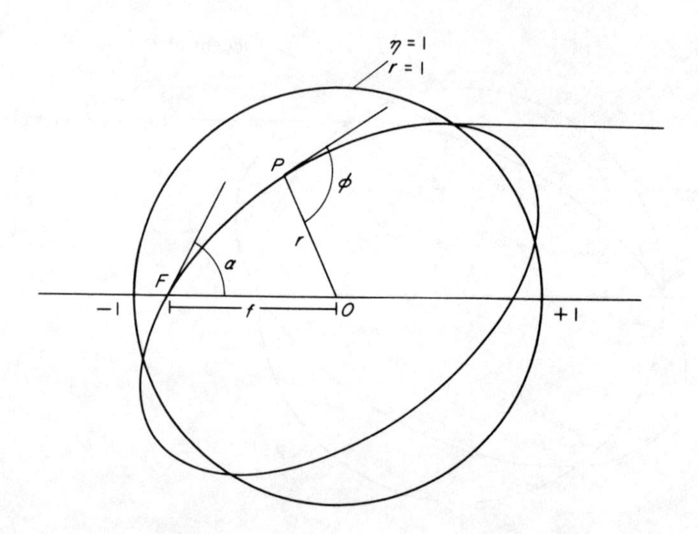

Figure 4.45: Geometry of elliptical ray Gutman lens

The general solution is a solution of the *functional* relation obtained by combining Equation 4.8.1 with the result $r_P r_Q = a^2$ to give

$$\frac{r_P}{a}\eta\left(\frac{r_P}{a}\right) = \frac{a}{r_P}\eta\left(\frac{a}{r_P}\right) \tag{4.8.10}$$

We can without loss of generality take the constant to be unity and the general solution of Equation 4.8.10 is seen to be

$$r\eta(r) = f\left\{\phi(r) \otimes \phi\left(\frac{1}{r}\right)\right\} \tag{4.8.11}$$

where f and ϕ are arbitrary functions of their arguments and $\otimes$ implies an associative law of combination. Taking this to be summation as the simplest instance then

$$r\eta(r) = f\left\{\phi(r) + \phi\left(\frac{1}{r}\right)\right\} \tag{4.8.12}$$

Maxwell's fish-eye is given by $\phi(r) = r$; $f(x) = A/x$ (Luneburg [8, p179]) gives a generalization of Maxwell's fish-eye obtained by a direct conformal mapping of the cross-section of the problem described above. This has the refractive index law

$$\eta(r) = \frac{2r^{\gamma-1}}{1 + r^{2\gamma}} \qquad \text{cf Equation 4.7.11}$$

This, in terms of Equation 4.8.12 can be seen to be obtained from the function

$$f(x) = \frac{1}{x} \qquad \phi(x) = r^{\gamma}$$

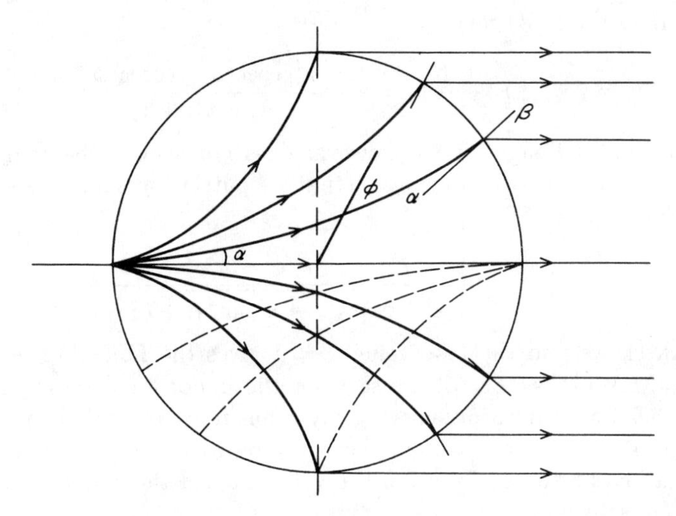

Figure 4.46: Lens with hyperbolic ray paths

No similar general solution can be obtained for Equation 4.8.2 other than the trivial $r\eta(r) = f(r)$ and hence no generalizations of the Luneburg lens can be obtained from this equation.

As a consequence of this theory the existence of a spherically symmetrical medium in which the ray paths are hyperbolae, the remaining form of central conic, can be postulated. Since these are outward curving as shown in Figure 4.46 the refractive index law will be an increasing function of the radius r and a real lens, η always larger than unity, will thus require a refraction at the surface $r = 1$. The refractive index there will be taken to have the value $\eta(1) = \kappa$. If this refraction is taken to turn each ray to the horizontal, the lens would have a focusing property similar to that of the Luneburg lenses, as is shown.

For central hyperbolae intersecting the axis at points $x = \pm 1$ and with the refractive property prescribed at the circle $r = 1$ the equations will be

$$x^2 + y^2 \left[1 + \frac{2\cot\alpha\sqrt{1 - \kappa^2\sin^2\alpha}}{\kappa\sin\alpha}\right] - 2xy\cot\alpha = 1 \tag{4.8.13}$$

With ϕ the angle between the radius vector and the tangent to the ray paths as in previous studies (Figure 4.46) we can obtain

$$2xyp + y^2 - x^2 = \tan\alpha / \tan\phi$$

and

$$\begin{aligned} r^2\cot\alpha &= 2x^2\cot\alpha - 2xyp\cot\alpha + \cot\phi \\ p &= \sqrt{1 - k^2\sin^2\alpha}/\kappa\sin\alpha \end{aligned} \tag{4.8.14}$$

resulting in the (r, ϕ) relation

$$\left(r^2 - \frac{1}{2}\right)^2 = \frac{1}{4} + \frac{r^4 \operatorname{cosec}^2\alpha - \operatorname{cosec}^2\phi}{2(p\cot\alpha + 1)} \tag{4.8.15}$$

Substitution of Bouguer's law however does not lead to the complete elimination of α and ϕ simultaneously, that is, putting $\eta r \sin\phi = \kappa \sin\alpha$ results finally in

$$\left(r^2 - \frac{1}{2}\right)^2 = \frac{1}{4} + \frac{\kappa^2 r^4 - \eta^2 r^2}{2[\kappa\cos\alpha\sqrt{1 - \kappa^2\sin^2\alpha} + \kappa^2\sin^2\alpha]} \tag{4.8.16}$$

As a check we find that the value $\kappa = 1$ gives the Luneburg lens with the *elliptical* ray paths. As will be seen for the uniform sphere for a linear ray $\kappa = 2$ thus for an inwardly curving ray κ has to be greater than 2. For these values of κ and over a small angular range of α the complicated denominator in Equation 4.8.16 can be given the value 2κ and the resulting *approximate* law of the lens is

$$[\eta(r)]^2 = 2\kappa + (\kappa^2 - 2\kappa)r^2 \tag{4.8.17}$$

which still incorporates the Luneburg result. For larger values of κ the approximation is valid over a decreasing range of α, but due to the increasing curvature of the rays this can still create an exit aperture of a considerable fraction of the total lens diameter. With the rays given by Equation 4.8.13 this can be seen to be

$$2y_{max} = 2\kappa \sin\alpha_{max}$$

The substitution of this refractive index law into the ray path integral of Equation 4.7.6 shows that to the first order in $\kappa^2\sin^2\alpha$, that is for the conditions of our approximate solution, the rays do follow the required trajectory.

That an approximation has to be resorted to is only an indication that the solution to this problem is unlikely to exist for an algebraic form of refractive index law. Solving this problem by the use of Abel's integration formula as shown in Appendix III results in a refractive index law involving elliptic integrals. A similar situation arises with the axially defocused Luneburg lens and computational processes have to be invoked.

A similar lens in which the *angles* α and β in Figure 4.46 are related by the factor κ, instead of their *sines* as required by refraction, has been investigated by Toraldo di Francia [38] (Appendix III).

The law of proportionality between these angles is taken to be

$$\kappa = 2(1 - p)$$

and the refractive index law of the lens is found to be given by

$$\eta^2 r^2 = \eta^{1/p}(2 - \eta^{1/p}) \tag{4.8.18}$$

When $p = 1/2$ we obtain Luneburg's result and when $p = 1$ we obtain Maxwell's fish-eye. This is then another generalization of this form of lens.

THE BENDING OF LIGHT RAYS NEAR THE SUN

The presence of a centre of gravitational attraction distorts the local space time geometry so that the increment of path length of a world line changes from the Lorentz line-element,

$$\mathrm{d}s^2 = c^2\,\mathrm{d}r^2 - \mathrm{d}r^2 - r^2\,\mathrm{d}\theta - r^2\sin^2\theta\,\mathrm{d}\phi^2 \tag{4.8.19}$$

to that of the Schwarzschild element,

$$\mathrm{d}s^2 = \left(1 - \frac{2Km}{c^2 r}\right) c^2\,\mathrm{d}r^2 - \frac{\mathrm{d}r^2}{1-(2Km)/(c^2 r)} - r^2\,\mathrm{d}\theta^2 - r^2\sin^2\phi\,\mathrm{d}\phi^2 \tag{4.8.20}$$

where K is the universal gravitational constant, m the mass of the centre of gravitational attraction and c is the velocity of light. The effect of this change is to cause a curvature of the geodesics of the geometry. If these are light rays, this curvature is most apparent when they just graze a massive body such as the sun. The observation of this effect is a well documented procedure during eclipses of the sun.

The coordinate velocity of light is given by $\mathrm{d}s = 0$ and in the radial direction ($\mathrm{d}\theta = \mathrm{d}\phi = 0$) this is $\mathrm{d}r/\mathrm{d}t = c$ for Lorentz space and

$$\frac{\mathrm{d}r}{\mathrm{d}t} \simeq \left(1 - \frac{2Km}{c^2 r}\right) c \equiv c'$$

for Schwarzschild space for $r \neq 0$ and $2Km/(c^2 r) \ll 1$.

Thus in the presence of a gravitational field we can specify a radial variation of refractive index given by [39]

$$c/c' = \eta(r) = 1 + \frac{2Km}{c^2 r} \tag{4.8.21}$$

To the same degree of approximation this is of the form

$$\eta^2 = 1 + \frac{A}{r}$$

which Luneburg [8] again shows to give ray trajectories which are hyperbolae with focus at the origin. Putting $2Km/c^2 = \phi$ for brevity, then in accordance with Equation 4.7.6 the angle turned through by the radius vector deflected from the straight line by an amount δ is

$$\pi + \delta = 2\int_{\infty}^{r_m} \frac{-h\,\mathrm{d}r}{r\sqrt{\eta^2 r^2 - h^2}} \tag{4.8.22}$$

where r_m is the radial distance of the nearest approach of the ray to the origin. This we take to be $r_\odot$ the radius of the sun since we consider the ray to be just grazing the surface. The ray constant is h so that at the closest approach to the sun

$$h = \eta(r) r \sin\alpha$$

where $r = r_\odot$ and $\alpha = \pi/2$, and

$$h = r_\odot\left(1 + \frac{2Km}{c^2 r_\odot}\right) \quad \text{or} \quad r_\odot = h - \phi$$

Transforming the integral by the substitution $v \to 1/r$ gives

$$\begin{aligned}
\pi + \delta &= 2\int_0^{1/(h-\phi)} \frac{h\,dv}{\sqrt{1 + 2\phi v + (\phi^2 - h^2)v^2}} \\
&= \frac{2h}{\sqrt{h^2 - \phi^2}}\left[\sin^{-1}\left(\frac{v(h^2 - \phi^2) - \phi}{h}\right)\right]_0^{1/(h-\phi)} \\
&= \frac{2h}{\sqrt{h^2 - \phi^2}}\left[\frac{\pi}{2} + \sin^{-1}\frac{\phi}{h}\right] \\
&\simeq \pi + 2\phi/h \quad \text{in the limit } \phi \ll h
\end{aligned}$$

Hence the angle of deflection

$$\delta = \frac{2\phi}{h} = \frac{4Km}{c^2 r_\odot} \tag{4.8.23}$$

This is the classical result for the bending of light rays near the sun.

In a similar way refraction through a variable atmosphere could be assessed for its effects upon the required beam shape of a satellite antenna for example or for propagation beyond the optical horizon.

4.9 SPHERICAL SHELL LENSES

The construction of non-uniform spherical lenses normally takes the form of a series of hemispherical shells, each of a material with a constant refractive index, nested together to form a layered sphere. The refractive index variation is then obtained by approximating the smooth refractive index law required by a stepped approximation. It is usually taken that with a sufficiently large number of steps the approximate curve is an adequate approximation for the required curvature of the rays [45]. This procedure disregards two basic principles of antenna design. Firstly it implies that each and every ray from the source within the forward hemisphere is required for full illumination of the aperture. This is so since rays with commencement angles $\pm\alpha = 90^o$ are included in the design as can be seen from Figure 4.34. In any practical situation a finite sized microwave source will have an optimum angular distribution of less than this angle and so rays outside this range are not only ignorable, but if included can reduce the final efficiency of the system. This case would be best approached by a design such as that shown in Figure 4.38(a).

The second principle is that which has been established by the work of SP Morgan [35] and others that the presence of an outer spherical layer of constant refractive index changes the refractive index law for the entire interior of the lens inside this layer as was shown in Figure 4.38(b). This is so for even a thin layer, and the attempts to construct this layer with very light density materials has led to unnecessary complications such as surface coatings of actually very high density materials.

It would appear obvious therefore that a lens with these two a priori conditions should be designed as a layered structure with a specified angular width of rays issuing from the source. The analysis required turns out to be a relatively simple geometrical study and results in far fewer constant index layers being required than the arbitrary stepped version. In fact for lenses of medium apertures, approximately 10λ in diameter, 2 or 3 shells only are necessary.

We illustrate the geometry first by deriving Bouguer's theorem in differential form from such a layered medium [46]. If as in Figure 4.47 a ray is refracted at successive layers at the points P_1 P_2 etc. where the radii of the layers are respectively $r_1, r_2 \ldots$ and the incident angles of the rays $\psi_1, \psi_2 \ldots$ we have by virtue of Snell's law at each point P,

$$\eta_1 \sin\psi_1 = \eta_2 \sin\psi_1'$$

or generally

$$\eta_i \sin\psi_i = \eta_{i+1} \sin\psi_i' \tag{4.9.1}$$

From the triangle P_iOP_{i+1} we have

$$r_i \sin\psi_i' = r_{i+1} \sin\psi_{i+1}$$

which substituted into Equation 4.9.1 gives

$$\eta_i r_i \sin\psi_i = \eta_{i+1} r_{i+1} \sin\psi_{i+1}$$

that is

$$\eta r \sin\psi \text{ is a constant at each radial layer}$$

This for a continuous smooth distribution becomes Bouguer's law.

Turning to the related problem of the field angle of the rays from the source, we consider the source P (Figure 4.48) to be at a distance h from the centre of a sphere of unit radius. A ray from the source making an angle α with the axis is refracted twice in its passage through the sphere at points Q and R. If this single ray emerges from the sphere parallel to the axis then the line QR can be shown quite simply to make an angle $\alpha/2$ with the axis and the required refractive index at the points of refraction has to be

$$\eta = \frac{h \sin\alpha}{\sin\{\sin^{-1}(h \sin\alpha) - \alpha/2\}} \tag{4.9.2}$$

In effect the ray is refracted by equal amounts of $\alpha/2$ at each of two points. If the ray is confined to a uniform shell with this refractive index the ray will be "exact" in the sense of its parallelism with the axis on emergence.

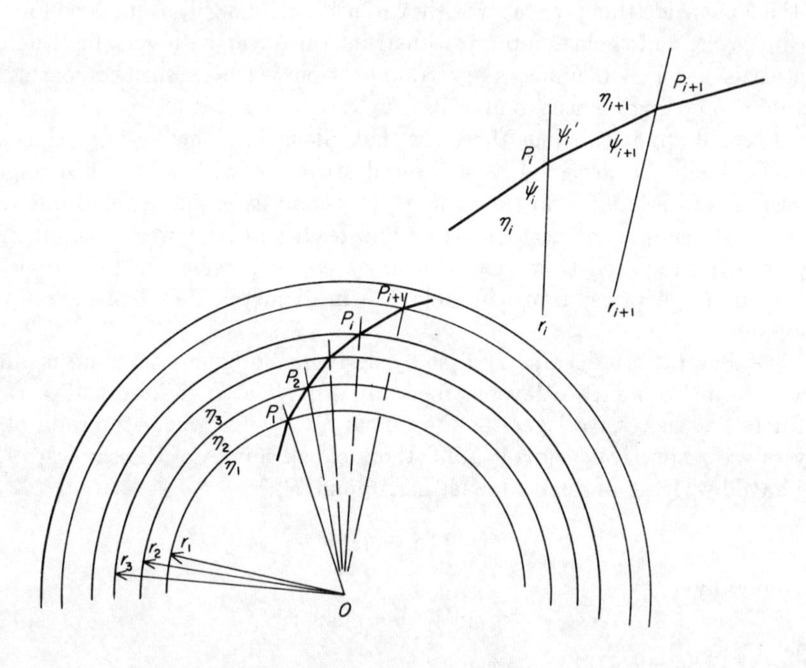

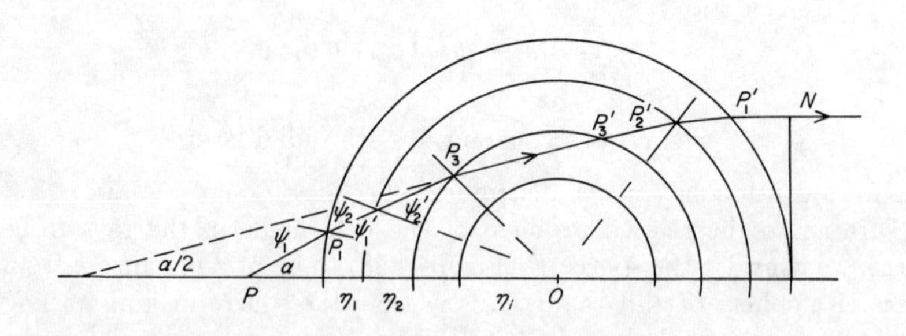

Figure 4.47: Ray paths in spherically stratified medium

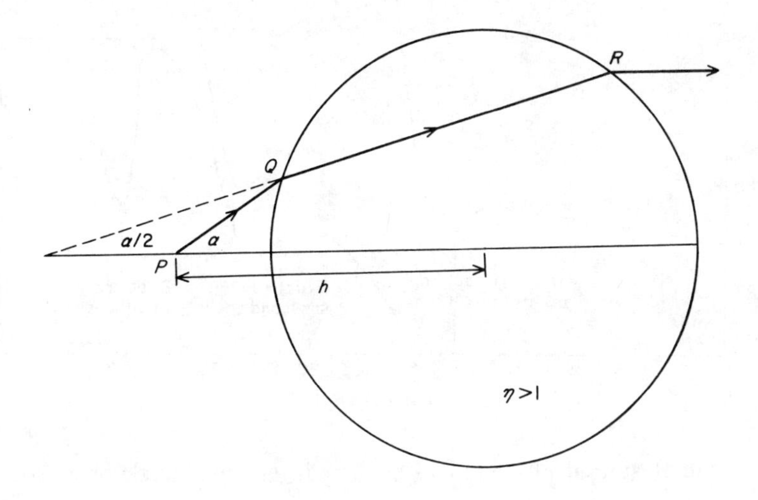

Figure 4.48: Single exact ray for homogeneous sphere

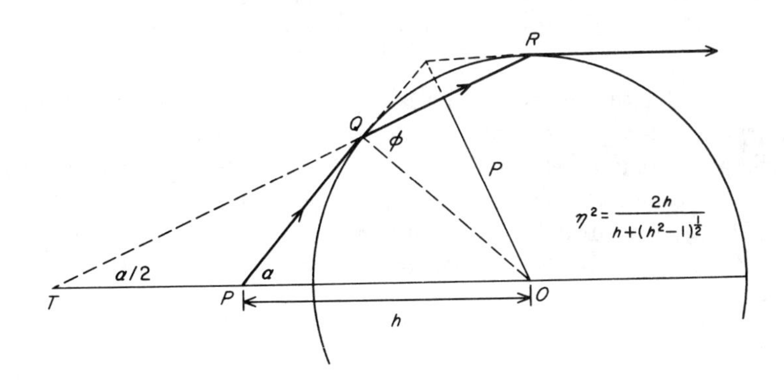

Figure 4.49: Limit ray for homogeneous sphere

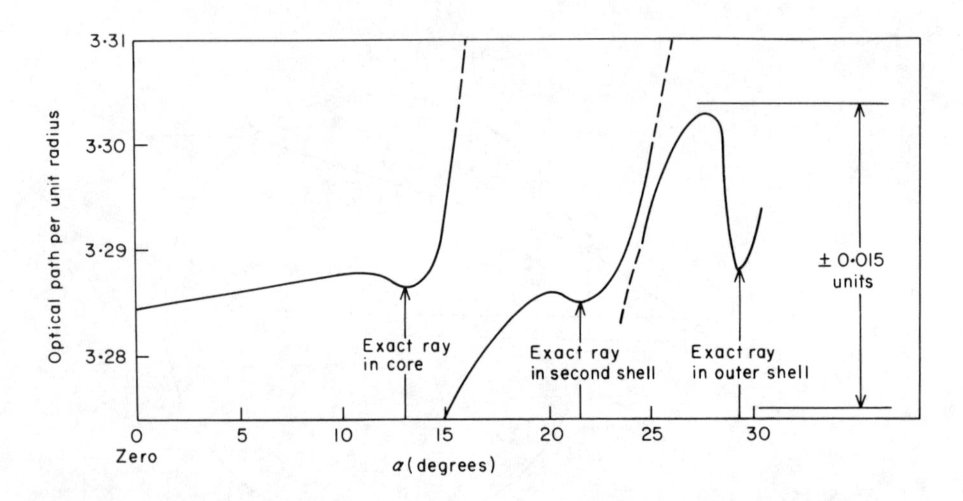

Figure 4.50: Resultant phase in lens of three layers; field angle of source $\pm 30^o$

In the limiting case, when the full aperture of the lens is required to be used the geometry is as shown in Figure 4.48. Then $PQ = PT = \sqrt{h^2 - 1}$ and since the ray is tangential both at Q and at R

$$\eta = \frac{1}{\sin\phi} = \frac{1}{p} \qquad \frac{p}{h + \sqrt{h^2 - 1}} = \sin\frac{\alpha}{2}$$

and hence

$$\eta = \sqrt{\frac{2h}{h + \sqrt{h^2 - 1}}} \tag{4.9.3}$$

For a source on the surface of the lens therefore an outer shell of refractive index $\eta = \sqrt{2}$ can be used with a source of angular width $\pm 45^o$.

The same analysis shows that when a ray is refracted at more than one spherical interface as in Figure 4.47, it will be transmitted horizontally if the sum of the angles turned through over *half* of its passage through the lens is half the angle of emission at the source. That is for a ray with angle α

$$\psi_1 - \psi_1' + \psi_2 - \psi_2' \ldots = \alpha/2$$

or

$$\sum_1^N \psi_i - \sum_1^N \psi_i' = \alpha/2$$

for a lens with N refracting surfaces (including the outer surface).

It is obvious therefore that all the rays cannot be exact [40] for a discretely layered spherical lens. In microwave practice it is not possible to establish a ray angle criterion up to which departures from "exactness" can be permitted.

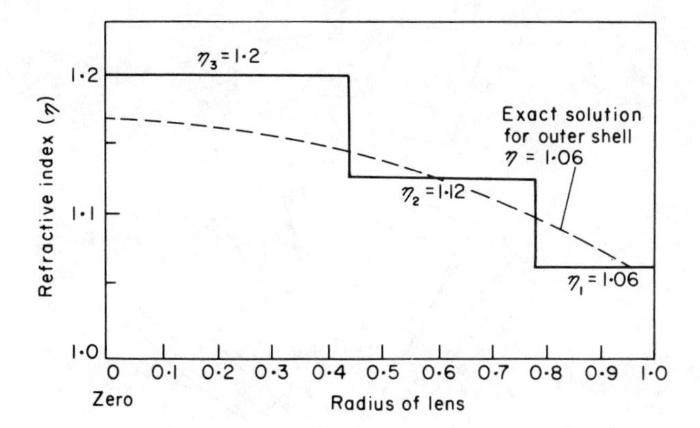

Figure 4.51: Refractive index of stepped lens

We have instead to calculate permitted departures from the planeness of the wave front. That is we require to calculate the optical path length

$$PP_1 + \eta_1 P_1 P_2 + \eta_2 P_2 P_3 \ldots \eta_1 P_2' P_1' + P_1' N$$

(Figure 4.47) and maintain its value by the choice of radii and refractive indices, within a prescribed tolerance. This can be done empirically with the concept of the exact ray, since such a ray will automatically have the correct optical path.

The design procedure is thus to establish the angular width at the source and thus the source position and the refractive index of the outer shell for an exact outer ray. The diameter of this shell is then taken to be that for which a tangential ray, now no longer exact, has optical path length within the limit of phase prescribed. A refractive index is then obtained for the next layer for which an exact ray occurs *near* to the interface and a check carried out that rays nearer to the interface are still within the phase tolerance. The next diameter can be obtained by calculation of ray phases until a further layer has to be introduced. The final phase distribution has the form shown in Figure 4.50 which was obtained for a lens of diameter 8λ constructed in three layers. Comparison of the resulting refractive index steps for this lens with the exact solution, given an outer shell of the same refractive index (Figure 4.51) shows how far a stepped solution departs from a presupposed smooth relation.

The same analysis can be used to determine the depth of focus for such a lens, which must, in principle be similar for all such non-uniform lenses. It is found that an axial defocusing movement of magnitude ε produces a phase error *differential* between the centre ray and a marginal one of 0.06ε. This allows a considerable source movement before noticeable deterioration of the radiated pattern is observed. Some measured results are shown in Figure 4.52.

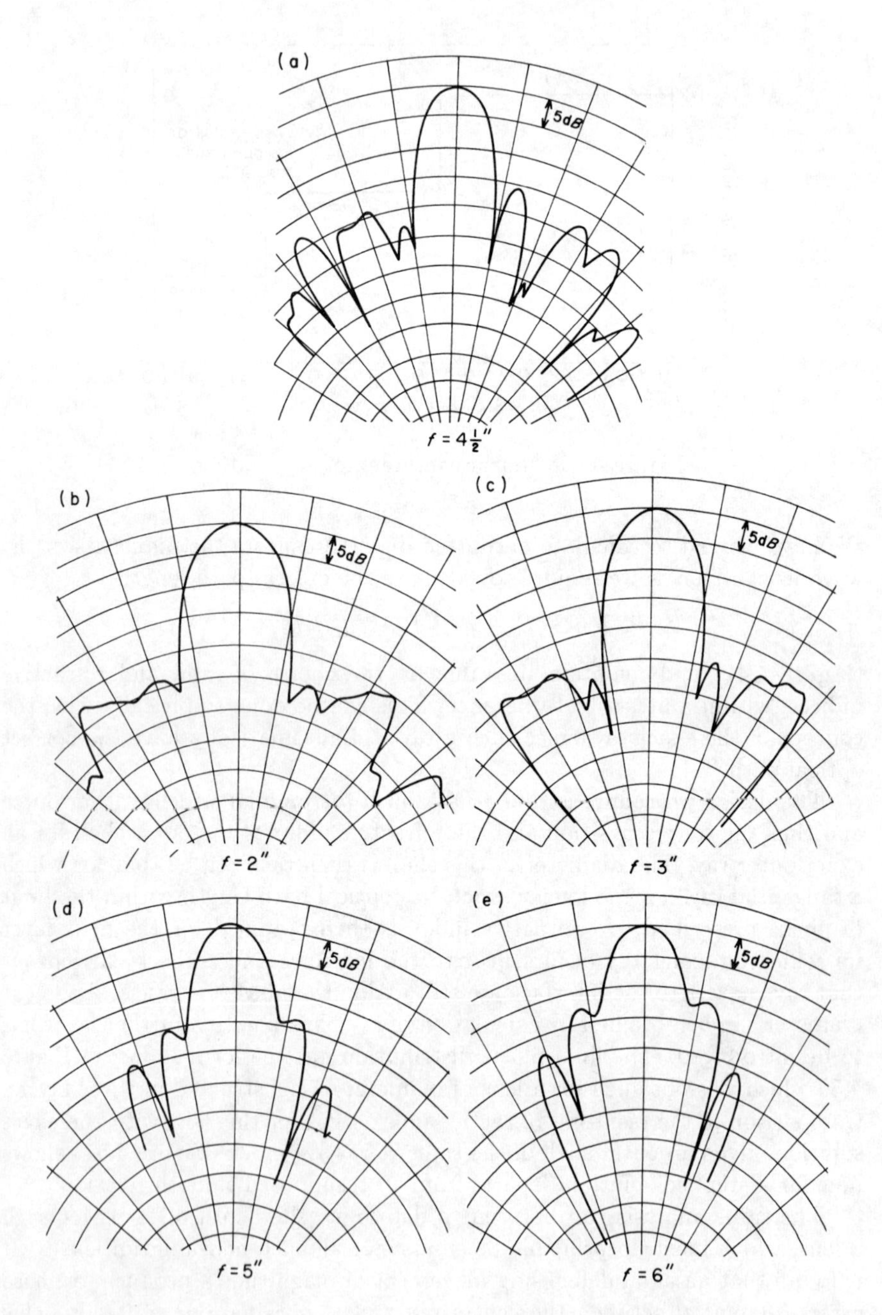

Figure 4.52: Patterns of lens for varying focal position (a) Design focus $f = 4.5$ (b) $f = 2$ (c) $f = 3$ (d) $f = 5$ (e) $f = 6$

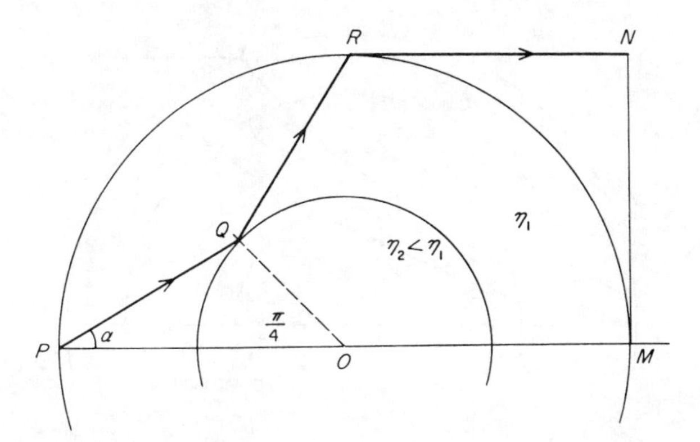

Figure 4.53: Ray paths in lens of two layers

A similar design procedure can be followed for lenses with refractive index law that increases outwardly from the centre. Some complexity occurs in that the field angle of the source rays has to be confined to a much smaller angle to prevent stray anomalous rays from interfering with the pattern. The dependence upon total internal reflection is also a hazardous procedure with regard to phase, but in this design, Figure 4.53, can be confined to the single most marginal ray. Such a lens, first suggested by Toraldo di Francia and Zoli [41] again has apparently a greater aperture efficiency for even fewer layers than the previous design. Both in this reference and in a later study [42] a two layer lens is given with very similar characteristics. That is an outer layer of refractive index 3.4 [41] (3.236 in [42]) and inner core of refractive index 2.665 [41] (2.618 in [42]).

If the full aperture is required to be used then by the geometry of Figure 4.54 such a lens can be designed upon the basic criterion that the optical length of the marginal ray be the same as that of the central ray.

From Figure 4.53 and the refraction of the ray at R we find $\eta_1 = 1/\sin\alpha$ for a marginal ray making an angle α with the axis at the source.

Then

$$PQ = \eta_1 r/\sqrt{2}$$

and hence

$$r = \sqrt{2}/(1 + \sqrt{\eta_1^2 - 1}) \tag{4.9.4}$$

Thus r and α are determined by the specification of the refractive index η_1. For the ray $PQRN$ to be of the same optical length as the ray POM we have simply (for a lens of unit radius)

$$2\eta_1 PQ + 1 = 2\eta_2 r + 2\eta_1(1 - r) \tag{4.9.5}$$

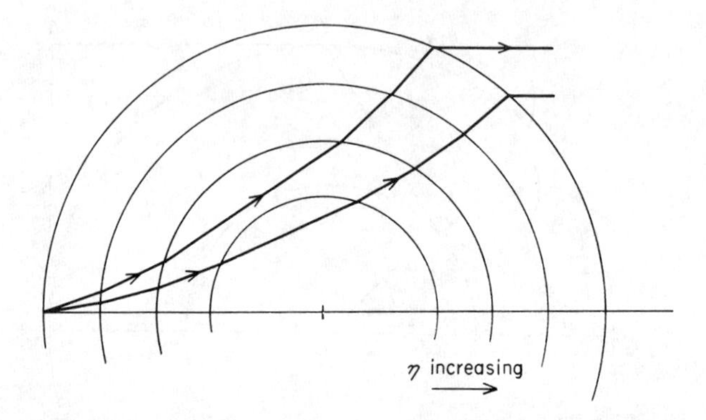

Figure 4.54: Extrapolation toward hyperbolic rays

which together with Equation 4.9.4 gives

$$\eta_2 = \frac{\eta_1^2}{\sqrt{2}} + \eta_1 - \frac{(2\eta_1 - 1)(1 + \sqrt{\eta_1^2 - 1})}{2\sqrt{2}} \tag{4.9.6}$$

This is only one of the many criteria that could be used however, and others may optimize the planeness of the complete wavefront.

The interest in this design of lens lies mainly in the extrapolation of it to a lens of many thin layers. It will be seen then that the rays will become smooth outwardly bending curves of precisely the kind hypothesized as the hyperbolae given by Equation 4.8.13.

4.10 TRANSFORMATIONS OF SPHERICAL LENSES

THE LEGENDRE TRANSFORMATION

One obvious connection between the designs of the spherical non-uniform lenses is their general reliance on the use of the Abel integral method of Appendix III. It is therefore of interest to see what transformations exist that connect the lens designs or that can extend them.

The first such transformation is that relating to the phase fronts given by Equation 4.1.6. Suppose as in Equation 4.1.6 the surfaces $S(x, y, z)$ = constant are wavefronts in a refractive space $\eta(x, y, z)$, and $T(\lambda, \mu, \nu)$ =constant are wavefronts in a medium with refractive index $n(\lambda, \mu, \nu)$. These surfaces are connected by a Legendre transformation if

$$\begin{array}{ccc} \frac{\partial S}{\partial x} = \lambda & \frac{\partial S}{\partial y} = \mu & \frac{\partial S}{\partial z} = \nu \\ \frac{\partial T}{\partial \lambda} = x & \frac{\partial T}{\partial \mu} = y & \frac{\partial T}{\partial \nu} = z \end{array} \tag{4.10.1}$$

These can be collectively written as $S + T = x\lambda + y\mu + z\nu$.

Since S and T are wavefronts, we have from Equation 4.1.8

$$\left(\frac{\partial S}{\partial x}\right)^2 + \left(\frac{\partial S}{\partial y}\right)^2 + \left(\frac{\partial S}{\partial z}\right)^2 = \eta^2$$

$$\left(\frac{\partial T}{\partial \lambda}\right)^2 + \left(\frac{\partial T}{\partial \mu}\right)^2 + \left(\frac{\partial T}{\partial \nu}\right)^2 = n^2 \tag{4.10.2}$$

We apply this to the fish-eye for which

$$\eta = \frac{2}{1+r^2} = \frac{2}{1+x^2+y^2+2^2} = \sqrt{\left(\frac{\partial S}{\partial x}\right)^2 + \left(\frac{\partial S}{\partial y}\right)^2 + \left(\frac{\partial S}{\partial z}\right)^2}$$

Then on transformation

$$\sqrt{\lambda^2+\mu^2+\nu^2} = 2 \Bigg/ \left[1 + \sqrt{\left(\frac{\partial T}{\partial \lambda}\right)^2 + \left(\frac{\partial T}{\partial \mu}\right)^2 + \left(\frac{\partial T}{\partial \nu}\right)^2}\right]$$

or

$$n^2 = \frac{2}{\rho} - 1$$

which is the refractive index law, Equation 4.7.20 of the lens of Eaton.

It is also to be noted that this same result can be effected by the transformation $\eta \to r$ and $r \to \eta$, leaving $r\eta(r)$ invariant. Thus the Luneburg lens, for which $\eta^2 = 2 - r^2$, is an invariant of the transformation. It does not always result in an explicit form for the refractive index, and once effected the rays have to be recalculated with the new refractive law.

THE INVERSION OF SPHERICAL LENSES

The second transformation is of a more fundamental nature. It could be anticipated, from the fact that the only completely conformal mapping of a three-dimensional space onto itself (apart from the elementary translation rotation or magnification) is an inversion [43], that this would be the transformation most relevant to spherically symmetrical problems.

As inversion leaves angles invariant, Bouguer's theorem for a spherical medium

$$r\eta(r)\sin\alpha = r_0\eta(r_0)\sin\alpha_0$$

(the suffix zero indicating the source condition for a ray with tangent making an angle α with the radius vector from the origin), would imply that the proper function for the transformation process is not the refractive index itself, but the function

$$r\eta(r) = f(r)$$

and this accords with the invariance of this function under the Legendre transformation and with the concept of refractive space $r\eta(r)$.

We therefore define a generalized inversion [44]

$$(r, \phi) \rightarrow (R, \Phi)$$

such that the refractive index transforms by

$$r\eta(r) = f(r) \rightarrow r\bar{\eta}(r) = f(1/r^n) \tag{4.10.3}$$

That is, in general, $r \rightarrow 1/R^n$ in the coordinate space, whereas the refractive index alone transforms by

$$\eta(r) = f(r)/r \rightarrow \bar{\eta}(R) = f(1/R^n)/R$$

For general values of n, the optical line element is scaled by a factor n. For a ray confined to the equatorial plane of the medium

$$\eta^2(r)\,\mathrm{d}s^2 = f^2(r)[\,\mathrm{d}r^2/r^2 + \mathrm{d}\phi^2]$$

which transforms to

$$f^2(1/R^n)(n^2\,\mathrm{d}R^2/R^2 + \mathrm{d}\phi^2)$$

Thus, if $\phi \rightarrow n\Phi$, we obtain

$$\begin{aligned}\eta^2(r)\,\mathrm{d}s^2 &= n^2 f^2(1/R^n)[\,\mathrm{d}R^2 + R^2\,\mathrm{d}\phi^2]/R^2 \\ &= n^2\bar{\eta}^2(R)\,\bar{\mathrm{d}s}^2\end{aligned}$$

We can now show that this formulation transforms the rays in one medium directly into rays in the transformed medium. If a ray at a fixed point (r_0, ϕ_0) makes an angle α_0 with the radius vector at that point, the ray constant κ is given by Bouguer's theorem to be

$$\kappa = r_0\eta(r_0)\sin\alpha_0 = f(r_0)\sin\alpha_0$$

the ray trajectory will then be given by Equation 4.7.6

$$\phi - \phi_0 = \int_r^{r_0} \frac{-\kappa\,\mathrm{d}r}{r\sqrt{f^2(r) - \kappa^2}}$$

This will transform by the method given to

$$n\Phi - n\Phi_0 = \int_{1/R^n}^{1/R_0^n} \frac{n\bar{\kappa}\,\mathrm{d}R}{R\sqrt{f^2(1/R^n) - \bar{\kappa}^2}}$$

and hence

$$\Phi - \Phi_0 = \int_{1/R^n}^{1/R_0^n} \frac{\bar{\kappa}\,\mathrm{d}R}{R\sqrt{R^2\bar{\eta}^2(R) - \bar{\kappa}^2}} \tag{4.10.4}$$

where $\bar{\kappa} = f(1/R_0^n)\sin(\pi - \alpha_0)$ is the transform of κ. (Note the factor $\sin(\pi - \alpha_0)$ for future reference.

Equation 4.10.4 is the ray equation for rays in the medium $\bar{\eta}(R)$ connecting point with coordinate $1/R_0^n$ and $1/R^n$.

Thus, points (r, ϕ) on rays in the first medium transform directly into points $(1/r^n, n\phi)$ in the second medium. The change of sign in Equation 4.10.4 arises because a ray in one medium, in which r increases with ϕ, will transform into a ray in the new medium in which $1/r^n$ decreases with increasing ϕ, or vice versa.

Only the special case of the above for $n = 1$ extends the argument from the equatorial plane to any ray trajectory in a three-dimensional spherically symmetrical space, and leaves the optical line element invariant.

We then have

$$\eta^2(r)\,\mathrm{d}s^2 = f^2(r)(\,\mathrm{d}r^2/r^2 + \mathrm{d}\phi^2 + \sin^2\phi\,\mathrm{d}\psi^2)$$

With $r \to 1/R$ and $\phi \to \Phi$ this gives (ψ is invariant)

$$\begin{aligned}\eta^2(r)\,\mathrm{d}s^2 &= f^2(1/R)(\,\mathrm{d}R^2/R^2 + \mathrm{d}\Phi^2 + \sin^2\Phi\,\mathrm{d}\psi^2)\\ &= \bar{\eta}^2(R)\,\bar{\mathrm{d}s}^2\end{aligned}$$

Applying this transformation to the spherical lenses discussed earlier gives an entire new class of refractive media in which the rays are the true geometrical inversions of the rays of the original lenses.

THE MEDIUM WITH $r\eta(r)$ CONSTANT

This law is obviously self-inverse. In the medium with $\eta(r) = 1/r$ the rays are equiangular spirals, with pole at the origin, which are likewise self-inverse with respect to an inversion in the pole (see Appendix III).

THE MAXWELL FISH-EYE

From the refractive law $\eta(r) = 2/(1+r^2)$ we obtain $r\bar{\eta}(r) = (2/r)(1+1/r^2)$ or $\bar{\eta}(r) = 2/(1+r^2)$. Hence this law is also invariant. In this case the inversion transforms a circle of the coaxal system in Figure 4.28 into its reflection in the axis - that is, a ray with source angle α into the ray with source angle $\pi - \alpha$.

THE LUNEBURG LENS

The refractive law is given by

$$\eta(r) = \sqrt{2 - r^2}$$

which on inversion becomes

$$\bar{\eta}(r) = \sqrt{2r^2 - 1}/r^3$$

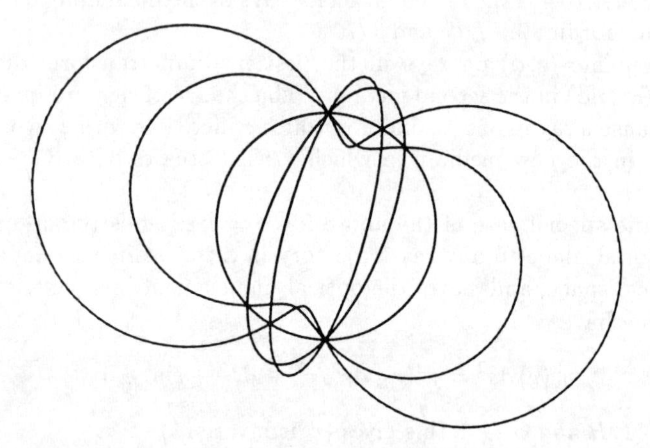

Figure 4.55: The inversion of the elliptical rays of the Luneburg lens into Cassinian ovals

and hence the ellipses given by Equation 4.7.15 become the Cassinian ovals (Figure 4.55)

$$r^2 = [1 - \cos\alpha\cos(2\phi + \alpha)]/\sin^2\alpha$$

THE EATON LENS

The refractive law for this is

$$\eta(r) = \sqrt{\frac{2}{r} - 1}$$

and hence on inversion we obtain

$$\bar{\eta}(r) = \sqrt{2r - 1}/r^2$$

The elliptical rays of Equation 4.7.19 thus invert to become limaçons of Pascal with equations

$$r = (1 - \cos\alpha\cos\phi)/\sin^2\alpha$$

as shown in Figure 4.56

THE UNIFORM MEDIUM

This has rays which are straight lines. The law $\eta(r)$ =constant transforms to $\bar{\eta}(r) = 1/r^2$, which therefore has rays which are circles passing through the centre (for a source not at the origin). This can simply be confirmed.

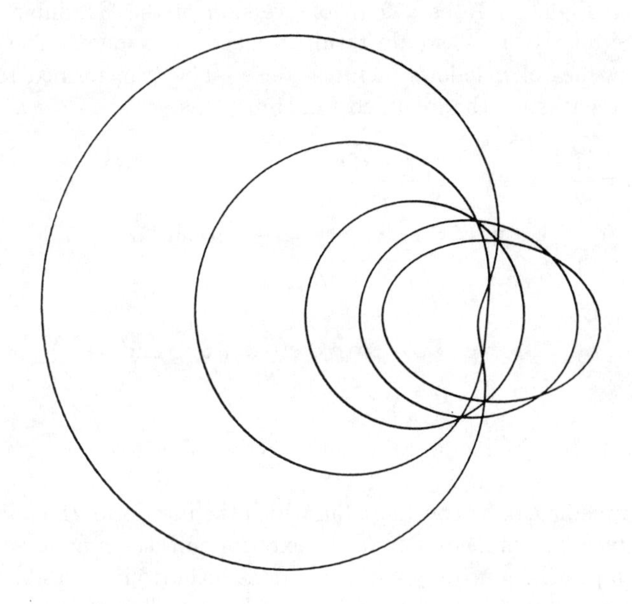

Figure 4.56: Inversion of the elliptical rays of the Eaton lens into limaçons of Pascal

It is elementary to show that, the repeat of an inversion reproduces the original medium.

The effect of an inversion with $n \neq 1$ is complicated by the factor multiplying ϕ.

Thus, the ray in one medium can be "compressed" in angle for $n > 1$ or extended, even "wrapped around" the origin in a new medium with $n < 1$.

For example, with $n = 2$ the transformation is $r \rightarrow 1/r^2$, applied to the refractive index

$$r\eta(r) = r\sqrt{2 - r^2}$$

of the Luneburg lens results in

$$r\bar{\eta}(r) = \sqrt{2 - 1/r^4}/r^2$$

For a ray from the same source making the angle $(\pi - \alpha)$ with the axis, the trajectory will therefore be

$$\phi = \int_1^r \frac{-\sin\alpha \,\mathrm{d}r}{r\sqrt{\frac{1}{r^4}\left(2 - \frac{1}{r^4}\right) - \sin^2\alpha}}$$

which integrates now, with the substitution $r^2 = v$, to give the identical integrand as arises from Equation 4.7.14, with the result

$$r^4 = [1 - \cos\alpha\cos(4\phi + \alpha)]/\sin^2\alpha$$

This, as shown Figure 4.57, is a four-fold version of the Cassinian ovals of Figure 4.55. Similarly, $n = 3$ would result in a six-fold version. The analysis for fractional values of n follows identical lines. The transformed refractive indices and ray paths for the modified Luneburg lens are

$$n = \frac{1}{2} \quad \bar{\eta}(r) = \sqrt{\frac{2 - 1/r}{r^3}}$$

$$r = [1 - \cos\alpha \cos(\phi + \alpha)]/\sin^2\alpha$$

$$n = \frac{1}{3} \quad \bar{\eta}(r) = \sqrt{2 - 1/r^{2/3}}/r^{4/3}$$

$$r = \{[1 - \cos\alpha\cos(\frac{2}{3}\phi + \alpha)]/\sin^2\alpha\}^{3/2}$$

$$0 < \phi < 4\pi$$

It can be shown that the reciprocal of an inversion of degree n is an inversion of degree $1/n$.

Composite media can be considered in which the interior of the unit sphere consists of a given known medium and the exterior consists of an inverse, both media having a priori defined ray paths. Regions hitherto inaccessible to rays, for instance $r > \sqrt{2}$ in the Luneburg lens medium, will in this way become usable as lens media.

The transformation provides a direct proof of the refocusing properties of the rays in the generalized fish-eye medium given by Equation 4.7.11. We have for the Maxwell fish-eye

$$r\eta(r) = \frac{2r}{1 + r^2}$$

and hence for an inversion of degree n

$$r\bar{\eta}(r) = \frac{2/r^n}{1 + 1/r^{2n}}$$

or

$$\bar{\eta}(r) = \frac{2r^{n-1}}{r^{2n} + 1} \qquad \text{the generalized medium}$$

The rays in Maxwell's fish-eye are the system of coaxal circles with parameter α given by (Equation 4.7.10)

$$r = \sqrt{1 + \sin^2\phi\cot^2\alpha} + \sin\phi\cot\alpha$$

Hence the rays in the generalized medium will be, as in Equation 4.7.12

$$r^{-n} = \sqrt{1 + \sin^2 n\phi\cot^2\alpha} + \sin n\phi\cot\alpha$$

and this gives the $2n$-fold refocusing at points around the unit circle shown in Figure 4.32, forming a ring resonator.

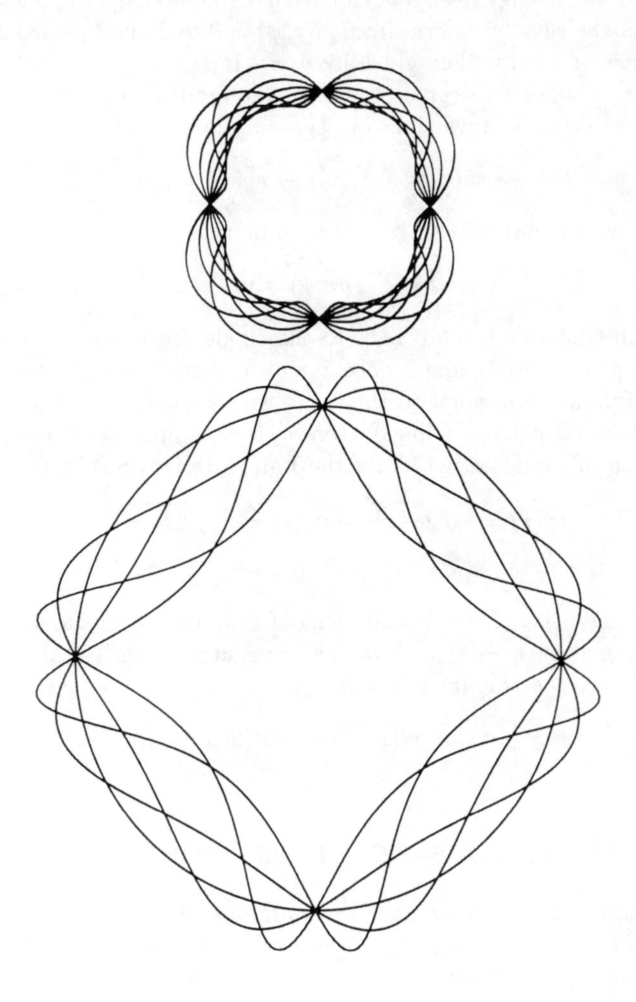

Figure 4.57: (a) Inversion of the rays of the Luneburg lens for $n = 2$
(b) Inversion of rays in the Luneburg lens for $t = 2$

Inversions for negative values of n are permissible, except that the condition governing the change in sign of Equation 4.10.4 no longer applies. Hence, for negative n the rays transform from $F(r, \phi) = 0$ to $F(1/r^n, -n\phi) = 0$. The identity transformation is then given by $n = -1$.

Inversions of different degree may be compounded. In general, we then have for two consecutive inversions of degree n and m

$$r\eta(r) = f(r) \to r\bar{\eta}(r) = f(1/r^n) \to r\bar{\bar{\eta}}(r) = f(r^{nm}) \tag{4.10.5}$$

and the rays with equation $F(r, \phi) = 0$ become rays

$$F(r^{nm}, nm\phi) = 0 \tag{4.10.6}$$

It can be seen that this is equivalent to the single inversion of degree $-nm$. This process produces a medium once again "realistic" in the interior of the unit circle. Putting nm equal to the single transformation t, say, results in a further set of refractive scaling factors. For example, with $nm = t$, the transformation of Equation 4.10.5 for the Luneburg lens results in

$$r\eta(r) = r\sqrt{2 - r^2} \to r\bar{\eta}(r) = r^t\sqrt{2 - r^{2t}}$$

$$r^2\eta^2(r) = r^{2t}(2 - r^{2t}) \tag{4.10.7}$$

which is precisely that of the Toraldo lens of Equation 4.8.18 under the Legendre transformation $\eta \to r$, $r \to \eta$. The rays can be derived directly from Equation 4.7.15 to be (Figure 4.56(b))

$$r^{2t} = \sin^2\alpha/[1 - \cos\alpha\cos(2t\phi + \alpha)] \tag{4.10.8}$$

4.11 RAYS IN AN AXISYMMETRIC COORDINATE SYSTEM

The ray equations for the axisymmetric coordinate system were given in Equations 4.2.13 and 4.2.14. A special case, as we shall see, is the spherical system again with the rays confined to the surface of a sphere. The problem is of interest more in the dynamical sense where the refractive index is analogous to a potential field and the rays to particle trajectories. The resulting trajectories have elements in common with the winds and tides prevailing, where the rotation of the Earth can be assumed to produce the potential field and thus to the equivalence of the problem to that of geodesic flow on the sphere.

Equations 4.2.14 are too general for complete solution. Assuming the rays to be confined to surfaces α = constant allows us to put $\mathrm{d}\alpha/\mathrm{d}s = 0$ and obtain

$$-\eta\frac{K}{H_\beta}\sqrt{F_\beta^2 + H_\beta^2}\frac{\mathrm{d}\beta}{\mathrm{d}s}\frac{\mathrm{d}}{\mathrm{d}s}\left[\frac{H_\alpha\sqrt{F_\beta^2 + H_\beta^2}}{K}\right] - \eta F F_\alpha\left(\frac{\mathrm{d}\psi}{\mathrm{d}s}\right)^2 = \frac{\partial\eta}{\partial\alpha} \tag{4.11.1}$$

$$\frac{\mathrm{d}}{\mathrm{d}s}\left[\eta\sqrt{F_\beta^2+H_\beta^2}\frac{\mathrm{d}\beta}{\mathrm{d}s}\right]-\eta\frac{FF_\alpha\left(\frac{\mathrm{d}\psi}{\mathrm{d}s}\right)^2}{\sqrt{F_\beta^2+H_\beta^2}}=\frac{1}{\sqrt{F_\beta^2+H_\beta^2}}\frac{\partial\eta}{\partial\beta} \tag{4.11.2}$$

$$\frac{\mathrm{d}}{\mathrm{d}s}\left(\eta F\frac{\mathrm{d}\psi}{\mathrm{d}s}\right)+\eta F_\beta\frac{\mathrm{d}\beta}{\mathrm{d}s}\frac{\mathrm{d}\psi}{\mathrm{d}s}=0 \tag{4.11.3}$$

Since F H and K and related derivatives as well as η are themselves ψ-independent

$$\frac{\mathrm{d}}{\mathrm{d}s}\equiv\frac{\partial}{\partial\beta}\frac{\mathrm{d}\beta}{\mathrm{d}s} \tag{4.11.4}$$

$$\mathrm{d}s^2=(F_\beta^2+H_\beta^2)\,\mathrm{d}\beta^2+F^2\,\mathrm{d}\psi^2 \tag{4.11.5}$$

Substituting for $\mathrm{d}s^2$ in Equation 4.11.1, we obtain

$$-\eta\left\{\frac{K}{H_\beta}\sqrt{F_\beta^2+H_\beta^2}\frac{\partial}{\partial\beta}\left[\frac{H_\alpha\sqrt{F_\beta^2+H_\beta^2}}{K}\right]\mathrm{d}\beta^2\right.$$
$$\left.+FF_\alpha\,\mathrm{d}\psi^2\right\}=\frac{\partial\eta}{\partial\alpha}[(F_\beta^2+H_\beta^2)\,\mathrm{d}\beta^2+F^2\,\mathrm{d}\psi^2] \tag{4.11.6}$$

The coefficients of $\mathrm{d}\beta^2$ and $\mathrm{d}\psi^2$ on each side of this equation can be equated, that is

$$-\eta FF_\alpha=F^2\frac{\partial\eta}{\partial\alpha} \tag{4.11.7}$$

and

$$-\eta\frac{K}{H_\beta}\sqrt{F_\beta^2+H_\beta^2}\frac{\partial}{\partial\beta}\left[\frac{H_\alpha\sqrt{F_\beta^2+H_\beta^2}}{K}\right]=\frac{\partial\eta}{\partial\alpha}(F_\beta^2+H_\beta^2) \tag{4.11.8}$$

Equation 4.11.6 is integrable and gives

$$\eta=\frac{A(\beta)}{F(\alpha,\beta)} \tag{4.11.9}$$

where $A(\beta)$ is an arbitrary function of the integration. Dividing Equation 4.11.7 by Equation 4.11.6, and performing the differentiations, results, after some manipulation, in the solubility condition

$$F[F_{\alpha\beta}H_\alpha-H_{\alpha\beta}F_\alpha]+H_\beta[F_\alpha^2+H_\alpha^2]=0 \tag{4.11.10}$$

It can be noted that, had the ray been assumed to be confined to a surface of constant β, the analysis above, and the solubility condition would have been arrived at in a similar manner subject to the substitutions $\alpha\leftrightarrow\beta$.

The relation for η in Equation 4.11.9 acts as an integrating factor for Equation 4.11.3 with the result

$$\frac{\mathrm{d}}{\mathrm{d}s}\left[A(\beta)F(\alpha,\beta)\frac{\mathrm{d}\psi}{\mathrm{d}s}\right]=0$$

or

$$A(\beta)F(\alpha,\beta)\frac{\mathrm{d}\psi}{\mathrm{d}s} = D \tag{4.11.11}$$

a constant for the ray trajectory. This is a three-dimensional generalization of Snell's law and contains all the previously known generalizations such as Bouguer's theorem.

Substitution of this result, and that for η, in Equation 4.11.2 results in

$$\frac{\mathrm{d}^2\beta}{\mathrm{d}s^2} + \left(\frac{\mathrm{d}\beta}{\mathrm{d}s}\right)^2 \frac{F_\beta F_{\beta\beta} + H_\beta H_{\beta\beta}}{F_\beta^2 + H_\beta^2} - \frac{D^2 A'(\beta)}{A^2(\beta)(F_\beta^2 + H_\beta^2)} = 0$$

or (by inspection!)

$$\frac{\mathrm{d}\beta}{\mathrm{d}s} = \sqrt{\frac{1 - D^2/[A(\beta)]^2}{F_\beta^2 + H_\beta^2}} \tag{4.11.12}$$

The condition for solubility appears to be highly restrictive, in that nearly all the common forms of axisymmetric and toroidal coordinate systems available [47] do not satisfy the criterion of Equation 4.11.10.

Quite general orthogonal axisymmetric coordinates can be obtained from the complex mapping

$$G(x + iz) = F(\alpha,\beta) + iH(\alpha,\beta)$$

and subsequent rotation about the z axis. These can be generalized still further into toroidal coordinate systems by translation along the positive x axis before rotation.

The simplest coordinate system so far found to satisfy Equation 4.11.10 is that defined by

$$\begin{aligned} F(\alpha,\beta) &= f(\alpha)\cos[g(\beta)] \\ H(\alpha,\beta) &= f(\alpha)\sin[g(\beta)] \end{aligned}$$

for general functions f and g. In this, and in most of the following, sine and cosine functions are found to be interchangeable, provided consistency is maintained. However, surfaces of constant α are spheres centred on the origin with radius $f(\alpha)$ and, consequently, the simplest permissible solutions are

$$\begin{aligned} F(\alpha,\beta) &= f(\alpha)\cos\beta \\ H(\alpha,\beta) &= f(\alpha)\sin\beta \end{aligned}$$

With this definition the factor $F_\beta F_{\beta\beta} + H_\beta H_{\beta\beta}$ in Equation 4.11.2 becomes zero. (This would also have been the case if, for $g(\beta)$, we had had $g''(\beta) = 0$ giving $g(\beta) = n\beta$, which is only a slight generalization and will be included subsequently.)

The final versions of Equations 4.11.11 and 4.11.12 thus become

$$A(\beta)f(\alpha)\cos\beta\frac{\mathrm{d}\psi}{\mathrm{d}s} = D \tag{4.11.13}$$

$$\frac{\mathrm{d}\beta}{\mathrm{d}s} = \sqrt{1 - \frac{D^2}{[A(\beta)]^2}} \tag{4.11.14}$$

with

$$\eta = \frac{A(\beta)}{f(\alpha)\cos\beta} \tag{4.11.15}$$

Equation 4.11.1 now becomes

$$-\eta f(\alpha)f'(\alpha)\left(\frac{\mathrm{d}\beta}{\mathrm{d}s}\right)^2 - \eta f(\alpha)f'(\alpha)\cos^2\beta\left(\frac{\mathrm{d}\psi}{\mathrm{d}s}\right)^2 = \frac{\partial\eta}{\partial\alpha} \tag{4.11.16}$$

with a solution $F'(\alpha) = 0$; $\eta'(\alpha) = 0$, so that the refractive index in the radial direction is uniform, as required for a ray to be confined to a spherical surface.

There remains at our disposal the arbitrary function $A(\beta)$. Equations 4.11.13 & 4.11.14 constitute the equations for the ray path in this medium and, thus the choice of the function $A(\beta)$ must be such that it contains no zeros in the range of β required, or for which $\mathrm{d}\beta/\mathrm{d}s$ is real. We will take β to be the angle of latitude on the sphere and, hence, its range will be $-\pi/2 \leq \beta \leq \pi/2$.

In the cases where $A(\beta)$ becomes infinite, the refractive index does too and we resort to the convention that rays terminate (or originate) there in a direction at right-angles to the local stratification of the refractive index. For real physical situations, regions of "realistic" refractive index only will be considered, and thus ray paths near singularities such as infinities of the refractive index or at the poles of the sphere will be purely hypothetical.

Equation 4.11.12 can now be integrated to give the angle of travel about the axis, as follows

$$\mathrm{d}\psi = \int \frac{D}{A(\beta)\cos\beta}\,\mathrm{d}s$$

on a sphere $\alpha =$ constant, and hence

$$\psi_1 - \psi_0 = \int_{\beta_0}^{\beta_1} \frac{D\,\mathrm{d}\beta}{\cos\beta\sqrt{[A(\beta)]^2 - D^2}} \tag{4.11.17}$$

where β_1 and β_0 are the values of the polar latitude of the end-points of a ray.

We consider the pencil of rays from a point source on the equator confined to the surface of the sphere. This situation produces the boundary condition for the solution of the ray integral equation.

In order to contain the ray within a zone about the equator, it is intuitively obvious that the refractive index should have a maximum value there, and so we can anticipate a function $A(\beta)$ of the form.

$$A(\beta) = AB(\beta) \tag{4.11.18}$$

where $B(\beta)$ has a maximum value of unity when $\beta = 0$.

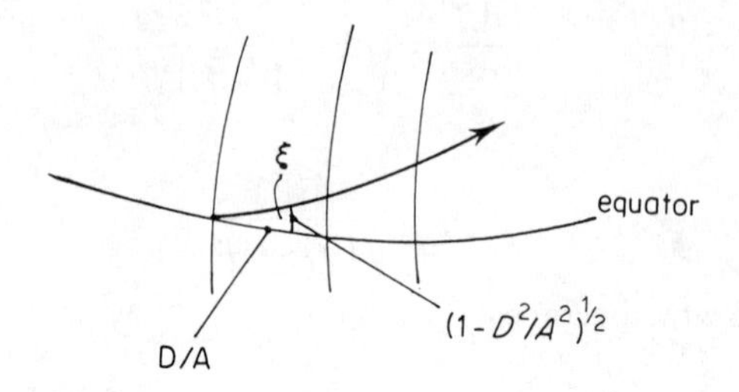

Figure 4.58: Rays on the surface of a sphere with angular variation of refractive index

At the source, therefore (Figure 4.58)

$$\left.\frac{\mathrm{d}\psi}{\mathrm{d}s}\right|_{\beta=0} = \frac{D}{A}$$

$$\left.\frac{\mathrm{d}\beta}{\mathrm{d}s}\right|_{\beta=0} = \sqrt{1-\frac{D^2}{A^2}}$$

That is, $D \leq A$ and the ray makes an angle with the equator at the source of $D/A = \cos\xi$. The ray becomes parallel to the equator at angles (ψ, β), where

$$\frac{\mathrm{d}\beta}{\mathrm{d}s} = 0 = 1 - \left[\frac{D^2}{A^2 B^2(\beta)}\right]$$

and turns to a line of longitude where $\mathrm{d}\psi/\,\mathrm{d}s = 0$.

From Equation 4.11.12 this occurs at values for which $A(\beta)$ is infinite, that is the refractive index also becomes infinite.

From Equation 4.11.17, for a ray commencing at the equator making an angle ξ there, at longitude $\psi_0 = 0$, the ray will be parallel to the equator at longitude

$$\begin{aligned}\psi_p &= \int_0^{\beta_0} \frac{D/A\,\mathrm{d}\beta}{\cos\beta\sqrt{B(\beta)^2 - D^2/A^2}} \\ &= \int_0^{\beta_0} \frac{\cos\xi\,\mathrm{d}\beta}{\cos\beta\sqrt{B(\beta)^2 - \cos^2\xi}} \qquad (4.11.19)\end{aligned}$$

where β_0 is the solution of $\mathrm{d}\beta/\,\mathrm{d}s = 0$ or, from Equation 4.11.14

$$B(\beta_0) = \cos\xi$$

making the denominator in Equation 4.11.19 zero.

The ray will reach the equator again at longitude $2\psi_p$. Thus, if ψ_p can be made independent of ξ, all the rays will come to a focus at the same point

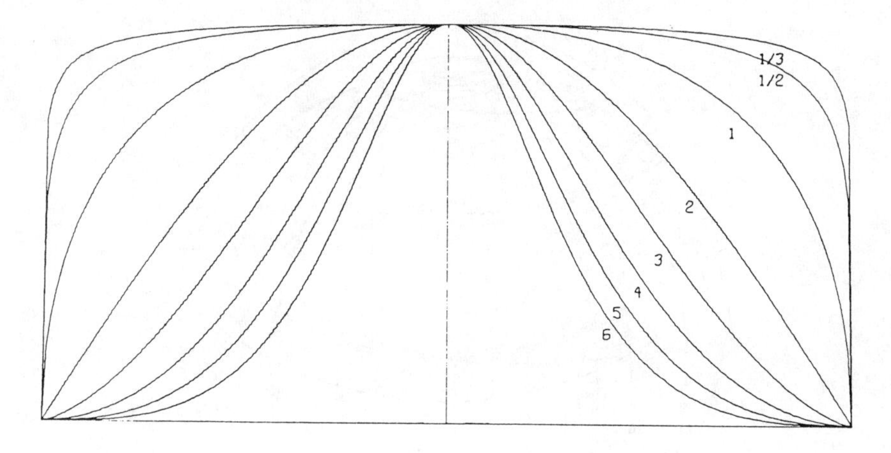

Figure 4.59: The law of Equation 4.11.19 for values of ν

on the equator. In particular, if $2\psi_p$ is a whole-number fraction of 2π, one of these foci will coincide with the original source. For this to occur, therefore, we require a solution of the integral equation

$$\psi_p = \int_0^{\beta_0} \frac{\cos\xi \, d\beta}{\cos\beta\sqrt{B^2(\beta) - \cos^2\xi}} = \frac{\pi}{\nu} \tag{4.11.20}$$

with ν preferably, but not essentially, an integer. Not surprisingly, the solution can be obtained by identical methods to the standard treatment of rays in a purely spherically symmetrical medium, given in Appendix III. As shown there this results in

$$B(\beta) = \frac{2\cos^{\nu/2}\beta}{(1+\sin\beta)^{\nu/2} + (1-\sin\beta)^{\nu/2}} \tag{4.11.21}$$

The function for various values of ν is shown in Figure 4.59

The resulting distribution of refractive index obtained is

$$\cos\beta\eta(\beta) = \frac{2A\cos^{\nu/2}\beta}{(1+\sin\beta)^{\nu/2} + (1-\sin\beta)^{\nu/2}}\eta(\alpha)$$

With $\nu = 0$ we obtain the elementary solution $A(\beta) =$ constant. This implies a variation in refractive index $\eta = \sec\beta$ and $\cos\beta(\,d\psi/\,ds) =$ constant as well as $d\beta/\,ds =$ constant. These are respectively the sine and cosine of the angle ξ made by the ray at line of longitude, and, when both remain constant, give the loxodrome path on the sphere (Figure 4.60). On a Mercator's projection these are rhumb lines, that is straight lines crossing the meridians at a fixed angle.

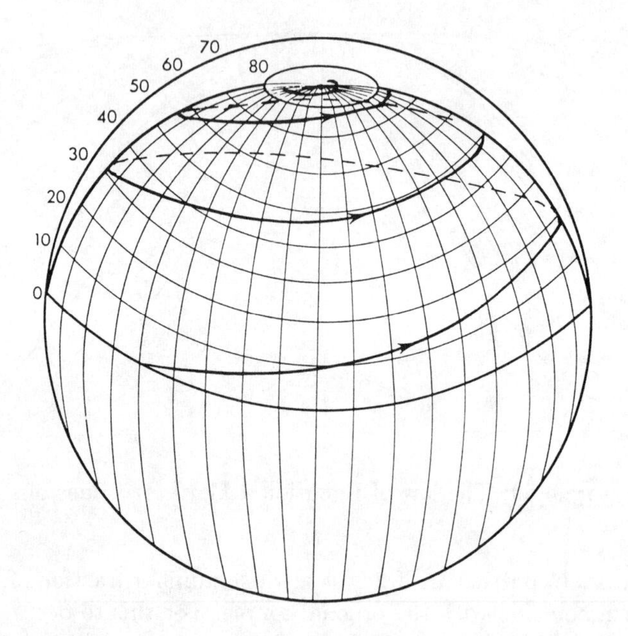

Figure 4.60: Loxodrome ($\nu = 0$) rays on the sphere with refractive index $\sec\beta$

With $\nu = 1$ the rays are parallel to the equator at the longitude diametrically opposed to the source, and hence circle the poles once (Figure 4.61).

With $\nu = 2$, $B(\beta) = \cos\beta$ and the result confirms the standard result (see also Equation III.3)

$$\int_0^{\xi} \frac{\cos\xi \, d\beta}{\cos\beta\sqrt{\cos^2\beta - \cos^2\xi}} = \frac{\pi}{2}$$

The rays are then great circles coming to a focus at the point at the equator diametrically opposite the source (Figure 4.62). The refractive index in this case is uniform. $\nu = 4$ gives the result

$$\int_0^{\sin^{-1}(\tan\xi/2)} \frac{\cos\xi \, d\beta}{\cos\beta\sqrt{\frac{\cos^4\beta}{1+\sin^2\beta} - \cos^2\xi}} = \frac{\pi}{4}$$

which is capable of confirmation by the application of complete elliptic integrals of the first and third kinds, but higher values of ν require hyperelliptic techniques or computational methods. Rays for other values of ν are shown in Figure 4.63

All other functions of $A(\beta)$ give diffuse ray systems: some, particularly

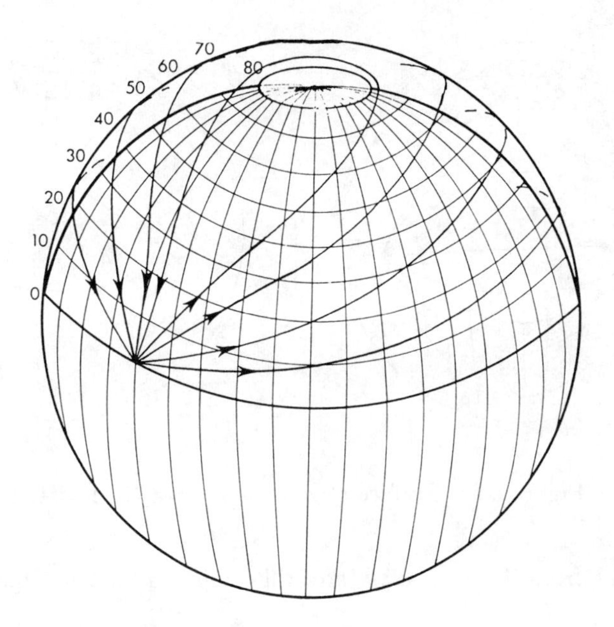

Figure 4.61: $\nu = 1$ rays are parallel to the equator at the longitude directly opposite the source

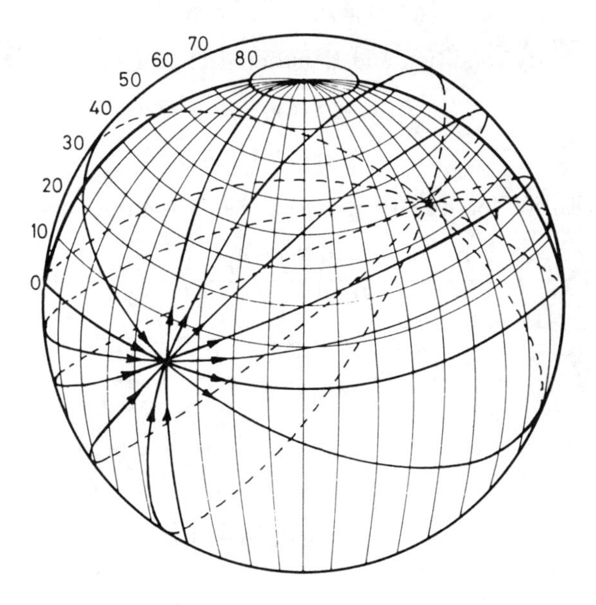

Figure 4.62: $\nu = 2$ rays are great circles

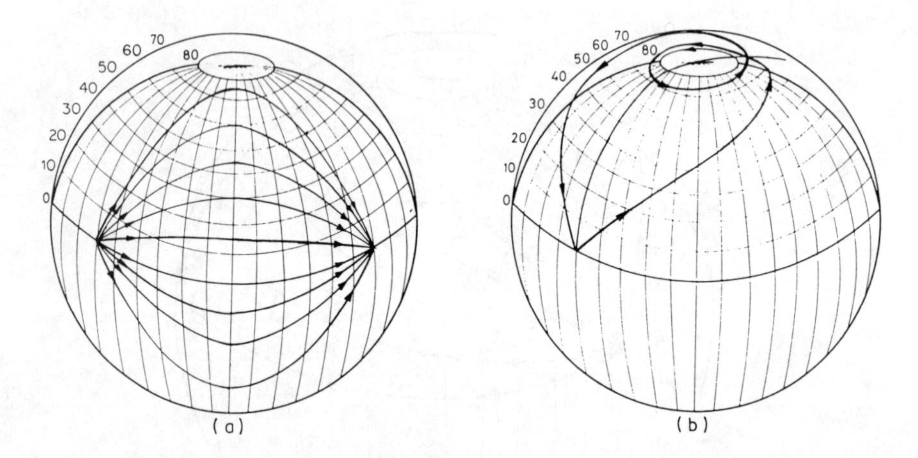

Figure 4.63: Surface rays for (a) $\nu = 3$ (b) $\nu = 1/2$

those arising from $A(\beta) = \cos^n \beta$, are still confined to a zone about the equator.

The results of the previous section can be generalized for separable solutions. That is, letting

$$\begin{aligned} F(\alpha, \beta) &= f_1(\alpha) f(\beta) \\ H(\alpha, \beta) &= h_1(\alpha) h(\beta) \end{aligned} \tag{4.11.22}$$

the orthogonality condition and Equation 4.11.10 requires

$$f_1(\alpha) \equiv h_1(\alpha) = \text{constant}$$

$$f^2(\beta) + h^2(\beta) = 1$$

After substitution for $\mathrm{d}\psi/\mathrm{d}s$, Equation 4.11.2 then becomes

$$\frac{\mathrm{d}^2\beta}{\mathrm{d}s^2} + \left(\frac{\mathrm{d}\beta}{\mathrm{d}s}\right)^2 \left(\frac{f''(\beta)}{f'(\beta)} + \frac{f'(\beta) f^2(\beta)}{f(\beta)[1 - f^2(\beta)]}\right) - \frac{A'(\beta)}{A^3(\beta)} D^2 \frac{1 - f^2(\beta)}{f'^2(\beta)} = 0$$

$$\frac{\mathrm{d}\beta}{\mathrm{d}s} = \frac{\sqrt{1 - f^2(\beta)}}{f'(\beta)} \sqrt{1 - \frac{D^2}{A^2(\beta)}}$$

The resultant functions $B(\beta)$ for repeated focusing are then given by

$$\int_0^{\beta_0} \frac{f'(\beta)}{f(\beta)\sqrt{1 - f^2(\beta)}} \frac{(D/A)\,\mathrm{d}\beta}{\sqrt{B^2(\beta) - D^2/A^2}} = \frac{\pi}{\nu}$$

or

$$B(\beta) = \frac{2[f(\beta)]^{\nu/2}}{[1 - \sqrt{1 - f^2(\beta)}]^{\nu/2} + [1 + \sqrt{1 - f^2(\beta)}]^{\nu/2}}$$

In fact these functions are not as general as first appears but are scale transformations of the β axis. For example, if $f(\beta) = \text{sech}\beta$, then β has the range $(-\infty, \infty)$, where ∞ refers to one pole of the sphere and $-\infty$ the other, but the functional form of the refractive index is unaltered.

STEREOGRAPHIC PROJECTION OF NON-UNIFORM SPHERICAL LENSES

Luneburg [8], showed that the ray trajectories associated with the Maxwell fish-eye could be derived from the stereographic projection of the great circle geodesics of the uniform unit sphere. The scale factor of the transformation then gives the requisite radial variation of the refractive index to create the fish-eye focusing property.

Luneburg continued the analysis to give the generalization of this refractive index law which, as stated, no longer applies to the spherical medium but is valid for axisymmetric media only. We have shown that the generalized law of refractive index is such that repeated focusing occurs at points around the periphery which, if spaced at whole number fractions of 2π, create a standing wave mode around the cylindrical system (Figure 4.32). The ray paths in the general axisymmetric medium, concludes that, for rays confined to the surface of a sphere, angular surface refractive index variations can be derived for which the rays would have the same repetitive focusing property around the equatorial plane of the sphere (Figure 4.63)

We now show that, in fact, the generalized refractive index of Luneburg is the stereographic projection of the *non-uniform* unit sphere. Furthermore, since the stereographic projection of any function which is symmetric about the equator is in the form of a curve for which the part interior to the unit circle is the geometrical inverse of the part exterior to it, this too is a property of the self-inverse nature of the generalized law of refractive index. Finally, the process can be reversed to give the surface refractive index appropriate to other well known radially variable media, such as the Luneburg lens itself and its generalisations which creates an endless variety of distributions on the sphere with the repeating focus property around the equator.

STEREOGRAPHIC PROJECTION

The stereographic projection of the unit sphere is shown in Figure 4.64. The sphere has a diameter equal to unity and the projection is made from the north pole N onto the plane tangent to the sphere at the south pole. Consequently the upper hemisphere projects into the region exterior to the unit circle and the lower hemisphere into the interior in the projection plane. A point $P(\beta, \phi)$ on the sphere, where ϕ is the angle of longitude and β the latitude (note: not the usual polar angle θ measured from the pole), projects into the point Q in the plane with polar coordinates (R, ϕ) where ϕ remains invariant (since the

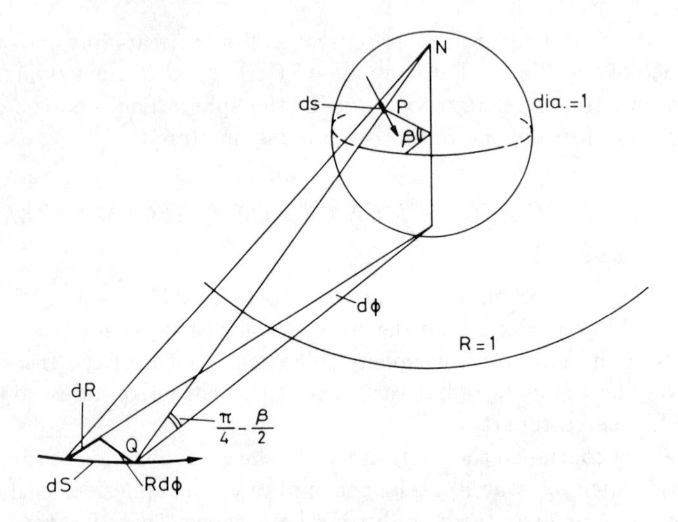

Figure 4.64: Stereographic projection

projection is conformal) and R is given by

$$R = (1 + \sin\beta)/\cos\beta \equiv \cot\theta/2 \tag{4.11.23}$$

Thus

$$\mathrm{d}R = (1 + \sin\beta)\,\mathrm{d}\beta/\cos^2\beta \tag{4.11.24}$$

It is this factor of $\cos\beta$, the ratio between R and $\mathrm{d}R$, that proves to be vital in the ensuing analysis.

Solving Equation 4.11.22 for $\sin\beta$ and $\cos\beta$ gives

$$\cos\beta = 2R/(R^2 + 1)$$

and

$$\sin\beta = (R^2 - 1)/(R^2 + 1) \tag{4.11.25}$$

The necessary refractive index law for repeated foci at points with longitude $2\pi/\nu$ is

$$\eta(\beta) = \frac{2\cos^{(\nu/2)-1}\beta}{(1 + \sin\beta)^{\nu/2} + (1 - \sin\beta)^{\nu/2}} \tag{4.11.26}$$

The stereographic projection of this law is made by equating the optical line elements on the sphere and in the plane. For the former we have

$$\eta^2\,\mathrm{d}s^2 = \mu^2(\,\mathrm{d}\beta^2 + \cos^2\beta\,\mathrm{d}\phi^2)$$

with η given by Equation 4.11.26.

We equate this with $N^2\,\mathrm{d}S^2$ in the plane, where

$$\mathrm{d}S^2 = \mathrm{d}R^2 + R^2\,\mathrm{d}\phi^2$$

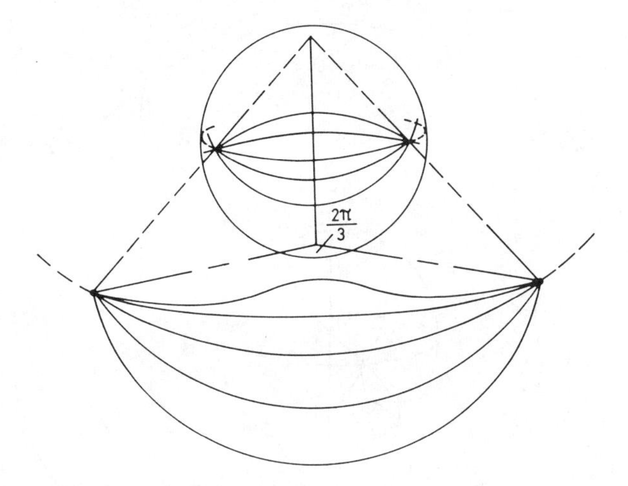

Figure 4.65: Projection of generalized law for $\nu = 3$

and R and dR are given by Equations 4.11.23 and 4.11.24

Consequently, we find

$$N = \frac{A\cos\beta\cos^{\nu/2}\beta}{2(1+\sin\beta)[(1+\sin\beta)^{\nu/2}+(1-\sin\beta)^{\nu/2}]} \tag{4.11.27}$$

Substituting for $\sin\beta$ and $\cos\beta$ from Equation 4.11.25 we obtain

$$N = AR^{(\nu/2)-1}/(1+R^{\nu}) \tag{4.11.28}$$

Putting A, which is an arbitrary constant anyway, equal to 2 gives the generalized Maxwell law of refractive index. The ray patterns for $\nu = 3$ and $\nu = 1$ are shown in Figures 4.65 and 4.66. $\nu = 2$ corresponds to Luneburg's result for the fish-eye lens.

The ray paths in the plane are given by Equation 4.7.12 by the substitution of the law of Equation 4.11.28 into the standard ray path integral as

$$R^{-\nu} = \sqrt{1+\sin^2\nu\phi\cot^2\alpha} + \sin\nu\phi\cot\alpha \tag{4.11.29}$$

where α is the angle made by the ray at the source with the unit circle, and thus also the angle made by the projected ray with the equator on the sphere.

It can be seen that they are self-inverse with respect to the unit circle since a change of sign of ν turns R^{ν} into $R^{-\nu}$. It follows that the refractive index law also is self inverse and this is easily demonstrated by the inversion

$$RN(R) = f(R) \rightarrow f(1/R) = R\overline{N(R)}$$

or

$$RN(R) = \frac{2R^{\nu/2}}{1+R^{\nu}} \rightarrow \frac{2R^{-\nu/2}}{1+R^{-\nu}} = \frac{2R^{\nu/2}}{1+R^{\nu}} = R\overline{N(R)}$$

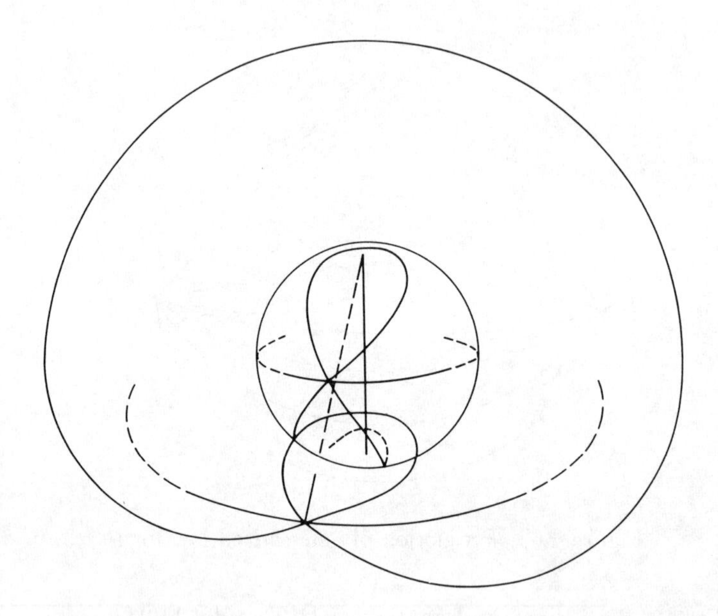

Figure 4.66: Projection of generalized law for $\nu = 1$

whence

$$\overline{N(R)} = 2R^{(\nu/2)-1}/(1 + R^{\nu})$$

Final confirmation of these inverse properties arises from the symmetry of the ray patterns on the sphere about the equator. Under stereographic projection the upper hemisphere pattern projects into the exterior of the unit circle and the lower hemisphere into the interior, and hence any pattern with symmetry about the equator forms projected curves in which the interior portion is the inverse of the exterior portion.

NONSYMMETRICAL RAYS ON A SPHERE

The process can be reversed. We can apply the projection of any given refractive index law with radial variation only into a spherical surface law varying in latitude only and in which the known ray paths in the plane will stereographically project into the ray paths on the sphere. The result is surprising in its simplicity

We have in the plane

$$\begin{aligned} N^2\,dS^2 &= N^2(dR^2 + R^2\,d\phi^2) && (4.11.30)\\ &= N^2[(1+\sin\beta)/\cos^2\beta]^2\,d\beta^2 + [(1+\sin\beta)/\cos\beta]^2\,d\phi^2 \\ &= N^2[(1+\sin\beta)/\cos^2\beta]^2[d\beta^2 + \cos^2\beta\,d\phi^2] \\ &= \eta^2(\beta)\,ds^2 \quad \text{as before} && (4.11.31)\end{aligned}$$

and the effect of the factor $\cos\beta$ can be clearly seen.

When $N(R)$ is replaced by $N(\beta)$ through Equation 4.11.27

$$\eta(\beta) = N(\beta)(1+\sin\beta)/\cos^2\beta \tag{4.11.32}$$

is the required surface refractive law for *any* refractive law $N(R)$.

As an example we project the Luneburg lens law

$$N^2(R) = 2 - R^2$$

into

$$\begin{aligned}\eta^2(\beta) &= [2-(1+\sin\beta)^2/\cos^2\beta](1+\sin\beta)^2/\cos^4\beta \\ &= \frac{(1-3\sin\beta)(1+\sin\beta)^3}{\cos^6\beta}\end{aligned} \tag{4.11.33}$$

The rays, which are complete ellipses in the plane, all refocus at the diametrically opposite point. Hence they must do so on the sphere.

The rays are confined in latitude for angles for which

$$3\sin\beta \leq 1 \tag{4.11.34}$$

corresponding to $R^2 = 2$, but even though part of the complete ellipse is exterior to the unit circle and thus in the region of refractive index less than unity, on the sphere the entire projected curve is viable.

The ray path can then be plotted simply as shown in Figure 4.67. For the narrower ellipses the ray paths approach the south pole while still confined in the upper hemisphere by the condition of Equation 4.11.33. Further generalizations of the Luneburg lens give multiple-refocusing around the unit circle and can be treated similarly. The rays then become confined to a *nonsymmetrical* zone about the equator of the sphere. Nonfocusing systems such as the Eaton lens can be projected in the same manner.

It is well known that when a system of rays undergoes endless refocusing there is an underlying resonant mode in the electromagnetic field. The foregoing therefore implies that in the cavity formed by a narrow annular zone of a sphere, resonances exist with mode number equal to the number of refocusing points around the equator. It follows that this would be the more natural form of torus for confining plasmas by replacing the variable refractive index with the equivalent variable field potential.

Similarly, the analogy between rays and particle trajectories suggest that stable fluctuating orbits can occur in a zone about the equator, should the potential of the nucleus have the latitudinal variation that the equivalent refractive index does.

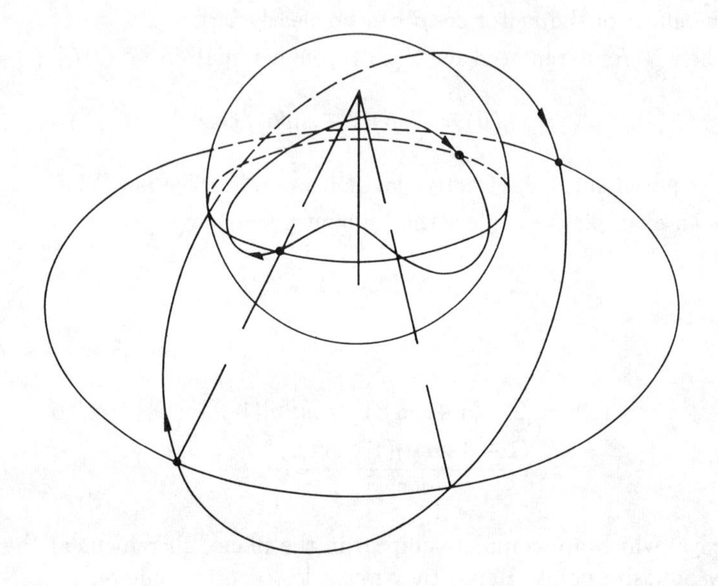

Figure 4.67: Projection of Luneburg's lens onto a sphere

4.12 OTHER COORDINATE SYSTEMS

SEPARABILITY

It can be seen from the foregoing discussion that there exist only a few laws of refractive index which lead to integrable forms of the ray path equation in terms of elementary functions, and that these laws are themselves algebraic and of a fairly elementary nature. In considering, in particular the elementary parabolic law giving rise to helical rays as in Section 4.6, Buchdahl [48] was lead to investigate separability conditions that could be applied to the eikonal equation for rays in a generally varying medium

$$|\nabla s|^2 = \eta^2$$

in different coordinate systems which would lead to algebraically simple results. By this method he formulated the simplest laws of refractive index that would arise in these coordinate systems. What is most remarkable is that this appears to be the first consideration of some of the more unusual coordinate systems that can support this form of lens construction. Any one of these could lead to the design of lenses with desirable qualities by the methods previously formulated, that of determining the ray paths and a suitable delimiting surface at which refraction, if it occurs, can be used to focus the rays in the required manner. The results of Buchdahl's investigations are outlined here, with a necessary change of notation for conformity with the previous sections. For the method the reader is referred to the original work.

CARTESIAN COORDINATES

$$\eta^2 = f(x) + g(y) + h(z)$$

where for symmetry about the z axis

$$f(x) + g(y) \equiv a(x^2 + y^2) \qquad a \text{ constant}$$

$$h(z) = \alpha + \beta \exp^{-\gamma z} \qquad \alpha, \beta, \gamma \text{ constant} \tag{4.12.1}$$

CYLINDRICAL POLAR COORDINATES

$$\eta^2 = f(\rho) + \rho^{-2} g(\phi) + h(z)$$

Axial symmetry requires $g(\phi)$ to be constant and for regularity along the axis (not essential however) this constant would be zero. $h(z)$ is also taken to be constant and the result for $f(\rho)$ is either

$$f(\rho) = \alpha\rho^2 + \beta + \frac{\gamma}{\rho^2} \quad \text{or} \quad \alpha + \frac{\beta}{\rho} + \frac{\gamma}{\rho^2} \qquad \alpha, \beta, \gamma \text{ constant} \tag{4.12.2}$$

This contains the parabolic law for helical rays, but not the sech $a\rho$ law of Equation 4.3.3 which essentially applies to rays in meridional planes alone and not general rays as in the separable solutions.

PARABOLIC COORDINATES

Parabolic coordinates defined by

$$x = \mu\nu\cos\theta \qquad y = \mu\nu\sin\theta \qquad z = \frac{1}{2}(\mu^2 - \nu^2)$$

gives the separable condition for the refractive index

$$\eta^2 = \frac{f(\mu) + g(\nu)}{\mu^2 + \nu^2} + \frac{h(\theta)}{(\mu\nu)^2}$$

Axial symmetry makes $h(\theta)$ constant and since $\mu^2\nu^2 = x^2 + y^2 = \rho^2$ the problem reverts to cylindrical symmetry with

$$\eta^2 = a + \frac{b + cz}{\sqrt{\rho^2 + z^2}} \qquad a, b, c \text{ constant} \tag{4.12.3}$$

SPHERICAL POLAR COORDINATES

$$\eta^2 = f(r) + g(\theta)/r \tag{4.12.4}$$

where

$$g(\theta) = \frac{\alpha + \beta \sin^2\theta + \gamma \sin^4\theta}{\sin^2\theta \cos^2\theta}$$

which if regular becomes

$$g(\theta) = \beta_1 + \beta_2 \sec^2\theta$$

$$f(r) = \frac{\alpha_1 + \alpha_2 r^\kappa + \alpha_3 r^{2\kappa}}{r^2(1 + \alpha_5 r^\kappa)^2} \tag{4.12.5}$$

all Greek letters are constant and κ is an arbitrary constant. Maxwell's fish-eye is given by

$$\kappa = 2 \qquad \beta_1 = \beta_2 = \alpha_1 = \alpha_2 = 0$$

It is interesting to observe that this result can be made to agree with the general solution given in Equation 4.8.12 by putting $\alpha_1 = \alpha_3$ in the above and

$$f(x) = \sqrt{A + B/x^2} \qquad \phi(r) = r^{\kappa/2}$$

in Equation 4.8.12

CARDIOID COORDINATES

Cardioid coordinates are defined by

$$x = \frac{\mu\nu\cos\theta}{(\mu^2+\nu^2)^2} \qquad y = \frac{\mu\nu\sin\theta}{(\mu^2+\nu^2)^2} \qquad z = \frac{\mu^2-\nu^2}{2(\mu^2+\nu^2)^2}$$

then, omitting refractive laws giving rise to singularities in the medium

$$\eta^2 = (\mu^2 + \nu^2)^3[f(\mu) + g(\nu)] \tag{4.12.6}$$

and $f(\mu) + g(\nu) = \alpha + \beta\mu^2 + \gamma\nu^2$ for simple solutions.

If $r^2 = x^2 + y^2 + z^2$ the result is

$$\eta^2 = (ar^2 + br + cz)/r^5 \qquad a, b, c \text{ constant} \tag{4.12.7}$$

For each of these coordinate systems the fundamental law

$$\frac{\mathrm{d}}{\mathrm{d}s}\left(\eta\frac{\mathrm{d}r}{\mathrm{d}s}\right) = \nabla\eta \tag{4.12.8}$$

can be reduced to ray equations and conditions of symmetry, and regularity applied. For given constant coordinate surfaces the ray paths can be determined and the above method of Buchdahl gives those refractive index laws which can be expected to yield analytically integrable results from which ray paths and focusing systems can be determined.

RAYS IN AN ANGULAR VARIABLE MEDIUM

Referring again to the equations for the spherical polar coordinate system for rays confined to a diametral plane $\mathrm{d}\phi/\mathrm{d}s = 0$ and for a refractive index independent of both the r and ϕ coordinates. They reduce to

$$\frac{\mathrm{d}}{\mathrm{d}s}\left(\eta\frac{\mathrm{d}r}{\mathrm{d}s}\right) - \eta r\left(\frac{\mathrm{d}\theta}{\mathrm{d}s}\right)^2 = 0 \tag{4.12.9}$$

$$\frac{\mathrm{d}}{\mathrm{d}s}\left(\eta r\frac{\mathrm{d}\theta}{\mathrm{d}s}\right) + \frac{\mathrm{d}r}{\mathrm{d}s}\frac{\mathrm{d}\theta}{\mathrm{d}s} = \frac{1}{r}\frac{\partial\eta}{\partial\theta} \tag{4.12.10}$$

The left-hand side of Equation 4.12.10 is

$$\frac{1}{r}\frac{\mathrm{d}}{\mathrm{d}s}\left(\eta r^2\frac{\mathrm{d}\theta}{\mathrm{d}s}\right)$$

giving

$$\frac{\mathrm{d}}{\mathrm{d}s}\left(\eta r^2\frac{\mathrm{d}\theta}{\mathrm{d}s}\right) = \frac{\mathrm{d}\eta}{\mathrm{d}\theta} \tag{4.12.11}$$

For a ray confined to the diametral plane

$$\frac{1}{r}\frac{\mathrm{d}r}{\mathrm{d}t} = \tan\alpha \qquad r\frac{\mathrm{d}\theta}{\mathrm{d}s} = \cos\alpha \qquad \frac{\mathrm{d}r}{\mathrm{d}s} = \sin\alpha$$

Then

$$\frac{\mathrm{d}}{\mathrm{d}s} \equiv \frac{\mathrm{d}}{\mathrm{d}\theta}\left(\frac{\mathrm{d}\theta}{\mathrm{d}s}\right) \equiv \frac{1}{r}\cos\alpha\frac{\mathrm{d}}{\mathrm{d}\theta} \tag{4.12.12}$$

and hence from Equation 4.12.9

$$\frac{\mathrm{d}}{\mathrm{d}s}(\eta\sin\alpha) - \frac{\eta}{r}\cos^2\alpha = 0$$

or

$$\frac{\tan\alpha}{\eta}\frac{\mathrm{d}\eta}{\mathrm{d}\theta} = 1 - \frac{\mathrm{d}\alpha}{\mathrm{d}\theta} \tag{4.12.13}$$

Equation 4.12.10 produces the same result

Putting $r'/r = \tan\alpha = v$ we obtain

$$\frac{\eta'}{\eta} = \frac{1}{v}\left[1 - \frac{v'}{1+v^2}\right] \tag{4.12.14}$$

the prime referring to differentiation with respect to θ.

Expanding the second term in Equation 4.12.14 by partial fractions and integrating, we obtain

$$\eta = \frac{C\sqrt{1+v^2}}{v}\exp\left(\int\frac{1}{v}\,\mathrm{d}\theta\right) \qquad v = \frac{r'}{r} \tag{4.12.15}$$

Thus given any one parameter pencil of rays, $r = r(a, \theta)$, the medium refractive index given by Equation 4.12.15 is a solution, only if the parameter a disappears in the derivation. On the other hand, the definition of the ray trajectory for a specified angular refractive medium depends on the solution of Equation 4.12.15 for $r(\theta)$ given $\eta(\theta)$. As can be seen, this is a very complicated problem.

It is important in what follows to remember that the θ coordinate is measured from the pole in the analysis, and thus the usual polar coordinate expressions for the curves we are to consider will require to be modified by the substitution of sine for cosine, and vice versa. The importance of this arises mainly in the effect it has on the sign of exponent in Equation 4.12.15 which inverts the solution.

The general rules of rays in a variable refractive medium also apply in this case. That is, the rays can never become orthogonal to the stratification of the medium without an infinite value of the refractive index occurring. In particular in this case, and contrary to expectations, circular rays with centre at the origin are impossible to achieve. This creates a severe limitation to those ray patterns that can be created by a realistic refractive medium, that is a medium with $\infty \geq \eta \geq 1$. Since the solutions are given in terms of $v = r'/r$, the ray patterns are form-invariant for scale changes in the radial coordinate. As a result, most of the examples will deal with spiral forms of rays, although for other forms finite-valued regions of the refractive medium can be delineated as boundary conditions in the solution of Equation 4.12.15.

SPIRAL RAYS

with $v =$ constant $= a$

$$r' = ar \quad \text{and thus} \quad r = \exp^{a\theta} \tag{4.12.16}$$

The rays are thus equiangular spirals and direct substitution into Equation 4.12.15 provides the refractive index

$$\eta = A \exp^{\theta/a} \tag{4.12.17}$$

the constant A containing all the other constant factors.

CIRCULAR RAYS PASSING THROUGH THE ORIGIN

With

$$r = a \sin\theta \quad (0 \leq \theta \leq \pi) \tag{4.12.18}$$

and therefore $r' = a\cos\theta$, we have $v = \cot\theta$. This gives

$$\eta = \frac{C\sqrt{1+\cot^2\theta}}{\cot\theta} \exp\left(\int \tan\theta \, \mathrm{d}\theta\right) = C/\cos^2\theta \tag{4.12.19}$$

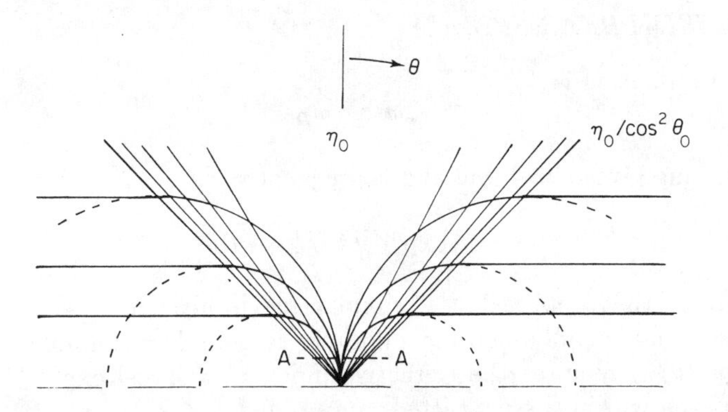

Figure 4.68: Circular rays in the conical medium. With suitable choice of parameters a completely azimuthal cone of rays is obtained

This function becomes infinite on the axis $\theta = \pi/2$ where the rays are perpendicular to the stratification.

If the medium is terminated at a realistic value of the refractive index, at a value θ_0 say, then the rays will all be incident upon the surface $\theta = \theta_0$ at the same angle. Consequently they will all refract by the same amount into free space, giving a collimated pencil of parallel rays. The angle of emission can be chosen quite arbitrarily to create any required cone of rays, upon rotation of the system about the axis $\theta = 0$. This is illustrated in Figure 4.68. With horizontal rays as shown, full azimuthal coverage is obtained from the now conical angularly non-uniform lens. The singularity at the tip can be removed by cutting the lens with a plane section AA in Figure 4.68, leaving a circular aperture in which a line at infinity is imaged into a horizontal radial line. This is the same effect as the camera obscura.

THE SINUSOIDAL SPIRALS

These have the general equation

$$r^n = a^n \sin n\theta \tag{4.12.20}$$

and $v = r'/r = \cot n\theta$, and hence from Equation 4.12.15

$$\eta = A(\cos n\theta)^{-(1+n)/n} \tag{4.12.21}$$

The straight line ray $n = -1$ thus requires a uniform refractive medium. Similar results can be obtained with the multi-folium curves

$$r = b \sin m\theta$$

Then $v = m \cot m$, and therefore

$$\eta = \frac{C\sqrt{\sin^2 m\theta + m^2 \cos^2 m\theta}}{m(\cos m\theta)^{1+1/m^2}} \tag{4.12.22}$$

THE ARCHIMEDEAN SPIRALS

The general relation is

$$r^m = a^m \theta$$

Differentiating gives $v = 1/m\theta$ and consequently

$$\eta = A \exp^{m\theta^2/2} \sqrt{1 + m^2\theta^2} \tag{4.12.23}$$

Spiral curves that approach the circular inevitably give rise to refractive index laws that rapidly become infinitely large. For example, the curve $r = a \tanh(\theta/2)$ requires the refractive index $\eta = A \cosh \exp^{\cosh\theta}$ which is excessive even within the permitted range of $0 \leq \theta \leq 2\pi$.

Numerous other curves can be illustrated in this manner, including the curves $r = a \sin^n \theta$ and the cardioids. For final illustration we show those curves which tend to and eventually pass beyond parallelism with the stratification. The designs are suitable for beam expanders or spot size reduction.

PARABOLIC RAYS

The appropriate polar equation for the parabola is

$$r = 4a \cos\theta / \sin^2\theta$$

and differentiation gives $v = -(\sin^2\theta + 2\cos^2\theta)/(\sin\theta\cos\theta)$. The exponent in Equation 4.12.15 is thus

$$-\int \frac{\sin\theta\cos\theta}{\sin^2\theta + 2\cos^2\theta}\, d\theta = \frac{1}{6}\log(\sin^2\theta + 2\cos^2\theta)$$

and hence

$$\eta = \frac{C\sqrt{1+v^2}}{v}(\sin^2\theta + 2\cos^2\theta)^{1/6} \tag{4.12.24}$$

with v given above.

This is one of many instances that occur in which both r'/r and r/r' can be derived in terms of elementary integrals.

THE KAPPA CURVES (Figure 4.69)

These are similar to parabolic curves with polar equation

$$r = a \cot\theta$$

Hence (Figure 4.70)

$$v = -2/\sin 2\theta \qquad \eta = \frac{C}{2}\sqrt{4 + \sin^2 2\theta}\, \exp^{\cos 2\theta/4} \tag{4.12.25}$$

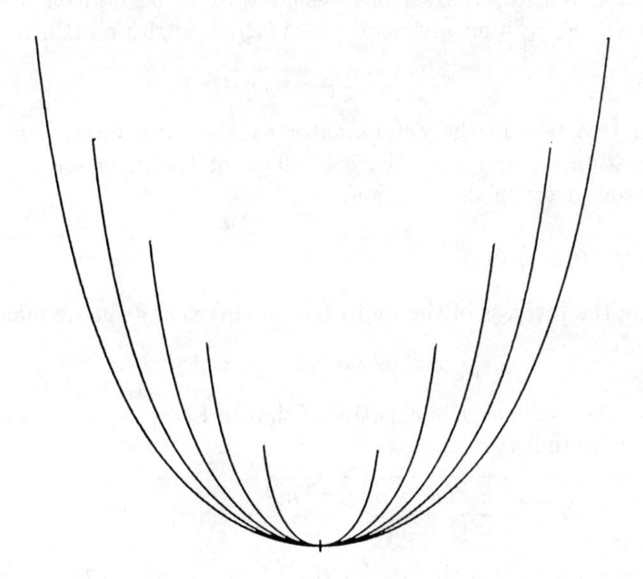

Figure 4.69: Rays in the conical medium with refractive law of Equation 4.12.25 - the kappa curves $r = a \cot \theta$

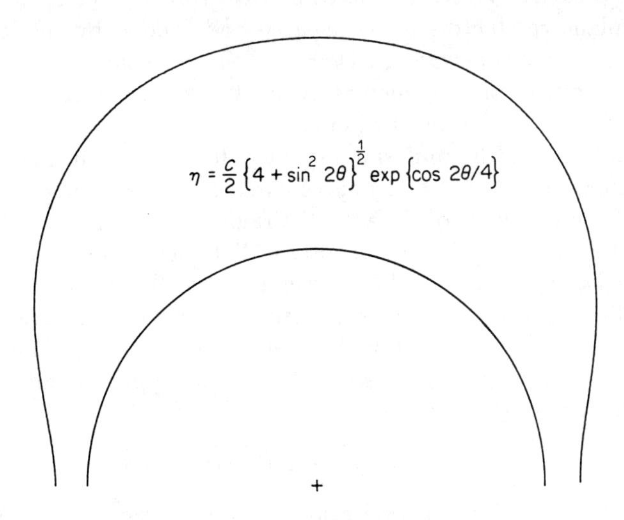

Figure 4.70: Refractive index variation for kappa curve rays

Since the entire analysis has been shown to be derived in terms of the variable $v = r'/r$, in an inversion $r \to 1/R$, all angles remain invariant and

$$R'/R = -r'/r = -v$$

The negative sign in the denominator can be incorporated in the constant of integration C, and thus the sole effect of the inversion is to invert the exponential factor in the solution.

THE EPI SPIRALS

These are the inverses of the multi-folium curves and their equations are thus

$$r = a/\sin m\theta \qquad m = 1, 2, 3$$

Hence $v = -m \cot m\theta$, the negative of that in Equation 4.12.22 and results in the refractive index

$$\eta = \frac{C\sqrt{\sin^2 m\theta + m^2 \cos^2 m\theta}}{-m} (\cos m\theta)^{(1/m)-1} \qquad (4.12.26)$$

giving the same m-fold singularity conditions at $m\theta = (2n+1)\pi/2$.

4.13 GEODESIC LENSES

There are numerous ways in which a microwave or an optical field can be confined to a surface layer. In the case of electromagnetic radiation, a pair of infinite plane conductors closely spaced and excited by the TEM mode, that is by a totally transverse electric field vector, confines the field to the planar space between the conductors. A point source of radiation between the planes would radiate isotropically in the space between the conductors and the refractive index would be uniform and unity. If, as shown in Figure 4.71, the two conductors were to be formed into curved surfaces remaining "parallel" in the sense of the geometry of surfaces, the radiation would proceed along rays which are the geodesics of a surface intermediate between the two conducting surfaces (and usually taken to be midway between the two.)

Two infinite plane conductors excited by the dominant H_{01} mode still contain an isotropic space but with a refractive index less than unity and dependent upon the spacing between them. The apparent index for planes separated by a distance h is given by

$$\eta = \sqrt{1 - \lambda^2/(2h)^2} \qquad (4.13.1)$$

λ being the wavelength of the excitation. In coordinates of η and $\lambda/2h$ this is a circle of unity radius. The space between the conductors could be filled if required with a lossless dielectric. With this mode of propagation a variable medium can be constructed by making the separation h a function of position.

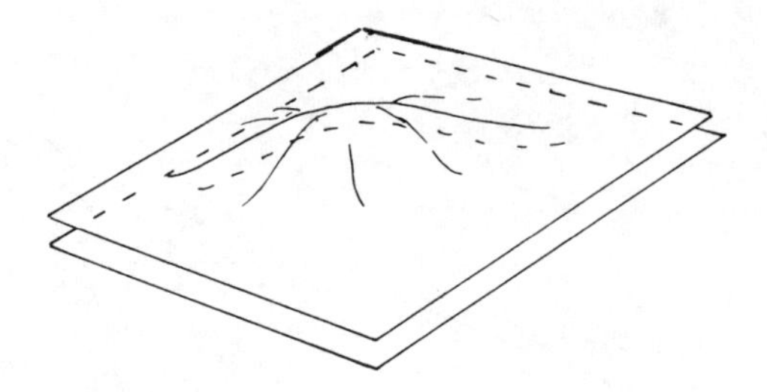

Figure 4.71: Geodesic surface: TEM propagation confined between two parallel curved conducting plates

This is ideally suited to the practical realization of the angular medium described in the previous section. We take, for example, the rays in the form of kappa curves $r = a\cot\theta$ shown in Figure 4.69; after a short distance the rays become virtually parallel, and thus a horn with curved faces that produces this ray configuration would be corrected for the phase curvature that occurs in the ordinary pyramidal horn. The profile is then obtained by substituting the refractive index law of Equation 4.12.25 into Equation 4.13.1 and solving for the separation h as a function of θ. That is

$$\sqrt{1-\frac{\lambda^2}{4h^2}} = \frac{C}{2}\sqrt{\sin^2 2\theta + 4}\exp^{\cos 2\theta/4}$$

and hence

$$h = \frac{\lambda}{2\sqrt{1-(C^2/4)(\sin^2 2\theta + 4)\exp^{\cos 2\theta/2}}} \tag{4.13.2}$$

an illustration of which is shown in Figure 4.72

Light, too, can be guided isotropically by a thin layer of uniform refractive material over a substrate of lower refractive index and confined, as in the optical fibre, by total internal reflection. In the previous chapters we have discussed the effects of making such a layer non-uniform in its medium properties, or equivalently of non-uniform thickness. However, we can retain uniformity both of material and thickness and obtain ray divergence by distorting the surface from the plane (Figure 4.73). The rays then follow the geodesics of these surfaces. Naturally for mechanical reasons such layers are supported by a substrate of lower refractive index, and hence the surface curvatures can be machined into such substrates.

Alternatively, the surface layer can be maintained plane and the required refractive index profile created by some method similar to ion implantation. With plane surfaces the entrapment of the field can best be met by placing the layer between conducting plates, with perpendicular polarisation giving

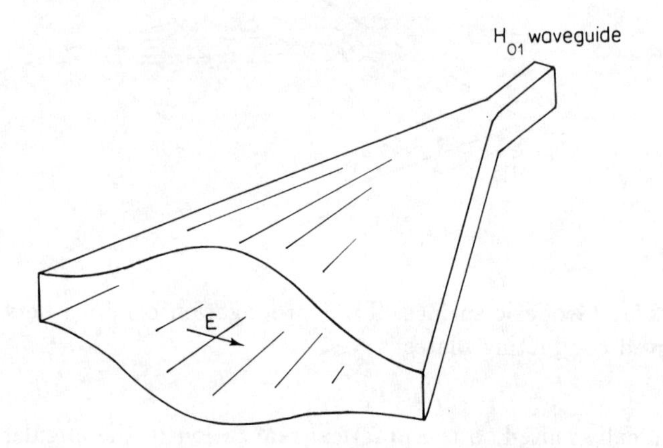

Figure 4.72: Non-uniform angular medium waveguide analogue

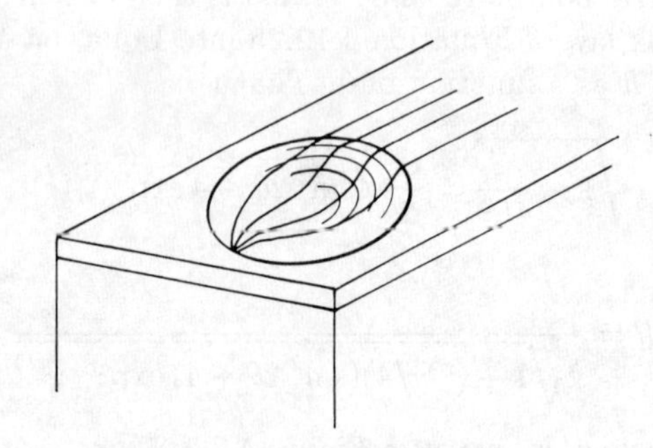

Figure 4.73: Geodesic lens, rays confined to a thin shaped dielectric coated indentation

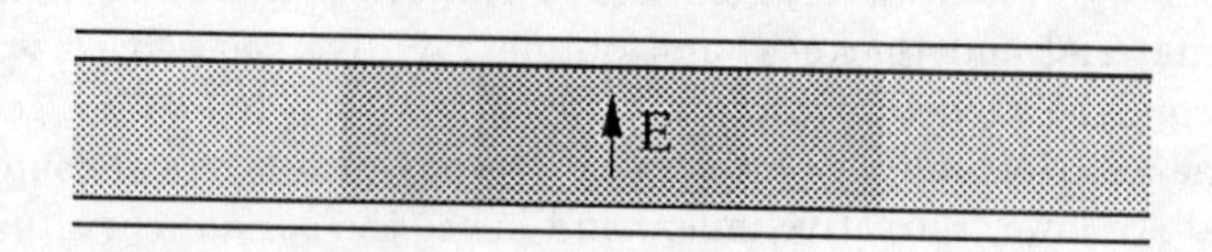

Figure 4.74: Variable density medium between parallel conducting plates - TEM mode

Figure 4.75: Variable height of dielectric TEM mode - pseudo-refractive index

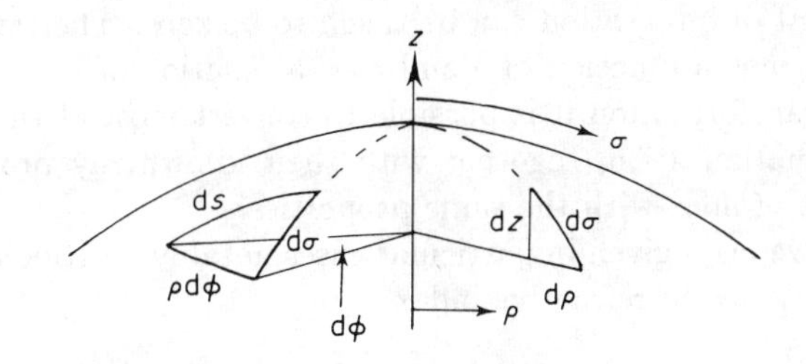

Figure 4.76: Geodesics on a surface of revolution

TEM propagation (Figure 4.74) There is also the possibility to use this mode between parallel plates in which a shaped layer of dielectric is inserted. If, as is usually the case, the surface form will be only gradually curved, the perturbation to the TEM mode will be insignificant and the height of the dielectric can create a pseudo-refractive index to the profile required (Figure 4.75) [?] [?] [?].

GEODESICS ON A SURFACE OF REVOLUTION

There is a one-to-one relation between the spherical non-uniform lens designs of the previous chapter and the circularly symmetrical surface, a surface of revolution, based upon the comparison between the rays as geodesics in the two systems. The resulting surfaces are collectively known as geodesic lenses and their optical properties are identical to the non-uniform spherical lenses (in a diametral cross-section). Such designs have long been applied to microwave antennas [49] [50] [51].

A fact of some importance that emerges from the analysis is that a refractive medium with a singularity in the refractive index can still be represented by a real and practicable geodesic surface. This is illustrated by the simple case of the equivalent refractive index for the geodesics on a right circular cone.

For a circularly symmetrical geodesic surface the optical increment of the geodesic path is, as shown in Figure 4.76

$$\eta^2 \, \mathrm{d}s = \mathrm{d}\sigma^2 + \rho^2 \, \mathrm{d}\phi^2 \tag{4.13.3}$$

since the refractive index is unity for TEM propagation.

The optical distance in the nonhomogeneous spherical refracting lenses is

$$\eta^2\,\mathrm{d}s^2 = \eta^2(r)(\,\mathrm{d}r^2 + r^2\,\mathrm{d}\theta^2) \tag{4.13.4}$$

For equivalence it can be seen by inspection that

$$\mathrm{d}\theta = \mathrm{d}\phi \quad \rho(\sigma) = r\eta(r) \quad \mathrm{d}\sigma = \eta(r)\,\mathrm{d}r \tag{4.13.5}$$

From the first of these $\theta = \phi$, and since both systems are circularly symmetrical the constant of integration can be taken to be zero. The transformation is then between η as a function of r and σ as a function of ρ.

With this transformation it is possible to convert most of the known lens solutions of Equation 4.7.6, together with their known ray properties, into geodesic surface profiles with the same properties.

Solutions have been given in particular cases notably by Rinehart [50] [51]. For the Luneburg law of refractive index

$$\eta(r)\sqrt{2 - r^2}$$

we obtain

$$\mathrm{d}\sigma = \sqrt{2 - r^2}\,\mathrm{d}r \quad \text{and} \quad (\rho(\sigma))^2 = 2r^2 - r^4$$

i.e.

$$r^2 = 1 - \sqrt{1 - \rho^2}$$

Elimination of r and $\mathrm{d}r$ leads to the differential equation relating to σ and ρ which is

$$2\mathrm{d}\sigma = \left[1 + \frac{1}{\sqrt{1 - \rho^2}}\right]\mathrm{d}\rho \tag{4.13.6}$$

or

$$\sigma = \frac{1}{2}(\rho + \sin^{-1}\rho) \tag{4.13.7}$$

the constant again being zero on the assumption that $\rho = 0$ when $\sigma = 0$. When this is done for the other known lenses, as is shown in Table 4.1, all the $\sigma(\rho)$ formulations can be seen to be included in a completely general formulation

$$\sigma = A\rho + B\sin^{-1} C\rho \tag{4.13.8}$$

characterized by the three apparently arbitrary parameters A B and C.

Consequently we use this general formulation to derive, by the reverse of the procedure, a general formulation for non-uniform refractive lenses using the same three parameters [52]. Thus from Equation 4.13.5

$$\mathrm{d}\sigma = A\,\mathrm{d}\rho + \frac{BC}{\sqrt{1 - C^2\rho^2}}\,\mathrm{d}\rho = \eta\,\mathrm{d}r = \frac{\rho}{r}\,\mathrm{d}r$$

$$\frac{\mathrm{d}r}{r} = \frac{\mathrm{d}\rho}{\rho}\left[A + \frac{BC}{\sqrt{1 - C^2\rho^2}}\right]$$

Lens form	$\eta(r)$	A	B	C	Ref
Maxwell's fish-eye	$\dfrac{2}{(1+r^2)}$	0	1	1	Eqn 4.7.9
Generalized fish-eye (resonator ring)	$\dfrac{2r^{\lambda-1}}{1+r^{2\lambda}}$	0	$\dfrac{1}{\lambda}$	1	Eqn 4.7.11
Luneburg	$\sqrt{2-r^2}$	$\dfrac{1}{2}$	$\dfrac{1}{2}$	1	Eqn 4.7.14
Eaton (transform of fish-eye $\eta \leftrightarrow r$)	$\sqrt{\dfrac{2}{r}-1}$	1	1	1	Eqn 4.7.20
Gutman	$\dfrac{\sqrt{1+f^2-r^2}}{f}$	$\dfrac{1}{2}$	$\dfrac{1+f^2}{4f}$	$\dfrac{2f}{1+f^2}$	Eqn 4.7.16
Toraldo	$\eta^2 r^2 = \eta^{1/p}(2-\eta^{1/p})$	1	p	1	Eqn 4.8.18
Transformed Toraldo ($\eta \leftrightarrow r$)	$\dfrac{\sqrt{2r^{1/p}-r^{2/p}}}{r}$	α	p	1	
Cone (semi-vertex angle ψ: $A =$ cosec ψ)	$r^{1/A-1}$	A	0	0	Eqn 4.13.16
Plane	1	1	0	0	
Lens with hyperbolic rays (including complex values $K > 2$)	$[(K^2-2K)r^2 + 2K]^{1/2}$	$\dfrac{1}{2}$	$\dfrac{K/2}{\sqrt{2K-K^2}}$	$\dfrac{\sqrt{2K-K^2}}{K}$	Eqn 4.8.17
Beam divider (point source) through angles $\pm p\pi/2$	Eqn 4.7.13	$\dfrac{1}{2}$	$-\dfrac{p+1}{2}$	1	Fig 4.32
Beam divider (parallel ray source) through angles $\pm q\pi/2$	Eqn 4.7.19	1	$\dfrac{2-q}{2}$	1	Fig 4.39

Table 4.2: Generalized law of refractive index

$$Cr^{2/BC} - 2r^{1/BC}(\eta r)^{(A/BC)-1} + C(\eta r)^{2A/BC} = 0$$

Putting $C\rho = \sin\gamma$, we have

$$\log r = A\log\rho + BC\log\tan(\gamma/2)$$

or

$$r = \rho^A \left[\frac{1-\sqrt{1-C^2\rho^2}}{1+\sqrt{1-C^2\rho^2}}\right]^{BC/2} \tag{4.13.9}$$

giving, since $\rho = \eta r$

$$C\eta r[r^{2/BC} + (\eta r)^{(2A/BC)}] = 2r^{1/BC}(\eta r)^{A/BC}$$

or

$$Cr^{2/BC} - 2r^{1/BC}(\eta r)^{(A/BC)-1} + C(\eta r)^{2A/BC} = 0 \tag{4.13.10}$$

It can readily be confirmed that the appropriate values of the parameters result in the refractive index laws shown in Table 4.2.

Some points of interest arise immediately from this result. First, a change in sign of the parameter B leaves the result unchanged. This can only mean that there are *two* geodesic surfaces (in some instances) lying on each side of the cone $\sigma = A\rho$, which have the same ray properties.

Secondly, the transformation $\eta \to r$ and $r \to \eta$ transform pairs of lenses with different properties as shown in Table 4.1. This transformation, also the result of a Legendre transformation (Equation 4.10.1), applied to Equations 4.13.9 and 4.13.10 shows it to remain form-invariant if A is replaced by $1 - A$. This simple rule is thus the equivalent of a Legendre transformation.

In the form given by Equation 4.13.8 it is a simple matter to establish a condition among the parameters that will give the geodesic surface the highly desirable property of becoming tangentially flat at the periphery. There will then be no edge giving ray diffraction effects. This requires $d\sigma/d\rho = 1$ when $\rho = 1$, or

$$(1 - C^2)(1 - A)^2 = B^2C^2 \tag{4.13.11}$$

which eliminates all the practical lenses in the table. In the same way a geodesic lens that is smooth and flat about the axis has $d\sigma/d\rho = 1$ at $\rho = 0$, or

$$A + BC = 1 \tag{4.13.12}$$

These conditions obviously cannot be met simultaneously as the combined result gives $C = 0$, which corresponds once more to a cone $\sigma = A\rho$ which clearly meets neither criterion.

In order to use the ray trace equation it is necessary to derive η explicitly as a function of r from the relations in Equations 4.13.9 or 4.13.10. Only particular values of the parameters make this possible and these include the complete range of known refractive devices. Obvious conditions for an explicit solution to Equation 4.13.10 are

1. $A = 0$

2. $A = BC$

3. $BC = 0$

4. $A = -BC/3$

to which can be added the result of a change of sign in B. The other transformation $A \rightarrow 1 - A$ does not necessarily transform an explicit solution into another explicit solution as can be seen from Table 4.1 when applied to the Toraldo lens.

With condition (1) we obtain

$$\eta = \frac{2}{C} \frac{r^{1/(BC)-1}}{1 + r^{2/(BC)}}$$

the generalized fish-eye solutions. In particular, if $BC = 1$ we have the original fish-eye for which the geodesic analogue is $\sigma = \sin \rho$, which is the equation of a circle. On rotation this is therefore the sphere, the stereographic projection of the great circles being the coaxal system of rays shown in Figure 4.28. This was the method first used by Luneburg to derive this result.

With condition (3)

$$A = \pm BC \qquad \eta^2 \frac{2}{r^2}(r^{1/BC} - Cr^{2/BC})$$

the Luneburg solutions which are non-singular if $BC \geq 2$. With $A = 1/2$ the conditions of Equation 4.13.12 are satisfied.

Condition (4) is highly unusual. With $A = -BC/3$ we obtain from Equation 4.13.10

$$2r^{-1/3A}(\eta r)^{-4/3} - C(\eta r)^{-2/3} - Cr^{-2/3A} = 0$$

This result has not been investigated further.

Condition (3), referred to Equation 4.13.10, results in the refractive index law

$$\eta = r^{\nu - 1} \tag{4.13.13}$$

and the geodesic surface

$$\sigma = 1/\nu \tag{4.13.14}$$

This is a right circular cone of semi-vertex angle ψ, where $\nu = \sin \psi$. Putting $A = 1/\nu$, then Equation 4.13.13 represents a refractive index with an infinite singularity at the origin, corresponding to the real geodesic surface, the cone.

Substituting $\eta = r^{(1/A)-1}$ into the ray equation (Equation 4.7.6) we have

$$\phi - \phi_0 = \int_{r_0}^{r} \frac{K\,\mathrm{d}r}{r\sqrt{r^{2/A} - K^2}} \tag{4.13.15}$$

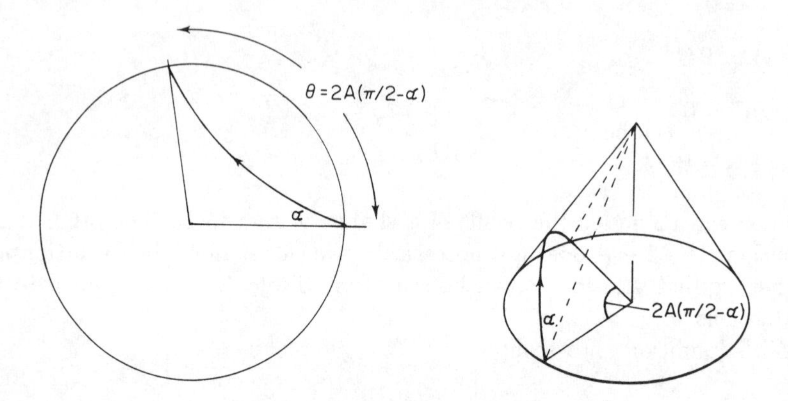

Figure 4.77: (a) Ray trajectory in the non-uniform medium analogous to the cone (b) Ray trajectory as the geodesic on a cone

the sign depending on the parts of the ray for which r increases or decreases with ϕ. This changes at the point r_m, where the ray is at its closest to the centre and consequently perpendicular to the radius vector. The ray starts from the periphery, that is $r_0 = 1$, making an angle α with the diameter (Figure 4.77(a)). Consequently $K = \sin\alpha$, and from the invariance of $\eta r \sin\psi = K$.

$$\eta_m r_m = K \qquad r_m = (\sin\alpha)^A$$

Substitution of these values into Equation 4.13.15 gives

$$\phi - \phi_0 = 2\int_1^{r_m} \frac{\sin\alpha\, dr}{r\sqrt{r^{2/A} - \sin^2\alpha}} \qquad r_m = \sin^A\alpha$$

The substitution $r^{1/4} = \sin\alpha \text{cosec}\beta$ results in

$$\phi - \phi_0 = A\text{cosec}^{-1}\left(\frac{r^{1/A}}{\sin\alpha}\right)\Bigg|_1^{\sin^4\alpha}$$

Taking ϕ_0 to be the zero angle of reference

$$\phi = 2A\left(\frac{\pi}{2} - \alpha\right) \tag{4.13.16}$$

the cone, being a developable surface has geodesics which are straight lines on the developed surface. Figure 4.77(b) shows that such a geodesic will have the result given in Equation 4.13.16. As a consequence of this result a refractive lens constructed of n uniform concentric shells with refractive index laws $r^{1/A_n - 1}$ has a geodesic analogue consisting of n continuous cone frustra with cone angles $A_n = \text{cosec}\psi_n$.

It should be noted that the reverse procedure, that is specification a priori of the ray pattern, even where it gives rise to implicit functions of r and $\eta(r)$, have all been reducible to the form of Equation 4.13.10.

The results given so far are in the form of a (σ, ρ) equation. For machining purposes this has to be converted to a (z, ρ) formulation. The methods inevitably require a computational procedure.

Combining the two results

$$\mathrm{d}\sigma^2 = \mathrm{d}z^2 + \mathrm{d}\rho^2$$

and

$$\mathrm{d}\sigma = \left(A + \frac{BC}{1 - C^2\rho^2}\right) \mathrm{d}\rho$$

then

$$\mathrm{d}z^2 = \mathrm{d}\rho^2 \left[A^2 + \frac{B^2C^2}{1 - C^2\rho^2} + \frac{2ABC}{\sqrt{1 - C^2\rho^2}} - 1\right] \tag{4.13.17}$$

This formula can be put into a finite-difference form and computed directly for incremental steps $\Delta\rho$. A check on the accuracy can be obtained by inserting the conditions for the Maxwell fish-eye and comparing the result with the true circle that should be obtained (no. 1 of Table 4.1). On substituting, as before, $C\rho = \sin\gamma$, we obtain

$$z = \frac{1}{C} \int_0^{\sin^{-1} C\rho} \sqrt{(A^2 - 1)\cos^2\gamma + 2ABC\cos\gamma + B^2C^2}\, \mathrm{d}\gamma \tag{4.13.18}$$

For the geodesic form of Luneburg lens, Equation 4.13.7, this gives, with $A = B = 1/2$ $C = 1$ [50]

$$z = \frac{2}{2\sqrt{3}}\kappa - \sqrt{3} - \log\left(\frac{\kappa + 3\sqrt{1 - \rho^2} + 2}{2 + \sqrt{3}}\right)$$

$$\kappa = \sqrt{12 - 9\rho^2 + 12\sqrt{1 - \rho^2}} \tag{4.13.19}$$

The profile computed for the beam divider given by Equation 4.7.19 is shown in Figure 4.78

In all these lenses, as in the non-uniform lenses, the central ray is a geodesic in its own right. In the devices which deflect the main body of rays, the central ray would require independent treatment to be defined in a manner compatible with the ray pattern. This could take the form of a simple reflecting prism or wedge placed in the medium at its centre

For those devices not meeting the criterion of Equation 4.13.11, the edge discontinuity could cause scattering and loss effects. A "rounded edge" procedure has been outlined [53] [54] [55] for this condition, and its incorporation into the lenses included here has to be carried out early in the design to prevent the creation of additional aberrations.

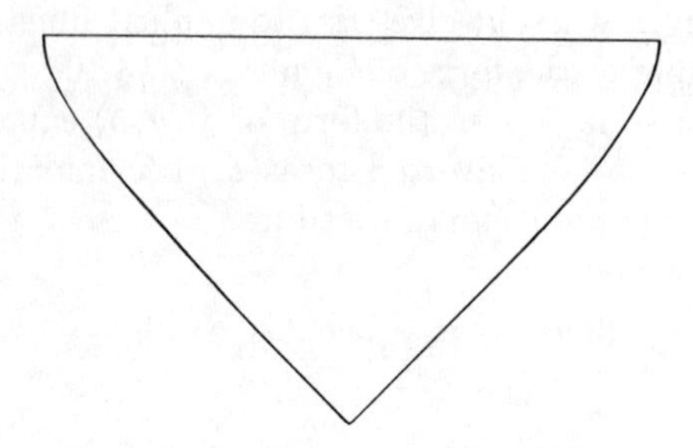

Figure 4.78: The geodesic surface for the beam divider of Figure 4.40

One method would be to determine the geodesic relation describing such a "rounded edge" and converting this to a non-uniform refractive layer surrounding the analogous non-uniform lens. Performing a ray trace operation on this combination by means of the standard integral will give the magnitude of the aberration created by the edge rounding process. Alternatively, a 45° chamfered edge could be used. This would be a frustrum of a cone and its equivalent refractive index would apply to a shell surrounding the original lens.

Surfaces of revolution can be "isometrically" deformed in such a way that geodesics on one surface transform into geodesics on the deformed surface. We can perform this operation on the lenses so far discussed and arrive at a new class of spherically symmetrical lens media.

We have as before $\mathrm{d}\sigma^2 = \mathrm{d}\rho^2 + \mathrm{d}z^2$ and $\mathrm{d}s^2 = \mathrm{d}\sigma^2 + \rho^2\,\mathrm{d}\phi^2$. This can be deformed to $\mathrm{d}s^2 = \mathrm{d}\sigma'^2 + \rho'^2\,\mathrm{d}\phi'^2$ and all arcs on one surface will correspond to arcs on the new surface if

$$\rho'\,\mathrm{d}\phi' = \rho\,\mathrm{d}\phi \quad \text{and} \quad \mathrm{d}\sigma' = \mathrm{d}\sigma$$

This is satisfied by

$$\rho' = k\rho \qquad \phi' = \phi/k \qquad \text{and} \qquad \mathrm{d}z'^2 + \mathrm{d}\rho'^2 = \mathrm{d}z^2 + \mathrm{d}\rho^2$$

or

$$\mathrm{d}z'^2 = \mathrm{d}z^2 + (1-k^2)\,\mathrm{d}\rho^2 \tag{4.13.20}$$

The required surfaces of revolution become

$$\rho' = k\rho \qquad z' = \int \sqrt{\mathrm{d}z^2 + (1-k^2)\,\mathrm{d}\rho^2}$$

For example the unit sphere $z = \sin\theta$ $\rho = \cos\theta$ becomes

$$\rho' = k\cos\theta \qquad z' = \int \sqrt{1 - k^2\sin^2\theta}\,\mathrm{d}\theta$$

Applying this to the general geodesic form $\sigma = A\rho + B\sin^{-1} C\rho$ gives $\rho' = k\rho$ and $\sigma' = A\rho'/k + B\sin^{-1}(C\rho'/k)$ that is simply to replace A with A/k and C with C/k. The condition for flatness at the centre then becomes $A + BC = k$.

REFERENCES

1. Synge JL (1937) *Geometrical Optics* Cambridge University Press

2. Born M & Wolf E (1959) *Principles of Optics* Pergamon Press

3. Budden KG (1961) *Radio Waves in the Ionosphere* Cambridge University Press

4. Atwater HA (1974) *Introduction to General Relativity* Pergamon Press p69

5. Born M & Wolf E (1959) *Principles of Optics* Pergamon Press p674

6. Damien R (1955) *Thèorém sur les Surfaces d'Onde en Optique Géometrique* Paris: Gauthier Villars

7. Brown J (1953) *Microwave Lenses* Methuen

8. Luneburg RK (1964) *The Mathematical Theory of Optics* University of California Press

9. Silver S (1949) *Microwave Antenna Theory and Design* McGraw Hill: MIT Radiation Laboratory Series **12** p522

10. Cornbleet S (1967) Waveguide sandwich electromagnetic window *Electronics Letters* **3** (12), p540

11. Kaplun VA (1964) Nomograms for determining the parameters of plane dielectric layers of various structure with optimum radio characteristics *Radioteknikha i Elektronika* **9** p81

12. Young PA (1970) Extension of Herpin's theorem *Jour Opt Soc Amer* **60** (10), p1422

13. Jones SSD (1947) A wide angle microwave radiator *Proc IEE* **97** Part III, p255

14. Herpin A (1947) Sur une nouvelle methode d'introduction des polynomes de Lucas (Chevychev polynomials) *Comptes Rendus* **225** (1) p17

15. Abeles F (1948) Transmission de la lumiere a travers un systeme de lames minces alternees *Comptes Rendus* **226** (22) p1809

16. Epstein LI (1952) The design of optical filters *Jour Opt Soc Amer* **42** (11) p806

17. Young L (1967) Multilayer interference filters with narrow stop-bands *Applied Optics* **6** (2), p267 (and refs therein)

18. Reed S (March, 1961) A note on loaded line synthesis *Trans IRE PGMTT* p201

19. Hansen RC (Ed) (1964) *Microwave Scanning Antennas* **Vol 1** Chapter 3. Optical scanners by Johnson RC London & New York: Academic Press p 224 and refs therein

20. Beauchamp KG (1976) *Walsh Functions and Their Applications* London & New York: Academic Press

21. Jones DS (1964) *Theory of Electromagnetism* Pergamon Press Section 6.21, p351

22. Heading J (1962) *Phase Integral Methods* Methuen Monographs

23. Eckersley TL (1932) Radio transmission problems treated by phase integral methods *Proc Roy Soc A* **136** (830), p499

24. Budden (1961) *Radio Waves in the Ionosphere* Cambridge University Press

25. Jacobsson R (1975) Inhomogeneous and co-evaporated films for optical applications, from *Physics of Thin Films* London & New York: Academic Press 8 p51

26. Jacobsson R (1966) Light reflection from thin films of continuously varying refractive index, from *Progress in Optics* (ed) Wolf E **V** p249

27. Richmond JH (1962) The WKB approximation for transmission through inhomogeneous plane layers *Trans IRE* **AP10** p472

28. Herzberger (1958) *Modern Geometrical Optics* New York: Interscience Publishers p30

29. Wood RW (1905) *Physical Optics* Macmillan p72

30. Byrd PF & Friedman MD (1971) *Handbook of Elliptic Integrals for Engineers and Scientists* Springer Verlag

31. Exner S (1891) *Die Physiologie der facettirten Augen von Krebsen und Insekten* Leipzig & Wien: Deutike

32. Fletcher A, Murphy T & Young A (1954) Solution to two optical problems *Proc Roy Soc* **223A** p216

33. Luneburg RK (1964) *The Mathematical Theory of Optics* University of California Press

34. Lenz W (1928) Theory of optical images from *Probleme der Modernen Physik* (ed) Debye P, Leipzig: Hirsel Press p198

35. Morgan SP (1958) General solution of the Luneburg lens problem *Jour Appl Physics* **29** (9) p1358

36. Gutman AS (1954) Modified Luneburg lens *Jour Appl Physics* **23** (7) p855

37. Eaton KE (1952) On spherically symmetric lenses *Trans IRE* **AP-4** p66

38. Toraldo di Francia G (1955) A family of perfect configuration lenses of revolution *Optica Acta* **1** (4) p157

39. Donoghue JF & Holstein BR (1986) Quantum mechanics in curved space *Am Jour Phys* **54** pp927-831

40. Mathis HF (1960) Checking the design of stepped Luneburg lens *Trans IRE* **AP8** p342

41. Toraldo di Francia G & Zoli MT (1958) *Perfect Concentric Systems with an Outer Shell of Constant Refractive Index* Firenze: Pubblicazione dell' Istituto Nazionale di Attica, Serie II, No 827

42. Ap Rhys TL (1970) The design of radially symmetric lenses *Trans IEEE* **AP18** (4) p497

43. Maxwell JC (1871-3) On the condition that, in the transformation of any figure ... in three dimensions, every angle in the new figure shall be equal to the corresponding angle in the original figure *Proc London Math Soc* **4** p117

44. Cornbleet S & Jones MC (1982) The transformation of spherical nonuniform lenses *Proc IEE* (pt H) **129** p321

45. Collin RE & Zucker FJ (1969) Antenna Theory, Part II, Chapter 18.6 from *Lens Antennas* Brown J, McGraw Hill, p133

46. Prache P (1961) Lenses and dielectric reflectors of homogeneous spherical layers *Annales de Telecomm* **16** (3-4) p85

47. Moon P & Spencer DE (1961) *Field Theory for Engineers* Van Nostrand p348

48. Buchdahl HA (1973) Rays in gradient index media: separable systems *Jour Opt Soc Amer* **63** (1) p46

49. Myers SP (1947) Parallel plate optics for rapid scanning *Jour Appl Phys* **18** p221

50. Rinehart RF (June 1952) A family of designs for rapid scanning radar antennas *Proc IRE* p686

51. Rinehart RF (1948) A solution of the problem of rapid scanning for radar antennae *Jour Appl Phys* **19** p80

52. Cornbleet S & Rinous PJ (1981) Generalised formulas for equivalent geodesic and nonuniform lenses *Proc IEE* (Pt H) **128** p95

53. Kassai & Marom E (1979) Aberration corrected rounded edge geodesic lenses *J Opt Soc Amer* **69** p1242

54. Southwell WH (1977) Geodesic optical waveguide analysis *J Opt Soc Amer* **67** p1293

55. Sottinin S, Russo V & Righini (1979) General solution of the problem of perfect geodesic lenses for integrated optics *J Opt Soc Amer* **69** p1248

5

SCALAR DIFFRACTION THEORY OF THE CIRCULAR APERTURE

The ray optical methods of the preceding chapters that have been used to determine the basic focusing properties of the reflector and lens antennas give a qualitative description of the resultant field intensity over the exit aperture, from the presumed or known field distribution of the source. They can no longer apply to the final problem of the determination of the radiated field in space from the antenna, the radiation pattern, without considering the major effect of the finiteness of the aperture, that is, diffraction. The effect is far more noticeable at microwave frequencies by virtue of the smallness (in wavelength number) of the apertures concerned. It is compensated in some measure by requiring the consideration of fewer aberration effects than does the optical system. It also differs from the optical problem in that amplitude distributions can be varied in a very general way as has been shown, with resultant variation in the radiation patterns.

The exact solution for this problem requires a knowledge of the total vector field distribution over the infinite surface containing the radiating apertures, and the application of Green's theorem to the half-space, otherwise source free, into which the antenna radiates. Green's theorem, applying as it does to a closed surface, requires the integration over the entire plane containing the sources, that is the radiating aperture and any fields occurring locally because of its presence, and a closing surface which is generally taken to be an infinite hemisphere. By invoking a "radiation condition at infinity" the integration over this hemisphere is shown to tend to zero as its radius becomes infinite.

Even where the sources of the radiating field have been precisely determined, the ensuing integration has to be approximated in some way, and therefore the effects of two approximations have to be taken into account, that of the definition of the source field, and that of its resulting radiation pattern. The procedure has become increasingly refined in recent years. The major problem in both cases is the precise effect of the edges and boundaries that must exist to define the finite radiating optical device, or its aperture.

An attempt to cover the various theories and their solutions is made in the accompanying table. Apart from the actual references given, all the names occurring in this chart are to be found in the bibliography to this chapter in the section relating to diffraction theory. In the following survey of these theories the references (A) (B) (C) etc. are to the equations given in the chart.

Diffraction

G is Green's function for freespace $= e^{ik|\mathbf{r}-\mathbf{r}'|}/4\pi|\mathbf{r}-\mathbf{r}'|$
S encloses volume V, A is an aperture in S, $S - A$ is a conducting surface

Symmetrised Maxwell Equations

Papas, p7 — *Retarded Potentials* (Volume, Surface)

$$\nabla\times\mathbf{E} = -\mathbf{j}_m - \frac{\partial \mathbf{B}}{\partial t} \qquad \mathbf{j}_m = -\mathbf{n}\times\mathbf{E} \qquad \mathbf{F} = \varepsilon\int_V \mathbf{j}_m G\,\mathrm{d}v = -\varepsilon\int_S \hat{\mathbf{n}}\times\mathbf{E}G\,\mathrm{d}s$$

$$\nabla\cdot\mathbf{B} = \rho_m \qquad \rho_m = \mu\hat{\mathbf{n}}\cdot\mathbf{H} \qquad X = -\frac{1}{\mu}\int_V \rho_m G\,\mathrm{d}v = -\int_S \hat{\mathbf{n}}\cdot\mathbf{H}G\,\mathrm{d}s$$

Transformation $\mathbf{E}\to\mathbf{H}$ $\quad \mathbf{H}\to-\mathbf{E}$ $\quad \mu\leftrightarrow\varepsilon$ $\quad \mathbf{j}\to\mathbf{j}_m$
then $\mathbf{A}\to\mathbf{F}$; $\mathbf{F}\to-\mathbf{A}$; $\Phi\to X$

Wave Equation

$$\nabla^2\mathbf{E} - \mu\varepsilon\frac{\partial^2\mathbf{E}}{\partial t^2} = \mu\frac{\partial\mathbf{j}}{\partial t} + \nabla\times\mathbf{j}_m + \frac{1}{\varepsilon}\nabla\rho = f(x', y', z', t')$$

Solution for time harmonic fields $e^{-i\omega t}$

$$u_p = -\frac{1}{4\pi}\int_V \frac{[f_{ret}]}{|\mathbf{r}-\mathbf{r}'|}\,\mathrm{d}v +$$

§

$$\mathbf{E}_p = i\omega\mathbf{A} - \frac{1}{\varepsilon}\mathrm{curl}\mathbf{F} - \mathrm{grad}\Phi$$

Silver (p83 complex conjugate) **Volume Source Solution +**

$$\mathbf{E}_p = \int_V [i\omega\mu\mathbf{j}G - \mathbf{j}_m\times\nabla G + \frac{\rho}{\varepsilon}\nabla G]\,\mathrm{d}v \qquad +$$

Far Field Approximation Silver p88

$$\mathbf{E}_p = \frac{i}{4\pi\omega\varepsilon}\int_V [k^2\mathbf{j} - k^2(\mathbf{j}\cdot\mathbf{R})\mathbf{R} + k\omega\varepsilon\mathbf{j}_m\times\mathbf{R}]\frac{e^{ikr(R-\mathbf{r}'\cdot\mathbf{R})}}{R}\,\mathrm{d}v$$

Physical Optics Approximation Rusch 1963 Ⓐ
$\mathbf{j} = 2\hat{\mathbf{n}}\times\mathbf{H}_{inc}$ $\qquad \mathbf{j}_m = \rho_m = 0$ $\qquad \hat{\mathbf{n}}\times\mathbf{E} = \hat{\mathbf{n}}\cdot\mathbf{H} = 0$

Aperture integrals $S\to A$ Silver p165, Stratton p469

$$\mathbf{E}_p = \int_A \left[G\frac{\partial\mathbf{E}}{\partial n} - \mathbf{E}\frac{\partial G}{\partial n}\right]\mathrm{d}a + \oint_{edge}(\mathbf{E}\times\hat{\tau})G\,\mathrm{d}l - \frac{i}{\omega\varepsilon}\oint_{edge}(\hat{\tau}\cdot\mathbf{H})\nabla G\,\mathrm{d}l =$$

Ⓑ Ⓒ additional edge current
Stratton and Chu

Theory

R is a vector from source point $\mathbf{r}'$ to field point $\mathbf{r}$ $\quad \mathbf{R} = \mathbf{r} - \mathbf{r}'$
of infinite plane screen. $\hat{\mathbf{n}}$ is an *inward* pointing normal.

Retarded Potentials

Volume Surface

$$\nabla \times \mathbf{H} = \mathbf{j} + \frac{\partial \mathbf{D}}{\partial t} \qquad \mathbf{j} = \hat{\mathbf{n}} \times \mathbf{H} \qquad \mathbf{A} = \mu \int_V \mathbf{j} G \, \mathrm{d}v \quad = \mu \int_S \hat{\mathbf{n}} \times \mathbf{H} G \, \mathrm{d}s$$

$$\nabla \cdot \mathbf{D} = \rho \qquad \rho = \varepsilon \hat{\mathbf{n}} \cdot \mathbf{E} \qquad \Phi = -\frac{1}{\varepsilon} \int_V \rho G \, \mathrm{d}v \quad = -\int_S \hat{\mathbf{n}} \cdot \mathbf{E} G \, \mathrm{d}s$$

$\mathbf{j}_m \to -\mathbf{j} \quad \rho \to \rho_m \quad \rho_m \to -\rho$
$X \to -\Phi$ and for all solutions $\mathbf{E}_p \to \mathbf{H}_p$ etc

Scalar Wave Equation

$$\left(\nabla^2 - \mu\varepsilon \frac{\partial^2}{\partial^2 t}\right) u = f(x', y', z', t')$$

Solution:
Stratton p470, Jackson p283

$$+ \qquad \int_S [G(\hat{\mathbf{n}} \cdot \nabla) u - u(\hat{\mathbf{n}} \cdot \nabla G)] \, \mathrm{d}s$$

$$u_p = \int_S \left[G \frac{\partial u}{\partial n} - u \frac{\partial G}{\partial n} \right] \mathrm{d}s$$

+ Surface Source Solution

by components

Stratton p466, Jackson p285
Ⓘ

$$\mathbf{E}_p = \int_S \left[G \frac{\partial \mathbf{E}}{\partial n} - \mathbf{E} \frac{\partial G}{\partial n} \right] \mathrm{d}s$$

Silver p108

$$+ \int_S [i\omega(\hat{\mathbf{n}} \times \mathbf{B}) G + (\hat{\mathbf{n}} \times \mathbf{E}) \times \nabla G + (\hat{\mathbf{n}} \cdot \mathbf{E}) \nabla G] \, \mathrm{d}s$$

Franz, Sommerfeld p325
CT Tai 1972

$$\mathbf{E}_p = \frac{i}{\omega\varepsilon} \nabla \times \nabla \times \int_S (\hat{\mathbf{n}} \times \mathbf{H}) G \, \mathrm{d}s + \nabla \times \int_S (\hat{\mathbf{n}} \times \mathbf{E}) G \, \mathrm{d}s \quad Ⓓ$$

$$= \frac{i}{\omega\mu\varepsilon} \nabla \times \nabla \times \mathbf{A} \; - \; \frac{1}{\varepsilon} \nabla \times \mathbf{F}$$

from § with div$\mathbf{A} = i\omega\Phi$

Kottler Luneburg Bouwkamp Vasseur Severin

Aperture integrals $S \to A$

$$= \int_A [i\omega(\hat{\mathbf{n}} \times \mathbf{B}) G + (\hat{\mathbf{n}} \cdot \mathbf{E}) \nabla G] \mathrm{d}a + \int_A (\hat{\mathbf{n}} \times \mathbf{E}) \times \nabla G \mathrm{d}a - \frac{i}{\omega\varepsilon} \oint_{edge} (\hat{\tau} \cdot \mathbf{H}) \nabla G \, \mathrm{d}l$$

$\hat{\mathbf{n}} \times \mathbf{E} = 0$ on $S - A$ $\qquad$ $\hat{\mathbf{n}} \times \mathbf{H} = 0$ on $S - A$

Vector Rayleigh Sommerfeld Ⓙ

Diffraction Theory

$$\mathbf{E}_p = \int_A \left[G\frac{\partial \mathbf{E}}{\partial n} - \mathbf{E}\frac{\partial G}{\partial n} \right] \mathrm{d}a \xrightarrow{\text{Far Field}} \frac{i}{2\pi}\frac{e^{ikR}}{R}\hat{\mathbf{k}} \times \int_A (\hat{\mathbf{n}} \times \mathbf{E}(\mathbf{r}'))e^{-i\mathbf{k}\cdot\mathbf{r}'} da$$

Ⓔ $\mathbf{r}'$ is a vector from the origin to element da $\hat{\boldsymbol{k}} = \mathbf{R}/|\mathbf{R}|$ (Jackson p292)

Kirchoff approximation Ⓕ

$$u_p \simeq \int_A \left[G\frac{\partial u}{\partial n} - u\frac{\partial G}{\partial n} \right] \mathrm{d}a$$

$$u_p = \int_S G\frac{\partial u}{\partial n}\,\mathrm{d}s - \int_S u\frac{\partial G}{\partial n}\,\mathrm{d}s$$

Heurtley Bouwkamp Ⓗ

$$u = 0 \text{ on } S{-}A \qquad \frac{\partial u}{\partial n} = 0 \text{ on } S{-}A$$

Rayleigh-Sommerfeld

Jackson p282
Silver p109, 166

$$u_p \simeq \frac{1}{4\pi}\int_A \frac{e^{ikR}}{R}\hat{\mathbf{n}} \cdot \left[\nabla u + ik\left(1 + \frac{i}{kR}\right)\frac{\mathbf{R}u}{R} \right] \mathrm{d}a$$

Far Field Silver p118

$$u_p = \frac{i}{2\lambda}\int_A u\left(1 - \frac{\cos\hat{\mathbf{n}}\cdot\mathbf{R}}{R}\right)\frac{e^{ikR}}{R}\mathrm{d}a$$

Silver p167 Born & Wolf p435

$$u_p = \frac{-ik}{4\pi}\frac{e^{ikR}}{R}(1 + \cos\theta)\int_A u e^{ik\mathbf{r}'\cdot\mathbf{R}}\mathrm{d}a$$

Silver p173

$$u_p = (1 + \cos\theta)\int_A F(\xi, \eta)e^{ik\sin\theta(\xi\cos\phi + \eta\sin\phi)}\,\mathrm{d}\xi\,\mathrm{d}\eta$$

Circular Aperture Jackson p293 Silver p192 Ⓖ

$$u_p = \int_0^{2\pi}\int_0^1 f(r, \phi')e^{ika\sin\theta\cos(\phi - \phi')}r\,\mathrm{d}r\,\mathrm{d}\phi' \equiv \text{Equation 5.1.1}$$

Diffraction of waves by apertures

Rubinowicz-Miyamoto-Wolf $u_{\text{Kirchoff}} = u_{\text{GeomOptics}} +$ Boundary Wave

Marchand $= u_G + u_B$ Sommerfeld p263

Ⓚ Born & Wolf p562

Miyamoto-Wolf-Sancer $u_K = u_G + u_{B_1} + u_{B_2}$

Geometrical Theory of Diffraction

Keller Sherman $E_p =$ Geom optics + Edge diffracted wave

Luneburg-Debye expansion

$$\frac{e^{ikr}}{r} = \frac{ik}{2\pi}\int_\Omega \exp\{ik[\alpha(x - x') + \beta(y - y') + \gamma(z - z')]\}\mathrm{d}\Omega$$

Plane Wave Spectrum $\quad \Omega = \sin\theta\,\mathrm{d}\theta\,\mathrm{d}\psi \quad \psi : 0 \to 2\pi \quad \theta : 0 \to i\infty$

Shafer Booker & Clemnow Boukwamp p41

In the case where the radiating system is based solely upon reflecting surfaces of presumed infinite conductivity, the most fruitful method appears to be that of the physical optics procedure (A), that is the determination of the actual currents arising on the surfaces concerned from the known incident field and their use as the source for the radiated field. The assumption is made that the current distribution arising at any point on the surface from a known value of the incident magnetic field, is the current that would be created by an infinite plane wave, of the same amplitude and polarization, incident upon an infinite conducting plane. The theory is thus most accurate for large reflectors with necessarily large radii of curvature and the contribution from the locality of the edges can still be expected to cause errors. As in what follows, this concern for the edge effect is the major problem in the whole of diffraction theory. In the physical optics method it can be shown to be small and the agreement with the best of observed measurements is the closest of the many theories available.

Unfortunately, it results in a procedure that is basically computational and is applied only to reflector systems. Hence most studies have been applied exclusively to the two reflector Cassegrain antennas [1] subsequent to their design by geometrical optics methods. Since this system is capable of an endless variation of the geometrical parameters and illumination functions, a great deal of work has to be done in order to provide sufficient data for general assessments to be made. It is difficult as yet to see how the method would apply to optical lenses but a case could be made out for its application to constrained waveguide lenses. In the case of the more complex geometries involved in reflector designs, as shown in Chapter 2, an entirely new computational programme would be required for each design and this fact may inhibit active development in these areas.

Much the same can be said of another recent method adopted specifically to tackle the problem of the effects of the edges in antenna diffraction. This is the geometrical theory of diffraction originated by Keller [2]. This adds to the geometrical optics field an "edge diffracted field" calculated at each point of the periphery as if from the diffraction of a plane wave incident there, with the same amplitude and polarization upon an infinite half-plane. The standard result of Sommerfeld is used for this half-plane effect. Modifications have been found to be necessary to this theory in the most vulnerable regions of the field, the light-shadow boundary or in the region of caustic and foci [3]. Again it remains essentially a computational procedure and no feed-back to the antenna design stage is possible without the collection of a great deal of data.

Earlier work on the subject is to a large extent unsurpassed. In general the integral solution to the problem has to be reduced from that covering the entire half-plane as used in the original application of Green's theorem, to the region covered by the antenna aperture alone and over which the total vector field is presumed known. A complete discussion is to be found in Bouwkamp

[4] (along with 500 relevant references). Silver [5] shows that the reduction of the infinite integral to a finite integral (B) results in a residual integral which by application of Stokes' theorem can be converted to a line integral around the periphery of the aperture (C). In addition there has to be a contribution from a line integral around the edge of the aperture from a line source of charge necessary to give continuity of the field between the illuminated aperture and the "dark" side of the plane. This same contour integral arises directly if Kirchhoff approximations (to be discussed) are made to the surface retarded potentials Φ and X (D).

The combined contributions of the line integrals are shown by Silver to be small enough to be ignored. The approximation that results is in effect that the integrand of the aperture integral reverts to the identical form that it had for the infinite plane (E). This is the vector Kirchhoff approximation. Interpretation of the result shows that it implies that the field components over the region of the infinite plane outside the aperture are considered to be zero. In the case of diffraction of waves by an aperture in an infinite plane, the field in the aperture in this approximation is the field that would have existed there in the absence of the plane. The problem is allied to the problem of antenna diffraction and has received far greater attention. It requires adaptation however before it can be applied to the antenna problem with a non-uniform amplitude distribution. For uniformly illuminated apertures both theories can be compared and this is sometimes used as a test in either case.

Other forms of Kirchhoff approximations occur whenever the infinite integral in whatever its form, is applied in that form to the area of the radiating aperture alone, ignoring in effect the residual integrals that result from such a transition. As was shown above, this can apply as far back in the theory as the surface retarded potentials themselves. It is tantamount to a Huygens' construction since it refers to radiating sources of spherical waves, the free-space Green's functions, at points within the illuminated aperture alone.

The scalar Kirchhoff approximation then arises by virtue of the fact that with the omission of the line edge integrals, the vector radiated field can be obtained from the addition of the contributions from each of the components of the aperture field separately. These then obey the scalar form of the approximation (F) which conforms with the earlier optical theory of scalar aperture illumination functions. The requirements of the antennas which are the main consideration of this study are for a circular aperture to which form the scalar diffraction integral is reduced in equation (G). Concerning the allied problem of the diffraction of an incident wave by an aperture in a plane screen, and in particular of a plane wave by a circular aperture other approximate solutions have been obtained. These have a relevance in suggesting a similar treatment for the radiating aperture study.

Transformations of the original infinite integrals both the scalar and the vector forms, can be made which separates them into two parts. Separate boundary conditions can be applied to each part which, when combined over the area of the aperture, gives the Kirchhoff approximation.

In the scalar case, for the diffraction of a plane wave, the two resultant integrals are the Rayleigh-Sommerfeld integrals (H).

In the case of the vector integrals for the radiating antenna a transformation by Franz gives a similar separability (I). Applying boundary conditions to each part does not give the vector form of the Kirchhoff approximation since an edge integral is included. The result could be termed the vector Rayleigh-Sommerfeld integrals (J). The boundary condition chosen closely agrees with the assumptions made in the physical optics approximation which, as stated, is among the best with regard to experimental observation. The resulting integral can thus be expected to give a similar order of agreement.

In considering the diffraction of waves by an aperture in a screen, a physical meaning to the difference that exists between the geometrical optics and the Kirchhoff approximate solution has been sought by Wolf and his co-workers, in the form of a "boundary wave" arising at the edge (K). The first and second orders of this boundary wave have been derived for plane wave illumination. This effect differs both from the edge current integrals and from the edge diffraction effect of the geometrical theory of diffraction, although it has some aspects in common with the latter.

In all these methods the Green's function for free-space arising from the original solution figures prominently. Methods to associate the non-uniform aperture fields with the plane or spherical wave diffraction theory involve expansion of this function as an infinite spectrum of the type of wave concerned. The expansion for plane waves is given (L) but the final results from their application have been shown to be less in agreement with measurements than are other approximations considered.

Other as yet untried methods to "improve" upon the Kirchhoff approximation can be conjectured. It is known from the experimental results of Andrews and Andrejewski [56] [57] that the smooth illumination function that would exist in the aperture if the Kirchhoff assumption were indeed correct, is perturbed by a standing wave. This is obviously created by the interaction between the edge currents by whatever means they are assumed to be caused. The magnitude of this standing wave in the aperture is a function of the aperture diameter and the strength of the illumination function at the edge. In principle the standing wave will increase as the aperture decreases and the illumination becomes more uniform until for the very small apertures it becomes the actual mode distribution of a waveguide with the same cross-section. In reverse, for apertures of ten wavelengths or over it forms but a slight ripple on the assumed smooth distribution. The diffraction pattern of this standing wave, *in addition* to that of the normal Kirchhoff diffraction pattern, should form the second order correction to the radiation patterns, in the form of the well known high period lobe patterns in the lower field strength regions.

For purely analytical procedures we find that only the scalar Kirchhoff approximation gives a closed form solution which enables an a priori assessment of the antenna radiation pattern to be used as a design instruction for the antenna itself. In this manner it applies to diffraction theory in the same way as geometrical optics applies to optical design. Where comparisons with more exact solutions have been made the Kirchhoff solution has always been found to be in close agreement and with experimental results. This is true as we shall show, even in disputed regions such as the neighbourhood of foci and caustics and for apertures of comparatively small dimension. Having regard to all the other contributions to errors of measurement that exist, such as spillover and scattering from support structures, most of which are greater in magnitude than can be attributed to the Kirchhoff approximation alone, it is evident that for a qualitative assessment of antenna performance under quite general conditions of illumination function, the result will be satisfactory.

Further justification for the use of this approximation for design and analysis arises in the sequel. It is found that the closed form series solution that arises from its use can be applied to the converse problem, that of pattern synthesis, the design of aperture functions to give prescribed radiation patterns. This will be dealt with subsequently.

5.1 THE SCALAR INTEGRAL TRANSFORMS

Considering mainly the circular aperture the Huygens-Green integral reduces as shown in Table 5.1 (G), to [5]

$$g(u,\phi) = \int_0^{2\pi} \int_0^1 f(r,\phi') e^{iur\cos(\phi-\phi')} r \, dr \, d\phi' \tag{5.1.1}$$

in which (Figure 5.1) r ϕ' are normalized polar coordinates in the plane of the circular aperture with origin at the centre and θ ϕ are polar coordinates of a field point at a sufficiently large distance with respect to an axis perpendicular to the plane of the aperture and passing through its centre. The parameter $u = 2\pi a \sin\theta/\lambda$, where a is the radius of the aperture and λ the wavelength of the operating radiation, is a generalized coordinate whose nominal maximum value for real values of θ is $2\pi a/\lambda$. At some stage however u will be considered as taking all possible values on the real axis.

The expression given in Equation 5.1.1 is in fact simplified by the exclusion of multiplying constants and more importantly by the factor $(1 + \cos\theta)$. The former are ignored since we are going to be concerned more with pattern shape than with the absolute values of the field in the radiation zone. It is customary to exclude the "obliquity factor" in those cases where the radiated pattern over only small values of θ is of interest. The main beam and side lobes of the focused antennas of a size for which Equation 5.1.1 is an adequate approximation, usually occur for values of $u < 5$.

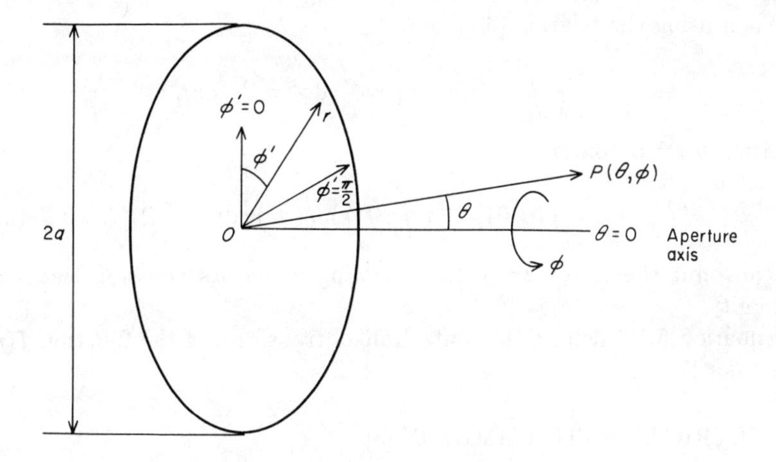

Figure 5.1: Circular aperture coordinate geometry

In subsequent sections we shall deal analytically with some aspects involving much larger values of u and hence of θ. In these cases, the theory can be modified to the extent that the pattern being considered is not the specific $g(u)$ of Equation 5.1.1 but

$$g(u) \to \frac{g(u)}{1+\cos\theta}$$

This will be indicated in those situations where it is relevant.

The function $f(r,\phi')$ represents the complex amplitude distribution of field in the aperture. This could be taken to be the separate components of a vector field quantity and added in the result to give a vector solution for the field at the far point. The derivation of Equation 5.1.1 makes certain conditions on $f(r,\phi')$ and in particular on the continuity of the imaginary part and on its differential coefficient. These add up to the limitation that sudden changes in amplitude or in phase are not allowable. In what follows however we take the function to be piecewise continuous with no limitation on the variation in amplitudes or phase apart from those which would be impossible in a practical situation.

We can then apply symmetry conditions to the illumination function $f(r,\phi')$ which will reduce the double integration to a single integral over the radial component only.

CIRCULAR SYMMETRY

For a perfectly centred circularly symmetric aperture distribution, $f(r,\phi')$ is independent of ϕ' and the integral independent of ϕ, which without loss of generality can be taken to be zero.

Then using the relation [6]

$$\frac{1}{2\pi}\int_0^{2\pi} \exp(iur\cos\phi')\,\mathrm{d}\phi' = J_0(ur) \tag{5.1.2}$$

Equation 5.1.1 becomes

$$g(u,0) = 2\pi \int_0^1 f(r)J_0(ur)r\,\mathrm{d}r \tag{5.1.3}$$

We can omit the factor 2π in the following where its omission leads to no ambiguity.

Equation 5.1.3 defines the finite Hankel transform of the function $f(r)$ of zero order [7].

DIAMETRICAL ANTISYMMETRY [8]

Dividing the aperture into two halves by the diameter $\phi' = 0$ and $\phi' = \pi$, we consider each half to be totally in antiphase. The complex field can then be expressed by

$$\begin{aligned} f(r,\phi') &= f(r) \qquad & 0 < \phi' < \pi \\ &= -f(r) \qquad & \pi < \phi' < 2\pi \end{aligned}$$

In a practical sense this distribution can only be achieved by the insertion of conducting surfaces into the aperture or by a similar discontinuity created artificially by phase delay media covering half of the aperture.

Equation 5.1.1 then gives

$$\begin{aligned} g(u,\phi) &= \int_0^{\pi}\int_0^1 f(r)\exp[iur\cos(\phi-\phi')]r\,\mathrm{d}r\,\mathrm{d}\phi' \\ &\quad - \int_0^{\pi}\int_0^1 f(r)\exp[-iur\cos(\phi-\phi')]r\,\mathrm{d}r\,\mathrm{d}\phi' \\ &= 2i\int_0^{\pi}\int_0^1 f(r)\sin[ur\cos(\phi-\phi')]r\,\mathrm{d}r\,\mathrm{d}\phi' \end{aligned} \tag{5.1.4}$$

In most cases the radiation pattern of most interest is that in the plane where the maximum effect occurs, that is, in the plane perpendicular to the diameter bisecting the aperture. Putting $\phi = \pi/2$ therefore in Equation 5.1.4 results in the separable integral

$$g\left(u,\pm\frac{\pi}{2}\right) = \pm 2i\int_0^1 f(r)\,\mathrm{d}r\int_0^{\pi} \sin(ur\sin\phi')\,\mathrm{d}\phi' \tag{5.1.5}$$

The second of these integrals defines the Lommel-Weber function Ω_0 that is from [6, p308]

$$\int_0^{\pi} \sin(ur\sin\phi')\,\mathrm{d}\phi' = \Omega_0(ur) \tag{5.1.6}$$

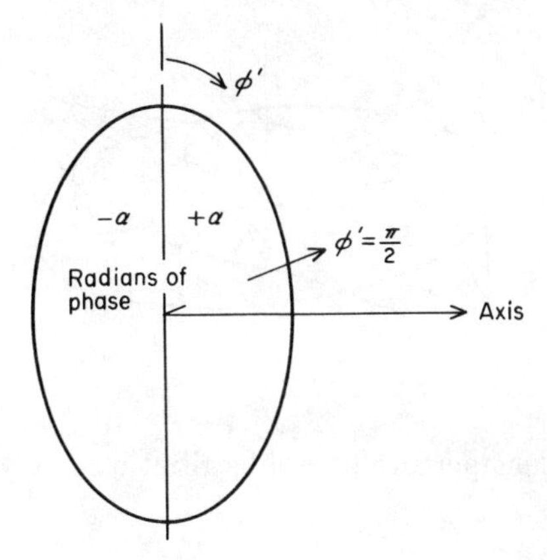

Figure 5.2: Circular aperture phase distributions. Diametral division of phase

It is identical to the zero order Struve function $H_0(ur)$ but higher orders of the two functions differ by a polynomial in the variable [6, p327].

If we allow positive and negative values of u to correpsond with values of ϕ separated by π then Equation 5.1.5. gives the result

$$g(u, \pm\pi/2) \equiv g(\pm u, \pi/2) = \pm 2\pi \int_0^1 f(r)\Omega_0(ur)r\,\mathrm{d}r \tag{5.1.7}$$

This defines the zero order Lommel transform of $f(r)$.

In a situation where two halves are not completely in antiphase the radiation pattern in the plane $\phi = \pi/2$ is expressible as a combination of the results of circular symmetry and diametrical antisymmetry. Thus if two halves are respectively $+\alpha$ and $-\alpha$ radians in phase (Figure 5.2)

$$\begin{aligned} g(u,\phi) = {} & \mathrm{e}^{-i\alpha} \int_0^\pi \int_0^1 f(r) \exp[iur\cos(\phi - \phi')]r\,\mathrm{d}r\,\mathrm{d}\phi' \\ & + \mathrm{e}^{i\alpha} \int_0^\pi \int_0^1 f(r) \exp[-iur\cos(\phi - \phi')]r\,\mathrm{d}r\,\mathrm{d}\phi' \end{aligned} \tag{5.1.8}$$

which in the plane $\phi = \pi/2$ reduces to

$$g\left(\pm u, \frac{\pi}{2}\right) = 2\pi\cos\alpha \int_0^1 f(r)J_0(ur)r\,\mathrm{d}r \pm 2\pi\sin\alpha \int_0^1 f(r)\Omega_0(ur)r\,\mathrm{d}r \tag{5.1.9}$$

CYCLIC PHASE VARIATION

The phase variation termed "cyclic" has p cycles of phase variation in any fixed circular path about the origin where p is an integer. The illumination

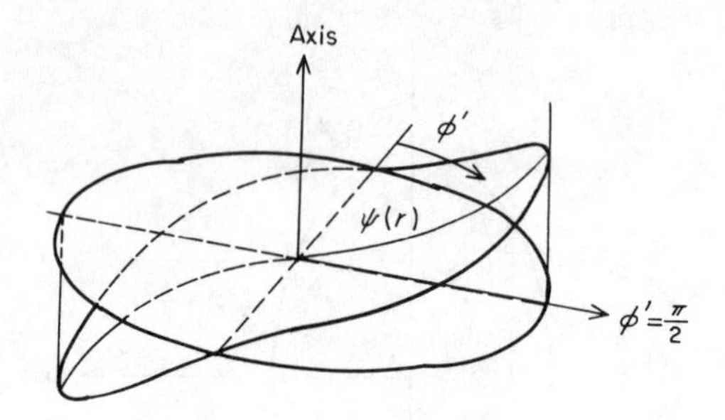

Figure 5.3: Circular aperture phase distributions. Asymmetric phase surface

function can be expressed as

$$f(r,\phi') = f(r)\exp(-ip\phi') \tag{5.1.10}$$

and

$$g(u,\phi) = \int_0^1 f(r)r\,\mathrm{d}r \int_0^{2\pi} \exp\{i[ur\cos(\phi-\phi') - p\phi']\}\,\mathrm{d}\phi' \tag{5.1.11}$$

The angular integral defines the higher order Bessel function [6, p20] leaving

$$g(u,\phi) = 2\pi \exp\left[ip\left(\frac{\pi}{2}-\phi\right)\right]\int_0^1 f(r)J_p(ur)r\,\mathrm{d}r \tag{5.1.12}$$

The factor outside the integral demonstrates that the radiation pattern exhibits the same cyclic variation in phase. The integral in Equation 5.1.12 defines the higher order Hankel transforms. This example is included to demonstrate a situation in which such higher order transforms can occur.

We can similarly define the higher order Lommel transforms of a radial function $f(r)$ analogously

$$\mathcal{L}_p[f(r)] \equiv \int_0^1 f(r)\Omega_p(ur)r\,\mathrm{d}r \tag{5.1.13}$$

SINUSOIDAL PHASE VARIATION

A sinusoidal phase variation is a phase surface obtained by rotating a function of the radial coordinate about the origin while varying its magnitude by a factor proportional to $\sin\phi'$. In practical antenna systems a phase front of this type arises when the source is displaced from the focus transversely to the axis of symmetry. In most instances then the phase variation is approximately cubic in the radial coordinate as shown in Figure 5.3 and gives rise to the aberration of primary coma.

The complex aperture function for this case is

$$f(r,\phi') = f(r)\exp[-i\psi(r)\sin\phi'] \tag{5.1.14}$$

in which $\psi(r)$ describes the shape of the phase surface in its cross-section in the plane $\phi' = \pi/2$. This would then be the curve given in Equation 1.2.16 as the focal curve of a parabolic cylinder with a displaced source. The scalar diffraction integral is then

$$g(u,\phi) = \int_0^{2\pi}\int_0^1 f(r)\exp\{i[ur\cos(\phi-\phi') - \psi(r)\sin\phi']\}r\,\mathrm{d}r\,\mathrm{d}\phi' \tag{5.1.15}$$

The plane showing the major effects of this perturbation is the plane $\phi = \pi/2$ hence

$$g(u,\pm\pi/2) \equiv g(\pm u,\pi/2) = 2\pi\int_0^1 f(r)J_0(ur - \psi(r))r\,\mathrm{d}r \tag{5.1.16}$$

Using the addition formula for Bessel functions [6, p30]

$$g(\pm u,\pi/2) = 4\pi\sum_{s=0}^{\infty}{}' (-1)^s\int_0^1 J_s(\mp\psi(r))J_s(ur)f(r)r\,\mathrm{d}r \tag{5.1.17}$$

where the prime in the summation indicates that the *first* term (s=0) is taken at only *half* its value.

For the particular case $\psi(r) = \alpha =$ constant the first Bessel function can be removed from the integration leaving

$$g(\pm u,\pi/2) = 4\pi\sum_{s=0}^{\infty}(-1)^s J_s(\mp\alpha)\int_0^1 f(r)J_s(ur)r\,\mathrm{d}r \tag{5.1.18}$$

For most cases of practical interest the first Bessel function in Equation 5.1.17 can be expanded as a series of powers of r, which can then be included with $f(r)$ in a second series of higher order Hankel transforms.

RADIAL PHASE VARIATION

All the phase conditions described have had variation with the angular position in the aperture. A major phase perturbation that has to be considered is that which is constant in the angular coordinate but varies with distance from the centre of the aperture. This condition arises in optical systems when the source is displaced axially from the focus and gives rise to the second major aberration that we need to consider in microwave systems that of spherical aberration (Figure 5.4). The illumination function is now given by

$$f(r,\phi') = f(r)\exp[-i\psi(r)] \tag{5.1.19}$$

where most usually $\psi(r) \sim br^2$ approximately.

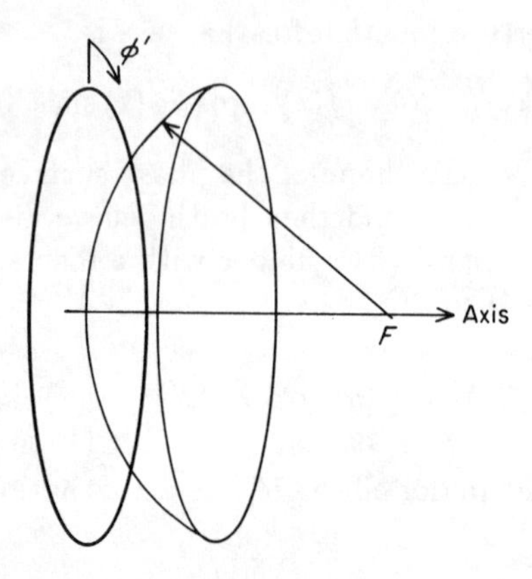

Figure 5.4: Circular aperture phase distributions. Axisymmetric phase surface

The methods to be obtained for the integrals of the previous section are found to apply to this phase variation, but are only of a simple enough form for the elementary condition $f(r) \equiv 1$. Other treatments, extend the method to general amplitude distributions but require separate considerations, which will be given in a later section.

AMPLITUDE VARIATION WITH ANGULAR POSITION

The higher order finite Hankel transforms also occur in the scalar integration when the illumination function has an amplitude variation which is periodic in the angular coordinate although the phase distribution remains constant. This condition can be shown to apply to a particular component of the vector field that is created when a paraboloid is illuminated from a linearly polarized element, a dipole or a waveguide horn, at its focus [10] [11]. This component gives rise to the well known cross polarized lobe patterns of this particular kind of reflector [5, p419].

This amplitude distribution of the cross polarized component can be described by a function of the kind

$$f(r, \phi') = f(r) \sin 2\phi' \tag{5.1.20}$$

The maximum perturbations, in this case the maximum values of the cross polarized lobes, occur in the quadrantal planes $\phi = \pm\pi/4$ and $3\pi/4$.

Substitution into the scalar integral gives

$$\begin{aligned} g(u, \pi/4) &= \int_0^1 \int_0^{2\pi} f(r) \exp[iur\cos(\phi' - \pi/4)] \sin 2\phi' \, \mathrm{d}\phi' r \, \mathrm{d}r \\ &\propto \int_0^1 f(r) J_2(ur) r \, \mathrm{d}r \end{aligned} \tag{5.1.21}$$

From these examples we see that most of the conditions that can be expected to arise in a practical antenna system, affecting the radiation pattern through diffraction or phase and amplitude perturbation can be assessed in the Kirchhoff approximation by the evaluation of integrals of the higher order finite Hankel and Lommel transforms defined by

$$g_n(u) = \int_0^1 f(r) J_n(ur) r \, dr$$
$$l_n(u) = \int_0^1 f(r) \Omega_n(ur) r \, dr \qquad (5.1.22)$$

for general real functions of $f(r)$.

5.2 ZERO ORDER TRANSFORMS

The most common requirement is for evaluation of the unperturbed aperture radiation patterns given by

$$g_0(u) = \int_0^1 f(r) J_0(ur) r \, dr \qquad (5.2.1)$$

The functions $f(r)$ to be considered are in most practical cases smooth monotonically decreasing functions with an edge value at $r = 1$ of between 0.1 and 0.3 (assuming $f(0) = 1$). They can readily be approximated with very close agreement by a few terms of the series

$$f(r) = \sum_{p=0} a_p (1 - r^2)^p \qquad (5.2.2)$$

A further degree of flexibility can be obtained by considering binomial expansions of such functions as

$$\begin{aligned} f(r) &= (1 - kr^2)^p \quad (k < 1, p \text{ integer}) \\ &= (1 + kr^2)^p \quad (k > 1, p > 0) \\ &= (1 + kr^2)^{-1/2} \end{aligned}$$

The resultant series are similar in kind to that in Equation 5.2.2. Hence for the result when substituted into Equation 5.2.1 we require the relations

$$\int_0^1 (1 - r^2)^p J_0(ur) r \, dr = 2^p p! \frac{J_{p+1}(u)}{u^{p+1}} \equiv \frac{\Lambda_{p+1}(u)}{2(p+1)} \quad (p = 0, 1, 2 \ldots) \quad (5.2.3)$$

Equation 5.2.3 is the defining relation for the Lambda functions [12]. The nature of the Bessel functions for the larger values of the argument coupled with the power of the argument in the denominator means that only a few terms of any resultant series of the form

$$g_0(u) = \sum_{p=0} a_p \frac{\Lambda_{p+1}(u)}{2(p+1)} \qquad (5.2.4)$$

need to be considered to reach the degree of approximation to which we are already committed. The results are given in the short table of transforms in Section 5.4

The same quadratic functions have a simple integral transform in the case of the zero order Lommel transforms. This arises from the relation [7, p160, no12]

$$\int_0^1 (1-r^2)^p \Omega_0(ur) r \, \mathrm{d}r = 2^p \frac{p! H_{p+1}(u)}{u^{p+1}} \tag{5.2.5}$$

where $H_p(x)$ is the Struve function.

The quadratic functions $(1-r^2)^p$ are not simply integrable in the same way for the higher order transforms and their use cannot be extended beyond the unperturbed aperture with circular symmetry or antisymmetry as given above. They do not form an orthogonal set in the interval and so cannot be used for the series expansion of a general function $f(r)$ or for an attack on the synthesis problem. Orthogonalization by the Schmidt method runs into excessive algebraic complexity after a very few terms. We examine means therefore of obtaining a series expansion in terms of more suitable polynomials.

5.3 THE CIRCLE POLYNOMIALS

The Lambda functions defined in Equation 5.2.3 contain the argument u in the denominator raised to the same power as the order of the accompanying Bessel function. In nearly all microwave antennas having pencil beam radiation patterns the main radiation falls within the range $-5 < u < 5$ approximately and the first diffraction lobe maxima occur within the adjacent region $5 < |u| < 10$. The effects of the denominator in even the second term of the series of Equation 5.2.4 will thus totally mask any pattern shaping effect of the Bessel function in the numerator. This fact was pointed out by McCormick [13] who suggested that a more appropriate series in which the effect of the denominator would be greatly reduced would be the Neumann series [6, Chapter XVI]

$$g(u) = \sum_{s=0}^{\infty} \alpha_{2s+1} \frac{J_{2s+1}(u)}{u} \tag{5.3.1}$$

with a similar series of Struve functions for the Lommel transforms.

We can consider the more general case

$$g(u) = \int_0^1 f(r) J_0(ur) r \, \mathrm{d}r = \sum_{s=0}^{\infty} \alpha(a)_{2s+1} \frac{J_{2s+1}(au)}{au} \tag{5.3.2}$$

where a is an arbitrary constant which can later be made equal to unity.

Taking the Laplace transform of Equation 5.3.2 with respect to u then

$$\int_0^1 \frac{r f(r) \, \mathrm{d}r}{\sqrt{t^2+r^2}} = \sum_{s=0}^{\infty} \alpha(a)_{2s+1} \frac{[\sqrt{a^2+t^2}-t]^{2s+1}}{(2s+1)a^{2s+2}} \tag{5.3.3}$$

On putting

$$\sqrt{a^2+t^2}-t=\sqrt{v} \quad \text{or} \quad t=\frac{a^2-v}{\sqrt{v}}$$

and

$$\sqrt{a^2+t^2}+t=\frac{a^2}{\sqrt{v}}$$

then

$$\int_0^1 \frac{2rf(r)\sqrt{v}\,dr}{\sqrt{a^4-2a^2v+v^2+4vr^2}}=\sum_{s=0}^{\infty}\alpha(a)_{2s+1}\frac{v^{s+1/2}}{(2s+1)a^{2s+2}} \tag{5.3.4}$$

The integrand can be expanded as a series of Legendre polynomials, that is

$$\left[\left(\frac{v}{a^2}\right)^2-2v\left(1-\frac{2r^2}{a^2}\right)+1\right]^{-1/2}=\sum_{s=0}^{\infty}P_s\left(1-\frac{2r^2}{a^2}\right)\left(\frac{v}{a^2}\right)^s$$

then comparing coefficients of powers of v/a^2 in the two series gives

$$\alpha(a)_{2s+1}=2(2s+1)\int_0^1 P_s\left(1-\frac{2r^2}{a^2}\right)f(r)r\,dr \tag{5.3.5}$$

We find therefore that for a radiation pattern to be described as a Neumann series of the kind in Equation 5.3.1 the coefficients of that series can be obtained directly from the aperture illumination function $f(r)$ by the relation in Equation 5.3.5

Putting $a=1$ then

$$\alpha_{2s+1}=2(2s+1)\int_0^1 P_s(1-2r^2)f(r)r\,dr \tag{5.3.6}$$

The functions $P_s(1-2r^2)$ where P is a Legendre polynomial are the circle polynomials first used by Zernike [14]. They are defined by

$$P_s(1-2r^2)=(-1)^s R_{2s}^0(r)$$

A table of polynomials and their properties is included in Appendix II.1.

With this terminology Equation 5.3.6 becomes

$$\alpha_{2s+1}=2(2s+1)\int_0^1 R_{2s}^0(r)f(r)r\,dr \tag{5.3.7}$$

The higher order circle polynomials can best be described as Jacobi polynomials [15] of which the Legendre polynomials are a subset. That is

$$R_n^m(r)=P_{(n-m)/2}^{(m,0)}(1-2r^2)(-1)^{(n-m)/2}r^m \tag{5.3.8}$$

$(n-m)$ a positive even integer or zero (Appendix II). As a consequence of this connection with the orthogonal Legendre and Jacobi polynomials, we have the orthogonality property

$$\int_0^1 R_n^m(r)R_p^m(r)r\,\mathrm{d}r = \frac{1}{2(n+1)}\delta_{n,p} \tag{5.3.9}$$

$\delta_{n,p}$ being the Kronecker symbol, and also

$$\int_0^1 R_m^m(r)J_m(ur)r\,\mathrm{d}r = (-1)^{(n-m)/2}\frac{J_{n+1}(u)}{u} \tag{5.3.10}$$

$(n-m)$ an even positive integer or zero.

This last result indicates the form of the appropriate expansion for the aperture distribution $f(r)$ suitable for all orders of Hankel (and Lommel) transforms.

That is we put

$$f(r) = \sum_{t=0}^{\infty} b_{2t+1}R_{2t}^m(r) \tag{5.3.11}$$

Confining ourselves temporarily to zero order transforms we now have

$$\begin{aligned} f(r) &= \sum_{t=0}^{\infty} b_{2t+1}R_{2t}^0(r) \quad \text{from Equation 5.3.11} \\ g(u) &= \sum_{s=0}^{\infty} \alpha_{2s+1}\frac{J_{2s+1}(u)}{u} \quad \text{from Equation 5.3.1} \end{aligned} \tag{5.3.12}$$

where from an application of Equation 5.3.9

$$b_{2t+1} = 2(2t+1)\int_0^1 R_{2t}^0(r)f(r)r\,\mathrm{d}r$$

and from Equation 5.3.7

$$\alpha_{2s+1} = (-1)^s 2(2s+1)\int_0^1 R_{2s}^0(r)f(r)r\,\mathrm{d}r$$

and hence

$$a_{2s+1} = (-1)^s b_{2s+1} \tag{5.3.13}$$

This recasts Equation 5.3.12 into the complete set

$$\begin{aligned} f(r) &= \sum_{s=0}^{\infty}(-1)^s\alpha_{2s+1}R_{2s}^0(r) \\ g(u) &= \sum_{s=0}^{\infty}\alpha_{2s+1}\frac{J_{2s+1}(u)}{u} \end{aligned}$$

and

$$\alpha_{2s+1} = (-1)^s 2(2s+1) \int_0^1 R_{2s}^0(r) f(r) r \, \mathrm{d}r \tag{5.3.14}$$

We have therefore a remarkable duality whereby both the aperture illumination function and the resultant radiation pattern can be described by two different series with (apart from an alternation in sign) identical coefficients.

The results can be readily extended to the higher order transforms in which case we have

$$\begin{aligned}
f(r) &= \sum_{n=m}^{\infty} a_{n,m} R_m^n(r) \qquad (n-m) \text{ an even positive integer or zero} \\
g(u) &= \sum_{n=m}^{\infty} a_{n,m} (-1)^{(n-m)/2} \frac{J_{n+1}(u)}{u} \\
a_{n,m} &= 2(n+1) \int_0^1 f(r) R_n^m(r) r \, \mathrm{d}r
\end{aligned} \tag{5.3.15}$$

Further interesting properties of the zero order transform now arise. Under the transformation

$$r \rightarrow \sqrt{1-r^2}$$

we have

$$1 - 2r^2 \rightarrow 2r^2 - 1$$

The circle polynomials being basically Legendre polynomials are alternatively odd and even functions of their arguments, that is

$$P_s(1-2r^2) = (-1)^s P_s(2r^2-1)$$

The following results of the transformation $r \rightarrow \sqrt{1-r^2}$ then become obvious with

$$\int_0^1 f(r) J_0(ur) r \, \mathrm{d}r = \sum_{s=0}^{\infty} \alpha_{2s+1} \frac{J_{2s+1}(u)}{u}$$

then

$$\int_0^1 f(\sqrt{1-r^2}) J_0(ur) r \, \mathrm{d}r = \sum_{s=0}^{\infty} (-1)^s \alpha_{2s+1} \frac{J_{2s+1}(u)}{u}$$

$$\int_0^1 f(r) J_0(u\sqrt{1-r^2}) r \, \mathrm{d}r = \sum_{s=0}^{\infty} (-1)^s \alpha_{2s+1} \frac{J_{2s+1}(u)}{u}$$

and

$$\int_0^1 f(\sqrt{1-r^2}) J_0(u\sqrt{1-r^2}) r \, \mathrm{d}r = \sum_{s=0}^{\infty} \alpha_{2s+1} \frac{J_{2s+1}(u)}{u} \tag{5.3.16}$$

all with the identical coefficients defined in Equation 5.3.14. This particular property does not extend to the higher order transforms.

The circle polynomials have the further property that

$$R^0_{2s}(0) = (-1)^s \quad \text{and} \quad R^0_{2s}(1) = 1$$

Substitution of these into Equation 5.3.14 gives

$$f(0) = \sum_{s=0}^{\infty} \alpha_{2s+1} \quad \text{and} \quad f(1) = \sum_{s=0}^{\infty} (-1)^s \alpha_{2s+1} \tag{5.3.17}$$

These are the values of the illumination function at the centre and the periphery of the aperture respectively.

If $f(r)$ is only defined in a circular region of radius $a \leq 1$ and is zero outside, then by the transformation $r' = r/a$

$$g(u) = \sum_{s=0}^{\infty} \alpha(a)_{2s+1} \frac{J_{2s+1}(au)}{au} \tag{5.3.18}$$

where

$$\begin{aligned} \alpha(a)_{2s+1} &= 2(2s+1) \int_0^a f(r) R^0_{2s}\left(\frac{r}{a}\right) r \, \mathrm{d}r \\ &= 2(2s+1) \int_0^1 f(r) R^0_{2s}\left(\frac{r}{a}\right) r \, \mathrm{d}r \end{aligned}$$

as in Equation 5.3.5 and since outside the range $x = \pm 1$

$$R^0_{2s}(x) = P_s(1 - 2x^2) \quad \text{is zero.}$$

Equation 5.3.18 is nothing more than the scaling of the radiation pattern by a factor equal to the reduction in size of the aperture and represents the obvious technique for producing patterns of greater width when required.

Analogous results to the above can be obtained in the case of the Lommel transforms. The key relation which parallels that of Equation 5.3.10 is [16].

$$\int_0^1 R^m_n(r) \Omega_m(ur) r \, \mathrm{d}r = (-1)^{(n-m)/2} \left[\frac{\Omega_{n+1}(u)}{u} + \frac{2\cos^2(n\pi/2)}{(n+1)\pi u} \right] \tag{5.3.19}$$

The Neumann series then becomes a series with the Bessel functions replaced by the Lommel-Weber functions of Equation 5.3.19 but with the identical coefficients α_{2s+1} derived in the same manner as for Equation 5.3.14

For the zero order transform this gives the *odd* function

$$g(u) = \sum_{s=0}^{\infty} \alpha_{2s+1} \left[\frac{\Omega_{2s+1}(u)}{u} + \frac{2}{(2s+1)\pi u} \right] \tag{5.3.20}$$

The original integral Equation 5.1.1 is basically a two dimensional Fourier transform, as would have been more apparent had we dealt with a rectangular coordinate system. In that case the integration would have separated into two single dimension Fourier transforms [5]. In Fourier transform theory a cosine transform can be transformed into a sine transform (and vice versa) through the application of a Hilbert transform. The equivalent even and odd transforms for the circular coordinate system are the Hankel and Lommel transforms. Thus the patterns of the two can be transformed into each other by the same Hilbert transformation. This was first observed by Moss [16].

Hence if by any method the radiation pattern of a circular aperture antenna can be described by the Neumann series (zero order transform, circular symmetric illumination),

$$g(u) = \sum_{s=0}^{\infty} \alpha_{2s+1} \frac{J_{2s+1}(u)}{u}$$

then the Hilbert transform of this relation gives the radiation in the plane perpendicular to the dividing diameter of the same aperture divided into two antiphase halves. For we have

$$ug(u) = \sum_{s=0}^{\infty} \alpha_{2s+1} J_{2s+1}(u)$$

the Hilbert transform of which [7, p243], is

$$yh(y) + \frac{1}{\pi} \int_{-\infty}^{\infty} g(y)\, \mathrm{d}y = -\sum_{s=0}^{\infty} \alpha_{2s+1} \Omega_{2s+1}(u)$$

Substituting $g(y)$ into the integral

$$\sum_{s=0}^{\infty} \int_{-\infty}^{\infty} \alpha_{2s+1} \frac{J_{2s+1}(y)}{y}\, \mathrm{d}y = 2 \sum_{s=0}^{\infty} \frac{\alpha_{2s+1}}{2s+1}$$

giving

$$h(y) = -\sum_{s=0}^{\infty} \alpha_{2s+1} \left[\frac{\Omega_{2s+1}(y)}{y} + \frac{2}{\pi y (2s+1)} \right]$$

which is identical to Equation 5.3.20 (the sign is ambiguous as is the definition of the phase in the aperture). The same transformation can be shown to transform the result of Equation 5.2.3 into that of Equation 5.2.5.

5.4 TRANSFORMS OF APERTURE DISTRIBUTIONS

UNIFORM PHASE - ZERO ORDER TRANSFORMS

To serve as an illustration of the method, we find the radiation patterns of uniform phase circularly symmetrical distributions with radial amplitude variation of the form which applies to the commonest types of illumination law found in practice.

Binomial Expansions

1. $f(r) = (1 - kr)^p \quad p \text{ integer} > 0 \quad k < 1$

$$g(u) = \sum_{m=0}^{p} \frac{p!(1-k)^{p-m}}{m!(p-m)!} k^m \frac{\Lambda_{m+1}(u)}{2(m+1)} \tag{5.4.1}$$

2. $f(r) = (1 + kr^2)^{-n} \quad n > 0 \quad k > 1$

$$g(u) = \frac{1}{(1+k)^n} \left[\frac{J_1(u)}{u} + \frac{2nk}{1+k} \frac{J_2(u)}{u^2} + 4n(n+1) \left(\frac{k}{1+k} \right)^2 \frac{J_3(u)}{u^3} + \dots \right] \tag{5.4.2}$$

Neumann Series

$$g(u) = \sum_{s=0}^{\infty} \alpha_{2s+1} J_{2s+1}(u)/u$$

3. $f(r) = 1$

Putting $f(r) = 1 = R_0^0(r)$ then $\alpha_1 = 1$; $s = 0$ and all

$$\alpha_{2s+1} = 0 \qquad s > 0$$

therefore

$$g(u) = \frac{J_1(u)}{u}$$

4. $f(r) = r^n$

From [7, p278], No 22 (corrected)

$$\alpha_{2s+1} = \frac{(-1)^s 2(2s+1)[(n/2)!]^2}{[(n/2)+s+1]![(n/2)-s]!} \tag{5.4.3}$$

where $(n/2)!$ can be replaced by the Gamma function $\Gamma[(n/2)+1]$ when n is odd. The result

$$g(u) = \sum_{s=0}^{\infty} \alpha_{2s+1} \frac{J_{2s+1}(u)}{u}$$

is also given in [7, p22], the form (corrected)

$$g(u) = \frac{1}{n+2} \, {}_1F_2 \left\{ \frac{n}{2} + 1; 1, \frac{n}{2} + 2; \frac{-u^2}{4} \right\} \qquad n > 0$$

where ${}_1F_2$ is the hypergeometric function.

This key function can now be used for any radial distribution which has a series expansion in powers of r or a good polynomial approximation. For example,

5. $f(r) = 1 - r$
   The "gabled" distribution

$$g(u) = \frac{1}{u}\left\{\frac{1}{3}J_1(u) + \frac{2}{5}J_3(u) + \frac{2}{21}J_5(u) + \frac{2}{45}J_7(u)\ldots\right\} \tag{5.4.4}$$

6. $f(r) = \cos(\pi r/2)$

$$g(u) \doteq 0.426267\frac{J_1(u)}{u} + 0.49904\frac{J_3(u)}{u} + 0.3733\frac{J_5(u)}{u} + 0.00095\frac{J_7(u)}{u} \tag{5.4.5}$$

7. $f(r) = \cos^2(\pi r/2)$

$$g(u) \doteq 0.297\frac{J_1(u)}{u} + 0.473\frac{J_3(u)}{u} + 0.205\frac{J_5(u)}{u} + 0.022\frac{J_7(u)}{u} + 0.002\frac{J_9(u)}{u} \tag{5.4.6}$$

The principle of linear superposition allows for linear combinations of these illumination laws $f(r)$ to give the same addition to the radiation patterns $g(u)$.

8. $f(r) = \exp(-2br^2)$
   For this function we find

$$\begin{aligned}\alpha_1 &= \frac{1}{2b}(1 - e^{-2b})\\ \alpha_3 &= \frac{-3}{2b}\left[\left(1 - \frac{1}{b}\right) + \left(1 + \frac{1}{b}\right)e^{-2b}\right]\end{aligned}$$

and the recurrence relation

$$\frac{\alpha_{2s+1}}{2s+1} = \frac{\alpha_{2s-1}}{b} + \frac{\alpha_{2s-3}}{2s-3}$$

For $b = 1/2$ we have

$$\begin{aligned}\alpha_1 &= 1 - e^{-1}\\ \alpha_3 &= 3(1 - 3e^{-1})\\ \alpha_5 &= 5(7 - 19e^{-1})\\ \alpha_7 &= 7(71 - 193e^{-1})\\ \alpha_9 &= 9(1001 - 2721e^{-1})\\ \alpha_{11} &= 11(18089 - 49171e^{-1})\end{aligned}$$

Since the coefficients decrease monotonically to zero each successive bracket contains a continuously improving rational approximation to e. Similar approximations to powers of e may be derived by inserting other values of b into the recurrence formula. The recurrence formula itself is related to the simple continued fraction of Gauss for the exponential function.

9. $f(r) = 1/\sqrt{1 - r^2/a^2} \qquad r < a$
   $= 0 \qquad a < r1$

   An "impossible" aperture distribution due to the divergence to infinity at $r = a$ [7, p278, No 23]

$$\alpha_{2s+1} = \int_0^1 \frac{R_{2s}^0(r)r}{\sqrt{1 - r^2/a^2}} \, \mathrm{d}r = \frac{2}{a} \sin[(2s+1)\sin^{-1} a] \qquad a < 1$$

   then $(-1)^s \alpha_{2s+1} = T_{2s+1}(a)$ the Chebychev polynomial of the first kind. Then

$$f(r) = \frac{1}{\sqrt{1 - r^2/a^2}} = \sum_{s=0}^{\infty} \frac{2}{a} T_{2s+1}(a) R_{2s}^0(r)$$

   and hence

$$g(u) = \sum_{s=0}^{\infty} 2(-1)^s T_{2s+1}(a) \frac{J_{2s+1}(u)}{u}$$

   Since we know the result for $a = 1$ [7, p7]

$$f(r) = \frac{1}{\sqrt{1 - r^2}} \qquad g(u) = \frac{\sin u}{u}$$

   we obtain the "scaled" result

$$f(r) = \frac{1}{\sqrt{1 - r^2/a^2}} \qquad g(u) = \frac{\sin au}{au}$$

   Hence

$$\frac{\sin au}{au} = 2 \sum_{s=0}^{\infty} 2(-1)^s T_{2s+1}(a) \frac{J_{2s+1}(u)}{u} \tag{5.4.7}$$

   possibly an original formula

10. $f(r) = 1/r$

   We apply the transformation rule given in Equation 5.3.16 to the last result to obtain

$$g(u) = 2 \sum_{s=0}^{\infty} 2 \frac{J_{2s+1}(u)}{u} \tag{5.4.8}$$

   (Since $T_{2s+1}(1) = 1$ and the factor $(-1)^s$ in Equation 5.4.7 is cancelled). In this case $g(u)$ is also known [7, p7] to be

$$g(u) = J_0(u) + \frac{\pi}{2}[J_1(u)H_0(u) - J_0(u)H_1(u)]$$

11. $f(r) = (1 - r^2)^p$

   The quadratic functions from which were derived the Lambda function in Equation 5.2.3 [7, p230, No 32]

$$\alpha_{2s+1} = \frac{(-1)^s (2s+1)(p!)^2}{(p+s+1)!(p-s)!} \tag{5.4.9}$$

The resultant series can be shown to be equivalent to a Lambda function by the application of the reduction formulae for Bessel functions [6, p143]. Non-integral values of p can now be considered by replacing the factorials in Equation 5.4.9 by their gamma function equivalents.

To this list may be added the considerable number of finite transforms of functions, many of which apply to special cases of amplitude distributions to be found in Erdelyi [7, Sections 8.2 8.3 8.5 and 8.7].

APERTURE DISTRIBUTION - TRANSFORMATION OF SERIES DESCRIPTION

The derivation of the radial in-phase amplitude distribution for the circular aperture from a radiation pattern given in the form of a Neumann series, entails the evaluation of the series given by Equation 5.3.14 from the known coefficients α_{2s+1}. This procedure can be simplified through an application of the recurrence relation for these coefficients based upon the recurrence relations which exist for the circle polynomials themselves, (Appendix II). For the required distribution $f(r)$ we then have

$$\begin{aligned}
\alpha_1 &= 2\int_0^1 f(r) r\,\mathrm{d}r \\
\alpha_3 &= 12\int_0^1 f(r) r^3\,\mathrm{d}r - 3\alpha_1 \\
\alpha_5 &= 60\int_0^1 f(r) r^5\,\mathrm{d}r - 5\alpha_3 - 10\alpha_1 \\
\alpha_7 &= 280\int_0^1 f(r) r^7\,\mathrm{d}r - 7\alpha_5 - 21\alpha_3 - 35\alpha_1 \qquad (5.4.10)
\end{aligned}$$

where α_i are all known.

If now $f(r)$ is assumed to be in the form

$$f(r) = \sum_{p=0} b_p (1 - r^2)^p$$

substitution into Equation 5.4.10 gives

$$\begin{aligned}
\alpha_1 &= \sum_p b_p/(p+1) \\
\alpha_3 &= -\sum_p b_p 3p/(p+1)(p+2) \\
\alpha_5 &= \sum_p b_p 5p(p-1)/(p+1)(p+2)(p+3) \\
\alpha_7 &= -\sum_p b_p 7p(p-1)(p-2)/(p+1)(p+2)(p+3)(p+4) \qquad (5.4.11)
\end{aligned}$$

and the continuation of the terms is now obvious.

The number of terms in each of the series descriptions must be made the same, making the summation in Equation 5.4.11 finite and hence the solution by a simple matrix inversion possible.

By this means, a superposition of two radiation patterns, the subsidiary lobes of which were substantially out of phase, and each being expressible as a Neumann series, the radiation pattern shown in Figure 5.5 can be obtained. The combination is found to be the series

$$g(u) = 0.34J_1(u)/u - 0.44J_3(u)/u + 0.198J_5(u)/u - 0.002J_7(u)/u$$

which has residual side lobes less than -47 dB everywhere in the range $u > 7$ and has not resulted in a greatly increased 3 dB beamwidth. Using the above procedure it is found that the amplitude distribution required to produce this highly desirable pattern is the function

$$f(r) = 0.076 - 0.0441(1 - r^2) + 0.528(1 - r^2)^2 + 0.44(1 - r^2)^3$$

which is shown in this figure.

5.5 PHASE ERRORS

There are two aspects to the inclusion of phase errors into an otherwise aberration-free optical system which are those applicable to microwave systems and those of purely optical systems. In the latter the main concern is with the purity of the image formed over a small area at the nominal focus of the system. For this purpose the object illumination is usually that of an incoming plane wave which is approximately uniform in amplitude over the final aperture of the system. The effects of numerous high order phase errors for uniform illumination are studied in greater detail in the literature of optics for example in Linfoot [18] [19] and it is to be noted that this study forms one of the most important applications of the circle polynomials.

In a microwave antenna we are less concerned with the focal distribution than with the radiated pattern and in either case with only the lowest orders of aberration. This could be put simply down to the fact of the much poorer definition that the focal spot has at microwave frequencies than at light frequencies. In the microwave case the non-uniformity of amplitude is the prerequisite study of the system which is the reason for "switching" the circle polynomial treatment from the phase distribution as in optical theory to the amplitude distributions of the theory presented above. In practice only the two primary aberrations, spherical aberration and coma, cause any concern in a microwave system and in general only make an appearance when the feed is displaced from the focus by a considerable fraction of the wavelength. To this extent it is possible to include in the circle polynomial method for amplitude distributions an additional circle polynomial method for these fundamental

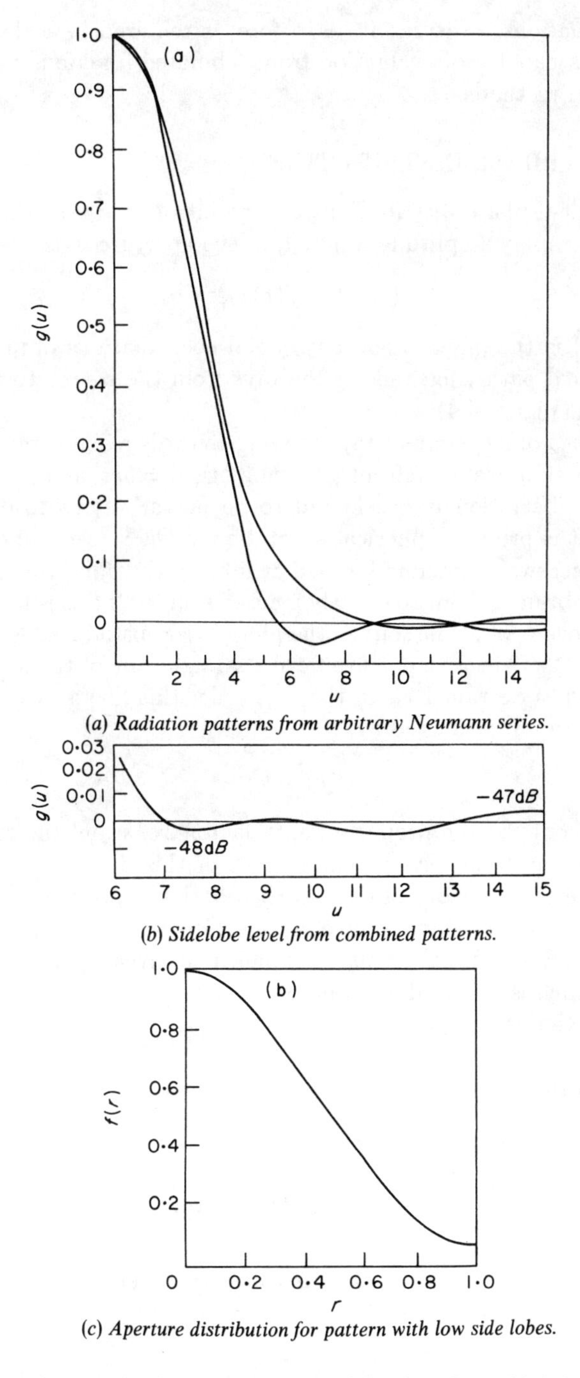

Figure 5.5: Radiation pattern with ultra-low sidelobes

phase perturbations. The result, as before, is required in a closed form series expression capable of evaluation from tabulated functions by elementary computational methods.

QUADRATIC PHASE DISTRIBUTION

The symmetrical phase distributions can be introduced into the analysis by putting the complex amplitude function in the aperture in the form

$$f(r, \phi') = f(r)\mathrm{e}^{-i\psi(r)}$$

in which $\psi(r)$ is the phase along any radial coordinate as obtained directly from the optical path length along the rays from the source to the aperture of the system (Figure 5.4).

In the optics of a system of rays tending towards a single point focus, the phase over the spherical wavefront is given by $\psi(r) = br^2$, and in the literature of optics the aberration is considered to be perturbations from the purely spherical, that is primary spherical aberration is $O(r^4)$, secondary $O(r^6)$ and so on. In a microwave antenna focused at infinity the unperturbed wavefront is plane, and obtained from an exactly focused source. If this is then perturbed by an axial movement of the source, the phase error introduced is proportional to (by a factor $2\pi/\lambda$) the departure from a straight line of the focal line given by the method in Section 1.2. To the first order this line is circular and thus $\psi(r)$ again can be represented by

$$\psi(r) = br^2$$

where b now contains the proportionality factor $2\pi/\lambda$ and the radius of curvature of the wavefront. As a consequence of this dual description we are able to use the entire theory of unperturbed optics to investigate the effect of spherical aberration in a microwave system. However, the optical literature rarely includes the non-uniformity of amplitude given by the function $f(r)$ above, and which is essential in a microwave system. This, therefore, requires extensions to the usual optical theory of aberrations.

UNIFORM AMPLITUDE DISTRIBUTION

Putting $f(r) = 1$ we require the evaluation of

$$g(u) = \int_0^1 \mathrm{e}^{-ibr^2} J_0(ur)r\,\mathrm{d}r \tag{5.5.1}$$

In Linfoot [18, p54] [19] (in which it is called the aberration-free case) use is made of Bauer's formula [6, p128]

$$\exp(iz\cos\theta) = \sqrt{\frac{\pi}{2z}}\sum_{n=0}^{\infty}(2n+1)i^n J_{n+1/2}(z)P_n(\cos\theta) \tag{5.5.2}$$

then

$$\exp(-ibr^2) = \exp(-ib/2)\exp[-ib(2r^2-1)/2] \tag{5.5.3}$$
$$= \exp(-ib/2)\sqrt{\frac{\pi}{b}}\sum_{n=0}^{\infty}(2n+1)(-i)^n J_{n+1/2}\left(\frac{b}{2}\right) R_{2n}^0(r)$$

This is then a complex radial distribution already put in the form of the series of circle polynomials required by the theory of Section 5.3 and the resulting Equations 5.3.14. Hence we have directly the coefficients α_{2s+1} and

$$g(u) = \exp(-ib/2)\sqrt{\frac{\pi}{b}}\sum_{n=0}^{\infty}(2n+1)i^n J_{n+1/2}\left(\frac{b}{2}\right) J_{2n+1}(u)/u \tag{5.5.4}$$

the appropriate Neumann series.

It is also to be noted that by operating on the integrand of Equation 5.5.1 by the transformation given in Equation 5.3.16 that is $r^2 \to 1 - r^2$ transforms both Equation 5.5.1 and the result in Equation 5.5.4 into their respective complex conjugates. The absolute radiated power distribution which is $P(u) = \sqrt{g(u)g^*(u)}$ is then the same for both cases. That is the radiative power pattern is distorted to the same degree by positive or negative values of the parameter b, that is (approximately) by motions of the feed source away from the focus in either a positive or negative axial direction.

Being a situation involving uniform amplitude distribution the major application of this result is in the focal distribution of a circular aperture antenna, which is receiving a plane wave normally incident upon it. This requires a re-interpretation of the parameters u and b. The parameter b is the sphericity of the wavefront contracting to the focus (Figure 5.6) and is given by [18] [19]

$$b = \frac{2\pi a^2 z}{\lambda f^2}$$

where a is the radius of the aperture, f the (geometrical) focal length and z the axial coordinate with respect to an origin at the focus. In this coordinate system u becomes

$$u = \frac{2\pi}{\lambda}\cdot\frac{a\rho}{f}$$

where, because of the circular symmetry, (ρ, z) are the polar coordinates required at this origin. This definition of u is not remote from the previous one since if ψ is the angle subtended by an off-axis point in the plane of the focus, at the centre of the aperture, the above definition of u becomes approximately

$$u = \frac{2\pi a}{\lambda}\sin\psi$$

The complete description of the field in the region of the focus is now obtained from Equation 5.5.4. In planes of constant z the pattern is a Neumann

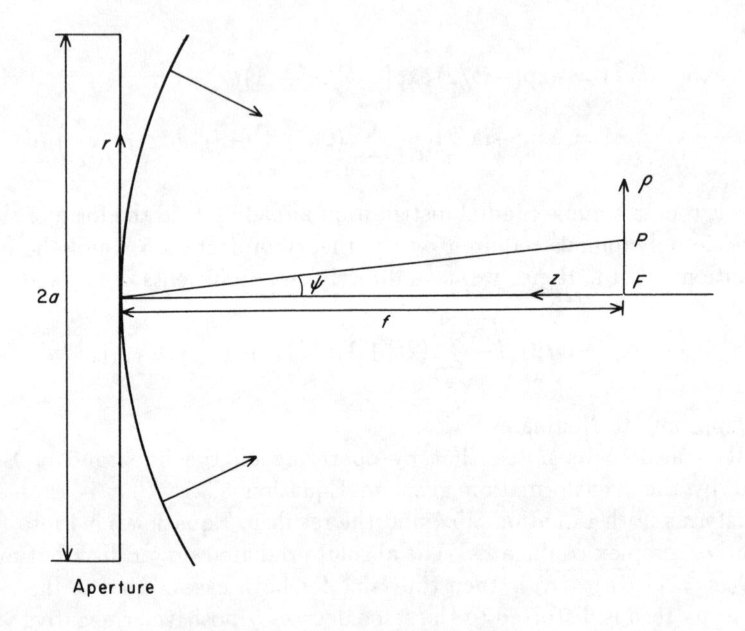

Figure 5.6: Spherical phase front

series and thus has the appearance of the usual radiation pattern, that of a central peak and decreasing amplitude side lobes. This has the usual appearance of a bright spot surrounded by rings of decreasing illumination. In the particular plane $z = 0$ this pattern is the usual Airy ring pattern proportional to $J_1(u)/u$.

In the axial direction we can put $u = 0$ and the result is proportional to $\sin(b/2)/(b/2)$.

Thus the entire ring system pattern obeys a sin X/X variation along the axis resulting in a system of elongated bright spots surrounded by similarly elongated but decreasingly bright spheroidal-like surfaces (Figure 5.7). The centre of the brightest region is the nominal focus, and is the obvious position for a receiving element. Further contributions to the received energy could be obtained by matching the receiving element to the right pattern. Identical results are obtained from the complete field analysis of receiving antennas [21].

It can be seen from the above that, whereas the additional contribution to be expected from the first subsidiary ring of the pattern is -17 dB on the centre illumination, since it is a $J_1(X)/X$ function, the additional energy that could be expected from a subsidiary feed placed at the secondary maximum along the axis would be only -13 dB since it derives from a $\sin X/X$ function. This amounts to only a slight increase in the total of the received energy of

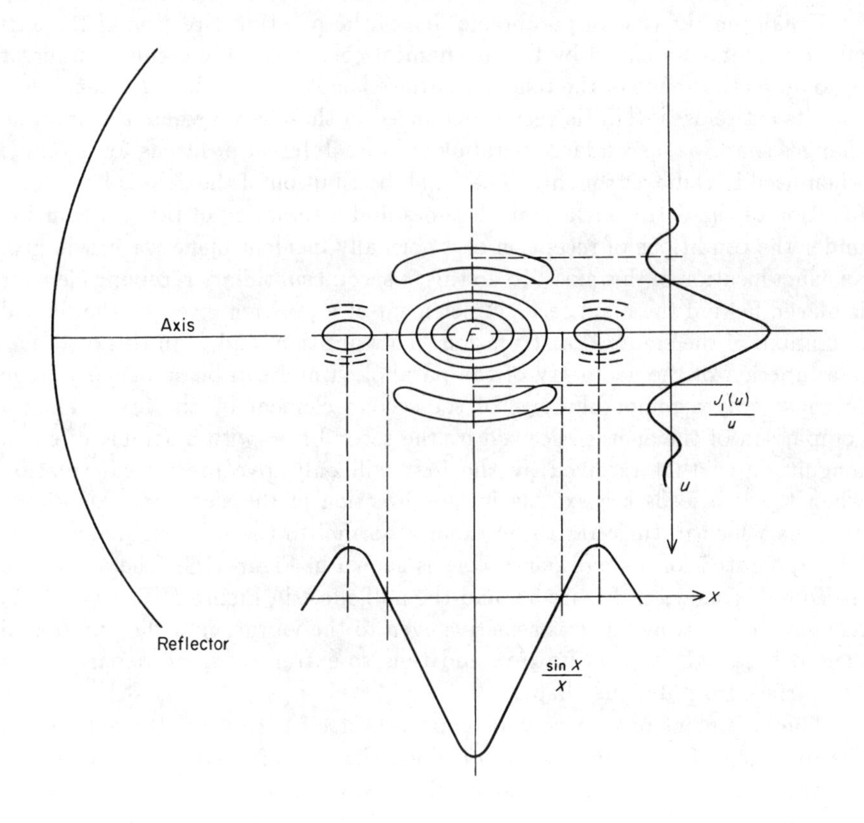

Figure 5.7: Field in focal region of reflector with incident axial plane wave

approximately 0.25 dB but the axial distribution can be used to play another important part in the requirements of large antennas, that of defining with great precision the operational axis of the antenna [22].

Usually in the case of paraboloid dishes the pointing direction of the axis of the antenna is defined by the mechanical construction and can be in error in so far as the shape of the reflector surface is not known with great accuracy over its entire area. This is very pronounced in those cases where the antenna changes shape as does a large paraboloid in its different positions for example when used in radio astronomy. The axial distribution of the field is however a function of the entire surface at all times and a sampling of this distribution under the conditions of reception of a normally incident plane wave, will give an exact location of this axis. To do this, a second subsidiary receiving element is placed behind the main receiving element at a position given by the second maximum of the $\sin X/X$ distribution. This position, and X in this instance, is a function of the geometry of the paraboloid and can be sufficiently large to cause only a minor blockage of the second element by the first. Then a comparison of the energy received by the second feed with a sample of equal magnitude to that received by the first will only give precise cancellation when the two feeds are exactly in the direction of the electrical axis of the antenna which in turn has to be exactly normal to the incident plane wave. The apparatus for a small paraboloid is shown in Figure 5.8. The axial field measured is shown in Figure 5.9 and the null effect in Figure 5.10. This null, as indicated on a meter, was sensitive even to the vibration of the paraboloid caused by a light tap on its rim, and thus to extremely small departures of the surface from its true shape.

The alignment procedure which suggests itself is to focus the antenna in the direction of a distant source, or along the axis of any source sufficiently distant to give an assumed incident plane wave, and to adjust the direction until the null position of the two receiving elements is exactly in line with the distant source. The line defined by the two receivers will then be the exact axis of the optical system and the first of these will be at the exact focus. The procedure being extremely sensitive as illustrated above requires the use of fine adjustment mechanisms.

Subsequently, a change in phase can be made which will allow the addition of the energy in the second receiving element to that of the first instead of the subtraction to give the additional 0.25 dB of gain.

NEAR AND FAR FIELD PATTERNS

The two situations in which the quadratic phase error has to be considered when applied to a *non-uniform* amplitude distribution are those of the far field radiation pattern of an axially defocused antenna and also the near field pattern of a correctly focused antenna.

In the latter case, when the measurement of the field is made in the near

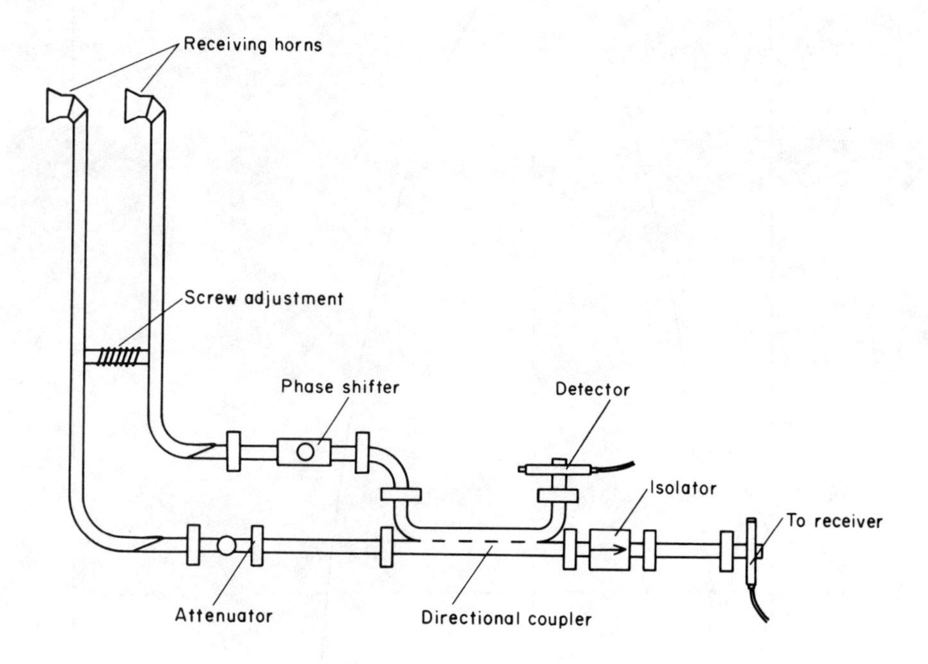

Figure 5.8: Feed arrangement for receiving secondary maximum for determination of reflector

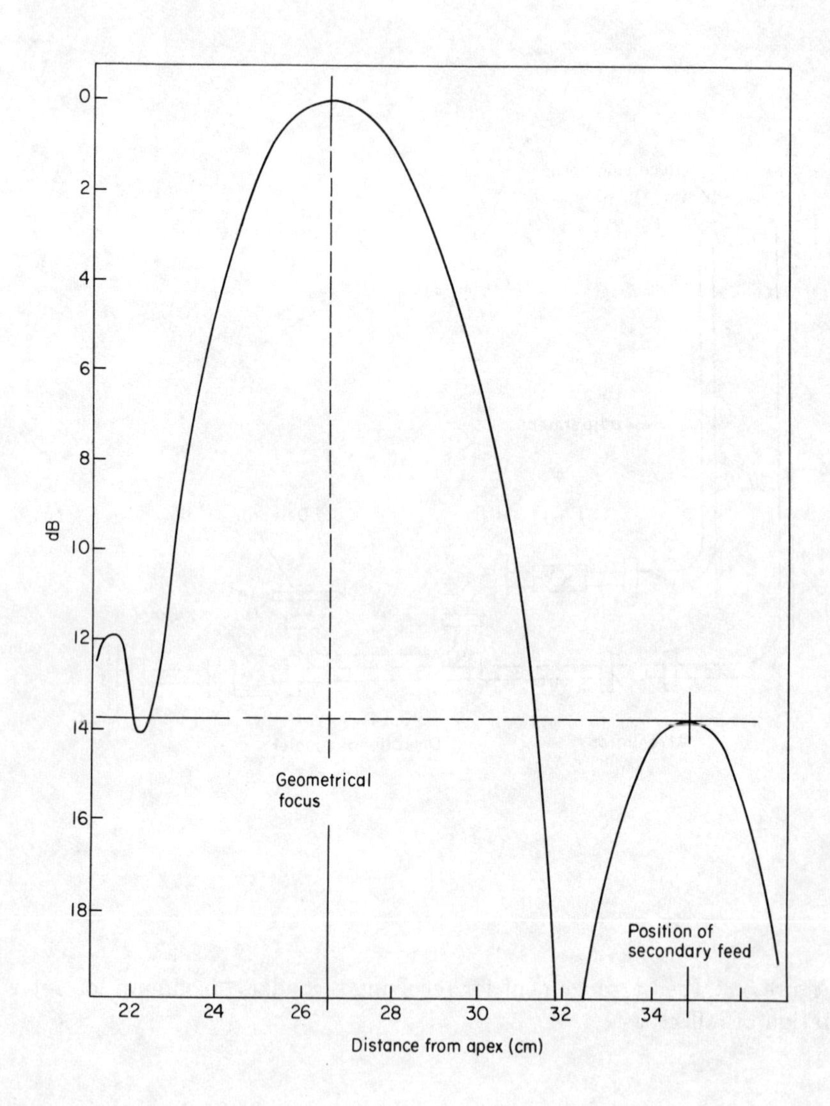

Figure 5.9: Measured field along paraboloid axis

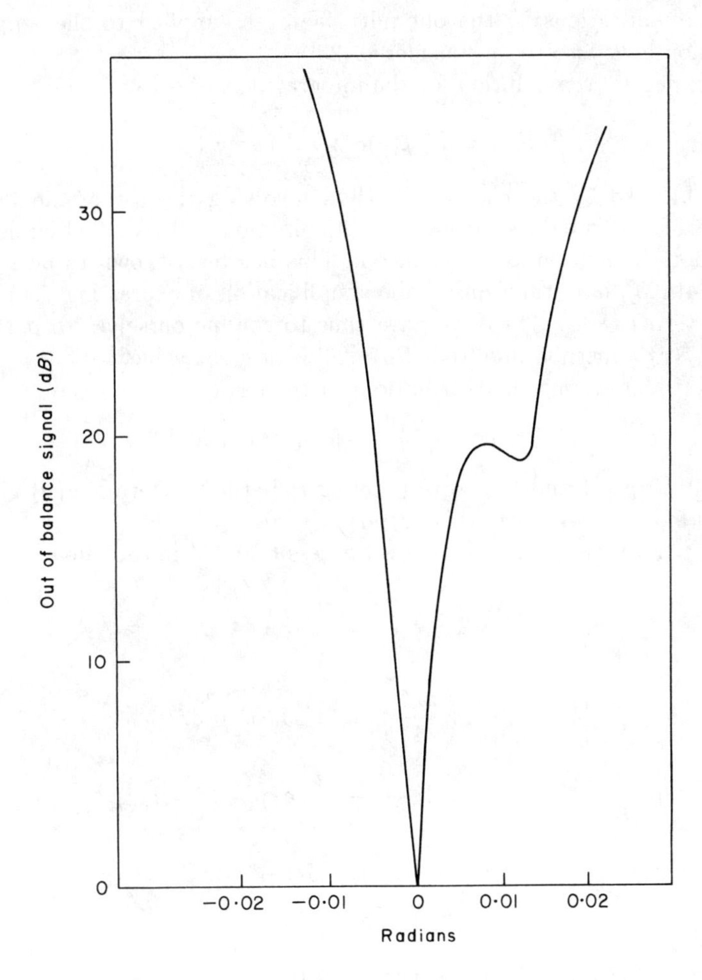

Figure 5.10: Error signal of alignment of two feed system in paraboloid receiver

field, that is up to approximately $a^2/2\lambda$, the error in phase that is created has the same parabolic form as in the former situation. At such close-in points however modifications to the obliquity factor as applied to the amplitude distribution have also to be considered [23].

The general series solution for the integral

$$g(u) = \int_0^1 f(r)\mathrm{e}^{-ibr^2} J_0(ur) r \,\mathrm{d}r$$

can be obtained by the classical method involving the use of the Lommel functions of two variables for uniform illumination and modified circle polynomials for the non-uniform extension. This has been shown to be possible by Hu [24] [25], but still requires the simplification of expressing $f(r)$ as the series of terms $(1 - r^2)^p$. If we have thus to confine ourselves to particular forms of $f(r)$ a method due to de Size [26] is of great value.

We consider amplitude distributions of the form

$$f(r) = \left(\cos\frac{\pi r}{2}\right)^p \qquad \text{for } p = 0, 1 \text{ and } 2$$

For $p = 0$ $f(r) = 1$ and the corresponding radiation pattern is $g(u)$ given in Equation 5.5.4 we will term this pattern $g(u, b)$.

For $p = 1$ we can put, with an error of less than 2% in the range $0 < r < 1$

$$\cos\left(\frac{\pi r}{2}\right) \simeq 1 - \sqrt{2}\sin\left(\frac{\pi r^2}{4}\right)$$

Then

$$\begin{aligned} g(u)_{p=1} &= \int_0^1 \left[1 - \sqrt{2}\sin\left(\frac{\pi r^2}{4}\right)\right] \mathrm{e}^{-ibr^2} J_0(ur) r \,\mathrm{d}r \\ &= \int_0^1 \mathrm{e}^{-ibr^2} J_0(ur) r \,\mathrm{d}r + \frac{i}{\sqrt{2}} \int_0^1 \exp[-i(b - \pi/4)r^2] J_0(ur) r \,\mathrm{d}r \\ &\quad - \frac{i}{\sqrt{2}} \int_0^1 \exp[-i(b + \pi/4)r^2] J_0(ur) r \,\mathrm{d}r \end{aligned} \tag{5.5.5}$$

and hence

$$g(u) = g(u, b) + \frac{i}{\sqrt{2}} g(u, b - \pi/4) - \frac{i}{\sqrt{2}} g(u, b + \pi/4)$$

Repeating the procedure for

$$f(r) = \cos^2\left(\frac{\pi r}{2}\right)$$

gives

$$\begin{aligned} g(u)_{p=2} &= g(u, b) + \frac{i}{\sqrt{2}} g(u, b - \pi/4) - \frac{i}{\sqrt{2}} g(u, b + \pi/4) \\ &\quad - \frac{1}{4} g(u, b - \pi/2) + \frac{1}{4} g(u, b + \pi/2) \end{aligned} \tag{5.5.6}$$

This method applies of course to any derivative of $g(u, b)$ that can be made and can be extended if necessary to higher orders of p.

ASYMMETRIC PHASE ERROR

The general description of the phase surface that is created when the source is moved away from its focal position in the plane transverse to the axis, can be given in simplified form as

$$f(r, \phi') = f(r)\exp(-ia_n r^n \sin\phi')$$

where for highly complex shapes more than one value of n may be required and a superposition of terms used. In this expression a_n is a phase angle in radian measure and the phase surface has the appearance given in Figure 5.3

The lowest order of these terms that is generally needed in a microwave antenna analysis is that given by the function representing primary coma,

$$f(r, \phi') = f(r)\exp(-iar^3 \sin\phi')$$

which approximates the focal line derived in Equation 1.2.16.

Substitution into the scalar integral and evaluation in the plane containing the offset source gives

$$g\left(u, \frac{\pi}{2}\right) = \int_0^{2\pi}\int_0^1 f(r)\exp(iur\sin\phi')\exp(-iar^3\sin\phi')r\,\mathrm{d}r\,\mathrm{d}\phi' \qquad (5.5.7)$$

The procedure adopted in optical theory [27] is to expand the second exponential term as a series of sines and cosines of multiples of ϕ', thus giving rise to a series of higher order Hankel transforms of the function $f(r)$. For uniform amplitude distributions $f(r) = 1$ this series simplifies greatly if the phase function ar^3 is replaced by a circle polynomial $R_3^1(r) = 3r^3 - 2r$ [18, p57] [19].

The method to be given here is to combine the two exponential terms and expand the resultant Bessel function by the addition theorem. Each term of the resultant series can then be evaluated by the method of circle polynomials individually. This can be done for the more general phase function $a_n r^n \sin\phi'$.

Then we have [28]

$$g\left(u, \frac{\pi}{2}\right) = 4\pi \sum_{m=0}^{\infty}{}' (-1)^m \int_0^1 f(r) J_m(a_n r^n) J_m(ur) r\,\mathrm{d}r \qquad (5.5.8)$$

$$= 4\pi \sum_{m=0}^{\infty}{}' \sum_{s=0}^{\infty} \frac{(-1)^s}{s!(m+s)!}\left(\frac{\pm a_n}{2}\right)^{m+2s}$$

$$\times \int_0^1 f(r) r^{n(m+2s)} J_m(ur) r\,\mathrm{d}r \qquad (5.5.9)$$

As always $\sum'$ refers to the sum when the first term ($m = 0$) is multiplied by the factor 0.5.

Choosing a short series of terms like $(1-r^2)^p$ to describe $f(r)$ or describing $f(r)$ by a power series makes the evaluation of these integrals elementary to give a Neumann series solution.

For those situations where the amount of the perturbation is small and the function more accurately a cubic curve, $n = 3$ in the above and the effect on the pattern is mainly that of a shift of the maximum by a small amount from the axial direction $u = 0$. The angular amount of this shift does not agree precisely with the angular offset of the source which creates the cubic wavefront. This difference or "squint" is of prime interest in those antennas where exact pointing accuracy is a requisite and which uses more than one system of sources dispersed about the true focus of the antenna. The amount of squint can be derived directly from Equation 5.5.9 by differentiation to obtain the value of u for which $g(u)$ is maximum.

HIGHER ORDER SYMMETRIC ERRORS

An expression has been derived for the evaluation of a uniformly illuminated circular aperture pattern in which the phase error is a symmetrical function of the radial coordinate alone, that is

$$f(r, \phi') = \exp(-i\beta_n r^n)$$

Substitution is therefore possible into the symmetric integral

$$g(u) = \int_0^1 \exp(-i\beta_n r^n) J_0(ur) r \, dr$$

Evaluation is obtained in closed form [29] by the expansion of both functions in the integrand and integration term by term of the double summation resulting. One of the ensuing summations then takes the form of a confluent hypergeometric function [6, p100]. The result is given by

$$g(u) = \sum_{m=0}^{\infty} \frac{(-1)^m}{(m!)^2}(m+1)\left(\frac{u}{2}\right)^{2m} {}_1F_1(a_m, c_m, -i\beta_n) \tag{5.5.10}$$

where ${}_1F_1$ is Kummer's confluent hypergeometric function

$${}_1F_1(a_m, c_m, X) = 1 + \frac{a_m}{c_m} X + \frac{a_m(a_m+1)}{c_m(c_m+1)} \frac{X^2}{2!} + \dots \tag{5.5.11}$$

In the analysis given

$$a_m = \frac{2(m+1)}{m} \quad \text{and} \quad c_m = 1 + a_m$$

leading to a considerable simplification of this result.

Results are illustrated in the reference for powers of r up to $n = 5$ and thus include the important quadratic and quartic terms of spherical aberration.

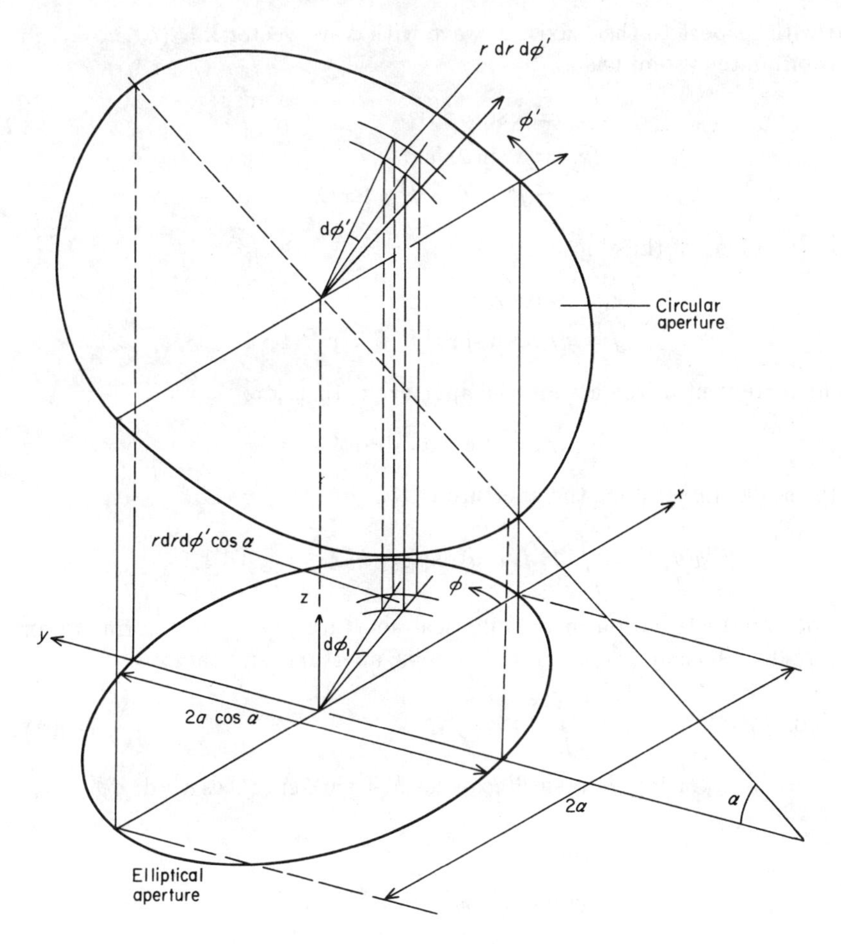

Figure 5.11: Projection into elliptical aperture

5.6 TRANSFORMATION TO ELLIPTICAL APERTURES

A common modification of the purely circular paraboloid antenna is a section of the paraboloid having an elliptical aperture. All of the foregoing analysis can be made directly applicable to elliptical distributions by the use of a projection transformation which converts the elliptical aperture into a circular one. As illustrated in Figure 5.11, the elliptical aperture may be taken to lie in the xy plane with z axis normal at its centre and with major axis, $2a$ along the x axis. The minor axis will then be $2a\cos\alpha$ where α is the angle between the plane containing the circular projection of the ellipse and the xy plane. In the plane containing the circle we use the coordinates (r, ϕ') as in the previous analysis and the field point at infinity has angular coordinates

(θ, ϕ) with respect to the z axis. A wave with wave vector $\mathbf{k} = (k_x, k_y, k_z)$ in this coordinate system has

$$\begin{aligned} k_x &= k \sin\theta \cos\phi \\ k_y &= k \sin\theta \sin\phi \\ k_z &= k \cos\theta \qquad k = 2\pi/\lambda \end{aligned}$$

and in the plane of the ellipse

$$\begin{aligned} x &= ar\cos\phi' \\ y &= ar\cos\alpha\sin\phi' \qquad 0 < r < 1 \end{aligned}$$

The increment of the area in the aperture is therefore

$$\mathrm{d}A = a^2 \cos\alpha\, r\, \mathrm{d}r\, \mathrm{d}\phi'$$

and the scalar integral for the aperture is

$$g(\theta, \phi) = \int\int_A f(x, y) \exp[-i(k_x x + k_y y)]\, \mathrm{d}A$$

The aperture distribution in the elliptical aperture $f(x, y)$ transforms to an aperture distribution $F(r, \phi')$ in the circular aperture, and thus

$$\begin{aligned} g(\theta, \phi) = a^2 \cos\alpha & \int_0^1 \int_0^{2\pi} F(r, \phi') \\ & \times \exp[-ikar\sin\theta(\cos\phi\cos\phi' + \sin\phi\sin\phi'\cos\alpha]r\,\mathrm{d}r\,\mathrm{d}\phi' \end{aligned} \tag{5.6.1}$$

Putting

$$\begin{aligned} u &= ka\sin\theta\sqrt{\cos^2\phi + \cos^2\alpha\sin^2\rho} \\ &= ka\sin\theta\sqrt{1 - \sin^2\alpha\sin^2\phi} \end{aligned}$$

and

$$\phi_1 = \tan^{-1}(\cos\alpha\tan\phi)$$

we find

$$g(\theta, \phi) = a^2\cos\alpha \int_0^1 \int_0^{2\pi} F(r, \phi')\exp[-iur\cos(\phi_1 - \phi')]r\,\mathrm{d}r\,\mathrm{d}\phi' \tag{5.6.2}$$

In this form all the previous theory can be adapted to the elliptical case.

5.7 THE MICROWAVE AXICON

As an illustrative example of the foregoing procedures, we investigate the effect of a conical phase distribution upon the radiated pattern of an already

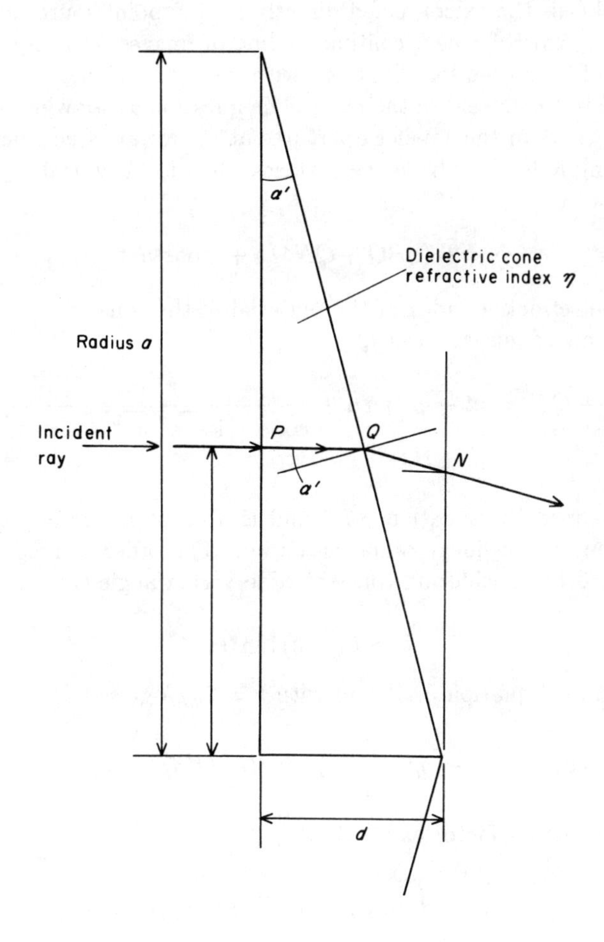

Figure 5.12: Geometry of microwave axicon

focused antenna. In practice this could be obtained by the insertion of a cone of wide angle (Figure 5.12) into the near field of a paraboloid or lens. It has analogies with, but is not identical to, the optical axicons of McLeod [30] [31]. In the optical case the axicon is fed directly from a point source and consists of a glass cone which forms a continuous line of images from a point source. This effect had been used for alignment with great precision over considerable distances before the advent of the laser. The question arises whether an effect akin to this occurs in the smaller apertures at microwave frequencies [32].

Discounting reflections from the surfaces, then for a typical ray PQN the phase is given by

$$\Phi = 2\pi(\eta PQ + QN)/\lambda + \text{ constant} \tag{5.7.1}$$

where η is the refractive index of the material of the cone.
With Snell's law of refraction at Q

$$\begin{aligned} \eta PQ + QN &= \eta t - \eta r \tan\alpha + \frac{r\tan\alpha}{\cos\alpha\sqrt{1-\eta^2\sin^2\alpha}+\eta\sin^2\alpha} \\ &= \eta t + b'r \end{aligned} \tag{5.7.2}$$

where b' is defined by Equation 5.7.2 and is a constant for any given cone. The phase thus has a linear radial variation. The order of magnitude of b' can be assessed by considering cones of wide vertex angle (i.e. small α), and therefore

$$b' \simeq (1-\eta)\tan\alpha$$

Substitution into Equation 5.1.1 and with $k = 2\pi/\lambda$ gives

$$g(u) = \exp(-ik\eta t)\int_0^1 a(\rho)\exp(-ikb'a\rho)J_0(u\rho)\rho\,\mathrm{d}\rho$$

Ignoring the constant factor $\exp(-ik\eta t)$

$$g(u) = \int_0^1 a(\rho)\exp(-ib\rho)J_0(u\rho)\rho\,\mathrm{d}\rho \tag{5.7.3}$$

where

$$b \simeq \frac{2\pi}{\lambda}a(1-\eta)\tan\alpha$$

and is therefore negative for dielectric materials with $\eta > 1$. The analysis, however, is even with respect to b, and the sign can be either positive or negative. The exact value of b can be obtained by inserting into Equation 5.7.3 the more complicated expression given by Equation 5.4.2. This is immaterial as b is a constant and only its order of magnitude is required. For cones of the usual plastic material, e.g. polythene or polystyrene, $\eta = 1.6$. With $\tan\alpha$ of the order of 0.25, b has a magnitude of a/λ (i.e. the ratio of aperture radius to wavelength).

The integral in Equation 5.7.3 can thus be evaluated by the technique of expansion in terms of the circle polynomials

$$g(u) = \int_0^1 a(\rho)\mathrm{e}^{-ib\rho} J_0(u\rho)\rho\,\mathrm{d}\rho = \sum_{n\ \mathrm{even}} \alpha_n(-1)^{n/2}\frac{J_{n+1}(u)}{u} \tag{5.7.4}$$

where

$$\alpha_n = 2(n+1)\int_0^1 \mathrm{e}^{-ib\rho}a(\rho)R_n^0(\rho)\rho\,\mathrm{d}\rho \tag{5.7.5}$$

In this equation $a(\rho)$ is the real amplitude distribution.

We shall only be considering amplitude distributions $a(\rho)$, which are simple quadratic functions of ρ such as $(1-\rho^2)^p$ for $p = 0$ and 1. Consequently, Equation 5.7.5 can be expanded and integrated term by term. This requires the summation of terms involving powers of ρ of the form

$$I_n = \int_0^1 \mathrm{e}^{-ib\rho}\rho^n\,\mathrm{d}\rho \tag{5.7.6}$$

which has a reduction formula

$$\begin{aligned} I_n &= -\frac{\mathrm{e}^{-ib}}{ib} + \frac{n}{ib}I_{n-1} \\ I_0 &= -\frac{\mathrm{e}^{-ib}}{ib} + \frac{ib}{1} \end{aligned} \tag{5.7.7}$$

hence

$$I_n = -\mathrm{e}^{-ib}\left[\frac{1}{ib} + \frac{n}{(ib)^2} + \frac{n(n-1)}{(ib)^3} + \ldots + \frac{n!}{(ib)^n} + \frac{n!}{(ib)^{n+1}}\right] + \frac{n!}{(ib)^{n+1}} \tag{5.7.8}$$

THEORETICAL FAR FIELD PATTERN

UNIFORM ILLUMINATION

We have, from Equations 5.7.4 and 5.7.5

$$g(u) = \alpha_0\frac{J_1(u)}{u} - \alpha_2\frac{J_3(u)}{u} + \alpha_4\frac{J_5(u)}{u} \tag{5.7.9}$$

for the first three terms. This confines the range of validity of u to those values for which $J_7(u)$ can be ignored with respect to $J_5(u)$ and for values of b for which the coefficients are either decreasing or remaining constant. Small values of b are therefore not suitable, since they would require an extension of the series of coefficients and correspondingly higher orders of J_n, possibly without convergence. Values of b greater than 5, however, are suitable for the first few coefficients, and hence for the series given by Equation 5.7.9. The range of u that can be then justified is approximately $u < 6$. This covers the region of the main lobe ($u < 4$) and the first side lobe ($u \simeq 5.5$) adequately. With $a(\rho) = 1$ the illumination is uniform with the result shown in Figure 5.13.

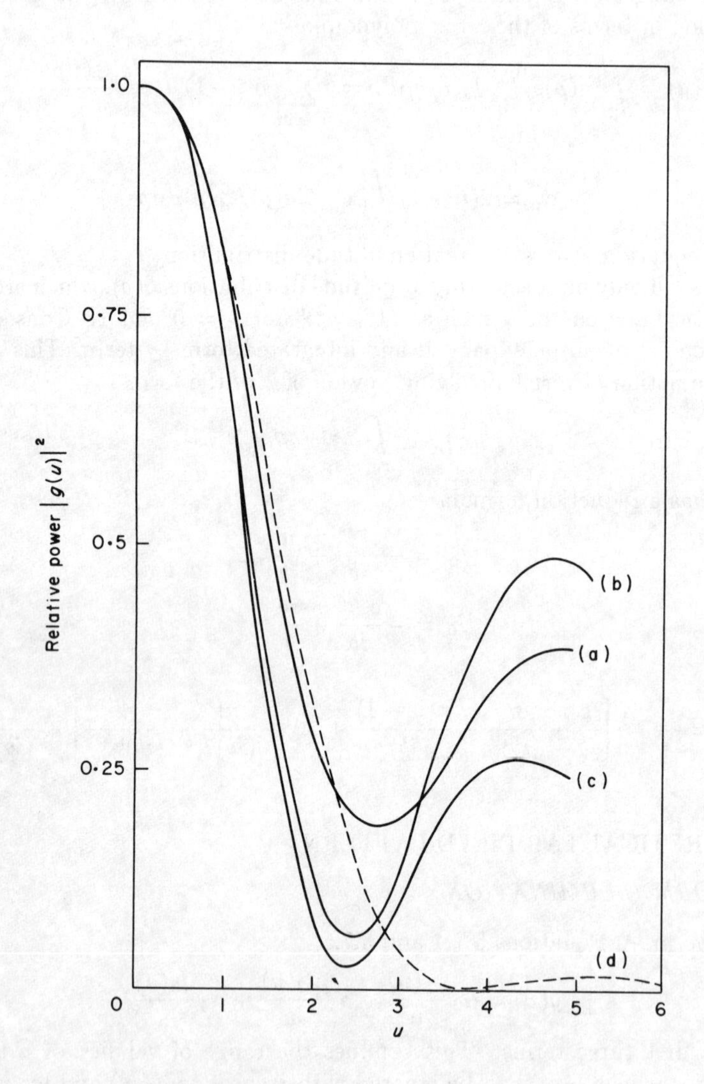

Figure 5.13: Microwave axicon. Uniform illumination. Theoretical patterns: (a) $b_m = 1.5\pi$ (b) $b_m = 3.5\pi$ (c) $b_m = 5.5\pi$ (d) $g(u) \propto J_1(u)/u$

TAPERED ILLUMINATION

With the illumination law $a(\rho) = 1 - \rho^2$ and the same considerations, we obtain the result shown in Figure 5.14.

Again the leading terms in α_0 and α_2 are in the ratio 3:1, giving the narrowing forced minimum.

Other illumination laws can be contrived to cause this effect. It appears to be general for functions $a(\rho) = (1 - \rho^2)^p$, and it occurs for $a(\rho) = 1 + \rho$, but not, strangely, for $a(\rho) = 1 - \rho$.

EXPERIMENTAL RESULTS AND CONCLUSIONS

Two experiments to confirm these effects have been carried out: namely one in the near field and one in the far field. The radiation patterns evaluated above are essentially far field patterns. The simple experiment to be performed is therefore to measure the radiation pattern of a focused antenna, and to repeat the measurement with a cone of dielectric supported in front of the antenna and in its near field. This experiment was carried out in x-band with a cone of 45 cm diameter and semivertex angle of approximately 76°, giving a b value of approximately 9. Although the result shown in Figure 5.15 confirms the theory precisely in the reduction of the beamwidth factor and forced minimum at $u = 2.5$, the side lobes are smaller than expected, and a decrease of gain of nearly 5 dB occurred. The final radiation pattern had a beamwidth at half power, which is much smaller than that which could have been achieved with an aperture of the largest dimension of the experimental apparatus, and with uniform inphase illumination.

This result is of some interest. In the system considered, none of the rays is travelling in the axial direction, yet the field has a maximum there. This bears out the principle of the Huygens lens discussed in Chapter 3. Superdirectivity has been achieved at the cost of a considerable reduction in forward gain.

5.8 THE RADIATION PATTERNS OF LUNEBURG LENSES

The Luneburg lens forms an ideal transformer between a source with angular power distribution proportional to powers of $\cos\alpha$ and radiation patterns in the far field proportional to a corresponding order of the lambda function.

With the refractive index law

$$\eta = \sqrt{2 - r^2}$$

the rays have trajectories

$$x^2 - 2xy\cot\alpha + y^2(1 + 2\cot^2\alpha) = R^2$$

Hence the ray intersects the unit sphere at the radial distance ρ from the axis given by

$$\rho = R\sin\alpha$$

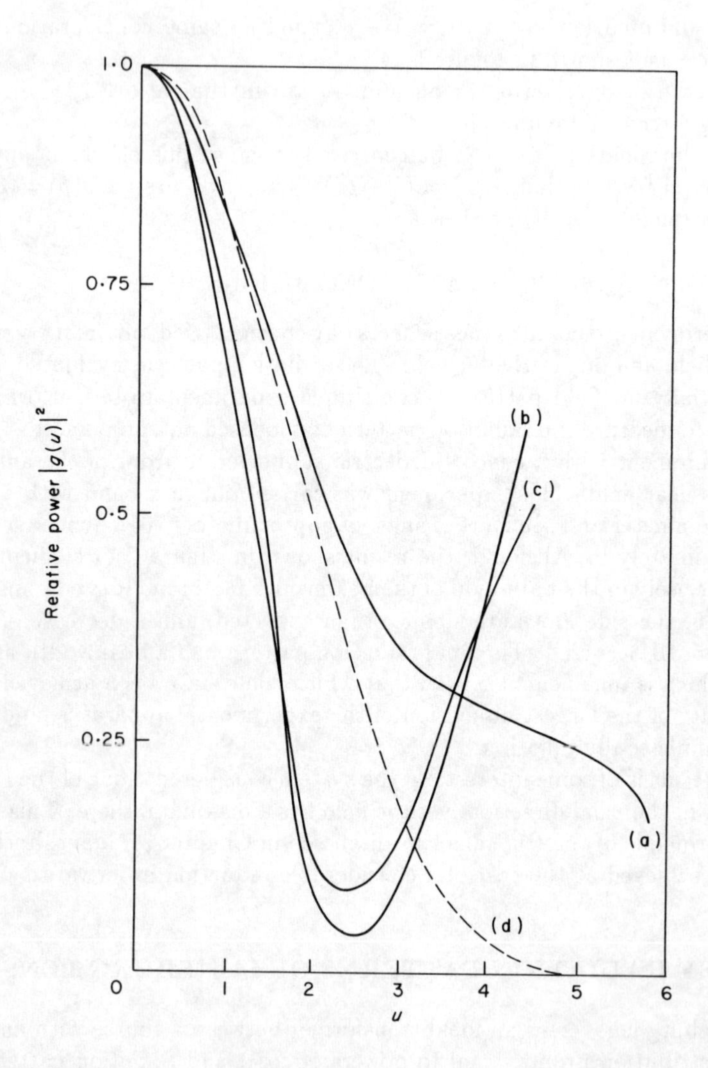

Figure 5.14: Microwave axicon with tapered illumination $[f(r) = 1 - r^2]$: (a) $b_m = 1.5\pi$ (b) $b_m = 3.5\pi$ (c) $b_m = 5.5\pi$ (d) $g(u) \propto J_2(u)/u^2$

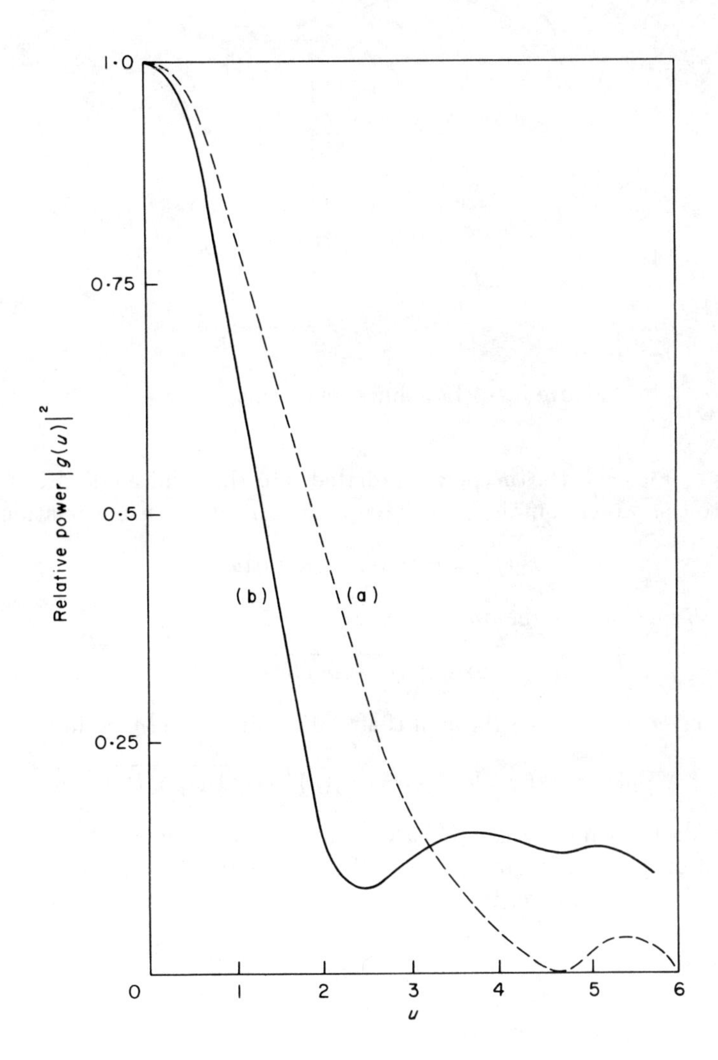

Figure 5.15: Microwave axicon experimental results. (a) without axicon (b) with axicon

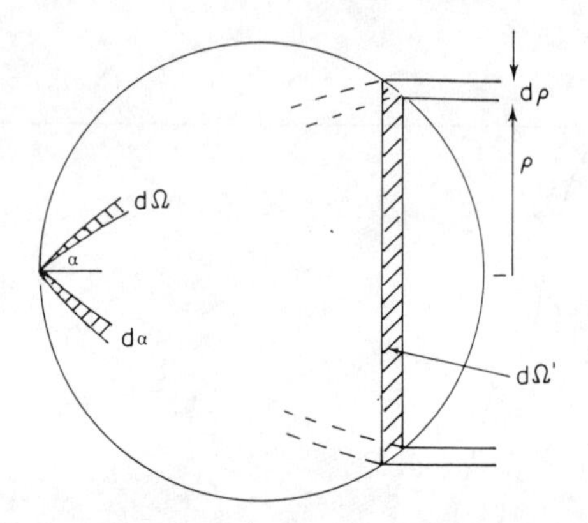

Figure 5.16: Luneburg lens transformer

Referring to Figure 5.16, the power radiated into the solid angle $\mathrm{d}\Omega$ is converted into the power contained in the annulus $\mathrm{d}\Omega'$ by the usual relation

$$P_2(\rho) = P_1(\alpha)\sin\alpha\,\mathrm{d}\alpha/\rho\,\mathrm{d}\rho$$

From $\rho = R\sin\alpha$ we obtain $\mathrm{d}\rho = R\cos\alpha\,\mathrm{d}\alpha$, and since

$$\cos\alpha = \sqrt{1-\rho^2/R^2}$$

$P_1(\alpha)$ is converted to $P_1(\rho)$ through the relation $\sin\alpha = R/\rho$, giving

$$P_2(\rho) = P_1(\alpha)/R^2\cos\alpha = P_1(\rho)/R^2\sqrt{1-\rho^2/R^2}$$

For a circularly symmetrical aperture amplitude distribution $f(\rho)$ the far-field radiation pattern is given, in the Huygens-Kirchoff approximation, by the zero-order finite Hankel transform (normalized to unit radius)

$$g(u) = \int_0^1 f(\rho)J_0(u\rho)\rho\,\mathrm{d}\rho \qquad u = \frac{2\pi R\sin\Theta}{\lambda}$$

λ being the wavelength of excitation, Θ the far-field polar coordinate and constants and scaling factors have been omitted.

Since $f(\rho)$ is the *amplitude* distribution it is proportional to $\sqrt{P_2(\rho)}$, with the result

$$g(u) = \int_0^1 \sqrt{\frac{P_1(\rho)}{\sqrt{1-\rho^2}}}J_0(u\rho)\rho\,\mathrm{d}\rho$$

There are several functions of $P_1(\rho)$ for which this integral is standard. A common description of the radiation from a microwave feed horn is to use a single function or a series of functions of the form

$$P_1(\alpha) = (\cos\alpha)^{2n+1} \rightarrow (1-\rho^2)^{n+1/2}$$

Then $n = 0$ gives $P_1(\alpha) = \cos\alpha$ and

$$g(u) = \int_0^1 J_0(u\rho)\rho\,\mathrm{d}\rho = \frac{J_1(u)}{u}$$

that is, a uniformly illuminated exit pupil and the diffraction-limited far-field pattern.

Generally for $P_1(\rho) = (1-\rho^2)^{n+1/2}$

$$\begin{aligned} g(u) &= \int_0^1 (1-\rho^2)^{n/2} J_0(u\rho)\rho\,\mathrm{d}\rho \\ &= \frac{J_{n/2+1}(u)}{u^{n/2+1}} = \Lambda_{n/2+1}(u) \end{aligned}$$

The result for $n = -1$ is of some interest. The primary pattern is $P_1(\alpha) = \sec\alpha$ and the far-field pattern is

$$g(u) = \int_0^1 \frac{J_0(u\rho)\rho\,\mathrm{d}\rho}{\sqrt{1-\rho^2}} = \frac{\sin u}{u}$$

This is narrower than the diffraction limit and gives evidence of some super directivity. The same primary distribution, $\sec\alpha$, applied to a paraboloid reflector gives the uniform illumination in the aperture and hence the far-field pattern, $J_1(u)/u$.

Source patterns of the form $P_1(\rho) = [R_n(1-2\rho^2/R^2)]^2\sqrt{1-\rho^2}$, where $R_n(1-2\rho^2/R^2)$ is a circle polynomial of Zernike (Appendix III), give radiation patterns

$$g(u) = \frac{J_{2n+1}u}{u}$$

and thus generate the terms of the Neumann series directly. Most source patterns can be generated by a series of such terms.

5.9 THE SYNTHESIS OF FAR FIELD RADIATION PATTERNS

One of the attractions of the Neumann series description of the far field radiation pattern is the insight it gives into the problem of constructing an aperture illumination function that will create a required or pre-determined radiation pattern. If, as in Popovkin [33], we regard the fundamental equation for radiation patterns, Equation 5.1.1, as the operational relation

$$Af = g \tag{5.9.1}$$

where g is the given function to be derived by the operation A on the unknown distribution f, then we require for the synthesis problem the inverse operator A^{-1}. This, as is stated in the reference, exists but is not continuous and then additional restrictions have to be made on the distribution $f(r)$. This function then becomes unstable and difficult of realization. In the classical sense the problem of antenna pattern synthesis is "incorrect". These aspects can all be demonstrated by the results of the method that has to be adopted in order to invert the series form of Equation 5.3.1. This will come as no surprise to experienced workers on antennas who find that even large variations in the amplitude distribution have little effect on the fundamental *shape* of a radiated beam, such as, for example, turning it from a pencil beam into a conical beam. The major effects are on the scale of the pattern, that is the beamwidth, and on the side lobe configuration. Thus, to synthesize a completely arbitrary beam shape, even a fairly simple one, turns out to be a very difficult proposition. This is basically so since we have to make a priori assumptions regarding the phase distribution in the aperture. Without these however the problem as a whole is indeterminate.

SYMMETRICAL PATTERNS - METHOD OF WEBB KAPTEYN

For symmetrical patterns and with a uniform phase distribution in the aperture, the operation required for the inversion in Equation 5.9.1 is the inverse of the finite Hankel transform of zero order.

Using the series description (Equation 5.3.1)

$$g(u) = \sum_{s=0}^{\infty} \alpha_{2s+1} \frac{J_{2s+1}(u)}{u}$$

then we only require to find the coefficients α_{2s+1} in order to obtain the aperture distribution from

$$f(r) = \sum_{s=0}^{\infty} (-1)^s \alpha_{2s+1} R_{2s}^0(r)$$

To do this we apply the theory [6] of Webb-Kapteyn which is based on the result

$$\int_0^{\infty} J_{2m+1}(t) J_{2n+1}(t) \frac{\mathrm{d}t}{t} = \frac{\delta_{m,n}}{2(2n+1)} \tag{5.9.2}$$

Then if an *odd* function $F(x)$ admits an expansion of the type

$$F(x) = \sum_{n=0}^{\infty} \alpha_{2n+1} J_{2n+1}(x) \tag{5.9.3}$$

the coefficients required are given by

$$\alpha_{2n+1} = (4n+2) \int_0^{\infty} F(t) J_{2n+1}(t) \frac{\mathrm{d}t}{t} \tag{5.9.4}$$

Unfortunately, the validity of this Fourier-like expansion is restrained by severe restrictions on the functions $F(x)$ to which it can be applied. These are [6, p535]

1. the integral $\int_0^\infty F(t)\,dt$ exists and is absolutely convergent
2. $F(t)$ has a continuous differential coefficient for all positive values of the variable which do not exceed x
3. the function satisfies the equation

$$F(t) = \frac{1}{2}\int_0^\infty \frac{J_1(v)}{v}[F(v+t) + F(v-t)]\,dv \qquad (5.9.5)$$

   when $t \leq x$.

Applying this to the *even* function

$$g(u) = \sum \alpha_{2s+1}\frac{J_{2s+1}(u)}{u}$$

we have the more stringent conditions

1. the integral $\int_0^\infty ug(u)\,du$ exists and is absolutely convergent
2. $g(u)$ has a continuous differential coefficient, and
3. the function $g(v)$ satisfies the relation

$$tg'(t) + g(t) = \frac{1}{2}\int_0^\infty \frac{J_1(v)}{v}[(v+t)g(v+t) + (v-t)g(v-t)]\,dv \quad (5.9.6)$$

   for all t in the range $[0, u]$.

As stated in the reference, no simple criteria have been established for functions which satisfy Equations 5.9.5 and 5.9.6.

We find that the ability to satisfy the latter is intimately bound up with the physical realizability of the pattern chosen. There are known functions with Neumann series expansions and there are other means of deriving Neumann series for arbitrary functions. But in every case not found to satisfy Equation 5.9.6 the introduction of α coefficients so obtained into the aperture distribution series (Equation 5.3.1) leads to the divergence of that series. In the absence of a general proof to this effect we give some illustrations

1.

$$g(u) = \cos u = 2\left[1^2\frac{J_1(u)}{u} - 3^2\frac{J_3(u)}{u} + 5^2\frac{J_5(u)}{u}\ldots\right]$$

   then with these coefficients the summation rule (Equation 5.3.17) for coefficients gives

$$f(1) = \sum_{s=0}^{\infty}(2s+1)^2$$

   which is clearly divergent.

2.

$$g(u) = \frac{\sin au}{au}$$

substitution into Equation 5.9.6 requires

$$\cos au = \frac{1}{2a}\int_0^\infty \frac{J_1(t)}{t}\sin au \cos au \, dt$$

which is true for $a \leq 1$ only. Therefore $\sin(au)/au$ satisfies the relation 5.9.6 but not the first condition on the absolute convergence of $\int_0^\infty \sin(au)\, du$. Hence the coefficients, obtained from Equation 5.9.4 namely

$$\begin{aligned} \alpha_{2s+1} &= 2(2s+1)\int_0^\infty \frac{\sin au}{au} J_{2s+1}(u)\, du \\ &= \frac{2}{a}\sin[(2s+1)\sin^{-1} a] = \frac{2}{a}T_s(a) \end{aligned}$$

lead to an amplitude distribution with an infinite singularity at $r = a$ as shown in Figure 5.17.

3. $g(u) = J_{2n+1}(au)/au$ gives equality when introduced into the relation 5.9.6 provided $a \leq 1$ and satisfies the two other criteria and is thus a viable function. From Equation 5.9.4 the coefficients are then

$$\begin{aligned} \alpha_{2s+2} &= 2(2s+1)\int_0^\infty J_{2n+1}(au)J_{2s+1}(u)\frac{du}{au} \\ &= \frac{1}{2}\sum_{m=0}^{s-n} \frac{(-1)^m a^{2n+2m}(s+m+n)!}{m!(2n+m+1)!(s-n-m)!} \qquad s \geq n \end{aligned} \qquad (5.9.7)$$

Comparing this with the alternative derivations of these coefficients gives us the two results

$$\begin{aligned} &\frac{1}{2}\sum_{s\geq n}^{\infty}(-1)^s \sum_{m=0}^{s-n}(-1)^m \frac{a^{2n+2m}}{m!}\frac{(s+m+n)!}{(2n+m+1)!(s-n-m)!}R_{2s}^0(r) \\ &= (-1)^n R_{2n}^0\left(\frac{r}{a}\right) \qquad 0 \leq r \leq a \\ &= 0 \qquad a < r \leq 1 \end{aligned} \qquad (5.9.8)$$

and

$$\frac{1}{2}\sum_{s\geq n}^{\infty}\sum_{m=0}^{s-n}(-1)^m \frac{a^{2n+2m}}{m!(2n+m+1)!}\frac{(s+m+n)!}{(s-n-m)!}\frac{J_{2s+1}(u)}{u} = \frac{J_{2n+1}(au)}{au} \qquad (5.9.9)$$

neither of which appears to be in the literature to date. The second of these is a particularly useful scaling theorem for Bessel functions.

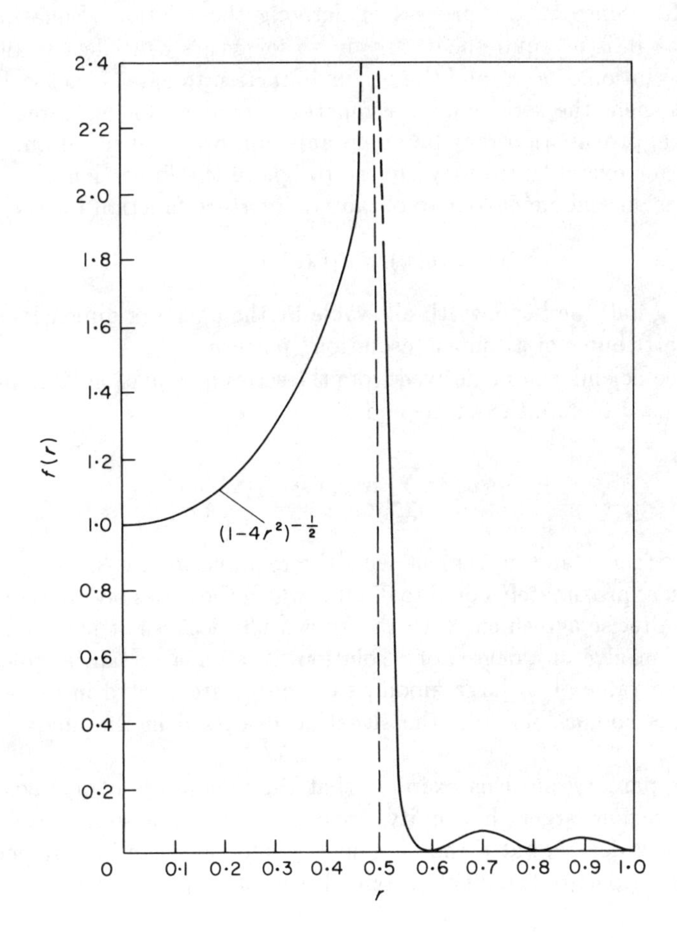

Figure 5.17: Series solution for aperture distribution required for a pattern $g(u) = \sin(u/2)/u$

$$\text{Exact solution } 1/\sqrt{1-4r^2}$$

$$\text{Series solution } \sum_{s=0}^{\infty} 4T_{2s+1}\left(\frac{1}{2}\right) R_{2s}^{0}(r)$$

This last example gives rise to the concept of "approximately valid functions". Reference to the process of deriving the relation (Equation 5.9.5) shows that it is a requirement only in so far as its cancellation allows the series description of $F(x)$ and the original function to agree exactly. If it were not exact then, the series and the function would no longer agree, but the degree of approximation may be acceptable in a practical situation.

Thus, for example, we may choose to ignore the limitation $a \leq 1$ in the last expansion and endeavour to obtain an aperture function for the pattern

$$g(u) = J_1(2u)/u$$

This, being half the beamwidth allowable by the aperture dimensions, would have the attributes of a "super resolution" pattern.

The coefficients can be derived from the series in Equation 5.9.7 but when these are used to sum the expression

$$g(u) = \sum_{s=0}^{\infty} \alpha_{2s+1} J_{2s+1}(u)/u$$

gives convergence and agreement with the required $J_1(2u)/u$ function up to values of u approximately equal to 8, after which the series becomes divergent. This is in precise agreement with the known physical result [34] that a finite aperture can give any degree of resolution, if, after a certain angular width (that is the value of u) large amounts of energy are wasted in radiated side lobes. It is compatible with the situation discussed in the analysis of the axicon.

It is apparent from this example that the magnitude of a, taken in the forbidden region larger than unity, creates this form of super resolution to an increasing degree, but if the high energy side lobes can be retained in the non-radiating part of the u field , some degree of "super resolution" would be achieved.

FUNCTIONS DESCRIBED BY THEIR MACLAURIN EXPANSION

Similar effects are observed when the Neumann series for a given radiation pattern function is derived from the Maclaurin expansion of the function. Incidentally many interesting Bessel function expansions can be obtained from the method independently of the practical consequences, by comparing two separate descriptions of the radiated field.

Neumann's own treatment for the expansion of an arbitrary function [6, p523] can be summed up by the statement that:

if the Maclaurin expansion of $F(z)$ is

$$F(z) = \sum_{n=0}^{\infty} b_n z^n$$

then

$$F(z) = \sum_{n=0}^{\infty} a_n J_n(z) \tag{5.9.10}$$

where a_n and b_n are related by

$$\begin{aligned} a_0 &= b_0 \\ a_n &= n \sum_{m=0}^{\leq n/2} 2^{n-2m} \frac{(n-m-1)!}{m!} b_{n-2m} \end{aligned} \tag{5.9.11}$$

The way in which the α coefficients of the Neumann series can be derived is obvious and hence the series description of the aperture distribution. No criteria have been established for the resultant convergence of these series and no success has been obtained in applying the method to such patterns as [35] [36] [37]

$$g(u) = \left[\frac{J_1(u)}{u}\right]^2 \qquad g(u) = \frac{\cos\sqrt{u^2 - A^2}}{\cosh A}$$

We are led therefore to make the conjecture that the only radiation patterns which can be synthesized from a circular aperture with a continuous inphase field distribution are those obeying the criteria of convergence, continuity and the relations given in Equation 5.9.6

Any attempt to obtain a meaningful physical understanding of Equation 5.9.6 must likewise fall into the category of pure conjecture. Integrating Equation 5.9.6 between the limits $[0, u]$ with respect to t results in the integral equation (after reversing the order of integration on the right-hand side)

$$ug(u) = \frac{1}{2}\int_0^{\infty} \frac{J_1(v)}{v}\,\mathrm{d}v \int_0^u [(v+t)g(v+t) + (v-t)g(v-t)]\,\mathrm{d}t \tag{5.9.12}$$

and the two cases, cases 2 and 3 given above are the simplest solutions for which the inner integral on the right-hand side gives a separable solution. That is

$$\int_0^u [(v+t)g(v+t) + (v-t)g(v-t)]\,\mathrm{d}t = F_1(v)F_2(u)$$

and then $F_1(v)$ must equal a constant $= b$ and so

$$F_2(u) = \frac{1}{b}ug(u)$$

Equation 5.9.12 can be said to have the appearance of a far field Huygens principle in which the kernel function is the uniform circular aperture function $J_1(v)/v$ in place of the free space function $\exp(ikr)/r$ and the inner integral is a symmetrized aperture weighting function, with all the appearance of being the average of a retarded and an advanced form. But beyond that our speculation should not go.

APERTURE EXPANSION AS A BESSEL SERIES

Other methods of pattern synthesis based on the scalar Kirchhoff formulation have been presented [38] [39]. The method is based upon a different expansion for the radial amplitude distribution $f(r)$, this time in terms of scaled Bessel functions of order zero. That is

$$\begin{aligned} f(r) &= \sum_{n=0}^{N} a_n J_0(u_n r) \qquad 0 < r < 1 \\ &= 0 \qquad\qquad\qquad\qquad r > 1 \end{aligned} \tag{5.9.13}$$

Substitution into the scalar integral for the condition of circular symmetry and uniform phase, require the application of Lommel's formula

$$\int J_0(u_n r) J_0(ur) r \, \mathrm{d}r = \frac{r}{u_n^2 - u^2}[u J_0(u_n r) J_0'(ur) - u_n J_0'(u_n r) J_0(ur)] \tag{5.9.14}$$

giving

$$g(u) = \sum_{n=0}^{N} \frac{a_n}{u_n^2 - u^2}[u J_0(u_n) J_0'(u) - u_n J_0'(u_n) J_0(u)] \tag{5.9.15}$$

The right-hand side can be simplified through the choice of N values of u_n to satisfy

$$J_0(u_n) \text{ or } J_0'(u_n) = \text{zero}$$

or more generally

$$u_n J_0'(u_n) + h J_0(u_n) = 0 \tag{5.9.16}$$

a form which we shall be meeting again later.

With $g(u)$ so fitted at these N values of u_n the coefficients a_n become

$$a_n = \frac{2u_n^2}{h^2 + u_n^2} J_0^2(u_n) g(u_n) \tag{5.9.17}$$

The first reference applies the theory to the positioning and the magnitudes of the side lobe zero and maxima, and thus is not strictly a beam shaping application. Ruze applies the same analysis to the production of a flat topped pattern. It is shown that the resultant series of Equation 5.9.15 does represent such a pattern but the aperture distribution arising from the substitution of the same coefficients into the series in Equation 5.9.13 is not illustrated. Asymmetric patterns can be derived by the application of the cyclic phase factor $\exp(-ip\phi')$ (Equation 5.1.10) for $p = 1$ only. Higher values of p would give patterns with higher periodicity in the ϕ coordinate and which have not had a practical application as yet. The analysis can then be repeated, but with first order Bessel function (since $p = 1$) and gives basically a pencil beam shape as before, but linearly shifted along the u axis. Thus again general pattern shaping cannot be said to have been achieved even though we have entered the area of non-uniform phases.

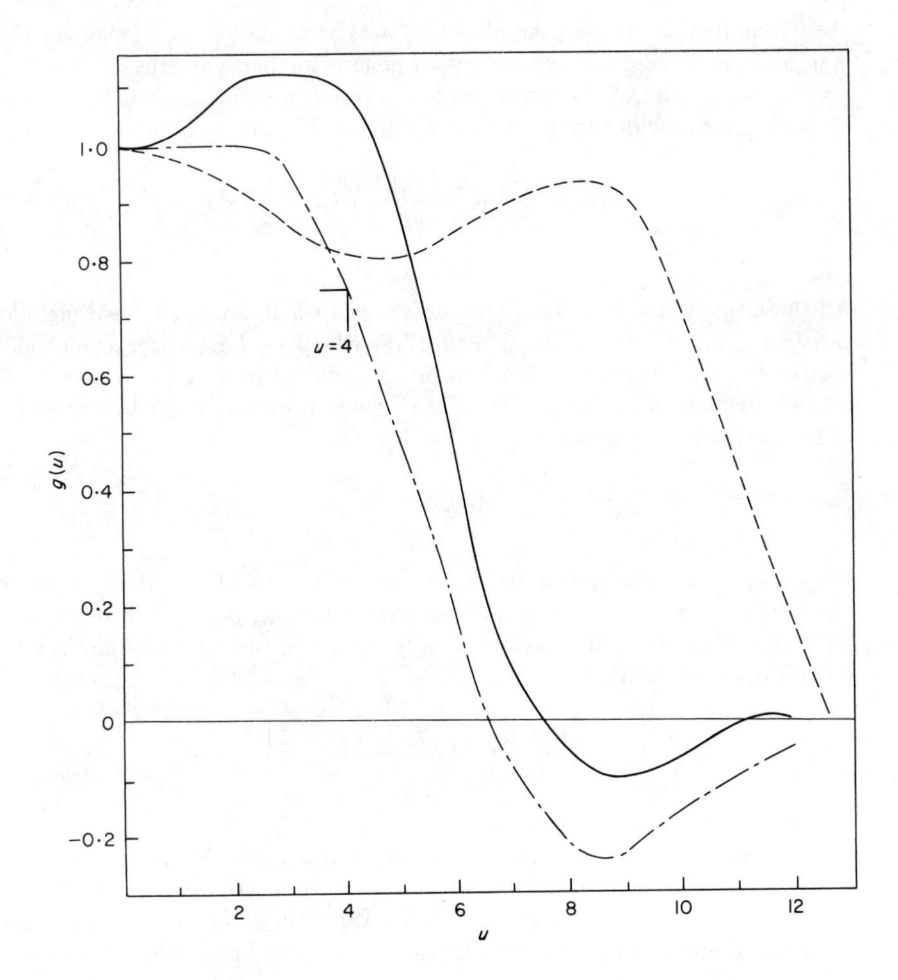

Figure 5.18: Patterns constructed from the functions $g(u) = J_n(\omega u)/u$

—— $\dfrac{2J_1(u)}{u} + \dfrac{2J_3(u)}{u} + \dfrac{J_3(u/2)}{u}$

- - - - $\dfrac{J_1(3u/4)}{u} + \dfrac{7J_3(u/2)}{u}$ (Normalized)

–·–·–·– Three term solution of Equation 5.9.19 with $\omega_2 = 0$ (for square topped pattern)

AN EXTENSION TO THE METHOD OF WEBB-KAPTEYN

We conclude with a final example which acts as an indication of the way that is needed to proceed in order to obtain general shaped patterns.

We use instead of the usual Neumann series, a series of scaled functions, each necessarily wider than its unscaled form. That is we assume

$$g(u) = \sum_{n=0}^{N} a_n \frac{J_1(\omega_n u)}{u} + b_n \frac{J_3(\mu_n u)}{u} \tag{5.9.18}$$

$$\omega_n, \mu_n \leq 1$$

All these functions, and the higher order ones which have not been included satisfy the criteria for physical realization which we have conjectured. The variety of shaped patterns that can be constructed from a few terms of this series is illustrated in Figure 5.18. The single condition that exists is the value of the function $g(u)$ at the origin, that is

$$g(0) = \sum_{n=1}^{N} \frac{\omega_n}{2} a_n$$

For the aperture distribution to be obtained by the use of the circle polynomials, the coefficients a_n and b_n and the parameters ω_n μ_n need to be derived. This can be performed computationally by a routine which minimizes the mean square difference

$$\left| \left[\sum_{n=1}^{N} a_n \frac{J_1(\omega_n u)}{u} + b_n \frac{J_3(\mu_n u)}{u} + \ldots \right] - g(u) \right|^2$$

over a finite range of values of u.

Now each term in the series is of the form $J_{2s+1}(au)/au$ for $a \leq 1$ and is therefore derived from an incomplete aperture distribution in which $f(r)$ is zero outside the value a (Equation 5.3.18). Thus unless the appropriate condition is applied, the aperture distribution required to produce the pattern of Equation 5.9.19, will be *discontinuous* at each value of ω_n used. It is soon to be found that when this further condition is applied and continuity of the aperture distribution (but not necessarily of its derivative) is insisted upon, then the patterns no longer are able to develop the shapes that the completely free choice of these coefficients gave.

Consequently, the next consideration must be that of the aperture distribution with deliberately placed discontinuities in the illumination function.

5.10 THE ZONED CIRCULAR APERTURE

Many practical microwave antennas and optical devices are known in which the aperture is divided into discrete zones. This includes all the zoned mirrors

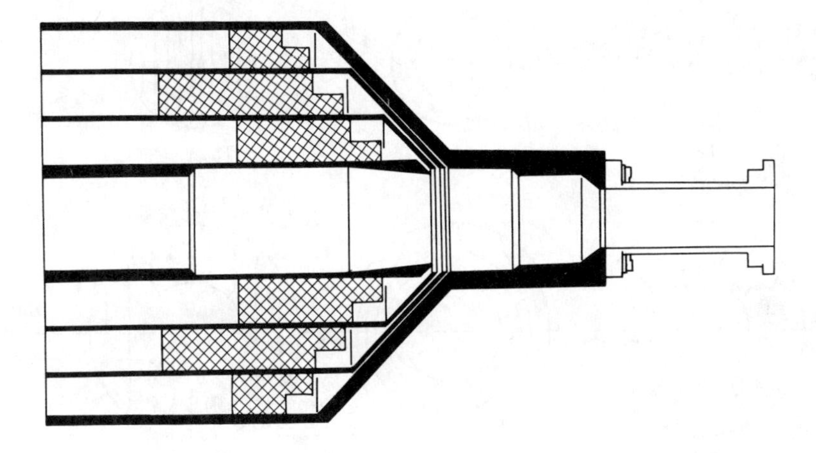

Figure 5.19: Cross-section of multi-zone antenna [37]

of the parageometrical optics design and the Fresnel zone plates. Antennas constructed from coaxial annuli, each with its own source and thus with independent control of the amplitude and phase, have been designed by Koch [40] [41] (Figure 5.19). By an ingenious coupling method between the coaxial elements he arranged the correct modal system in each annulus to give a manifestly uniform amplitude distribution in that region. This is a combination of the TE_{11} and TM_{11} modes for the central region, and of the TE_{11} and TE_{12} modes for each annular region is shown in Figure 5.20. Much of the current work on antenna feeds or horns which by themselves create shaped beam patterns now centres on the creation of these modes by mode transformers at the throat of the horn [42] or by the addition of hybrid modes such as would be created by impedance surfaces, corrugations or dielectric layers in the interior of the flare of the horn. The essence of the design is to create the correct mode in the correct amplitude and phase to create what in practice becomes a zoned distribution. It can be noted from Figure 5.20 that it is possible to insert an intermediate diametral conductor to divide each zone into semi-circular halves without upsetting the mode pattern, and that subsequently each semi-circular zone can be considered to have independence of amplitude and phase. This degree of freedom, we shall see, can be of great assistance in the creation of asymmetric patterns.

Larger antennas, if not constructed by the zonal method noted, can be constructed from individual circular arrays of discrete elements, although the construction itself is intricate and would only be used for antennas of a few wavelengths in diameter [43] (Figure 5.21). The particular antenna illustrated, has the additional capability of being fed directly from a planar hybrid combi-

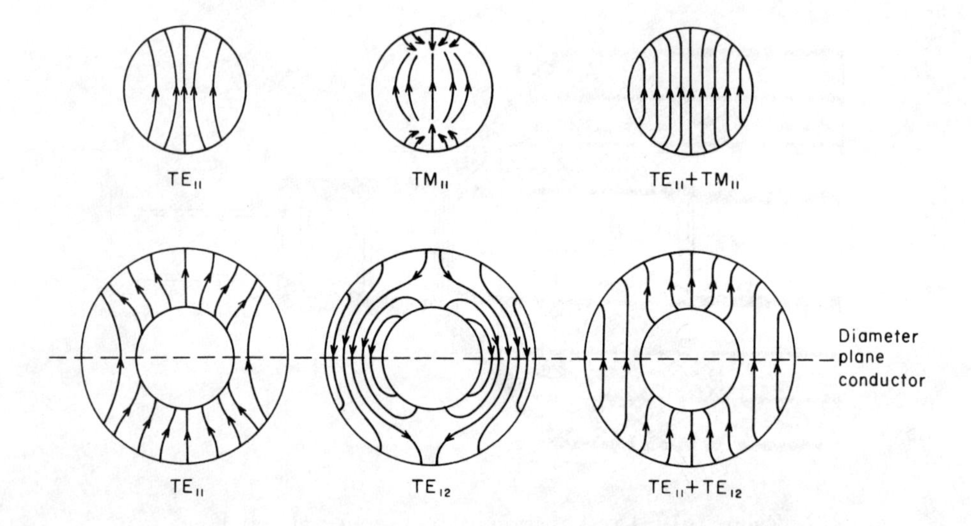

Figure 5.20: Mode summation to give zones of constant amplitude

nation in a way that enables monopulse operation among the four quadrants of the antenna.

In the study of optics the Fresnel zone plate and phase zone plate are known, and the analysis encompasses these designs. Similar devices have possible applications in microwave antenna designs. The difference as before is the additional phase factor $\exp(-ibr^2)$ in the integrand for rays focusing on the axis at a local point. With its omission for the far field pattern of a microwave zoned aperture, the analysis becomes considerably simpler, but leads to many interesting effects applicable to these apertures and to multimode horns. The radii of the zones are arbitrary in the analysis of the free aperture. In the case of the horns these radii have to be chosen to be compatible with the various TE or TM mode combinations which create the zones.

CIRCULAR SYMMETRIC PATTERNS: FOURIER-BESSEL AND DINI SERIES

Applying the scalar integral to the construction of the radiation pattern of a single zone with inner radius r_{n-1} and outer radius r_n with a constant complex illumination of magnitude a_{n+1}, gives

$$g_n(u) = \int_{r_{n-1}}^{r_n} a_{n+1} J_0(ur) r \,\mathrm{d}r = a_n \left[r_n \frac{J_1(ur_n)}{u} - r_{n-1} \frac{J_1(ur_{n-1})}{u} \right] \quad (5.10.1)$$

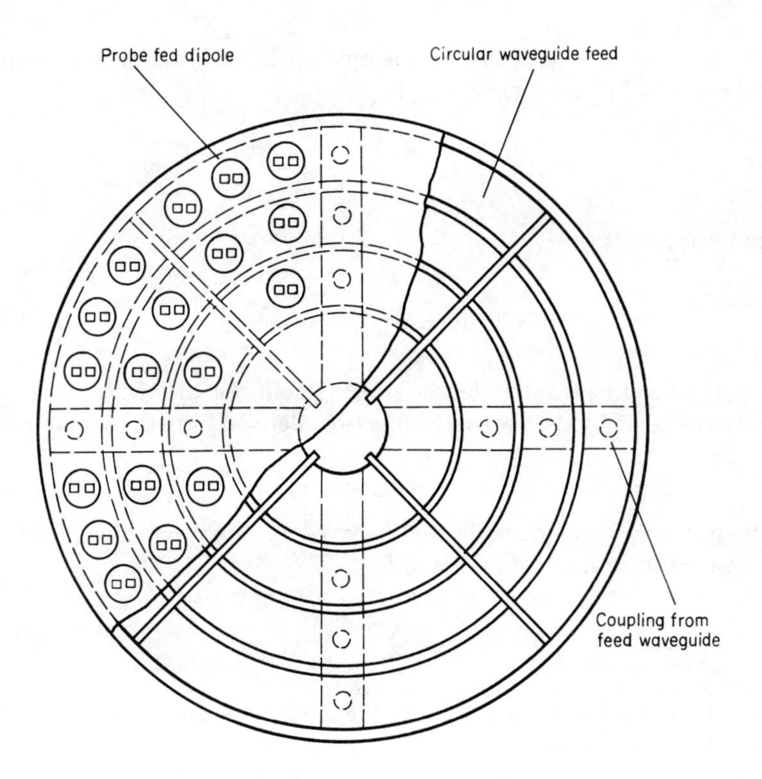

Figure 5.21: Zoned array of discrete elements

Summing these for an aperture of unit radius divided into N zones gives

$$g(u) = \sum_{n=1}^{N} r_n \frac{J_1(ur_n)}{u}(a_n - a_{n+1}) \tag{5.10.2}$$

where $a_{N+1} = 0$

Putting $A_n = a_n - a_{n+1}$ and r_n the outer radius of the nth zone equal to $\lambda_n r_1$ where $r_1 = 1/m$ and $r_N = 1$

$$g(u) = \sum_{n=1}^{N} \frac{A_n \lambda_n J_1(\lambda_n u/m)}{um} \tag{5.10.3}$$

Finally putting $u = mv$

$$m^2 vg(mv) = \sum_{n=1}^{N} A_n \lambda_n J_1(\lambda_n v) \tag{5.10.4}$$

This equation can be taken to be the starting point for the synthesis of radiation patterns given by the associated function, derived from the given function $g(u)$

$$F(v) = m^2 vg(mv) \tag{5.10.5}$$

$F(v)$ then has to be a generally well behaved function and $\lim_{v\to 0}(F(v)/v)$ must exist, or in terms of Equation 5.10.3

$$g(0) = \sum_{n=1}^{N} \frac{A_n \lambda_n^2}{2m} \tag{5.10.6}$$

This is a basic condition applicable to the amplitude coefficients $A_n = a_n - a_{n+1}$ by whatever means that are to be determined.

The procedure is to compare the series of Equation 5.10.4 with the infinite series of the same kind

$$F(v) = \sum_{n=1}^{\infty} b_n J_\nu(\lambda_n v) \tag{5.10.7}$$

there being several ways in which this latter expression can be used for the description of quite general functions $F(v)$. The choice depends entirely upon the radii of zones decided upon and hence upon the parameters λ_n.

The first choice is to take the $\lambda_n = \lambda_1, \lambda_2, \lambda_3 \ldots$ proportional to the positive zeros in ascending order of magnitude of the function previously met in Equation 5.9.16 [6, p577], namely

$$x^{-\nu}[xJ'_\nu(x) + hJ_\nu(x)] \tag{5.10.8}$$

where in the present case $\nu = 1$, h is any given constant. The choice $h = 0$ or $h = 1$ then makes λ_n the zeros of $J'_1(x)$ and $J_0(x)$ respectively. The choice of infinite h gives the zeros of $J_1(x)$. These are all in accord with the multi-mode requirements for the correct boundary conditions for TE or TM propagation in circular horns. The coefficients of the series in Equation 5.10.7 are then

$$b_n = \frac{2\lambda_n^2}{(\lambda_n^2 - 1)J_1^2(\lambda_n) + \lambda_n^2 J_1'(\lambda_n)^2} \int_0^1 vF(v)J_1(\lambda_n v)\,dv \tag{5.10.9}$$

and hence for the finite series the amplitude coefficients are given by

$$A_n = \frac{2m^2\lambda_n}{(\lambda_n^2 - 1)J_1^2(\lambda_n) + \lambda_n^2 J_1'(\lambda_n)^2} \int_0^1 v^2 g(mv)J_1(\lambda_n v)\,dv \tag{5.10.10}$$

The procedure being a Fourier-Bessel expansion it can be shown that with these coefficients in the *finite* series the resultant patterns will be the best mean-square fit to the required function $g(u)$.

The integrals on the right-hand side of Equation 5.10.10 can now be evaluated by the circle polynomial method of Section 5.3. That is

$$\int_0^1 v^2 g(mv)J_1(\lambda_n v)\,dv = \sum_{s=0}^{\infty} \alpha_{2s+1}(-1)^s \frac{J_{2s+2}(\lambda_n)}{\lambda_n} \tag{5.10.11}$$

where

$$\alpha_{2s+1} = 2(2s+2)\int_0^1 v^2 g(mv) R_{2s+1}^1(v)\,dv \tag{5.10.12}$$

whence

$$A_n = \frac{2m^2}{(\lambda_n^2 - 1)J_1^2(\lambda_n) + \lambda_n^2 J_1'(\lambda_n)^2} \sum_{s=0}^{\infty} \alpha_{2s+1}(-1)^s J_{2s+2}(\lambda_n) \tag{5.10.13}$$

The three cases of major interest then give the following results

1. h infinite; λ_n are proportional to the zeros of $J_1(x)$ and the series is the Fourier-Bessel series

$$A_n = \frac{2m^2}{\lambda_n[J_2(\lambda_n)]^2} \sum_{s=0}^{\infty} \alpha_{2s+1}(-1)^s J_{2s+2}(\lambda_n) \tag{5.10.14}$$

2. $h = 0$; λ_n are proportional to the zeros of $J_1'(x)$, the series is a Dini series [6, p577]

$$A_n = \frac{2m^2}{(\lambda_n^2 - 1)[J_1(\lambda_n)]^2} \sum_{s=0}^{\infty} \alpha_{2s+1}(-1)^s J_{2s+2}(\lambda_n) \tag{5.10.15}$$

3. $h = 1$; λ_n are proportional to the roots of $J_0(x)$

$$A_n = \frac{2m^2}{(\lambda_n^2 - 1)[J_1(\lambda_n)]^2 + \lambda_n^2[J_2(\lambda_n)]/4} \sum_{s=0}^{\infty} \alpha_{2s+1}(-1)^s J_{2s+2}(\lambda_n) \tag{5.10.16}$$

All the solutions are subject to the summation condition of Equation 5.10.6. If, for a given $g(u)$ pattern this summation is carried out there results an oscillatory function which therefore only crosses the prescribed value $g(0)$ at a discrete set of values. This limits the total freedom with which $g(u)$ can be selected in dimension but not in shape.

The effect can best be illustrated by an example and the synthesis of a flat topped pattern from an aperture of eight zones will be given [44]. The pattern is to be described by

$$\begin{aligned} g(u) &= 1 \text{ for } -u_0 < u < u_0 \\ &\quad 0 \qquad u_0 < |u| < N\pi \end{aligned}$$

where $N =$ the number of zones in this case 8. That is

$$\begin{aligned} m^2 v g(mv) &= m^2 v \quad |v| < v_0 \\ &= 0 \qquad |v| < v_0 \qquad v_0 < 1 \end{aligned} \tag{5.10.17}$$

From Equation 5.10.10 or 5.10.14

$$A_n \lambda_n = \frac{2}{[J_2(\lambda_n)]^2} \int_0^{v_0} m^2 v^2 J_1(\lambda_n v)\, dv \tag{5.10.18}$$

and $m = \lambda_N$, where the integrals are

$$\begin{aligned} I &= \int_0^{v_0} v^2 J_1(\lambda_n v)\, dv = v_0^3 \int_0^1 t^2 J_1(\lambda_n v_0 t)\, dt \\ &= v_0^3 \frac{J_2(\lambda_n v_0)}{\lambda_n v_0} \end{aligned}$$

either by the method of circle polynomials or Sneddon [45]. Hence

$$A_n \lambda_n = \frac{2u_0^2 \lambda_1^2}{[J_2(\lambda_n)]^2} \frac{J_2(\lambda_n u/\lambda_N)}{\lambda_n} \tag{5.10.19}$$

These coefficients have first to be substituted into the summation condition of Equation 5.10.6. This gives a curve of

$$\sum_{n=0}^{N} \frac{A_n \lambda_n^2}{2\lambda_N}$$

as a function of u_0 as shown in Figure 5.22(a).

Only those values of u_0 which equal the required value of $g(0)$ can then give patterns with correct value at the origin. Other selections of u_0 would have other values of $g(0)$ and hence different shaped patterns would result. These discrete crossings show the relationship between the summation function and the proper set of radiation patterns as illustrated in the figure. The narrowest pattern is that of an ordinary pencil beam with a monotonically decreasing in-phase amplitude distribution among the zones. The widest fills

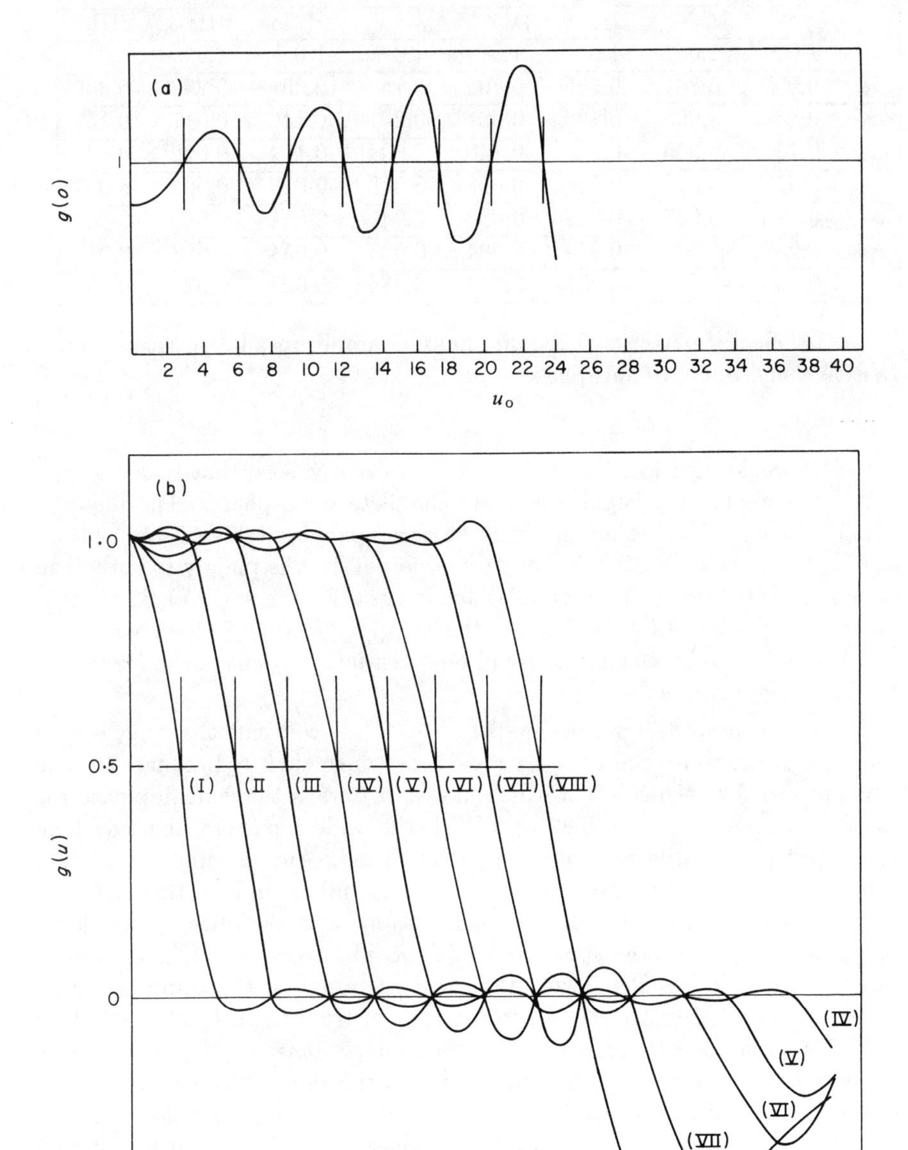

Figure 5.22: (a)Summation function for proper beam widths (b) Sector patterns - Fourier-Bessel solution

	I	II	III	IV	V	VI	VII	VIII
a_1	1.0	1.0	1.0	1.0	1.0	1.0	1.0	1.0
a_2	0.95	0.84	0.605	0.402	0.15	-0.05	-0.257	-0.345
a_3	0.86	0.60	0.186	-0.08	-0.20	-0.11	0.017	0.162
a_4	0.74	0.335	-0.102	-0.142	0.004	0.114	0.022	-0.10
a_5	0.67	0.07	-0.156	0.018	0.078	-0.025	-0.08	0.051
a_6	0.423	-0.057	-0.057	0.083	-0.04	-0.015	0.039	-0.038
a_7	0.285	-0.10	0.044	0.004	-0.033	0.054	-0.058	0.012
a_8	0.136	-0.067	0.057	-0.044	0.036	-0.025	0.02	-0.006

Table 5.1: Sector patterns. Zone illumination amplitudes for sector patterns. Negative sign refers to anti-phase.

the entire region for which $u = 2a \sin\theta/\lambda$ has real values, that is $u < 2\pi a/\lambda$ and corresponds to a beam filling the complete hemisphere. The required amplitudes for the zones are given in the accompanying table (Table 5.1). Of course, the physical realization of such wide patterns is mainly hypothetical since it is based upon the original scalar integral for the small angle approximation (Equation 4.1.1) and omits the obliquity factor of $(1 + \cos\theta)$. For closer approximation this factor could be taken into account in the manner of the following example.

A further interesting result manifests itself if we continue the range of u in the computation into the complex region for which $\sin\theta > 1$. A far out side lobe appears for each solution which moves in toward the widening patterns as can be seen in Figure 5.22(b). With the widest pattern the side lobe is greatest and contiguous with the pattern itself. Any attempt to create a super-wide pattern, the reverse of super directive in principle, is then defeated by the mutual destruction of the main pattern and the out-of-phase lobe. Mathematically it has the appearance of the Gibbs overshoot of linear Fourier series. Physically, as is known, "radiation patterns" in the complex region represent stored inductive energy in the form of trapped surface waves. This would be created if there were interaction effects between the zones. This concept is in accord with the amplitude distributions shown in the table, where for the widest pattern, that with the highest lobe and therefore greatest stored energy, the zones are alternative in phase, a condition which would be expected to give rise to such inductive coupling between the zones. This is remarkable in so far as it has been produced by a purely scalar theory, yet has the final result that would be expected to arise from a full field and mode expansion. Other effects of interest to be observed from the patterns shown are the half amplitude points which occur at values of u corresponding to the crossing values of the summation curve. This is analogous to the geometrical shadow of an edge which crosses the diffraction shadow at the same level. The common set of zeros for all the patterns occurs in similar solutions and these

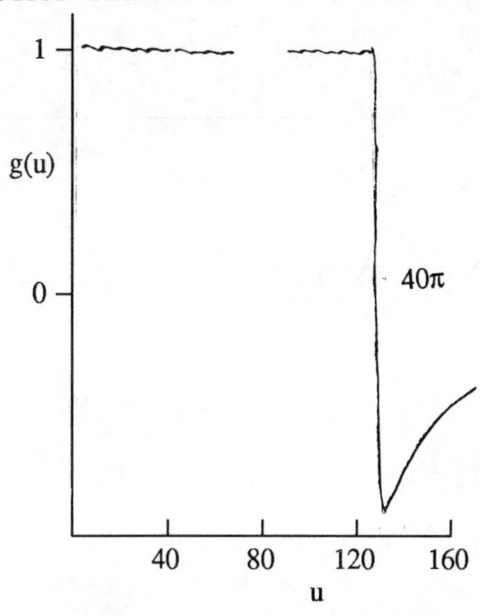

Figure 5.23: Pattern from aperture of 40 zones

zeros occur at the turning values of the summation function.

To investigate the overshoot lobe the result of the analysis for a 40 zone aperture is shown in Figure 5.23. The cut off for the beam is now much sharper and the resultant overshoot much greater. It would appear that in the limit of an exactly square pattern the overshoot would have the same magnitude as the pattern itself.

In an identical manner the derivation of a shaped symmetrical pattern of an increasingly useful kind is shown in the next example. These are for "parabolic" patterns of the geometric type

$$g(u) = 1 + p^2u^2 \tag{5.10.20}$$

where p can be chosen to accommodate such patterns (or their binomial approximations) as $g(\theta) = \sec^2\theta$, $\sec^2\theta/2$ or to take into account the obliquity factor $(1 + \cos\theta)$.

Such patterns have increasing amplitude with increasing u and are generally desirable when the *uniform* illumination of a curved surface is required. These encompass the illumination of the surface of the earth from a stationary satellite antenna and the more efficient illumination of a paraboloid from a source at its focus [46] as observed previously (Equation 1.7.8).

The patterns have similar properties to the square topped beam and one of the set is shown in Figure 5.24. The zone illumination amplitudes are obtained from the result

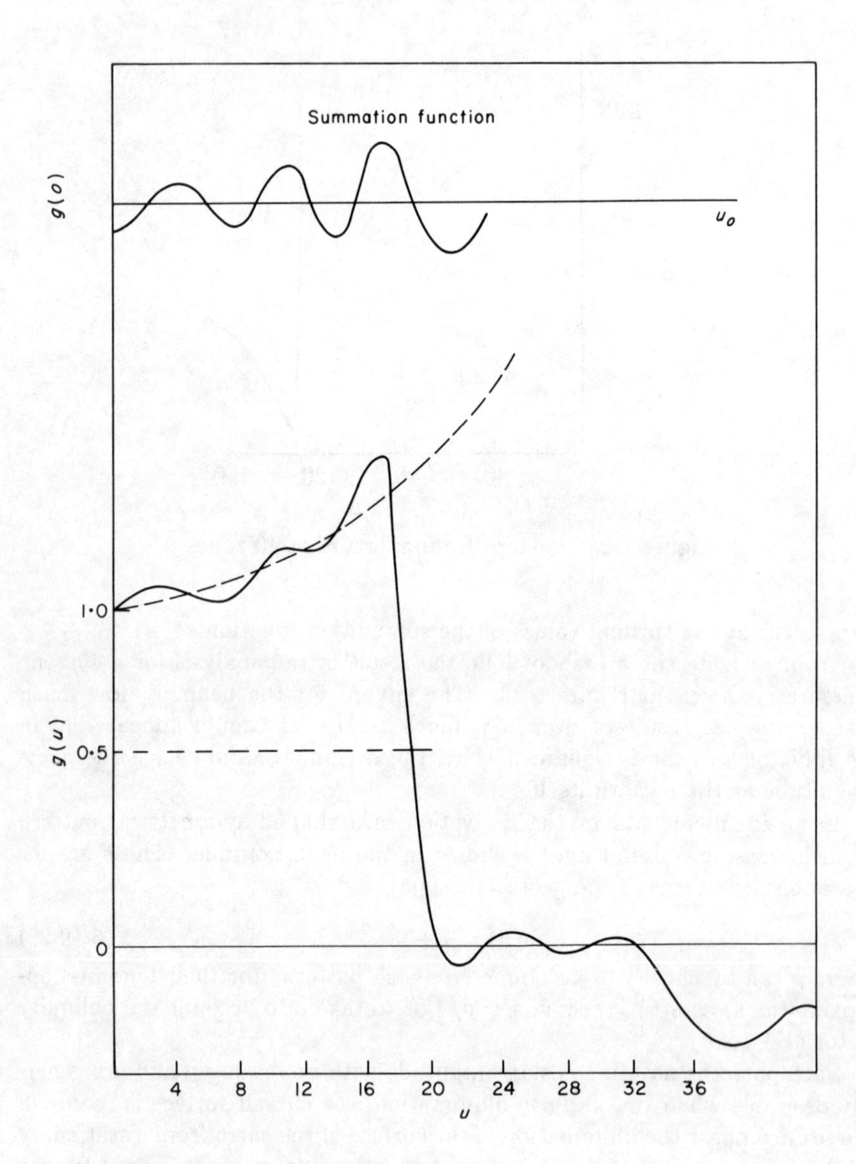

Figure 5.24: Secant squared pattern - Fourier-Bessel solution

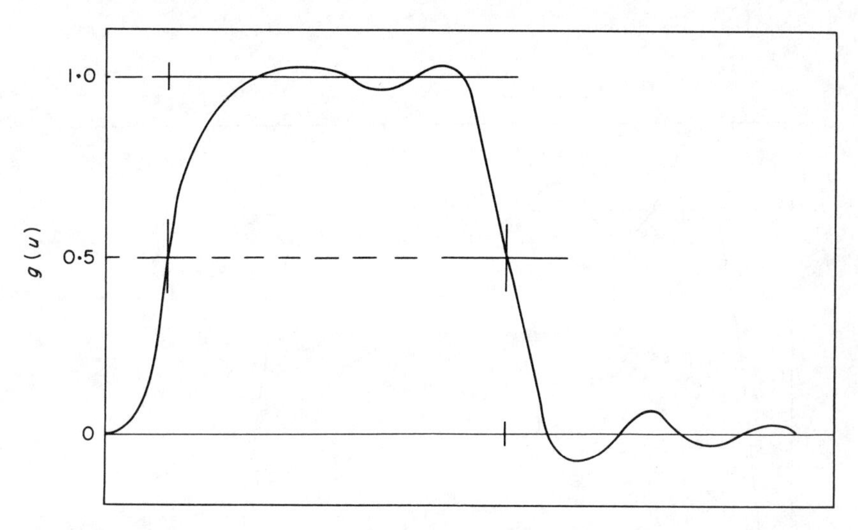

Figure 5.25: Conical pattern by superposition. Pattern created by the subtraction of pattern No. I from pattern No. V in Figure 5.22(b)

$$A_n\lambda_n = \frac{2}{\lambda_n[J_2(\lambda_n)]^2}\left[(u_0^2+\frac{2}{3}p^2u_0^4)\frac{J_2(\lambda_n u/\lambda_N)}{\lambda_n} - \frac{p^2u_0^4}{3}\frac{J_4(\lambda_n u/\lambda_N)}{\lambda_n}\right] \tag{5.10.21}$$

A consequence of this analysis allows the generation of patterns with a central hollow of any given depth, including conical beam shapes with a central null of any width. More than one method is available for this derivation.

1. Superposition
   If any two of the square topped patterns are superimposed in anti-phase and with due regard to their absolute gains, a pattern with a central null of the width of the smaller of the two patterns can be obtained. The required zone amplitudes are then the appropriate subtraction of the zone amplitudes of the two patterns concerned (Figure 5.25).

2. By direct application to the function

$$\begin{aligned} g(u) &= 1 \; u_0 < u < u_1 \\ &= 0 \text{ all other values of } u \end{aligned}$$

   Then

$$\begin{aligned} A_n\lambda_n &= \frac{2m^2}{[J_2(\lambda_n)]^2}\int_{v_0}^{v_1} v^2 J_1(\lambda_n v)\,\mathrm{d}v \qquad (5.10.22) \\ &= \frac{2}{\lambda_n[J_2(\lambda_n)]^2}\left[u_1^2 J_2\left(\frac{\lambda_n u_1}{\lambda_N}\right) - u_0^2 J_2\left(\frac{\lambda_n u_0}{\lambda_n}\right)\right] \end{aligned}$$

   for those values of u_0 and u_1 which make $g(0) = 0$.

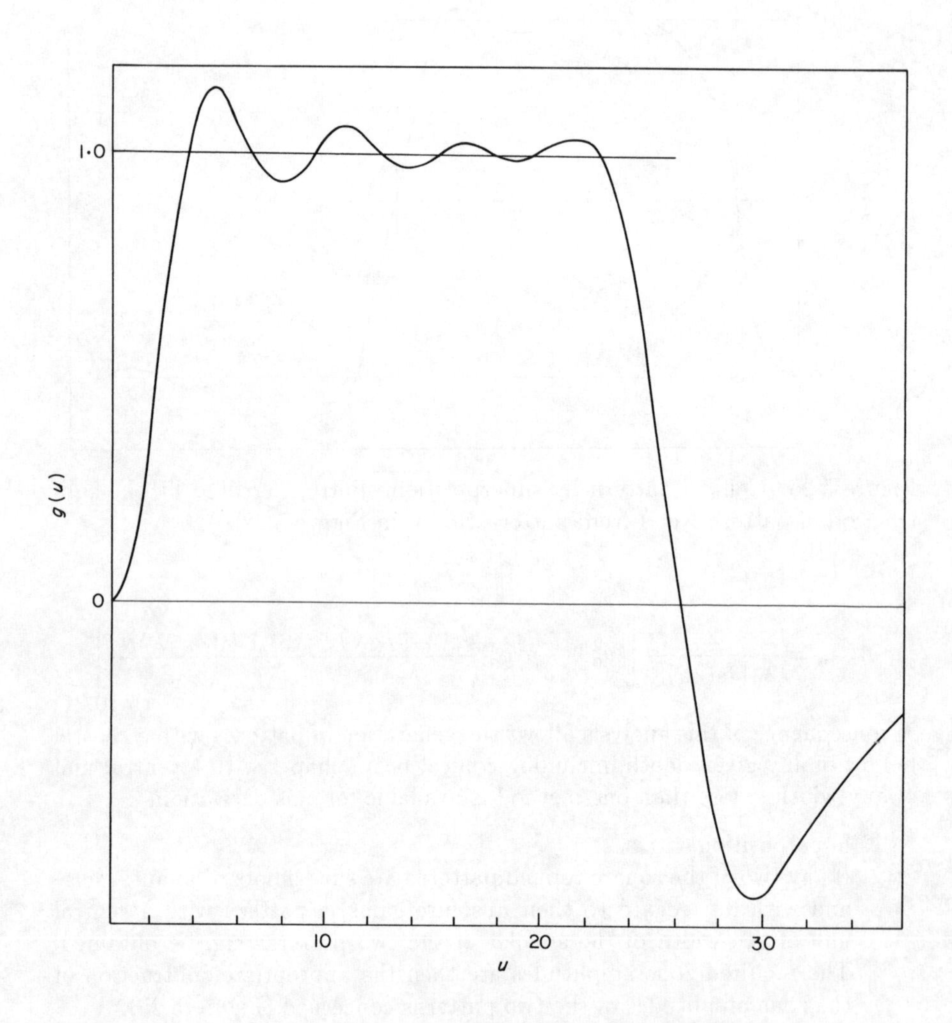

Figure 5.26: Conical pattern created by $g(u) = 0$

3. By the choice of a value of u for which the summation function Equation 5.10.6 gives the value $g(0) = 0$ instead of its "proper" value $g(0) = 1$.

   This procedure compels the pattern to start off at a zero value after which it recovers to give the remaining flat topped portion. A pattern based on this process is illustrated in Figure 5.26.

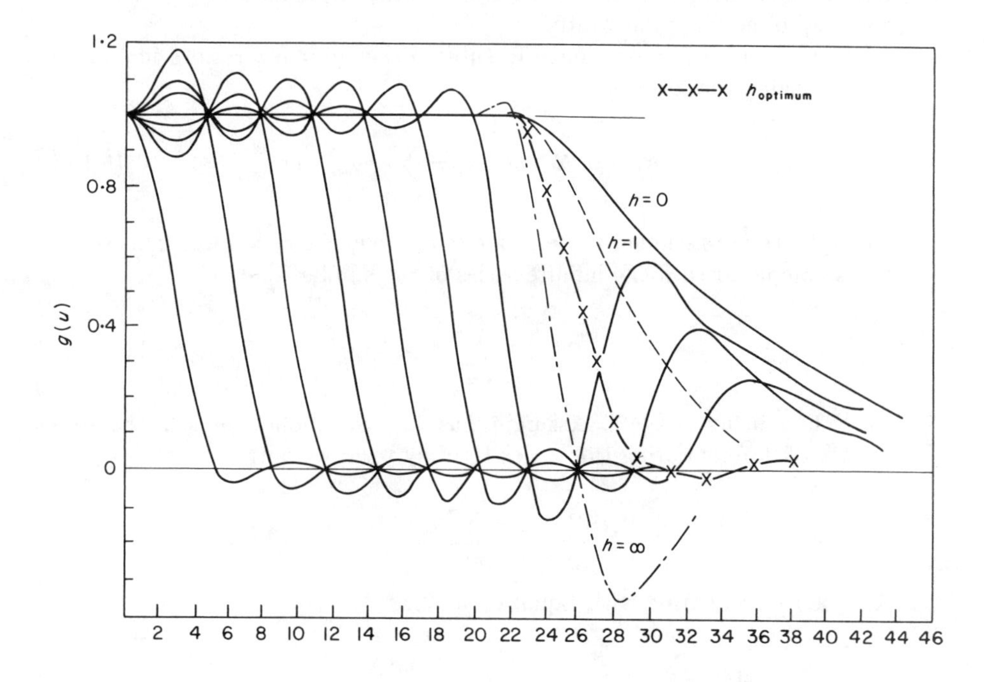

Figure 5.27: Sector patterns. Dini series solution

The patterns derived by using a Dini series instead of the Fourier-Bessel series have only two fundamental differences from the patterns so far described. Comparing the patterns for a flat topped beam from an aperture of eight zones by each method, shows that apart from the central pencil beam pattern, the approximation to a flat topped beam in the other members of the set is not as good as are the Fourier-Bessel patterns. This is to be expected since the latter is theoretically the best mean-square approximation. The result is higher side lobe levels for the Dini patterns and a less uniform level over the flat region of the pattern. However, the really important effect occurs for the widest patterns. These are all virtually identical except for the cut-off of the square beam. The effect is shown in Figure 5.27. The value of h gives the kind of Dini series from the expressions in Equations 5.10.14 to 5.10.16. As shown the effect of increasing h from zero to infinity is to increase the sharpness of the cut-off from a gradual slope to the maximum at which level the overshoot lobe is at its greatest. Obviously, a choice of value for h can be made which optimizes the slope of the cut-off of the beam and the level of the overshoot. One such possibility is guessed at in the figure.

CIRCULARLY SYMMETRIC PATTERNS - SCHLÖMILCH SERIES

An entirely different form of series results with the choice of zone radii to give zones all of equal radial width.

In this case $r_n = nr_1$ and the substitution $u = Nv$ results in Equation 5.10.4 having the form

$$F(v) = N^2 v g(Nv) = \sum_{n=1}^{N} nA_n J_1(nv) \tag{5.10.23}$$

where there are assumed to be N zones and whence $r_1 = 1/N$. This series is to be compared with the infinite series of the similar kind

$$S(v) = \sum_{n=1}^{\infty} A_n J_0(nv)$$

the Schlmölich series [47]. Assuming, as does Schlmölich himself, the admissibility of differentiating this series term by term we obtain

$$S'(v) = -\sum_{n=1}^{\infty} nA_n J_1(nv) \tag{5.10.24}$$

And hence comparing with Equation 5.10.23

$$S(v) = -\int^{v} N^2 v g(Nv)\, \mathrm{d}v \tag{5.10.25}$$

Then [6, p630] the coefficients A_n of Equation 5.10.23 are given by

$$A_n = \frac{1}{\pi}\int_{-\pi}^{\pi}\int_{0}^{\pi/2} \sec\phi \frac{\mathrm{d}}{\mathrm{d}\phi}[S(v\sin\phi)] \cos nv \,\mathrm{d}\phi\,\mathrm{d}v \tag{5.10.26}$$

where again the form of Equation 5.10.23 insists that

$$g(0) = \sum_{n=1}^{N} \frac{n^2 A_n}{N^2 2} \tag{5.10.27}$$

This replaces the summation condition of the previous section. For a flat topped beam with

$$g(u) = 1 \quad u < u_0$$
$$g(u) = 0 \quad u_0 < u < N\pi$$

we find

$$A_n = \frac{-2N^2}{\pi}\left[\frac{u_0^2}{N^2 n}\sin\left(\frac{nu_0}{N}\right) + \frac{2u_0}{Nn^2}\cos\left(\frac{nu_0}{N}\right) - \frac{2}{n^3}\sin\left(\frac{nu_0}{N}\right)\right] \tag{5.10.28}$$

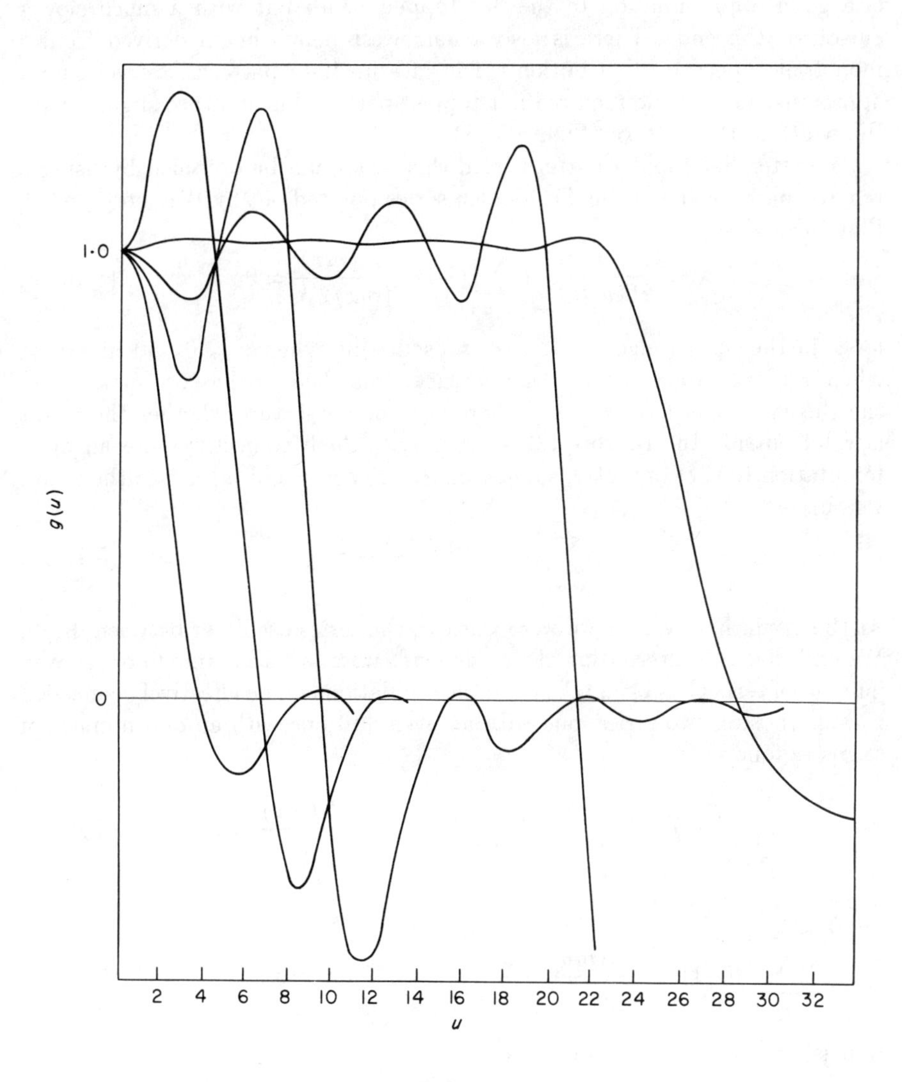

Figure 5.28: Sector patterns. Schlömilch solution

the only complication being integration by parts. The widest of these forms is a good approximation to the flat topped beam but with a much slower cut-off at the end. There is also a narrowest pencil beam derived from a monotonic aperture distribution. The intermediate patterns are very poor approximations to the required flat topped pattern but improve slightly with the width of the pattern (Figure 5.28)

A better flat topped pattern from this series can be obtained by using a remarkable property of the Schlömilch series pointed out by Watson [6, p636] that the series

$$\frac{1}{2\Gamma(\nu+1)} + \sum_{m=1}^{\infty}(-1)^m \frac{J_\nu(mx)}{(mx/2)^\nu} = 0 \tag{5.10.29}$$

for x in the *open* range $-\pi < x < \pi$ oscillating when $x = 0$ and diverging when $x = \pi$. As stated in the reference, this theorem has no analogue in the theory of Fourier series and there is some conjecture whether the result is valid outside the range $-1/2 < \nu < 1/2$. Such conjectures are an open temptation to the computer, so putting $\nu = 1$, $x = v$ and $|v| < \pi$ in the above we obtain

$$2\sum_{m=0}^{\infty}(-1)^{m-1}\frac{J_1(mv)}{mv} = \frac{1}{2} \tag{5.10.30}$$

At the origin however, as we were warned, the series oscillates between $\pm 1/2$. We find that this alternation of sign depends upon whether an odd or an even number of terms has been taken. So the oscillation can be effectively cancelled by superposing two series, one with an even and one with an odd number of terms as follows

$$\sum_{m=1}^{8} 2(-1)^{m-1}\frac{J_1(mv)}{mv} + \sum_{m=1}^{7} 2(-1)^{m-1}\frac{J_1(mv)}{nv} = 1 \tag{5.10.31}$$

That is

$$\sum_{m=1}^{7} 4(-1)^{m-1}\frac{J_1(mv)}{mv} - \frac{J_1(8v)}{4v} = 1 \quad \text{for } 0 < v < \pi \tag{5.10.32}$$

that is for $u < N\pi$ with in this case $N = 8$.

The A_n coefficients and hence the zone amplitudes are already written in Equation 5.10.32 that is

$$\frac{nA_n}{N^2} = \frac{4(-1)^{n-1}}{n} \qquad 1 < n < 7$$

and

$$\frac{A_8}{N} = -\frac{1}{4}$$

The resultant pattern is as shown in Figure 5.29.

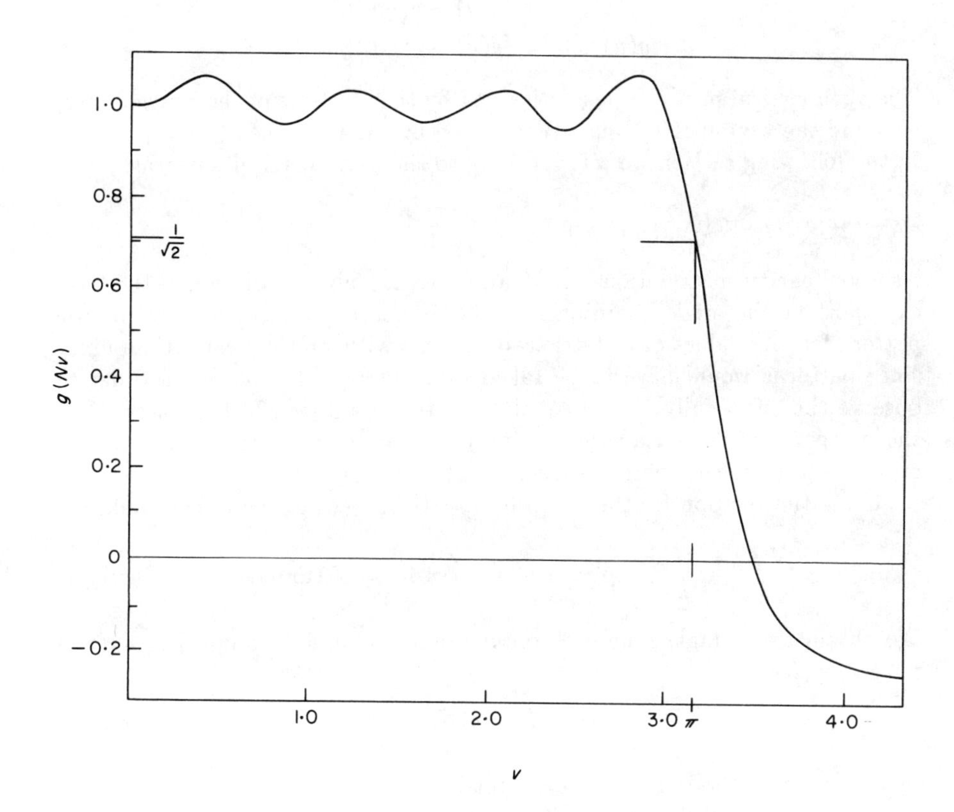

Figure 5.29: Sector pattern - null function solution

5.11 NON-SYMMETRICAL PATTERNS

Any general pattern $g(u)$ can be separated into its symmetrical and asymmetrical parts by the usual method

$$\begin{aligned} g(u)_{sym} &= [g(u) + g(-u)]/2 \\ g(u)_{asym} &= [g(u) - g(-u)]/2 \end{aligned}$$

The symmetrical part can be synthesized by the methods of the previous section and the asymmetrical part then added by the principle of superposition. In the following section we refer entirely to the asymmetrical patterns.

CYCLIC PHASE VARIATION

As noted earlier in Equations 5.1.10 and 5.1.12, a phase variation of the form $\exp(ip\phi')$ in the angle coordinate of the aperture produces an asymmetric pattern for odd values of p. Little can be done with values greater than unity since patterns would have to be taken in a number of ϕ planes in order to observe the effect fully. For p equal to unity the major effect is observed in the plane $\phi = \pi/2$ and a series definition for patterns in that plane can be obtained from zoned apertures also.

Using the relation for the integration with respect to the ϕ' coordinate

$$\int_0^{2\pi} \exp[iur\sin(\phi' - \phi)]\, d\phi' = J_1(ur) \tag{5.11.1}$$

We obtain for a single zone with outer radius r_n uniform amplitude a_n and phase $\exp(i\phi')$

$$g_n(u) = \int_{r_{n-1}}^{r_n} a_n J_1(ur) r\, dr \tag{5.11.2}$$

Invoking the method of circle polynomials to evaluate this integral

$$\int_0^1 J_1(ur) r\, dr = \sum_{s=1,3,5} \frac{2(s+1)}{s(s+2)} \frac{J_{s+1}(u)}{u} \tag{5.11.3}$$

then for a single zone

$$g_n(u) = \sum_{s=1,3,5} \frac{2(s+1)}{s(s+2)} \left[r_n^2 \frac{J_{s+1}(r_n u)}{r_n u} - r_{n-1}^2 \frac{J_{s+1}(r_{n-1} u)}{r_{n-1} u} \right] \tag{5.11.4}$$

Summing as before for N zones gives

$$g(u) = \sum_{n=1}^{N} \frac{A_n r_n}{u} \left[\frac{4}{3} J_2(ur_n) + \frac{8}{15} J_4(ur_n) + \frac{12}{35} J_6(ur_n) + \ldots \right] \tag{5.11.5}$$

where $A_n = a_n - a_{n+1}$ and $a_{N+1} = 0$ as before.

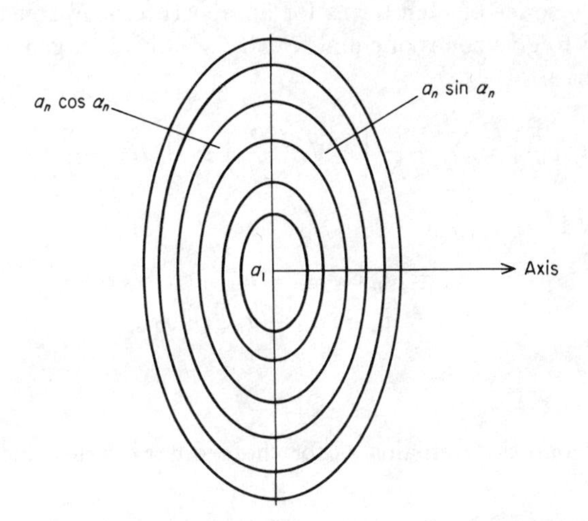

Figure 5.30: Aperture of phased semi-circular zones

With the same transformations as Equations 5.10.2 and 5.10.3

$$m^2vg(mv) = \sum_{n=1}^{N} A_n\lambda_n \left[\frac{4}{3}J_2(\lambda_n r_n) + \frac{8}{15}J_4(\lambda_n r_n) + \ldots\right] \tag{5.11.6}$$

This series does not lend itself in any obvious way to the synthesis process but could of course be applied to a trial and error procedure. Equation 5.11.5 being a completely odd function of u provides the asymmetrical part of the pattern as required.

THE GENERALIZED SCHLÖMILCH SERIES

Consider the aperture of N zones to be divided by a diameter, and that the excitation phase of the opposite halves of each zone be $\pm\alpha_n$ radians with respect to a nominal zero of phase which is the same for all zones (Figure 5.30). The real part of the amplitude of each zone a_n is considered to be constant. Then from the relation given in Equation 5.1.9 for a complete aperture divided in this way, we have for a *single* zone

$$\begin{aligned} g_n\left(u, \pm\frac{\pi}{2}\right) = a_n \Bigg[& \cos\alpha_n \left(r_n^2 \frac{J_1(ur_n)}{ur_n} - r_{n-1}^2 \frac{J_1(ur_{n-1})}{ur_{n-1}} \right) \\ & \pm \sin\alpha_n \left(r_n^2 \frac{H_1(ur_n)}{ur_n} - r_{n-1}^2 \frac{H_1(ur_{n-1})}{ur_{n-1}} \right) \Bigg] \end{aligned} \tag{5.11.7}$$

where $H_1(x)$ is the Struve function.

Summing a series of such terms for an aperture of N zones, and choosing the zone radii to give zones of equal width * we obtain a generalization of the Schlömilch series

$$g\left(u, \pm\frac{\pi}{2}\right) = \sum_{n=1}^{N} \frac{n}{uN}\left[A_n J_1\left(\frac{un}{N}\right) \pm B_n H_1\left(\frac{un}{N}\right)\right] \tag{5.11.8}$$

where we now have

$$\begin{aligned} A_n &= a_n \cos\alpha_n - a_{n+1}\cos\alpha_{n+1} \\ B_n &= a_n \sin\alpha_n - a_{n+1}\sin\alpha_{n+1} \end{aligned}$$

and

$$a_{N+1} = 0 \tag{5.11.9}$$

One limiting condition remains as for the ordinary series since $H(x)/x \to 0$ as $x \to 0$

$$g(0) = \sum_{n=1}^{N} \frac{n^2 A_n}{N^2 2}$$

Putting $u = Nv$, $-\pi < v < \pi$, we obtain the associated function

$$N^2 v g(Nv) = \sum_{n=1}^{N} n A_n J_1(nv) \pm B_n H_1(nv) \tag{5.11.10}$$

In this relation, the first series represents the symmetrical part of the pattern and the second series the anti-symmetrical part.

We compare the series given in Equation 5.11.10 with the complete Schlömilch series [6, Chapter XIX]

$$S(v) + R = \sum_{n=1}^{\infty} A_n J_0(nv) + B_n H_0(nv) \tag{5.11.11}$$

where R is a constant to be determined.

Integrating Equation 5.11.10 we have

$$-\int^{v} N^2 v g(Nv)\,\mathrm{d}v = \sum_{n=1}^{N} A_n J_0(nv) + B_n H_0(nv) - \frac{2B_n}{\pi} \tag{5.11.12}$$

Then letting

$$S(v) = -\int^{v} N^2 v g(Nv)\,\mathrm{d}v$$

requires

$$\sum_{n=1}^{N} \frac{2B_n}{\pi} = 0 \tag{5.11.13}$$

---

*This condition is essential since $H_1(x)$ has no roots outside $x = 0$ and hence a Fourier-Struve series would not be possible for the odd functional part of this expansion

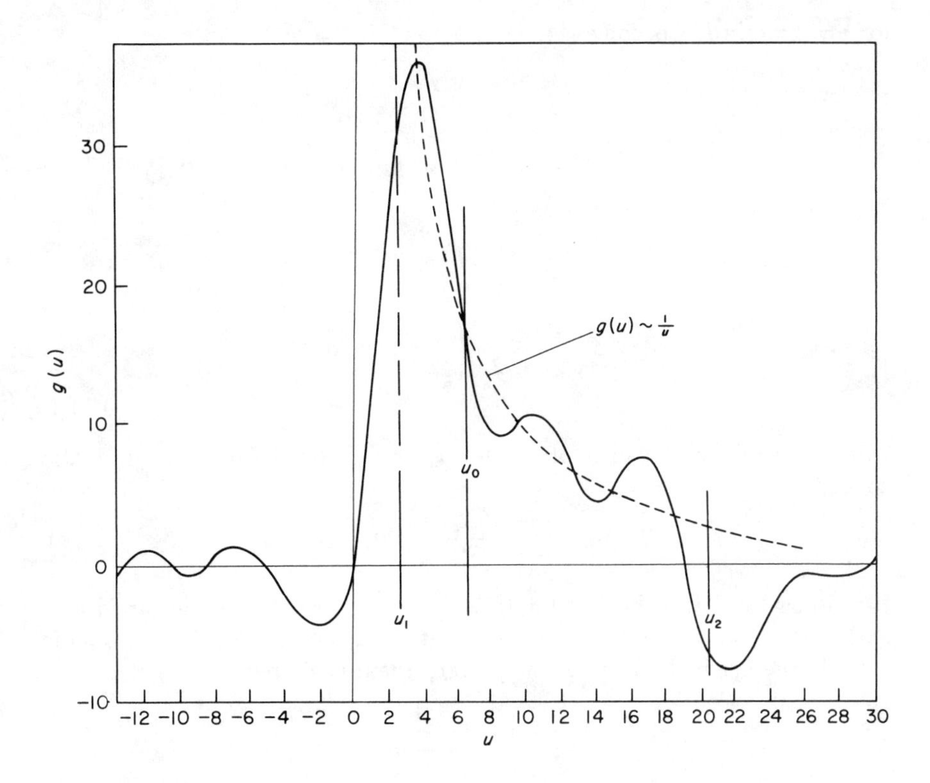

Figure 5.31: Theoretical cosecant pattern Schlömlich series

the finite series can be compared with an infinite series to which it forms an approximation. Then, provided $\sum_{n=1}^{\infty} B_n = 0$ we have from [6, p630]

$$A_n = \frac{1}{\pi}\int_{-\pi}^{\pi}\int_{0}^{2\pi} \sec\phi \frac{\mathrm{d}}{\mathrm{d}\phi}[S(v\sin\phi)]\cos nv\,\mathrm{d}\phi\,\mathrm{d}v$$

and

$$B_n = \frac{1}{\pi}\int_{-\pi}^{\pi}\int_{0}^{2\pi} \sec\phi \frac{\mathrm{d}}{\mathrm{d}\phi}[S(v\sin\phi)]\sin nv\,\mathrm{d}\phi\,\mathrm{d}v \tag{5.11.14}$$

where the constant R disappears in the internal differentiations.

From A_n and B_n we now obtain the amplitudes and phases of the illumination functions of the individual zones from Equation 5.11.9.

For illustration we can generate the often required cosecant^2 pattern [48] [5] given by

$$|g(u)|^2 = \operatorname{cosec}^2\theta$$

for the range $\theta_1 < \theta < \theta_2$ not containing the axis $\theta = 0$ (Figure 5.31). This, for $u = 2\pi a \sin\theta/\lambda$ becomes

$$g(u) = 1/u \text{ for } u_1 < u < u_2$$

and hence

$$g(Nv) = 1/(Nv) \text{ for } v_1 < v < v_2$$

We have directly from Equation 5.11.14

$$\begin{aligned} A_n &= -\frac{N}{2}\int_{v_1}^{v_2} v\cos nv\,dv \\ B_n &= -\frac{N}{2}\int_{v_1}^{v_2} v\sin nv\,dv \end{aligned} \tag{5.11.15}$$

where "permissible" values of v_1 and v_2 are obtained from the summation condition

$$\sum_{n=1}^{N}\frac{n^2 A_n}{N^2 2} = g(0)$$

It can now be shown that the limitation on the range of θ can be varied by an axis shift to a mean value of the pattern at u_0 say (Figure 5.31) corresponding to θ_0 in the original pattern. The pattern then, given by

$$g(u) = \frac{1}{u - u_0}$$

results in the *same* relations for A_n and B_n but now of course v_1 and v_2 can be on opposite of the *new* axis and the value of $g(0)$ in the summation condition becomes

$$g(0) = -\frac{1}{u_0}$$

Some computational procedure is required to obtain the values of v_1 and v_2 and the results for one such pattern are shown in Figure 5.31.

5.12 ZONE PLATES

AMPLITUDE ZONE PLATES

One obvious application of the zone theory is in the construction and application of the Fresnel zone plate. These as used in optics are essentially for the focusing of rays at a local point on the axis and thus the optical theory invariably incorporates the focusing factor $\exp(ibr^2)$. The theory given above is essentially simplified by the omission of such a factor for the far field of a microwave antenna. Unfortunately, most of the published optical results do not allow for the simple limit $b \to 0$ to be made which would allow for a transition to be made between the two theoretical approaches.

This problem does not arise if the zone plate is considered to be an *addition* to an already collimated beam of rays. In practical terms this implies the addition of the zone plate to the existing aperture of, for example, a focused paraboloid. For narrow enough zones (the minimum being $\lambda/2$) the amplitude distribution can be considered to be constant over the zone and the foregoing analysis applied directly.

Used in the reverse mode as a collimating device in its own right, the zone plate has major drawbacks mainly in relation to its aperture efficiency. When illuminated by a point source, the amplitude at each zone could still be assumed to be constant, but an additional "spherical" phase term has to be incorporated making the a_n coefficients complex.

The simplicity of the series resulting in the relation in Equation 5.10.3 was the objective for its use in a pattern synthesis procedure. It, of course, is directly applicable to the more simple problem of presenting the major aspects of a radiated pattern from a *known* distribution of zone amplitudes and phases, and hence to the Fresnel zone plate whether used as a collimating device or as a pattern shaping device added to an already collimated pattern.

Several quite interesting results can be achieved by inserting quite arbitrary distributions of zone radii, amplitudes and phases, into the coefficients and obtaining resultant radiation patterns.

In a standard form of Fresnel zone plate the zones are alternately transparent and opaque and the radii are proportional to the square root of the zone number [49]. That is with

$$r_n = \beta\sqrt{n} \qquad \beta = \sqrt{\lambda f}$$

f = focal distance

$$g(u) = \sum_{n=1}^{N} \beta A_n \sqrt{n} \frac{J_1(\beta u \sqrt{n})}{u} \tag{5.12.1}$$

where A_n are alternatively positive and negative. The actual form of A_n depends on the illumination function creating the appropriate a_n and upon whether the central zone is transparent or opaque.

PHASE ZONE PLATES

Zone plates which are totally transparent do not have the drawback of inefficiency associated with the alternative opaque zones of the amplitude zone plate. Such a plate relies for its action in creating phase shift zones, and has both a collimating and beam shaping action as does the amplitude zone plate. They are a natural form of Huygens lens.

The action can again be directly evaluated from the series form of Equation 5.10.3 when added to an already collimated aperture. In a practical form this would appear as a sheet of dielectric material with zones of various depths acting as phase shifters over those regions as shown in Figure 5.32. It is a

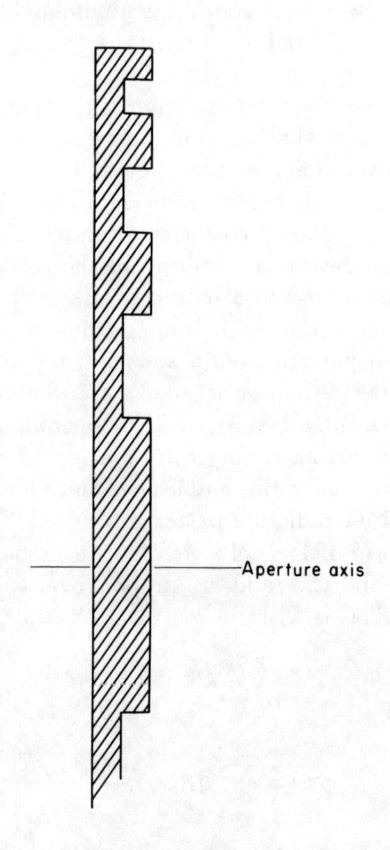

Figure 5.32: Microwave phase zone plate

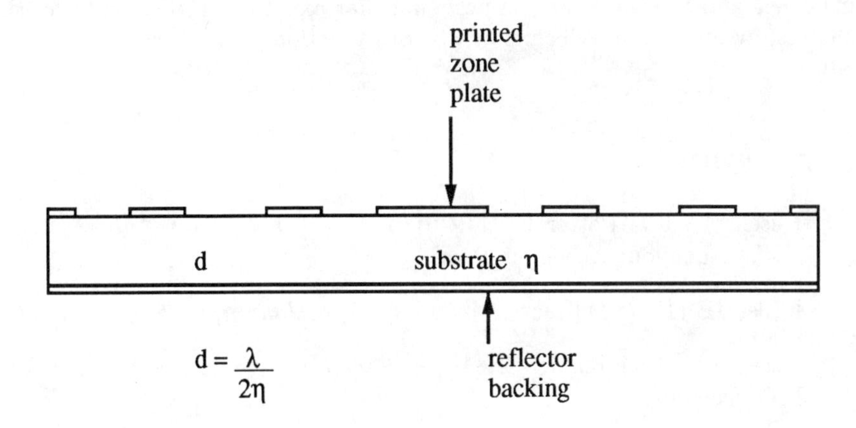

Figure 5.33: Reflector backed zone plate

close approximation to a stepped lens [50] but with an additional degree of freedom not available to a simple collimating device. It, in fact, provides the additional surface, which could be applied to any one optical advantage. The design procedure given above applies directly to obtain the pattern from an antenna incorporating such a plate.

MODIFIED ZONE PLATES

There are several modern adaptations of the zone plate as a receiving antenna, suitable for use at millimeter wavelengths in systems where low efficiency can be tolerated [52]. The first was the noticeable increase in efficiency, from about 10% to nearer 40%, that is obtained by converting the standard Fresnel zone plate of alternative opaque and transparent circular zones, into a reflector by backing it with a half-wave distance (in the medium of the supporting substrate) plane reflector [51] [52] [53] (Figure 5.33).

Continuing this concept to the phase correcting zone plate reproduces the Huygens reflector of Section 3.10 with square stepped zones instead of the tapered ones shown there (Figure 3.52).

The great advantage of zone plates as focusing devices or lens reflectors, is their flat construction and use of lightweight materials. However there is an advantage to be gained by considering zone plates printed on non-planar surfaces such as spheres or paraboloids [54]. This arises mainly because a larger number of zones can be accommodated within a specified aperture radius. In this respect the paraboloid surface presents the greatest advantage.

Finally an original concept of Wright [55] shows that all the previous designs based on circular concentric zones, can be converted without loss to confocal elliptical zones. In that case the focusing occurs at offset angles of incidence and receiver from the perpendicular axis through the centre of the ellipses, by an amount which is an obvious function of the eccentricity of the ellipses.

REFERENCES

1. Rusch WVT & Potter PD (1970) *Analysis of Reflector Antennas* New York & London: Academic Press

2. Keller JB (1957) Diffraction by an aperture *Jour App Phys* **28** p426

3. Felsen LB & Marcuvitz N (1973) *Radiation and Scattering of Waves* Prentice Hall

4. Bouwkamp CJ (1954) Diffraction theory *Reports in Progress in Physics* **Vol XVII** p35

5. Silver S (1949) *Microwave Antenna Theory and Design* MIT Series **Vol12** McGraw Hill p192

6. Watson GN (1948) *Bessel Functions* Cambridge University Press

7. Erdelyi A et al (1954) *Tables of Integral Transforms* **Vol 2** McGraw Hill

8. Cornbleet S (1959) The diffraction fields of a non-uniform circular aperture *Symposium on EM Theory and Antennas* ed Jordan EC Pergamon Press p157

9. Lansraux G (1953) Conditions functionelles de la diffraction instrumentalle cas particulier des zeros d'amplitude de figure de diffraction de revolution *Cahier de Physique* **45** p29

10. Pinney E (1946) Laguerre functions in the mathematical foundations of the electromagnetic theory of the paraboloid reflector *Jour Maths and Phys* **25** 49 and **26**

11. Afifi MS (1967) Radiation from a paraboloid of revolution, Electromagnetic wave theory **Part 2** ed Brown J *Delft Symposium 1965* Pergamon Press p669

12. Jahnke E & Emde F (1945) *Tables of Functions* Dover p180

13. McCormick GC (1959) *McGill Symposium on Microwave Optics* **Part 2** p363

14. Zernike F (1934) Beugungstheorie des Schneidenvefahrens *Physica* **Vol 1** p689

15. Chako N (1959) Characteristic curves in image space *McGill Symposium on Microwave Optics* **Part 1** p67

16. Moss SH (1964) Lommel transforms in diffraction theory *Trans IEEE* **AP12** p777

17. Cornbleet S (1966) Circular aperture pattern with ultra-low side-lobes *Electronics Letters* **2**(2) p79

18. Linfoot EH (1955) *Recent Advances in Optics* Oxford University Press p51

19. Born M & Wolf E (1959) *Principles of Optics* Pergamon Press p436

20. Minnett HC & Thomas BM (1968) Fields in the image space of symmetrical focusing antennas *Proc IEE* **115** (10) p1419

21. Matthews PA & Cullen AL (July 1956) *A Study of the Field Distribution at an Axial Focus of a Square Microwave Lens* IEE monograph no 186R

22. Cornbleet S (1973) Feed arrangement for the axis definition of a paraboloid reflector *Electronics Letters* **9** (3) p66

23. Hansen RC & Bailin IL (1960) A new method of near field analysis *Trans IRE* **AP7** Special Supplement pS458

24. Ming Kwei Hu (1960) Fresnel region field distributions of circular aperture antennas *Trans IRE* **AP8** p344

25. Ming Kwei Hu (1961) *Jour Res Nat Bur Stand Sect D* **65** (2) p137

26. de Size LK (December 1957) *Uniform, Cosine and Cosine Squared Illumination with a Curved Phase Front* AIL Report No 3585-4

27. Nijboer BRA (1947) The diffraction theory of optical aberrations *Physica* **13** (10) p605

28. Cornbleet S (1964) Asymmetric phase effects in the circular aperture *Symposium on Quasi-Optics* Polytechnic Institute of Brooklyn p487

29. McElvery RM & Smerczynski JE (1964-5) The gain of a defocused circular aperture *Quart Jour Appl Maths* **29** p319

30. McLeod JH (1954) The axicon: a new type of optical element *Jour Opt Soc Amer* **44** p592

31. McLeod JH (1960) Axicons and their uses *Jour Opt Soc Amer* **50** (2) p166

32. Cornbleet S (1970) Superdirectivity property of the microwave axicon *Proc IEE* **117** (5) p869

33. Popovkin VI, Shcherbakov GI & Yelumeyev VI (1969) Optimum solutions of problems in antenna synthesis *Rad Eng & Electronic Phys* **14** (7) p1025

34. Toraldo di Francia G (1952) Super gain antennas and optical resolving power *Supplement Nuovo Cimento* **IX** Series IX (3) p426

35. Hansen RC (1964) *Microwave Scanning Antennas* London & New York: Academic Press Vol 1 Chap 1 p67

36. Taylor TT (1960) Design of circular apertures for narrow beam width and low side lobes *Trans IRE* **AP8** (1) p17

37. Sinnott J (1966) Patterns for out of phase semi-circular apertures *Trans IEEE* **AP14** (3) p390

38. Ishimaru A & Held G (1960) Analysis and synthesis of radiation patterns from circular apertures *Can J Phys* **39** p78

39. Ruze J (1964) Circular aperture synthesis *Trans IEEE* **AP12** (6) p691

40. Koch GF (January 1968) Koaxialstrahler als Erreger fur Rauscharme Parabolantenn *FTZ*

41. Rebhan W *Theoretical Analysis of Antennas with Sector Shaped Radiation Patterns for Communications Satellites* Zentral Laboratorium fur Nachrichtentechnik Siemens and Halske AG Munich

42. Potter PD (1963) A new horn antenna with suppressed sidelobes and equal beamwidths *Microwave Journal* **6** p71

43. Cornbleet S & Brown J (9 June 1971) Circular array antenna, Brit Patent no 1234751

44. Cornbleet S (1966) Pattern synthesis from zoned circular apertures *Trans IEEE* **AP14** p646

45. Sneddon IN (1966) *Mixed Boundary Value Problems in Potential Theory* North Holland Publishing Co p27

46. Ajioka JS (1970) Shaped beam antenna for earth coverage from a stabilized satellite *Trans IEEE* **AP18** (3) p323

47. Schlömilch O (1857) On Bessel's function *Zeitschrift für Math und Phys* **2** p155

48. Silver S (1949) Microwave antenna theory and design *MIT Series* **Vol 12** p469

49. van Buskirk LF & Hendrix CE (1961) The zone plate as a radio frequency focusing element *Trans IRE* **AP9** p319

50. Sobel F, Wentworth FL & Wiltse JC (1961) Quasi-optical surface waveguide and other components for 100-300 GHz region *Trans IRE* **MTT9** p512

51. Wright TMB & Collinge G (23 January 1992) International Patent no W092/01319

52. Garrett JE & Wiltse JC (1991) Fresnel zone plate antennas at millimetre wavelengths *Int Jour IR & MM Waves* **12** no 3 p195

53. Guo YJ & Barton SK (1992) Flat printed zone plate antennas for DBS reception *International Broadcasting Conference, Amsterdam*

54. Khastgir P, Chabravorty JN & Dey KK (1973) Microwave paraboloid, spherical and plane zone plate antennas *Indian Jour Radio & Space Phys* **2** no 1, p47

55. Wright TMB Mawzones Developments Ltd GB

Diffraction Theory Table

56. Adrejewski W (1951) Rigorous theory of diffraction of plane EM waves at a perfectly conducting circular disc and a circular aperture in a perfectly conducting plane screen *Naturwissenschaften* **38** p406

57. Andrews CL (1960) *The Optics of the Electromagnetic Spectrum* Prentice Hall

58. Baker BB & Copson ET (1960) *The Mathematical Theory of Huygens' Principle* Oxford University Press

59. Booker HG & Clemmow PC (1950) The concept of an angular spectrum of plane waves and its relation to that of polar diagram and aperture distribution *Proc IEE* **97** (3) p11

60. Born M & Wolf E *Principles of Optics* Chapter 8 Section 3 p374 Pergamon Press

61. Bouwkamp CJ (1954) Diffraction Theory in *Reports on Progress in Physics* **XVII** p35 (including 500 refs)

62. Franz W (ref in Bouwkamp)

63. Heurtley JC (1973) Scalar Rayleigh-Sommerfeld and Kirchoff diffraction integrals. A comparison of exact evaluations at axial points *Jour Opt Soc Amer* **63** (8) p1003

64. Jackson JD (1962) *Classical Electrodynamics* John Wiley

65. Keller JB (1957) Diffraction by an aperture *Jour App Phys* **28** p426

66. Kline M & Kay IW (1965) *Electromagnetic Theory and Geometrical Optics* Interscience

67. Kottler F (1923) *Ann Phys* **71** p457

68. Kottler F (1967) Diffraction at a black screen in *Progress in Optics* Vol VI p333

69. Levine H & Schwinger J (1951) On the theory of electromagnetic wave diffraction by an aperture in an infinite conducting screen, in *Theory of EM Waves* Washington Square Symposium, Interscience

70. Marchand EW & Wolf E (1962) Boundary diffraction wave in the domain of the Rayleigh-Kirchoff diffraction theory *Jour Opt Soc Amer* **52** p76

71. Marchand EW & Wolf E (1970) Transmission cross section for small apertures in black screens *Jour Opt Soc Amer* **60** (11) p1501

72. Miyamoto K & Wolf E (1962) Generalization of the Maggi-Rubinowicz theory of the boundary diffraction wave *Jour Opt Soc Amer* **52** p615

73. Papas CH (1965) *The Theory of Electromagnetic Wave Propagation* McGraw Hill

74. Rubinowicz A (1965) The Miyamoto-Wolf diffraction wave, in *Progress in Optics* Wolf E (ed) **IV** p201 North Holland

75. Rusch WVT (1963) Scattering from a hyperboloid reflector in a Cassegrain feed system *Trans IEEE* **AP11** p414

76. Sancer MI (1968) An analysis of the vector Kirchoff equations and the associated boundary line charge *Radio Science* **3** (2) (new series) p141

77. Schelkunoff SA (1951) Kirchoff's formula, its vector analogue and other field equivalence theorems, in *Theory of EM Waves* Washington Square Symposium, Interscience p107

78. Severin H (1952) Methods of light optics for the calculation of the diffraction phenomena within the range of centimeter waves. Supplemento al vol IX serie IX *Nuovo Comento* no 3 p381

79. Shafer AB (1967) Hamilton's mixed and angle characteristic functions and diffraction aberration theory *Jour Opt Soc Amer* **57** (5) p630

80. Sherman GC (1969) Diffracted wave fields expressible by plane-wave expansions containing only homogeneous waves *Jour Opt Soc Amer* **59** (6) p697 also (1967) Integral transform formulation of diffraction theory *Jour Opt Soc Amer* **57** (12) p1490

81. Silver S (1949) *Microwave Antenna Theory and Design* MIT Radiation Laboratory Series **12**

82. Sommerfeld A (1964) *Optics* Academic Press, Chapter V, p179

83. Stratton JA & Chu LJ (1939) Diffraction theory of electromagnetic waves *Physics Review* **56** p99

84. Tai CT (1972) Kirchoff Theory: scalar vector or dyadic? *Trans IEEE* **AP20** p114

85. Vasseur JP (1952) Diffraction of electromagnetic waves by apertures in a plane conducting screen *L'Onde Electrique* **32** pp3, 55, 97

6
POLARIZATION

One of the subjects that clearly demonstrates the unity of physics is the polarization property of the electromagnetic field. Most of the fundamental properties of the polarization of light or of radio waves are long established and can be found in the classical literature. In recent years studies being made of the polarization of light, have produced a representation which shows the deeper connection between the common description of polarization as an elliptically rotating field vector and the spin properties of particles and the entire group theoretic approach to theoretical physics.

Many more applications of this representation have been used in optics than in microwave studies, with the result that the beauty of this particular unifying theory has not been appreciated by antenna designers, who may be able to apply these concepts to problems of immediate practical concern.

The first part of the chapter is concerned with the complete definition of the polarization ellipse. In this we derive in full some of the fundamental relations often merely quoted in the standard literature, but which are required in subsequent analysis. The newer optical representation is introduced and applied to some problems of interest to antenna designers and microwave engineers. In the final part, an indication is given of some of the intimate connections which exist between polarization and general theoretical physics with the expectation that some concepts in the latter could be introduced to assist in the problems of polarized antennas. This is most marked in the study of radar polarimetry.

6.1 THE POLARIZATION ELLIPSE

We can express the total electric field E in terms of its vector components along fixed x and y directions denoted by the unit vectors $\hat{e}_x$ and $\hat{e}_y$. Propagation is assumed to be in the positive z direction. Then

$$\begin{aligned} \mathbf{E}_x &= E(x)\hat{e}_x \exp i(\omega t - kz) \\ \mathbf{E}_y &= E(y)\hat{e}_y \exp i(\omega t - kz) \end{aligned} \tag{6.1.1}$$

$E(x)$ and $E(y)$ being complex and $k = 2\pi/\lambda$.

Putting $E(x) = a_x \exp(i\delta_x)$ and $E(y) = a_y \exp(i\delta_y)$, then at any fixed plane, say $z = 0$, the *real* part of the amplitudes are

$$\begin{aligned} E_x &= a_x \cos(\omega t + \delta_x) \\ E_y &= a_y \cos(\omega t + \delta_y) \end{aligned} \tag{6.1.2}$$

Elimination of the ωt factor results in the relation

$$\frac{E_x^2}{a_x^2} + \frac{E_y^2}{a_y^2} - \frac{2E_xE_y}{a_xa_y}\cos\delta = \sin^2\delta \tag{6.1.3}$$

where $\delta = \delta_y - \delta_x$. This is the equation of an ellipse which, because of the presence of the cross product term, has not been referred to its principle axes. We assume that the major axis of the ellipse makes an angle θ with the x axis. An ellipse with major axis $2a$ and minor axis $2b$ with centre at the origin and whose major axis makes an angle θ with the x axis (Figure 6.1) has the equation (obtained by rotating the ellipse $x^2/a^2 + y^2/b^2 = 1$ through a positive angle θ)

$$x^2\left(\frac{\cos^2\theta}{a^2} + \frac{\sin^2\theta}{b^2}\right) + y^2\left(\frac{\sin^2\theta}{a^2} + \frac{\cos^2\theta}{b^2}\right) - 2xy\left(\frac{1}{b^2} - \frac{1}{a^2}\right)\sin\theta\cos\theta = 1 \tag{6.1.4}$$

Taking lines parallel to the x and y axes such that their intersections with Equation 6.14 have equal roots we find that this ellipse fits into the rectangle

$$\begin{aligned} x &= \pm\sqrt{a^2\cos^2\theta + b^2\sin^2\theta} = \pm c \\ y &= \pm\sqrt{a^2\sin^2\theta + b^2\cos^2\theta} = \pm d \end{aligned} \tag{6.1.5}$$

The ellipse of Equation 6.1.4 intersects the axes at

$$\begin{aligned} x &= \pm a' \quad \text{where} \quad a'^2 = \frac{a^2b^2}{b^2\cos^2\theta + a^2\sin^2\theta} \\ y &= \pm b' \quad \text{where} \quad b'^2 = \frac{a^2b^2}{b^2\sin^2\theta + a^2\cos^2\theta} \end{aligned} \tag{6.1.6}$$

Solving these we find

$$a^2 = \frac{a'^2b'^2\cos^2\theta}{b'^2\cos^2\theta - a'^2\sin^2\theta} \qquad b^2 = \frac{a'^2b'^2\cos^2\theta}{a'^2\cos^2\theta - b'^2\sin^2\theta}$$

and thus

$$\frac{1}{a'^2} + \frac{1}{b'^2} = \frac{1}{a^2} + \frac{1}{b^2}$$

and

$$\frac{1}{b^2} - \frac{1}{a^2} = \frac{1}{\cos 2\theta}\left(\frac{1}{b'^2} - \frac{1}{a'^2}\right) \tag{6.1.7}$$

The ellipse of Equation 6.1.4 with this substitution becomes

$$\frac{x^2}{a'^2} + \frac{y^2}{b'^2} - \frac{2xy\sin\theta\cos\theta}{\cos 2\theta}\left(\frac{1}{b'^2} - \frac{1}{a'^2}\right) = 1 \tag{6.1.8}$$

Comparing this with Equation 6.1.3 in the form

$$\frac{E_x^2}{(a_x\sin\delta)^2} + \frac{E_y^2}{(a_y\sin\delta)^2} - \frac{2E_xE_y\cos\delta}{a_xa_y\sin^2\delta} = 1$$

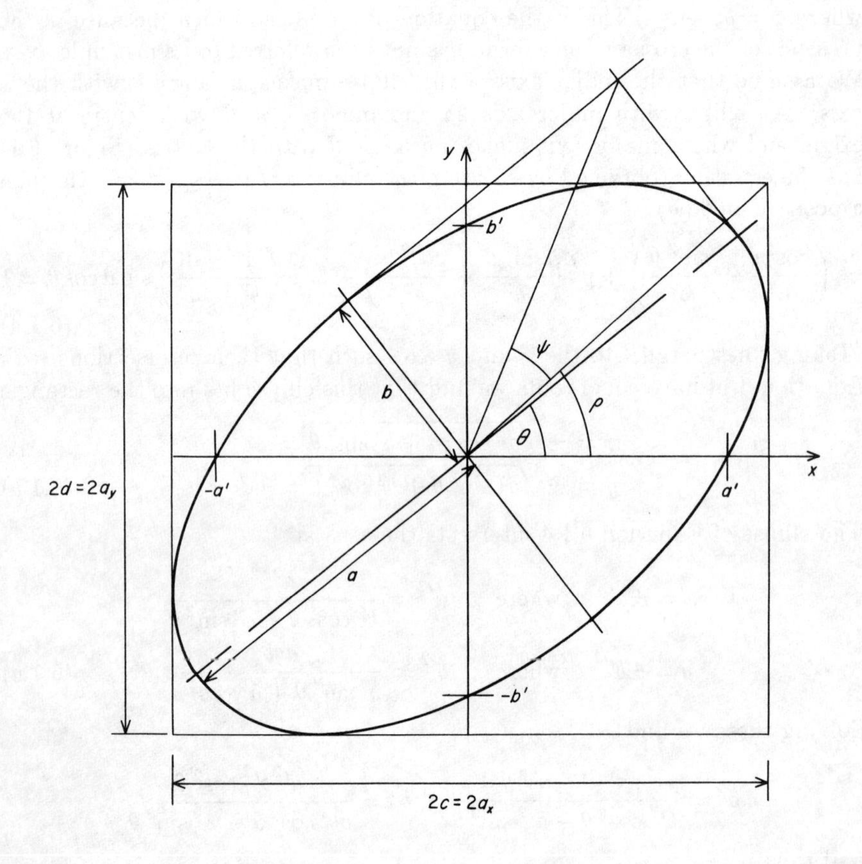

Figure 6.1: Geometry of polarization ellipse

we find that the tip of the E vector describes the same ellipse, in a rectangle $\pm a_x$ and $\pm a_y$ in magnitude, if

$$a' = a_x \sin\delta \qquad b' = a_y \sin\delta$$

and

$$(a_x^2 - a_y^2)\tan 2\theta = 2a_x a_y \cos\delta \tag{6.1.9}$$

From the enclosing rectangle given by Equation 6.1.5 we have

$$c^2 + d^2 = a^2 + b^2 = a_x^2 + a_y^2$$

and from Equation 6.1.6

$$a' = ab/d$$

Therefore

$$a' = ab/a_y \quad \text{or} \quad ab = a_x a_y \sin\delta \tag{6.1.10}$$

Hence

$$\frac{2ab}{a^2 + b^2} = \frac{2a_x a_y \sin\delta}{a_x^2 + a_y^2} \tag{6.1.11}$$

The ellipticity is defined by

$$\varepsilon = 1 - b/a$$

Putting $b/a = \tan\psi$ Equations 6.1.9 and 6.1.11 become

$$\tan 2\theta = \frac{2a_x a_y \cos\delta}{a_x^2 - a_y^2} \qquad \sin 2\psi = \frac{2a_x a_y \sin\delta}{a_x^2 + a_y^2} \tag{6.1.12}$$

The numerical value of $\tan\psi$ gives the axial ratio b/a and the sign differentiates the two senses for the description of the ellipse by the convention

$$0 < \delta < \pi \begin{cases} \text{the polarization is right-handed and} \\ \tan\psi = +b/a \end{cases}$$

$$-\pi < \delta < 0 \begin{cases} \text{the polarization is left-handed and} \\ \tan\psi = -b/a \end{cases} \tag{6.1.13}$$

STOKES PARAMETERS

Four new parameters may be derived from Equation 6.1.12. These are usually denoted by the symbols

$$\begin{aligned} I &= a_x^2 + a_y^2 = |E(x)|^2 + |E(y)|^2 \\ Q &= a_x^2 - a_y^2 = |E(x)|^2 - |E(y)|^2 \\ U &= 2a_x a_y \cos\delta = 2\mathrm{Re}\{E(x)E(y)^*\} \\ V &= 2a_x a_y \sin\delta = 2\mathrm{Im}\{E(x)E(y)^*\} \end{aligned} \tag{6.1.14}$$

We note that I is the "intensity" in optics or the "power" in electromagnetic field theory. Only three of the four parameters are independent since they are related by

$$I^2 = Q^2 + U^2 + V^2 \tag{6.1.15}$$

We denote the set given by Equation 6.1.14 as the Stokes vector $\{I, Q, U, V\}_{lin}$ and then

$$\tan 2\theta = \left(\frac{U}{Q}\right)_{lin} \qquad \sin 2\psi = \left(\frac{V}{I}\right)_{lin} \tag{6.1.16}$$

This set of parameters is capable of an interesting physical interpretation [1] [2] [3]. Consider a set of hypothetical filters F_1 F_2 F_3 and F_4 with the following properties

Each has a transmittance of 50%
Each is normal to the incident beam
F_1 is independent of the incident polarization
F_2 only transmits horizontally polarized radiation (a vertical wire grating)
F_3 only transmits 45° linearly polarized radiation
F_4 only transmits right-handed circular polarization

A power detector is assumed which is independent of the incident polarization. Then if each of the filters is placed in turn between the source and the detector and the readings multiplied by 2 are V_1 V_2 V_3 and V_4, then

$$I \equiv V_1 \qquad Q \equiv V_2 - V_1 \qquad U \equiv V_3 - V_1 \qquad V \equiv V_4 - V_1$$

will give the Stokes parameters of the radiation. The Stokes parameters of a mixture of independent waves (as in partial coherence) are the sums of the Stokes parameters of the separate waves.

Elliptical polarization can also be defined in terms of right and left-handed circularly polarized waves travelling in the same (positive z) direction.

The basis vectors are then

$$\hat{\alpha}_r = \frac{1}{\sqrt{2}}(\hat{\mathbf{e}}_x - i\hat{\mathbf{e}}_y) \qquad \hat{\alpha}_l = \frac{1}{\sqrt{2}}(\hat{\mathbf{e}}_x + i\hat{\mathbf{e}}_y) \tag{6.1.17}$$

then if

$$\mathbf{E} = E(r)\hat{\alpha}_r + E(l)\hat{\alpha}_l$$

the Stokes parameters are

$$\begin{aligned} I &= |E(r)|^2 + |E(l)|^2 \\ Q &= -2\text{Re}\{E(l)E(r)^*\} \\ U &= 2\text{Im}\{E(l)E(r)^*\} \\ V &= |E(r)|^2 - |E(l)|^2 \end{aligned} \tag{6.1.18}$$

This Stokes vector will be denoted by $\{I, Q, U, V\}_{circ}$.

Putting **E** of Equation 6.1.1 in the form

$$\begin{aligned}\mathbf{E} &= E(x)\hat{\mathbf{e}}_x + E(y)\hat{\mathbf{e}}_y \\ &= \frac{1}{2}[E(x) + iE(y)][\hat{\mathbf{e}}_x - i\hat{\mathbf{e}}_y] + \frac{1}{2}[E(x) - iE(y)][\hat{\mathbf{e}}_x + i\hat{\mathbf{e}}_y] \quad (6.1.19)\end{aligned}$$

and comparing with Equations 6.1.16 and 6.1.17 gives

$$\begin{aligned}E(r) &= \frac{1}{\sqrt{2}}(E(x) + iE(y)) \\ E(l) &= \frac{1}{\sqrt{2}}(E(x) - iE(y)) \quad (6.1.20)\end{aligned}$$

Equation 6.1.17 can be put in the equivalent form

$$\begin{pmatrix}\hat{\alpha}_r \\ \hat{\alpha}_l\end{pmatrix} = \frac{1}{\sqrt{2}}\begin{pmatrix}1 & -i \\ 1 & i\end{pmatrix}\begin{pmatrix}\hat{\mathbf{e}}_x \\ \hat{\mathbf{e}}_y\end{pmatrix} \quad (6.1.21)$$

If $E(r) = a_r \exp(i\delta_r)$ and $E(l) = a_l \exp(i\delta_l)$ Equation 6.1.20 is

$$\begin{pmatrix}a_r \exp(i\delta_r) \\ a_l \exp(i\delta_l)\end{pmatrix} = \frac{1}{\sqrt{2}}\begin{pmatrix}1 & i \\ 1 & -i\end{pmatrix}\begin{pmatrix}a_x \exp(i\delta_x) \\ a_y \exp(i\delta_y)\end{pmatrix} \quad (6.1.22)$$

and its inverse is

$$\begin{pmatrix}a_x \exp(i\delta_x) \\ a_y \exp(i\delta_y)\end{pmatrix} = \frac{1}{\sqrt{2}}\begin{pmatrix}1 & 1 \\ -i & i\end{pmatrix}\begin{pmatrix}a_r \exp(i\delta_r) \\ a_l \exp(i\delta_l)\end{pmatrix} \quad (6.1.23)$$

ROTATION OF REFERENCE FRAME

If the (x, y) frame is rotated through an angle γ in a right-handed screw sense with respect to the propagation vector **k**, linear fields are changed by a rotation matrix but the circularly polarized fields are only changed in phase. If primes denote the rotated system we have for the linear system

$$\begin{pmatrix}a'_x \exp(i\delta'_x) \\ a'_y \exp(i\delta'_y)\end{pmatrix} = \begin{pmatrix}\cos\gamma & -\sin\gamma \\ \sin\gamma & \cos\gamma\end{pmatrix}\begin{pmatrix}a_x \exp(i\delta_x) \\ a_y \exp(i\delta_y)\end{pmatrix} \quad (6.1.24)$$

and for the circular system

$$\begin{pmatrix}a'_r \exp(i\delta'_r) \\ a'_l \exp(i\delta'_l)\end{pmatrix} = \begin{pmatrix}\exp(i\gamma) & 0 \\ 0 & \exp(-i\gamma)\end{pmatrix}\begin{pmatrix}a_r \exp(i\delta_r) \\ a_l \exp(i\delta_l)\end{pmatrix} \quad (6.1.25)$$

It can be simply shown that Equations 6.1.22 and 6.1.23 transform by these rules.

The column vectors (I, Q, U, V) for the same rotation have the same transformation law in both the linear and the circular representations. This is [1] [2] [3]

$$\begin{pmatrix}I' \\ Q' \\ U' \\ V'\end{pmatrix} = \begin{pmatrix}1 & 0 & 0 & 0 \\ 0 & \cos(2\gamma) & -\sin(2\gamma) & 0 \\ 0 & \sin(2\gamma) & \cos(2\gamma) & 0 \\ 0 & 0 & 0 & 1\end{pmatrix}\begin{pmatrix}I \\ Q \\ U \\ V\end{pmatrix} \quad (6.1.26)$$

JONES VECTORS

The parameters I Q U and V all have the dimensions of power and require a 4x4 matrix for transformation an illustration of which is Equation 6.1.26. Such matrices are called Mueller matrices and the transformation laws the Mueller calculus. On the other hand Equations 6.1.24 and 6.1.25 represent the polarization by a two (complex) element column vector and their transformation by 2x2 matrices. This method was developed by Jones [4]. There are nine major differences between the Mueller calculus and the Jones calculus listed in Shurcliff [1, p122]. The outstanding one, from the microwave point of view, is that the Jones calculus retains the phase information of the individual components. The Mueller calculus is able to deal with a completely depolarized wave which of course, is not normally encountered in microwave practice. Consequently Mueller matrices can be derived from Jones matrices in much the same way that power can be derived from complex field components, but the reverse is in general not possible. Such a derivation can be found in the work of Schmeider [5] which gives the method of obtaining a Mueller matrix from the equivalent Jones matrix. For ease of handling, Jones vectors can be "normalized" by the division by any appropriate factor which leaves the vector in its simplest normalized state.

The interpretation of a normalized Jones vector is then made as follows

1. convert the vector to its full form

$$\begin{pmatrix} a_x \exp(i\delta_y) \\ a_y \exp(i\delta_y) \end{pmatrix}$$

2. compute $\rho = |\tan^{-1}(a_y/a_x)|$ and $\delta = \delta_y - \delta_x$

3. the azimuth of the ellipse is then given by $\tan 2\theta = \tan 2\rho \cos \delta$ (see Equation 6.3.8)

4. the axial ratio is given by $\tan \psi$ where $\sin \psi = \sin 2\rho |\sin \delta|$

5. the handedness is given by sgn($\sin\delta$) that is if
   $\sin \delta > 0$ ellipse is right-handed
   $\sin \delta < 0$ ellipse is left-handed.

In particular (see Section 6.2) if a normalized Jones vector is represented by $\begin{pmatrix} m \\ n \end{pmatrix}$ the Jones vector of the orthogonally polarized state is $\begin{pmatrix} -n^* \\ m^* \end{pmatrix}$ (* refers to complex conjugate).

The transformation of the Jones vector of a wave, as it is transmitted through a system of optical components, is made through the application of the Jones matrix appropriate to those components. These can be derived heuristically by analysing the properties of each component individually. The matrices in the table are those given in the original work of Jones (loc.cit.)

Horizontal polarization	$\begin{pmatrix} a_x \exp(i\delta_x) \\ 0 \end{pmatrix}$	$\begin{pmatrix} 1 \\ 0 \end{pmatrix}$
Vertical polarization	$\begin{pmatrix} 0 \\ a_y \exp(i\delta_y) \end{pmatrix}$	$\begin{pmatrix} 0 \\ 1 \end{pmatrix}$
$\pm 45^o$ linear polarization	$\begin{pmatrix} a_x \exp(i\delta_x) \\ \pm a_x \exp(i\delta_x) \end{pmatrix}$	$\frac{1}{\sqrt{2}} \begin{pmatrix} 1 \\ \pm 1 \end{pmatrix}$
Linear polarization at angle $\rho = \lvert \tan^{-1}(a_y/a_x) \rvert$	$\begin{pmatrix} a_x \exp(i\delta_x) \\ \pm a_y \exp(i\delta_x) \end{pmatrix}$	$\begin{pmatrix} \cos\rho \\ \pm \sin\rho \end{pmatrix}$
RH Circular polarization	$\begin{pmatrix} a_x \exp(i\delta_x) \\ \pm a_x \exp[i(\delta_x + \pi/2)] \end{pmatrix}$	$\frac{1}{\sqrt{2}} \begin{pmatrix} -i \\ \pm 1 \end{pmatrix}$
LH Circular polarization	$\begin{pmatrix} a_x \exp(i\delta_x) \\ \pm a_x \exp[i(\delta_x - \pi/2)] \end{pmatrix}$	$\frac{1}{\sqrt{2}} \begin{pmatrix} i \\ \pm 1 \end{pmatrix}$
Elliptical polarization $\delta = \delta_y - \delta_x$	$\begin{pmatrix} a_x \exp(i\delta_x) \\ a_y \exp(i\delta_y) \end{pmatrix}$	$\begin{pmatrix} \cos\rho \exp(-i\delta/2) \\ \sin\rho \exp(i\delta/2) \end{pmatrix}$

Table 6.1: Jones Vectors

6.2 REFLECTED WAVES

As in most optical processes the analysis so far has been confined to a progressive wave through a system of optical components and the surface reflections at each component have been ignored. In microwave practice considerable loss of power by internal reflections can occur and accumulate in a system which contains many internal surfaces. This is one of the facts limiting the number of such surfaces to be found in a microwave optical system. The reflected power, if not scattered into generally undesirable directions, will be returned to the source and give rise to impedance mismatch. This for any high power source is deleterious to performance. In communications systems, internal reflections can cause serious inter-channel interference and, in general, have to be kept to the minimum possible.

A necessary extension to the above theory therefore is to include in the matrix description the effects of reflected waves.

We can consider S, in Figure 6.2, as an intermediate plane surface in an optical system with a normally incident beam giving rise to a reflected wave in addition to the transmitted wave. The Jones vectors for waves in both directions can be written as a column vector for four components, the upper two for propagation in the positive z direction and the lower two for the negative z direction. These vectors are then related by a 4 x 4 matrix, not to be confused with the Mueller matrix, which for this problem would require doubling to an 8 x 8 matrix.

Considering both reflection and transmission the components of Figure 6.2

1	Free space	$\begin{pmatrix} 1 & 0 \\ 0 & 1 \end{pmatrix}$
2	Ideal isotropic absorber with transmission (voltage) coefficient τ (This could be separated into different τ_x τ_y values)	$\begin{pmatrix} \tau & 0 \\ 0 & \tau \end{pmatrix}$
3	Horizontal linear polarizer	$\begin{pmatrix} 1 & 0 \\ 0 & 0 \end{pmatrix}$
4	Vertical linear polarizer	$\begin{pmatrix} 0 & 0 \\ 0 & 1 \end{pmatrix}$
5	$\pm 45^o$ linear polarizer	$\frac{1}{2}\begin{pmatrix} 1 & \pm 1 \\ \pm 1 & 1 \end{pmatrix}$
6	Polarization rotator to angle α	$\begin{pmatrix} \cos^2\alpha & \cos\alpha\sin\alpha \\ \cos\alpha\sin\alpha & \sin^2\alpha \end{pmatrix}$
7	Linear phase shifter, free space distance d	$\begin{pmatrix} \exp(i2\pi d/\lambda) & 0 \\ 0 & \exp(i2\pi d/\lambda) \end{pmatrix}$
8	Phase shift of 90°, azimuth of fast axis $= 0^o$	$\begin{pmatrix} \exp(i\pi/4) & 0 \\ 0 & \exp(-i\pi/4) \end{pmatrix}$
9	Phase shift of 90°, azimuth of fast axis $= 90^o$	$\begin{pmatrix} \exp(-i\pi/4) & 0 \\ 0 & \exp(i\pi/4) \end{pmatrix}$
10	Phase shift of 90°, azimuth of fast axis $= \pm 45^o$	$\frac{1}{\sqrt{2}}\begin{pmatrix} 1 & \pm i \\ \pm i & 1 \end{pmatrix}$
11	Phase shift of 90°, azimuth of fast axis at general angle γ	$\begin{pmatrix} \cos^2\gamma\exp(i\pi/4) + \sin^2\gamma\exp(-i\pi/4) & i\sqrt{2}\cos\gamma\sin\gamma \\ i\sqrt{2}\cos\gamma\sin\gamma & \cos^2\gamma\exp(-i\pi/4) + \sin^2\gamma\exp(i\pi/4) \end{pmatrix}$
12	Phase shift of 180°, fast axis at angle γ	$\begin{pmatrix} \cos 2\gamma & \sin 2\gamma \\ \sin 2\gamma & -\cos 2\gamma \end{pmatrix}$
13	General phase shift ϕ, fast axis at angle γ	$\begin{pmatrix} \cos^2\gamma\exp(i\phi/2) + \sin^2\gamma\exp(-i\phi/2) & 2i\cos\gamma\sin\gamma\sin(\phi/2) \\ 2i\cos\gamma\sin\gamma\sin(\phi/2) & \cos^2\gamma\exp(-i\phi/2) + \sin^2\gamma\exp(i\phi/2) \end{pmatrix}$
14	Transmitter of RH circular polarization only i.e. total reflection of LH circular polarization	$\frac{1}{2}\begin{pmatrix} 1 & -i \\ i & 1 \end{pmatrix}$
15	Transmitter of LH circular polarization	$\frac{1}{2}\begin{pmatrix} 1 & i \\ -i & 1 \end{pmatrix}$

Table 6.2: Jones transmission matrices

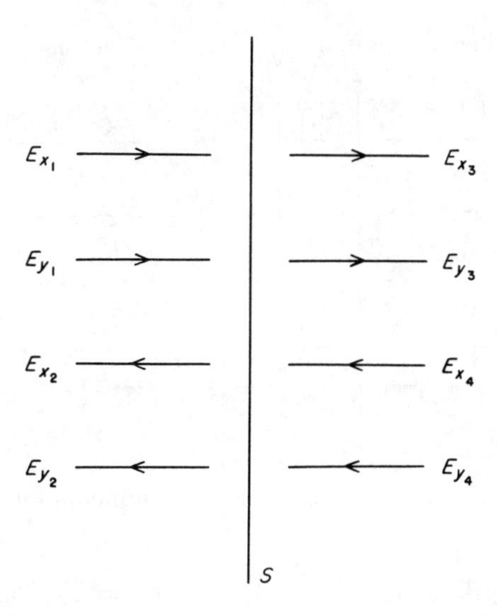

Figure 6.2: Intermediate surface: reflection and transmission components

are related by

$$\begin{pmatrix} E_{x2} \\ E_{y2} \\ E_{x3} \\ E_{y3} \end{pmatrix} = \begin{pmatrix} R_{1\,1} & R_{1\,2} & T_{1\,1} & T_{1\,2} \\ R_{2\,1} & R_{2\,2} & T_{2\,1} & T_{2\,2} \\ R_{1\,1} & R_{1\,2} & T_{1\,1} & T_{1\,2} \\ R_{2\,1} & R_{2\,2} & T_{2\,1} & T_{2\,2} \end{pmatrix} \begin{pmatrix} E_{x1} \\ E_{y1} \\ E_{x4} \\ E_{y4} \end{pmatrix} \qquad (6.2.1)$$

E_x and E_y being, as before complex field components parallel to fixed x and y axes.

The matrix in Equation 6.2.1 can be rearranged to give a form applicable to a succession of surfaces, that is a cascade matrix [6] [7], so that components on the transmission side may be obtained from the components on the incident side. This results in

$$\begin{pmatrix} E_{x3} \\ E_{y3} \\ E_{x4} \\ E_{y4} \end{pmatrix} = A \begin{pmatrix} E_{x1} \\ E_{y1} \\ E_{x2} \\ E_{y2} \end{pmatrix}$$

where A has elements $A_{ij}/(T_{2\,2}T_{1\,1} - T_{1\,2}T_{2\,1})$ and where

grating 1 grating 2 grating r grating n

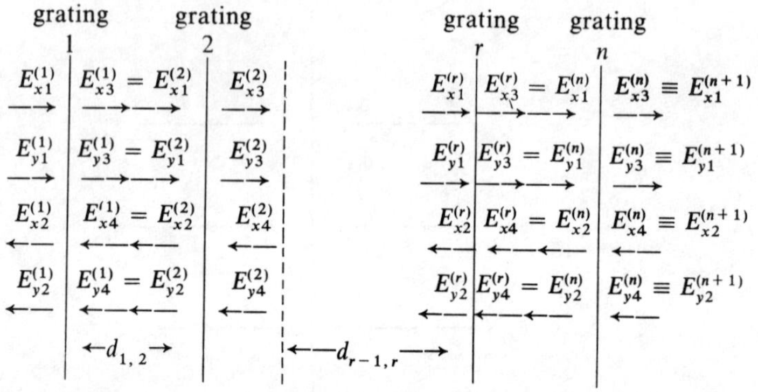

$E_{x1}^{(1)}$ $E_{x3}^{(1)} = E_{x1}^{(2)}$ $E_{x3}^{(2)}$ $E_{x1}^{(r)}$ $E_{x3}^{(r)} = E_{x1}^{(n)}$ $E_{x3}^{(n)} \equiv E_{x1}^{(n+1)}$

$E_{y1}^{(1)}$ $E_{y3}^{(1)} = E_{y1}^{(2)}$ $E_{y3}^{(2)}$ $E_{y1}^{(r)}$ $E_{y3}^{(r)} = E_{y1}^{(n)}$ $E_{y3}^{(n)} \equiv E_{y1}^{(n+1)}$

$E_{x2}^{(1)}$ $E_{x4}^{(1)} = E_{x2}^{(2)}$ $E_{x4}^{(2)}$ $E_{x2}^{(r)}$ $E_{x4}^{(r)} = E_{x2}^{(n)}$ $E_{x4}^{(n)} \equiv E_{x2}^{(n+1)}$

$E_{y2}^{(1)}$ $E_{y4}^{(1)} = E_{y2}^{(2)}$ $E_{y4}^{(2)}$ $E_{y2}^{(r)}$ $E_{y4}^{(r)} = E_{y2}^{(n)}$ $E_{y4}^{(n)} \equiv E_{y2}^{(n+1)}$

$d_{1,2}$ $d_{r-1,r}$

Figure 6.3: Reflection and transmission components for a system of n surfaces

$$
\begin{aligned}
A_{11} &= T_{11}(T_{11}T_{22} - T_{12}T_{21}) & A_{21} &= T_{21}(T_{11}T_{22} - T_{12}T_{21}) \\
&\quad + R_{11}(T_{12}R_{21} - T_{22}R_{11}) & &\quad + R_{21}(T_{12}R_{21} - T_{22}R_{11}) \\
&\quad + R_{12}(T_{21}R_{11} - T_{11}R_{21}) & &\quad + R_{22}(T_{21}R_{11} - T_{11}R_{21}) \\
A_{12} &= T_{12}(T_{11}T_{22} - T_{12}T_{21}) & A_{22} &= T_{22}(T_{11}T_{22} - T_{12}T_{21}) \\
&\quad + R_{11}(T_{12}R_{22} - T_{22}R_{12}) & &\quad + R_{21}(T_{12}R_{22} - T_{22}R_{12}) \\
&\quad + R_{12}(T_{21}R_{12} - T_{11}R_{22}) & &\quad + R_{22}(T_{21}R_{12} - T_{11}R_{22}) \\
A_{13} &= T_{22}R_{11} - T_{21}R_{12} & A_{23} &= T_{22}R_{21} - T_{21}R_{22} \\
A_{14} &= T_{11}R_{12} - T_{12}R_{11} & A_{24} &= T_{11}R_{22} - T_{12}R_{21} \\
A_{31} &= T_{12}R_{21} - T_{22}R_{11} & A_{41} &= T_{21}R_{11} - T_{11}R_{21} \\
A_{32} &= T_{12}R_{22} - T_{22}R_{12} & A_{42} &= T_{21}R_{12} - T_{11}R_{22} \\
A_{33} &= T_{22} & A_{43} &= -T_{12} \\
A_{34} &= -T_{21} & A_{44} &= T_{11}
\end{aligned}
\tag{6.2.2}
$$

The space between the surfaces is represented by the space matrix shown in the table for a linear phase shift in free space of magnitude $2\pi d/\lambda$, where d is the distance between the surfaces. For the forward travelling wave this phase shift is positive and for the reflected wave it is negative. The spacing matrix is therefore a diagonal matrix $D^{r,s}$ with elements

$$
\begin{aligned}
D_{11} &= D_{22} = \exp(i2\pi d^{r,s}/\lambda) \\
D_{33} &= D_{44} = \exp(-i2\pi d^{r,s}/\lambda)
\end{aligned}
\tag{6.2.3}
$$

where $d^{r,s}$ is the free space distance between the r and s surfaces. Since the transmitted fields at one surface are the incident fields upon the next as shown in Figure 6.3, for an overall system of n surfaces, therefore

$$\begin{pmatrix} E_{x1} \\ E_{y1} \\ E_{x2} \\ E_{y2} \end{pmatrix}_{n+1} = A_n D^{n,n-1} A_{n-1} \dots D^{2,1} A_1 \begin{pmatrix} E_{x1} \\ E_{y1} \\ E_{x2} \\ E_{y2} \end{pmatrix}_{\text{incident}} \tag{6.2.4}$$

This gives the transmitted field $\begin{pmatrix} E_{x1}^{n+1} \\ E_{y1}^{n+1} \end{pmatrix}$ and the reflected field $\begin{pmatrix} E_{x2}^{\text{inc}} \\ E_{y2}^{\text{inc}} \end{pmatrix}$ in terms of the incident field $\begin{pmatrix} E_{x1}^{\text{inc}} \\ E_{y1}^{\text{inc}} \end{pmatrix}$ with the field $\begin{pmatrix} E_{x2}^{n+1} \\ E_{y2}^{n+1} \end{pmatrix}$ identically zero, it being the reflection from the non-existent $(n+1)$th surface. In those cases where the medium between the surfaces is not free space, the phase factor in Equation 6.2.3 can be modified by the inclusion of the appropriate refractive index. This, for normal incidence, oblique parallel incidence or oblique perpendicular incidence, is given by Equation 4.4.3.

6.3 TRANSMISSION AND RECEPTION OF ELLIPTICALLY POLARIZED WAVES

We consider the power received by an antenna whose elliptically polarized transmitted radiation is the ellipse $\begin{pmatrix} a_x \exp(i\delta_x) \\ a_y \exp(i\delta_y) \end{pmatrix}$ with major axis $2a$ and minor axis $2b$ and whose major axis makes an angle θ with the x axis. The relations of Section 6.1, Equations 6.1.1 to 6.1.12 therefore apply to this ellipse.

A second elliptically polarized wave is incident upon this antenna, and this incident ellipse can, without loss of generality be referred to the principle x and y axes.

If we designate the properties of the incident elliptical wave by capital letters we therefore have $A_x = A$ $A_y = B$ and the axial ratio is given by $\tan \psi_i = B/A$.

The field received by the first antenna is then obtained from the product (note the transposition)

$$E = \begin{pmatrix} A_x & iA_y \end{pmatrix} \begin{pmatrix} a_y \exp(-i\delta_y) \\ a_x \exp(-i\delta_x) \end{pmatrix} \tag{6.3.1}$$

and hence the received power

$$P = EE^* = A_x^2 a_y^2 + A_y^2 a_x^2 + 2A_x A_y a_x a_y \sin \delta \tag{6.3.2}$$

Use of the relations (Equations 6.1.10 and 6.1.6)

$$a_x a_y \sin \delta = ab$$

$$a'b' = a_x a_y \sin^2 \delta$$

allows for the substitution for $a_x a_y$ in terms of a and b in Equation 6.3.2 to give the result

$$P = EE^* = (A^2a^2 + B^2b^2)\cos^2\theta + (B^2a^2 + A^2b^2)\sin^2\theta \pm 2ABab \tag{6.3.3}$$

where the positive or negative sign is taken when the ellipses have the same handedness or opposed handedness.

The same result is obtained by rotating the receiving ellipse to the axes of the incident ellipse. That is if

$$E = \begin{pmatrix} A & iB \end{pmatrix} \begin{pmatrix} \cos\theta & -\sin\theta \\ \sin\theta & \cos\theta \end{pmatrix} \begin{pmatrix} a \\ \pm ib \end{pmatrix}$$

then $P = EE^*$ gives the result of Equation 6.3.3.

The power in the incident field is $P_i = A^2 + B^2$ and the power of the receiving antenna is $P_r = a^2 + b^2$.

Equation 6.3.3 can then be put in the form

$$P = P_i P_r \eta$$

where η is an efficiency factor. Putting $B/A = r_i$, $b/a = r_r$ the axial ratio of the respective ellipses

$$\eta = \frac{(1 + r_i^2)(1 + r_r^2) + (1 - r_i^2)(1 - r_r^2)\cos 2\theta \pm 4r_i r_r}{2(1 + r_i^2)(1 + r_r^2)} \tag{6.3.4}$$

the same result as Kales [8] [9] [10].

THE POINCARÉ SPHERE

From the definitions of Section 6.1, Equations 6.1.9 to 6.1.12 we can derive

$$\tan\psi = \frac{b}{a} \qquad \frac{a_y}{a_x} = \tan\rho \tag{6.3.5}$$

$$\begin{aligned} a_x^2 + a_y^2 &= a^2 + b^2 \\ a_x^2 - a_y^2 &= (a^2 - b^2)\cos 2\theta \qquad \text{from Equation 6.1.7} \\ a_x a_y \sin\delta &= \pm ab \\ 2a_x a_y \cos\delta &= (a^2 - b^2)\sin 2\theta \end{aligned} \tag{6.3.6}$$

and hence

$$\cos 2\psi = \cos 2\rho \cos 2\theta + \sin 2\rho \sin 2\theta \cos\delta \tag{6.3.7}$$

Equation 6.3.7 immediately suggests a spherical triangle, one internal angle being $\pi/2$ and another δ.

Other relations which can be derived from the set in Equation 6.3.6 and the spherical triangle are then

$$\begin{aligned} \pm\sin 2\psi &= \sin 2\rho \sin\delta \\ \tan 2\theta &= \tan 2\rho \cos\delta \\ \cos 2\rho &= \cos 2\psi \cos 2\theta \\ \pm\tan 2\psi &= \sin 2\theta \tan\delta \end{aligned} \tag{6.3.8}$$

These are identical (with change in notation only) to those given in Rumsey [8].

If we now put

$$\frac{a_y e^{i\delta_x}}{a_x e^{i\delta_y}} = \frac{a_y}{a_x} e^{i\delta} \equiv u + iv$$

then

$$u^2 + v^2 + 2u \cot 2\theta - 1 = 0 \tag{6.3.9}$$

and

$$u^2 + v^2 - 2v\,\mathrm{cosec}2\psi + 1 = 0 \tag{6.3.10}$$

Constant values of θ in Equation 6.3.9 give circles of radius cosec 2θ centred on the point $(-\cot 2\theta, 0)$ in the u, v plane.

In Equation 6.3.10 constant values of ψ give circles of radius cot 2ψ centred at (0, cosec 2ψ). These circles are orthogonal in the u, v plane and their point of intersection has polar angle δ.

Equations 6.3.9 and 6.3.10 can be recognized as analogous to equations of constant resistance and reactance that comprise the system of orthogonal circles in the Cartesian form of the impedance chart [11]. The connection between Equations 6.3.7 6.3.9 and 6.3.10 shows that the impedance chart is a stereographic projection of the Poincaré sphere.

We consider a sphere of radius 1/2 centred at the origin (Figure 6.4) which is tangential to the u, v plane at its origin, the point (-1/2, 0, 0). The point m in the plane is the stereographic projection of the point M on the sphere from the point(1/2, 0, 0) the opposite end of the diameter perpendicular to the u, v plane at its origin. If l is the latitude and k the longitude of M with respect to the polar axis P_1P_2 as shown then

1. the equation of the sphere is $x^2 + y^2 + z^2 = 1/4$
2. M has coordinates (x, y, z) and m has coordinates $(-1/2, \tan\rho\cos\delta$, $\tan\rho\sin\delta)$

which are connected by

$$\begin{aligned} 2x &= \cos 2\psi \cos 2\theta \\ 2y &= \cos 2\psi \sin 2\theta \\ 2z &= \sin 2\psi \end{aligned} \tag{6.3.11}$$

Since $2z = \sin l$, $l = 2\psi$ and since $\tan k = y/x = \tan 2\theta$; $k = 2\theta$.

From the equation for the spherical triangle given by Equation 6.3.7 the final side of the triangle is the angle 2ρ as shown.

The upper half of the projection being for $\delta > 0$ is for right-handed elliptical polarizations and the lower half for left-handed. The projection of the poles are then the points corresponding to right-handed circular polarization in the lower half.

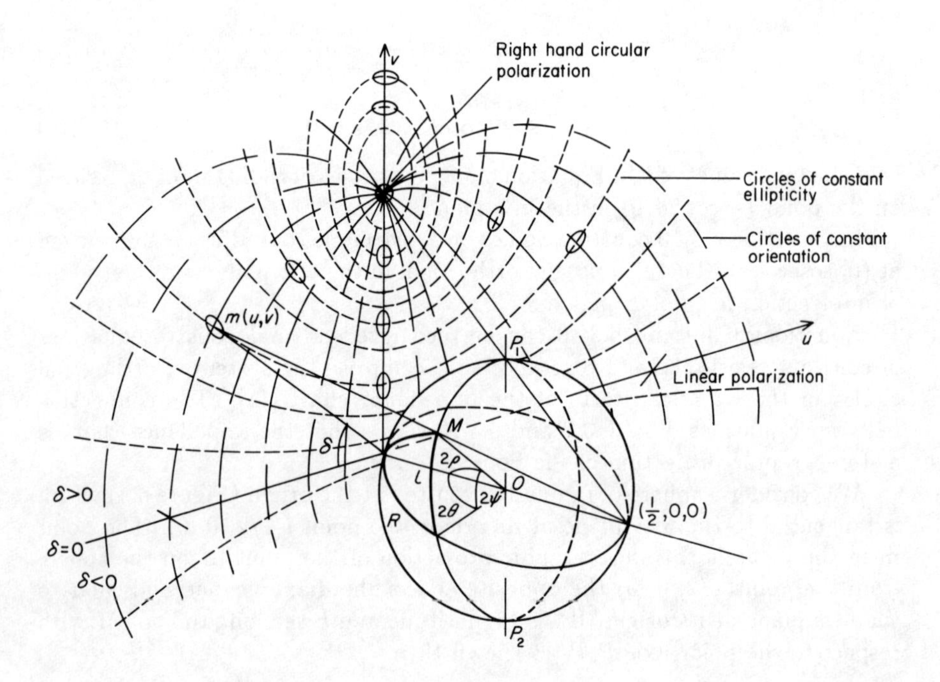

Figure 6.4: Stereographic projection of the Poincare sphere

In order to use the Poincaré sphere the effects of the various devices listed in the table for Jones matrices have to be converted to rotations on the sphere. Other projections of the sphere could be considered, for example orthographic projection, which would give rise to other forms of impedance chart. This has been considered fully by Deschamps [8]. Conformal mappings of the impedance chart, for example the Smith chart, could by stereographic projection give other forms of the polarization sphere. Rotations or other automorphisms which map the sphere onto itself are then representations of polarization changes of state and continuous automorphisms represent a continuous change in polarization state as a wave progresses through an optically active medium. This analogy between polarization and impedance has, as a practical consequence, the application of a circular polarizer as a standing wave impedance indicator. If as shown in Figure 6.5 the coupling from rectangular waveguide to circular waveguide is made through a correctly positioned slot in the broad wall of the waveguide, positively travelling and negatively travelling waves in the rectangular waveguide will induce right-handed and left-handed circularly polarized components in the circular guide. The amplitudes of these will be proportional to the amplitudes of the waves in the rectangular waveguide and thus the elliptical wave which they combine to produce, will be of the axial ratio equal to the ratio of the two waves in the rectangular waveguide and hence to the standing wave ratio. A measurement of this ellipticity can be made by rotating a probe in the circular waveguide. The position of this on the polarization chart is then identical to its position on the equivalent impedance chart and the orientation of the ellipse determines the phase.

6.4 APPLICATIONS TO MICROWAVE ANTENNAS

CIRCULAR POLARIZATION

A standard form of microwave circular polarizer for incident linear polarization consists of a set of thin parallel plates (Figure 6.6) set at an azimuth of 45°. Propagation through the plate medium consists of two modes, the TEM at a wavelength equal to the free space wavelength λ_0 for which the **E** field vector is perpendicular to the plates, and the TE_{01} at a wavelength of λ_g for which the **E** field vector is parallel to the plates. The two wavelengths are related by the waveguide relation.

$$\lambda_g = \lambda_0/\sqrt{1-(\lambda_0/2a)^2}$$

where a is the separation between the plates. In most practical applications the ratio λ_0/λ_g lies in the range 0.5 to 0.8. The two modes can then be separated by any prescribed phase value ϕ by adjusting the width d to suit

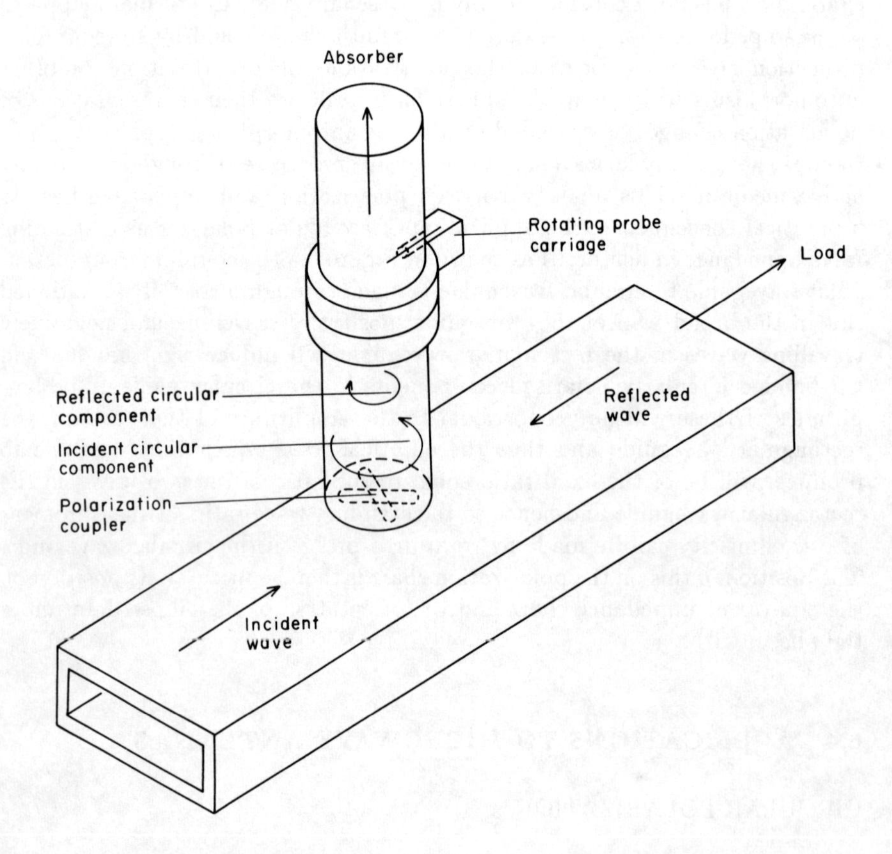

Figure 6.5: Polarization impedance meter

the relation

$$\frac{2\pi d}{\lambda_0} - \frac{2\pi d}{\lambda_g} = \phi \tag{6.4.1}$$

The direction parallel to the edges of the plate is then the azimuth γ of the "fast" axis in the terminology of Table 6.2. For a phase shift $\phi = \pi/2$ and fast axis azimuth $\gamma = \pi/4$ the appropriate Jones matrix, number 10 of Table 6.2 is $\frac{1}{\sqrt{2}}\begin{pmatrix} 1 & i \\ i & 1 \end{pmatrix}$. Applied to an incident x polarized wave this gives right-handed circular polarization through the relation

$$\frac{1}{\sqrt{2}}\begin{pmatrix} 1 & i \\ i & 1 \end{pmatrix}\begin{pmatrix} 1 \\ 0 \end{pmatrix} = \frac{1}{\sqrt{2}}\begin{pmatrix} 1 \\ i \end{pmatrix}$$

With the more general form of phase shift matrix, number 13 of Table 6.2, we can now extend this analysis to determine the effect of the same polarizer operating over a band of frequencies containing the frequency for which ϕ of Equation 6.3.12 is exactly $\pi/2$. This gives the operating bandwidth of the polarizer, subject to prescribed limits on the permissible ellipticity of the transmitted wave.

The same matrix can be used to design a metal plate polarizer (and obtain the operating bandwidth) that will create any given ellipticity of polarization from an incident linearly polarized beam. Without loss of generality this can be taken to be x polarized. Then to create a given ellipticity defined by the Jones vector $\begin{pmatrix} me^{i\theta} \\ ne^{i\pi/2} \end{pmatrix}$ (since the absolute phase is arbitrary) we require the solution for γ and ϕ of

$$\begin{pmatrix} \cos^2\gamma e^{i\phi/2} + \sin^2\gamma e^{-i\phi/2} & 2i\cos\gamma\sin\gamma\sin(\phi/2) \\ 2i\cos\gamma\sin\gamma\sin(\phi/2) & \cos^2\gamma e^{-i\phi/2} + \sin^2\gamma e^{i\phi/2} \end{pmatrix}\begin{pmatrix} 1 \\ 0 \end{pmatrix} = \begin{pmatrix} me^{i\theta} \\ in \end{pmatrix}$$

This is readily found to be

$$\begin{aligned} \tan 2\gamma &= \frac{n}{m\sin\theta} \\ \sin(\phi/2) &= m^2\sin^2\theta + n^2 \end{aligned}$$

or

$$\cos(\phi/2) = m\cos\theta \qquad \text{since } m^2 + n^2 = 1$$

POLARIZATION ROTATORS

The application of a series of phase shifters of the metal plate type illustrated in Figure 6.6 can be shown to have the property of rotating the direction of polarization of any linearly polarized wave through a prescribed angle, which, for simplicity, we illustrate by a rotation through 90°.

We consider first two 180° phase shifters in series whose fast axes make an angle of 45°. We choose in the first instance the azimuths to be 22.5^o and 67.5^o and show later that this occasions no loss of generality.

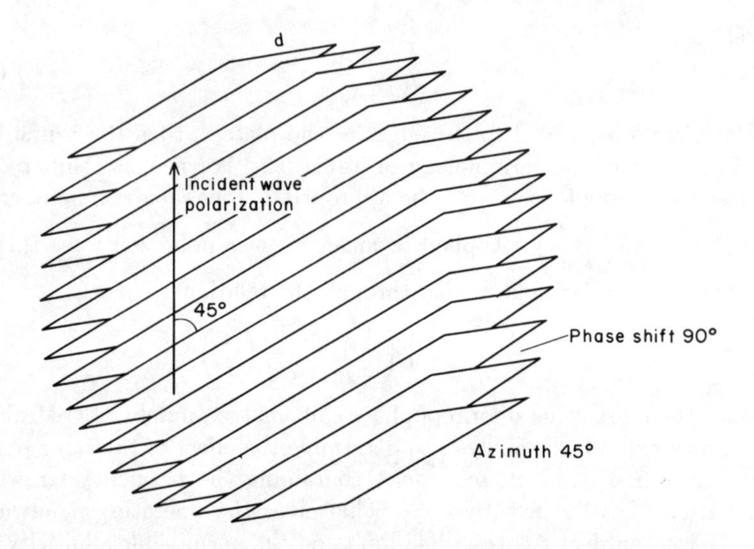

Figure 6.6: Metal plate circular polarizer

The Jones matrices are then (from Number 12 of Table 6.2)

$$\begin{pmatrix} \cos 2\gamma & \sin 2\gamma \\ \sin 2\gamma & -\cos 2\gamma \end{pmatrix}$$

with $\gamma = 22.5^o$ for the first and 67.5^o for the second metal plate system.

Applying these to an incident linearly polarized wave polarized in the arbitrary direction α results in

$$\begin{pmatrix} -\frac{1}{\sqrt{2}} & \frac{1}{\sqrt{2}} \\ \frac{1}{\sqrt{2}} & \frac{1}{\sqrt{2}} \end{pmatrix} \begin{pmatrix} \frac{1}{\sqrt{2}} & \frac{1}{\sqrt{2}} \\ \frac{1}{\sqrt{2}} & -\frac{1}{\sqrt{2}} \end{pmatrix} \begin{pmatrix} \cos\alpha \\ \sin\alpha \end{pmatrix} = \begin{pmatrix} 0 & -1 \\ 1 & 0 \end{pmatrix} \begin{pmatrix} \cos\alpha \\ \sin\alpha \end{pmatrix} = \begin{pmatrix} -\sin\alpha \\ \cos\alpha \end{pmatrix}$$

That is the final polarization, $\begin{pmatrix} -\sin\alpha \\ \cos\alpha \end{pmatrix}$ is at right angles to the incident polarization $\begin{pmatrix} \cos\alpha \\ \sin\alpha \end{pmatrix}$.

The independence of this result to the choice of azimuths can be seen from the arbitrary choice of the incident polarization direction α.

It can also be shown by rotating each of the phase shifters through an arbitrary azimuthal angle β while retaining the 45° separation. The Jones matrix for the combination becomes

$$\frac{1}{2} \begin{pmatrix} \cos\beta & \sin\beta \\ -\sin\beta & \cos\beta \end{pmatrix} \begin{pmatrix} -1 & 1 \\ 1 & 1 \end{pmatrix} \begin{pmatrix} \cos\beta & \sin\beta \\ -\sin\beta & \cos\beta \end{pmatrix} \begin{pmatrix} 1 & 1 \\ 1 & -1 \end{pmatrix}$$

$$= \begin{pmatrix} 0 & -1 \\ 1 & 0 \end{pmatrix}$$

which rotates the polarization through 90° as before.

Application of the more general form of the matrix Number 12 now shows that the polarization rotation caused by a combination of phase shifters of this kind is simply twice the angle between the azimuths of the two individual elements.

Another series combination with the property of rotation of the polarization of a linearly polarized wave through 90°, is the sequence that consists of

1. a 90° degree phase shifter with azimuth at 0°

2. a 180° phase shifter with azimuth 45°

3. a 90° phase shifter with azimuth at 90°

The result by the product of the respective matrices is

$$\frac{1}{2}\begin{pmatrix} 1-i & 0 \\ 0 & 1+i \end{pmatrix}\begin{pmatrix} 0 & 1 \\ 1 & 0 \end{pmatrix}\begin{pmatrix} 1+i & 0 \\ 0 & 1-i \end{pmatrix} = i\begin{pmatrix} 0 & -1 \\ 1 & 0 \end{pmatrix}$$

this being the matrix appropriate to a 90° rotation and with a shift in the phase of $\pi/2$.

This series is also capable of generalization and leads to the following combination

1. a 90° degree phase shifter with azimuth γ

2. a 180° phase shifter with azimuth zero

3. a 90° phase shifter with azimuth $-\gamma$

This rotates the angle of a linearly polarized incident wave through the angle $\pi - 2\gamma$. In its most symmetric form $\gamma = 60^o$ and the polarization is rotated through 60°.

Although the devices illustrated here have been given in a free-space form, that is the effect upon an infinite normally incident plane wave, identical results can be obtained with analogues using the dominant mode in a circular waveguide. The differential phase required can be obtained by the insertion of a vane of dielectric along a diameter of the guide. The "fast" axis is the axis perpendicular to this vane and the length of the vane, including its matching sections, determines the value of ϕ. The matrix theory can then be applied in full to these waveguide components. Similar electronically actuated devices using ferrite materials will have the same design principles.

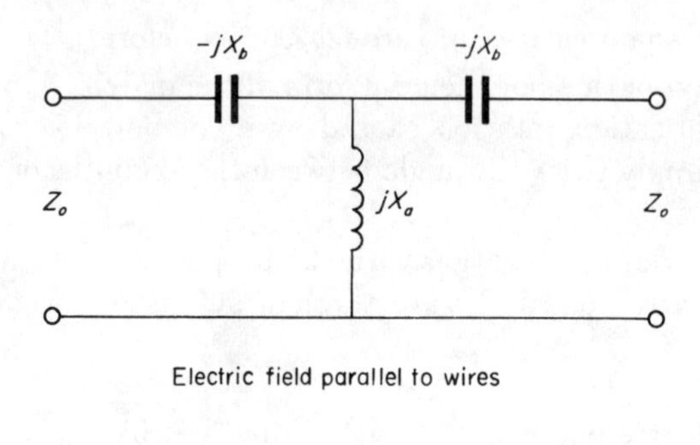

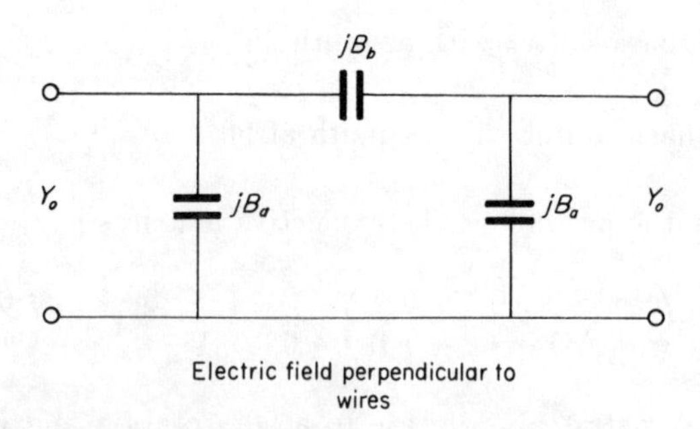

Figure 6.7: Equivalent circuits for parallel wire gratings

6.5 MULTIPLE INCLINED GRATINGS

Major disadvantages of the polarization rotators discussed in the previous sections are the common microwave ones of large size and complexity in manufacture. A polarization rotator will now be described that will rotate linear polarization from any predetermined direction into any other fixed direction without loss. This is done by a series of closely spaced parallel wire gratings each of which has a slightly different azimuth from the preceding grating. The incident polarization is then perpendicular to the wires of the first grating and the transmitted polarization is perpendicular to the wires of the final grating. Analysis shows that this can be done for gratings which are as close as 1/16 of a wavelength and hence a rotation through as much as 90° can be made within a structure less than half a wavelength in depth. This compares very favourably with the metal plate structures. The analysis is performed for plane gratings and for normal incidence.

Each grating consists of parallel fine wires of circular cross-section, the spacing between the wires, the diameter of the wires and the separation between the gratings are all small fractions of the wavelength of the incident radiation.

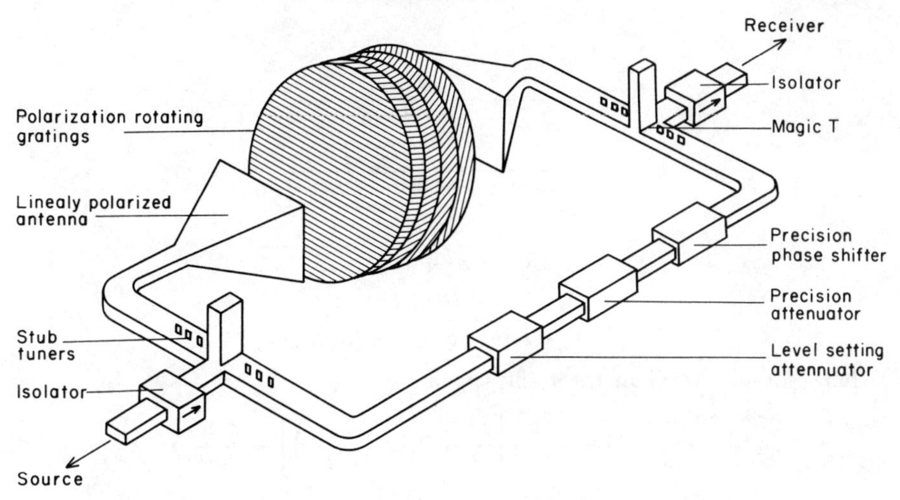

Figure 6.8: Experimental apparatus for polarizing grating measurement (Schematic)

The problem becomes one of determining how all the incremental reflected components which are parallel to the wires at each intermediate surface, combine through reflection, and re-reflection through the previous gratings in such a way that the overall reflection coefficient is minimized. To do this we employ the cascade matrix set up in Equation 6.2.4.

For any individual grating, the complex transmission and reflection coefficients parallel and perpendicular to the grating wires will be designated $T_{\|}$ $T_{\perp}$ $R_{\|}$ and $R_{\perp}$.

Then the elements T_{ij} and R_{ij} that go to make up the cascade matrix A_{ij} in Equation 6.2.2 are

$$\begin{aligned} T_{11} &= T_{\|} \cos^2\theta_r + T_{\perp} \sin^2\theta_r \\ T_{12} &= T_{21} = \frac{1}{2}(T_{\|} - T_{\perp}) \sin 2\theta_r \\ T_{22} &= T_{\perp} \cos^2\theta_r + T_{\|} \sin^2\theta_r \\ R_{11} &= R_{\|} \cos^2\theta_r + R_{\perp} \sin^2\theta_r \\ R_{12} &= R_{21} = \frac{1}{2}(R_{\|} - R_{\perp}) \sin 2\theta_r \\ R_{22} &= R_{\perp} \cos^2\theta_r + R_{\|} \sin^2\theta_r \end{aligned} \tag{6.5.1}$$

where the azimuth of the individual grating is the (variable) angle θ_r made by the wires with respect to the fixed x axis.

For a close grating of wires with circular cross-section the coefficients $T_{\|}$ $T_{\perp}$ $R_{\|}$ and $R_{\perp}$ can be obtained from the equivalent circuits of Figure 6.7. They are [12].

$$\begin{aligned}
R_{\parallel} &= \frac{2iX}{2Xx - x^2 + 1 + 2i(X - x)} \\
T_{\perp} &= \frac{2iB}{-2Bb - b^2 + 1 + 2i(B + b)} \\
T_{\parallel} &= \frac{2Xx - x^2 - 1}{2Xx - x^2 + 1 + 2i(X - x)} \\
R_{\perp} &= \frac{2Bb + b^2 + 1}{-2Bb - b^2 + 1 + 2i(B + b)}
\end{aligned} \tag{6.5.2}$$

where, for wire of diameter a and spacing d

$$\begin{aligned}
B_a &= \frac{2\pi^2 a^2}{\lambda d}\left\{1 - \frac{\pi^2 a^2}{\lambda^2}\left[\frac{11}{2} + 2\log\left(\frac{2\pi a}{d}\right)\right] - \frac{\pi^2 a^2}{6d^2}\right\} \\
B_b &= \frac{d}{\lambda}\left[\frac{3}{4} - \log\left(\frac{2\pi a}{d}\right)\right] + \frac{d\lambda}{2\pi^2 a^2} \\
X_a &= \frac{d}{\lambda}\left\{\log\left(\frac{d}{2\pi a}\right) + \frac{1}{2}\sum_{m=-\infty}^{\infty}\left[1 \Big/ \sqrt{m^2 - \frac{d^2}{\lambda^2}} - \frac{1}{|m|}\right]\right\} \\
X_b &= \frac{d}{\lambda}\left(\frac{2\pi a}{d}\right)^2
\end{aligned} \tag{6.5.3}$$

and

$$X = X_a/Z_0 \quad x = X_b/Z_0 \quad B = B_b/Y_0 \quad \text{and} \quad b = B_a/Y_0 \tag{6.5.4}$$

The analysis need by no means be confined to wires of circular cross-section. The reflection and transmission coefficients for gratings of elements with other cross-sections, including thin strips and elliptical cross-sections, can be found in the literature, in a form suitable for direct substitution into either the equivalent circuit Equations 6.5.2 or Equation 6.5.3 [13].

In the practical case where the gratings are only separated by a fraction ($\simeq$ 1/16) of a wavelength, the question of grating interaction has to be considered. The analysis has only considered plane TEM wave propagation and ignored evanescent or higher modes. This has been investigated experimentally in a microwave bridge by the method shown diagrammatically in Figure 6.8. A set of five fine wire gratings was used to rotate an incident linear polarization through 90°. The results agreed with the analysis for a range of spacings between the individual gratings which indicates a bandwidth of 3:1 for almost 100% transmission. In fact the arrangement can be used at any frequency for which the spacing between gratings is well below the critical value of $\lambda/2$. At this value, a resonant reflection effect occurs (Figure 6.9) and this is found to increase in sharpness with increasing number of gratings. Such a reflection resonance indicates the possible use of this grating array as a polarization frequency filter.

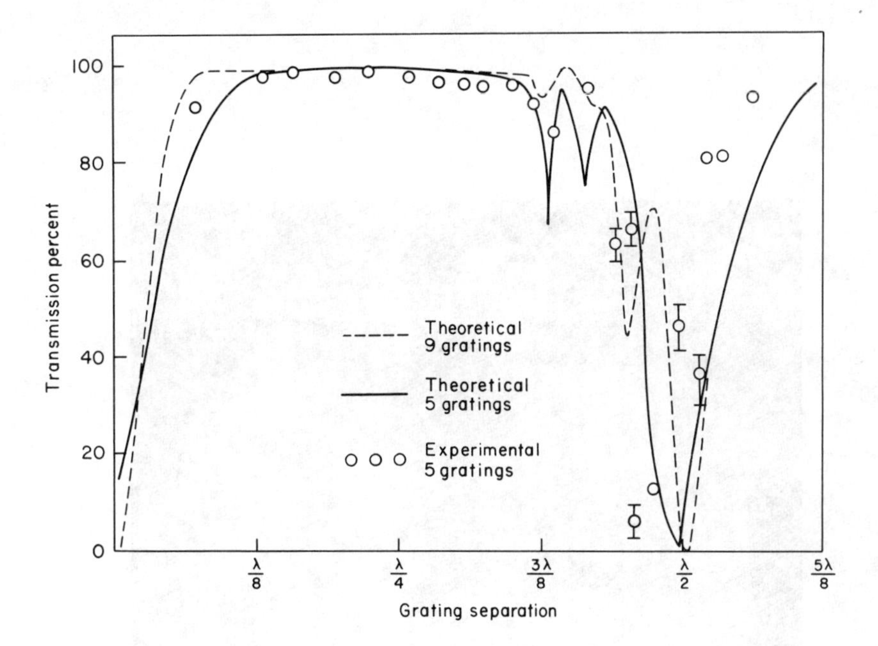

Figure 6.9: Polarization rotation through 90° with closely spaced gratings

This would include systems such as circular polarizers and analysers. The method can also be applied to the operation of any of these systems at other than normal incidence by suitably modifying the grating coefficients in Equations 6.5.1 and 6.5.2 for oblique incidence and adjusting the transmission and spacing matrices accordingly.

MOIRÉ FRINGE PLANE WAVE PHASE SHIFTER

When two parallel gratings of thin strips are contra-rotated about an axis perpendicular to the surface containing them, Moiré fringes are created which take the form of a parallel grid of black and white stripes, perpendicular to the original gratings and whose width and separation are a function of the angle of rotation (Figure 6.10). When the two gratings are both parallel and initially perpendicular to the polarization of a plane incident wave, these fringes will be parallel to that polarization. On rotation therefore, the Moiré fringes act as a variable inductance to the incident field, whose effect is to change the phase of the transmitted wave. For the parallel gratings and for a considerable degree of rotation of the gratings, the transmission is virtually lossless, and the structure then acts as a (nearly) lossless free space phase shifter.

Transmission is through the diamond shaped regions between the fringes, and, provided the long axis of these diamond areas is well above cut-off for the

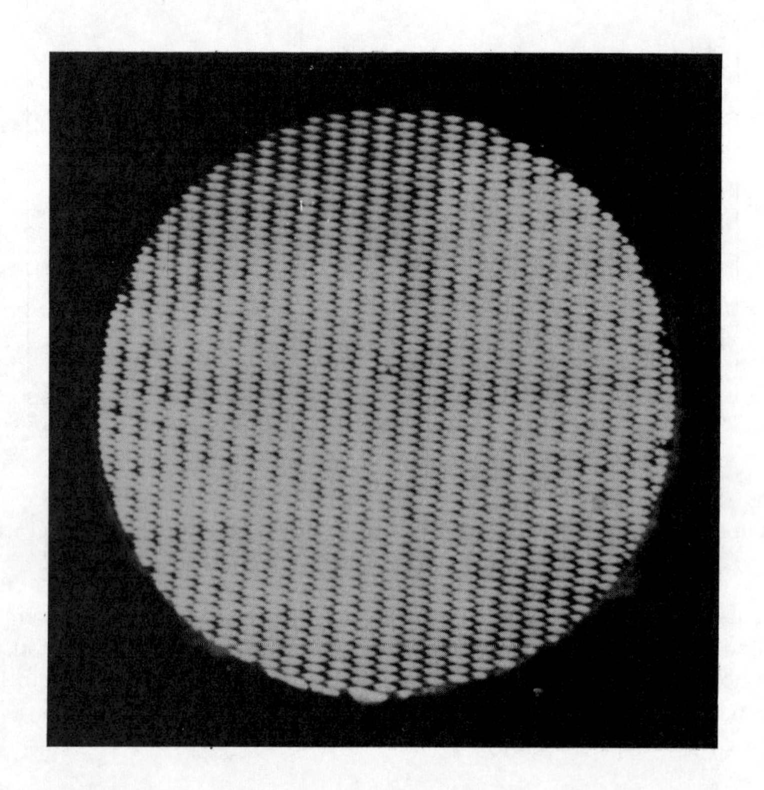

Figure 6.10: Moiré fringes of rotated gratings

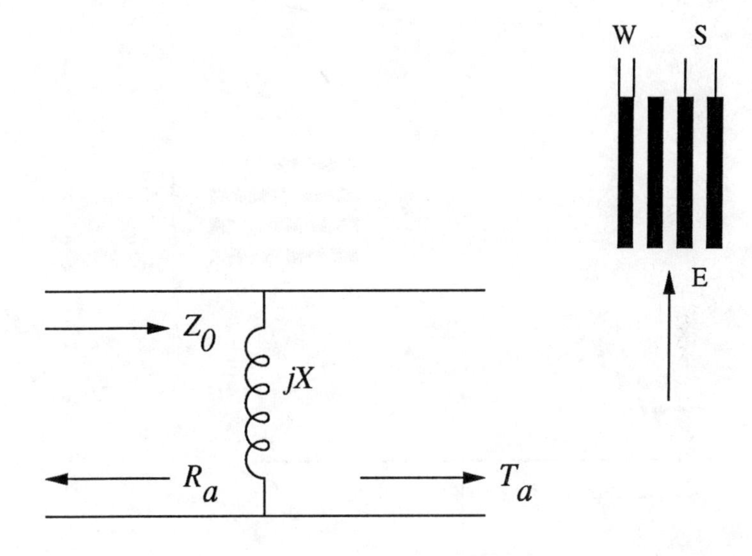

$$R_a = \frac{-1}{1 + 2jX_0} \qquad T_a = \frac{2jX_0}{1 + 2jX_0}$$

Figure 6.11: Inductive grating

operational wavelength, virtually the same degree of transmission is obtained as when the pair are both parallel and perpendicular to the incident polarization. For the strip dimensions being considered this is virtually 100%. This allows for a considerable amount of rotation during which there is (nearly) total transmission, while the fringes move closer together and decrease in width (and in definition) thus varying the inductance presented to the incoming wave and hence the phase of the transmitted wave.

Transmission is then calculated as for a pair of closely spaced inclined gratings, using the cascade matrix with elements derived from the equivalent circuits of a grating of thin metal strips.

The matrix elements are derived from the equivalent circuits (simplified), shown in Figure 6.12 for grating parallel and perpendicular to the electric field vector. For thin plane parallel strips of width w and centre to centre spacing S, these are [12]

$$B_0 = \frac{B}{Y_0} = \frac{4S}{\lambda}\left[\ln\frac{1}{c} + \frac{(1-c^2)^2(2P(1-c^2/4)+4c^2P^2)}{2[1-c^2/4+2Pc^2(1+c^2/2-c^4/8)+2c^6P^2]}\right] \tag{6.5.5}$$

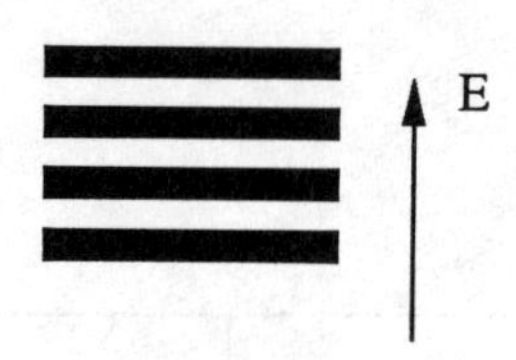

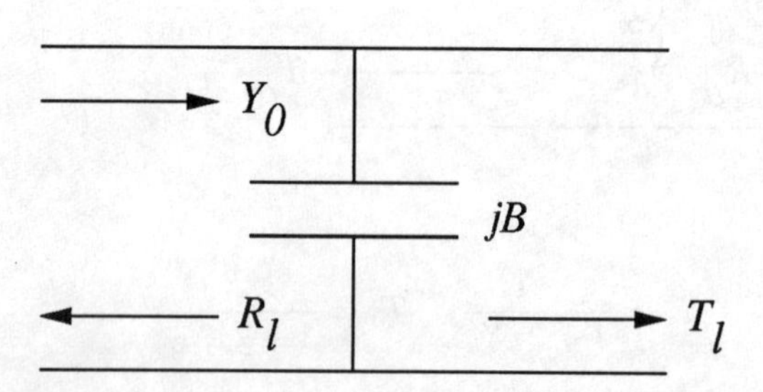

$$R_l = \frac{-jB_0}{1+2jB_0} \qquad T_l = \frac{2}{1+2jB_0}$$

Figure 6.12: Capacitative grating

where

$$c = \sin\frac{\pi}{2}(1 - \frac{w}{S}) \qquad P = \frac{1}{\sqrt{1 - S^2/\lambda^2}} - 1$$

$$X_0 = \frac{X}{Z_0} = \frac{S}{\lambda}\left[\ln\frac{1}{d} + \frac{(1-d^2)^2(2P(1-d^2/4) + 4d^2P^2)}{2[1 - d^2/4 + 2Pd^2(1 + d^2/2 - d^4/8) + 2d^6P^2]}\right] \tag{6.5.6}$$

where

$$d = \sin\frac{\pi}{2}\frac{w}{S}$$

The cascade matrix is then applied to two such gratings inclined at angles $\pm\theta$ with close ($< \lambda/10$) air spacing. The computer model uses complex matrix theory to determine the transmitted and reflected wave coefficients by resolving the incident wave into components parallel and perpendicular to the rotated gratings. The complex values of the components of the transmitted wave then give both the transmitted power and phase.

The results are shown in Figure 6.13 for gratings with mark space ratios of 1:1 (3mm. 3mm.). Cut-off is approached when the grids are at 80° (± 40 degrees) angle with each other as can be seen. Over the range ± 20 degrees very little attenuation occurs with a phase shift of nearly 30 degrees. Over a considerable intermediate range, ± 10 degrees to ± 25 degrees, this phase change is virtually linear, and the transmission loss less than 1db. With increased attenuation the full range of phase shift is over 90°. The experimental method used the same bridge circuit as in Figure 6.8 and was carried out in X-band. The results are shown in Figure 6.13. *

6.6 POLARIZATION TWIST REFLECTORS

Any transmission system which is *symmetrical* about a mid-plane can be converted into a reflection system with identical properties by the insertion of a conducting surface at the mid-plane. This is true in optical filter theory as it is in electric circuit theory and thus applies to the previous examples of this chapter.

In a most elementary application of this principle a polarization rotator which consists of a single stack of plates can be converted into a reflector with the same polarization rotation property by a conducting surface at the centre of the plates bisecting the stack. In the case of a 90° polarization rotator the resultant structure has the simple form of a set of parallel plane fins on a plane conducting backing surface, in which a differential phase of π radians is obtained between the TEM mode perpendicularly polarized to the fins and the TE_{01} mode of the parallel component (Figure 6.14(a)).

---

*The author is indebted to Miss J Beale MA MSc of the Mullard Science Laboratory for carrying out this analysis and computation.

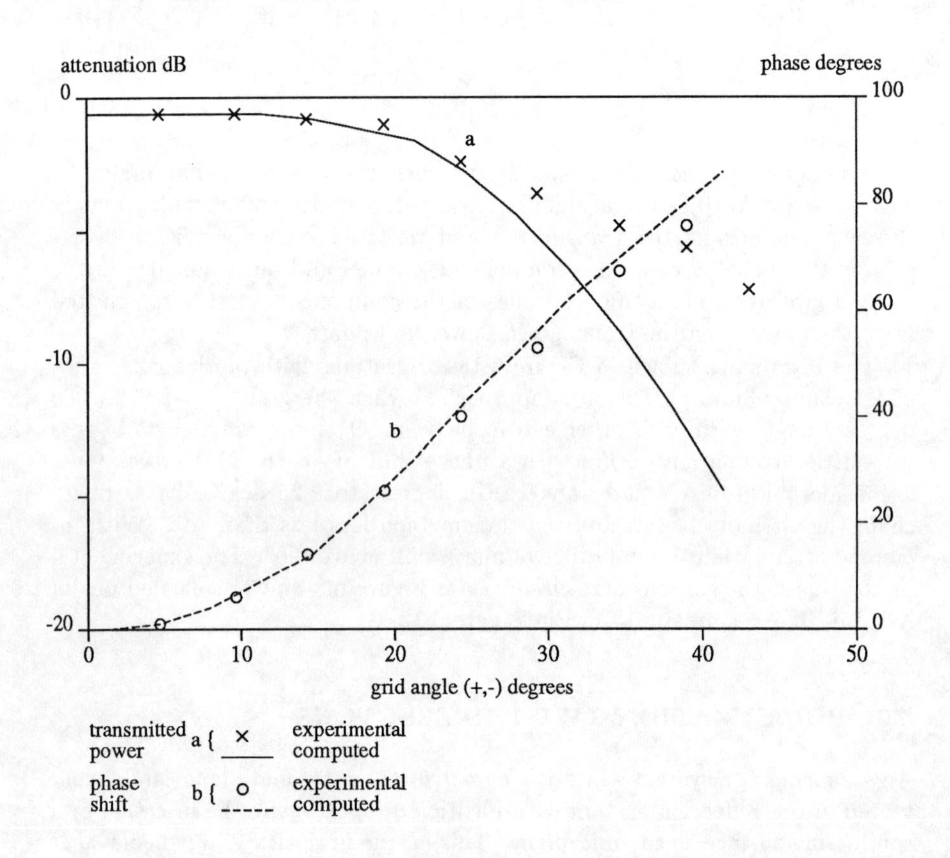

Figure 6.13: Moiré fringe phase shifter, comparison of results

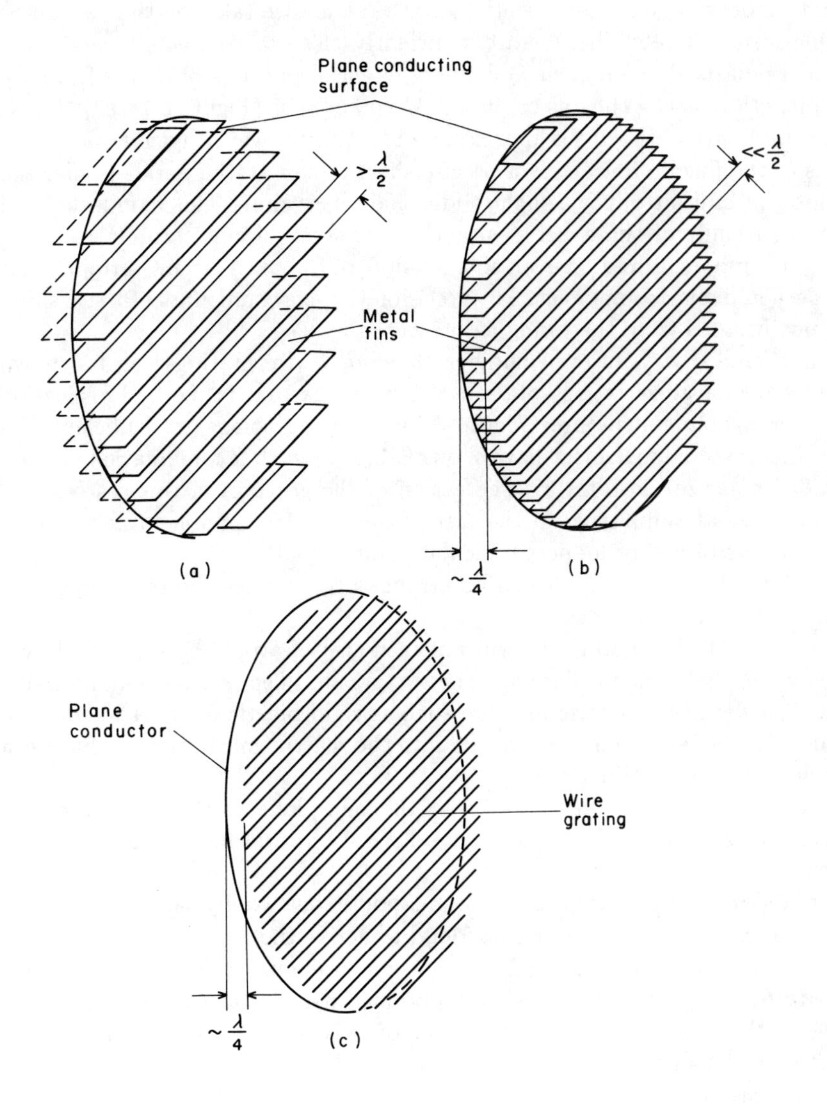

Figure 6.14: Polarization rotating reflectors (a) Bisected polarization rotator (b) and thin version with closely spaced fins (c) bisected grating polarization rotator

Since the device is no longer active as a transmitting system, propagation of both modes is no longer essential and the structure takes on the even more simple form whereby the fins are sufficiently close to completely reflect the parallel polarized component and the differential phase is obtained from the path length between the fins of the TEM component (Figure 6.14(b)). In the main this depth of the fins is approximately a quarter wavelength, however the edges of the fins do not act as a precise short circuit to the parallel polarized component and introduce a slight additional phase shift. This manifests itself as a small shift of the exact position of the short circuit terminals for the parallel component and an even smaller shift of the open circuit terminals for the perpendicular component. For precision the exact amounts of these shifts can be obtained from the formulae presented in [12].

In a similar way an even number of *parallel* wire gratings can be shown to introduce a polarization rotation of 90° by the matrix method of Section 6.4. For a simple structure (Figure 6.14(c)) two gratings are sufficient and both theory and experiment have shown [13] that such a combination can act as a 90° polarization rotator. The action of the gratings in this case can be visualised as an inductance in the case of the parallel component and a (very small) capacitance to the perpendicular component.

The resulting phase shift can be arranged to produce the required polarization rotation.

In a practical structure the wires of each grating would be supported by a transparent dielectric medium and the two gratings kept separate and parallel by a light density dielectric medium such as rigid plastic foam. The addition of the effects of such refractive media into the matrix calculation, as has been shown, presents no difficulty.

The insertion of a metal wall into the mid-plane of this structure now creates a particularly effective form of polarization rotating reflector, or twist-reflector.

As a design procedure the matrix method is not wholly satisfactory as no obvious iterative procedure can be found to derive the matrix elements which alone define the system parameters. Hence a large number of a priori systems have to be analysed and an optimum obtained usually by graphical methods.

A method for the design of this type of twist-reflector has been given by Hannan [14] using the circuit analogue, which has been found to produce accurate results for most applications.

If, as in Figure 6.15, the wires are of radius a and have a grating spacing of d the approximate admittance from Equation 6.5.3

$$\frac{1}{X_a} = B_w = \frac{\lambda}{d \log(d/2\pi a)}$$

at normal incidence. At oblique incidence at an angle θ, this becomes modified to

$$B_w = \frac{\lambda}{d \cos\theta \log(d/2\pi a)}$$

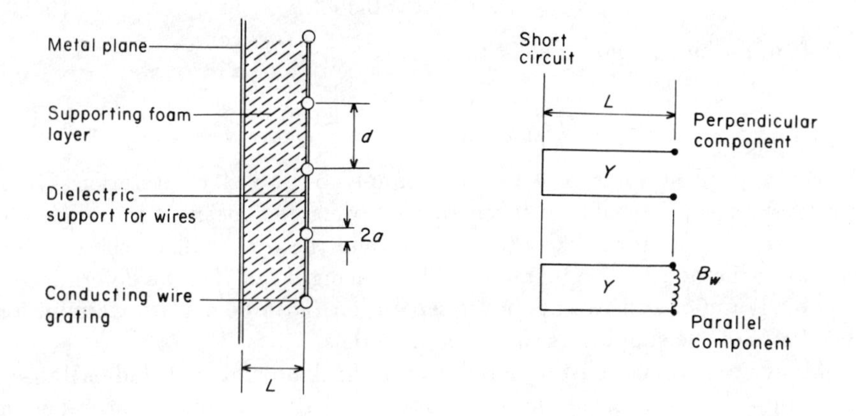

Figure 6.15: Polarization twist reflector and equivalent circuits

If the wires are spaced a distance L from the conducting surface, then, referred to a terminal plane at the grating, the perpendicular component sees a short circuited length of transmission line with admittance Y of length L and the parallel component sees a similar short circuited line but with a parallel admittance B_w at the terminal plane.

For operation as a 90° polarization rotator the phase differential is contained in the condition

$$Y_\perp = 1/Y_\| \tag{6.6.1}$$

where

$$Y_\perp = i\cot(2\pi L\cos\theta/\lambda)$$

and

$$Y_\| = i\cot(2\pi L\cos\theta/\lambda) - iB_w$$

the solution of which gives

$$B_w = 2\text{cosec}(4\pi L\cos\theta/\lambda)$$

or

$$2\text{cosec}(4\pi L\cos\theta/\lambda) = \frac{\lambda}{d\cos\theta\log(d/2\pi a)} \tag{6.6.2}$$

Optimum performance is obtained if this relation holds to the second order for changes in λ or θ. Differentiating therefore with respect to θ we obtain

$$\frac{4\pi L\cos\theta}{\lambda} = \tan\frac{4\pi L\cos\theta}{\lambda}$$

and hence

$$(L\cos\theta)/\lambda = 0.358 \tag{6.6.3}$$

Substitution into Equation 6.6.2 then gives

$$\frac{\lambda}{d\cos\theta\log(d/2\pi a)} = 2.046 \tag{6.6.4}$$

For any given angle of θ this can simply be solved by iteration. In a practical design θ would be taken to be a weighted mean of the angles of incidence the reflector is likely to intercept and the wire radius a chosen from the catalogue of available materials. The spacing d can then be determined.

Finally minor adjustments in the separation L can be made to account for the effects of the supporting dielectric materials.

Many applications exist for a reflector of this kind and are listed in Hansen [15]. In the optical sense the main interest lies in the many two reflector arrangements that become possible through its application as described in Chapter 2. If the first reflector is constructed as a grating of closely spaced wires which reflects the parallel polarized incident radiation entirely, the second reflector can be a twist reflector which turns the polarization through 90° and thus makes it pass through the first reflector without loss. This principle is illustrated in Figure 2.50 of Section 2.8 and was first used in a paraboloid-plane combination as shown [16] [17]. In this form it can scan a pencil beam over the entire forward hemisphere. As always other arrangements are possible which, while limiting the scan through the introduction of aberrations, make possible other desirable effects such as varying the illumination function.

The properties of these twist reflectors have been found to be constant over the wide range of incident angles that such systems and those illustrated in Chapter 2 require.

IMPOSSIBLE POLARIZERS

There are few polarizing systems whose application to microwaves would be highly desirable but whose design can be shown to be impossible by the nature of the Jones matrix required to describe them. For example, in nearly all open antenna systems and particularly in communications systems over long terrestrial paths or through the atmosphere to and from satellites, signal fading occurs at random. This can be caused by absorption and scattering mechanisms. Quite often however, and not always taken into consideration, it can be caused fully or in part by polarization rotation. Polarization diversity reception at the receiving antenna could be an unwarranted technical complication. Should, however, a device with total polarization agility be possible, the fade could be adjusted instantaneously, as opposed to having a form of servo-driven polarization seeker that would be otherwise required. Such a device if achievable, would be described by the matrix number 6 in Table 6.2, which rotates an arbitrary incident linear polarization into a given prescribed direction, and without reflection loss of any kind. Such a device has not, so far, been shown possible of practical achievement.

Similarly we observe that polarizers corresponding to matrices numbered 14 and 15 in the table have as yet no physical counterparts. These refer to the total reflectors of one hand of circular polarization and hence total transmitters of the opposite hand.

The feature that distinguishes these matrices from the others in the table is that they are singular.

They cannot therefore be constructed from a series product of the other non-singular matrices each of which corresponds to a realizable practical device.

We are led therefore to the assumption that a lossless, which includes loss by reflection, polarizing device is only capable of practical realization if the Jones matrix representing it is non-singular.

6.7 BI-VECTORS AND QUATERNIONS IN POLARIZATION ANALYSIS

The theory so far given in this chapter has been concerned exclusively with wave propagation in the positive or negative z axis directions. For these cases the polarization state can be uniquely defined by the two component Jones vector or by a combination of the component circular polarizations to give the representations of Equations 6.1.27 and 6.1.22.

We now wish to describe the propagation of an elliptically polarized wave in any general direction in a three dimensional space, in terms of what may be called "basis polarizations" along orthogonal Cartesian axes.

Polarization studies in the field of optics and microwave antennas have involved concepts which, by themselves, would have led to many of the mathematical methods now applied to the study of particle spin. As long ago as 1843 Hamilton himself had found in his discovery of quaternions

> "there seems to be something analogous to polarized intensity in the purely imaginary part and to the unpolarized energy (indifferent of direction) in the real part, of a quaternion, and thus we have some slight glimpse of a future calculus of polarities ..." (letter to Graves, 17th October 1843 [18].

In this discovery he had in fact anticipated the advent of the spin matrices by approximately a century. Spin matrix methods are now current in the study of optical polarization and concurrently quaternion methods are coming into regular usage [5] [19] [20] [21].

The application in optics has been mainly to the fields of reflection polarimetry and ellipsometry [22] [23]. These are studies of imperfect or irregular conducting surfaces or of the properties of dielectric materials in the form of thin films backed by perfect conductors through the measurement of the ellipticity of a reflected wave arising from an obliquely illuminating linearly polarized wave. This procedure could well be adapted to the measurement of the dielectric properties of materials at microwave frequencies, for example, such properties at elevated temperatures, or the thickness of transparent layers.

The method by which polarization studies as applied to microwave antennas derive all the properties attributed to quaternions is fascinating. In the classic set of papers published in 1951 [8] dealing with the polarization of a generally directed plane wave Kales used the concept of a rotating vector in the plane normal to the propagation direction and the theory of bi-vectors to describe it. The results agree with the results obtained through the spinor method as discussed in the final section of this chapter.

Both bi-vectors and quaternions have been used in the fundamental study of the relativistic invariance of Maxwell's equations [24] (see Chapter 7). Rumsey and later Crout [25] [26] in considering the polarization of the radiation from n-arm spiral antennas introduced a bi-complex notation which only falls short of the quaternions themselves in a single particular. A similar consideration and the same result occurs in an entirely different context in a paper by Lewin [27] in which he discusses the wave equation in orthogonal curvilinear coordinates. In brief, the bi-vector notation in calling for a complex combination of the form $\mathbf{E}+iZ\mathbf{H}$ of the electric and magnetic field vectors requires to distinguish between the complex $\sqrt{-1}$ as used there with the $\sqrt{-1}$ used in the angular frequency of the temporal component $\exp(j\omega t)$. Such a bi-complex notation is the essence of the quaternion method. Commutation properties are required of course and these are given as $i^2 = j^2 = -1$ but the essential non-commutative property $ij = -ji$ that distinguishes the quaternions from all that had gone before, is not included.

The full relationship between bi-vectors and quaternions is given by Ehlers et al [28]. We shall demonstrate this by a physical quaternion description of a polarized wave.

We specify (Appendix IV) three hypercomplex numbers α β γ with the algebraic properties

$$\alpha^2 = \beta^2 = \gamma^2 = -1$$

$$\alpha\beta = \gamma \qquad \beta\gamma = \alpha \qquad \gamma\alpha = \beta$$

from which we find fundamentally $\alpha\beta = -\beta\alpha$ (since for example $\alpha\beta\beta\gamma = \gamma\alpha$ but $= -\alpha\gamma$). This last commutation rule must be adhered to in all products of quaternions.

The "ordinary" complex number $e^{i\theta}$ now refers to a circular motion in the Argand plane that is to a rotation about the z axis. It is compounded of the two harmonic motions $\cos\theta$ $\sin\theta$, which are in phase quadrature and orthogonal in space. If we let each of the complex numbers α β and γ refer similarly to rotations about the three orthogonal Cartesian axes then say

$$e^{\alpha\theta} \text{ refers to a rotation about the } z \text{ axis,}$$
$$e^{\beta\theta} \text{ refers to a rotation about the } y \text{ axis,}$$

and

$$e^{\gamma\theta} \text{ refers to a rotation about the } x \text{ axis,}$$

This notation is the only variant between quaternion notation and the spin matrices.

The required circularity is a real circle $e^{i\psi}$ in a plane perpendicular to the direction (θ, ϕ) of wave propagation. That is ψ is a real rotation angle and i the ordinary $\sqrt{-1}$. This means in principle that $i = \sqrt{-1}$ has to be projected into the three basis coordinate directions and each component has to represent a basis circularity of the form $e^{\alpha\theta}$....

This can be done by the projection (Appendix IV)

$$i = \alpha\cos\theta + \beta\sin\theta\sin\phi + \gamma\sin\theta\cos\phi \qquad (6.7.1)$$

The validity of Equation 6.7.1 can readily be tested by the direct square product of both sides which gives -1 when the rules of quaternion multiplication are used. Then the rotation in the plane normal to the θ, ϕ direction becomes

$$\cos\psi + i\sin\psi = \cos\psi + \sin\psi[\alpha\cos\theta + \beta\sin\theta\sin\phi + \gamma\sin\theta\cos\phi] \qquad (6.7.2)$$

which relation is established algebraically in the appendix.

This is then of the form $\cos\psi+$ (complex number)$\sin\psi$ for *any* choice of θ and ϕ. For example rotations about the x y and z axes are given by

$$\begin{array}{llll} x \text{ axis} & \theta = \frac{\pi}{2} & \phi = 0 & e^{i\psi} = \cos\psi + \gamma\sin\psi \\ y \text{ axis} & \theta = \frac{\pi}{2} & \phi = \frac{\pi}{2} & e^{i\psi} = \cos\psi + \beta\sin\psi \\ z \text{ axis} & \theta = 0 & & e^{i\psi} = \cos\psi + \alpha\sin\psi \end{array} \qquad (6.7.3)$$

The connection with particle spin is now made through the fact that *complex* spin matrices have the same algebra as the complex α β γ given above and so we find [2].

$$\alpha = i\sigma_3 \qquad \beta = i\sigma_2 \qquad \gamma = i\sigma_1$$

where

$$\sigma_1 = \begin{pmatrix} 0 & 1 \\ 1 & 0 \end{pmatrix} \qquad \sigma_2 = \begin{pmatrix} 0 & -i \\ i & 0 \end{pmatrix} \qquad \sigma_3 = \begin{pmatrix} 1 & 0 \\ 0 & -1 \end{pmatrix}$$

are the Pauli spin matrices.

With this decomposition and the definition of a coherency matrix

$$J \equiv \begin{vmatrix} |E_x|^2 & E_x^* E_y \\ E_x E_y^* & |E_y|^2 \end{vmatrix} \qquad (6.7.4)$$

$$= \frac{E_x^2 + E_y^2}{2}\begin{pmatrix} 1 & 0 \\ 0 & 1 \end{pmatrix} + \frac{E_x^2 - E_y^2}{2}\begin{pmatrix} 1 & 0 \\ 0 & -1 \end{pmatrix}$$

$$+ \frac{E_x E_y^* + E_x^* E_y}{2}\begin{pmatrix} 0 & 1 \\ 1 & 0 \end{pmatrix} + \frac{E_x^* E_y - E_x E_y^*}{2} i \begin{pmatrix} 0 & i \\ -i & 0 \end{pmatrix} \qquad (6.7.5)$$

then the coefficients of the Pauli matrices are the $I\ Q\ U\ V$ of the Stokes vector 6.1.15 [29, p550] or

$$J = J^{\mu}\sigma\mu \tag{6.7.6}$$

The notation is suitable for partially coherent waves.

6.8 RADAR POLARIMETRY

One subject that has made distinct progress in the past decade is the theory and application of the polarization properties of target returns. A survey of this topic [31] contains 67 references, only twelve of which predate 1980 and most of those are concerned with standard texts on group and algebraic theory. It is as obvious therefore, that only a covering view of the subject can be given here, but that becomes possible since the basic notation and properties of the polarized wave have been given already in this chapter. This alone is a sufficient preparation for the study of most of the references in the survey.

There is one major difference, however, the resolution of which will take us deep into the methods of general theoretical physics, and is worth pursuing for that reason alone. So far in this chapter we have been exclusively considering a wave propagating in the positive z axis direction and where reflected waves are considered, only their linearly polarized components are dealt with. Radar polarimetry, as in optical polarimetry, deals with both transmitted and received waves of general polarization. At the very outset of the study in this chapter, the time component was eliminated, and the system referred to the plane $z = 0$ for the description of the polarization ellipse. The essential fact that this ellipse is swept out by the tip of the rotating **E** vector due to the progression of the wave along the axis and *not* by a rotating vector in the plane is lost, together with all the information concerning the direction of travel of the wave.

The problem can be simply demonstrated. We consider an infinite conducting plane and require it to have a Jones matrix that will describe the reflection of a wave of *any* polarity in the manner which is known from experimental evidence. That is plane polarized waves are reflected with the same plane of polarization and circularly polarized waves are reflected with the "handedness" reversed. We take the incident waves from Table 6.1 of Jones vectors. Then it is found quite readily that the reflecting matrix

$$\begin{bmatrix} -1 & 0 \\ 0 & -1 \end{bmatrix}$$

which leaves linearly polarized Jones vectors linearly polarized (with if necessary a 180° phase change) will *not* convert right-hand circular polarization $\begin{bmatrix} -i \\ 1 \end{bmatrix}$ into left-hand $\begin{bmatrix} i \\ 1 \end{bmatrix}$. Contrarily, a matrix that does perform this conversion can be constructed, for example $\begin{bmatrix} 1 & 0 \\ 0 & -1 \end{bmatrix}$ which also leaves the two major (x and y) linear polarizations unchanged. When applied to a 45° linear polarization however it is reflected at -45°.

This illustration serves to show that whereas matrix multiplication in the usual sense can be applied to progressive waves and devices, the algebra is inadequate to deal with reflected waves and the scattering from objects that creates them.

Hence a descriptor of either the handedness of the polarization or its equivalent the direction of travel of the wave, has to be included, either in the definitions of the wave polarity, or in the properties of the reflecting matrix. It is also a requirement in general polarimetry that the notation and methodology arrived at can be extended to include complex scattering from objects, and, although not to be considered in this volume, bistatic scattering and the description and scattering of partially coherent waves.

One of the first to recognise this was Graves [32], who attributed a + or − suffix to the Jones vector to imply a wave travelling in the positive z direction or in the negative z direction. The following analysis is taken mainly from this reference. Thus in any specified basis, (say linear at this stage),

$$E_+ = \begin{pmatrix} E_1 \\ E_2 \end{pmatrix}_+ = \begin{pmatrix} a_x e^{i\delta x} \\ a_y e^{i\delta y} \end{pmatrix}_+ \quad \text{or better} \quad \begin{pmatrix} E_{1+} \\ E_{2+} \end{pmatrix} \tag{6.8.1}$$

$$E_- = \begin{pmatrix} E_1 \\ E_2 \end{pmatrix}_- = \begin{pmatrix} a_x e^{i\delta x} \\ a_y e^{i\delta y} \end{pmatrix}_- \quad \equiv \quad \begin{pmatrix} E_{1-} \\ E_{2-} \end{pmatrix} \tag{6.8.2}$$

(the reason will be made apparent later).

Change of *basis* (say to circular) requires a unitary change-of-basis transformation matrix which, operating on a positive direction vector E_+ performs the transformation

$$E'_+ = Q_{++} E_+$$

and negatively

$$E'_- =^* Q_{--} E_- \tag{6.8.3}$$

Here the equal sign means that, when viewed from the same coordinate system, a positive basis change Q_{++} on the positive direction vector E_+ has the same effect as a negative transformation $^*Q_{--}$ has on a negative direction vector E_-. (* is complex conjugate).

Finally there is the requirement for a matrix operation which transforms the components and reverses the direction of propagation. This operator will be designated M_{+-} or M_{-+} according to its action in transforming a +vector into a −vector or vice versa. Hence

$$E'_- = M_{-+} E_+ \quad \text{and} \quad E'_+ = M_{+-} E_- \tag{6.8.4}$$

It is important to note that the second subscript on the operator matrix M has to conform with the type of vector being transformed. Transformations such as $Q_{++} E_-$ or $M_{+-} E_+$ are not defined.

THE SCATTERING MATRIX

What is now required is the incorporation of the +, − notation into the Jones matrices of reflection. This will be termed a back-scattering matrix with complex elements

$$S = \begin{pmatrix} s_{11} & s_{12} \\ s_{21} & s_{22} \end{pmatrix}$$

where by reciprocity $s_{12} = s_{21}$ and therefore S is symmetric for monostatic reflection.

Scattering is then defined by the operation of S on the transmitted vector $\begin{pmatrix} E_x^T \\ E_y^T \end{pmatrix}$

$$\begin{pmatrix} E_x^S \\ E_y^S \end{pmatrix} = \begin{pmatrix} s_{11} & s_{12} \\ s_{21} & s_{22} \end{pmatrix} \begin{pmatrix} E_x^T \\ E_y^T \end{pmatrix} \tag{6.8.5}$$

or

$$E^S = SE^T$$

This has to be modified now to take into account the fact that the transmitted and reflected waves are travelling in opposite directions. This means the inclusion of direction symbols in the scattering matrix and so Equation 6.8.5 becomes

$$E_-^S = S_{-+}E_+^T \tag{6.8.6}$$

Consequently a transmitter of the elliptically polarized wave $E_+^T = \begin{pmatrix} E_1^T \\ E_2^T \end{pmatrix}_+$ will receive the component of the backscattered field $E_-^S = \begin{pmatrix} E_1^S \\ E_2^S \end{pmatrix}_-$ by the direct scalar product $E_+^T \cdot E_-^S$. Similarly another antenna with (transmit) polarization E^R will receive $E_+^R \cdot E_-^S$.

The change of basis, from say linear polarized components to circularly polarized components, cannot have any effect on the end result. So the result of applying the transformation of 6.8.3 to the relation 6.8.6 must leave the result the same. So from 6.8.3

$$\begin{aligned} E_+^T &= Q_{++}E_+^{T'} \\ E_-^S &= {}^*Q_{--}E_+^{S'} \end{aligned} \tag{6.8.7}$$

where the primes refer to the transformed basis.

Substituting in 6.8.8 from 6.8.6 therefore

$$S_{-+}Q_{++}E_+^{T'} = {}^*Q_{--}E_+^{S'}$$

and hence

$$E_-^{S'} = \tilde{Q}_{--}S_{-+}Q_{++}E_+^{T'} \tag{6.8.8}$$

where for unitary transformations $\tilde{Q} = {}^*Q^{-1}$ is the transpose of Q.

Comparing 6.8.8 with 6.8.6 shows that going from an unprimed basis to a primed one involves the *congruent* transformation of S.

$$S' = \tilde{Q}SQ$$

that is the left multiplication is with the transpose and not as is usual in a *similarity* transformation, with the inverse.

We now see that there has been an introduction of a second index to each entity. The outer one labels the direction as + or − and the inner one labels the element in the vector. Both these symbols can only take on one of two values since, fortunately, we are only dealing with two component elements. Hence we can replace them with the capital symbols A B C etc, each of which is two valued 1 or 2 for the vector element and + or − for the direction symbol (currently 0 or 1 is the preferred symbolism in keeping with binary notation). Thus the vectors in Equations 6.8.1 and 6.8.2 can now be defined as E_{AA} or E_{AB} etc.

A reflected elliptical polarization from a plane reflector is the same ellipse with other handedness. As can be seen from the Poincaré sphere, (Figure 6.4), this implies the change from δ to $-\delta$. This is then equivalent to taking the complex conjugate of the original incident polarization, and it is to be noted, *not* the same as the orthogonal polarization. Thus, for example, an incident right-hand circular polarization in a forward direction namely $\begin{pmatrix} -i \\ 1 \end{pmatrix}_+$ is reflected as left-handed in the opposite direction or $\begin{pmatrix} i \\ 1 \end{pmatrix}_-$ and a matrix transformation that uses regular matrix formulations with the conjugation addition is the requirement. The method now in use [33] is to imply the taking of complex conjugate by dotting the appropriate index † To enable this dotting process to be performed by the same procedure as other transformation matrix processes, a system of unit metric matrices has been devised and is basic to the entire spinor formulation. While adding in a (manageable) complication to the algebra of matrices, it does present a coherent formulation for the study of polarity and spin. The final justification is that this formalism connects homeographically with the group theory of SU(2) and the Lie algebra of quaternions as outlined below [34].

SPINORS AND GROUPS

The two complex-element vectors used to define the polarization state in Equations 6.8.1 and 6.8.2 are examples of "spin-vectors". In the same manner that the ordinary vectors can be generalized to vector fields, spin-vectors generalise to spinor fields. The higher continuation of this concept, which we have

---

†This was the original notation of Infeld and van der Waerden [35] and was superseded for typographical reasons (in the pre-processor era) by a prime or overbar.

no need to consider in the context of radar polarimetry, is for further indices to be applied taking the vector analogy to higher order tensors and tensor fields. The general analysis calls all such entities spinors. Tensors also have a matrix analogue, and commonly their spinor equivalents, also matrices, are included in the terminology as spinors. In what follows we only deal with two component two index spinors and 2 x 2 matrices needed to transform them. The indices in this case, as we have noted, are binary numbers, but although fundamental in form, have to conform with the general rules of the algebra of abstract indices. This is similar in most respects to that of tensor analysis. Second rank spinors therefore have to have an index system implying multiplication which involves both an upper or contravariant index and a lower or covariant one, and most importantly, a process to transform between them.

In tensor analysis, this transformation consists of a matrix multiplication which raises or lowers the index. The matrices are called metric tensors, which in the present context become metric spinors. In the physical world described by tensors, the metric matrices have a physical meaning, usually relating the incremental distance along a world line to the coordinate space being considered. As can be seen by the derivation, the "metric" for polarization states is a purely algebraic process, but still one that accords with tensor analysis. Hence we define metric spinors

$$\varepsilon_{AB} \qquad \varepsilon^{AB} \qquad \varepsilon_{\dot{A}\dot{B}} \qquad \varepsilon^{\dot{A}\dot{B}} \qquad \varepsilon_{AB} = -\varepsilon_{BA} \tag{6.8.9}$$

with the properties that they are independent of the basis in which they are applied and invariant under the conjugation process. In matrix form

$$\varepsilon_{AB} = \varepsilon^{AB} = \begin{bmatrix} 0 & 1 \\ -1 & 0 \end{bmatrix} \tag{6.8.10}$$

and their operation is given by

$$\Psi^{A} = \varepsilon^{BA}\Psi_{B} \tag{6.8.11}$$

Raising (or lowering) an index on an ε matrix means multiplying two matrices of the type in 6.8.10 hence mixed index ε matrices are negative unit matrices.

Conjugation has to be a transform by an identity matrix and this is formed by the product of two mixed index ε matrices as

$$\eta'_{\dot{A}A} = \varepsilon_{\dot{A}}{}^{B}\,\varepsilon_{A}{}^{C}\,\eta_{BC} \tag{6.8.12}$$

If we take the dotting process to automatically include the change in direction of the wave, it can be seen that the A and C symbols become redundant and can be omitted.

Thus for the scattering process we replace the unit matrices in 6.8.12 by a scattering matrix $S_{\dot{A}A}{}^{BC}$ with action defined by 6.8.15 and reduced to $S_{\dot{A}}{}^{B}$. Simple backscattering is then given by [36]

$$\eta'_{\dot{A}} = S_{\dot{A}}{}^{B}\,\eta_{B} \tag{6.8.13}$$

where η is the transmitted spinor and η' the reflected one. S therefore has the same action as M_{+-} in Equation 6.8.4.

The essential difference between Equation 6.8.13 and ordinary matrix transformation lies in the fact that the latter maps a state into another with the same basis, whereas Equation 6.8.13 maps a state into one in the conjugate basis.

The basis spinors are the quaternions α β and γ and their analogues the spin matrices of Pauli given in Equation 6.7.3. As shown in Appendix IV the general quaternion

$$Q = t + \alpha x + \beta y + \gamma z$$

can be represented by the matrix

$$Q = \begin{bmatrix} t + iz & -y + ix \\ y + ix & t - iz \end{bmatrix} = \begin{bmatrix} Z_1 & Z_2 \\ -Z_2^* & Z_1^* \end{bmatrix} \tag{6.8.14}$$

and all the spin matrices concerned, including the scattering matrix of Equation 6.8.5 and the metric spinors, conform to the representation.

The relation of these matrices and the basis α β γ to the algebra su(2) of the group SU(2) and rotations in general, is outlined in Appendix IV.

In the context of polarimetry, polarization operations are illustrated by operations on the Poincaré sphere. A transformation of polarization occurs through a scattering matrix in the case where a complex figure is illuminated by a polarized wave and by a transmission matrix when considering continuous propagation through an optically active medium. The former, which is our main concern, can then be demonstrated by a fixed rotation on the Poincaré sphere (the latter by a continuous rotation). Rotations in three dimensions are actions of the group SO(3).

However this is insufficient to describe the total action of the scatterer omitting as it does the phase of the scattered wave. This omission is identical to the neglect of the time component in the basic definition of polarity that was mentioned at the outset. To include it moves us from three dimensional rotations to four, categorised by the SU(2) group. It could be said heuristically that the inclusion of time, that is the specification of the phase, arises from the fact that in a general polarization state the i of the space quadrature of two linear components is of a different nature to the i of the phase (or time) quadrature. This immediately necessitates a bicomplex notation leading directly to the quaternion algebra.

An imaginary rotation about the time axis is a definition of a Lorentz transformation in relativity theory and there is therefore an association between scattering and Lorentz transformations.

The relationship between the Lie groups and wave propagation and scattering, polarization and spin states including those of the fundamental particles, which have been indicated, forms a theoretical study which is too large to go into here in any further detail. The interplay of the applications in the various branches of physics to which the analysis applies, can form a rich field for ongoing research. The interested reader can find a basic introduction and full referencing in such papers as [37] [38] and others.

Graves [32] defines the "polarization power scattering matrix" $P^2 = \tilde{S}^* S$

$$P^2 = \begin{pmatrix} s_{11}^* s_{11} + s_{12}^* s_{12} & s_{11}^* s_{12} + s_{12}^* s_{22} \\ s_{12}^* s_{11} + s_{22}^* s_{12} & s_{12}^* s_{12} + s_{22}^* s_{22} \end{pmatrix} \tag{6.8.15}$$

If P^2 is referred to a linear basis polarized along the x and y axes, then it can be completely derived from the transmission of a horizontally polarized wave, a vertically polarized wave and waves polarized at $\pm 45^o$, and measuring the total backscattering power in each case. This can be compared with the procedure for deriving the Stokes vector given in Section 6.1.

TARGET VECTOR REPRESENTATION

A vector representation in the spinor basis takes the form of an expansion of the relation in Equation 6.7.6 to include spinor indices. A general theorem states that one of the indices has to be dotted, and thus the notation derived is

$$J^{\dot{A}B} = J^{\mu} \sigma_{\mu}{}^{\dot{A}B} \tag{6.8.16}$$

The backscattering matrix in 6.8.13 is of mixed index type, so to use this vector representation we require the steps

$$\begin{aligned} S^{\dot{A}B} &= \varepsilon^{CB} S_C{}^{\dot{A}} \\ S_\mu &= \frac{1}{2} \sigma_{\mu \dot{A} B} S^{\dot{A}B} \qquad \mu = 0, 1, 2, 3 \end{aligned} \tag{6.8.17}$$

This is an expansion of the scattering matrix in terms of the basic spinors. Other expansions are possible which aim to separate the target return into components representing a pure state, a mixed state and an unpolarized state. These states have more practical significance than the basis states in 6.8.17. One such [39] shows that (after transformation by a rotation operator) the target return can be uniquely decomposed into three components which are respectively the return from a sphere, the return from an orientated dihedral and that from a helix.

The basic analysis of target returns was that of Huynen [40]. There the scattering process is considered as producing components which are "receivable" by the illuminating antenna and orthogonally "non-receivable", giving rise to what has become termed "Huynen's polarization fork" consisting of the co-polar and cross-polar components of each as represented on the Poincaré sphere.

Elementary illustrations of this concept can be given. Firstly a sphere (or infinite plane), if illuminated by a pure circularly polarized wave, will give zero return and thus the co-polar nulls will be located at the north and south poles of the Poincaré sphere. For maximum return any linear polarization will suffice, and hence the cross-polar nulls can be any point on the equator of the sphere. For this case

$$[S] = \begin{bmatrix} 1 & 0 \\ 0 & 1 \end{bmatrix}$$

Alternatively, a dihedral corner, say two planes intersecting at right angles, would give complete return to circular polarization by virtue of the double reflection and zero return for linearly polarized waves perpendicular to the line of intersection of the planes, since the reflections will reverse the polarization. Hence the co-polar nulls are diametrically opposite points on the equator and the cross-polar nulls the circles of points giving circular polarization. This can be derived from the scattering matrix

$$[S] = \begin{bmatrix} 1 & 0 \\ 0 & -1 \end{bmatrix}$$

Helices are described by

$$[S] = \begin{bmatrix} 1 & -i \\ -i & -1 \end{bmatrix}$$

for right-handed and

$$[S] = \begin{bmatrix} 1 & i \\ i & -1 \end{bmatrix}$$

for left-handed.[‡] Methods for deriving representations from a general scattering matrix and their association with other descriptions, are still a current item of research.

PRACTICAL APPLICATIONS

There is far more information about the form or nature of a scattering target to be derived from the analysis of its action on a polarized wave, than from illumination by a simple linearly polarized wave as in general current radar systems. There is, of course, considerable added complexity required in the transmitting and receiving apparatus for the detection of the full data comprising as it does the three measurements, two polarizations and a phase, plus a fourth if amplitude is included. Thus there is a time consideration involved for the necessary gathering of this information. This is greatly dependent on the application involved. For remote sensing, radar meteorology and inverse

[‡]This author has some misgivings about the planar helix in this respect. While it will have an action along the axis where the spiral curves are below a wavelength in dimension, the outer arms of the spiral become almost parallel and could not have any action other than as a grating. It is possible to propose an active medium in which elemental spirals act on locally plane waves. Such is the chiral medium which, along with other active or non-isotropic media, are outside the scope of this study. However it is conceivable that such a medium if electronically activated would be of great use in polarization diversity transmission or reception.

scattering [41] [42], for example, long time scales for data reduction can be used together with slow changes in aspect of the scatterer. In targeted radar systems, short pulses are normally used, requiring the development of ultra wide bandwidth radars. In addition a polarization agile system is necessary for the optimum illumination, and the target return can only be expected to be aspect independent at long ranges.

The objective of the foregoing analysis is to provide algorithms for on-board computer systems to allow the rapid reduction of the received polarization data to enable a first appreciation of the target form, say the difference between a missile shape and an aeroplane or that between real targets and decoys, to be made with a fair degree of certainty. The same data becomes available for the choice of optimizing the illumination polarization. A discussion is included in [31].

REFERENCES

1. Shurcliff WA (1962) *Polarized Light* Harvard University Press

2. Newton RC (1966) *Scattering Theory of Waves and Particles* McGraw Hill pp4-10

3. Beckman P (1968) *The Depolarization of Electromagnetic Waves* Golem Press

4. Jones RC (1941) A new calculus for the treatment of optical systems, Parts I-VIII *Jour Opt Soc Amer* **31** p488; (1941) **31** p493, with Hurwitz; (1941) **31** p500; (1942) **32** p486; (1947) **37** p107; (1947) **37** p110; (1948) **38** p671; (1956) **46** p126

5. Schmeider RW (1969) Stokes algebra formalism *Jour Opt Soc Amer* **59** p297

6. Hill N & Cornbleet S (1973) Microwave transmission through a series of inclined gratings *IEE* **12** (4) p407

7. Groves WE (1953) Transmission of electromagnetic waves through pairs of parallel wire grids *Jour App Phys* **24** p845

8. (1951) Techniques for handling elliptically polarized waves with special reference to antennas *Proc IRE* **39** pp533-556
Booker HG. Introduction p533
Rumsey VH. Transmission between elliptically polarized antennae p535
Deschamps GA. Geometrical representation of the polarizations of a plane electromagnetic wave p540
Kales ML. Elliptically polarized waves and antennas p554
Bohnert JI. Measurement of elliptically polarized waves p549

Morgan MG & Evans WR Jnr. Synthesis and analysis of elliptic polarization loci in terms of space-quadrature sinusoidal components p552

9. Azzam RMA & Bishara NM (1972) Polarization transfer function of an optical system as a bilinear transformation *Jour Opt Soc Amer* **62** (7) p222

10. Kaplan LJ (1962) Bilinear transformation of polarization *Jour Opt Soc Amer* **62** (10) p1239

11. Barlow HM & Cullen AL (1950) *Microwave Measurements* Constable. Appendix II p376

12. Marcuvitz N (1951) *Waveguide Handbook* MIT Series **vol 10** McGraw Hill

13. Skwirzynski JK & Thackray JC (1959) Transmission of electromagnetic waves through wire gratings *Marconi Review* **22** p77

14. Hannan PW (1961) Microwave antennas derived from the Cassegrain telescope *Trans IRE* **AP9** p140

15. de Size LK & Ramsey JF (1964) Reflecting systems in "Microwave scanning antennas" Hansen RC (ed) London & New York: Academic Press, Vol 1 p128

16. Mariner PF & Cochrane CA (August 1953) High frequency radio aerials, British Patent no 716939

17. Mariner PF (1958) Microwave aerials with full hemispherical scanning *L'Onde Electrique* Supplement August No 376 Proceedings of the International Congress on Ultra High Frequency Circuits and Antennas, Paris 21-26 October 1957, Vol II p767

18. Halberstam H & Ingram RE (1967) *Collected Papers of W R Hamilton* vol 3 **Algebra** Cambridge University Press

19. Marathay AS (1971) Matrix operator description of the propagation of polarized light *Jour Opt Soc Amer* **61** (10) p1363

20. Whitney C (1971) Pauli-algebraic operators in polarization optics *Jour Opt Soc Amer* **61** (9) p1207

21. Eichmann (1971) Complex polarization variable description of polarizing instruments *Jour Opt Soc Amer* **61** (11) p1585

22. Azzam RMA & Bishara NM (1972) Ellipsometric measurement of the polarization transfer function of an optical system *Jour Opt Soc Amer* **62** (3) p336

23. Ghezzo M (1968) Thickness calculations for a transparent film from ellipsometric measurements *Jour Opt Soc Amer* **58** (3) p368

24. Silberstein L (1924) *The Theory of Relativity* MacMillan

25. Rumsey VH (1966) *Frequency Independent Antennas* London & New York: Academic Press

26. Crout PD (1970) The determination of antenna patterns of n-arm antennas by means of bicomplex functions *Trans IEEE* **AP-18** p686

27. Lewin L (1972) A decoupled formulation of the vector wave equation in orthogonal curvilinear coordinates *Trans IEEE* **MTT-20** (5) p339

28. Ehlers J, Rindler W & Robinsom I (1966) Quaternions, bivectors and the Lorentz group in *Perspectives in Geometry and Relativity* Hoffmann B (Ed) Indiana University Press p134

29. Born M & Wolf E (1954) *Principles of Optics* Pergamon Press p550

30. Altman SL (1986) *Rotations, Quaternions and Double Groups* Oxford: Clarendon Press

31. Cloude SR (1992) Recent developments in radar polarimetry, a review *Proc IEEE* Asia Pacific Microwave Conference, Adelaide p955 (includes 67 refs)

32. Graves CD (1956) Radar polarization power scattering matrix *Proc IRE* **44** p248

33. Penrose R & Rindler W (1984) *Spinors and Space-Time* Cambridge University Press p106

34. Helgason S (1962) *Differential Geometry and Symmetric Spaces* London & New York: Academic Press

35. van der Waerden BL (1929) *Spinoranalyse* Nackr Akad Wiss. Gottingen math-Physik Kl p100-109

36. Bebbington DHO (1992) Target vectors - spinorial concepts *Proceedings of the Second International Workshop on Radar Polarimetry, University of Nantes* p2

37. Cloude SR (1992) Lie groups in electromagnetic wave propagation and scattering *Jour Elect Waves & Applications* **6** no 8 pp947-974

38. Wolf KB (1985) *Group Theoretical Methods in Lie Optics* Rep Department of Mathematics, Universidad Autónoma Metripolitana, Itztapalapa DF, Mexico **IV** no 4

39. Krogager E (1990) New decomposition of the radar target scattering matrix *Electronic Letters* **26** no 18 p1525

40. Huynen JR (1970) *Phenomenological Theory of Radar Targets* PhD Thesis, Technical University of Delft, The Netherlands

41. Boerner WM, El-Arini MB, Chan CY & Mastiris PM (1981) Polarization dependence in electromagnetic inverse problems *IEEE Trans Ant & Prop* **AP-29** no 2 p262

42. Kong JA (1990) *Polarimetric Remote Sensing* New York: Elsevier Press

7
FIELDS RAYS AND TRAJECTORIES

7.1 TRANSFORMATIONS IN OPTICAL GEOMETRY

A recurring theme in this volume has been the ability to transform one optical system into another by purely geometrical methods. This implies an underlying discipline which can suitably be termed "Optical Geometry". In this context, optical rays are no longer time evolving trajectories, but are fixed curves in a space determined by the refractive contents of the elements in it. This space is the "refractive space" and is obtained by the product of the coordinates at any point with the refractive index of the medium at that point. In such a space the line increment is $\eta^2\,\mathrm{d}s^2$ as defined in Section 4.10 and rays are the geodesics of the metric (see Equation 4.1.12)

$$\mathrm{d}\kappa^2 = \eta^2\,\mathrm{d}s^2 = \eta^2\,\mathrm{d}r^2 + \eta^2 r^2\,\mathrm{d}\theta^2 + \eta^2 r^2 \sin^2\theta\,\mathrm{d}\phi^2 \tag{7.1.1}$$

If reflecting and refracting surfaces are present, the geometry is basically Euclidean with the inclusion of the laws of reflection and refraction at those surfaces.

We find that in this refractive space, certain transformations can be made that transform one optical system into another. These have been discussed in previous chapters, and are notably the inversion theorem of Damien in Chapter 3 and the inversions of spherically non-uniform media relating to the lenses of Chapter 4. To these can be added the following

BUDDEN'S RECIPROCITY [1]

In an inhomogeneous, and not necessarily isotropic, medium vectors tangent to a curved ray can be mapped onto an alternative space, such that they all derive from a common origin and are in the same direction as the tangent vectors to the real rays in the refractive space, and of length equal to the refractive index at the point of tangency of the ray in the medium. The end points of these vectors then define a surface termed by Budden the "refractive surface", but previously considered by Hamilton as "the surface of components" and by Fresnel as "the surface of normal slowness". The real rays in the refractive medium have associated with them a phase front, to which, in an isotropic medium, they are normal. In a non-isotropic medium there can be defined instead a "ray surface" which becomes a phase front when isotropy sets in.

This is the Fresnel wave surface. Budden then proves that the normals to the ray surface are parallel to the radius vectors of the refractive surface (by construction) and that the normals to the refractive surface are parallel to the radius vectors of the ray surface. An illustration of this effect was given at the end of Section 4.1 (Figure 4.2).

THE LENS TRANSFORMATION HYPOTHESIS

One is hesitant to include this unproven transformation given in detail in Section 3.12, but should it prove to be viable it also would have to be considered among the corpus of geometrical optical transformations, which only arise in the field of optical geometry.

It is the very existence of such a comprehensive body of geometrical theories applying only to optical systems of a variety of types, that poses a fundamental question. Light rays are the high frequency limit to solutions of Maxwell's equations of electromagnetism. The electromagnetic treatment of fields in completely general media [2] rarely consider this limit. As a consequence there is little or no correlation between the geometrical properties in the foregoing and the electromagnetic field. In principle there should be equivalent transformations of the electromagnetic field that would give the appropriate transformations of the optical rays in the infinite frequency limit. It is pertinent to note that nearly all of the geometrical theorems involved, and further to come, have at their basis some form of inversion. Inversion is a transformation in the conformal group so the connection would be made through the action of this group on the electromagnetic field. In fact Maxwell himself has proved that there is only one conformal mapping that maps three-dimensional space into itself (apart from "trivial" mappings such as reflection translation and magnification) and that is the inversion [3]. This was extended to a real (Euclidean) space of four dimensions by Forsyth [4] and can be shown to extend further to a complex space of four dimensions (Minkowski space). This would be the necessary space for radiating systems and optical rays. (The algebra is extensive, 56 pages at the last count and thus cannot be included here).

There are two aspects to be considered. The well-known conformal mapping in two dimensions is made through a function of a complex variable. The Cauchy-Riemann equations of the transformation automatically ensure that the potentials obey transverse (two dimensional) Laplace equations. Attempts to apply conformal mapping to three dimensional *static* problems involve the use of multi-valued potentials [5]. For example the field of a line source over a conducting semi-infinite plane, uses the method of images in a Riemann surface. Sommerfeld [6] continues the process for a point source over a semi-infinite plane, by proposing a Riemann space, having the same relation to real space that the Riemann surface has to the uncut complex plane. Similar considerations arise in the fundamental analysis of Huygens' Principle by Baker and Copson [7]. However functions of a complex variable have been used in one archetypal radiation problem, by Smirnov [8], where the diffraction of a plane wave by a semi-infinite plane, the Sommerfeld diffraction problem, is solved by that method. Even in that case the four dimensionality is reduced to three through the choice of the geometry of the situation.

To proceed to mappings of the electromagnetic field and thence to optical problems requires the general extension to four dimensional analysis. Thus we either have the extension of Riemann connected surfaces to Riemann connected spaces (a branch cut in the surface becoming a branch surface in the higher space [9]), or to advance to a higher form of complex algebra.

There are several reasons why the extension of complex variable theory to hypercomplex theory is to be preferred, not least the elegant way it completes the cycle of Hamilton's studies. This, as with Newton, consisted of mechanics optics and the necessary underlying algebra. The universal algebra that Hamilton devised was, of course, the theory of quaternions that was met with in Chapter 6. Hamilton showed that it was the only algebra able to perform the essential step from two dimensions to any higher dimension. It is nice to note in this context that the quaternion algebra predicted the cone of rays from an obliquely illuminated thin wire or sharp edge now used as a canonical scatterer in current GTD theories [10] [11].

The optical requirement for a bicomplex notation can be demonstrated by an extension to the standard treatment for the derivation of rays from the (reduced or Helmholtz) wave equation. This is the substitution of the Sommerfeld-Runge ansatz

$$u = A(x, y, z) \exp(ik_0 S(x, y, z)) \tag{7.1.2}$$

into the wave equation

$$(\nabla^2 + k^2)u = 0 \quad \text{(source free)} \tag{7.1.3}$$

from which arises the eikonal equation

$$\left(\frac{\partial S}{\partial x}\right)^2 + \left(\frac{\partial S}{\partial y}\right)^2 + \left(\frac{\partial S}{\partial z}\right)^2 = \left(\frac{k}{k_0}\right)^2$$

which, being identical to that of Equation 4.1.8 with k/k_0 in place of η, is taken to signify the optical ray field.

"Complex rays" [12] [13] can then be introduced by making the eikonal function S complex. However, if this were to be the same "complex" as i in Equation 7.1.2 only a k dependent amplitude term results. Thus S has to be $R + jI$ and there are four terms to equate to zero after substitution into the Helmholtz equation

1. real terms

$$\frac{\nabla^2 \mathbf{A}}{A} - k_0^2[|\nabla R|^2 - |\nabla I|^2] + k^2 = 0$$

2. i terms

$$\frac{2k_0}{A}\nabla R\cdot\nabla A - k_0\nabla^2 R = 0$$

3. j terms

$$2k_0^2\nabla R\cdot\nabla I = 0$$

4. ij terms

$$\frac{2k_0}{A}\nabla I\cdot\nabla A - k_0\nabla^2 I = 0$$

(retaining the factor k_0 for illustration) the result given in the reference. If we relabel i j and ij α β γ we have the quaternion notation, but without the non-commutative property. Treating $\nabla^2 A$ as ignorably small (medium changes slowly on the wavelength scale). The "refractive index" is given by

$$\eta^2 = (k/k_0)^2 = |\nabla R|^2 - |\nabla I|^2 \qquad (7.1.4)$$

which can therefore give complex values.

It has to be noted that the wave Equation (7.1.3), into which the asymptotic series of all forms has to be substituted, is already subject to the condition indicating a slowly varying refractive index function. The insertion $\mathbf{B} = \mu\mathbf{H}$ and $\mathbf{E} = \varepsilon\mathbf{D}$ into Maxwell's (source free) equations with varying ε (with at least non-varying μ), results in the equations

$$\nabla^2\mathbf{E} - \mu\varepsilon\frac{\partial^2\mathbf{E}}{\partial t^2} = -\nabla\left(\frac{\mathbf{E}\cdot\nabla\varepsilon}{\varepsilon}\right) \simeq \text{grad }\log\varepsilon\times\nabla\times\mathbf{E} \qquad (7.1.5)$$

$$\nabla^2\mathbf{H} - \mu\varepsilon\frac{\partial^2\mathbf{H}}{\partial t^2} = -\frac{\nabla\varepsilon}{\varepsilon}\times\nabla\times\mathbf{H} = \text{grad }\log\varepsilon\times\nabla\times\mathbf{H} \qquad (7.1.6)$$

These complete equations appear to have been treated in only a few instances and then only for a one-dimensional (Cartesian) variation in ε. However, even in Equation 7.1.5, second differentials of ε have been ignored. Budden [1, Chapter 9] gives this omission for the one-dimensional problem as

$$\frac{1}{k^2}\left|\frac{3}{4}\left(\frac{1}{\eta^2}\frac{\partial\eta}{\partial z}\right)^2 - \frac{1}{2\eta^3}\frac{\partial^2\eta}{\partial z^2}\right| \ll 1 \qquad \eta = \sqrt{\varepsilon}$$

which can be extrapolated to

$$\frac{1}{k^2}\left|\frac{3}{4}\left(\frac{1}{\eta^2}\nabla\eta\right)^2 - \frac{1}{2\eta^3}\nabla^2\eta\right| \ll 1 \qquad (7.1.7)$$

This is the WKB approximation for this problem [14, p13]. Its worst consequence is that it prevents the extension of the theory to rapidly varying refractive indices, such as would permit the "derivation" of Snell's law of refraction from the electromagnetic field equations. At such a discontinuity, the derivatives are infinite.

Budden shows that this approximation is valid up to within about one (free space) wavelength of the sharp refracting region. This accords exactly with the effect which we observe with the highly curved rays in the Eaton lens. Since this approximation is determined by the gradient of the refractive index, the homogeneous Helmholtz equation is itself deficient for the case of the non-uniform medium if gradient terms on the right-hand side of Equation 7.1.5 are ignored. They are *not* recovered by the expedient of substituting a space variable $\eta^2 k_0^2$ for k^2 in the homogeneous equations, thus

$$(\nabla^2 + \eta^2 k_0^2)u = 0 \tag{7.1.8}$$

The eikonal too has to be re-interpreted as an optical distance, it has to be replaced by

$$S(x, y, z) = \int^{\tau} \eta \, ds \tag{7.1.9}$$

where η is variable and parametrized by τ as

$$\eta = \eta[x(\tau), y(\tau), z(\tau)]$$

and ds is an increment of the geometrically determined path with the same parametrization. These are the "phase integrals" [15] [16] and the integral in Equation 7.1.9 is thus an action integral.

The Helmholtz equation for the slowly varying medium with solution $u(x, y, z)$, can be converted to the Schrödinger equation by the substitution

$$u(x, y, z) = Bk_0 v(\xi, \eta, \zeta)$$

where v is a new variable of characteristic size unity, while B tending to infinity will be the geometrical optics limit. We follow the working as in [17]. Substitution into Equation 7.1.8 gives

$$\left(\frac{1}{B^2} \overline{\nabla}_v^2 + \eta^2 \right) v = 0 \tag{7.1.10}$$

where

$$\overline{\nabla}_v^2 = \frac{\partial^2}{\partial \xi^2} + \frac{\partial^2}{\partial \eta^2} + \frac{\partial^2}{\partial \zeta^2}$$

Assuming $\eta(x, y, z) \to E$ as $r \to \infty$ ($E = 1$ for vacuum) and

$$V(\xi, \eta, \zeta) = E - \eta^2 \tag{7.1.11}$$

Equation 7.1.10 becomes

$$\frac{1}{B^2} \overline{\nabla}_v^2 - Vv = -Ev$$

Putting

$$\psi(v, \tau) = ve^{-iE\tau B}$$

then

$$\frac{1}{B^2}\overline{\nabla}^2_{\psi} - V\psi = \frac{-i}{B}\frac{\partial\psi}{\partial t} \qquad (7.1.12)$$

and

$$v(\xi,\eta,\zeta) = \int_{-\infty}^{\infty} \mathrm{d}\tau \exp(iE\tau B)\psi(\xi,\eta,\zeta,\tau) \qquad (7.1.13)$$

Equation 7.1.12 parallels the time-dependent Schrödinger equation and the limits $B \to \infty$ and $\hbar \to 0$ are formally identical. The latter is the "semi-classical approximation" [18] [19] in wave mechanics. For trajectories of constant energy the relation $E = \hbar\nu$ gives the same limit $\hbar \to 0$ $\nu \to \infty$. The Green's function for Equation 7.1.13 is the "propagator", which, for large values of B, gives a stationary phase condition, equivalent to the "classical" path. This is the stationarity of the action integral,

$$\delta \int \sqrt{2m(E - V)}\,\mathrm{d}s = 0 \qquad (7.1.14)$$

in physical space, formally identifying the refractive index of Equation 4.1.1 with the function $\sqrt{E - V}$ as in Equation 7.1.11.

The archetypal problems of both disciplines are also closely related to each other. Most specifically this has been put by Buchdahl [20] [21] in comparing the equivalent problems of the rays in the Maxwell fish-eye with Kepler orbits. He shows that constant vectors can be derived for the non-uniform medium totally analogous to the constant angular momentum vector giving Bouguer's relation and Runge-Lenz vector of the orbit in a $1/r$ potential. He later discusses the same problem in the light of conformal transformations and the invariance of optical systems. Klein [22] in discussing the classical deflection function (Rutherford scattering), derives a new deflection formula, which on inspection, is the ray trajectory integral of Equation 4.7.7 in terms of a scalar unit V/E. More recently [23], there has been derived a transformation of the potential function $V(r)$ which gives the trajectories of particles in the transformed potential in the same way that we have just derived for rays in a spherical medium. This too is an inversion and it acts on the phase-space of the mechanical motion. The reference includes the observation that "it is amusing to recall the intriguing fact that the potential outside a uniform spherically symmetrical mass is the Kepler while inside we have the harmonic oscillator." The Kepler orbit itself is "simple harmonic" on a circle, in that for every ellipse there exists a circle centred on the focus for which the time spent outside the circle equals the time spent inside, as derived from Kepler laws of equal areas.

However, there are some lingering discontinuities in this otherwise happy relationship. The connection established by Buchdahl is between the Kepler problem and the fish-eye, but the former has elliptical orbits with a centre of potential at a focus of the ellipse, whereas the latter has circular orbits (rays). The connection can be made in two different ways. As was shown in Chapter

4 (Section 4.10) a Legendre transformation of the Maxwell fish-eye transforms the coaxal system of rays into the elliptical rays in the Eaton lens. These are exactly analogous to the constant energy trajectories of the Kepler problem.

Alternatively the hodographs of the constant energy trajectories are the (r, v) curves in the position/momentum phase space (Figure 7.1). It is comparatively simple to show that for an elliptical orbit, the "fast" velocity v_f at the perihelion, the "slow" velocity v_s at the aphelion and the median velocity v_m at points at the ends of the minor axis satisfy the relation

$$v_f v_s = v_m^2$$

and that v_m is the same for all equal energy orbits. Hence the hodograph of the constant energy orbits form a set of *coaxal* circles identical to those of the rays in the fish-eye lens. The question arises whether the transformations of the optical rays for other lenses can provide equivalent transformations of the phase-space trajectories for transformed potentials. A particularly relevant example comes with the demonstration that the Kepler problem can be transformed into geodesic flow on the unit sphere [24]. The Kepler problem was shown to be analogous to the elliptical rays of the Eaton lens (Section 4.7) which in turn was a Legendre transform of the Maxwell fish-eye (Section 4.10) which stereographically projected into the geodesic flow on the sphere (Section 4.11). This is stated to be ultimately related to Fock's treatment of the quantum theoretical problem for the hydrogen atom in which the Schrödinger equation is transformed by the stereographic projection of the momentum variables into a form that is invariant under the four dimensional rotation group on a hypersphere.

Rays in lenses have also been shown to have a counterpart in the rays confined geodesically to a surface of revolution (Section 4.13). Geodesics on surfaces of revolution have transformation properties of their own. These include stereographic or other projection, conformal mapping and isometric transformation. Geodesic mappings exist which map the geodesics of one surface onto the geodesics of another. If a mapping is both geodesic and conformal it is isometric [25]. Surfaces can be deformed continuously into other surfaces (allowing no stretching or tearing) and such surfaces are termed "applicable surfaces". Hence any geodesic transformation of this kind can be applied to lenses through its analogy to the refractive index law of a spherical non-uniform refractive medium.

Geodesics on surfaces of revolution are also intimately bound up with the trajectories of particles in potential distributions, since they, like the optical rays, have been obtained from similar variational principles. These have been studied for a considerable period, but their applications to lens or trajectory transformations have not been considered [26].

The natural families of trajectories, of which the optical ray is one class, consist of all those problems governed by the extremum of an action integral, for example, catenaries or brachistochrones etc. It has been proved that the

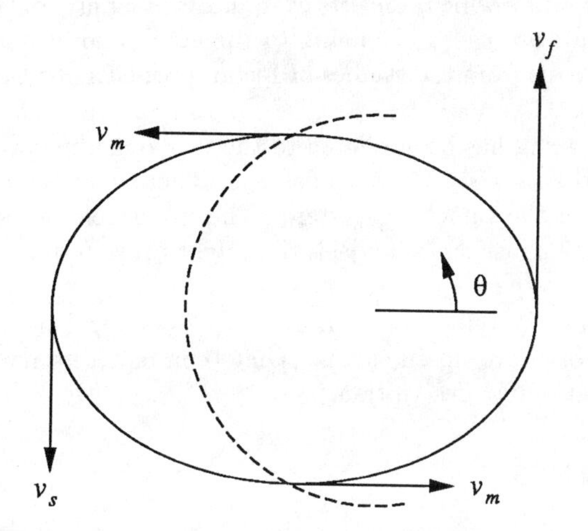

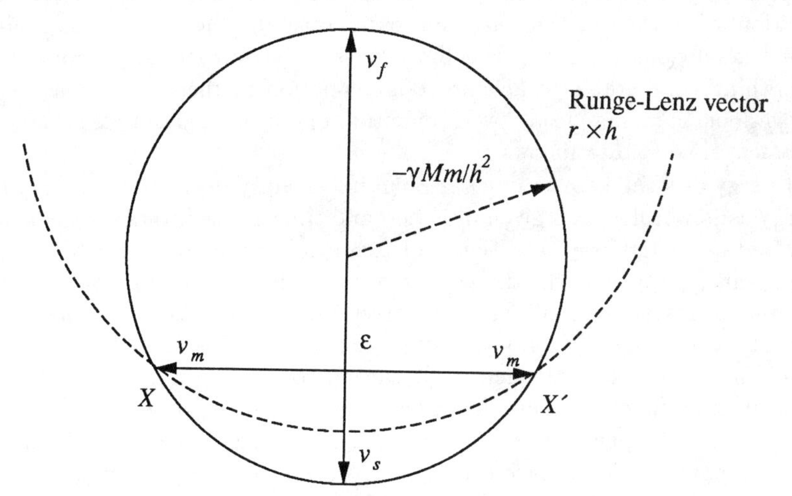

Figure 7.1: Hodograph of Kepler orbits

$$v_f = \frac{h}{l}(1+\varepsilon) \quad v_s = \frac{h}{l}(1-\varepsilon) \quad v_m = \frac{h^2}{l^2}(1-\varepsilon^2)$$

h = areal constant, l = semi-latus rectum
v_m is constant and so all curves pass through XX' for equal energy orbits

only point transformation which converts every natural family (of trajectories) into a natural family, are those belonging to the conformal group. This line of investigation extends from the studies of Lie and Scheffers to Bateman and Fueter [27] [28] [29].

Most of the foregoing has been related to the two archetype problems, the Kepler orbits and the rays in Maxwell's fish-eye. There is the second form in optical rays - that of the Luneburg system. The nature of the potential for which these would be phase-space trajectories might prove to be equally basic in trajectory theory.

Perfectly focusing systems are analogous to dynamical systems whose phase-space trajectories are closed curves. They then represent normal modes in the electromagnetic field description.

STATISTICAL OPTICS

There are other aspects of geometrical optics as the limiting process or first-order approximation to the solution of a physical problem. As already noted, the analogy between geometrical optics and classical mechanics, and the wave equation with that of the harmonic oscillator, has the semi-classical limit $\hbar \to 0$ analogous to the optical limit $k \to \infty$. However, there appears to be nothing comparable to an uncertainty relation in the optical context. If we overlook the difference between quantum operators and optical scalars, an argument might go as follows.

Energy and time are conjugates in an uncertainty law, [30] thus regarding energy as power flow in a given direction and time as the inverse of frequency, it would seem that ray direction and frequency would similarly be subject to an uncertainty law. In the context of ultra-high-speed microwave beam scanning or instantaneous frequency detection, this uncertainty becomes very real [31]. Consequently, a ray is only precisely determined in direction in the optical limit. At other frequencies it can be considered to be a disturbance with an error function in its direction. Assuming this to be Gaussian with a width which depends on k, we have the effect that every ray in a system has a probability of influence throughout the whole space. There would be a similar uncertainty about the direction of a ray, its momentum and the point at which it intersected a given surface, its location.

If we were to use three such uncertainty rays as the orthogonal geodesic axes of a Cartesian coordinate system, there arises an uncertainty of the origin (the mathematical or unobservable origin) with respect to the observable (the physical) origin. The transformation between these frames forms the basis of one of the last studies by Eddington [32] concerning the connection between relativity theory and the quantum theory.

A similar probability theory arises if the ray from a given point *towards* a second point is not the determinate Fermat path joining the two, but the most probable of all the possible paths between them. If all paths are in this manner putative rays with a small probability attached, the concept is that of the path integrals [33] the summation of such rays. Although this has been mainly applied to particle mechanics, the "propagator" for the particle is analogous to the ansatz function for the wave equation, and the path integral with the angular spectrum of plane waves with amplitude proportional to the ray direction.

Besides the direction of rays, the other aspect, the transport equations, have a statistical interpretation. This arises in the theories of radiative transfer, where sources and fields have a frequency distribution. Here an ensemble average is required both for the amount of energy transported and its direction of transport. These can be obtained either from the correlation tensors of the Maxwell field equations, or the mutual coherence functions in the optics of partially coherent light [34]. The geometrical optics limit for this case is that obtained as the limit of total coherence. The next order term in this solution, which reduces to zero in the limit, has the nature of a small circulatory motion which extends over a distance of the order of a wavelength [35]. No physical interpretation is given for this, but it is to be noted that Stavroudis also observes a rotatory motion about the binormal vector of a ray in an orthotomic system (a normal congruence) [36, p196].

Approaching the problem from the alternative direction Stavroudis [36, p194] shows that from the pure geometry of a curve in refractive space and the inclusion of the variational integral

$$I = \int \eta \, \mathrm{d}s$$

the Frenet-Seret formulae for curves in space themselves produce what are termed "pseudo-Maxwell" equations with $\eta\hat{\mathbf{n}} = -\mathbf{E}$ and $\eta\hat{\mathbf{b}} = \mathbf{H}$ ($\hat{\mathbf{n}}$ the principal normal and $\hat{\mathbf{b}}$ the binormal to the curve). Later, Appendix IV, we shall show that Maxwell's equations can also be derived purely algebraically from holomorphic properties in a space with quaternion structure.

7.2 BATEMAN'S TRANSFORMATION OF OPTICAL SYSTEMS

The surprise then occurs to find that problems of this nature were tackled about 1910 in a series of papers by Bateman [37] [38] based, as in his own title, on the conformal transformations of a space of *four* dimensions.

Bateman bases his study upon the requirement for transformations to leave the world line increment invariant

$$\mathrm{d}x^2 + \mathrm{d}y^2 + \mathrm{d}z^2 - c^2\,\mathrm{d}t^2 = 0 \tag{7.2.1}$$

in itself a definition of conformality.

A transformation that does this is extremely well-known, namely the Lorentz transformation of the restricted theory of relativity. But even in that context it is to be applied in a free-space situation. To include the effects of refractive pondero-motive materials is still a highly complicated even questionable, procedure and not one that could be easily adapted to the shapes that constitute lens systems. However, Bateman [39] finds that many other transformations leave both the wave equation and the eikonal equation invariant and that the Lorentz transformation is only a particular version of these more general possibilities. Some in fact refer back to earlier work by Bianchi in differential geometry and of Maxwell himself.

He singles out for particular interest the transformation

$$X = \frac{x}{r^2 - c^2t^2} \quad Y = \frac{y}{r^2 - c^2t^2} \quad r^2 = x^2 + y^2 + z^2$$
$$Z = \frac{z}{r^2 - c^2t^2} \quad T = \frac{t}{r^2 - c^2t^2} \tag{7.2.2}$$

and applies it directly to an optical lens (spherical surfaces paraxial case) to obtain a transformed lens.

The form of the transformations are inversions, but whereas the origin normally transforms by inversion into the point at infinity, in these cases the entire circle $r = ct$ (or the point $z = ct$) does. In a further discussion on more general transformations of the form

$$\lambda[\mathrm{d}x^2 + \mathrm{d}y^2 + \mathrm{d}z^2] + \mu[(s_x\,\mathrm{d}x + s_y\,\mathrm{d}y + s_z\,\mathrm{d}z)^2 - s^2\,\mathrm{d}t^2]$$
$$= \mathrm{d}x'^2 + \mathrm{d}y'^2 + \mathrm{d}z'^2 - \mathrm{d}t'^2 \tag{7.2.3}$$

Bateman derives those transformations "that can be used to solve problems in the reflection or refraction of light when the orthotomic surfaces before or after incidence are known". As applied to the paraboloid, one example of a transformation is of the form

$$x' = x = a - r \quad t' = t$$
$$y' = \frac{ay}{r + x} \quad z' = \frac{az}{r + x} \tag{7.2.4}$$

The fact that these are all inversions by nature illustrates a connection with the method of Damien in previous chapters. Rather revealingly the reference concludes with a note much in accord with this author's own investigations: "It should be remarked that the present method cannot be used to solve any problem in reflection; so far transformations have only been found in cases where a single plane or spherical wave is transformed into a single plane or spherical wave ..." This is however sufficient for our microwave antenna purpose as can now be illustrated.

A transformation that matches the inversion of Equation 7.2.4 can be found that transforms the elliptical *single* refracting surface into the hyperbolic. We have for the elliptical profile

$$r = \frac{f(\eta - 1)}{\eta - \cos\theta}$$

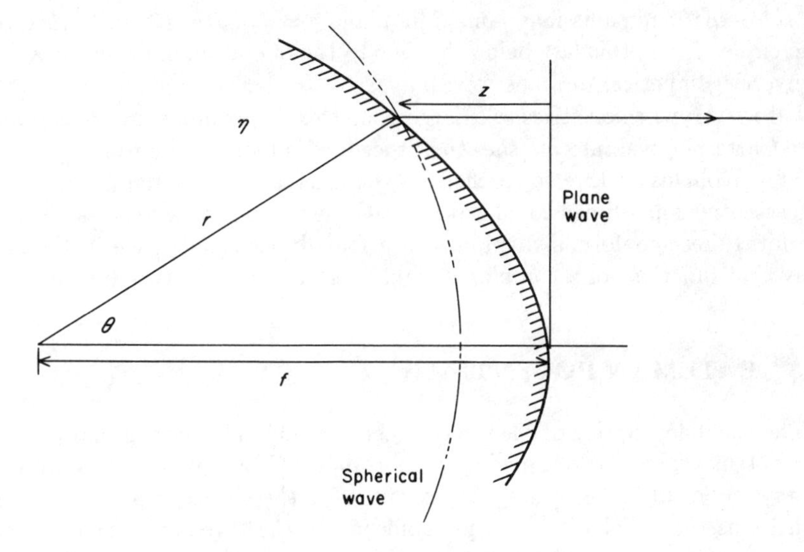

Figure 7.2: Elliptical refractor

to which we apply the transformation $\theta' = \theta$ (from conformality)

$$R = \frac{kr}{r + x} \qquad \text{(from } y = r\sin\theta\text{)}$$

Making the radius of inversion k equal to the focal length f, we choose (fairly arbitrarily) x to be the path length along the axis between the point where the wave front is still spherical up to the point where it becomes plane (Figure 7.2) that is $x = r\cos\theta - f$.

The transformation then becomes

$$R = \frac{fr}{r - f + r\cos\theta}$$

and its inverse by direct solution

$$r = \frac{fR}{R - f + R\cos\theta}$$

Substitution of

$$r = \frac{f(\eta - 1)}{\eta - \cos\theta}$$

results in

$$R = \frac{f(\eta - 1)}{\eta\cos\theta - 1}$$

and inversely.

That is the transformation converts the elliptical profile to the hyperbolic and vice versa. With $\eta = -1$ the transformation shows that the parabola

$2f/(1+\cos\theta)$ remains invariant. The plane $r = f/\cos\theta$ is transformed into the circle $R = f$, this last being the clue to the operation of these surfaces in converting spherical to plane waves.

However, when one tries to enlarge upon this, for example, by transforming one Cartesian oval into another the procedure is found to be wanting.

One conclusion that is reached by this exercise is that the problem is of an essential four dimensional nature. We next look for a process that can perform these conformal mappings in a four dimensional space in the same way that functions of a complex variable perform them in two dimensions.

7.3 BATEMAN POTENTIALS

* The usual description of the electromagnetic field in terms of scalar and vector potentials has proved inadequate for applications involving the inversion of rays in geometrical optics. An extension to the theory is proposed which can be used to deal with such phenomena. In the three dimensional vector calculus the electromagnetic field is derived from a vector potential **A** such that $\mathbf{B} = \nabla\times\mathbf{A}$ is the well-known solution to $\nabla\cdot\mathbf{B} = 0$ of the electromagnetic field equations. A second, less widely known solution exists, namely

$$\mathbf{B} = \nabla\sigma\times\nabla\tau \tag{7.3.1}$$

where σ and τ are completely arbitrary scalar functions of position and time. These scalar potentials were first investigated by Bateman [37] when seeking transformations of the electromagnetic field which leave the eikonal and wave equations invariant, up to a scale factor. These scalar potentials are therefore termed Bateman potentials, and their relevance to the current discussion is apparent.

The complete solution to $\nabla\cdot\mathbf{B} = 0$ is then given by

$$\mathbf{B} = \nabla\times\mathbf{A} + \lambda(\nabla\sigma\times\nabla\tau)$$

for λ constant, from which it is observed to conform to the three dimensional version of the Clebsch potential [40] which is given by

$$B_{\mu\nu} = \left[\frac{\partial A_\mu}{\partial x^\nu} - \frac{\partial A_\nu}{\partial x^\mu}\right] + \lambda\left[\frac{\partial\sigma}{\partial x^\mu}\frac{\partial\tau}{\partial x^\nu} - \frac{\partial\tau}{\partial x^\mu}\frac{\partial\sigma}{\partial x^\nu}\right] \qquad \mu,\nu = 0,1,2,3 \tag{7.3.2}$$

and derived by Clebsh in 1857, when he observed that an arbitrary differentiable vector field on a three-dimensional Euclidean space could always be decomposed into the sum of a gradient and scalar multiple of another gradient.

*Much of the work in this and the following sections has been done in collaboration with Dr CC Sullivan during the preparation of his thesis "The application of bi-quaternion analysis to the transformation of the electromagnetic field and geometrical optics" (1993) presented at the University of Surrey.

From Equation 7.3.1 it is possible to derive the usual scalar potential ϕ and vector potential $\mathbf{A}$ in the form

$$\mathbf{A} = \frac{1}{2}[\sigma\nabla\tau - \tau\nabla\sigma] \tag{7.3.3}$$

$$\phi = \frac{1}{2}\left[\tau\frac{\partial\sigma}{\partial t} - \sigma\frac{\partial\tau}{\partial t}\right] \tag{7.3.4}$$

Since

$$\mathbf{B} = \frac{1}{2}\nabla\times[\sigma\nabla\tau - \tau\nabla\sigma] = \nabla\sigma\times\nabla\tau$$

and $\nabla\cdot\mathbf{B} = 0$ as required.

We also have

$$\begin{aligned}\mathbf{E} &= -\frac{\partial\mathbf{A}}{\partial t} - \nabla\phi \\ &= \frac{1}{2}\frac{\partial}{\partial t}[\sigma\nabla\tau - \tau\nabla\sigma] - \frac{1}{2}\nabla\left[\tau\frac{\partial\sigma}{\partial t} - \sigma\frac{\partial\tau}{\partial t}\right] \\ &= \nabla\sigma\frac{\partial\tau}{\partial t} - \nabla\tau\frac{\partial\sigma}{\partial t}\end{aligned} \tag{7.3.5}$$

Maxwell's free space equations are satisfied directly since

$$\begin{aligned}\nabla\times\mathbf{E} &= \nabla\times\left[\nabla\sigma\frac{\partial\tau}{\partial t} - \nabla\tau\frac{\partial\sigma}{\partial t}\right] \\ &= \nabla\left[\frac{\partial\tau}{\partial t}\right]\times\nabla\sigma - \nabla\left[\frac{\partial\sigma}{\partial t}\right]\times\nabla\tau \\ &= -\frac{\partial\mathbf{B}}{\partial t}\end{aligned}$$

GAUGE CONDITIONS

As with all other potential descriptions of the electromagnetic field, gauge conditions give separate wave or Laplace equations for the separate potentials. This is true for the classical scalar and vector potentials, ϕ and $\mathbf{A}$, the Whittaker potentials, Hertz potentials and so on

Lorentz gauge

$$\nabla\cdot\mathbf{A} + \frac{1}{c^2}\frac{\partial\phi}{\partial t} = 0$$

is satisfied provided σ and τ satisfy separately the wave equations

$$\begin{aligned}\Box^2\sigma &\equiv \nabla^2\sigma - \frac{1}{c^2}\frac{\partial^2\sigma}{\partial t^2} = 0 \\ \Box^2\tau &\equiv \nabla^2\tau - \frac{1}{c^2}\frac{\partial^2\tau}{\partial t^2} = 0\end{aligned} \tag{7.3.6}$$

Coulomb gauge

$$\nabla\cdot\mathbf{A} = \frac{1}{2}[\sigma\nabla^2\tau - \tau\nabla^2\sigma] = 0$$

and both potentials obey separate Laplace equations.

Such conditions are most suited to boundary value problems and will only arise incidentally in the infinite medium we shall be dealing with. In addition the addendum to $\mathbf{B} = \nabla \times \mathbf{A}$ can be $\mathbf{B} = \nabla \times \mathbf{A}'$ where

$$\mathbf{A}' = \mathbf{A} + \frac{1}{2}[\sigma\nabla\tau - \tau\nabla\sigma]$$

from Equation 7.3.3. This is therefore a new gauge transformation which can be suitably entitled Bateman gauge.

A problem arises with the inclusion of the constitutive relations and the second pair of Maxwell equations

$$\text{curl}\mathbf{H} = \frac{\partial \mathbf{D}}{\partial t} \qquad \text{div}\mathbf{D} = 0 \qquad \text{and} \qquad \mathbf{D} = \varepsilon\mathbf{E} \tag{7.3.7}$$

(we shall use $\mu = 1$; $\mathbf{H} = \mathbf{B}$ throughout)

In *all* standard treatments of electromagnetic field theory, a second pair of potentials has to be defined. The symmetrised potentials shown on page 364, the dual Hertz potentials and dual electromagnetic field tensors are examples. Sommerfeld points out that if one considers the components of the electric field vector $\mathbf{E}$ to be point coordinates, then the components of $\mathbf{D}$ are plane coordinates [4]. This is a Legendre transformation and a transformation to a "refractive space". The logical step would be to satisfy Equation 7.3.7 with two further potentials defined from

$$\mathbf{D} = \nabla u \times \nabla v$$

Then the constitutive relations give cross relationships between σ, τ and u, v as follows

$$\nabla u \times \nabla v = \varepsilon\frac{\partial \tau}{\partial t}\nabla\sigma$$
$$\nabla\sigma \times \nabla\tau = -\frac{\partial v}{\partial t}\nabla u \tag{7.3.8}$$

on the assumption that σ and u are both independent of time.

Thus we require from the divergence equation

$$\frac{\partial \tau}{\partial t}\nabla\varepsilon \cdot \nabla\sigma + \varepsilon\nabla\sigma \cdot \frac{\partial \nabla\tau}{\partial t} + \varepsilon\frac{\partial \tau}{\partial t}\nabla^2\sigma = 0 \tag{7.3.9}$$

Assuming time harmonic potentials, then $\partial/\partial t = -i\omega$ and Equation 7.3.9 reduces to a wave equation

$$\tau\nabla\varepsilon \cdot \nabla\sigma + \varepsilon\nabla\sigma \cdot \nabla\tau + \varepsilon\tau\nabla^2\sigma = 0$$

As stated originally , these potentials as defined by Bateman are analogous to orthogonal stream and velocity potentials. Hence the relations $\nabla\sigma \cdot \nabla\tau = 0$ and $\nabla u \cdot \nabla v = 0$ are implicit in the analysis. Taking the divergence of

Equations 7.3.8 we obtain wave equations for the potentials σ and u in the form

$$\begin{aligned} \nabla^2 u &= 0 \\ \nabla^2 \sigma &= -\nabla\varepsilon \cdot \nabla\sigma/\varepsilon = \text{grad} \log \varepsilon \cdot \text{grad}\ \sigma \end{aligned} \tag{7.3.10}$$

Taking dot products of Equations 7.3.8 gives further connections

$$\nabla\sigma \cdot \nabla u = \nabla\sigma \cdot \nabla v = \nabla\tau \cdot \nabla u = 0 \tag{7.3.11}$$

Hence $\nabla\sigma$, $\nabla\tau$, ∇u, ∇v form a tetrad of orthogonal vectors with $\nabla\tau$ co-parallel to ∇v, and $\nabla\sigma$ and ∇u transverse to them. Since $\nabla\sigma$ is parallel to the **E** polarization it follows, as it does from the basic definitions, that **B** is parallel to ∇u. Thus ∇v is also a definition of the ray tangent vector since $\nabla\tau$ is parallel to $\mathbf{E} \times \mathbf{B}$ (Equation 7.3.15). In free space $\varepsilon = 1$ and $\nabla^2\sigma = 0$.

These conditions and the orthogonality condition can be met quite simply in source free vacuum if the gradients concerned are independent of each other's coordinates, for example in Cartesian coordinates $\nabla\sigma$ would be $\hat{\imath}$ directed and $\nabla\tau$ $\hat{\jmath}$ directed.

PLANE WAVE SOLUTION

For plane waves, say, along the positive z axis we have

$$\begin{aligned} \mathbf{E} &= E_0 \exp[-i(\omega t - kz)]\hat{\imath} \\ \mathbf{B} &= B_0 \exp[-i(\omega t - kz)]\hat{\jmath} \end{aligned} \tag{7.3.12}$$

Rays of a congruence are defined by different values of the constant in the relation $\sigma =$ constant. Parallel rays in the z axis direction arise from $\sigma = f(x)$ the function itself expressing the amplitude distribution among the rays. For a uniform wave, from $\nabla^2\sigma = 0$; $\sigma = ax + b$. This method of defining ray constants is a consequence of the theory. $\nabla\tau$ is then the ray tangent at points on the ray whose constant value specifies it.

Thus the field in Equation 7.3.12 can be derived from

$$\sigma = E_0 x \qquad \tau = T \exp[-i(\omega t - kz)]$$

T constant.

THE DIPOLE (FREE SPACE - FAR FIELD)

The optical field of the radiating dipole can only truly be considered in the extreme far field where the rays are then radially outward pointing. The appropriate form for $\sigma =$ constant is then $\sigma = \cos\theta$ giving individual rays for different values of θ. That this only satisfies $\nabla^2\sigma = 0$ in the same limit is a consequence of the optical as opposed to the total solution of the field equations. (The attempt to use the harmonic function $\sigma = ar\cos\theta$ fails on the grounds that $\sigma =$ constant cannot define a radially outward ray).

We then require $\tau = \exp[-i(\omega t - kr)]$ to derive the field

$$\mathbf{E} = i\omega a \sin\theta \frac{\exp[-i(\omega t - kr])}{r}\hat{\boldsymbol{\theta}}$$

$$\mathbf{B} = -aik \sin\theta \frac{\exp[-i(\omega t - kr])}{r}\hat{\boldsymbol{\phi}}$$

which are the correct descriptions of the field where, as is customary, terms in $1/r^2$ and $1/r^3$ have been ignored, and the factor $\sin\theta/r$ comes from $\nabla\sigma$.

THE POYNTING VECTOR

The Poynting vector is the electromagnetic equivalent of the ray tangent (in isotropic media). With the designation above we obtain

$$\mathbf{P} = \mathbf{E} \times \mathbf{B} = \left(\nabla\sigma\frac{\partial\tau}{\partial t} - \nabla\tau\frac{\partial\sigma}{\partial t}\right) \times (\nabla\sigma \times \nabla\tau) \tag{7.3.13}$$

which reduces, when all the conditions are taken into consideration, to

$$\mathbf{P} = -\frac{\partial\tau}{\partial t}|\nabla\sigma|^2\nabla\tau \tag{7.3.14}$$

and thus "ray tangents" are parallel to $\nabla\tau$ and $\nabla\sigma \cdot \mathbf{P} = 0$.

It is perhaps worth noting in passing that in free space σ and τ are invariant scalars, and with the field given in terms of their differentials, the Lorentz transformation of the field vectors has a very elementary derivation from the Lorentz transformation of the space-time coordinate system alone. This gives the most simple and direct way for deriving the well-known transformation laws for the field vectors. In performing this operation σ and τ have both to be taken as time dependent, since time independence in one frame cannot guarantee time independence in a co-moving frame.

The Bateman potentials are truly "optical" in that, as shown above, they are found only to apply to fields that have a ray optical description. Their use in deriving complex vector and scalar potentials in the Bateman gauge has been illustrated.

7.4 FIELD BI-QUATERNIONS

The fundamentals of the analysis advancing the theory of quaternions to bi-quaternions is given in Appendix IV. We now define "physical bi-quaternions" which act in principle exactly as Minkowski four-vectors, but in the $\alpha\ \beta\ \gamma$ space of the quaternion algebra instead of $x\ y\ z$ space. Physical bi-quaternions will be defined as they arise, but the most basic is the coordinate bi-quaternion $\mathbf{X} = \{ict, x, y, z\}$ alternatively written as $\mathbf{X} = \{ict, \mathbf{r}\}$ implying $\mathbf{X} = \{ict + \alpha x + \beta y + \gamma z\}$.

As is stated in the appendix, all the properties of vector analysis in $\hat{\imath}\hat{\jmath}\hat{k}$ space, such as grad div curl the relations between them and double products etc. are expressed identically in $\alpha\beta\gamma$ notation.

We also define the bi-quaternion differential operator

$$\Diamond = \left(\frac{1}{ic}\frac{\partial}{\partial t}, \frac{\partial}{\partial x}, \frac{\partial}{\partial y}, \frac{\partial}{\partial z}\right) \tag{7.4.1}$$

(the positioning of the i is crucial in obtaining consistency and the result of a long investigation.)

Then for any bi-quaternion $\mathbf{Q} = \{iq_0, \alpha q_1, \beta q_2, \gamma q_3\}$

$$\begin{aligned} \Diamond\mathbf{Q} &= \Diamond\{iq_0, \mathbf{q}\} \\ &= \left\{\frac{1}{c}\frac{\partial q_0}{\partial t} - i\nabla \odot \mathbf{q}, \frac{1}{ic}\frac{\partial \mathbf{q}}{\partial t} + \nabla q_0 + \nabla \otimes \mathbf{q}\right\} \end{aligned} \tag{7.4.2}$$

where

$$\nabla q_o = \alpha\frac{\partial q_0}{\partial x} + \beta\frac{\partial q_0}{\partial y} + \gamma\frac{\partial q_0}{\partial z} \tag{7.4.3}$$

$$\nabla \odot \mathbf{q} = \frac{\partial q_1}{\partial x} + \frac{\partial q_2}{\partial y} + \frac{\partial q_3}{\partial z}$$

$$\nabla \otimes \mathbf{q} = \alpha\left(\frac{\partial q_3}{\partial y} - \frac{\partial q_2}{\partial z}\right) + \beta\left(\frac{\partial q_1}{\partial z} - \frac{\partial q_3}{\partial y}\right) + \gamma\left(\frac{\partial q_2}{\partial z} - \frac{\partial q_1}{\partial y}\right)$$

Hence all the major first order derivatives are incorporated in the single algebraic operation.

The D'Alembertian is obtained as usual as the norm of the Laplace-Beltrami operator

$$\Diamond\overline{\Diamond} = \frac{-1}{c^2}\frac{\partial^2}{\partial t^2} + \frac{\partial^2}{\partial x^2} + \frac{\partial^2}{\partial y^2} + \frac{\partial^2}{\partial z^2} \tag{7.4.4}$$

We shall be mainly concerned with space time transformations of the coordinate bi-quaternion

$$\mathbf{X} = \{ict, \mathbf{r}\} = \{ict, x, y, z\}$$

(using real (x, y, z) rather than involve unnecessary suffices) and the field four-potential

$$A = \left\{\frac{\phi}{ic}, \mathbf{A}\right\}$$

where ϕ and $\mathbf{A}$ are the quaternion "scalar" and vector potentials.

A more comprehensive treatment applied to general physics theory would include the bi-quaternion representations of velocity momentum force and wave vectors. These will be obvious when they arise.

Applying the operator $\Diamond$ to the potential $A = \{\phi/(ic), \mathbf{A}\}$

$$\Diamond A = \left\{\frac{-1}{c^2}\frac{\partial\phi}{\partial t} - \nabla \odot \mathbf{A}, \frac{1}{ic}\frac{\partial \mathbf{A}}{\partial t} + \frac{1}{ic}\nabla\phi + \nabla \otimes \mathbf{A}\right\} \tag{7.4.5}$$

and if the field satisfies the Lorentz condition

$$\frac{1}{c^2}\frac{\partial \phi}{\partial t} + \nabla \odot \mathbf{A} = 0$$

then

$$\Box A = \left\{ 0, \frac{i\mathbf{E}}{c} + \mathbf{B} \right\} \tag{7.4.6}$$

where

$$\mathbf{B} = \nabla \otimes \mathbf{A}$$

and

$$\mathbf{E} = -\frac{\partial \mathbf{A}}{\partial t} - \nabla \phi$$

We define $\{0, i\mathbf{E}/c + \mathbf{B}\}$ to be the field bi-quaternion $\mathcal{M}$.

ALGEBRAIC MAXWELL EQUATIONS

Specify a bi-quaternion $\Phi = \{i\phi_0, \phi_1, \phi_2, \phi_3\}$ and

$$\Box = \left(\frac{-i}{c}\frac{\partial}{\partial t}, \frac{\partial}{\partial x}, \frac{\partial}{\partial y}, \frac{\partial}{\partial z} \right)$$

then $\Box \Phi = 0$ results in

$$\begin{aligned}
&\frac{1}{c}\frac{\partial \phi_0}{\partial t} - \frac{\partial \phi_1}{\partial x} - \frac{\partial \phi_2}{\partial y} - \frac{\partial \phi_3}{\partial z} \\
&+ \alpha \left\{ i\frac{\partial \phi_0}{\partial x} - \frac{i}{c}\frac{\partial \phi_1}{\partial t} + \frac{\partial \phi_3}{\partial y} - \frac{\partial \phi_2}{\partial z} \right\} \\
&+ \beta \left\{ i\frac{\partial \phi_0}{\partial y} - \frac{i}{c}\frac{\partial \phi_2}{\partial t} - \frac{\partial \phi_3}{\partial x} + \frac{\partial \phi_1}{\partial z} \right\} \\
&+ \gamma \left\{ i\frac{\partial \phi_0}{\partial z} - \frac{i}{c}\frac{\partial \phi_3}{\partial t} + \frac{\partial \phi_2}{\partial x} - \frac{\partial \phi_1}{\partial y} \right\} = 0
\end{aligned} \tag{7.4.7}$$

This condition will be referred to as a "left regularity" ("right regularity" is the condition $\Phi \Box \equiv \Box \bar{\Phi} = 0$) [42] [43].

We put

$$\begin{aligned}
\Phi &= A + iB = \{0, a_1, a_2, a_3\} + i\{0, b_1, b_2, b_3\} \\
&\equiv \alpha(a_1 + ib_1) + \beta(a_2 + ib_2) + \gamma(a_3 + ib_3)
\end{aligned}$$

Then Equations 7.4.7 yield

$$\nabla \odot \mathbf{a} = 0 \qquad \nabla \odot \mathbf{b} = 0$$

$$-\frac{1}{c}\frac{\partial \mathbf{a}}{\partial t} + \nabla \otimes \mathbf{b} = 0 \qquad \frac{1}{c}\frac{\partial \mathbf{b}}{\partial t} + \nabla \otimes \mathbf{a} = 0$$

These are source free Maxwell equations and the derivation is purely algebraic and due to the skew symmetric or symplectic nature of the algebra.

Thus putting $\mathbf{a} \equiv \mathbf{E}$ and $\mathbf{b} \equiv c\mathbf{B}$ we obtain

$$\text{curl } \mathbf{E} = -\frac{\partial \mathbf{B}}{\partial t} \qquad \text{curl } \mathbf{B} = \frac{1}{c^2}\frac{\partial \mathbf{E}}{\partial t}$$

(curl being the quaternion operator $\nabla\otimes$)

Sources can be included by equating $\Box\Phi$ to a source bi-quaternion $\mathcal{S}$

$$\mathcal{S} = \{\frac{\rho}{\varepsilon}, \mathbf{j}_m\} + ic\{\rho_m, \mu\mathbf{j}\} \tag{7.4.8}$$

ρ ρ_m $\mathbf{j}$ and $\mathbf{j}_m$ being electric and magnetic charge and current densities.

The complete set of symmetrised Maxwell equations result

$$\nabla \otimes \mathbf{E} = -\frac{\partial \mathbf{B}}{\partial t} - \mu\mathbf{j}_m \qquad \nabla \otimes \mathbf{B} = \mu\mathbf{j} + \varepsilon\frac{\partial \mathbf{E}}{\partial t}$$

$$\nabla \odot \mathbf{B} = \rho_m \qquad \nabla \odot \mathbf{E} = \frac{\rho}{\varepsilon}$$

(Note that A B above are specified to have zero scalar components. Imaeda [42] terms this a "vector condition" on the free choice of such bi-quaternions.) $A + iB$ is then an electromagnetic field $\mathbf{E} + ic\mathbf{B}$.

Finally, since $\Box\Phi = \mathcal{S}$ does not uniquely determine Φ any left regular bi-quaternion K can be added. That is with $\Box K = 0$: $\Box\Phi \equiv \Box(\Phi + K) \equiv \mathcal{S}$ constitutes a gauge transformation. An extension can be made with the inclusion of the electric vector potential $\mathbf{A}_e$ and scalar magnetic potential ϕ_m.

Hence a total field with

$$A = \{\frac{-i\phi}{c}, \mathbf{A}\} + i\{-i\phi_m, c\mathbf{A}_e\}$$

then $\Box A = \{0, \mathbf{B} + i/c\mathbf{E}\} \equiv \mathcal{M}$ results in

$$\text{div } \mathbf{A} + \frac{1}{c^2}\frac{\partial \phi}{\partial t} = 0 \qquad \text{div } \mathbf{A}_e + \frac{1}{c^2}\frac{\partial \phi_m}{\partial t} = 0 \tag{7.4.9}$$

$$\mathbf{E} = -\frac{\partial \mathbf{A}}{\partial t} - \nabla\phi - \nabla \otimes \mathbf{A}_e$$

$$\mathbf{B} = \nabla \otimes \mathbf{A} + \nabla\phi_m + \frac{\partial \mathbf{A}_e}{\partial t}$$

the Lorentz condition and symmetrised potential description of the electromagnetic field. The apparently arbitrary definition of the regularity condition of Equation 7.4.7 is given full justification by Imaeda [42] in terms of functional derivatives in a four dimensional complex space.

Let $\mathcal{F}$ be a force bi-quaternion

$$\mathcal{F} = \left\{-\frac{if_0}{c}, \frac{\mathbf{f}}{c}\right\}$$

and $\mathcal{U}$ a velocity bi-quaternion $\mathcal{U} = \{-iu_o, \mathbf{u}\}$ then for the field bi-quaternion $\mathcal{M} = \{0, \mathbf{B} + \frac{i\mathbf{E}}{c}\}$

$$\mathcal{F} = -\frac{1}{2}\{\mathcal{M}\mathcal{U}^* - \mathcal{U}\mathcal{M}^*\} \quad (\mathcal{U}^* \equiv \mathcal{U})$$

gives

$$f_0 = \frac{\mathbf{u}\cdot\mathbf{E}}{c} \quad \mathbf{f} = \left\{\frac{\mathbf{E}}{c} + \mathbf{u}\times\mathbf{B}\right\}$$

that is the Lorentz force and temporal component of the four-force.

7.5 FOUR-SPACE REPRESENTATION OF BATEMAN POTENTIALS

We analytically continue into the complex four-space by operating with the bi-quaternion operator

$$\Box = \left\{\frac{i}{c}\frac{\partial}{\partial t}, \nabla\right\}$$

Equations 7.3.3 and 7.3.4 are combined in

$$A = \frac{1}{2}(\tau\Box\sigma - \sigma\Box\tau) \equiv \left\{\frac{\phi}{ic}, \mathbf{A}\right\}$$

where

$$\mathbf{A} = \frac{1}{2}(\tau\nabla\sigma - \sigma\nabla\tau) \quad \text{and} \quad \phi = \frac{1}{2}\left(\tau\frac{\partial\sigma}{\partial t} - \sigma\frac{\partial\tau}{\partial t}\right)$$

Then $\frac{1}{2}(\Box\sigma\overline{\Box}\tau - \Box\tau\overline{\Box}\sigma)$ (or the direct bi-quaternion product $\Box\sigma\Box\tau$ both) $= \left\{0, \frac{i\mathbf{E}}{c} + \mathbf{B}\right\}$ where

$$\mathbf{E} = \frac{\partial\sigma}{\partial t}\nabla\tau - \frac{\partial\tau}{\partial t}\nabla\sigma \quad \text{and} \quad \mathbf{B} = \nabla\sigma\otimes\nabla\tau \tag{7.5.1}$$

and $\nabla\sigma\cdot\nabla\tau = 0$ if the Lorentz condition applies (Equation 7.4.5). $\Box\sigma\Box\tau$ is an antisymmetric tensor (cf $\nabla\sigma\times\nabla\tau$ and Equation 7.2.2) defined by

$$F^{\mu\nu} = \frac{\partial\sigma}{\partial x}\mu\frac{\partial\tau}{\partial x}\nu - \frac{\partial\tau}{\partial x}\mu\frac{\partial\sigma}{\partial x}\nu$$

(cf Equation 7.3.2)

$$F^{\mu\nu} = \begin{bmatrix} 0 & \frac{i}{c}\left(\frac{\partial\sigma}{\partial t}\frac{\partial\tau}{\partial x} - \frac{\partial\tau}{\partial t}\frac{\partial\sigma}{\partial x}\right) & \frac{i}{c}\left(\frac{\partial\sigma}{\partial t}\frac{\partial\tau}{\partial y} - \frac{\partial\tau}{\partial t}\frac{\partial\sigma}{\partial y}\right) & \frac{i}{c}\left(\frac{\partial\sigma}{\partial t}\frac{\partial\tau}{\partial z} - \frac{\partial\tau}{\partial t}\frac{\partial\sigma}{\partial z}\right) \\ & 0 & \left(\frac{\partial\sigma}{\partial x}\frac{\partial\tau}{\partial y} - \frac{\partial\tau}{\partial x}\frac{\partial\sigma}{\partial y}\right) & \left(\frac{\partial\sigma}{\partial x}\frac{\partial\tau}{\partial z} - \frac{\partial\tau}{\partial x}\frac{\partial\sigma}{\partial z}\right) \\ & & 0 & \left(\frac{\partial\sigma}{\partial y}\frac{\partial\tau}{\partial z} - \frac{\partial\tau}{\partial y}\frac{\partial\sigma}{\partial z}\right) \\ & & & 0 \end{bmatrix} \tag{7.5.2}$$

$$= \begin{bmatrix} 0 & \frac{i}{c}E_x & \frac{i}{c}E_y & \frac{i}{c}E_z \\ & 0 & B_z & -B_y \\ & & 0 & B_x \\ & & & 0 \end{bmatrix} \tag{7.5.3}$$

which is the classical electromagnetic field tensor.

Forming the dual tensor $^*F^{\mu\nu}$ either by the transformation $E \to H$, $H \to E$, $\mu \leftrightarrow \varepsilon$ or the product with the antisymmetric tensor density $\varepsilon_{ij\mu\nu}$ results as usual in the form of Maxwell's equations given by

$$F^{\mu\nu},\nu = 0 \qquad {}^*F^{\mu\nu},\nu = -J^{\mu}$$

If

$$\Box^2 A = \Box\overline{\Box} A = J = \frac{1}{ic}\left\{\frac{\rho}{\varepsilon}, ic\mu\mathbf{j}\right\}$$

and both $\mathbf{A}$ and ϕ obey the separated wave equations

$$\nabla^2\phi - \frac{1}{c^2}\frac{\partial^2\phi}{\partial t^2} = -\frac{\rho}{\varepsilon} \qquad \nabla^2\mathbf{A} - \frac{1}{c^2}\frac{\partial^2\mathbf{A}}{\partial t^2} = \mu\mathbf{j} \tag{7.5.4}$$

Maxwell's equations have the bi-quaternion form

$$\begin{aligned} \nabla \otimes \mathbf{E} &= -\frac{\partial \mathbf{B}}{\partial t} \quad \nabla \odot \mathbf{E} = \frac{\rho}{\varepsilon} \\ \nabla \otimes \mathbf{B} &= \frac{1}{c^2}\frac{\partial \mathbf{E}}{\partial t} + \mu\mathbf{j} \quad \nabla \odot \mathbf{B} = 0 \\ \nabla \odot \mathbf{j} &+ \frac{\partial \sigma}{\partial t} = 0 \end{aligned} \tag{7.5.5}$$

7.6 TRANSFORMATION OF BI-QUATERNIONS

Physical bi-quaternions (four-vectors) transform under similarity transformations of the kind

$$\mathcal{X}' = \mathcal{Q}\mathcal{X}\mathcal{Q} \tag{7.6.1}$$

where $\mathcal{Q}$ is a bi-quaternion. We list without proofs the following

1.

$$\mathcal{Q} = \frac{a_0 + i\mathbf{b}}{\sqrt{a_0^2 + \mathbf{b}\cdot\mathbf{b}}}$$

then $\mathcal{Q}$ has unit norm and Equation 7.6.1 represents a spatial rotation with axis $\mathbf{b}$ through the origin of coordinates and magnitude

$$\omega = 2\tan^{-1}\left[\frac{|\mathbf{b}|}{a_0}\right]$$

2.

$$\mathcal{Q} = \frac{1}{\sqrt{2}}\left(\sqrt{\lambda+1} - i\frac{\mathbf{v}}{v}\sqrt{\lambda-1}\right) \tag{7.6.2}$$

$$\mathbf{v} = \alpha v_x + \beta v_y + \gamma v_z$$

$$\lambda = \frac{1}{\sqrt{1-v^2/c^2}} \qquad \text{norm } \mathcal{Q} = 1$$

Equation 7.6.1 then represents a pure Lorentz transformation [44]. That is if

$$\begin{aligned}
\mathcal{X}' &= ix_0' + \mathbf{x}' \\
x_0' &= \lambda\left[x_0 - \frac{\mathbf{x}\cdot\mathbf{v}}{c}\right] \\
\mathbf{x}' &= \mathbf{x} + \frac{\mathbf{v}}{c}[\mathbf{x}\cdot\mathbf{v}(\lambda-1) + \lambda x_0]
\end{aligned}$$

the standard form.

BI-QUATERNION TRANSFORMATIONS

A polynomial function of a bi-quaternion variable $Q = \{iq_0, \mathbf{q}\}$ can be put into the form

$$f(Q) = [f_1(q_0+q) + f_2(q_0-q)] \;\; ; \;\; \frac{i\mathbf{q}}{2q}[f_1(q_0+q) - f_2(q_0-q)] \tag{7.6.3}$$

Hence choosing Q to be the basic coordinate bi-quaternion $Q = \{ict, \mathbf{r}\}$ we can regard Equation 7.6.3 as a transformation and obtain

$$\begin{aligned}
ct' &= \frac{1}{2}[f_1(ct+r) + f_2(ct-r)] \\
x' &= \frac{x}{2r}[f_1(ct+r) - f_2(ct-r)] \\
y' &= \frac{y}{2r}[f_1(ct+r) - f_2(ct-r)] \\
z' &= \frac{z}{2r}[f_1(ct+r) - f_2(ct-r)]
\end{aligned}$$

For example choosing $f_1(\xi) \equiv f_2(\xi) = \frac{1}{\xi}$ we have [45]

$$t' = \frac{t}{s^2} \quad x' = \frac{x}{s^2} \quad y' = \frac{y}{s^2} \quad z' = \frac{z}{s^2} \tag{7.6.4}$$

(cf Equation 7.1.5) where $s^2 = c^2t^2 - r^2$.

It can readily be verified that if $f(x, y, z, t)$ is a wave function then

$$\frac{1}{s^2} f\left[\frac{x}{s^2}, \frac{y}{s^2}, \frac{z}{s^2}, \frac{t}{s^2},\right]$$

is also a wave function. For $t = 0$ this becomes a pure space inversion.

If we apply a pure inversion $r' = -1/r$ to the outgoing spherical wave function $\exp(-ikr)/r$ we obtain $r'^2 \exp(ik'r)/r'$ and the outgoing wave with amplitude decreasing as $1/r$ is inverted to an incoming spherical wave with amplitude proportional to r' in a medium with pseudo refractive index given by $k' = k/r^2$ or $\eta = 1/r^2$. This, as shown in Chapter 4, is one example of a more general theorem.

It is well known that the transformation $r \Rightarrow 1/R$; $\phi = R\phi'$ leaves Laplace's equation invariant. Applying it to the wave equation results in

$$\nabla^2\phi' = \frac{1}{R^4c^2}\frac{\partial^2\phi'}{\partial t^2}$$

implying a new velocity $c' = R^2c$ or a new refractive index $\eta = c/c' = 1/R^2$.

The transformation which was used in Equation 7.2.4 to convert the spherical/elliptical lens into a hyperbolic/plane lens, can be obtained from

$$f_1 = ct + \frac{ar}{a+x} \qquad f_2 = ct - \frac{ar}{a+x} \tag{7.6.5}$$

A further transformation given by Bateman is

$$t' = \frac{s^2+1}{2c(z-ct)} \quad x' = \frac{x}{z-ct}$$
$$y' = \frac{y}{z-ct} \quad z' = \frac{s^2-1}{2c(z-ct)} \tag{7.6.6}$$

and this is "equivalent to a conformal transformation of a space of four dimensions which was discovered by Cremona" [46]. It has not proved possible as yet to obtain this transformation by bi-quaternion methods.

7.7 INVERSION OF FIELDS AND RAYS IN A SPHERICALLY SYMMETRICAL MEDIUM

We begin by noting the following property of geometrically inverse curves. For any curve $r = F(\theta)$

$$r\frac{d\theta}{dr} = \frac{r}{F'(\theta)}$$

Then the tangent is given by

$$r\frac{d\theta}{dr} = \frac{r}{F'(\theta)} = \frac{F(\theta)}{F'(\theta)} \tag{7.7.1}$$

For the inverse curve $R = 1/F(\theta)$ $dR = -\frac{F'(\theta)}{F^2(\theta)}\,d\theta$ and thus

$$R\frac{d\theta}{dR} = -\frac{F^2(\theta)}{F'(\theta)}R = -\frac{F(\theta)}{F'(\theta)} \tag{7.7.2}$$

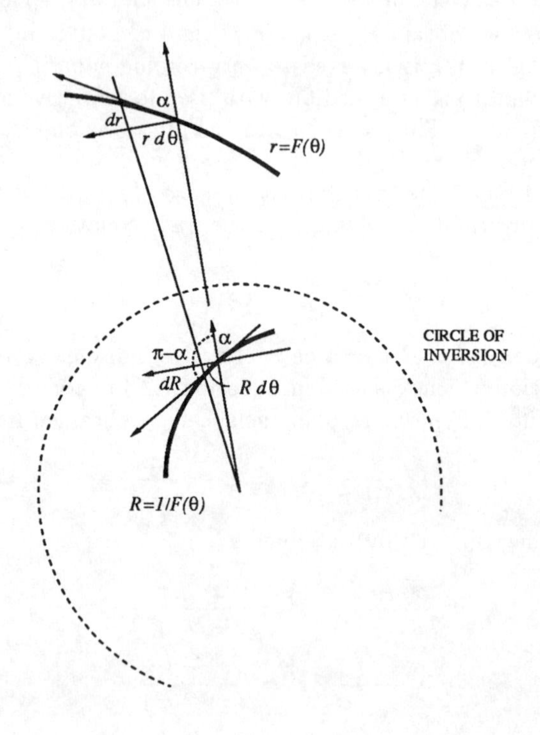

Figure 7.3: Geometrical inversion of a space ray

The minus sign in Equation 7.7.2 plays an all important role in the analysis leading to the inversion of rays and the derivation of the electromagnetic field equivalent. It manifests itself as a change of sign of either the r or θ component of the tangent as shown in Figure 7.2, whereby, on inversion the angle α becomes $\pi - \alpha$.

The inversion of the field bi-quaternion (a six-vector) is performed by the transformation

$$\mathcal{M}' = \mathcal{X}^c \mathcal{M} \mathcal{X} \tag{7.7.3}$$

where $\mathcal{M}$ is given by Equation 7.4.6 and normalised through division by the norm, $|\mathcal{X}^c \mathcal{X}| = c^2t^2 - r^2$. Hence writing out the expression in full gives

$$\left\{0, \frac{i}{c}\mathbf{E}' + \mathbf{B}'\right\} = \frac{\{-ict, \mathbf{r}\}\{0, (i/c)\mathbf{E} + \mathbf{B}\}\{ict, \mathbf{r}\}}{c^2t^2 - r^2} \tag{7.7.4}$$

Carrying out the multiplication of 7.7.4 with the rules of bi-quaternion algebra and equating real and imaginary parts gives the two relationships

$$\mathbf{E}' = \mathbf{E} + \frac{2c^2t\mathbf{r}\times\mathbf{B} - 2\mathbf{r}\times(\mathbf{r}\times\mathbf{E})}{c^2t^2 - r^2} \tag{7.7.5}$$

$$\mathbf{B}' = -\mathbf{B} - \frac{2t\mathbf{r}\times\mathbf{E} - 2\mathbf{r}\times(\mathbf{r}\times\mathbf{B})}{c^2t^2 - r^2} \tag{7.7.6}$$

which apart from changes in notation, is identical in form to the inversion presented without derivation by Cunningham † We assume the time independent solution, putting $t = 0$ in these relations. This presupposes that a ray is a fixed curve in the refractive space and not a time evolving trajectory. It is a common solution in dynamical problems where, for example, a planetary orbit can be considered to be a fixed curve in space. The inverted Poynting vector is therefore given by

$$\mathbf{E}'\times\mathbf{B}' = \mathbf{E}\times\mathbf{B} - 2\frac{(\mathbf{E}\times\mathbf{r})(\mathbf{B}\cdot\mathbf{r}) - (\mathbf{B}\times\mathbf{r})(\mathbf{E}\cdot\mathbf{r})}{r^2} \tag{7.7.7}$$

With either polarization condition $\mathbf{E}\cdot\mathbf{r} = 0$ or $\mathbf{B}\cdot\mathbf{r} = 0$ the term that remains in the numerator on the right-hand side of Equation 7.7.7 is found to reverse precisely the $\hat{\theta}$ component of the first term $\mathbf{E}\times\mathbf{B}$ thus reflecting the ray inversion shown in Figure 7.2 and by Equation 7.7.2.

In the spherically symmetrical medium, rays lie in planes through the origin. We can use Bouguer's theorem to establish a specific form of the σ scalar potential. Bouguer's theorem states that, for a spherically symmetric medium, the generalization of Snell's law is given by Equation 4.7.5

$$r\eta(r)\sin\alpha = \text{constant}$$

where α is the angle between the radius vector to a point P on the ray and the tangent to the ray at P, Figure 7.1. The angle α is a function of the spherical polar coordinates r and θ, $\alpha = \alpha(r,\theta)$. Along a ray

$$r\eta(r)\sin\alpha \equiv f(r)\sin\alpha = \text{constant} = p$$

For this case τ can be assumed to be totally arbitrary, so long as $\nabla\tau$ does not have a $\hat{\phi}$ component, that is $\nabla\tau$ is of the form

$$\nabla\tau = a\hat{\mathbf{r}} + b\hat{\boldsymbol{\theta}} \tag{7.7.8}$$

†For completeness, the field transformations of Cunningham [39] for the inversion are

$$\mathbf{E}' = \frac{r^2 - c^2t^2}{d^4}[(r^2 - c^2t^2)\mathbf{E} + 2\mathbf{r}\times(\mathbf{r}\times\mathbf{E}) - 2ct(\mathbf{r}\times\mathbf{H})]$$

$$\mathbf{H}' = \frac{r^2 - c^2t^2}{d^4}[-(r^2 - c^2t^2)\mathbf{H} - 2\mathbf{r}\times(\mathbf{r}\times\mathbf{H}) - 2ct(\mathbf{r}\times\mathbf{E})]$$

where d is the radius of inversion and $r^2 = x^2 + y^2 + z^2$. It can be seen that, for $t = 0$, ray directions $(\mathbf{E}'\times\mathbf{H}')\cdot\hat{\mathbf{r}} = (\mathbf{E}\times\mathbf{H})\cdot\hat{\mathbf{r}}$ since in an inversion $\hat{\mathbf{r}}' = \hat{\mathbf{r}}$. We have taken $d^2 = c^2t^2 - r^2$ to obtain Equations 7.7.5 and 7.7.6

(With these definitions of σ and $\nabla\tau$, it has not proved possible to guarantee agreement with the orthogonality condition or the conditions of Equation 7.3.11. The heuristic assumption is made that these are satisfied, on the grounds of the final result which shows full agreement with the previous analysis.)

Then using the form of σ and $\nabla\tau$ given above

$$\nabla\sigma = \left[\sin\alpha\frac{\partial f}{\partial r} + f\cos\alpha\frac{\partial\alpha}{\partial r}\right]\hat{\mathbf{r}} + \frac{f}{r}\cos\alpha\frac{\partial\alpha}{\partial\theta}\hat{\boldsymbol{\theta}}$$

$$\mathbf{E} \approx \left[\sin\alpha\frac{\partial f}{\partial r} + f\cos\alpha\frac{\partial\alpha}{\partial r}\right]\hat{\mathbf{r}} + \frac{f}{r}\cos\alpha\frac{\partial\alpha}{\partial\theta}\hat{\boldsymbol{\theta}}$$

$$\mathbf{B} = A\hat{\boldsymbol{\phi}} \tag{7.7.9}$$

where, for simplicity, all the multiplying factors are subsumed into A. The Poynting vector is obtained straightforwardly and is given by

$$\mathbf{E}\times\mathbf{B} \approx \frac{fA}{r}\cos\alpha\frac{\partial\alpha}{\partial\theta}\hat{\mathbf{r}} - A\left[\sin\alpha\frac{\partial f}{\partial r} + f\cos\alpha\frac{\partial\alpha}{\partial r}\right]\hat{\boldsymbol{\theta}} \tag{7.7.10}$$

Inversion of the Poynting vector is then achieved by changing the sign of the $\hat{\boldsymbol{\theta}}$ component as proved. This gives for the *inverted* Poynting vector $\mathbf{E}'\times\mathbf{B}'$

$$\mathbf{E}'\times\mathbf{B}' \approx \frac{fA}{r}\cos\alpha\frac{\partial\alpha}{\partial\theta}\hat{\mathbf{r}} + A\left[\sin\alpha\frac{\partial f}{\partial r} + f\cos\alpha\frac{\partial\alpha}{\partial r}\right]\hat{\boldsymbol{\theta}} \tag{7.7.11}$$

Consider a new medium, characterised by $F(R) \equiv R\bar{\eta}(R)$. Within this new medium the electromagnetic field components, $\mathbf{E}$ and $\mathbf{B}$ are given by

$$\mathbf{E} \approx \left[\sin\alpha\frac{\partial F}{\partial R} + F\cos\alpha\frac{\partial\alpha}{\partial R}\right]\widehat{\mathbf{R}} + \frac{F}{R}\cos\alpha\frac{\partial\alpha}{\partial\theta}\hat{\boldsymbol{\theta}}$$

$$\mathbf{B} = A'\hat{\boldsymbol{\phi}} \tag{7.7.12}$$

This gives the Poynting vector within the new medium $F(R) \equiv R\bar{\eta}(R)$, as

$$\mathbf{E}\times\mathbf{B} \approx \frac{FA'}{R}\cos\alpha\frac{\partial\alpha}{\partial\theta}\widehat{\mathbf{R}} - A'\left[\sin\alpha\frac{\partial F}{\partial R} + F\cos\alpha\frac{\partial\alpha}{\partial R}\right]\hat{\boldsymbol{\theta}}$$

Since $\hat{\mathbf{r}} = \widehat{\mathbf{R}}$ and α and $\partial\alpha/\partial\theta$ are invariant, the coefficients of the $\hat{\mathbf{r}}$ and $\widehat{\mathbf{R}}$ in Equations 7.7.11 and 7.7.12 respectively, are equivalent, giving

$$A' = A\frac{Rf}{rF}$$

and comparing the coefficients of the $\hat{\boldsymbol{\theta}}$ component, of the two Poynting vectors

$$-\frac{Rf}{rF}\frac{\partial F}{\partial R} = \frac{\partial f}{\partial r} \qquad -\frac{R}{r}\frac{\partial\alpha}{\partial R} = \frac{\partial\alpha}{\partial r}$$

the first equation from terms in $\cos\alpha$ and the second from terms in $\sin\alpha$.

The equations are satisfied by $rF = Rf$ and $A = A'$ and therefore

$$R = \frac{1}{r} \quad \text{and} \quad F(R) = f\left[\frac{1}{r}\right]$$

Hence the transformation of the refractive index is given by

$$r\eta(r) \equiv f(r) \Rightarrow f\left[\frac{1}{r}\right] \equiv r\bar{\eta}(r) \tag{7.7.13}$$

Thus rays within the medium with refractive index $\bar{\eta}(r)$ are the geometrical inverse of the rays within the medium with refractive index $\eta(r)$, as shown in Section 4.10, to be conformally transformed.

This result confirms that the inversion of the electromagnetic field and its Poynting vector is equivalent to the geometrical inversion of the optical rays provided the refractive space is inverted by the law of 7.7.13.

LORENTZ INVARIANCE

The Lorentz transformation of field bi-quaternions (six-vectors, tensors) is given by

$$\mathcal{M}' = \mathcal{Q}^* \mathcal{M} \mathcal{Q}$$

where * denotes complex conjugate. Hence

$$(0, \frac{i\mathbf{E}'}{c} + \mathbf{B}') = \frac{1}{2}\left(\sqrt{\lambda+1}\ \tfrac{i\mathbf{v}}{v}\sqrt{\lambda-1}\right)\left(0, \frac{i\mathbf{E}}{c} + \mathbf{B}\right)\left(\sqrt{\lambda+1}\ \tfrac{-i\mathbf{v}}{v}\sqrt{\lambda-1}\right)$$

then

$$\begin{aligned} \mathbf{E}' &= \mathbf{E} + \mathbf{v} \times \mathbf{B} \\ \mathbf{B}' &= \mathbf{B} + \frac{\mathbf{v} \times \mathbf{E}}{c^2} \end{aligned}$$

the standard result.

Lorentz invariance

ꝗ transforms as a physical bi-quaternion

$$ꝗ' = \mathcal{Q} ꝗ \mathcal{Q}$$

and from Equation 7.4.8 Maxwell's equations are

$$ꝗ \mathcal{M} = \mathcal{J} \qquad \mathcal{J} = \left(\frac{i\rho}{c\varepsilon}; \mu\mathbf{j}\right)$$

and $\mathcal{J}$ is a physical bi-quaternion. Then

$$\begin{aligned} ꝗ'\mathcal{M}' &= \mathcal{Q} ꝗ \mathcal{Q}\mathcal{Q}^* \mathcal{M} \mathcal{Q} \\ &= \mathcal{Q} ꝗ \mathcal{M} \mathcal{Q} = \mathcal{Q}\mathcal{J}\mathcal{Q} = \mathcal{J}' \end{aligned}$$

demonstrating the Lorentz invariance of Maxewll's equations.

7.8 SUMMARY AND CONCLUSIONS

There are several themes running through this volume, and an attempt has to be made to draw them together and show their relevance to each other and to physics in general.

The inclusion of the laws of reflection and refraction into the corpus of Euclidean geometry, sets up what has been termed "optical geometry". These at first referred to optical systems defined by general aspheric surfaces, that is discontinuous spaces connected by piecewise straight geometrical rays. Later, with the inclusion of isotropic non-uniform media, the rays become continuous curves in a "refractive space". There exists in this geometry entities and relationships with no counterpart in Euclidean geometry, most notably the concept of the orthogonal congruence, the existence of the ray system and its orthogonal surfaces the wave or phase front. Once included, this orthogonal congruence turns out to have a specific set of theorems of its own which has in this book been embodied in the phrase "optical geometry".

A complete description of these theorems and their practical applications to the design of optical devices and microwave antennas has been given in those chapters relevant to those designs. These include the inversion theorem of Damien and its associated concept of the zero distance phase front and the inversion of the refractive space and the geometrical ray system in a spherically non-uniform refractive medium. Other geometrical or analytical theorems which have a relevance have been discussed. These were the reciprocity of Budden and the transformations of Bateman. It is highly likely that many more geometrical theorems relating to rays, surfaces and phase fronts exist and may surface eventually. Hamilton too had a geometrical theorem of inversion, closely allied to that we have termed Damien's theorem.

One study has been to try to see in what way these theorems can be associated with other known relations that exists in optical lens systems. Stavroudis [47] proves the existence of a lens group defined in terms of the matrices of the Sp(4) group and a leading term which is a member of the quaternion group (Appendix IV). He also shows that in going to the paraxial approximation for propagation through such lenses, they can be represented exactly by a subgroup the Sp(2) group. It is probably too much to hope for at this stage that transformations in these groups could supply the proof required for the hypothesis relating to the geometrical transformation of lenses given by the inversion method at the end of Chapter 3. However this illustrates something of the attempt to derive relationships between different procedures in the field of optical geometry.

Optical rays in a continuous non-uniform space have been shown to be strongly analogous to particle trajectories. Buchdahl [48] has compared the equivalence of rays in the Maxwell fish-eye and the Kepler orbit through the analogy between Bouguer's theorem and a constant angular momentum and the existence of a new ray constant equivalent to the Runge-Lenz vector. Both problems of rays and trajectories in a non-uniform potential derive from an Euler-Lagrange variational problem. Thus all the theorems of optical geometry should have a dynamical analogue, and indeed this is found to be the case [49]. It should be observed that the comparison is made in the phase space of the dynamical problem and the refractive space of the optical one. Thus there is a direct compatibility between Herzberger's optical invariant [50] and the Poisson bracket.

The transformations considered, almost invariably involve geometrical inversion. Inversions are a key transformation of the conformal group and so it is to that group that attention has to be turned. It has been shown [51] that every continuous conformal transformation may be represented as the product of (a) a Lorentz transformation (b) a translation (c) a dilatation and T one of four special transformations which include the identity and the inversions of Bateman.

Thus the subject of geometrical optics has come to be closely associated with the action of the conformal group both on optical rays and on particle trajectories and there should be a close interaction in the future discoveries in either discipline.

A further recombination has to be considered however, between the optical ray and its derivation from the high frequency limit of the wave equation. The attempt to create such a connection through the intermediary of the scalar and vector potentials of electromagnetism, the field vectors themselves or other well-known potential derivations, proved unsuccessful. It was to be hoped that, in such a case, the conformal invariance of the electromagnetic field [52] would derive the inversions of the optical rays and provide the solid theoretical basis for optical geometry.

A solution was found in the definition of the field in terms of Bateman potentials. These are a gauge addition to the standard potentials of electromagnetic theory and strictly of relevance *only* to the field in the optical limit. Only one or two examples of their application have so far been obtained and these were demonstrated. With an amount of hypothesising, it was shown that this formalism did agree in principle with the inversion theory of the non-uniform medium previously found by conformal transformation. These potentials turn out to have well defined connections to the Clebsch potentials and the Debye potentials [53] [54] [55]. This approach therefore appears to give greater physical insight than does the usual (rather arbitrary but justified by the result) ansatz into the wave equation that gives rise to so-called rays.

Looked at a bit obliquely, we do have a somewhat tenuous connection between optical rays, their analogue particle trajectories and the fact that the former can be derived from electromagnetic theory and the latter from gravitation theory in their appropriate phase spaces. Thus an association between two of the fundamental forces is created. It is possible in that context to derive Maxwell-like equations with the gravitational and inertial fields in lieu of the electric and magnetic field strengths [56] [57], but the role of gravitational permittivity or permeability is undecided.

Unfortunately the confinement of the theory to conformal transformations is a severe limitation. Inversions, even with the addition of translations rotations or magnifications do not result in a large variety of practicable optical devices. Therefore an attempt has been made to widen transformation theory to more general applications. To do this, an extension from the conformal mapping in two dimensions, applicable to transverse electrostatic problems, has been pursued. The two dimensional problem depends on the conformal property of the function of a complex variable, so it was proposed to study the properties of a four-dimensional hypercomplex variable. In this manner it was hoped that quite general transformations of radiating or scattering problems could provide solutions to complicated problems from the transformation of simpler or known canonical problems. This involved the study of the (regarded as obvious) extension to the ordinary complex number the biquaternion. This once again made a connection with the theory of groups, and also with the higher algebras such as the Clifford algebra [58] [59] [60] [61]. The equivalent in the four-space of the Cauchy-Riemann equations of complex variable theory is termed "left-regularity" and the resultant equations are shown to be algebraic versions of Maxwell equations [62] [63].

However, nature is not to be denied, and left-regularity applies to only very few functions of the hypercomplex variable, in contrast to two dimensional electrostatics where conformal mapping can be performed with all functions. In fact the application to real radiating systems or real electromagnetic fields has yet to be discovered. Salingaros [64] shows that four dimensional complete simply connected Riemannian space with quaternionic structure can describe all four fundamental fields. Geometrodynamics is defined by Synge [65] and geometric optics has been the theme of this volume. What is required is an associated geometroelectrodynamics theory [66].

To repeat a statement from reference 47 where it is shown that geometrical Maxwell-like equations can be obtained from the geometry of twisted curves in a refractive space: "One certain implication is that the geometrical component in physical optics is greater than anyone thought and that the usual methods for extracting geometrical optics from Maxwell's equations may be too restrictive". What has been attempted is precisely that, finding unusual methods and showing where these methods agree with the unusual geometry that the optics in geometrical optics provides. It is hoped that some indication of the depth and value of such an approach has been given in this work, without departing too far from real life and practical device designing that microwave optics and geometrical optics are used for.

REFERENCES

1. Budden KG (1961) *Radio Waves in the Ionosphere* Cambridge University Press, UK

2. Kline M & Kay IW (1965) *Electromagnetic Theory and Geometrical Optics* Interscience Press

3. Maxwell JC (1871-1873) On the condition that, in the transformation of any figure ... in three dimensions, every angle in the new figure shall be equal to the corresponding angle in the original figure *Proc London Math Soc* **4** p117

4. Forsyth AR (1905) *Differential Geometry* Cambridge University Press, 5th edition

5. Jeans J (1933) *Electricity and Magnetism* Cambridge University Press, 5th edition p279

6. Sommerfeld A (1964) *Lectures on Theoretical Physics* **IV Optics** New York & London: Academic Press p150

7. Baker HF & Copson ET (1953) *The Mathematical Theory of Huygens' Principle* Oxford University Press, England

8. Smirnov VI (1964) *Course of Higher Mathematics* **3** pt 2 New York: Pergamon Press Chapter 11, pp120-123 & 202-217

9. Bateman H (1912) Some general theorems connected with Laplace's equation and the equation of wave motion *Amer Jour Math* **34** p325

10. Jones DS (1979) *Methods in Electromagnetic Wave Propagation* Oxford University Press Chapter 8

11. Ursell P (1958) On the short wave asymptotic expansion of the wave equation *Proc Camb Phil Soc* **53** p113

12. Choudhary S & Felsen LB (1973) Asymptotic theory for inhomogeneous waves *IEEE Trans Ant & Prop* **AP-21** p827

13. Weinberg S (1962) Eikonal methods in magnetohydrodynamics *Phys Rev* **126** no 6, p1899

14. Froman N & Froman PO (1965) *JWKB Approximation* Amsterdam: North Holland Press

15. Heading J (1962) *Phase Integral Methods* Methuen Press, England

16. Saxon DS (1959) Modified WKB methods for the propagation and scattering of electromagnetic waves *Proc Symposium on EM Theory, Toronto, Canada*

17. Schulman LS (1981) *Techniques and Applications of Path Integrations* Addison Wesley, Reading, Massachusetts

18. Berry MV & Mount KE (1972) Semi-classical approximation in wave mechanics *Rep Prog Phys* **35** p315

19. Schiff L (1969) *Quantum Mechanics* New York: McGraw Hill p268

20. Buchdahl HA (1978) Kepler problem and the Maxwell fish-eye *Amer Jour Phys* **46** p840

21. Collinson CD (1973) Investigation of planetary orbits using the Runge-Lenz vector *Jour of Inst Math & Applications* London p377

22. Klein A (1978) Alternative formula for the classical deflection function in a central field *Amer Jour Phys* **46** p1019

23. Collas P (1981) Equivalent potentials in classical mechanics *Jour Math Phys* **22** p2512

24. Moser J (1970) Regularization of Kepler's problem and the averaging method on a manifold *Comm Pure & App Math* **XXIII** pp609-636

25. Eisenhart LP (1909) *Differential Geometry* Ginn & Co. Boston, Massachusetts

26. Hartwell GW (1909) Plane fields of force whose trajectories are invariant under a projective group *Trans Amer Math Soc* **10** p220

27. Scheffers G (1893) Verallgemeinerung der Grundlagen der gewohnich complexen Functionen *Rep Trans Roy Soc Sci* **45** Leipzig, Germany p828

28. Fueter R (1934-35 1936-7) On the analytical representation of regular functions of a quaternion variable *Comm Math Helvetica* **7** p307 **8** p371

29. Deschamps GA (1970) *Exterior Differential Forms in Mathematics Applied to Physics* Roubine (ed) Berlin: Springer Chapter 3

30. Mizushima M (1972) *Theoretical Physics* New York: Wiley p327

31. Cornbleet S (1971) On the uncertainty limit of high speed electronic scanning *IEE Conf Publication* **77** p83

32. Eddington AS (1943) *The Combination of Relativity Theory and Quantum Theory* Dublin: Institute of Advanced Studies **Series A** no 2

33. Schulman L (1975) *Caustics and Multivaluedness: Two results of adding path amplitudes in functional integration and its application* Arthurs AM (ed) Oxford University Press Chapter 12 p144

34. Wolf E (1978) Coherence and radiometry *Jour Opt Soc Amer* **68** p6

35. Sudarshan ECG (1979) Quantum theory of radiative transfer *Tech Rep Center for Particle Theory, University of Texas, Austin and Phys Lett* **37A** p269

36. Stavroudis ON (1972) *The Optics of Rays Wavefronts and Caustics* New York & London: Academic Press p196

37. Bateman H (1908) The conformal transformations of a space of four dimensions and their applications to geometrical optics *Proc Lond Math Soc* **7** p70

38. Buchdahl HA (1975) Conformal transformations and conformal invariance of optical systems *Optik* **43** p259

39. Cunningham E (1910) The principle of relativity in electrodynamics and an extension thereof *Proc Lond Math Soc* **8** p70

40. Rund H (1977) Clebsch potentials in the theory of electromagnetic fields admitting electric and magnetic charge distributions *Jour Math Phys* **18** p84

41. Sommerfeld A (1964) *Lectures on Theoretical Physics* **III Electrodynamics** New York & London: Academic Press p150

42. Imaeda K (1983) Quaternionic formulation of classical electrodynamics and theory of functions of a biquaternion variable *Report FPL-1-83-1* Fundamental Physics Laboratory, Dept of Electronic Science, Okayama University of Science, Japan

43. Milner SR (1963) Generalized electrodynamics and the structure of matter *Technical reports of the University of Sheffield, England*

44. Ehlers J, Rindler W & Robinson I (1966) *Quaternions, Bi-vectors and the Lorentz Group in Perspectives in Geometry and Relativity* Hoffman B (ed) Indiana: Indiana University Press p134

45. Bateman H (1955) *The Mathematical Analysis of Electrical and Optical Wave Motion* Dover Paperback s **14** New York: Dover

46. Hudson HP (1978) *Cremona Transformations* Cambridge University Press

47. Stavroudis ON (1972) *The Optics of Rays Wavefronts and Caustics* New York & London: Academic Press p281-297

48. Buchdahl HA (1978) Kepler problem and Maxwell fish-eye *Amer Jour Phys* **46** p840

49. Collas P (1981) Equivalent potentials in classical mechanics *Jour Math Phys* **22** p2512

50. Herzberger M (1958) *Modern Geometrical Optics* New York: Wiley Interscience

51. Klotz FS (1974) Twistors and the conformal group *Jour Math Phys* **15** p2242

52. Fulton T, Rohrlich F & Witten L (1962) Conformal invariance in physics *Rev Modern Physics* **34** no 3 p442

53. Rund H (ed) (1976) *Generalized Clebsch Representations on Manifolds in Topics in Differential Geometry* New York & London: Academic Press

54. Weingarten D (1973) Complex symmetries of electrodynamics *Annals of Physics* **76** p510

55. Stephani H (1974) Debye potentials in Riemannian spaces *Jour Math Phys* **15** no 1 p14

56. Coster HGL & Shepanski JR (1969) Gravito-inertial fields and relativity *Jour Phys A (Gen Phys)* ser 2 **vol 2** p22

57. de Witt BS (1963) Dynamical theory of groups and fields, in *Relativity, Groups & Topology* (eds de Witt C & de Witt B) *Les Houches Lectures on Theoretical Physics* New York: Gordon & Breach p659

58. Hestenes D (1966) *Space-Time Algebra* New York: Gordon and Breach

59. Post EJ (1962) *Formal Structure of Electromagnetics* Amsterdam: North Holland

60. Penney R (1968) Octonions and the Dirac equation *Amer Jour Phys* **36** p871

61. Edmonds JD Jnr (1973 et seq) *Quaternion Electrodynamics* Foundations of Physics

62. Typaldos ZA & Porgozelski RJ (1975) Quaternion calculus and the solution of Maxwell's equations *Math Notes* **43** University of California

63. Evans DD (1976) Complex variable theory generalized to electromagnetics: the theory of functions of a quaternion variable *PhD Thesis* University of California

64. Salingaros N (1981) Electromagnetism and the holomorphic properties of space-time *Jour Math Phys* **22** p1919

65. Synge JL (1950) Electromagnetics without metric *Pub Amer Monthly Soc*

66. Cornbleet S (1984) Geometrical optics reviewed: a new light on an old subject *Proc IEEE* **71** no 4 p471 (including 261 refs)

I
APPENDIX

This appendix deals in the main with the fundamental laws of geometrical optics and the derivation of the particular forms used as the basis for reflector and lens design in the earlier chapters of the book. To do this one has of necessity to draw on the many and varied versions of these laws that have appeared in many and varied notations over the entire history of the subject. Many formulae will be given without derivation or proof where it can be more simply observed that the result is consistent with the generally accepted version. One or two applications will be introduced as illustrative examples as they are relevant to the general subject even if not exactly optical designs with direct application to microwave antennas. The resultant forms of the laws so obtained, besides being fundamental to optical design, do permit generalization and extension. For complete proofs of the formulae a standard text or the references should be consulted [1] [2].

I.1 HAMILTON'S DIFFERENTIAL FORM

The basic definition of the laws of reflection and refraction can be derived from the principle of least action (and vice versa) and are most simply given by the relation [3]

$$\eta_r \, \mathrm{d}r = \eta_\rho \, \mathrm{d}\rho \tag{I.1.1}$$

in which from Figure I.1(a) η_r and η_ρ are the refractive indices of the media on the incident side with coordinates (r, θ) and on the transmission side with coordinates (ρ, ϕ) of a refracting surface.

For a perfect reflector we put the ratio $\eta_\rho/\eta_r = -1$ and Equation I.1.1 becomes the even more simple relation

$$\mathrm{d}r = \pm \, \mathrm{d}\rho \tag{I.1.2}$$

The ambiguity in sign has been introduced to define the nature of the reflecting surface where the negative sign applies to a surface creating a real focus and the positive sign for a virtual focus.

From the elementary geometry of Figure I.1(b) a relation equivalent to that of Equation I.1.2 can be seen to be

$$r \, \mathrm{d}\theta = \pm \rho \, \mathrm{d}\phi \tag{I.1.3}$$

where the sign in this case relates to the positive or negative value of the angle

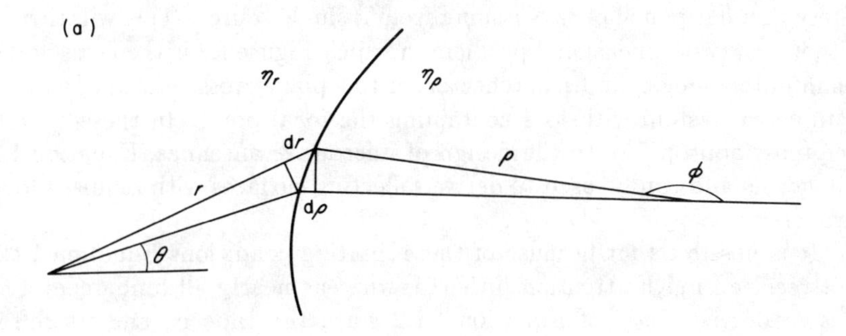

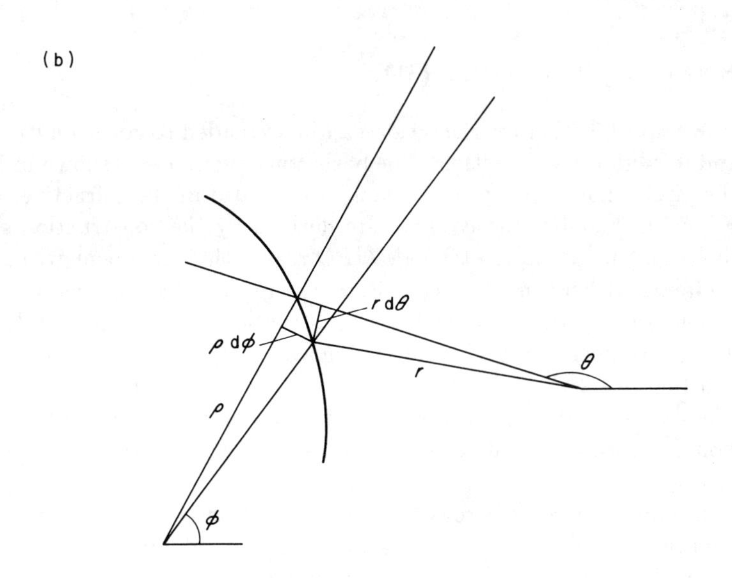

Figure I.1: (a) Refraction (b) Reflection

increment *. The familiar law of sines can be derived from Equation I.1.1 by division with the line increment ds of the surface.

We now consider the case where the surface concerned acts upon the entirety of a flat pencil of rays issuing from a single source. This will then apply either to a two-dimensional problem in which Figure I.2 is the cross-section of an infinitely long cylindrical reflector, or to a plane cross-section of a circularly symmetric system with axis containing the focal point. In these conditions, the most appropriate to the design of microwave antennas, Equation I.1.3 is integrable and can be used to derive reflecting surfaces with required focusing properties.

It is possible that because of these limiting conditions Equation I.1.3 has not received much attention hitherto, whereas nearly all fundamental optics has made much use of Equation I.1.2 since the time it, and its refractive counterpart Equation I.1.1, were presented by Hamilton.

Equation I.1.3 joins together the optics and the geometry of surfaces in a most intimate manner in which the geometrical properties of curves can be derived from known optical systems as well as the reverse.

I.2 SNELL'S CONSTRUCTION

Snell's own construction for refraction can be extended to cover multiple surfaces and total internal reflection. The basic construction is as shown in Figure I.2. The circles have radii proportional to the ratio of the refractive indices and the incident and refractive rays are derived by the construction shown. Applying the sine law to the triangle OPQ shows the agreement with Snell's law of refraction. For a multilayer system, the refracted ray in one medium is the incident ray for the next, and the construction can continue with piecewise straight segments and abutting triangles OP_1Q_1 OQ_1P_2 OP_2Q_2 etc. as shown in Figure I.2.

If the ray is incident from the side with the higher refractive index, the condition for total internal reflection, $\sin\theta = 1/\eta$, is derived immediately (Figure I.3)

We take now the case where two media with refractive indices η_1 and η_2 are separated by a small air gap as shown in Figure I.4. The ray is incident at an angle beyond the critical value and is totally internally reflected. That is, the normal ñ through P does not intersect the unit circle in real points. It does, however, do so at two conjugate imaginary points, given by the intersection of the line $y = \eta_1 \sin\theta$ with the circle

$$x^2 + y^2 = 1$$

that is, at points $x = \pm i(\eta_1^2 \sin^2\theta - 1)$, $y = \eta_1 \sin\theta$.

---

*The author is indebted to Mr SH Moss, Senior Mathematician, Marconi Space & Defence Systems Ltd for pointing out this vital relationship.

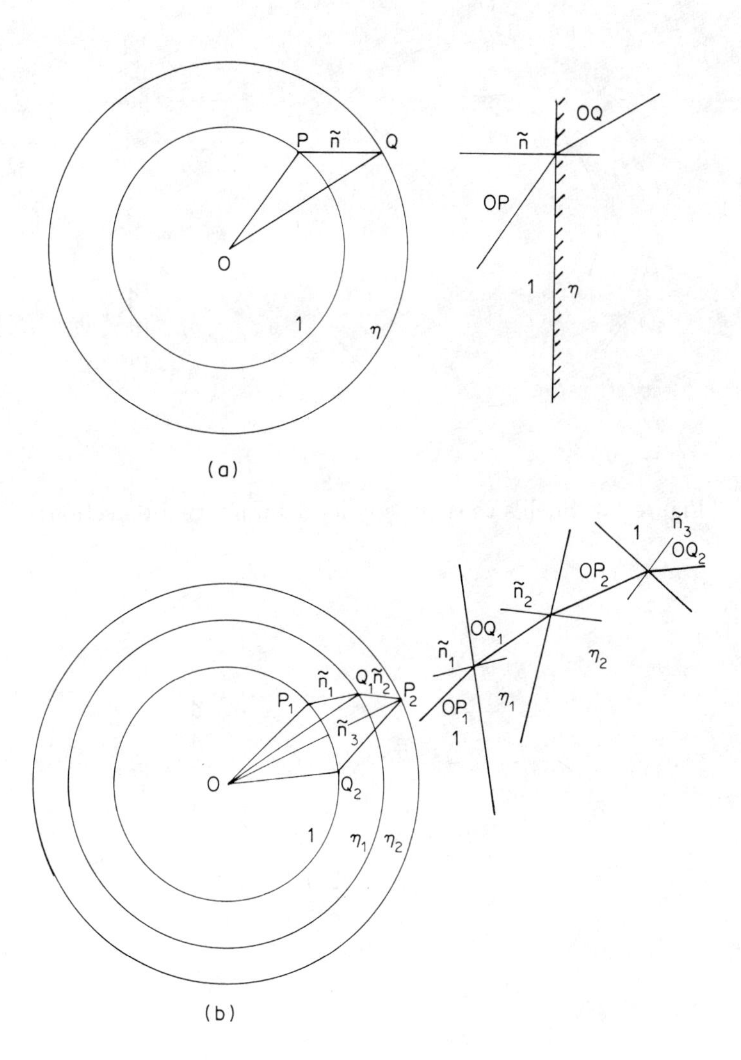

Figure I.2: (a) Snell's construction for refraction (b) Snell's construction for several surfaces

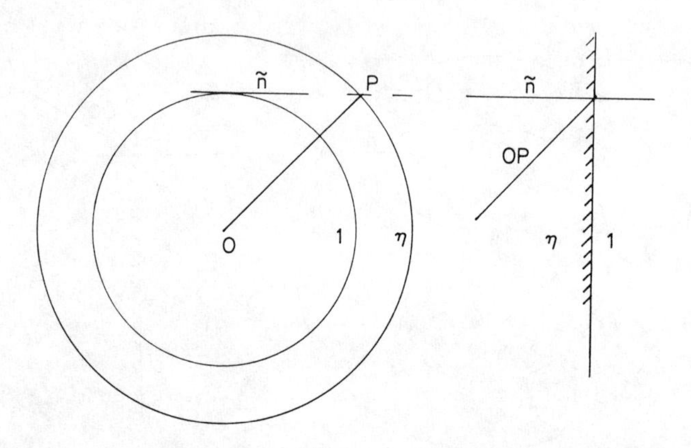

Figure I.3: Snell's construction for total internal reflection

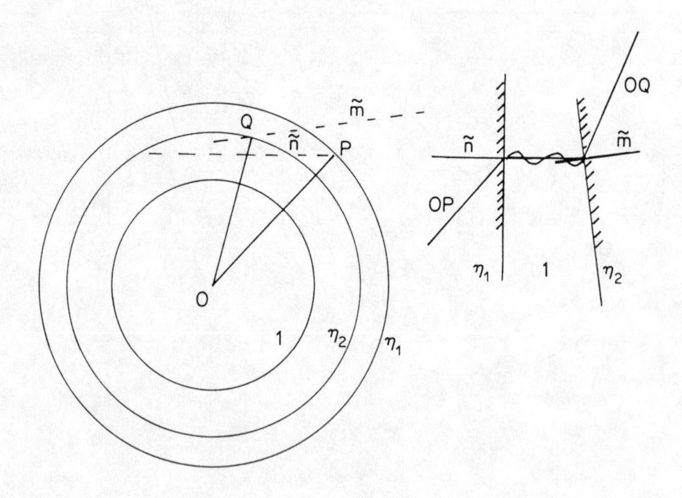

Figure I.4: Tunnelling rays in total internal reflection. ñ intersects the unit circle in complex conjugate points, and m̃ passes through this intersection and intersects the circle η_2 at the real point Q.

If the normal $\hat{\mathrm{m}}$ to the second surface passing through one of these points has a *real* intersection with the circle $x^2 + y^2 = \eta_2^2$, then a real ray will exist in the second medium as shown. However, its amplitude will be reduced by the exponential decay of evanescent fields along the path PQ, and this too can be obtained from the geometrical construction given.

I.3 CLASSICAL VECTOR FORMULATION

We use suffices i r and t throughout to denote the incident, reflected and transmitted (or refracted) sides of a surface of discontinuity between two media with refractive indices η_i and η_t accordingly. Unit vectors $\hat{\mathbf{s}}$ will denote the direction of rays, which, for homogeneous isotropic media, will be straight lines. For non-homogeneous media, where the rays are not taken as piecewise straight increments, $\hat{\mathbf{s}}$ will be tangential to a curved path at any point on that ray. The increment of path length in such a case will be ds. Position vectors from an origin will be r or $\boldsymbol{\rho}$. Where the ray itself is a radius vector from the origin, as in a source distribution, $\hat{\mathbf{s}}$ and ds may be replaced by $\hat{\mathbf{r}}$ and dr. It is in this form that they are used in Chapter 1. Care must be exercised in differentiating between $\hat{\mathbf{r}}$ and $\hat{\mathbf{s}}$ whenever these do not derive from the same origin.

The laws of reflection and refraction can be obtained rigorously from the solution of the boundary value problem for the incident plane electromagnetic wave upon an interface between two refracting media [4]. The geometrical optics approximation enables one to dispense with the infinities and treat a local region as if it were part of a tangential infinitely plane discontinuity. This is a good approximation for regions with radius of curvature that is a large number of wavelengths and consequently where a uniquely defined normal $\mathbf{n}$ exists. The alternative derivation of the same law by Fermat's principle does not at first sight appear to suffer from this limitation, but, as observed by Pauli [1] the fact that the reflected ray does not appear in this formulation shows that Fermat's principle is essentially an approximation to reality.

The result with the notation of Figure I.5 is

$$\hat{\mathbf{n}} \times \hat{\mathbf{s}}_i = \hat{\mathbf{n}} \times \hat{\mathbf{s}}_r \quad \text{and} \quad \hat{\mathbf{n}} \cdot \hat{\mathbf{s}}_i = -\hat{\mathbf{n}} \cdot \hat{\mathbf{s}}_r \tag{I.3.1}$$

$$\eta_i \hat{\mathbf{n}} \times \hat{\mathbf{s}}_i = \eta_t \hat{\mathbf{n}} \times \hat{\mathbf{s}}_t$$

which are Snell's laws of reflection, and the law of refraction.

The solution of the boundary value problem for a plane electromagnetic wave, shows, as does the above result, that the vectors are all co-planar.

Solving I.3.1 gives for a reflected ray

$$\hat{\mathbf{s}}_r = \hat{\mathbf{s}}_i - 2\hat{\mathbf{n}}(\hat{\mathbf{s}}_i \cdot \hat{\mathbf{n}}) \tag{I.3.2}$$

probably the most immediately useful of all the various forms of Snell's law of reflection.

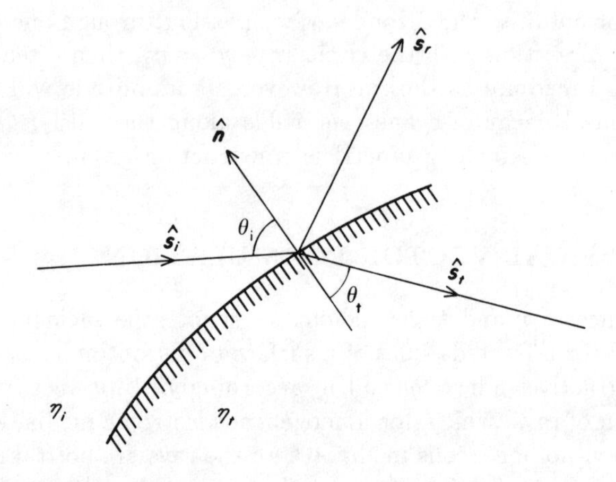

Figure I.5: The principle of reflection and refraction

The scalar form of Equation I.3.1 gives the familiar law of refraction

$$\eta_1 \sin\theta_i = \eta_t \sin\theta_t \tag{I.3.3}$$

To obtain a vector solution for the refracted ray we note that, since the incident and reflected rays are within the same medium, that is $\eta_i = \eta_r$ Equations I.3.1 combine to give

$$\hat{\mathbf{n}} \times \eta_i \hat{\mathbf{s}}_i = \hat{\mathbf{n}} \times \eta_r \hat{\mathbf{s}}_r = \hat{\mathbf{n}} \times \eta_t \hat{\mathbf{s}}_t \tag{I.3.4}$$

We define, in the light of this relation, the *directed* ray vector $\mathbf{t} = \eta\hat{\mathbf{s}}$ and thus relation I.3.4 states that the vector $\hat{\mathbf{n}} \times \mathbf{t}$ is an invariant of the refraction at a plane surface. With this notation Equation I.3.1 is

$$(\mathbf{t}_i - \mathbf{t}_t) \times \hat{\mathbf{n}} = 0 \tag{I.3.5}$$

which is the Sommerfeld-Runge relation [5]. Hence the vector $\mathbf{t}_i - \mathbf{t}_t$ is parallel to $\hat{\mathbf{n}}$ and there exists a scalar γ such that

$$\mathbf{t}_t = \mathbf{t}_i + \gamma\hat{\mathbf{n}} \tag{I.3.6}$$

Taking the cross product of Equation I.3.5 with $\hat{\mathbf{n}}$ gives

$$\mathbf{t}_t = \mathbf{t}_i + \hat{\mathbf{n}}(\mathbf{t}_t \cdot \hat{\mathbf{n}} - \mathbf{t}_i \cdot \hat{\mathbf{n}})$$

which on comparison with Equation I.3.6 shows that

$$\gamma = \mathbf{t}_t \cdot \hat{\mathbf{n}} - \mathbf{t}_i \cdot \hat{\mathbf{n}} = \eta_t \cos\theta_t - \eta_i \cos\theta_i \tag{I.3.7}$$

Using Equation I.3.3 to eliminate θ_t in Equation I.3.7 results in several equivalent forms for γ dependent only upon properties of the incident ray, which are

$$\begin{aligned}\gamma &= \sqrt{\eta_t^2 - \eta_i^2 \sin^2 \theta_i} - \eta_i \cos \theta_i \\ &= \sqrt{\eta_t^2 - \eta_i^2 (1 - \hat{\mathbf{n}} \cdot \hat{\mathbf{s}}_i^2)} - \eta_i \hat{\mathbf{n}} \cdot \hat{\mathbf{s}}_i \\ &= \sqrt{\eta_t^2 - \eta_i^2 (\hat{\mathbf{n}} \times \hat{\mathbf{s}}_i)^2} - \eta_i \hat{\mathbf{n}} \cdot \hat{\mathbf{s}}_i \\ &= \sqrt{\eta_t^2 - \eta_i^2 (\hat{\mathbf{n}} \times \hat{\mathbf{s}}_t)^2} - \eta_i \hat{\mathbf{n}} \cdot \hat{\mathbf{s}}_i \end{aligned} \tag{I.3.8}$$

Taking now the differential form of Equation I.3.6 we obtain

$$\mathrm{d}\mathbf{t}_t = \mathrm{d}\mathbf{t}_i + \gamma \mathrm{d}\hat{\mathbf{n}} + \hat{\mathbf{n}}\, \mathrm{d}\gamma \tag{I.3.9}$$

From Equation I.3.7 $\mathrm{d}\gamma = -\eta_t \sin \theta_t \, \mathrm{d}\theta_t + \eta_i \sin \theta_i \, \mathrm{d}\theta_i$ and from Equation I.1.3 $\eta_i \cos \theta_i \, \mathrm{d}\theta_i = \eta_t \cos \theta_t \, \mathrm{d}\theta_t$ so that θ_t and $\mathrm{d}\theta_t$ can be eliminated to give

$$\gamma = f(\theta_i)\, \mathrm{d}\theta_i$$

Since $\hat{\mathbf{n}}$ is a unit vector $\hat{\mathbf{n}} \cdot \mathrm{d}\hat{\mathbf{n}} = 0$ hence taking the dot product of Equation I.3.9 with $\hat{\mathbf{n}}$ there results

$$\hat{\mathbf{n}} \cdot \mathrm{d}\mathbf{t}_t - \hat{\mathbf{n}} \cdot \mathrm{d}\mathbf{t}_i = \mathrm{d}\gamma \tag{I.3.10}$$

Comparing this with the differential of Equation I.3.7 leaves

$$\mathbf{t}_t \cdot \mathrm{d}\hat{\mathbf{n}} - \mathbf{t}_t \cdot \mathrm{d}\hat{\mathbf{n}} = 0$$

which since $\mathrm{d}\hat{\mathbf{n}}$ is tangential to the surface of refraction is a restatement of the original Snell's law.

In the reference given [6] the following statement concludes this analysis:

> "In essence formula I.3.9 already contains the answer to all thinkable questions concerning the properties of the infinitesimal refracted pencil in terms of those of the incident one. It is enough to read it intelligently and to develop it appropriately in each particular case."

We can illustrate an application of Equation I.3.6 by the transmission of an oblique ray through a prism (Figure I.6). Consider two consecutive plane refractions at a prism with surface normals $\hat{\mathbf{n}}_1$ and $\hat{\mathbf{n}}_2$ and let the intersecting edge have direction $\hat{\mathbf{e}}$. Then since the refracted ray of the first surface becomes the incident ray of the second we have from Equation I.3.6

$$\mathbf{t}_t = \mathbf{t}_i + \gamma_1 \hat{\mathbf{n}}_1 \qquad \mathbf{t}_T = \mathbf{t}_t + \gamma_2 \hat{\mathbf{n}}_2$$

so that the final ray is given by

$$\mathbf{t}_T = \mathbf{t}_i + \gamma_1 \hat{\mathbf{n}}_1 + \gamma_2 \hat{\mathbf{n}}_2$$

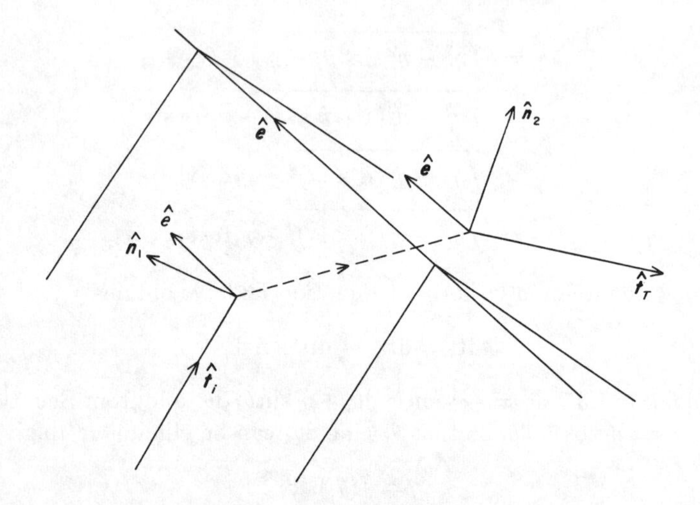

Figure I.6: Double refraction at a plane prism

Taking the scalar product with $\hat{\mathbf{e}}$

$$\mathbf{t}_T \cdot \hat{\mathbf{e}} = \mathbf{t}_i \cdot \hat{\mathbf{e}}$$

producing the well-known result that the incident and final rays make the same angle with the prism edge.

The same property can be shown to occur with a double mirror for a single reflection at each surface. In the case of a pure reflection we make $\eta_t = -\eta_i$ then from Equation I.3.7

$$\gamma = -2\eta_i \cos\theta_i$$

and hence from Equations I.3.2 or I.3.6 (with $\mathbf{t}_t$ replaced by $\mathbf{t}_r$)

$$\hat{\mathbf{s}}_r = \hat{\mathbf{s}}_i - 2\hat{\mathbf{n}}\cos\theta_i \tag{I.3.11}$$

Consider now two adjacent rays of a flat pencil of rays as in Figure 1.7 issuing from the point O. After refraction the rays (or their extensions) intersect again at the point F, and, letting the adjacent points PP' on the surface be a distance Δ apart then

$$\mathrm{d}s_i = \Delta\sin\theta_i \quad \text{and} \quad \mathrm{d}s_t = \Delta\sin\theta_t \tag{I.3.12}$$

and hence by virtue of Equation I.3.4

$$\eta_i\,\mathrm{d}s_i = \eta_t\,\mathrm{d}s_t \tag{I.3.13}$$

An ambiguity in sign has to be included in these results to relate the changes between $\mathrm{d}s_i$ and $\mathrm{d}s_t$ which will be seen to determine whether the image formed is virtual (on the same side as O as shown) or real (on the opposite side).

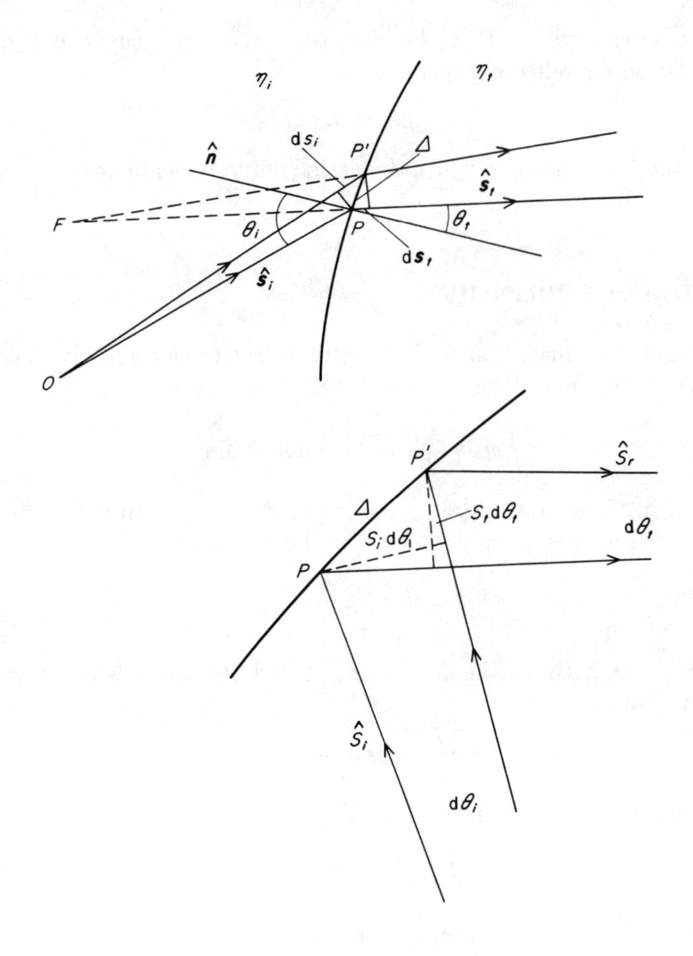

Figure I.7: Refraction and reflection of a ray pencil

Once again for a purely reflecting surface $\eta_i = -\eta_t$ and Equation I.3.13 becomes

$$\mathrm{d}s_i = \pm\,\mathrm{d}s_t \tag{I.3.14}$$

The two triangles with PP' as hypotenuse become congruent and give the further relation for reflecting surfaces

$$s_i\,\mathrm{d}\theta_i = s_t\,\mathrm{d}\theta_t \tag{I.3.15}$$

These are the relations upon which the design of fundamental focusing systems are based in Chapter 1.

I.4 FERMAT'S PRINCIPLE

The principle of Fermat, from which the above relations could also have been derived [7], states that along a true ray path the function

$$\int \eta\,\mathrm{ds} \qquad \text{is an extremum}$$

Consider as in Figure 1.8 several refracting surfaces separating discrete media with refractive indices η_1, η_2... Then from Fermat's principle

$$\delta \int_A^B \eta\,\mathrm{ds} = 0$$

for all ray paths between the end points A and B (one of which may be at infinity) we have

$$\int_A^B \eta\,\mathrm{ds} = \text{const}$$

which for the piecewise straight ray in Figure I.8 becomes

$$\sum_n \eta_n\,\mathrm{d}s_n = \text{const} \tag{I.4.1}$$

with the correct sign attributed to each term.

In the case of a sequence of reflectors we can use Equation I.3.15 to give

$$\sum_n s_n\,\mathrm{d}\theta_n = \text{const} \tag{I.4.2}$$

upon which the design of optical systems of more than one surface can be based, as for example, in Section 2.10.

The basic form of Equation I.3.6 is

$$\eta_t\hat{\mathbf{s}}_t - \eta_i\hat{\mathbf{s}}_i = \gamma\hat{\mathbf{n}} \tag{I.4.3}$$

at a single refracting surface. If this surface has the equation

$$\phi(x, y, z) = 0$$

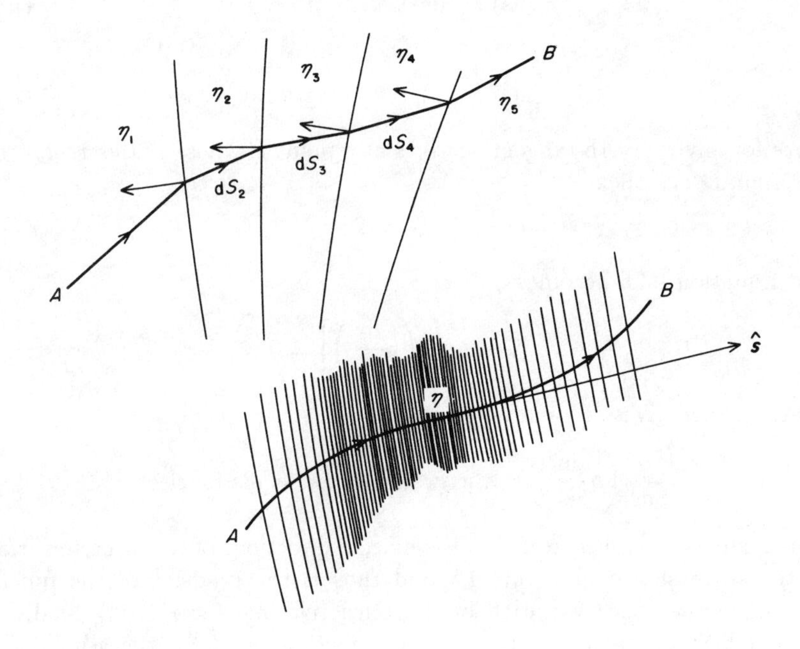

Figure I.8: Ray paths in discontinuous and continuous variable media

then $\hat{\mathbf{n}}$ is proportional to $\nabla\phi$ and hence the curl of Equation I.4.3 will be zero giving

$$\nabla \times (\eta_t \hat{\mathbf{s}}_t) = \nabla \times (\eta_i \hat{\mathbf{s}}_i) \qquad \text{(I.4.4)}$$

Continuing this, first for a system of closely packed refracting surfaces as in Figure I.8 and then for a continuously varying refracting medium will give

$$\nabla \times (\eta \hat{\mathbf{s}}) = \text{const along a ray path} \qquad \text{(I.4.5)}$$

or

$$\frac{\mathrm{d}}{\mathrm{d}s}(\nabla \times \eta \hat{\mathbf{s}}) = 0 \text{ along a ray} \qquad \text{(I.4.6)}$$

Since for any ray with tangent vector $\hat{\mathbf{s}}$ at a point P whose *radius vector* from an origin O is $\mathbf{r}$ then

$$\frac{\mathrm{d}\mathbf{r}}{\mathrm{d}s} = \hat{\mathbf{s}}$$

and Equation I.4.6 becomes

$$\nabla \times \left[\frac{\mathrm{d}}{\mathrm{d}s}\left(\eta \frac{\mathrm{d}\mathbf{r}}{\mathrm{d}s}\right)\right] = 0 \qquad \text{(I.4.7)}$$

which in turn gives

$$\frac{\mathrm{d}}{\mathrm{d}s}\left(\eta \frac{\mathrm{d}\mathbf{r}}{\mathrm{d}s}\right) = \text{the gradient of a space function}$$

This gradient must lie parallel to the normal directions of the discrete surfaces in the stage shown in Figure I.8 and thus is the gradient of the function describing the space variation of the refractive index, so giving finally the result which forms the basis of the ray method used for the study of the optics of non-homogeneous media in Chapter 4 namely

$$\frac{\mathrm{d}}{\mathrm{d}s}\left(\eta \frac{\mathrm{d}\mathbf{r}}{\mathrm{d}s}\right) = \text{grad } \eta \qquad \text{(I.4.8)}$$

A more rigorous derivation of this result can be found in many places in the literature of optics for example [2] or Section 4.1.

I.5 THE THEOREM OF MALUS AND DUPIN

This is given in [6, p130], as an application of Fermat's principle and in [7, p116], as a consequence of Equation I.4.4. That it derives from the latter is a point of great interest in the derivation of the four dimensional analysis. We will give only an adapted statement of the theorem here for reference purposes.

If a system of rays, issuing in the first instance from some source point and thus having a spherical wave front about that point, is refracted or reflected any number of times, and if equal optical paths are measured along each ray, from either the source itself or any of its original wave fronts, then the surface so formed by the end points of the equal optical path rays will be orthogonal to the system of rays and form a wave front.

This interprets the usual statement that a normal congruence (q.v.) (or orthotomic system of rays) will remain a normal congruence after repeated reflection or refraction.

I.6 THE EIKONAL EQUATION

We take as starting point Equation I.4.5 and choosing the arbitrary constant to be zero we have

$$\nabla \times (\eta \hat{\mathbf{s}}) = 0 \tag{I.6.1}$$

or

$$\eta \hat{\mathbf{s}} = \nabla S$$

where S is a function of the space coordinates, and also the consequence that the unit vectors $\hat{\mathbf{s}}$ are perpendicular to the surfaces $S(x, y, z) = \text{const.}$

Since $\hat{\mathbf{s}}$ is tangential to the ray at any point these surfaces are the normal surfaces referred to in the theorem of Malus and Dupin and are thus the wave fronts of the propagating field.

Thus from Equation I.6.1 we have (Equation 4.1.8)

$$\nabla S \cdot \nabla S = \eta^2 \quad \text{or} \quad |\nabla S| = \eta \tag{I.6.2}$$

From the integral definition of curl, Equation I.6.1 implies

$$\int \eta \hat{\mathbf{s}} \cdot \mathrm{d}\mathbf{s} = 0$$

for any closed path within the medium and hence

$$\int_A^B \eta \hat{\mathbf{s}} \cdot \mathrm{d}\mathbf{s} = S_B - S_A \tag{I.6.3}$$

independently of the path taken. This can be seen to be a restatement of the fundamental principle of Fermat $\delta \int \eta \, ds = 0$ along a true ray path. S is the eikonal function (for example [7, p131]) and its analogy with the potential function in the theory of conservative force fields demonstrated by Equation I.6.3 provides a further connection between the theory of ray paths and particle trajectories as discussed at more than one point in the text.

Equation I.6.2 in a Cartesian coordinate system becomes

$$\left(\frac{\partial S}{\partial x}\right)^2 + \left(\frac{\partial S}{\partial y}\right)^2 + \left(\frac{\partial S}{\partial z}\right)^2 = \eta^2 \tag{I.6.4}$$

The same relation can be obtained by the substitution of an assumed solution into the scalar wave equation

$$\nabla^2 u + k^2 u = 0$$

where $k = \omega\sqrt{\mu\varepsilon} = 2\pi/\lambda$ and μ ε are the permeability and the permittivity of the medium, λ the wavelength of the radiated wave in the medium. Using the suffix 0 for the same parameters in a vacuum, we substitute

$$u = A\exp(-ik_0 S)$$

then since $\eta = k/k_0$ there results

$$\nabla^2 u + k^2 u = -k_0^2 u\left[\left(\frac{\partial S}{\partial x}\right)^2 + \left(\frac{\partial S}{\partial y}\right)^2 + \left(\frac{\partial S}{\partial z}\right)^2 - \frac{k^2}{k_0^2}\right]$$
$$+2ik_0 u\left[\frac{1}{2}\nabla^2 S + \frac{1}{A}\text{grad } A \cdot \text{grad } S\right] + \ldots \qquad \text{(I.6.5)}$$

The remaining terms remain finite in the geometrical optics limit that is when $k_0 \to \infty$. We thus have an illustration of the approximate solution we have been using in that making the first term zero gives us Equation I.6.2 which derives from Fermat's principle. We further require the second term to be zero or

$$\frac{1}{A}\text{grad } A \cdot \text{grad } S = -\frac{1}{2}\nabla^2 S \qquad \text{(I.6.6)}$$

Since $\nabla S = \eta\hat{\mathbf{s}}$ and grad $A \cdot \hat{\mathbf{s}} = \partial A/\partial s$

$$\frac{\mathrm{d}A}{\mathrm{d}s} + \frac{1}{2}\frac{\nabla^2 S}{\eta}A = 0$$

or

$$A(s) = A(0)\exp\left(-\frac{1}{2}\int_0^s \frac{\nabla^2 S}{\eta}\,\mathrm{d}s\right) \qquad \text{(I.6.7)}$$

The exponential factor can be expressed in terms of the Gaussian curvature of the wave front at the two chosen points with parameters zero and s [8]. If the principle radii of curvature are R_1 and R_2 and $K = 1/R_1 R_2$ then

$$\exp\left(-\frac{1}{2}\int_0^s \frac{\nabla^2 S}{\eta}\,\mathrm{d}s\right) = \sqrt{\frac{K(s)}{K(0)}} \qquad \text{(I.6.8)}$$

These relations can be seen to be a statement of the conservation of energy. A full discussion is contained in [7] which concludes with the interesting result that the orthogonal triad of vectors associated with a curved ray in space, that is the tangent vector $\hat{\mathbf{s}}$, the normal $\hat{\mathbf{n}}$ and the bi-normal $\hat{\mathbf{b}}$, satisfy "pseudo-Maxwell" equations in which $\eta\hat{\mathbf{n}}$ and $\eta\hat{\mathbf{b}}$ play the part of $\mathbf{E}$ and $\mathbf{H}$ and $\hat{\mathbf{s}}$ is the Poynting vector $\mathbf{E} \times \mathbf{H}$.

I.7 THE REFLECTION DYADIC

The law of reflection given in Equation I.1.2 is

$$\hat{\mathbf{s}}_r = \hat{\mathbf{s}}_i - 2\hat{\mathbf{n}}(\hat{\mathbf{s}}_i \cdot \hat{\mathbf{n}})$$

and can be written with a change in notation as

$$\hat{\mathbf{s}}_r = [I - 2(\hat{\mathbf{n}}\hat{\mathbf{n}})]\hat{\mathbf{s}}_i \tag{I.7.1}$$

where I is an identity operator or idemfactor and $(\hat{\mathbf{n}}\hat{\mathbf{n}})$ is a dyadic. The properties of the reflection dyadic

$$\Omega = I - 2(\hat{\mathbf{n}}\hat{\mathbf{n}})$$

can be obtained by the comparison between these two equations. It can be shown that this operator has the necessary algebraic properties of distributivity and associativity required for the following but not commutativity [9]. Then to determine the direction of a final ray after a number of intermediate reflections we have

$$\hat{\mathbf{s}}_{r_n} = \Omega_n \Omega_{n-1} \ldots \Omega_1 \hat{\mathbf{s}}_i \tag{I.7.2}$$

By the associativity property these may be bracketed in any way (without disturbing the order of course) that affords simplification of the result.

The angle between the incident and final rays will then be ψ where

$$\cos \psi = \hat{\mathbf{s}}_i \Omega \hat{\mathbf{s}}_i \tag{I.7.3}$$

where Ω stands for the product of all the operators in Equation I.7.2. For example the double mirror system with normals $\hat{\mathbf{n}}_1$ and $\hat{\mathbf{n}}_2$ has

$$\begin{aligned} \Omega = \Omega_2\Omega_1 &= [I - 2(\hat{\mathbf{n}}_2\hat{\mathbf{n}}_2)][I - 2(\hat{\mathbf{n}}_1\hat{\mathbf{n}}_1)] \\ &= I - 2(\hat{\mathbf{n}}_2\hat{\mathbf{n}}_2) - 2(\hat{\mathbf{n}}_1\hat{\mathbf{n}}_1) + 4\hat{\mathbf{n}}_1 \cdot \hat{\mathbf{n}}_2(\hat{\mathbf{n}}_1\hat{\mathbf{n}}_2) \end{aligned}$$

so that

$$\hat{\mathbf{s}}_{r_2} = \hat{\mathbf{s}}_i - 2\hat{\mathbf{s}}_i \cdot \hat{\mathbf{n}}_2\hat{\mathbf{n}}_2 - 2\hat{\mathbf{s}}_i \cdot \hat{\mathbf{n}}_1\hat{\mathbf{n}}_1 + 4\hat{\mathbf{n}}_1 \cdot \hat{\mathbf{n}}_2\hat{\mathbf{s}}_i \cdot \hat{\mathbf{n}}_1\hat{\mathbf{n}}_2 \tag{I.7.4}$$

If the line of the intersection of the mirrors has direction $\hat{\mathbf{e}}$ then $\hat{\mathbf{e}} \cdot \hat{\mathbf{n}}_1 = \hat{\mathbf{e}} \cdot \hat{\mathbf{n}}_2 = 0$ and $\hat{\mathbf{e}} \cdot \hat{\mathbf{s}}_1 = \hat{\mathbf{e}} \cdot \hat{\mathbf{s}}_i$ and the ray undergoes a rigid rotation about $\hat{\mathbf{e}}$ as was the case for the prism.

Since $\Omega_1\Omega_2 \neq \Omega_2\Omega_1$ in general, a beam of rays wide enough to illuminate both sides will separate into two reflected beams. If the mirrors are at right angles the operator Ω becomes self-conjugate and the order of reflections is immaterial. However, no such double mirror can reverse the direction of *every* incident ray as is the case with (some) triple mirrors.

I.8 THE REFRACTION DYADIC [10]

With regard to refraction the situation is more complex. From Equations I.1.6 and I.1.8 we obtain

$$\hat{\mathbf{s}}_t = \frac{\eta_i}{\eta_t}\hat{\mathbf{s}}_i - \frac{\eta_i}{\eta_t}\hat{\mathbf{n}}(\hat{\mathbf{n}}\cdot\hat{\mathbf{s}}_i) + \hat{\mathbf{n}}\sqrt{1 - \frac{\eta_i^2}{\eta_t^2} + \frac{\eta_i^2}{\eta_t^2}(\hat{\mathbf{n}}\cdot\hat{\mathbf{s}}_i)^2} \tag{I.8.1}$$

which can be seen to be Snell's law by taking the scalar product with $\hat{\mathbf{n}}$ to give

$$\cos\theta_t = \sqrt{1 - \frac{\eta_i^2}{\eta_t^2}\sin^2\theta_i}$$

Putting

$$\frac{\eta_i}{\eta_t} = \beta \qquad \hat{\mathbf{n}}\cdot\hat{\mathbf{s}}_i = -p \qquad \text{and} \qquad \frac{\eta_i}{\eta_t}(\hat{\mathbf{n}}\cdot\hat{\mathbf{s}}_i - \cos\theta_t) = \alpha$$

Equation I.8.1 becomes

$$\alpha = \beta p\left[1 - \sqrt{1 + \frac{1-\beta^2}{\beta^2 p^2}}\right]$$

or

$$\alpha = \beta p q \tag{I.8.2}$$

Then Equation I.1.6 is simply

$$\hat{\mathbf{s}}_t = \beta[I - q\hat{\mathbf{n}}\hat{\mathbf{n}}]\hat{\mathbf{s}}_i \tag{I.8.3}$$

whence $R = \beta[I - q\hat{\mathbf{n}}\hat{\mathbf{n}}]$ is the refraction dyadic. As noted by Silberstein [6], q contains p explicitly and this fact prevents R from being distributive. It does however have a defined inverse which is found to be [10]

$$\frac{1}{R} = \frac{1}{\beta}\left[1 - \frac{q}{q-1}\hat{\mathbf{n}}\hat{\mathbf{n}}\right] \tag{I.8.4}$$

This is the reversal of the ray and hence is obtained from R by the transformation $\eta_i \leftrightarrow \eta_t$ and $\theta_i \leftrightarrow \theta_t$.

It is thus limited by the condition for total internal reflection since

$$\frac{q}{q-1} = 1 - \sqrt{1 + \frac{\beta^2 - 1}{1 - \beta^2 + \beta^2 p^2}}$$

becomes imaginary when $\eta_i/\eta_t < \sin\theta_t$.

I.9 THE REFLECTION QUATERNION

Reverting to the basic relations of Snell's law for reflection given in Equation I.1.1 which are (noting change of cross product)

$$\hat{\mathbf{n}} \cdot \hat{\mathbf{s}}_i = -\hat{\mathbf{n}} \cdot \hat{\mathbf{s}}_r \qquad \hat{\mathbf{n}} \times \hat{\mathbf{s}}_r = -\hat{\mathbf{s}}_i \times \hat{\mathbf{n}}$$

and adding we obtain

$$\begin{aligned} -\hat{\mathbf{n}} \cdot \hat{\mathbf{s}}_r + \hat{\mathbf{n}} \times \hat{\mathbf{s}}_r &= \hat{\mathbf{n}} \cdot \hat{\mathbf{s}}_i - \hat{\mathbf{s}}_i \times \hat{\mathbf{n}} \\ &= -[-\hat{\mathbf{s}}_i \cdot \hat{\mathbf{n}} + \hat{\mathbf{s}}_i \times \hat{\mathbf{n}}] \end{aligned} \tag{I.9.1}$$

The combination of dot and cross product typifies the product of two quaternions which have zero scalar elements (Appendix IV).

Calling them $N \equiv (0, \hat{\mathbf{n}})$ $R \equiv (0, \hat{\mathbf{s}}_r)$ and $I = (0, \hat{\mathbf{s}}_i)$ Equation I.9.1 is $NR = -IN$ where quaternion algebra has to be used.

Consequently $R = N^{-1}IN$ for a single reflection. Again in a system of multiple reflections we have after m reflections

$$R_m = (-1)^m N_m^{-1} N_{m-1}^{-1} \ldots N_1^{-1} I N_1 \ldots N_m \tag{I.9.2}$$

or

$$R_m = (-1)^m Q^{-1} I Q$$

where Q is the single quaternion, the product of $N_1 \ldots N_m$.

As always happens with such condensed versions of operator algebras and tensors, there is as much complexity in calculating the final result as in other longhand versions. The procedure is however straightforward though complicated vector algebra.

We give as illustration the result for the three mirror system illustrated in Figure I.9. The mirror in the x, y plane has a 90° degree corner and two mirrors intersect it along the x and y axes with inclinations θ and ϕ respectively. Due to the non-commutative property of quaternion products the order of the mirror reflections affects the final result (as is physically obvious). Taking a reflection first from the "θ" mirror, then the "ϕ" mirror and finally the horizontal mirror, we find for a ray with original direction cosines (l, m, n) that the reflected ray has direction cosines given by

$$\begin{aligned} \xi &= l\cos 2\phi + m\sin 2\theta \sin 2\phi - n\cos 2\theta \sin 2\phi \\ \eta &= m\cos 2\theta + n\sin 2\theta \\ \zeta &= -l\sin 2\phi + m\sin 2\theta \cos 2\phi - n\cos 2\theta \cos 2\phi \end{aligned} \tag{I.9.3}$$

For example with $\theta = \phi = \pi/2$ we have the usual 90^0 corner reflector and $(\xi, \eta, \zeta) = (-l, -m, -n)$ that is the ray is returned along its original direction.

Other interesting results can be obtained by putting $\theta = \pi/2$ $\phi = \pi/4$ and $\theta = \pi/4$ $\phi = \pi/2$ in turn in Equation I.9.3, giving reflections from the semi-octant corner reflector. For a full appraisal however the result is required

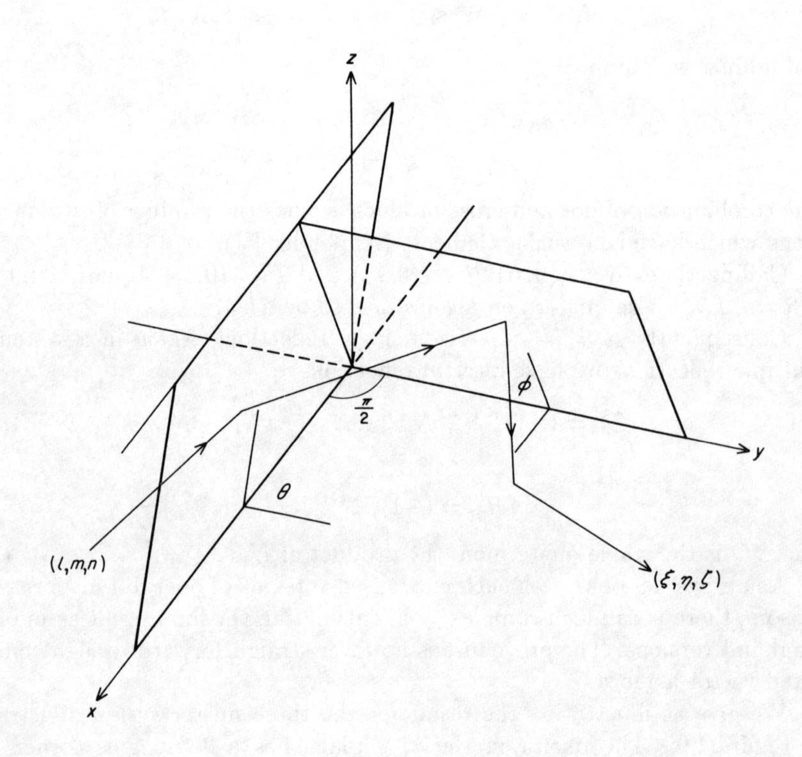

Figure I.9: Reflection in a triple mirror

for all combinations of the order of reflections. If, in the example given, the incidence was first upon the "ϕ" mirror and second upon the "θ" mirror the result is a transformation of Equations I.9.3 by $l \leftrightarrow m$; $\xi \leftrightarrow \eta$; $\theta \leftrightarrow \phi$.

A study of the more complex refraction quaternion has been made by Wagner [9].

I.10 THE REFLECTION MATRIX

A quaternion is an algebraic form of a matrix or tensor operator in much the same way as a complex number is an algebraic form of vector. Hence the results of the previous paragraph will have a matrix version and this is to be found in the work of Beggs [11].

For a plane mirror whose surface is given by the equation

$$Ax + By + Cz + D = 0$$

the direction cosines of the reflected ray are related to those of the incident ray by the matrix M given by

$$\begin{pmatrix} \xi \\ \eta \\ \zeta \end{pmatrix} = \begin{pmatrix} 1 - 2A^2/F^2 & -2AB/F^2 & -2AC/F^2 \\ -2AB/F^2 & 1 - 2B^2/F^2 & -2BC/F^2 \\ -2AC/F^2 & -2BC/F^2 & 1 - 2C^2/F^2 \end{pmatrix} \begin{pmatrix} l \\ m \\ n \end{pmatrix} \qquad (I.10.1)$$

where $F^2 = A^2 + B^2 + C^2$.

Successive reflections are then calculated via the successive multiplication of reflection matrices.

The main role played by this combination of plane mirrors in the field of microwaves is in the design of corner reflectors or clusters of corner reflectors [12] [13]. Such arrangements are necessary to smooth out some of the blind spots which may occur with the single 90° trihedral corner. The connection between such systems and the crystal symmetries is fairly obvious and thus it is of no surprise that some of the methods used here make their appearance in those subjects also and have been studied by authors more renowned in those fields [14].

By regarding the reflections of a ray from a system of mirrors as a single rigid rotation of the direction of the ray by some angle about an axis, both of which can be determined by the methods given, shows again the connection that exists between the quaternions for example, the rotations in three (and four) space and the symmetry groups [15] [16] [17].

The refractive matrix can be obtained in detail from the Mueller-Stokes formulations of Fresnel's equations in [18].

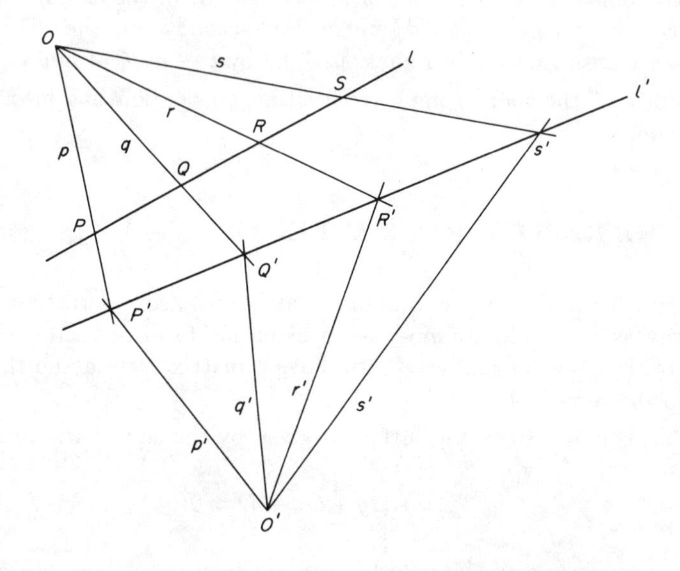

Figure I.10: Geometry of the projectivity

I.11 THE REFRACTIVE PROJECTIVITY

Four co-planar straight lines through a common point O, as shown in Figure I.10, having a transversal l which intersects them in points P Q R and S have the elementary geometrical property

$$PR/OP = \sin(pr)/\sin(ORP)$$

indicating by (pr) the angle between the orientated lines p and r (taken counter clockwise for positive angles). So for a single ratio

$$\frac{PR}{QR} = \frac{OP \sin(pr)}{OQ \sin(qr)}$$

Indicating this ratio by the symbol (PQR) then for a ratio of ratios

$$(PQRS) = \frac{PQR}{PQS} = \frac{\sin(pr)}{\sin(qr)} \bigg/ \frac{\sin(ps)}{\sin(qs)} = \frac{pqr}{pqs} = (pqrs) \qquad \text{(I.11.1)}$$

This expression defines the anharmonic or cross-ratio between the points $PQRS$ or the lines $pqrs$. As can be seen it is dependent only upon the angles between the lines and not upon the position of the transversal l. It forms a condition between the four entities, so that for a given cross-ratio *any* three points or lines, will determine the unique fourth member.

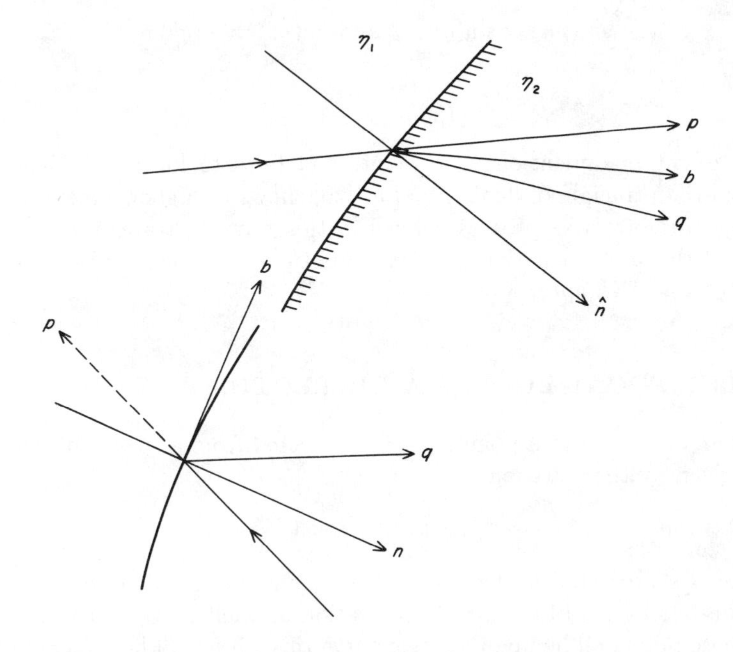

Figure I.11: Refraction and reflection projectivities

The property remains constant for any other transversal l' and by the uniqueness property for any other point of common origin O' whether in the same plane or not. The proof of this fundamental property comes from an application of Desargue's theorem [19]. The relation in Equation I.11.1 is said to be invariant under the projection from a point, or the points (or lines) are a projectivity.

In the case of refraction of a ray between two media with refractive indices η_1 and η_2 let the incident ray be p and the refracted ray q, the line bisecting the angle between them b and the normal to the surface n (Figure I.11) then as Cambi [20] shows

$$\frac{\sin(pn)}{\sin(qn)} = \frac{\eta_2}{\eta_1} \quad \text{and} \quad \frac{\sin(pb)}{\sin(qb)} = -1$$

and hence $(pqnb) = -\eta_2/\eta_1$.

Thus Snell's law defines a projectivity on the refracting surface which thus remains invariant under projective (and involutary) transformations of the plane (or line) containing the rays (or points) and its cross-section of the refracting surface.

In the case of a reflector

$$\sin(pn) = \sin(qn) \quad \text{and} \quad (pb) = -(qb)$$

thus

$$(pqnb) = -1$$

The ascent in dimension from points and lines to lines and planes follows from the principle of duality and as [20] shows to higher dimensionality. Therefore the refractive process defined in this way illustrates the relevance of some of the transformations, for example the inversions, conjectured as a design process in Chapter 3.

I.12 THE FOCAL LINE OF A REFLECTOR

If a system of rays from a point source is reflected from a specified reflecting surface given by the equation

$$f(x, y, z) = \text{const}$$

then the rays projected backwards will meet any specified surface at points which can be obtained by a direct application of Snell's law. For those cases where these points all lie upon a single curve this curve will be a focal line for the reflector and surface given. Such is the case in particular for cylindrical reflectors and the perpendicular plane through the source. The focal line in such cases is also a cross-section of the zero distance wave-front which is required for the application of Damien's theorem.

Normals to the surface at a point $P(a, b, c)$ have the direction

$$\hat{\mathbf{n}} = \frac{\text{grad } f_p}{|\text{grad } f|_p}$$

The incident ray $\hat{\mathbf{s}}_i$ from a point $(d, 0, 0)$ will have direction

$$\hat{\mathbf{s}}_i = \frac{(a-d)\mathbf{i} + b\mathbf{j} + c\mathbf{k}}{\sqrt{(a-d)^2 + b^2 + c^2}} \tag{I.12.1}$$

Hence the reflected ray will be in the direction

$$\hat{\mathbf{s}}_r = \frac{1}{\sqrt{(a-d)^2 + b^2 + c^2}} \left[(a-d)\mathbf{i} + b\mathbf{j} + c\mathbf{k} \right.$$
$$\left. - \frac{2\left(\frac{\partial f}{\partial x}\mathbf{i} + \frac{\partial f}{\partial y}\mathbf{j} + \frac{\partial f}{\partial z}\mathbf{k}\right)\left((a-d)\frac{\partial f}{\partial x} + b\frac{\partial f}{\partial y} + c\frac{\partial f}{\partial z}\right)}{\left(\frac{\partial f}{\partial x}\right)^2 + \left(\frac{\partial f}{\partial y}\right)^2 + \left(\frac{\partial f}{\partial z}\right)^2} \right] \tag{I.12.2}$$

where all the differentials are evaluated at the point P.

Letting

$$\left(\frac{\partial f}{\partial x}\right)^2 + \left(\frac{\partial f}{\partial y}\right)^2 + \left(\frac{\partial f}{\partial z}\right)^2 = D$$

and

$$(a-d)\frac{\partial f}{\partial x} + b\frac{\partial f}{\partial y} + c\frac{\partial f}{\partial z} = B$$

the direction cosines of $\hat{\mathbf{s}}_r$ are thus proportional to

$$\begin{aligned} l &\equiv (a-d)D - 2\frac{\partial f}{\partial x}B \\ m &\equiv bD - 2\frac{\partial f}{\partial y}B \\ n &\equiv cD - 2\frac{\partial f}{\partial z}B \end{aligned} \tag{I.12.3}$$

The equation of the ray is therefore

$$\frac{x-a}{l} = \frac{y-b}{m} = \frac{z-c}{n} \tag{I.12.4}$$

This line meets a specified surface in a point which varies as the point (a, b, c) is moved on the reflector. One variable may be eliminated in these relations therefore by the condition $f(a, b, c) = \text{const.}$ For certain surfaces and reflectors two of the variables can be eliminated resulting in a single parameter system of points which is the focal line.

I.13 REFLECTION AND REFRACTION AT A LOSSY MEDIUM

The boundary value solution giving rise to the form of Snell's law for refraction so far considered was expressly for the incidence of a wave from free space or medium with refractive index η_i upon a semi-infinite lossless medium with refractive index η_t. The following result due to Bell at al [21] generalizes this result for the situation where the second medium has a loss mechanism and is thus characterised by a complex refractive index

$$\eta_t = \eta_2 + iK \tag{I.13.1}$$

In the notation of this appendix the generalized Snell's law is given as

$$\sin\theta_t = [\eta_1/\eta_2 \sin\theta_i]\sqrt{F} \tag{I.13.2}$$

where F is given as

$$F = -(\eta_2^2 - K^2 + \eta_1^2 \sin^2\theta_i) + \frac{\sqrt{(\eta_2^2 - K^2 + \eta_1^2\sin^2\theta_i)^2 + 4[\eta_2^2K^2 - \eta_1^2\sin^2\theta_i(\eta_2^2 - K^2)]}}{2(K^2 - \eta_1^2\sin^2\theta_i(1 - K^2/\eta_2^2))} \tag{I.13.3}$$

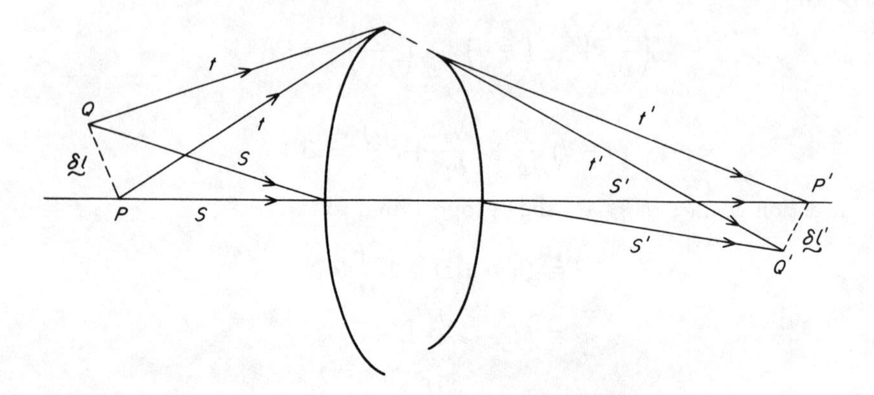

Figure I.12: Ray paths for small displacements of source

when $K = 0$ this reduces to the law for lossless media.

The Fresnel coefficients for this case for two conditions of reflection from a lossy interface are

1. for polarization perpendicular to the plane of incidence

$$R_{\perp} = \frac{\sin^2(\theta_i - \theta_t) + \left(\frac{\eta_2 K}{\eta_1}\right)^2 \left(\frac{\sin^2\theta_t}{\sin\theta_i \cos\theta_t}\right)^2}{\sin^2(\theta_i + \theta_t) + \left(\frac{\eta_2 K}{\eta_1}\right)^2 \left(\frac{\sin^2\theta_t}{\sin\theta_i \cos\theta_t}\right)^2} \tag{I.13.4}$$

2. for polarization parallel to the plane of incidence

$$R_{\parallel} = \frac{N \text{ with upper signs}}{N \text{ with lower signs}}$$

where

$$N = \tan^2(\theta_i \mp \theta_t) + \left[\frac{\sin^4\theta_t \sin^2 2\theta_i}{\eta_1^4 \sin^4\theta_i (\cos 2\theta_i + \cos 2\theta_t)^2}\right] G$$

and

$$\begin{aligned} G = {} & \eta_2^4 - \eta_1^4 \frac{\sin^4\theta_i}{\sin^4\theta_t} \mp 2\eta_1^2 \frac{\cos\theta_t \sin\theta_i}{\cos\theta_i \sin\theta_t} \left(\eta_2^2 - \eta_1^2 \frac{\sin^2\theta_i}{\sin^2\theta_t}\right) \pm K^4 \\ & \pm 2\eta_1^2 K^2 \frac{\sin\theta_i \cos\theta_t}{\cos\theta_i \sin\theta_t} + 2\eta_2^2 K^2 \left(1 + \frac{2\sin^2\theta_t}{\sin^2 2\theta_i \cos^2\theta_t} \mp \frac{4\sin\theta_t}{\sin 2\theta_i \cos\theta_t}\right) \end{aligned} \tag{I.13.5}$$

when $K = 0$ these reduce to the Fresnel equations for lossless media.

I.14 ABBE'S SINE CONDITION AND THE HERSCHEL CONDITION

In all optical systems with more than a single surface, the additional surfaces allow the application of a further condition for each surface, upon the properties of the focusing ray pencils. For a fundamental two surface axi-symmetric system one such condition is that the system of rays focuses a second focal source point into a second image point near to the first. Two basic situations exist, that where the second foci are displaced transversely to the optical axis of the system and where the displacement is axially. The situation is as shown in Figure I.12

If P and P' are the axial foci and Q and Q' the displaced foci and using primes to indicate properties of the image space, we consider rays $\hat{\mathbf{s}}$ and $\hat{\mathbf{t}}$ which are subtended by the semi-aperture first at P and then at Q. These appear in the image space as $\hat{\mathbf{s}}'$ and $\hat{\mathbf{t}}'$ subtended at P' and Q' respectively.

From Equation I.6.3 we have for any closed true ray path

$$I \equiv \oint \eta \hat{\mathbf{s}} \cdot \mathrm{d}\mathbf{s} = 0 \tag{I.14.1}$$

Taking this closed path to be the route $PP'Q'QP$ first along the s rays then along the t rays, and writing

$$\int_A^B \ldots \equiv \{AB\}$$

for simplicity, then

$$\begin{aligned} \delta I &\equiv \{PP' + \mathrm{d}l' + Q'Q + \mathrm{d}l\} \text{ along } s \text{ rays} \\ &= \{PP' + \mathrm{d}l' + Q'Q + \mathrm{d}l\} \text{ along } t \text{ rays} \end{aligned}$$

Since $\{PP'\}$ along s rays $= \{PP'\}$ along t rays and we require $\{QQ'\}$ along s rays to equal $\{QQ'\}$ along t rays, then this relation will be satisfied if for $\mathrm{d}l$ and $\mathrm{d}l'$

$$\int_P^Q \eta \hat{\mathbf{s}} \cdot \mathrm{d}\mathbf{s} = \int_{P'}^{Q'} \eta \hat{\mathbf{s}}' \cdot \mathrm{d}\mathbf{s}'$$

and

$$\int_P^Q \eta \hat{\mathbf{t}} \cdot \mathrm{d}\mathbf{s} = \int_{P'}^{Q'} \eta \hat{\mathbf{t}}' \cdot \mathrm{d}\mathbf{s}' \tag{I.14.2}$$

For *small* displacements $\mathrm{d}l$ Equation I.14.2 can be approximated by

$$\eta(\hat{\mathbf{s}} - \hat{\mathbf{t}}) \cdot \mathrm{d}\mathbf{l} = \eta'(\hat{\mathbf{s}}' - \hat{\mathbf{t}}') \cdot \mathrm{d}\mathbf{l}' \tag{I.14.3}$$

which is Brun's law [1, p11].

For a displacement $\mathrm{d}\mathbf{l}$ perpendicular to the axis of this system the second of the relations in Equation I.14.2 gives (approximately)

$$\eta \, \mathrm{d}l \sin\theta = \eta' \, \mathrm{d}l' \sin\theta' \tag{I.14.4}$$

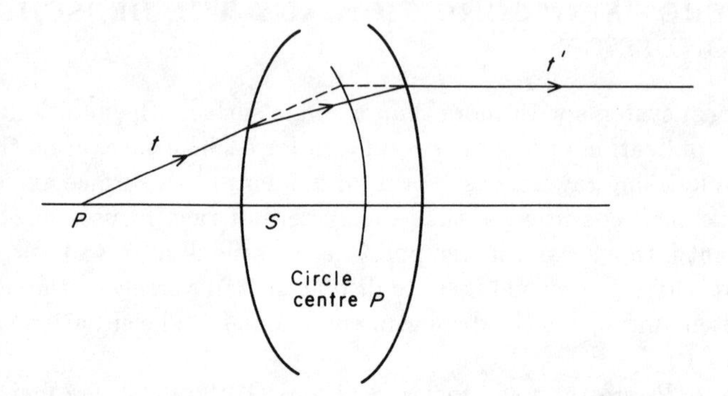

Figure I.13: The sine condition with focus at infinity

which is the Abbe sine condition [2, p165].

In the case where the image focus is at infinity the integral I along the s ray from P must equal the integral along any t ray from P up to the point where it intersects its corresponding t' ray. Thus the sine rule is said to be obeyed if the intersection of the incident rays and the finally transmitted rays all lie on a circle centred on the source of rays (Figure I.13). This can be derived as a limiting condition on Equation I.14.4 [2, p167]. If we now take $\mathbf{dr}$ along the axis of symmetry Equation I.14.3 gives the relation

$$\eta\, \mathrm{d}r(1 - \cos\theta) = \eta'\, \mathrm{d}r'(1 - \cos\theta') \tag{I.14.5}$$

This is Herschel's condition for an axial displacement of focus.

It is obviously incompatible with the sine condition to which it approximates only for small angles θ, that is for paraxial rays only. Thus no optical system can simultaneously produce perfect foci at local transversely and axially displaced points from a point source similarly displaced from its true position.

I.15 REFRACTION OF A NORMAL CONGRUENCE [22]

A congruence is a two-dimensional family of rays characterized by the fact that when intersected by a suitable arbitrary surface $\mathbf{w}(x, y, z)$ where $z = f(x, y)$, the *optical* direction cosines (u, v, w) are functions of the coordinates x y of the intersection of the ray with the surface. A congruence becomes a *normal* congruence if there exists a wave surface or phase front which is normal to all the rays of a congruence. That is, there exists a surface $\mathbf{a}(x, y, z)$ such that the vector $\mathbf{t}(u, v, w)$ (see Section I.3) equal in length to the refractive index η at the point (x, y, z) of the medium tangential to the ray satisfies

$$\mathbf{t} \cdot \mathbf{dw} = 0 \tag{I.15.1}$$

From Huygens' construction for a phase front τ, there exists a relation of the form

$$\mathbf{w} = \mathbf{a} + \lambda \mathbf{t} \tag{I.15.2}$$

Since $\mathbf{t} \cdot \mathrm{d}\mathbf{t} = 0$

$$\mathbf{t} \cdot \mathrm{d}\mathbf{a} = \mathbf{t} \cdot (\mathrm{d}\mathbf{w} - \lambda \mathrm{d}\mathbf{t} - \mathbf{t}\mathrm{d}\lambda) = -\mathrm{d}(\eta^2 \lambda) \tag{I.15.3}$$

is a total differential. The function g defined by $g = \eta^2 \lambda$ is one of the well-known characteristic functions, whose geometrical meaning is the optical distance from the wave surface back to the surface **a** measured along the ray. Equation I.15.3 determines g only to within an arbitrary constant, which means there are an infinite number of wave surfaces.

If we consider as independent variables the intersection coordinates x and y with the arbitrary surface (the only restriction is that one ray only goes through each point of the surface), the equation for which may be assumed to be given in the form $z = f(x, y)$, we have

$$\begin{aligned} \mathrm{d}g &= u\,\mathrm{d}x + v\,\mathrm{d}y + w\,\mathrm{d}z \\ \mathrm{d}z &= p\,\mathrm{d}x + q\,\mathrm{d}y \end{aligned} \tag{I.15.4}$$

where $p = \partial f / \partial x$ and $q = \partial f / \partial y$. It follows that

$$\begin{aligned} g_1 &= u + pw \\ g_2 &= v + qw \end{aligned} \tag{I.15.5}$$

where $g_1 = \partial g / \partial x$; $g_2 = \partial g / \partial y$. Equations I.15.5 and the relation

$$u^2 + v^2 + w^2 = \eta^2$$

can be used to determine u and v as functions of x and y.

If the surface **a** is one of the refracting surfaces, the refraction law has the form

$$(\mathbf{t}' - \mathbf{t}) \cdot \mathrm{d}\mathbf{a} = 0$$

showing that

$$\mathbf{t}' \cdot \mathrm{d}\mathbf{a} = \mathrm{d}g \tag{I.15.6}$$

From this we find that the same function $g(x, y)$ is connected to the congruence *after* refraction by the equations

$$\begin{aligned} g_1 &= u' + pw' \\ g_2 &= v' + qw' \end{aligned}$$

and taking note of the relation

$$u'^2 + v'^2 + w'^2 = \eta'^2$$

we can find the direction cosines u' and v' of the refracted congruence as functions of x and y.

Proof of the law of Malus and Dupin is obtained at once, since the refraction law gives

$$\mathbf{t}' \cdot d\mathbf{a} = \mathbf{t} \cdot d\mathbf{a} = dg \tag{I.15.7}$$

from which we easily verify that

$$\mathbf{w}' = \mathbf{a} - \frac{g}{\eta'^2}\mathbf{t}' \tag{I.15.8}$$

is a wave surface for the refracted congruence.

I.16 REFLECTION USING COMPLEX COORDINATES [23]†

Considerable simplification in the mathematics can be achieved by parametrising ray directions using a single complex coordinate rather than, for example, two (real) polar angles in a spherical polar coordinate system. The complex notation has two main advantages. Not only are the resulting mathematical formulas relatively simple, being rational algebraic expressions rather than complicated trigonometric expressions, but interesting implications are obtained by invoking theorems of complex variable theory.

The new formalism makes obvious the fact that the mapping between incident and reflected ray directions can be any analytic (in particular, conformal) mapping. Thus complex potential theory can be used in reflector design and offers interesting possibilities.

COMPLEX COORDINATES ON THE UNIT SPHERE

We begin by summarizing some standard notation associated with complex coordinates. A complex-valued function f of real variables x y can be written $f = u + iv$ where u v are real-valued functions and $i = \sqrt{-1}$. The partial derivatives are defined by $f_x = u_x + iv_x$ $f_y = u_y + iv_y$.

We can also regard f as a function of the complex variable $\eta = x + iy$. The relations $2x = \eta + \bar{\eta}$; $2iy = \eta - \bar{\eta}$ motivate the definitions

$$f_\eta = \frac{1}{2}(f_x - if_y) = \frac{1}{2}[u_x + v_y + i(v_x - u_y)] \tag{I.16.1}$$

$$f_{\bar{\eta}} = \frac{1}{2}(f_x + if_y) = \frac{1}{2}[u_x - v_y + i(v_x + u_y)] \tag{I.16.2}$$

The equation $f_{\bar{\eta}} = 0$ is equivalent to the Cauchy-Riemann equations $u_x = v_y$, $v_x = -u_y$. Consequently a function f satisfying this condition is an analytic function of η. Similarly, if $f_\eta = 0$ then f is an analytic function of $\bar{\eta}$.

We denote the conjugate complex function $u - iv$ by $\bar{f}$. Its derivatives satisfy the relations

$$\bar{f}_\eta = (\overline{f_{\bar{\eta}}}) \quad \bar{f}_{\bar{\eta}} = (\overline{f_\eta}) \tag{I.16.3}$$

†Section I.16 is a copy of Section 2.1 of Westcott's book [23] and is reproduced here with the kind permission of the author and the Research Studies Press, Letchworth England

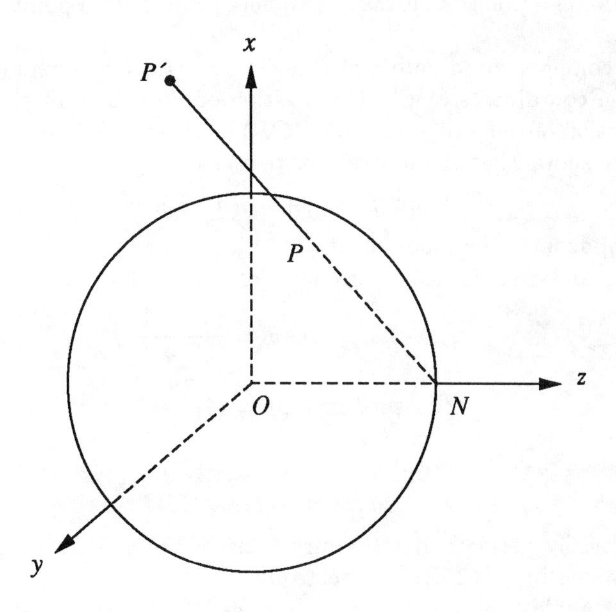

Figure I.14: Diagram showing stereographic projection

The commutativity relation $f_{\eta\bar{\eta}} = f_{\bar{\eta}\eta}$ follows from Equations I.16.1 and I.16.2, both of these derivatives being equal to $\Delta f/4$ where $\Delta f = f_{xx} + f_{yy}$ is the Laplacian of f.

The following lemma will be used in later work.

Lemma 1

Let f be a function with continuous derivatives of first order. There exists a real-valued function g such that $g_\eta = f$ if, and only if, $f_{\bar{\eta}}$ is real-valued.

Proof

If g exists then $f_{\bar{\eta}} = g_{\eta\bar{\eta}} = \Delta g/4$ and is therefore real-valued. Conversely, if $f_{\bar{\eta}}$ is real-valued then, from Equation I.16.2, $u_y = -v_x$. Consequently, there exists a real-valued function ϕ such that $u = \phi_x$, $v = -\phi_y$. The function $g = 2\phi$ satisfies $g_\eta = f$.

Let O (Figure I.14) denote the centre of the sphere and choose a rectangular set of axes OX OY OZ. Let (x, y, z) be the corresponding Cartesian coordinates. Under stereographic projection from the point N of the coordinates $(0, 0, 1)$ a point P on the unit sphere projects to a point P' in the plane $z = 0$.

The complex coordinate η of P is defined as $x + iy$ where $(x, y, 0)$ are the Cartesian coordinates of P'. The position of P of the unit sphere and hence the direction of the unit vector $\mathbf{p}$ $(= OP)$ is parametrised by η. In fact simple geometry shows that the unit vector $\mathbf{p}$ is given by

$$(1 + |\eta|^2)\mathbf{p} = (\eta + \bar{\eta}, i(\bar{\eta} - \eta), |\eta|^2 - 1) \tag{I.16.4}$$

where $|\eta|$ denotes the modulus of η.

The standard spherical coordinates of P are related to η by

$$\cos\theta = \frac{|\eta|^2 - 1}{|\eta|^2 + 1}$$

$$\sin\theta \exp(i\phi) = \frac{2\eta}{|\eta|^2 + 1} \tag{I.16.5}$$

(cf Equation 4.11.25) so that

$$\eta = \cot(\theta/2)\exp(i\phi) \tag{I.16.6}$$

It is easily verified that the directions of the axes Ox Oy and Oz have complex coordinates $1, i, \infty$ respectively.

The complex coordinate η is determined by the choice of a right-handed rectangular Cartesian coordinate system. If we take a different choice we get a different coordinate $\tilde{\eta}$. It can be shown that the relation between $\tilde{\eta}$ and η is necessarily of the form

$$\eta = \frac{\alpha_0\tilde{\eta} + \beta_0}{-\bar{\beta}_0\tilde{\eta} + \bar{\alpha}_o} \tag{I.16.7}$$

where α_0 and β_0 are constants. For example, in Figure I.15 axes $\tilde{X}$ $\tilde{Y}$ $\tilde{Z}$ are taken with $\tilde{Y} = -Y$ and the direction of the $\tilde{Z}$ axis is given by the complex coordinate η_0 when referred to the X Y Z axes. Then the relation between $\tilde{\eta}$ and η for an arbitrary direction is

$$\eta = \frac{\eta_0\tilde{\eta} + 1}{\tilde{\eta} - \eta_0} \tag{I.16.8}$$

corresponding to $\alpha_0 = i\eta_0$, $\beta_0 = i$ in the general formula. It is noted that since η_0 corresponds to a direction in the $Y = 0$ plane, it is real and related to θ_0 by $\eta_0 = \tan(\theta_0/2)$.

The metric on the sphere assumes a simple form in terms of the coordinate η. To show this we first differentiate Equation I.16.4 to obtain

$$(1 + |\eta|^2)^2\mathbf{p}_\eta = (1 - \bar{\eta}^2, -i(1 + \bar{\eta}^2), 2\bar{\eta}) \tag{I.16.9}$$

It follows from Equation I.16.3 that $\mathbf{p}_{\bar{\eta}} = \overline{(\mathbf{p}_\eta)}$. We can now calculate the scalar products.

$$\begin{aligned} \mathbf{p} \cdot \mathbf{p}_\eta &= \mathbf{p} \cdot \mathbf{p}_{\bar{\eta}} = 0 \\ \mathbf{p}_\eta \cdot \mathbf{p}_\eta &= \mathbf{p}_{\bar{\eta}} \cdot \mathbf{p}_{\bar{\eta}} = 0 \\ \mathbf{p}_\eta \cdot \mathbf{p}_{\bar{\eta}} &= 2/(1 + |\eta|^2)^2 \end{aligned} \tag{I.16.10}$$

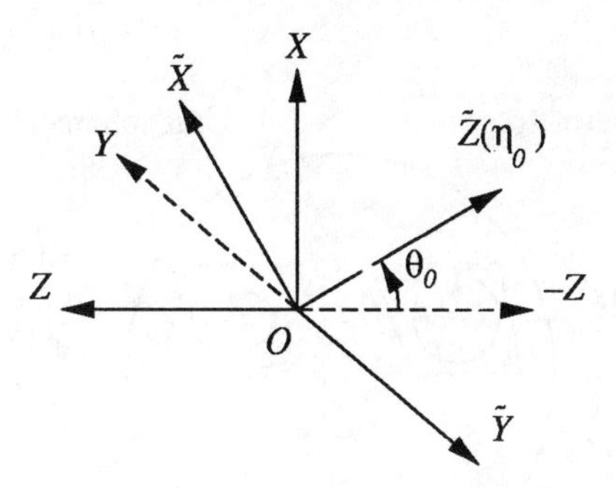

Figure I.15: Relationship between $X\ Y\ Z$ and $\tilde{X}\ \tilde{Y}\ \tilde{Z}$ axes

Consider a curve $\eta = \eta(t)$ on the sphere where t is a real parameter. This curve has a tangent vector given by

$$\frac{\mathrm{d}}{\mathrm{d}t}(\mathbf{p}\eta(t)) = \mathbf{p}_\eta \frac{\mathrm{d}\eta}{\mathrm{d}t} + \mathbf{p}_{\bar{\eta}} \frac{\mathrm{d}\bar{\eta}}{\mathrm{d}t}$$

and, using the Equations I.16.10 we obtain

$$\begin{aligned} \left|\frac{\mathrm{d}\mathbf{p}}{\mathrm{d}t}\right|^2 &= \left(\mathbf{p}_\eta \frac{\mathrm{d}\eta}{\mathrm{d}t} + \mathbf{p}_{\bar{\eta}} \frac{\mathrm{d}\bar{\eta}}{\mathrm{d}t}\right) \cdot \left(\mathbf{p}_{\bar{\eta}} \frac{\mathrm{d}\bar{\eta}}{\mathrm{d}t} + \mathbf{p}_\eta \frac{\mathrm{d}\eta}{\mathrm{d}t}\right) \\ &= \frac{4}{(1+|\eta|^2)^2} \cdot \left|\frac{\mathrm{d}\eta}{\mathrm{d}t}\right|^2 \end{aligned}$$

and hence the promised simple formula for the length of $\mathrm{d}\mathbf{p}/\mathrm{d}t$ is

$$\frac{2}{1+|\eta|^2}\left|\frac{\mathrm{d}\eta}{\mathrm{d}t}\right| \tag{I.16.11}$$

In the following sections we shall be concerned with a local mapping τ of the unit sphere into itself. Suppose that τ is given by $\zeta = \zeta(\eta)$ in terms of the complex coordinates η of P and ζ of $Q = \tau(P)$. The curve $\eta = \eta(t)$ transforms into the curve $\zeta = \zeta(\eta(t))$ and according to Equation I.16.11 the length of the tangent vector to this curve is

$$2\left|\zeta_\eta \frac{\mathrm{d}\eta}{\mathrm{d}t} + \zeta_{\bar{\eta}} \frac{\mathrm{d}\bar{\eta}}{\mathrm{d}t}\right| / (1+|\zeta|^2) \tag{I.16.12}$$

Using this formula, it can be shown that the circle of unit vectors tangent to the sphere at P transforms into an ellipse of tangent vectors at Q (see Figure I.16) whose major and minor axes are of lengths

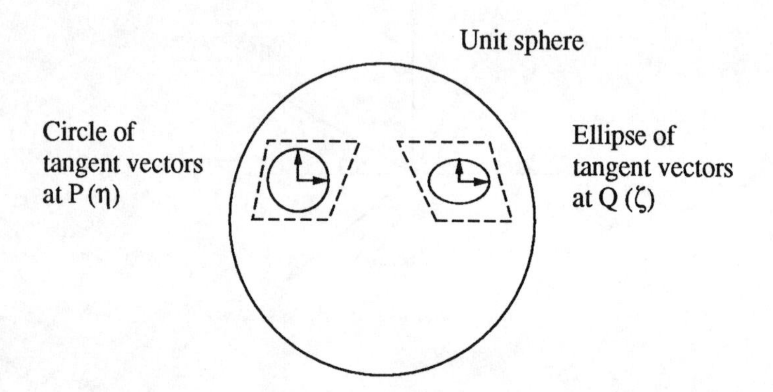

Figure I.16: Transformation of areas on the unit sphere under general mapping τ

$$\frac{1+|\eta|^2}{1+|\zeta|^2}(|\,|\zeta_\eta| \pm |\zeta_{\bar{\eta}}|\,|) \qquad \text{(I.16.13)}$$

These numbers measure the *distortion* of τ. We say that τ has *uniform distortion* if they are equal. It follows that the mappings of uniform distortion satisfy either $\zeta_{\bar{\eta}} = 0$ or $\zeta_\eta = 0$. In the first case ζ is an analytic function of η in the sense of complex variable theory and we say that τ is *analytic.* In the second case we say that τ is *anti-analytic.*

It is a consequence of the formula (Equation I.16.13) that τ alters elementary areas by the factor

$$\frac{(1+|\eta|^2)^2}{(1+|\zeta|^2)^2}|J(\tau)| \qquad \text{(I.16.14)}$$

where $J(\tau)$, the Jacobian of τ is given by

$$J(\tau) = |\zeta_\eta|^2 - |\zeta_{\bar{\eta}}|^2$$

BASIC EQUATIONS FOR A REFLECTION

In Figure I.17 a sphere of unit radius and centre O is drawn, where O is the point source of incident rays. The points P Q are the end-points of unit vectors $\mathbf{p}$ $\mathbf{q}$ drawn from O. The unit vector $\mathbf{p}$ is in the direction of the incident ray, $\mathbf{r} = r\mathbf{p}$ is the position vector of the point of reflection R, and $\mathbf{q}$ is the direction of the reflected ray at R. Our aim is to relate the far-field power density pattern $G(\mathbf{q})$ and the source power density pattern $I(\mathbf{p})$ to the geometry of the reflector under the laws of geometrical optics.

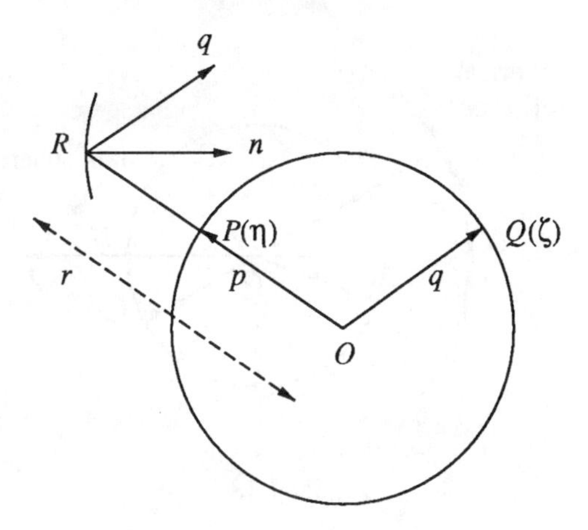

Figure I.17: Diagram showing incident and reflected ray directions

First of all we obtain the differential equations governing the mapping $\tau : P \rightarrow Q$. We shall use complex coordinates η, ζ to parametrize the incident and reflected directions respectively.

Under geometrical optics energy is conserved under reflection. This implies that the mapping τ has to alter elementary areas on the unit sphere by the factor I/G. Equation I.16.14 leads immediately to one of the differential equations

$$|\zeta_\eta|^2 - |\zeta_{\bar{\eta}}|^2 = \pm\frac{(1+|\zeta|^2)^2 I}{(1+|\eta|^2)^2 G} \qquad \text{(I.16.15)}$$

The differential Equation I.16.15 is not the only restriction on the mapping τ. We shall see that $\rho = \ln r$ satisfies a differential equation whose integrability condition gives a second condition. The law of reflection implies that the vector $\mathbf{r} - r\mathbf{q}$ is normal to the reflector surface. Since $\mathbf{r}_\eta$ is a tangential vector it follows that

$$\mathbf{r}_\eta \cdot (\mathbf{r} - r\mathbf{q}) = 0 \qquad \text{(I.16.16)}$$

This condition can be modified when we use $\rho = \ln r$ to give

$$\rho_\eta = (\mathbf{q} \cdot \mathbf{p}_\eta)/\Lambda \qquad \text{(I.16.17)}$$

where $\Lambda(\eta, \zeta) = 1 - \mathbf{p} \cdot \mathbf{q}$ is regarded as a function of the *two* variables η, ζ. We put $\psi(\eta, \zeta) = -\ln(\Lambda)$ so that we can write Equation I.16.17 as

$$\rho_\eta = \psi_\eta(\eta, \zeta(\eta)) \qquad \text{(I.16.18)}$$

A calculation based on Equation I.16.4 shows that

$$\Lambda(\eta, \zeta) = \frac{2|\zeta - \eta|^2}{(1+|\zeta|^2)(1+|\eta|^2)} \qquad \text{(I.16.19)}$$

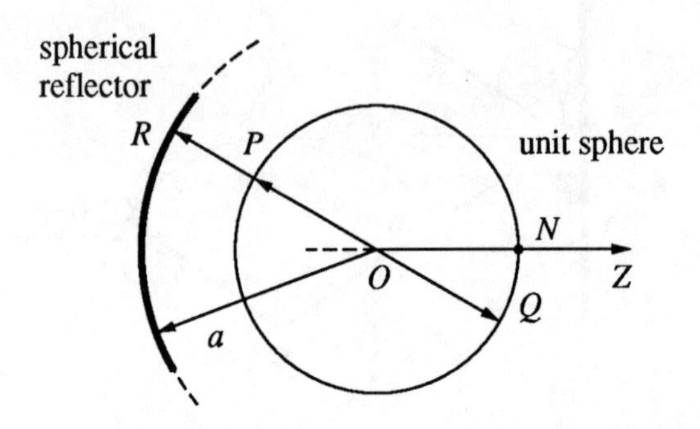

Figure I.18:

We differentiate to obtain ψ_η and so express Equation I.16.18 as

$$\rho_\eta = \frac{1}{\zeta - \eta} + \frac{\bar{\eta}}{1 + |\eta|^2}$$

It is convenient to write this result as

$$L_\eta = \frac{1}{\zeta - \eta} \tag{I.16.20}$$

where $L(\eta) = \ln(r/(1 + |\eta|^2))$. L is a real valued function and consequently Lemma 1 implies that

$$\frac{1}{(\zeta - \eta)^2}\zeta_{\bar{\eta}} \quad \text{is real valued} \tag{I.16.21}$$

This is the second condition on the mapping τ and is necessary if the mapping is to be derived from an integrable surface. Examples to show that this condition is satisfied by elementary mappings derived from simple reflecting surfaces are indicated below:

1. The sphere $r = a$ (with source at centre)
   We have $L = \ln(a/(1 + |\eta|^2))$

$$L_\eta = -\frac{\bar{\eta}}{1 + |\eta|^2}$$

   Hence $\zeta = -1/\bar{\eta}$ which satisfies condition I.16.21

2. The plane
   We have

$$r = -d \sec\theta = -\frac{d(|\eta|^2 + 1)}{|\eta|^2 - 1}$$

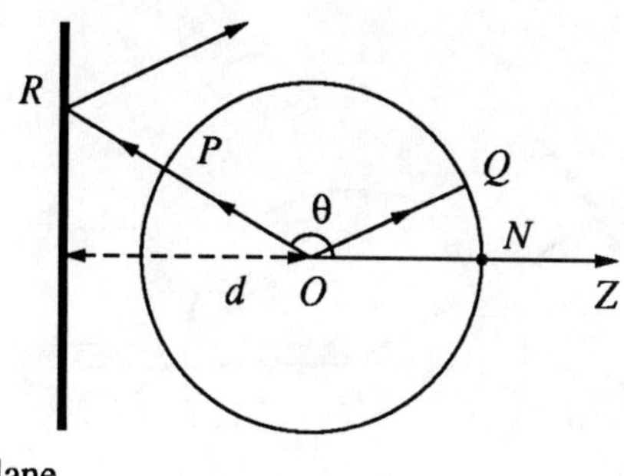

Figure I.19:

$$L = \ln\left(\frac{d}{1-|\eta|^2}\right)$$

$$L_\eta = \frac{\bar{\eta}}{1-|\eta|^2}$$

Hence $\zeta = 1/\bar{\eta}$ which satisfies condition I.16.21

3. The hyperboloid of revolution (with source at one focus)
   We have

$$r = -\frac{\ell}{e\cos\theta - 1} = -\frac{\ell(1+|\eta|^2)}{(e-1)|\eta|^2-(e+1)}$$

$$L = \ln\left(\frac{\ell}{e+1-(e-1)|\eta|^2}\right)$$

$$L_\eta = \frac{(e-1)\bar{\eta}}{e+1-(e-1)|\eta|^2}$$

   Hence

$$\zeta = \frac{e+1}{(e-1)\bar{\eta}}$$

   which satisfies condition I.16.21

4. The paraboloid of revolution (with source at focus)
   Since all rays derived from the focus are reflected parallel to the z-axis ζ = constant $= \infty$. Hence condition I.16.21 is satisfied trivially.

   Note that the mapping $\eta \to \zeta$ is singular (the Jacobian vanishes in Equation I.16.15.

It is noted that 1. 2. and 3. have the property that $\zeta_\eta = 0$ and are anti-analytic mappings; 4. gives a singular mapping and may be regarded as a degenerate member of either class.

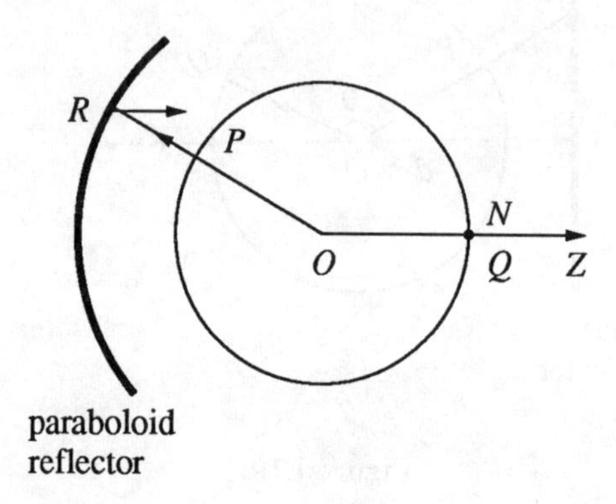

Figure I.20:

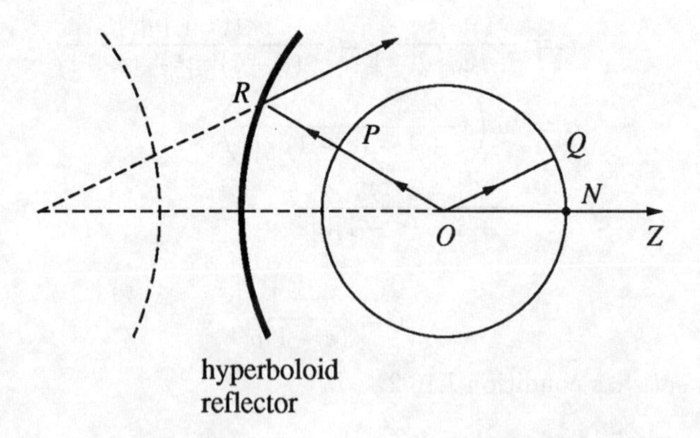

Figure I.21:

It is obvious that any analytic (and in particular conformal) mapping of the unit sphere into itself satisfies the condition I.16.21 and therefore belongs to a reflector. A corresponding ratio I/G can be obtained from Equation I.16.15 (with the + sign).

The class of anti-analytic mappings satisfying condition I.16.21 is restricted and it can be shown that the mappings are necessarily of the form

$$\zeta = \frac{a\bar{\eta} + b}{c\bar{\eta} + d}$$

where $a\ b\ c\ d$ are constants satisfying further restrictions. Examples 1. 2. 3. given above are particular members of this class. It can be proved that this class corresponds to reflectors which are all quadrics of revolution (or degenerate ones) with the source at a focus.

Of course to satisfy the condition I.16.21 the mapping does not necessarily have to be analytic or anti-analytic. For the synthesis problem it is sufficient to note that any local mapping τ of a sphere into itself, satisfying the conditions I.16.15 and I.16.21 will lead to a reflector. For, according to Lemma 1, there exists a real-valued function $L(\eta)$ satisfying Equation I.16.20 and the reflector is given by

$$r = (1 + |\eta|^2) \exp L(\eta) \tag{I.16.22}$$

For a given non-zero ratio I/G one can obtain two types of reflector depending on the choice of sign in Equation I.16.15. We refer to the choice of $+(-)$ sign as the hyperbolic (elliptic) case respectively.

Instead of finding the mapping function $\zeta(\eta)$ it is often more convenient to derive an equation to be solved for $L(\eta)$ directly. In fact we can deduce, by differentiating Equation I.16.20 that

$$L_{\eta\eta} - L_\eta^2 = -L_\eta^2 \zeta_\eta \qquad L_{\eta\bar{\eta}} = -L_\eta^2 \zeta_{\bar{\eta}}$$

Consequently we obtain from Equation I.16.15 that

$$|L_{\eta\eta} - L_\eta^2|^2 - |L_{\eta\bar{\eta}}|^2 = \pm |L_\eta|^4 \left(\frac{1 + |\zeta|^2}{1 + |\eta|^2} \right)^2 \frac{I(\eta)}{G(\zeta)} \tag{I.16.23}$$

When the variable ζ is replaced by $\eta + 1/L_\eta$ this equation becomes a partial differential equation of second order for L in terms of η. It is non-linear and is a Monge-Ampere equation of hyperbolic or elliptic type depending on the choice of sign.

I.17 LENS EQUATIONS [24]

The two surfaces considered are $S : z = z(x, y)$ and $S' : z' = z'(x', y')$ of a lens of refractive index η. A source at the primary focal point $F(x_0, y_0, z_0)$ is required to give a stigmatic image at the (symmetrically disposed) point

$F'(x_0', y_0', z_0')$. The two parameter system of rays from F are characterised by independent variables u and v which can be direction cosines of individual rays or the point of intersection x, y of the ray with a given surface usually the first lens surface. From Snell's law in the form of the cross products of the unit vectors in Equation I.3.4 at each surface we can derive the partial derivatives involved in the form

$$\frac{\partial z}{\partial x} = \frac{\partial(z,y)}{\partial(u,v)} \Big/ \frac{\partial(x,y)}{\partial(u,v)} \qquad \frac{\partial z'}{\partial x'} = \frac{\partial(z',y')}{\partial(u,v)} \Big/ \frac{\partial(x',y')}{\partial(u,v)}$$

$$\frac{\partial z}{\partial y} = \frac{\partial(z,x)}{\partial(u,v)} \Big/ \frac{\partial(y,x)}{\partial(u,v)} \qquad \frac{\partial z'}{\partial y'} = \frac{\partial(z',x')}{\partial(u,v)} \Big/ \frac{\partial(y',x')}{\partial(u,v)} \tag{I.17.1}$$

in order to express all of the derivatives in terms of differentiations with respect to the independent variables.

The optical path constant is $P = p + \eta p' + p'' = P_0$ along every ray (Figure I.22) from F to $F'(x_0', y_0', z_0')$.

Referring all surfaces and equations to a single set of coordinate axes results, after suitable rearrangement in the following:

for $\mathcal{S}$: for $\mathcal{S}'$:

$$\frac{\partial(z,y)}{\partial(u,v)} = F\frac{\partial(x,y)}{\partial(u,v)} \qquad \frac{\partial(z',y')}{\partial(u,v)} = F'\frac{\partial(x',y')}{\partial(u,v)}$$

$$\frac{\partial(z,x)}{\partial(u,v)} = G\frac{\partial(y,x)}{\partial(u,v)} \qquad \frac{\partial(z',x')}{\partial(u,v)} = G'\frac{\partial(y',x')}{\partial(u,v)} \tag{I.17.2}$$

where

$$F = \frac{\eta(x'-x)p - (x-x_0)p'}{(z-z_0)p' - \eta(z'-z)p} \qquad G = \frac{\eta(y'-y)p - (y-y_0)p'}{(z-z_0)p' - \eta(z'-z)p}$$

$$F' = \frac{(x_0'-x')p' - \eta(x'-x)p''}{\eta(z'-z)p'' - (z_0'-z')p'} \qquad G' = \frac{(y_0'-y')p' - \eta(y'-y)p''}{\eta(z'-z)p'' - (z_0'-z')p'} \tag{I.17.3}$$

and

$$p^2 = (x-x_0)^2 + (y-y_0)^2 + (z-z_0)^2$$
$$p'^2 = (x'-x)^2 + (y'-y)^2 + (z'-z)^2$$
$$p''^2 = (x_0'-x')^2 + (y_0'-y')^2 + (z_0'-z')^2 \tag{I.17.4}$$

are the lengths of the linear increments of the ray between F and F'.

Symmetry conditions take the form

$$z(-x,y) = z(x,y) \qquad z'(-x',y') = z'(x',y')$$
$$z(x,-y) = z(x,y) \qquad z'(x',-y') = z'(x',y') \tag{I.17.5}$$

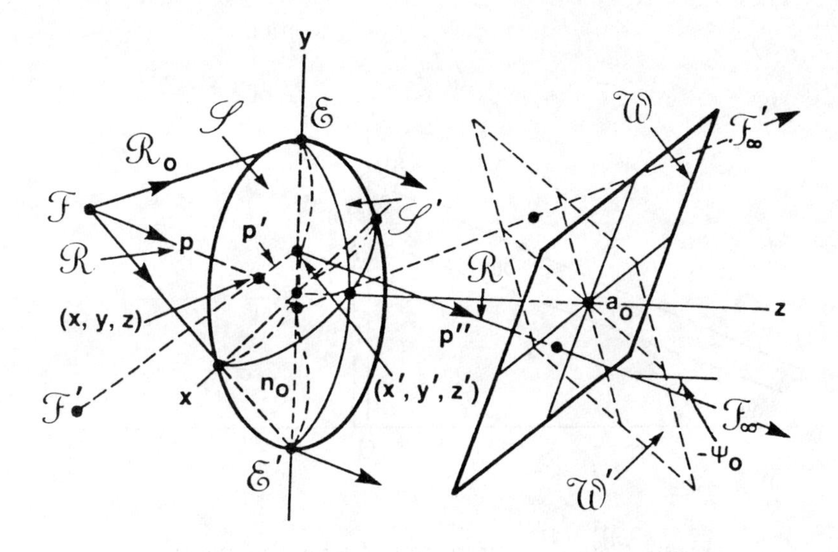

Figure I.22: Bifocal lens with surfaces $\mathcal{S}$ and $\mathcal{S}'$

and boundary conditions of the form

$$z(x, y) = 0 \quad z'(x, y) = 0 \tag{I.17.6}$$

The path length condition I.17.1 is not independent of the differential Equations I.17.2, but can be substituted for one set given that the other set is valid, in keeping with the general understanding that the problem is soluble using an eikonal relation and Snell's law at one surface.

Symmetry conditions on the surfaces, that is rotational symmetry about the axis, and the symmetrical disposition of the points F and F' simplify the relations. The parameters u and v can be replaced by x and y giving further simplification. For the collimating lens with two off-axis foci, F' goes to infinity in the required direction, and the eikonal relation for P in terms of $p\,p'$ and p'' is adjusted. The article itself is concerned with the numerical methods of solving such differential equations subject to two a priori conditions [25]. Numerical methods involve using a series expansion of generalized elliptical surfaces, comparable to the method of generic surfaces for lenses and reflectors given previously in this volume.

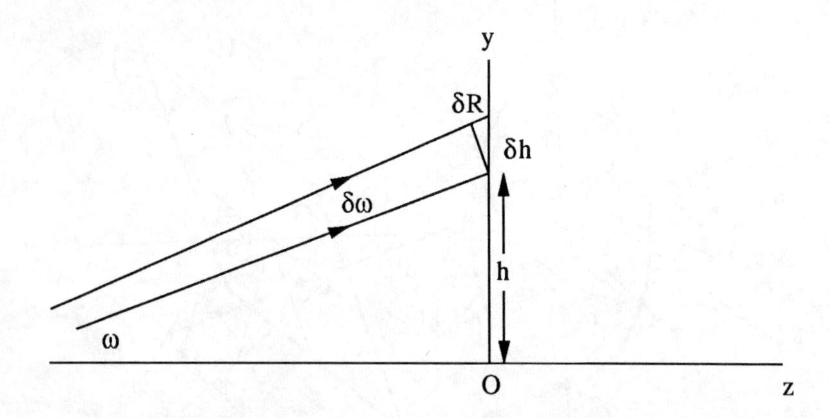

Figure I.23: ω, h parameters of a ray congruence

I.18 REFRACTIVE TRANSFORMATIONS BETWEEN NORMAL CONGRUENCES [26][27]

We choose rectangular axes with Oz along the axis of symmetry and Oy perpendicular to it, and denote by h the height at which a member of the ray congruence intersects Oy. The angle this ray makes with the z-axis is $\omega(h)$ and this function specifies the normal congruence (dealing only in the plane of symmetry.)

Hence for two closely spaced rays (Figure I.23) with parameters ω and $\omega + \delta\omega$ the path difference δR is

$$\delta R = \delta h \sin(\omega + \delta\omega)$$

Dividing by h and proceeding to the limit $\delta h = 0$ gives to the first order

$$\mathrm{d}R = \mathrm{d}h \sin\omega \tag{I.18.1}$$

In a medium with uniform refractive index η, this becomes

$$\eta \mathrm{d}R = \eta \int \sin\omega \mathrm{d}h \tag{I.18.2}$$

If two rays with parameters h_1 and h_2 intersect a wave front W of uniform phase front at P_1 and P_2 and the axis Oy in Q_1 and Q_2 (Figure I.24), their optical path differences

$$[P_2Q_2] - [P_1Q_1] = \eta \int_{h_1}^{h_2} \sin\omega \mathrm{d}h \tag{I.18.3}$$

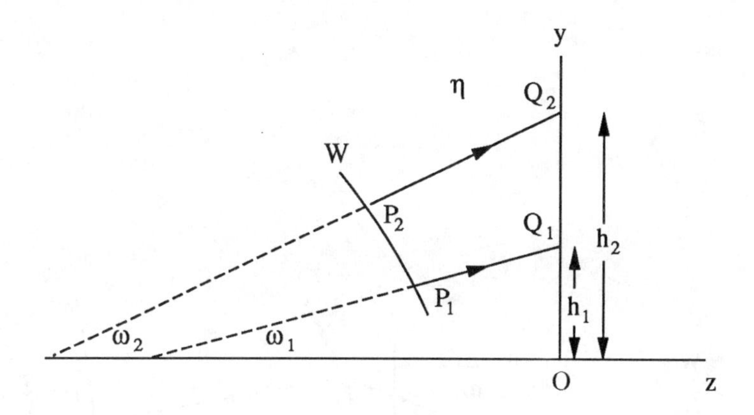

Figure I.24: Ray parameters of incident wave front

Consider now a curved refracting surface with the profile Σ intersecting the z-axis at O, with an illuminating phase front W as shown in Figure I.25. This will refract into (a normal congruence) the phase front W'. Let P and Q be points on rays of the incident congruence and P' Q' the corresponding points on the rays after refraction. P and Q are parametrised by ω_P and ω_Q and the "Q" ray meets the Oy axis at h_Q. On the refracted side the corresponding parameters are primed.

Then since the optical paths from P to P' and Q to Q' are equal

$$[QN] + [NQ'] = [PO] + [OP']$$

where N is the point of intersection of the "Q" ray with the surface Σ with coordinates (y, z). Then from I.18.3

$$[QN] = [Qh_Q] + [h_Q N] = [PO] + \eta \int_0^{h_Q} \sin\omega \mathrm{d}h + \eta z \sec\omega \tag{I.18.4}$$

and similarly

$$[NQ'] = -[PO] - \eta' \int_0^{h_{Q'}} \sin\omega' \mathrm{d}h' - \eta' z \sec\omega_{Q'} \tag{I.18.5}$$

Substituting I.18.4 into I.18.5 we obtain

$$z(\eta \sec\omega - \eta' \sec\omega') + \eta \int_0^{h_Q} \sin\omega \mathrm{d}h - \eta' \int_0^{h'_Q} \sin\omega' \mathrm{d}h' = 0 \tag{I.18.6}$$

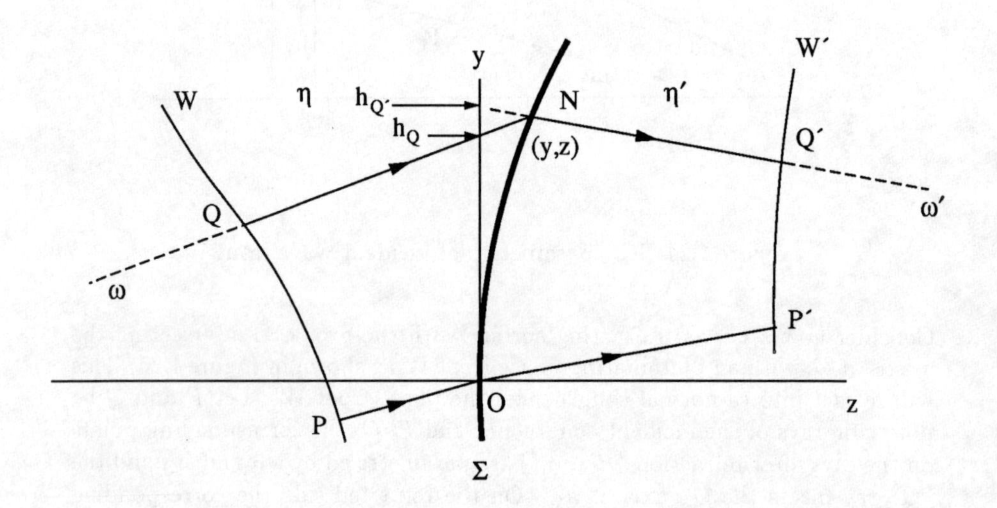

Figure I.25: Refraction at a curved interface

As seen in the figure

$$y = h_Q + z \tan \omega_Q \qquad \text{and} \qquad y = h_{Q'} + z \tan \omega_{Q'} \tag{I.18.7}$$

Eliminating y and z between Equations I.18.6 and I.18.7 we obtain a refractive invariant property of the transformation between the incident and the refracted congruences:-

$$\sin(\omega_Q - \omega'_Q) \left(\eta \int_0^{h_Q} \sin \omega \mathrm{d}h - \eta' \int_0^{h_{Q'}} \sin \omega' \mathrm{d}h' \right)$$
$$+ (\eta \cos \omega_{Q'} - \eta' \cos \omega_Q)(h_{Q'} - h_Q) = 0 \tag{I.18.8}$$

This relation enables the profile Σ to be computed. For if h_Q is on any particular ray of the incident congruence, the corresponding $h_{Q'}$ can be obtained by solving I.18.8 as a transcendental equation. The point (y, x) on Σ is then given by I.18.6 and I.18.7 (essentially by a numerical procedure needed to evaluate the integrals involved).

For the design of aspheric reflector profiles, the same procedure can be followed with the substitutions $\eta = 1$ and $\eta' = 1$, with some ensuing simplification.

REFERENCES

1. Enz CP (1972) *Optics and the Theory of Electrons* Pauli Lectures on Physics **2** MIT Press p4

2. Born M & Wolf E (1959) *Principle of Optics* Pergamon Press p36

3. Conway AW & Synge JL (eds) (1981) *The Collected Papers of WR Hamilton* Cambridge University Press p10

4. Lee SW, Sheshadri MS, Jamnejad V & Mittra R (1982) Reflection at a curved dielectric interface: geometrical optics solution *IEEE Trans Microwave Theory & Techniques* **vol MTT - 30** pp12-19

5. Stavroudis ON (1976) Simpler derivation of the formulas for generalized ray tracing *Jour Opt Soc Amer* **66** no 12 p1330

6. Silberstein L (1918) *Simplified Method of Tracing Rays Through Any Optical System of Lenses Prisms or Mirrors* London: Longmans Green & Co

7. Stavroudis ON (1972) *The Optics of Rays Wavefronts and Caustics* New York and London: Academic Press Chapter 2

8. Collin RE & Zucker FJ (1969) *Antenna Theory* **II** McGraw Hill Inter University Electronics Series, Vol 7 p5 & p31

9. Wagner H (1951) Zur mathematischen behandlung von Spiegelungen *Optik* **8** (10) p456

10. Bennett FD (1957) Refraction operators and ray tracing through cones of constant refractive index *Jour Opt Soc Amer* **47** (1) p85

11. Beggs JS (1960) Mirror image kinematics *Jour Opt Soc Amer* **50** (4) p388

12. Southworth GC (1966) *Principles and Applications of Waveguide Transmission* Van Nostrand p475

13. Synge JL (1946) Reflection in a corner formed by three plane mirrors *Quart J App Maths* **4** (2) p116

14. Coxeter HSM (1947) The product of three reflections *Quart J App Maths* **5** (2) p217

15. Montgomery AJ (1967) Analysis of two-tilt compensating interferometers *Jour Opt Soc Amer* **57** (9) p1121

16. Tuckerman LB (1947) Multiple reflection by plane mirrors *Quart J App Maths* **5** (2) p133

17. Polasek JC (1967) Matrix analysis of gimballed mirror and prism systems Jour Opt Soc Amer **57** (10) p1193

18. Collet E (1970) Mueller-Stokes formulation of Fresnel's equations *Amer Jour Phys* **vol 39** pp577-528

19. Baker HF (1929) *Principles of Geometry* Cambridge University Press Vol 1 pp6-12

20. Cambi E (1959) Projective formulation of the problems of geometrical optics *Jour Opt Soc Amer* **49** (1) pp 2 & 15

21. Bell RJ, Armstrong KR, Nichols CS & Bradley RW (1969) Generalized laws of refraction and reflection *Jour Opt Soc Amer* **59** (2) p187

22. Herzberger M & Marchand E (1954) Tracing a normal congruence through an optical system *Jour Opt Soc Amer* **44** (2) p146

23. Westcott BS (1983) *Shaped Reflector Antenna Design* Letchworth, England: Research Studies Press

24. Sternberg RL (1985) A survey of numerical methods for a new class of non-linear partial differential equations arising in non-spherical geometrical optics *Proceedings of 11th Congress of IMAC*

25. Sternberg RL, Grafton RB & Childs DR (1985) On a numerical minimization problem for a new class of non-linear partial differential equations arising in non-spherical geometrical optics and beamforming lenticular antennas *Jour of Computing & Applied Maths* **12 & 13** pp591-607

26. Wolf E (1949) On the designing of aspheric surfaces *Proc Phys Soc* **61** p494

27. Wolf E & Preddy WS (1947) On the determination of aspheric profiles *Proc Phys Soc* **59** p704

II
APPENDIX

II.1 THE CIRCLE POLYNOMIALS

The circle polynomials of Zernike [1] as applied to the theory of optical aberrations are the polynomials

$$R_n^m(r) = \sum_{k=0}^{(n-m)/2} (-1)^k \frac{(n-k)!}{[(n+m-2k)/2]![(n-m-2k)/2]!} \frac{r^{n-2k}}{k!} \qquad \text{(II.1.1)}$$

for $(n-m)$ an even positive integer ($R_n^m(r) = 0$ otherwise). Other descriptions include

$$R_n^m(r) = r^{-m} \frac{1}{\Gamma[(n-m+2)/2]} \left(\frac{\mathrm{d}}{\mathrm{d}(r^2)}\right)^{(n-m)/2} (r^{(m+n)/2}(r^2-1)^{(n-m)/2}) \qquad \text{(II.1.2)}$$

and more particularly

$$R_n^m(r) = (-1)^{(n-m)/2} \binom{(m+n)/2}{m} r^m \, {}_2F_1\left\{\frac{m+n+2}{2}, -\frac{n-m}{2}; m+1; r^2\right\} \qquad \text{(II.1.3)}$$

A full derivation of these polynomials including the differential equation satisfied by them can be found in Chako [2]. A table of $R_n^m(r)(n < 10; m < 10)$ is given in Table II.1.

Comparison of Equation II.1.3 with one of the possible definitions of the Jacobi polynomials [3]

$$P_s^{(\alpha,\beta)} = \binom{s+\alpha}{s} \, {}_2F_1\left\{s+\alpha+\beta+1, -s; \alpha+1; \frac{1-x}{2}\right\} \qquad \text{(II.1.4)}$$

shows that

$$R_n^m(r) = (-1)^{(n-m)/2} r^m P_{(n-m)/2}^{(m,0)}(1-2r^2) \qquad \text{(II.1.5)}$$

With $m = 0$ and n therefore an even positive integer the Jacobi polynomials reduce to the Legendre polynomials and Equation II.1.5 becomes

$$R_{2n}^0(r) = P_n(2r^2 - 1) \qquad \text{(II.1.6)}$$

From the orthogonality properties of the Jacobi and Legendre polynomials we obtain the orthogonality relation of the circle polynomials

$$\int_0^1 R_n^m(r) R_p^m(r) r \, \mathrm{d}r = \frac{\delta_{n,p}}{2n+2} \qquad \text{(II.1.7)}$$

where $\delta_{n,p}$ is the Kronecker symbol.

$R_0^0 = 1$ $R_1^1 = \mathrm{r}$ $R_0^2 = 2r^2 - 1$ $R_2^2 = r^2$ $R_3^1 = 3r^3 - 2r$ $R_3^3 = r^3$	$R_7^1 = 35r^7 - 60r^5 + 30r^3 - 4r$ $R_7^3 = 21r^7 - 30r^5 + 10r^3$ $R_7^5 = 7r^7 - 6r^5$ $R_7^7 = r^7$ $R_8^0 = 70r^8 - 140r^6 + 90r^4 - 20r^2 + 1$ $R_8^2 = 56r^8 - 105r^6 + 60r^4 - 10r^2$ $R_8^4 = 28r^8 - 42r^6 + 15r^4$ $R_8^6 = 8r^8 - 7r^6$ $R_8^8 = r^8$
$R_4^0 = 6r^4 - 6r^2 + 1$ $R_4^2 = 4r^4 - 3r^2$ $R_4^4 = r^4$	$R_9^1 = 126r^9 - 280r^7 + 210r^5 - 60r^3 + 5r$ $R_9^3 = 84r^9 - 168r^7 + 105r^5 - 20r^3$
$R_5^1 = 10r^5 - 12r^3 + 3r$ $R_5^3 = 5r^5 - 4r^3$ $R_5^5 = r^5$	$R_9^5 = 36r^9 - 56r^7 + 21r^5$ $R_9^7 = 9r^9 - 8r^7$ $R_9^9 = r^9$
$R_6^0 = 20r^6 - 30r^4 + 12r^2 - 1$ $R_6^2 = 15r^6 - 20r^4 + 6r^2$ $R_6^4 = 6r^6 - 5r^4$ $R_6^6 = r^6$	$R_{10}^0 = 252r^{10} - 630r^8 + 560r^6 - 210r^4 + 30r^2 - 1$ $R_{10}^2 = 210r^{10} - 504r^8 + 420r^6 - 140r^4 + 15r^2$ $R_{10}^4 = 120r^{10} - 252r^8 + 168r^6 - 35r^4$ $R_{10}^6 = 45r^{10} - 72r^8 + 28r^6$ $R_{10}^8 = 10r^{10} - 9r^8$ $R_{10}^{10} = r^{10}$

Table II.1: Zernike Polynomials $R_n^m(r)$

$R_n^m(r) = 0$ if $(n - m)$ is not an even integer

From the Hankel transform of the Jacobi polynomial as given in Erdelyi [4] we have

$$\int_0^a \frac{r^\nu}{a^{\nu+2}} P_n^{(\nu,0)}\left(1-\frac{2r^2}{a^2}\right) J_\nu(ur) r\,\mathrm{d}r = \frac{J_{\nu+2n+1}(au)}{au} \qquad \text{(II.1.8)}$$

In terms of the circle polynomials (with $a = 1$) we obtain Equation 5.3.10

$$\int_0^1 R_n^m(r) J_m(ur) r\,\mathrm{d}r = (-1)^{(n-m)/2} J_{n+1}(u)/u \qquad \text{(II.1.9)}$$

The Fourier-Bessel inverse of this is

$$(-1)^{(n-m)/2} R_n^m(r) = \int_0^\infty J_{n+1}(ru) J_m(u)\,\mathrm{d}u \qquad r < 1$$
$$= 0 \qquad r > 1 \qquad \text{(II.1.10)}$$

Recurrence relations can be obtained from the recurrence relations for Legendre and Jacobi polynomials [3, p782]. These include, from Chako [2]

$$2(n+1) r R_n^m(r) = (n+m+2) R_{n+1}^{m+1}(r) + (n-m) R_{n-1}^{m+1}(r) \qquad \text{(II.1.11)}$$

and from Myrick [5] (with correction in the case of Equation II.1.12)

$$(2n+4)(n+m)(n-m) R_{n-2}^m(r) + [2n(n-m+2)^2 + (2n+4)(n+m)^2$$
$$-2nr^2(2n+4)(2n+2)] R_n^m(r) + 2n(n-m+2)(n+m+2) R_{n+2}^m(r) = 0$$
$$\text{(II.1.12)}$$

and

$$(n-m+2)(n+m) R_n^{m-2}(r) - [2n^2 + 2m^2 - 2m(2m-2)/r^2] R_n^m(r)$$
$$+(n+m+2)(n-m) R_n^{m+2}(r) = 0 \qquad \text{(II.1.13)}$$

Chako also gives the relation (corrected)

$$\frac{\mathrm{d}}{\mathrm{dr}} R_n^0(r) = 2n R_{n-1}^1(r) + 2(n-2) R_{n-3}^1(r) + \cdots + 4 R_1^1(r) \qquad \text{(II.1.14)}$$

The coefficients a_n for *uniform* illumination are particularly elementary for all orders of Hankel transform. They are given by

$$a_{nm} = 2(n+1) \int_0^1 f(r) R_n^m(r) r\,\mathrm{d}r$$

which for uniform illumination, $f(r) = 1$, become simply

$$a_{nm} = (-1)^{(n-m)/2} \frac{2m(n+1)}{n(n+2)} \qquad \text{(II.1.15)}$$

For zero order Hankel transforms of polynomial functions $f(r)$ the following integral is useful in obtaining the coefficients α_{2s+1} in Equation 5.3.14

$$\int_0^1 r^{2(n-1)} R_{2s}^0(r)\,\mathrm{d}r = \frac{[\Gamma(n)]^2}{2\Gamma(n+s)\Gamma(n-s+1)} \qquad n > 0 \qquad \text{(II.1.16)}$$

[4, p278 corrected]

Finally, in the light of the discussion in Chapter 7, it is interesting to observe that both Chako and Myrick (quoting Zernike) consider as natural the extension of the circle polynomial theory to solutions of Laplace's equation in higher dimensional polar coordinates.

II.2 BESSEL FUNCTIONS, ASSOCIATED BESSEL FUNCTIONS AND SERIES

We list below basic series of Bessel and associated functions and fundamental relations which have been found to be of value in deriving Hankel and Lommel transforms in the form of infinite series by the method of circle polynomials. Page references are those of Watson [6]

$$J_n(z) = \sum_{m=0}^{\infty} \frac{(-1)^m (z/2)^{n+2m}}{m!(n+m)!} \tag{II.2.1}$$

$$J_{n-1}(z) + J_{n+1}(z) = \frac{2n}{z} J_n(z) \tag{II.2.2}$$

$$J_{n-1}(z) - J_{n+1}(z) = 2J_n'(z) \tag{II.2.3}$$

$$J_n(z) = \frac{1}{2\pi} \int_0^{2\pi} \cos(n\theta - z\sin\theta)\, \mathrm{d}\theta \tag{II.2.4}$$

$$J_n(y+z) = \sum_{m=-\infty}^{\infty} J_m(y) J_{n-m}(z) \tag{II.2.5}$$

$$\begin{aligned}
\cos(z\sin\theta) &= J_0(z) + 2\sum_{n=1}^{\infty} J_{2n}(z)\cos 2n\theta \\
\sin(z\sin\theta) &= 2\sum_{n=1}^{\infty} J_{2n+1}(z)\sin(2n+1)\theta \\
\cos(z\cos\theta) &= J_0(z) + 2\sum_{n=1}^{\infty} (-1)^n J_{2n}(z)\cos 2n\theta \\
\sin(z\cos\theta) &= 2\sum_{n=1}^{\infty} (-1)^n J_{2n+1}(z)\cos(2n+1)\theta
\end{aligned} \tag{II.2.6}$$

$$\begin{aligned}
\frac{\sin z}{z} &= \frac{2}{z^2}[2^2 J_2(z) - 4^2 J_4(z) + 6^2 J_6(z) \ldots] \\
\cos z &= \frac{2}{z}[1^2 J_1(z) - 3^2 J_3(z) + 5^2 J_5(z) \ldots]
\end{aligned} \tag{II.2.7}$$

Watson p34

$$\begin{aligned} 1 &= J_0(z) + 2\sum_{n=1}^{\infty} J_{2n}(z) \\ z^m &= 2^m \sum_{n=0}^{\infty} \frac{(m+2n)(m+n-1)!}{n!} J_{m+2n}(z) \end{aligned} \tag{II.2.8}$$

Watson p142

$$J_\nu(\lambda z) = \lambda^\nu \sum_{m=0}^{\infty} \frac{(-1)^m(\lambda^2-1)^m(z/2)^m}{m!} J_{\nu+m}(z) \tag{II.2.9}$$

The Anger-Weber function, Watson p308

$$E_\nu(z) = -\Omega_\nu(z) = \frac{1}{\pi}\int_0^\pi \sin(\nu\theta - z\sin\theta)\,\mathrm{d}\theta \tag{II.2.10}$$

$$\begin{aligned} \Omega_n(z) &= \frac{2}{\pi}\sum_{s=0}^{\infty} \frac{2(2s+1)}{(2s+1)^2-n^2} J_{2s+1}(z) \qquad n \text{ even} \\ &= \frac{2}{\pi}\left[-\frac{1}{n}J_0(z) + \sum_{s=1}^{\infty}\frac{2n}{4s^2-n^2}J_{2s}(z)\right] \qquad n \text{ odd} \end{aligned} \tag{II.2.11}$$

$$\begin{aligned} \Omega_0(z) + 2\Omega_2(z) + 2\Omega_4(z) + \cdots &= 0 \\ \Omega_0(z) - 2\Omega_2(z) + 2\Omega_4(z) - \cdots &= \sin z \\ 2[\Omega_1(z) - \Omega_3(z) + \Omega_5(z) - \cdots] &= -\cos z \end{aligned} \tag{II.2.12}$$

Watson p32

$$J_n^2(z) = \sum_{m=0}^{\infty} \frac{(-1)^m(2n+2m)!(z/2)^{2n+2m}}{m!(2n+m)![(n+m)!]^2} \tag{II.2.13}$$

Watson p634

$$\frac{1}{2} + \sum_{m=1}^{\infty}(-1)^m J_0(mx) = 0 \qquad x > 0 \tag{II.2.14}$$

$$\begin{aligned} \left(\frac{az}{2}\right)^{\mu-\nu} J_\nu(az) &= \frac{1}{\Gamma(\nu+1)}\sum_{t=0}^{\infty}\frac{(\mu+2t)\Gamma(\mu+t)}{t!} \\ &\quad \times {}_2F_1\{-t;\mu+t;\nu+1;a^2\}J_{\mu+2t}(z) \end{aligned}$$

(Erdelyi: "Higher Transcendental Functions" **Vol 2** 19 p64)

This relation can also be derived from a result by MacRobert [7] and agrees with that obtained in Chapter 5 Equation 5.7.9 Hence for $\nu = n$ and $\mu - \nu = 1$

$$\frac{1}{2} J_n(z) = \frac{1}{z} \sum_{t=0}^{\infty} (-1)^t (n + 2t + 1) J_{n+2t+1}(z)$$

and thus

$$J_0(z) = \frac{2}{z} \sum_{t=0}^{\infty} (-1)^t (2t + 1) J_{2t+1}(z)$$

$$J_1(z) = \frac{2}{z} \sum_{t=0}^{\infty} (-1)^t (2t + 2) J_{2t+2}(z) \qquad \text{etc.}$$

The modified Anger-Weber function [8]
From II.2.10 we define the modified Anger-Weber function

$$\Phi_\nu(z) = \frac{1 - \cos \nu\pi}{\nu\pi} + \Omega_\nu(z) \tag{II.2.15}$$

then for integer values

$$\Phi_{2n}(z) = \Omega_{2n}(z)$$

but

$$\Phi_{2n+1}(z) = \Omega_{2n+1}(z) + \frac{2}{(2n+1)\pi}$$

Most relevant properties can then be obtained from those of $E_\nu(z)$ and in particular the two expansions [9]

$$\begin{aligned} \Phi_{2n}(z) &= \frac{2}{\pi} \sum_{t=0}^{\infty} \frac{2(2t+1)}{(2t+1)^2 - (2n)^2} J_{2t+1}(z) \\ \Phi_{2n+1}(z) &= \frac{2}{(2n+1)\pi} [1 - J_0(z)] + \frac{1}{\pi} \sum_{t=0}^{\infty} \frac{2(2n+1)}{(2t)^2 - (2n+1)^2} J_{2t}(z) \end{aligned}$$

Moss [10]

$$\begin{aligned} \Omega_n(z) &= -\frac{1}{\pi} \int_{-\infty}^{\infty} \frac{J_n(u)\, du}{u - z} \\ &= -\frac{1}{\pi} \int_{-\infty}^{\infty} \frac{J_n(z+u)\, du}{u} \\ &= \sum_{s=-\infty}^{\infty} \Omega_{n-s}(0) J_s(z) \end{aligned} \tag{II.2.16}$$

$$\Omega_0(z) = \frac{4}{\pi} \left[J_1(z) + \frac{1}{3} J_3(z) + \frac{1}{5} J_5(z) + \cdots \right] \tag{II.2.17}$$

$$\Phi_{2n+1}(z) = (-1)^n \sum_{s=1}^{\infty} \frac{(-1)^s (z/2)^{2s}}{\Gamma(s-n-1/2)\Gamma(s+n+3/2)} \tag{II.2.18}$$

From

$$J_n(Nxy) = \sum_{s=0}^{\infty} c_s R_{n+2s}^n(y)$$

$$\int_0^1 J_n(Nxy) R_{n+2t}^n(y) y \, \mathrm{d}y = \frac{c_t}{2(n+2t+1)}$$

hence

$$J_n(Nxy) = \sum_{s=0}^{\infty} \frac{2(n+2s+1)}{Nx} (-1)^s J_{n+2s+1}(Nx) R_{n+2s}^n(y) \tag{II.2.19}$$

(cf II.2.9)
Similarly

$$\begin{aligned} &(-1)^t \int_0^1 R_{n+2s}^n(y) J_{n+2t+1}(Ny) \, \mathrm{d}y \\ &= (-1)^s \int_0^1 R_{n+2t}^n(y) J_{n+2s+1}(Ny) \, \mathrm{d}y \end{aligned} \tag{II.2.20}$$

$$\begin{aligned} &\int_0^{2\pi} \exp(i[-n\theta + \alpha r \cos\theta + \beta r \sin\theta]) \, \mathrm{d}\theta \\ &= 2\pi \exp[in \tan^{-1}(\alpha/\beta)] J_n(\sqrt{\alpha^2+\beta^2}\, r) \end{aligned} \tag{II.2.21}$$

$$\int J_1(ur) r \, \mathrm{d}r = \sum_{s \text{ odd}} \frac{2(s+1)}{s(s+2)} \frac{J_{s+1}(u)}{u} \tag{II.2.22}$$

$$\int J_2(ur) r \, \mathrm{d}r = \sum_{s \text{ even}} \frac{4(s+1)}{s(s+2)} \frac{J_{s+1}(u)}{u}$$

$$\int J_3(ur) r \, \mathrm{d}r = \frac{8}{5} \frac{J_4(u)}{u} + \frac{36}{35} \frac{J_6(u)}{u} + \frac{16}{21} \frac{J_8(u)}{u} + \frac{60}{99} \frac{J_{10}(u)}{u} \ldots$$

$$\int J_4(ur) r \, \mathrm{d}r = \frac{10}{6} \frac{J_5(u)}{u} + \frac{14}{12} \frac{J_7(u)}{u} + \frac{18}{20} \frac{J_9(u)}{u} + \frac{22}{30} \frac{J_{11}(u)}{u} \ldots$$

Watson p416

$$\frac{\exp(ikR)}{R} = \int_0^{\infty} \frac{u \, \mathrm{d}u}{\sqrt{u^2-k^2}} J_0(ur) \exp[\mp \sqrt{u^2-k^2}\, z] \tag{II.2.23}$$

upper sign for $+$ ve z

lower sign for $-$ ve z

$$R^2 = z^2 + r^2$$

REFERENCES

1. Zernike F (1934) *Physica* **I** p687
2. Chako N (1959) Characteristic curves in image space *The McGill Symposium on Microwave Optics* ASTIA No AD211499 p67
3. Abramowitz M & Stegun IA (eds) (1965) *Handbook of Mathematical Tables* Dover p779, equation (22.5.42)
4. Erdelyi A (ed) (1954) *Tables of Integral Transforms* **II** McGraw Hill p47
5. Myrick DR (1966) A generalization of the radial polynomials of F Zernike (and references therein) SIAM *Jour App Maths* **14** (2) p476
6. Watson GN (1941) *The Theory of Bessel Functions* Cambridge University Press
7. MacRobert TM (1961) Expression of an E-function as a finite series of E-functions *Math Rev* **22** (7A) p967 and (1966) *Math Ann* **140** p414
8. Bernard GD & Ishimaru A (1962) *Tables of Anger and Lommel-Weber Functions* AFCRL Report no 53 University of Washington Press
9. Brauer P & Brauer E (1941) Über unvollständiger Anger-Webersche Funktionen, Z *Agnew Math Mech* **21** (3)
10. Moss SM (1964) Lommel transforms in diffraction theory *Trans IEEE* **vol AP-12** p777

III
APPENDIX

III.1 CURVES AND THEIR FORMULAE*

1. Astroid - see under hypocycloid

2. Cardioid

   (a) $(x^2 + y^2 - 2ax)^2 = 4a^2(x^2 + y^2)$

   (b) $r = 2a(1 + \cos\theta)$

   (c) $x = 2a\cos t(1 + \cos t)$
   $y = 2a\sin t(1 + \cos t) \quad -\pi \leq t \leq \pi$

3. Catenary

   (a) $y = a\cosh(x/a)$

4. Cayley's sextic

   (a) $4(x^2 + y^2 - ax)^3 = 27a^2(x^2 + y^2)^2$

   (b) $r = a\cos^3(\theta/3)$

5. Cissoid (Diocles)

   (a) $y^2(a - x) = x^3$

   (b) $r = a\sin\theta\tan\theta$

   (c) $x = a\sin^2 t \quad y = a\tan t\sin^2 t \quad -\pi < t < \pi$

6. Circle

   (a) $(x - h)^2 + (y - k)^2 = a^2$

   (b) $r = a \quad r = a\cos\theta,\ 0 < \theta < \pi/2$

   (c) $x = a\cos t \quad t = a\sin t \quad -\pi \leq t \leq \pi$

7. Cubic (Tschirnhausen)

   (a) $27ay^2 = x^2(x + 9a)$

   (b) $r\cos^3(\theta/3) = a$

   (c) $x = 3a(t^2 - 3) \quad y = at(t^2 - 3) \quad -\infty < t < \infty$

*The formulae and derived curves of Appendix III Sections 1 to 5 are taken from Lawrence [1] and reproduced by kind permission of the publishers.

8. Semi-cubical parabola

   (a) $27ay^2 = 4x^3$
   (b) $4r = 27a \sin^2\theta \sec^3\theta$
   (c) $x = 3at^2 \quad y = 2at^3 \quad -\infty < t < \infty$

9. Cycloid

   (c) $x = at - h\sin t \quad y = a - h\cos t \quad -\infty < t < \infty$

10. Ellipse

   (a) $x^2/a^2 + y^2/b^2 = 1$
   (b) $r^2 = b^2/(1 - e^2\cos^2\theta); \quad ar = b^2/(1 + e\cos\theta)$
   (c) $x = a\cos t \quad y = b\sin t \quad -\pi \le t \le \pi$
   $b^2 = a^2(1 - e^2)$
   $e \equiv$ eccentricity

11. Epicycloid

   (c) $x = m\cos t - b\cos(mt/b)$
   $y = m\sin t - b\sin(mt/b) \quad -\pi \le t \le \pi$
   $m = a + b$

   for $a = b$ see cardioid; $a = 2b$ see nephroid

12. Hyperbola

   (a) $x^2/a^2 - y^2/b^2 = 1 \quad b^2 = a^2(e^2 - 1)$
   $xy = c^2$
   (b) $r^2 = b^2/(e^2\cos^2\theta - 1)$ (equilateral if $a = b$)
   $ar = b^2/(1 + e\cos\theta)$
   $2r^2\sin^2\theta = a^2e^2$
   (c) $x = a\sec t \quad y = b\tan t \quad -\pi \le t \le \pi$
   $x = ct \quad y = c/t$
   $x = a\cosh t \quad y = b\sinh t$
   $e \equiv$ eccentricity

13. Hypocycloid

   (c) $x = n\cos t + b\cos(nt/b)$
   $y = n\sin t - b\sin(nt/b) \quad -\pi \le t \le \pi$
   $n = a - b$
   $b = 0$ see circle

14. Astroid ($a = 4b$)

(a) $x^{2/3} + y^{2/3} = a^{2/3}$

(c) $x = 4a\cos^3 t \quad y = 4a\sin^3 t$

15. Lemniscate (Bernoulli)

(a) $(x^2 + y^2)^2 = a^2(x^2 - y^2)$

(b) $r^2 = a^2 \cos 2\theta$

(c) $x = a\cos t/(1 + \sin^2 t)$
$y = a\sin t\cos t/(1 + \sin^2 t) \quad -\pi \le t \le \pi$

16. Limaçon (Pascal) [2]

(a) $(x^2 + y^2 - 2ax)^2 = b^2(x^2 + y^2)$

(b) $b^2r^2 = (r^2 - 2ar\cos\theta)^2$

(c) $x = \cos t(2a\cos t + b)$
$y = \sin t(2a\cos t + b) \quad -\pi \le t \le \pi$

17. Nephroid (Huygens, Figure III.1)

(a) $(x^2 + y^2 - 4a^2)^3 = 180a^4y^2$

(b) $(r/2a)^{2/3} = (\sin\theta/2)^{2/3} + (\cos\theta/2)^{2/3}$

(c) $x = a(3\cos t - \cos 3t)$
$y = a(3\sin t - \sin 3t) \quad -\pi \le t \le \pi$

18. Oval (Descartes)

(a) $[b(x^2 + y^2 + 1) - 2cx]^2 = 2c(x^2 + y^2 + 1) - 4bx - 1$

19. Oval (Cassini)

(a) $(x^2 + y^2 + a^2)^2 = b^2 + 4a^2x^2$

(b) $r^4 - 2a^2r^2\cos 2\theta = b^4 - a^4$

20. Parabola

(a) $y^2 = 4ax$

(b) $r = 2a/(1 - \cos\theta)$

(c) $x = at^2 \quad y = 2at \quad -\infty < t < \infty$

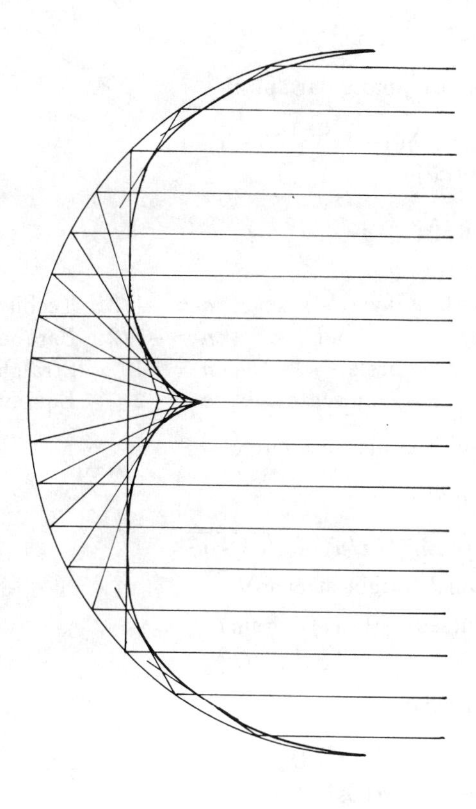

Figure III.1: the nephroid

21. Spiral (Archimedes)

(b) $r^m = a^m \theta$

$m = 1$	Archimedean
$m = 2$	Fermat
$m = -1$	Hyperbolic
$m = -2$	Lituus

22. Logarithmic or equiangular spiral

(a) $x^2 + y^2 = \exp[2a \tan^{-1}(y/x)]$

(b) $r = \exp(a\theta)$

23. Sinusoidal spirals

(b) $r^n = a^n \cos n\theta$

$n = 1/3$	Cayley's sextic	$n = -1/3$	Tschirnhausen's cubic
$n = 1/2$	Cardioid	$n = -1/2$	Parabola
$n = 1$	Circle	$n = -1$	Straight line
$n = 2$	Lemniscate	$n = -2$	Equilateral hyperbola

24. Tschirnhausen's cubic - see cubic

25. Tractrix (Huygens)

(a) $x = \pm a \cosh^{-1}(a/y) - \sqrt{a^2 - y^2}$

(b) $r = a \tan\theta$ (origin at cusp)

(c) $x = a \ln(\sec t + \tan t) - a \sin t$
$y = a \cos t \quad -\pi/2 \le t \le \pi/2$

26. Trisectrix (Maclaurin)

(a) $y^2(a - x) = x^2(x + 3a)$

(b) $r = a \sec\theta - 4a \cos\theta$

(c) $x = a(t^2 - 3)/(t^2 + 1) \quad y = at(t^2 - 3)/(t^2 + 1) \quad -\infty < t < \infty$

27. Kappa curve

(a) $(x^2 + y^2)y^2 = a^2 x^2$

(b) $r = a \cot\theta$

(c) $x = a \cos t \cot t \quad y = a \cos t \quad 0 < t < 2\pi$

28. Right strophoid

(a) $y^2(a - x) = x^2(a + x)$

(b) $r = a(\sec\theta - 2\cos\theta)$

(c) $x = a(1 - 2\cos^2 t)$
$y = a \tan t(1 - 2\cos^2 t) \quad -\pi \le t \le \pi$

III.2 CURVES AND THEIR INVERSES

Curve	*← Centre of Inversion →*		*Inverse*
straight line	point not on line		circle
circle	point on circumference		straight line
	point not on circumference		circle
parabola	focus	cusp	cardioid
	vertex	cusp	cissoid
central conic	focus	pole	limaçon
	centre	centre	oval, figure of eight
rectangular	centre	centre	lemniscate
hyperbola	vertex	node	right strophoid
right strophoid	pole		same strophoid
sinusoidal spiral	pole		sinusoidal spiral
logarithmic spiral	pole		logarithmic spiral
Archimedean spiral	pole		Archimedean spiral

III.3 EVOLUTES AND INVOLUTES

Involute	*Evolute*
parabola	semi-cubical parabola
limaçon of Pascal	catacaustic of the circle
cardioid, parameter a	cardioid, parameter $a/3$
epicycloid	epicycloid
nephroid, parameter a	nephroid, parameter $a/2$
hypocycloid	hypocycloid
logarithmic spiral	equal logarithmic spiral
cycloid	equal cycloid
tractrix	catenary

III.4 CATACAUSTICS

Curve	*Source*	*Catacaustic*
circle	on circumference	cardioid
	not on circumference	limaçon of Pascal
	parallel rays	nephroid
parabola	axial parallel rays	focal point
	rays not parallel to axis	Tschirnhausen's cubic [3]
cardioid	cusp	nephroid
logarthmic spiral	pole	equal spiral
cycloidal arch	rays perpendicular to axis	two equal cycloidal arches
$y = \ln x$	rays parallel to axis	catenary
Tschirnhausen's cubic	focus	semi-cubical parabola
cissoid of Diocles	focus	cardioid
quadrifolium	centre	astroid

III.5 PEDAL CURVES

Curve	*Pole*	*Pedal*
line	any point	point
circle	any point	limaçon of Pascal
	on circumference	cardioid
parabola	on directrix	strophoid
	foot of directrix	right strophoid
	reflection of focus in directrix	trisectrix of Maclaurin
	vertex	cissoid of Diocles
	focus	line
central conic	focus	circle
	centre	lemniscate of Bernoulli
Tschirnhausen's cubic	focus (of pedal)	parabola
cissoid	focus	cardioid
cardioid	cusp	Cayley's sextic
logarithmic spiral	pole	an equal spiral
involute of a circle	centre	Archimedes' spiral

III.6 APPLICATIONS OF ABEL'S INTEGRAL

The integral equation, occurring in nearly all of the ray-tracing solutions for non-uniform media can be put in the general form

$$I = \int_{x_1}^{x_2} \frac{\kappa f(x)\,\mathrm{d}x}{\sqrt{F^2(x) - \kappa^2}} \tag{III.6.1}$$

Here I is a function of x with given values at the limits x_1 and x_2, and $f(x)$ is known. It is required to derive $F(x)$ subject to the boundary conditions of I. κ is a constant of an individual ray whose value can be determined from the value of I at x_1. The solution for a pencil of rays to have a given specified behaviour thus requires $F(x)$ to be independent of κ. For the source conditions used in this volume we have $f(x_1) = 1$ and x_2 will be a turning value, that is a value when I changes from a function increasing with x to one decreasing. At this value $F(x_2) = \kappa$, and the denominator becomes zero.

With these conditions the formal method is to make the substitutions

$$w = \int f(x)\,\mathrm{d}x \qquad v = \kappa \qquad u = F(x)$$

which converts the integral into Abel's form

$$I = \int_1^v \frac{v \frac{\mathrm{d}w}{\mathrm{d}u}\,\mathrm{d}u}{\sqrt{u^2 - v^2}} \tag{III.6.2}$$

Multiplication by $\mathrm{d}v/\sqrt{v^2 - y^2}$ for an as yet unspecified function y, and integration in the range $[y, 1]$ gives

$$\int_y^1 \frac{v\,\mathrm{d}v}{\sqrt{v^2 - y^2}} \int_1^v \frac{\frac{\mathrm{d}w}{\mathrm{d}u}\,\mathrm{d}u}{\sqrt{u^2 - v^2}} = \int_y^1 \frac{I\,\mathrm{d}v}{\sqrt{v^2 - y^2}}$$

Reversing the order of integration on the left-hand side and using the result

$$\int_y^1 \frac{v}{\sqrt{u^2 - v^2}} \frac{\mathrm{d}v}{\sqrt{v^2 - y^2}} = \frac{\pi}{2} \tag{III.6.3}$$

gives

$$\int_y^1 \frac{\mathrm{d}w}{\mathrm{d}u}\,\mathrm{d}u \int_y^u \frac{v\,\mathrm{d}v}{\sqrt{u^2 - v^2}\sqrt{v^2 - y^2}} = \frac{\pi}{2}\int_y^1 \mathrm{d}w = \frac{\pi}{2}\int_y^1 f(x)\,\mathrm{d}x$$

and hence

$$\frac{\pi}{2}\int f(x)\,\mathrm{d}x = \int_y^1 \frac{I\,\mathrm{d}v}{\sqrt{v^2 - y^2}} \tag{III.6.4}$$

For those situations where I is independent of v (or κ), such as the conditions governing specified refocusing of rays, the right-hand side is integrable, giving

$$\frac{\pi}{2}\int^x f(x)\,\mathrm{d}x = I \log\left[\frac{1 + \sqrt{1 - y^2}}{y}\right] \tag{III.6.5}$$

Finally putting $y = F(x)$ gives the required solutions. The limits are such that the upper limit in the integral on the left-hand side gives zero value.

1. Cylindrical systems with repeated focusing (Section 4.3)

$$f(x) = 1 \quad F(x) = \eta(x) \quad I = a \tag{III.6.6}$$

$$\eta(x) = \operatorname{sech}\frac{\pi x}{2a} \quad \text{(Equation 4.3.4)}$$

2. Spherical systems (Section 4.7)

$$f(r) = \frac{1}{r} \quad F(r) = r\eta(r) \quad I = \frac{2\pi}{y}$$

$$\operatorname{sech}^{-1} r\eta(r) = -\frac{\nu}{2}\log r \tag{III.6.7}$$

or

$$r\eta(r) = \frac{2r^{\nu/2}}{r^{\nu}+1}$$

the generalized fish-eye solutions (Equations 4.7.11) $\nu = 2$ gives the original fish-eye.

3. Surface rays on the sphere (Section 4.11)

$$f(\beta) = \sec\beta \quad F(\beta) = \eta(\beta)\cos\beta \quad I = \frac{2\pi}{\nu}$$

$$\operatorname{sech}^{-1}[\eta(\beta)\cos\beta] = \nu\log\left(\sqrt{\frac{1+\sin\beta}{1-\sin\beta}}\right)$$

$$\cos\beta\eta(\beta) = \frac{2\cos^{\nu/2}\beta}{\sqrt{1+\sin\beta}+\sqrt{1-\sin\beta}} \quad \text{(Equation 4.11.9)} \tag{III.6.8}$$

Also, the general result (Equation 4.11.21) follows when $f(\beta)$ is replaced by

$$\frac{f'(\beta)}{f(\beta)}\sqrt{1-f^2(\beta)}$$

When I is not independent of v (or therefore κ) integration depends upon the integrability of the function as given in Equation III.6.4.

4. The Luneberg lens (Section 4.7)

$$f(r) = \frac{1}{r} \quad F(r) = r\eta(r) \quad I = \frac{\pi}{2} - \frac{\sin^{-1} v}{2}$$

The right-hand side of Equation III.6.4 now consists of two terms

$$\frac{\pi}{2}\log\left[\frac{1+\sqrt{1-y^2}}{y}\right] - \frac{1}{2}\int_y^1 \frac{\sin^{-1} v}{\sqrt{v^2-y^2}}\,dv \tag{III.6.9}$$

with the result

$$\log\left[\frac{1+\sqrt{1-y^2}}{y}\right]-\frac{1}{2}\log[1+\sqrt{1-y^2}]=-\log r \qquad \text{(III.6.10)}$$

or with $y = r\eta$; $\eta^2 = 2 - r^2$ (Equation 4.7.14)

5. Beam divider (point source)
   These are generalizations of the Luneburg lens with

$$I = p\pi/4 - \sin^{-1} v/2 + \pi/2$$

giving the result in (4) modified to

$$(2+p)\log\left[\frac{1+\sqrt{1-y^2}}{y}\right]-\log[1+\sqrt{1-y^2}]=-2\log r$$

or

$$[1+\sqrt{1-\eta^2r^2}]^{p+1}=\frac{(r\eta)^{p+2}}{r^2} \qquad \text{(Equation 4.7.13)} \qquad \text{(III.6.11)}$$

6. Beam divider (parallel beam)

$$I=\frac{q\pi}{4}-\sin^{-1} v$$

and

$$[1+\sqrt{1-\eta^2r^2}]^{q-2}=\frac{(r\eta)^{q}}{r^2} \qquad \text{(Equation 4.7.19)} \qquad \text{(III.6.12)}$$

From the form of Equation III.6.9 it can be seen that very few functions other than $\sin^{-1} v$ are integrable in the form needed to give the logarithmic function necessary for an algebraic solution. Hence for other focusing conditions, and in particular for two foci external to the spherical lens, the integral in Equation III.6.9, and its eventual exponential, require computational methods and have been tabulated.

REFERENCES

1. Lawrence JD (1972) *A Catalog of Special Plane Curves* New York: Dover Publications
2. Salmon G (1960) *Higher Plane Curves* New York: Chelsea Publishing
3. Scarborough JB (1964) The caustic curve of an off-axis parabola *Applied Optics* **Vol 3** p1445

IV
APPENDIX: QUATERNIONS AND BI-QUATERNIONS

"It's a long road that has no quaternions" JR Blakey*.

IV.1 QUATERNIONS

Quaternions have a long and interesting history, the most intriguing aspect of which being the manner whereby they have begun to make an immense impact in the field of modern theoretical physics [1] after being forgotten for nearly a hundred years since their discovery as a pure algebraic-geometrical concept. This history can be traced in many ways, by concentrating on the algebra leading back to Hamilton and Grassmann and even to Euler, the geometry of Clifford, Servois and Cayley and the physics in more recent times to Pauli and Dirac. Such a history and the recent advent of a spate of applications should be the concern of a separate volume on the subject. The basic algebra that is all that is required at this time for the material in this book can be discussed with comparative brevity. This unfortunately hides the enormous beauty and potency of the subject illustrated as it has been here by comparatively simple examples.

One illustrative example arises from the definition of a circularly polarized wave as shown in Section 6.8. Such a wave, as is well-known, comprises two components, orthogonally polarized in space and in time quadrature. Thus the i defining space orthogonality and the i defining time (or phase) quadrature have separate physical meanings. A more general form of space orthogonality - applicable to quite general directions (θ, ϕ) is required.

We require to label each of the three space coordinate axes with a complex number in the manner that $i = \sqrt{-1}$ labels the y-axis in the Argand diagram. In common with the concept that this complex i is also a rotation of $\pi/2$ about the z axis, the other complex labels will then give rotations about the other axes as well. We label a rotation of $\pi/2$ about the x axis as the complex number α with $\alpha^2 = -1$, about the y axis as β and about the z axis as γ. Thus the first elements of the algebra are $\alpha^2 = \beta^2 = \gamma^2 = -1$.

The next step is algebraic and is to find the equivalent in four dimensions of the product of scalar invariants of the two dimensional field. Given two complex numbers $z_1 = (a + ib)$; $z_2 = (c + id)$ one can find by the product of

*The author regrets that he was not responsible for this appalling pun.

the norms

$$|z_1|^2|z_2|^2 = |z_1 z_2|^2 \tag{IV.1.1}$$

that $(a^2+b^2)(c^2+d^2) = A^2+B^2$ where $A = ac - bd$ $B = ad + bc$. This result can be termed the "two square" problem.

Performing this for a "four square" problem (and proving that there is no solution in three dimensions) was Hamilton's great discovery. It takes the form [2].

$$(a^2+b^2+c^2+d^2)(t^2+x^2+y^2+z^2) = A^2+B^2+C^2+D^2 \tag{IV.1.2}$$

with

$$\begin{aligned} A &= at - bx - cy - dz \\ B &= ax + bt + cz - dy \\ C &= ay - bz + ct + dz \\ D &= az + by - cx + dt \end{aligned} \tag{IV.1.3}$$

This is now called a skew (or symplectic) field or a division ring.

To obtain the same product $|\mathcal{Q}_1|^2|\mathcal{Q}_2|^2 = |\mathcal{Q}_1\mathcal{Q}_2|^2$ requires hypercomplex numbers as in Equation IV.1.1 with

$$\begin{aligned} \mathcal{Q}_1 &= (a + \alpha b + \beta c + \gamma d) \\ \mathcal{Q}_2 &= (t + \alpha x + \beta y + \gamma z) \end{aligned}$$

and thus the additional non-commutative property

$$\alpha\beta = \gamma \qquad \beta\gamma = \alpha \qquad \gamma\alpha = \beta$$

and

$$\alpha\beta = -\beta\alpha \qquad \text{etc (cf Section 6.7)}$$

The numbers $\mathcal{Q}$ are the quaternions. Hamilton originally labelled the axes i j and k and this led on to vector analysis and the non-commutative cross product. The absorption of Gibbs-Heaviside [3] vector theory into the science of the time led to the neglect of quaternion theory. The essential three dimensionality of the former had greater appeal it seems than the four dimensionality of the latter. This is in the process of being corrected.

A unique inverse is defined by

$$\mathcal{Q}^{-1}\mathcal{Q} = \mathcal{Q}\mathcal{Q}^{-1}$$

whence we have $\alpha\alpha^{-1} = \alpha^{-1}\alpha = 1$ or $\alpha = -\alpha^{-1}$. In the quotient α/β we have the alternative of forming the left hand quotient $\beta^{-1}\alpha$ or the right hand quotient $\alpha\beta^{-1}$ and these are different (by -1). This ambiguity between $\alpha/\beta = \beta/\alpha$ and $\alpha/\beta = -\beta/\alpha$ is closely associated with the concept of left and right hand regularity given in Section 7.4.7. For consistency with later analysis (Equation IV.1.20) we will call $\alpha/\beta = \beta/\alpha$ left regular division and $\alpha/\beta = -\beta/\alpha$ right regular division.

A quaternion $\mathcal{Q} = (a + \alpha b + \beta c + \gamma d)$ has a conjugate, the quaternion

$$\overline{\mathcal{Q}} = (a - \alpha b - \beta c - \gamma d) \tag{IV.1.4}$$

We have the square

$$\begin{aligned} \mathcal{Q}^2 = &= (a + \alpha b + \beta c + \gamma d)^2 \\ &= \{a^2 - b^2 - c^2 - d^2, 2\alpha ab, 2\beta ac, 2\gamma ad\} \end{aligned}$$

and the positive definite norm

$$\begin{aligned} \mathcal{Q}\overline{\mathcal{Q}} &= (a + \alpha b + \beta c + \gamma d)(a - \alpha b - \beta c - \gamma d) \\ &= \{a^2 + b^2 + c^2 + d^2, 0, 0, 0\} \end{aligned}$$

From the rule of multiplication we find

$$\begin{aligned} \mathcal{Q}_1\mathcal{Q}_2 &= (a + \alpha b + \beta c + \gamma d)(t + \alpha x + \beta y + \gamma z) \\ &= A + \alpha B + \beta C + \gamma D \end{aligned}$$

with A B C and D defined above.

Then *if* b c d and x y z were *considered* as the component of three-space vectors $\mathbf{V}$ and $\mathbf{W}$

$$\mathcal{Q}_1 = (a, \mathbf{V}) \quad \mathcal{Q}_2 = (t, \mathbf{W})$$

and

$$\mathcal{Q}_1\mathcal{Q}_2 = (at - \mathbf{V} \circ \mathbf{W} \cdot a\mathbf{W} + t\mathbf{V} + \mathbf{V} \otimes \mathbf{W}) \tag{IV.1.5}$$

where $\circ$ and $\otimes$ refer to the inner product and outer product of $\mathbf{V}$ and $\mathbf{W}$ in the "ordinary" vector sense. This last result can be seen directly by expanding the terms and comparing with A B C and D. It follows that if

$$\mathcal{Q} = (a, \mathbf{V}) \quad \overline{\mathcal{Q}} = (a, -\mathbf{V})$$

The norm is $\mathcal{Q}\overline{\mathcal{Q}} = a^2 - \mathbf{V} \circ \mathbf{V}$, the norm of $(\mathcal{Q} - a)$ is

$$(\mathcal{Q} - a)(\overline{\mathcal{Q} - a}) = +\mathbf{V} \circ \mathbf{V} \tag{IV.1.6}$$

and the square of $(\mathcal{Q} - a)$ is

$$(\mathcal{Q} - a)(\mathcal{Q} - a) = -\mathbf{V} \circ \mathbf{V}$$

POLAR QUATERNIONS

Using the ordered notation we have

$$\begin{aligned} (a, b, c, d)^2 &= (a^2 - b^2 - c^2 - d^2, 2ab, 2ac, 2ad) \\ (0, x, y, z)^2 &= -(x^2 + y^2 + z^2) \\ (0, x, y, z)^3 &= -(0, x, y, z)(x^2 + y^2 + z^2) \\ (0, x, y, z)^4 &= (x^2 + y^2 + z^2)^2 \quad \text{etc} \end{aligned}$$

Hence

$$\begin{aligned}\exp\{0,x,y,z\} &= 1+(0,x,y,z)-\frac{x^2+y^2+z^2}{2!}-\frac{(0,x,y,z)(x^2+y^2+z^2)}{3!}\\ &= \cos\sqrt{x^2+y^2+z^2}+\frac{(0,x,y,z)}{\sqrt{x^2+y^2+z^2}}\sin\sqrt{x^2+y^2+z^2}\end{aligned} \tag{IV.1.7}$$

Therefore

$$|\exp\{0,x,y,z\}|=1$$

Putting $x^2+y^2+z^2=\rho^2$

$$\begin{aligned} x &= \rho\cos\phi\\ y &= \rho\sin\phi\cos\psi\\ z &= \rho\sin\phi\sin\psi \end{aligned}$$

polar coords w.r.t. the x–axis

then

$$\begin{aligned}&\exp[\rho(\alpha\cos\phi+\beta\sin\phi\cos\psi+\gamma\sin\phi\sin\psi)]\\ &\quad=\cos\rho+\sin\rho[\alpha\cos\phi+\beta\sin\phi\cos\psi+\gamma\sin\phi\sin\psi]\end{aligned} \tag{IV.1.8}$$

Hence

$$\alpha\cos\phi+\beta\sin\phi\cos\psi+\gamma\sin\phi\sin\psi$$

is the three dimensional analogue of $\sqrt{-1}$ as can be verified by the direct product.

In hyperspherical coordinates we can put

$$\begin{aligned} a &= \mu\cos\rho\\ b &= \mu\sin\rho\cos\phi\\ c &= \mu\sin\rho\sin\phi\cos\psi\\ d &= \mu\sin\rho\sin\phi\sin\psi \end{aligned}$$

then

$$\mu=\sqrt{a^2+b^2+c^2+d^2}$$

Any quaternion is then

$$\mathcal{Q}=\mu[\cos\rho+\sqrt{-1}\sin\rho] \tag{IV.1.9}$$

where $\sqrt{-1}$ is given above and a generalized De Moivre's theorem states

$$\begin{aligned}&[\mu\cos\rho+\mu\sin\rho(\alpha\cos\phi+\beta\sin\phi\cos\psi+\gamma\sin\phi\sin\psi)]^\nu\\ &\quad=\mu^\nu\cos(\nu\rho+2\nu n\pi)+\mu^\nu\sqrt{-1}\sin(\nu\rho+2\nu n\pi)\end{aligned}$$

The matrix representation of the complex number is

$$z = x + iy \quad \text{namely} \quad z = \begin{pmatrix} x & y \\ -y & x \end{pmatrix}$$

which can be confirmed by taking the product

$$z_1 z_2 = \begin{pmatrix} a & b \\ -b & a \end{pmatrix} \begin{pmatrix} c & d \\ -d & c \end{pmatrix}$$

We form

$$Q = \begin{pmatrix} t + ix & y + iz \\ -y + iz & t - ix \end{pmatrix} \tag{IV.1.10}$$

that is the complex matrix

$$Q = \begin{pmatrix} w & u \\ -\overline{u} & \overline{w} \end{pmatrix} \tag{IV.1.11}$$

with

$$w = t + ix \qquad u = y + iz$$

This matrix, it will be found, conforms in all respects with the product rules and the definitions of the product factors A B C and D required by a quaternion.

Thus we find

$$Q = t\begin{pmatrix} 1 & 0 \\ 0 & 1 \end{pmatrix} + x\begin{pmatrix} i & 0 \\ 0 & -i \end{pmatrix} + y\begin{pmatrix} 0 & 1 \\ -1 & 0 \end{pmatrix} + z\begin{pmatrix} 0 & i \\ i & 0 \end{pmatrix}$$

This is termed the "symplectic decomposition" of Q.

If Q also is $(t + \alpha x + \beta y + \gamma z)$ then

$$\alpha = \begin{pmatrix} i & 0 \\ 0 & -i \end{pmatrix} = i\sigma_z$$

$$\beta = \begin{pmatrix} 0 & 1 \\ -1 & 0 \end{pmatrix} = i\sigma_y$$

$$\gamma = \begin{pmatrix} 0 & i \\ i & 0 \end{pmatrix} = i\sigma_x \tag{IV.1.12}$$

where σ_x σ_y and σ_z are the Pauli spin matrices (cf Section 6.8). The reversal of the order of x y z is to be noted.

Putting each complex number into its 2 x 2 matrix form results in the 4 x 4 representation of the quaternion basis

$$E = \begin{pmatrix} 1 & 0 & 0 & 0 \\ 0 & 1 & 0 & 0 \\ 0 & 0 & 1 & 0 \\ 0 & 0 & 0 & 1 \end{pmatrix} \qquad I = \begin{pmatrix} 0 & 1 & 0 & 0 \\ -1 & 0 & 0 & 0 \\ 0 & 0 & 0 & -1 \\ 0 & 0 & 1 & 0 \end{pmatrix}$$

$$J = \begin{pmatrix} 0 & 0 & 1 & 0 \\ 0 & 0 & 0 & 1 \\ -1 & 0 & 0 & 0 \\ 0 & -1 & 0 & 0 \end{pmatrix} \qquad K = \begin{pmatrix} 0 & 0 & 0 & 1 \\ 0 & 0 & -1 & 0 \\ 0 & 1 & 0 & 0 \\ -1 & 0 & 0 & 0 \end{pmatrix} \tag{IV.1.13}$$

in which $E\ I\ J\ K$ has the identical algebra to $1\ \alpha\ \beta\ \gamma$. The matrix J in particular has significant applications in transformation theory (see Section IV.3). Complex forms of $E\ I\ J\ K$ are bi-quaternion and relate to the Clifford numbers.

The physical applications are as many and as varied. In some cases they require the even more general form the bi-quaternion $Q_1 + iQ_2$ (Argand i) as for example the transmission matrices of Chapter 4 and in Chapter 7.

A quaternion $Q = (a, \mathbf{V}) = (a, a_1, a_2, a_3)$ satisfies the relations IV.1.5

$$(Q - a)^2 = b^2 \qquad b^2 = -\mathbf{V} \circ \mathbf{V}$$

where since

$$\overline{Q} = (a - \mathbf{V}) \text{ and } N(Q) = Q \cdot \overline{Q} = a^2 + \mathbf{V} \circ \mathbf{V}$$
$$a = (q + \overline{q})/2 \text{ and } b^2 = -N(Q - a) \qquad \text{(IV.1.14)}$$

A *polynomial* function of a quaternion variable can then be expressed as

$$f(Q) = \left\{ \frac{f(a+b) + f(a-b)}{2}; (Q - a)\frac{f(a+b) - f(a-b)}{2b} \right\} \qquad \text{(IV.1.15)}$$

In the event that $b = 0$ this becomes

$$f(Q) = \{f(a); (Q - a)f'(a)\} \qquad \text{(IV.1.16)}$$

If we take as our example the basic coordinate quaternion $Q_c = (it, x, y, z)$ then

$$-b^2 = x^2 + y^2 + z^2 \qquad a = t$$

and

$$f(Q_c) = \left\{ \frac{1}{2}[f(t+r) + f(t-r)]; \frac{(x, y, z)}{2r}[f(t+r) - f(t-r)] \right\}$$

that is the transformation

$$\begin{aligned} t' &= \frac{1}{2}[f(t+r) + f(t-r)] \\ x' &= \frac{x}{2r}[f(t+r) - f(t-r)] \\ y' &= \frac{y}{2r}[f(t+r) - f(t-r)] \\ z' &= \frac{z}{2r}[f(t+r) - f(t-r)] \end{aligned} \qquad \text{(IV.1.17)}$$

This is virtually identical to the second transformation of Bateman [4] based on the invariance of Maxwell's equations (cf Section 7.6). The main difference is that, in the reference the functions on the right hand side are given with greater generality as the difference between two independent functions f_1 and f_2 as in Equation 7.6.4.

DIFFERENTIABILITY

Define a function of a quaternion variable, in analogy to the function of a complex variable $f(x+iy) = u + iv$ by

$$f(\mathcal{Q}) = W \text{ or } f(t,x,y,z) = (p,u,v,w) \tag{IV.1.18}$$

The differential can then be defined in the same way as

$$\begin{aligned} W' = \lim_{h\to 0}[\frac{1}{h}p(t+\delta t, x+\delta x, y+\delta y, z+\delta z), u(t+\delta t, x+\delta x, y+\delta y, z+\delta z), \\ v(t+\delta t, x+\delta x, y+\delta y, z+\delta z), w(t+\delta t, x+\delta x, y+\delta y, z+\delta z) \\ -p(t,x,y,z), u(t,x,y,z), v(t,x,y,z), w(t,x,y,z)] \end{aligned} \tag{IV.1.19}$$

For uniqueness W' has to have the same value in whatever manner the limit $h \to 0$ is taken. Therefore putting $h = \delta t$, $\alpha\delta x$, $\beta\delta y$ and $\gamma\delta z$ in turn and making the others zero, we approach the limit along each of the *four* coordinate axes independently and obtain

$$\begin{aligned} W' &= \frac{\partial p}{\partial t} + \alpha\frac{\partial u}{\partial t} + \beta\frac{\partial v}{\partial t} + \gamma\frac{\partial w}{\partial t} \\ &= \frac{1}{\alpha}\frac{\partial p}{\partial x} + \frac{\partial u}{\partial x} + \frac{\beta}{\alpha}\frac{\partial v}{\partial x} + \frac{\gamma}{\alpha}\frac{\partial w}{\partial x} \\ &= \frac{1}{\beta}\frac{\partial p}{\partial y} + \frac{\alpha}{\beta}\frac{\partial u}{\partial y} + \frac{\partial v}{\partial y} + \frac{\gamma}{\beta}\frac{\partial w}{\partial y} \\ &= \frac{1}{\gamma}\frac{\partial p}{\partial z} + \frac{\alpha}{\gamma}\frac{\partial u}{\partial z} + \frac{\beta}{\gamma}\frac{\partial v}{\partial z} + \frac{\partial w}{\partial z} \end{aligned} \tag{IV.1.20}$$

By the uniqueness of the inverse we have $\alpha = -1/\alpha$ etc. Then we have :-

$$\frac{\partial p}{\partial t} = \frac{\partial u}{\partial x} = \frac{\partial v}{\partial y} = \frac{\partial w}{\partial z} \tag{IV.1.21}$$

and

$$\frac{\partial p}{\partial x} = -\frac{\partial u}{\partial t} \quad \frac{\partial p}{\partial y} = -\frac{\partial v}{\partial t} \quad \frac{\partial p}{\partial z} = -\frac{\partial w}{\partial t} \tag{IV.1.22}$$

$$\frac{\partial u}{\partial y} = \pm\frac{\partial v}{\partial x} \quad \frac{\partial u}{\partial z} = \pm\frac{\partial w}{\partial x} \quad \frac{\partial v}{\partial z} = \pm\frac{\partial w}{\partial y} \tag{IV.1.23}$$

where the upper sign has been taken for the left regular division $\alpha/\beta = \beta/\alpha$ and the lower for the right regular division $\alpha/\beta = -\beta/\alpha$.

From these equations we find that u v and w each satisfy the wave equation in separated form.

$$\frac{\partial^2}{\partial x^2} + \frac{\partial^2}{\partial y^2} = 0 = \frac{\partial^2}{\partial t^2} - \frac{\partial^2}{\partial z^2} \tag{IV.1.24}$$

In addition the relations 21-23 give rise to three pair combinations of the kind

$$\frac{\partial u}{\partial x}\frac{\partial p}{\partial x} + \frac{\partial u}{\partial t}\frac{\partial p}{\partial t} = 0 = \frac{\partial v}{\partial y}\frac{\partial w}{\partial y} + \frac{\partial v}{\partial z}\frac{\partial w}{\partial z} \tag{IV.1.25}$$

showing the orthogonality of the respective coordinate surfaces and demonstrating the orthogonality of the transformation.

The left regularity condition in Equations 7.4.7 also provides a transformation equivalent to Bateman's first transformation. Replacing ϕ_0 ϕ_1 ϕ_2 ϕ_3 with p u v w we obtain

$$\frac{\partial p}{\partial t} - \frac{\partial u}{\partial x} - \frac{\partial v}{\partial y} - \frac{\partial w}{\partial z} = 0 \qquad \text{(IV.1.26)}$$

and

$$\frac{\partial p}{\partial x} = \frac{\partial u}{\partial t} \quad \frac{\partial p}{\partial y} = \frac{\partial v}{\partial t} \quad \frac{\partial p}{\partial z} = \frac{\partial w}{\partial t} \qquad \text{(IV.1.27)}$$

$$\frac{\partial w}{\partial y} = \frac{\partial v}{\partial z} \quad \frac{\partial w}{\partial x} = \frac{\partial u}{\partial z} \quad \frac{\partial v}{\partial x} = \frac{\partial u}{\partial y} \qquad \text{(IV.1.28)}$$

Thus all p u v and w obey the wave equation

$$\frac{\partial^2}{\partial t^2} = \nabla^2$$

If in addition any two of u v or w obey "original Cauchy-Riemann conditions" for example

$$\frac{\partial v}{\partial x} = \frac{\partial u}{\partial y} \quad \frac{\partial u}{\partial x} = -\frac{\partial v}{\partial y}$$

then

$$\frac{\partial p}{\partial t} - \frac{\partial w}{\partial z} = 0$$

then each obeys the wave equation in separated form that is

$$\underset{(\text{for } u, v)}{\frac{\partial^2}{\partial x^2} + \frac{\partial^2}{\partial y^2}} = 0 = \underset{(\text{for } p, w)}{\frac{\partial^2}{\partial t^2} - \frac{\partial^2}{\partial z^2}}$$

as in Equation IV.1.24

The left-hand side of this equation can be satisfied by the transformation of "ordinary" complex variable theory and the right-hand side is a one-dimensional wave equation. The result is the transformation

$$\begin{aligned} u + iv &= \Psi(x + iy) \\ u - iv &= \Psi(x - iy) \\ w &= F_1(z + t) + F_2(z - t) \\ p &= F_1(z + t) - F_2(z - t) \end{aligned}$$

for arbitrary functions Ψ F_1 and F_2, which is identical to the first of the Bateman transformations.

The comparison between Equations IV.1.21, IV.1.22 and Equations IV.1.26, IV.1.27 shows that differentiability and left regularity are essentially different, a point noted by Imaeda [5]. Due however to the derivation of Maxwell-like equations from the regularity conditions, these are taken to be the more relevant.

e_0	e_1	e_2	e_3	e_4	e_5	e_6	e_7
e_1	e_0	$-e_7$	e_6	$-e_5$	$-e_4$	e_3	$-e_2$
e_2	e_7	e_0	$-e_5$	$-e_6$	$-e_3$	$-e_4$	e_1
e_3	$-e_6$	e_5	e_0	$-e_7$	e_2	$-e_1$	$-e_4$
e_4	$-e_5$	$-e_6$	$-e_7$	$-e_0$	e_1	e_2	e_3
e_5	$-e_4$	e_3	$-e_2$	e_1	$-e_0$	e_7	$-e_6$
e_6	$-e_3$	$-e_4$	e_1	e_2	$-e_7$	$-e_0$	e_5
e_7	e_2	$-e_1$	$-e_4$	e_3	e_6	$-e_5$	$-e_0$

Table IV.1: Product table for the bi-quaternion basis

IV.2 BI-QUATERNIONS

The difficulties experienced in obtaining complete consistency between the standard derivations in Silberstein [6] and the algebraic derivation in Chapter 7 are due in the main to the problem of correctly positioning the "ordinary" complex i, which I have termed the Argand i. It may even be possible that Hamilton's concentration on the *real* octonions as the development of the quaternions, led to their failure to adequately describe the physical phenomena he anticipated. As is well known, the last twenty or so years of his life were devoted to such an outcome, in which he was continually disappointed.

We therefore introduce the Argand i in the form of an additional basis element in a complex eight dimensional space, one that commutes with the other basis elements α β and γ of the quaternion algebra. That is $i\alpha = \alpha i$ etc. and accordingly we establish the complete basis e_i $i = 0 \ldots 7$ where

$$\begin{array}{l} e_0 = 1 \;\; e_1 = i\alpha \;\; e_2 = i\beta \;\; e_3 = i\gamma \\ e_4 = i \;\; e_5 = \alpha \;\; e_6 = \beta \;\; e_7 = \gamma \end{array} \tag{IV.2.1}$$

With this commutative property of the Argand i we can define the bi-quaternion

$$\mathcal{Q} = Q + iP = \{q_0 + \alpha q_1 + \beta q_2 + \gamma q_3\} + i\{p_0 + \alpha p_1 + \beta p_2 + \gamma p_3\} \tag{IV.2.2}$$

In most practical cases this is reduced to Minkowski space by the simple expedient of putting q_0 p_1 p_2 p_3 equal to zero.

Thus there are some instances where $e_i e_j = e_j e_i$ and others where $e_i e_j = -e_j e_i$ and $e_0 - e_3$ square to $+1$ while $e_4 - e_7$ square to -1.

The resulting product rules are given in Table IV.1 for the product of elements of the left-hand column with elements of the top row. Consequently there exist three conjugations of the bi-quaternion $\mathcal{Q} = Q + iP$

$$\begin{array}{rl} \text{Quaternion conjugation} & \overline{\mathcal{Q}} = \overline{Q} + i\overline{P} \\ \text{Complex conjugation} & \mathcal{Q}^* = Q - iP \\ \text{Hermitian conjugation} & \mathcal{Q}^+ = \overline{Q} - i\overline{P} \end{array} \tag{IV.2.3}$$

($^-$ denoting quaternion conjugate) in which the application of any two consecutively will produce the third.

The general requirements of most physical systems need only the complex four-space of Minkowski and in this form $\overline{Q} = -Q$ and $\overline{P} = P$ and the conjugations are considerably simplified.

The "missing" coordinates form a space with three real "time-like" dimensions and one complex "space-like" dimension, the supposed conditions governing superluminal space, the space where the velocity of light has been exceeded. The complete bi-quaternion may therefore have these effects already included in it.

The reduced bi-quaternion $\mathcal{Q} = (ip_0; \mathbf{q})$ has norm

$$\mathcal{Q}\overline{\mathcal{Q}} = \overline{\mathcal{Q}}\mathcal{Q} = -p_0^2 + \mathbf{q}^2 = \{p_0^2 + q_1^2 + q_2^2 + q_3^2\} \stackrel{\text{def}}{=} \mathcal{Q}^2 \qquad \text{(IV.2.4)}$$

and then the inverse is

$$\mathcal{Q}^{-1} = \frac{\mathcal{Q}}{|\mathcal{Q}|^2}$$

for $\mathcal{Q}$ with non-zero norm.

THE CLIFFORD ALGEBRA

This notation is also a variant of that of the Clifford algebra [7]. The latter is an n-dimensional algebra with basis elements e_i $i = 1 \ldots n$ with $e_i e_j = -e_j e_i$ for all $i \neq j$; $e_i^2 = -1$. The Clifford algebra extends vector notation through the use of bi-vectors tri-vectors and generally multi-vectors. (Note *not* the bi-vectors of the polarization state of Chapter 6 defined as $\mathbf{u} + i\mathbf{v}$, but the bi-vector defined by the exterior product $\mathbf{u} \wedge \mathbf{v}$ and multi-vectors by continued exterior products [8]). The general Clifford number then consists of a series with terms consisting of a scalar, a one-form, a two-form and so on such as :-

$$a_0 + a_1\mathbf{u} + a_2\mathbf{v} \wedge \mathbf{w} + a_3\mathbf{x} \wedge \mathbf{y} \wedge \mathbf{z} \qquad \text{etc.}$$

Conjugation is performed by a process termed the Hodge star operator, which changes the signs of the basis products involved. Clifford algebra is being applied to general physics increasingly in recent years. The concept of the one-form being a directed line, a two-form a directed area and a three-form a directed volume, originated in the original work of Grassmann [9]. The work actually predated the discovery of quaternions by Hamilton but the volume outlining the theory was so obscure that its relevance was not appreciated for half a century (a well-known syndrome). There is increasing application in theoretical physics for Grassmann geometry also.

The exterior product rules can be expressed in terms of the basis elements e_i (different from the bi-quaternion basis) and gives rise to an n-dimensional algebra with 2^n elements. Thus for the three space $\mathcal{C}_3$ has eight elements corresponding to the requirements of the scalar the one-form the two-form etc. [10]

$$\begin{array}{llll} \{E_0\} & 1 & & \\ \{E_1\} & e_1 & e_2 & e_3 \\ \{E_2\} & e_2e_1 & e_3e_2 & e_1e_3 \\ \{E_3\} & e_1e_2e_3 & & \end{array} \qquad \text{(IV.2.5)}$$

the product rules applying to higher multiples of e_i.

If we now establish the mapping

$$\begin{array}{c} 1 \leftrightarrow 1 \quad \alpha \leftrightarrow e_2e_1 \\ \beta \leftrightarrow e_3e_2 \quad \gamma \leftrightarrow e_1e_3 \end{array} \qquad \text{(IV.2.6)}$$

we find that α β γ obey the rules of quaternion multiplication. Thus the *quaternion* algebra is a sub-algebra of the Clifford algebra, but the *bi-quaternion* algebra, with its introduction of the Argand complex i, obeys independent rules which could be termed a skew Clifford algebra.

THE BI-QUATERNION DIFFERENTIAL OPERATOR

The bi-quaternion differential operator and its conjugate are defined to be

$$\Box = \left[\frac{1}{i}\frac{\partial}{\partial x_0} + \alpha\frac{\partial}{\partial x_1} + \beta\frac{\partial}{\partial x_2} + \gamma\frac{\partial}{\partial x_3}\right] \equiv \left[\frac{1}{i}\frac{\partial}{\partial x_0}; \nabla\right] \qquad \text{(IV.2.7)}$$

$$\Box^c = \left[-\frac{1}{i}\frac{\partial}{\partial x_0} + \alpha\frac{\partial}{\partial x_1} + \beta\frac{\partial}{\partial x_2} + \gamma\frac{\partial}{\partial x_3}\right] \equiv \left[-\frac{1}{i}\frac{\partial}{\partial x_0}; \nabla\right] \qquad \text{(IV.2.8)}$$

respectively, where ∇ is the quaternion gradient.

The vector type gradient, divergence and curl operators have their bi-quaternion equivalents. All the vector relationships, such as triple products, differentials of scalar and vector products etc. in standard vector analysis are also valid in the quaternion basis.

$$\nabla \odot \mathbf{q} = \frac{\partial q_1}{\partial x_1} + \frac{\partial q_2}{\partial x_2} + \frac{\partial q_3}{\partial x_3}$$

$$\nabla \otimes \mathbf{q} = \alpha\left[\frac{\partial q_3}{\partial x_2} - \frac{\partial q_2}{\partial x_3}\right] + \beta\left[\frac{\partial q_1}{\partial x_3} - \frac{\partial q_3}{\partial x_1}\right] + \gamma\left[\frac{\partial q_2}{\partial x_1} - \frac{\partial q_1}{\partial x_2}\right]$$

Thus we can write in vector and scalar terms

$$\Box Q = \left[\frac{\partial q_0}{\partial x_0} - \nabla \odot \mathbf{q}; \frac{1}{i}\frac{\partial q}{\partial x_0} + i\nabla q_0 - \nabla \otimes \mathbf{q}\right] \qquad \text{(IV.2.9)}$$

Hence all the major first order derivatives are incorporated in a single algebraic bi-quaternion operation.

This incorporation of the Argand i makes such expressions doubly separable, into Argand real and imaginary parts and into quaternion scalar and vector parts. For example applying Equation IV.2.7 to the field bi-quaternion $\mathcal{M}$ of Chapter 7 and setting the result equal to zero, the condition for the left regularity of $\mathcal{M}$, gives

$$\Box\mathcal{M} = \Box\{(0;\mathbf{B}) + i(0;\mathbf{E}/c)\} = 0$$

$$\left\{-\nabla \odot \mathbf{B} - i\nabla \odot \frac{\mathbf{E}}{c}; \frac{\partial \mathbf{B}}{i\partial(ct)} + \frac{\partial \mathbf{E}/c}{\partial(ct)} + \nabla \otimes \mathbf{B} + i\nabla \otimes \frac{\mathbf{E}}{c}\right\} = 0$$

Equating real and Argand imaginary parts

$$\nabla \odot \mathbf{B} = 0 \qquad \nabla \odot \mathbf{E} = 0$$

$$\nabla \otimes \mathbf{B} = -\frac{\partial \mathbf{E}}{\partial t} \qquad \nabla \otimes \mathbf{E} = \frac{\partial \mathbf{B}}{\partial t}$$

Hence the single equation $\Box\mathcal{M} = 0$ incorporates all four Maxwell source free free-space equations. Sources and medium constituent relations can be included by an obvious extension (Section 7.4).

THE BI-QUATERNION D'ALEMBERTIAN

The d'Alembertian is obtained by taking the norm of the Laplace-Baltrami operator.

$$\begin{aligned}\Box^c\Box &= \left[-\frac{1}{ic}\frac{\partial}{\partial t} + \alpha\frac{\partial}{\partial x} + \beta\frac{\partial}{\partial y} + \gamma\frac{\partial}{\partial z}\right]\left[\frac{1}{ic}\frac{\partial}{\partial t} + \alpha\frac{\partial}{\partial x} + \beta\frac{\partial}{\partial y} + \gamma\frac{\partial}{\partial z}\right] \\ &= \left[\frac{1}{c^2}\frac{\partial^2}{\partial t^2} - \frac{\partial^2}{\partial x^2} - \frac{\partial^2}{\partial y^2} - \frac{\partial^2}{\partial z^2}\right] \qquad \text{(IV.2.10)}\end{aligned}$$

whence

$$\Box^c\Box\Phi = 0$$

is the wave equation.

IV.3 ROTATIONS AND GROUPS

In many places in the text of this book, we have used similar concepts in apparently differing contexts. Thus, for example, quaternions were used in the analysis of polarization effects as well as in the bi-quaternion derivation of Maxwell's equations and the wave equation. It can be shown that this is no coincidence that there is an underlying theory to which both belong, and the quaternion analysis in particular is the paradigm which demonstrates this most clearly.

We start therefore with the quaternion description of rotation based on the spin matrices of Section 6.8 and the definitions in Section IV.1

Rotations in two dimensions are categorised by the usual matrix

$$\begin{pmatrix} \cos\theta & \sin\theta \\ -\sin\theta & \cos\theta \end{pmatrix}$$

applied to the real vector $\begin{pmatrix} x \\ y \end{pmatrix}$. The vectors of polarization theory are complex and form four element entities as a result. Thus there needs to be a rotation operation that will accommodate complex vectors that will of necessity be four dimensional and thus either a four by four matrix of real elements or two by two with complex elements.

To extend the two dimensional rotation above, we first note that the simple rotation is the transformation

$$\begin{pmatrix} x' \\ y' \end{pmatrix} = \begin{pmatrix} \cos\theta & \sin\theta \\ -\sin\theta & \cos\theta \end{pmatrix} \begin{pmatrix} x \\ y \end{pmatrix}$$

which can equally be written

$$\begin{pmatrix} x' & y' \\ y' & -x' \end{pmatrix} = \begin{pmatrix} \cos\frac{\theta}{2} & \sin\frac{\theta}{2} \\ -\sin\frac{\theta}{2} & \cos\frac{\theta}{2} \end{pmatrix} \begin{pmatrix} x & y \\ y & -x \end{pmatrix} \begin{pmatrix} \cos\frac{\theta}{2} & -\sin\frac{\theta}{2} \\ \sin\frac{\theta}{2} & \cos\frac{\theta}{2} \end{pmatrix} \qquad \text{(IV.3.1)}$$

and the minus sign can be "shared" to give

$$\begin{pmatrix} ix' & y' \\ y' & ix' \end{pmatrix} = \begin{pmatrix} \cos\frac{\theta}{2} & i\sin\frac{\theta}{2} \\ i\sin\frac{\theta}{2} & \cos\frac{\theta}{2} \end{pmatrix} \begin{pmatrix} ix & y \\ y & ix \end{pmatrix} \begin{pmatrix} \cos\frac{\theta}{2} & i\sin\frac{\theta}{2} \\ i\sin\frac{\theta}{2} & \cos\frac{\theta}{2} \end{pmatrix} \qquad \text{(IV.3.2)}$$

Not only but also

$$\begin{pmatrix} x'+iy' & 0 \\ 0 & x'-iy' \end{pmatrix} = \begin{pmatrix} e^{i\frac{\theta}{2}} & 0 \\ 0 & e^{-i\frac{\theta}{2}} \end{pmatrix} \begin{pmatrix} x+iy & 0 \\ 0 & x-iy \end{pmatrix} \begin{pmatrix} e^{i\frac{\theta}{2}} & 0 \\ 0 & e^{-i\frac{\theta}{2}} \end{pmatrix} \qquad \text{(IV.3.3)}$$

(clockwise rotation)
In physical terms this means that the rotation through θ has been divided into two semi-rotations, with the right-hand product in one direction and the left in the other but about a reversed axis thus completing the rotation. The distinction is not trivial. It is connected to the "twoness" of rotations and hence to the concept of spin half values. It can be observed in action on the Poincaré sphere, where along the equator and its projection onto the stereographic plane, the linear polarized state reverses its direction on going from the extreme left to the extreme right as shown in Figure 6.4 corresponding to a rotation of 2π on the sphere itself. Consequently it requires a rotation of 4π to reproduce the original state, and so to the concept of the double group [10].

A vector with complex elements $X = \begin{pmatrix} x_1 \\ x_2 \end{pmatrix}$ that transforms by $X' = AX$ where A is the general unimodular quaternion

$$\begin{pmatrix} a & b \\ -b^* & a^* \end{pmatrix} = \cos\theta + \hat{\mathbf{n}}\sin\theta$$

is likewise a rotation of θ about the axis direction

$$\hat{\mathbf{n}} = \alpha n_1 + \beta n_2 + \gamma n_3 \tag{IV.3.4}$$

provided

$$|a|^2 + |b|^2 = 1$$

Separated in the same manner the rotation is composed of two semi-rotations

$$X' = A^{-1}XA$$

with A now given by $\cos(\theta/2) + \hat{\mathbf{n}}\sin(\theta/2)$.

Taking A in turn to be the three axis rotations †

1. $\theta/2$ about γ the z-axis

2. $\phi/2$ about β the y-axis

3. $\psi/2$ about the *z-axis*

the product will give the matrix for the general rotation in three dimensions in terms of the Euler angles θ ϕ ψ [11].

The A rotations are "screw" or "conical" rotations. Combined as above to give the general rotation on the sphere shows the connection between the group of A matrices and the orthogonal group of spherical rotations. The latter is the group $SO(3)$ and we shall see later that the group of A matrices is the group $SU(2)$. Hence the relation shown is the ***homomorphism*** between these groups. Equation IV.2.6 gave the relation between the quaternion basis and the Clifford algebra $\mathcal{C}_3$ and thus demonstrated that $\mathcal{C}_3$ is *realized* by $SO(3)$.

It has been shown, therefore, that "to every unitary matrix of the second order with determinant unity, there corresponds a rotation in three-dimensional space. Conversely each rotation corresponds to two such matrices u differing only in sign." The non-commutativity of rotations arises directly from the non-commutativity of the quaternion algebra.

The aggregate of all such matrices u constitutes the "special unitary" group usually denoted by $SU(2)$ which has the basis generally given as the Pauli spin matrices (plus the identity matrix) but has been used throughout this book as their complex conjugates the α β γ of the quaternion basis.

Rotations are also prime examples of *continuous* groups. This demands that a small change in any of the factors involved produces a small change in the result of their operation. Obviously rotations can be infinitesimal as can translations (but not reflections nor inversions). Those groups representing transformations of the infinitesimal kind are Lie groups and have their own complete bibliography (see for example [13]). They would apply in the context

---

† The alternative is to take rotations about the x the y and z axes in turn. The method given here is the classical form with rotations about the z the y and a repeat of the z axis. [12] discusses this point in detail.

of microwave polarization for example to propagation through a continuously perturbing medium such as a chiral medium or optically through a medium giving Faraday rotation, and systems with aberrations [14] [15].

Groups can also be designated by the type of invariant quadratic form that occurs when group matrix members are applied as a transformation. This designation is due to Cartan [16]. Thus transformations that leave $x_1y_1 + x_2y_2 + x_3y_3 + x_4y_4$ invariant are members of the $O(4)$ group, while for the invariance of the indefinite norm in relativity we require the invariance (in standard notation) $x_0y_0 - x_1y_1 - x_2y_2 - x_3y_3$. These matrices specify the Lorentz group. If a condition is applied to a group, the part of the group satisfying the condition becomes a sub-group. Thus applying the condition of symmetry in monostatic scattering as in Section 6.8 makes b in IV.3.3 pure imaginary and its A matrix symmetrical. Thus the scattering matrices are a sub-group of the general A matrix group.

Similarly the transformations of special relativity can be obtained from complex rotations about an imaginary axis. This adds to the three rotation matrices (recognisable from IV.3.2)

$$\begin{pmatrix} \cos\frac{\theta}{2} & i\sin\frac{\theta}{2} \\ i\sin\frac{\theta}{2} & \cos\frac{\theta}{2} \end{pmatrix} \begin{pmatrix} \cos\frac{\theta}{2} & -\sin\frac{\theta}{2} \\ \sin\frac{\theta}{2} & \cos\frac{\theta}{2} \end{pmatrix} \begin{pmatrix} \exp\frac{i\theta}{2} & 0 \\ 0 & \exp\frac{-i\theta}{2} \end{pmatrix} \tag{IV.3.5}$$

the complex rotations

$$\begin{pmatrix} \cosh\frac{\theta}{2} & \sinh\frac{\theta}{2} \\ \sinh\frac{\theta}{2} & \cosh\frac{\theta}{2} \end{pmatrix} \begin{pmatrix} \cosh\frac{\theta}{2} & i\sinh\frac{\theta}{2} \\ -i\sinh\frac{\theta}{2} & \cosh\frac{\theta}{2} \end{pmatrix} \begin{pmatrix} \exp\frac{\theta}{2} & 0 \\ 0 & \exp\frac{-\theta}{2} \end{pmatrix} \tag{IV.3.6}$$

to give the group $SL(2, C)$ a spinor representation of the Lorentz group [17].

The matrices of IV.3.4 are the same as the basis matrices of the $SU(2)$ group above. The involvement of these matrices in the applications of group theory to quantum and atomic theories is now basic to theoretical physics [18].

It was noted earlier, Equation IV.1.10 that the coordinate quaternion $\{t, x, y, z\}$ can have the order of axes transformed without affecting any result. This is so since the labelling of the axes is arbitrary, provided left-handed orthogonality is maintained. The transformations between them are then rotations of the basic x y z axes about axes passing through the origin. For example a rotation of 90° about the z-axis is the transformation $x - y$; $y - -x$. There are 24 possible rotations which leave a cube with the origin at the centre invariant. These are point transformations of the cube and form the octahedral group O. If inversion in the origin is included, there results a further 24 improper transformations, positions of the cube that are obtained by rotation alone. These correspond to the conjugate quaternions of the 24 original quaternions resulting from the x y z permutations. The complete group O_h results with applications in crystallography as with the NaCl cubic lattice [19].

Matrices relating to the transmission through a non-uniform dielectric parallel sided slab given in Chapter 4 (Equation 4.4.36) have the general form

$$T = \begin{pmatrix} \cos\Phi & (i/Y_N)\sin\Phi \\ iY_1\sin\Phi & (Y_1/Y_N)\cos\Phi \end{pmatrix}$$

It is not known whether matrices of this kind can form a group, but were they to do so, they would have interesting applications in non-uniform linear media and in non-uniform waveguide transformations. They are however not unitary and a group theory of non-unitary matrices has not been considered hitherto.

THE LENS AND SYMPLECTIC GROUPS

After a considerable amount of extremely complicated algebra Stavroudis [20] produces general lens equations giving a matrix description for the two refractions at the surfaces and a "transfer" matrix for propagation in between them. They are derived from the very basic fundamental optical invariant of Herzberger [21] which states that the function

$$\eta\left[\frac{\partial \mathbf{r}}{\partial u}\cdot\frac{\partial \mathbf{s}}{\partial v}-\frac{\partial \mathbf{r}}{\partial v}\cdot\frac{\partial \mathbf{s}}{\partial u}\right]$$

is invariant, where $\mathbf{r}$ is the radius vector to a point on a ray in a congruence specified by the parameters u and v where the ray has unit tangent vector $\mathbf{s}$ in the medium of refractive index η. If the family of rays all issue from a single source then at that point

$$\frac{\partial \mathbf{r}}{\partial u}=\frac{\partial \mathbf{r}}{\partial v}=0$$

and thus the invariant vanishes at the origin and hence is zero at all points in its path. It follows that the rays can always be brought to a point focus again and that this is a restatement of the theorem of Malus and Dupin.

The matrices involved in the lens equation are 4 x 4 and Jacobian in form. The piece de resistance of the work described here (author's term) is that these matrices are related to those of the symplectic group.

In the manner described above, the symplectic group consists of those matrices whose application leaves the skew quadratic form

$$x_1y_3 - x_3y_1 + x_2y_4 - x_4y_2 \tag{IV.3.7}$$

invariant and is termed the group $Sp(4)$ [22]. The matrix representation of this group consists of matrices $\mathcal{R}$ which satisfy

$$\mathcal{R}J\mathcal{R}^T = J \tag{IV.3.8}$$

(^{T} meaning transpose) where J is the matrix

$$J = \begin{pmatrix} 0 & 0 & 1 & 0 \\ 0 & 0 & 0 & 1 \\ -1 & 0 & 0 & 0 \\ 0 & -1 & 0 & 0 \end{pmatrix} \tag{IV.3.9}$$

This is instantly recognisable as one of the basic quaternion matrices, see Equation IV.1.13. The theory is applied mainly to the restricted problem of the circularly symmetrical lenses. In this case a sub-group of $Sp(4)$ is sufficient. As shown in [22], the generators of the symplectic group are

1. translations $\begin{pmatrix} 1 & 0 & a & c \\ 0 & 1 & c & b \\ 0 & 0 & 1 & 0 \\ 0 & 0 & 0 & 1 \end{pmatrix}$

2. rotations $\begin{pmatrix} U & \begin{matrix} 0 & 0 \\ 0 & 0 \end{matrix} \\ \begin{matrix} 0 & 0 \\ 0 & 0 \end{matrix} & U^{T^{-1}} \end{pmatrix} \qquad |U| = \pm 1$

3. the semi-involutions $\begin{pmatrix} Q & \mathbf{I} - Q \\ -\mathbf{I} + Q & Q \end{pmatrix}$ where Q is one of

$$\begin{pmatrix} 0 & 0 \\ 0 & 0 \end{pmatrix} \quad \begin{pmatrix} 1 & 0 \\ 0 & 0 \end{pmatrix} \quad \begin{pmatrix} 0 & 0 \\ 0 & 1 \end{pmatrix} \quad \begin{pmatrix} 1 & 0 \\ 0 & 1 \end{pmatrix}$$

Stavroudis then goes on to show that the transfer matrix is identical to the translation matrix even to the symmetric property of the (a, b, c) matrix included and the matrix for the refractions at the surfaces is composed of a rotation, a translation and *one* of the four possible semi-involutions. Hence the rotationally symmetric lenses form a proper sub-group of $Sp(4)$. The significance of the remaining semi-involutions in this context, has not been determined.

Finally, in considering the paraxial approximation to these lenses, discarding all but the first order linear terms, the resulting simplification is shown to reduce to a *proper* sub-group of $Sp(4)$ homomorphic with $Sp(2)$. That an arbitrary approximation should lead to a proper sub-group is of some significance.

This has of necessity to have been but a very brief encounter with the intricacies and the wealth of literature governing the applications of the theory of groups to physics generally and to geometrical optics in particular. There is an objective though. We have shown the general application of matrices and their groups to optical problems. They mostly involve the quaternion group either in polarization studies, electromagnetic theory or, as in the last instance, the lens group. We have also shown in our studies of lenses and non-uniform media the part played by inversions. These too can be expressed by an operator in a matrix form. At the end of Chapter 3 there was proposed a hypothesis regarding one method of transforming a given (circularly symmetrical) lens into another. It could be possible that the matrix definition of lenses could go together with a matrix description of inversion to actually provide a proof (or disproof) of that hypothesis. That would provide a transformation technique between members of the lens group that becomes transformations of actual lenses. The continuous involvement of the quaternions in these groups and other analysis points to one possible way to make the connection.

This then is another source for possible investigation which has to be left for the future.

REFERENCES

1. Kyrala A (1967) *Applications of Vectors Matrices Tensors and Quaternions* Philadelphia: WB Saunders Chapters 8 & 9

2. Halberstam H & Ingram RE (1967) *The Mathematical Papers of Sir William Rowan Hamilton* **III** Algebra, Cambridge University Press

3. Crowe MJ (1967) *A History of Vector Analysis* Notre Dame University Press

4. Bateman H (1909) The transformation of the electrodynamical equations *Proc Lond Math Soc* **vol 8** no 2 p223

5. Imaeda K (1983) *Quaternionic Formulation of Classical Electrodynamics and Theory of Functions of a Biquaternion Variable* Fundamental Physics Laboratory, Dept of Electronic Science, Okayama University of Science, Japan

6. Silberstein L (1924) *The Theory of Relativity* London: Macmillan & Co

7. Lounesto P (1979) Spinor valued regular functions in hypercomplex analysis Report HTKK-MAT-A154 *Institute of Mathematics, Helsinki University of Technology, Finland*

8. Jancewicz B (1989) *Multivectors and Clifford Algebras* World Scientific Books

9. Grassmann HG (1844) Die Ausdehnungslehre see [2]

10. Altmann SL (1986) *Rotations, Quaternions and Double Groups* Oxford: Clarendon Press p221

11. Gel'fand IM, Minloss RA & Shapiro Z Ya (1963) *Representations of the Rotation and Lorentz Groups* Pergamon Press p11

12. Goldstein H (1950 reprinted 1964) *Classical Mechanics* Addison-Wesley p108

13. Helgason S (1962) *Differential Geometry and Symmetric Spaces* London & New York: Academic Press

14. Saad MN & Wolf KB (1984) The group theoretical treatment of aberrating systems **Pt I** Rep no 363 *Institute of Mathematics, University of Mexico*

15. Wolf KB (1984) The group theoretical treatment of aberrating systems **Pt II** Rep no 366 *Institute of Mathematics, University of Mexico*

16. Cartan E La théorie des groupes finis et continuus et l'analysis situs *Memor Sci Math* **vol 42**

17. Carmeli M & Malin S (1976) *Representations of the Rotation and Lorentz Groups* New York: Marcel Dekker Inc p71

18. Mizushima M (1972) *Theoretical Physics* New York: Wiley & Sons Inc

19. Keown R (1975) *An Introduction to Group Representation Theory* New York & London: Academic Press p179

20. Stavroudis ON (1972) *The Optics of Rays Wavefronts and Caustics* New York & London: Academic Press pp281-297

21. Herzberger M (1958) *Modern Geometrical Optics* New York: Wiley (Interscience)

22. Stanek PFG (1961) Two element generation of the symplectic group *Bull Amer Math Soc* **vol 67** p225

INDEX